HYDROTHERMAL PROCESSES AT SEAFLOOR SPREADING CENTERS

NATO CONFERENCE SERIES

- I Ecology
- II Systems Science
- III Human Factors
- IV Marine Sciences
- V Air–Sea Interactions
- VI Materials Science

IV MARINE SCIENCES

Volume 1 Marine Natural Products Chemistry
edited by D. J. Faulkner and W. H. Fenical

Volume 2 Marine Organisms: Genetics, Ecology, and Evolution
edited by Bruno Battaglia and John A. Beardmore

Volume 3 Spatial Pattern in Plankton Communities
edited by John H. Steele

Volume 4 Fjord Oceanography
edited by Howard J. Freeland, David M. Farmer, and Colin D. Levings

Volume 5 Bottom-interacting Ocean Acoustics
edited by William A. Kuperman and Finn B. Jensen

Volume 6 Marine Slides and Other Mass Movements
edited by Svend Saxov and J. K. Nieuwenhuis

Volume 7 The Role of Solar Ultraviolet Radiation in Marine Ecosystems
edited by John Calkins

Volume 8 Structure and Development of the Greenland–Scotland Ridge
edited by Martin H. P. Bott, Svend Saxov, Manik Talwani, and Jörn Thiede

Volume 9 Trace Metals in Sea Water
edited by C. S. Wong, Edward Boyle, Kenneth W. Bruland, J. D. Burton, and Edward D. Goldberg

Volume 10A Coastal Upwelling: Its Sediment Record
Responses of the Sedimentary Regime to Present Coastal Upwelling
edited by Erwin Suess and Jörn Thiede

Volume 10B Coastal Upwelling: Its Sediment Record
Sedimentary Records of Ancient Coastal Upwelling
edited by Jörn Thiede and Erwin Suess

Volume 11 Coastal Oceanography
edited by Herman G. Gade, Anton Edwards, and Harald Svendsen

Volume 12 Hydrothermal Processes at Seafloor Spreading Centers
edited by Peter A. Rona, Kurt Boström, Lucien Laubier, and Kenneth L. Smith, Jr.

Photograph of mounds surmounted by elongate chimneys several meters high composed primarily of polymetallic sulfide minerals discharging high-temperature pure hydrothermal solutions as "black smokers." A white crab is present near the center of the photograph on one of the chimneys. The objects in the foreground are water sampling bottles. The photograph was taken from the Deep Submergence Research Vehicle *Alvin* at a water depth of about 2000 meters at the "Hanging Gardens" hydrothermal site near the axis of the East Pacific Rise at latitude 21° North (K. Kim, Isotope Laboratory, Scripps Institution of Oceanography).

Library of Congress Cataloging in Publication Data

Main entry under title:

Hydrothermal processes at seafloor spreading centers.

(NATO conference series. IV, Marine sciences; v. 12)
"Proceedings of a NATO Advanced Research Institute, held April 5–8, 1982, at the Department of Earth Sciences of Cambridge University, England"—T.p. verso.
Includes bibliographical references and index.
1. Sea-floor spreading—Congresses. 2. Hydrothermal deposits—Congresses. 3. Heat budget (Geophysics)—Congresses. 4. Chemical oceanography—Congresses. 5. Hydrothermal vent ecology—Congresses. I. Rona, Peter A. II. Series.
QE511.7.H93 1983 551.46′08 83-17747
ISBN 0-306-41482-1

Proceedings of a NATO Advanced Research Institute, held April 5–8, 1982, at the Department of Earth Sciences of Cambridge University, England

A Division of Plenum Publishing Corporation
233 Spring Street, New York, N.Y. 10013

Printed in the United States of America

PREFACE

During the past ten years, evidence has developed to indicate that seawater convects through oceanic crust driven by heat derived from creation of lithosphere at the Earth-encircling oceanic ridge-rift system of seafloor spreading centers. This has stimulated multiple lines of research with profound implications for the earth and life sciences. The lines of research comprise the role of hydrothermal convection at seafloor spreading centers in the Earth's thermal regime by cooling of newly formed lithosphere (oceanic crust and upper mantle); in global geochemical cycles and mass balances of certain elements by chemical exchange between circulating seawater and basaltic rocks of oceanic crust; in the concentration of metallic mineral deposits by ore-forming processes; and in adaptation of biological communities based on a previously unrecognized form of chemosynthesis. The first workshop devoted to interdisciplinary consideration of this field was organized by a committee consisting of the co-editors of this volume under the auspices of a NATO Advanced Research Institute (ARI) held 5-8 April 1982 at the Department of Earth Sciences of Cambridge University in England. This volume is a product of that workshop.

The papers were written by members of a pioneering research community of marine geologists, geophysicists, geochemists and biologists whose work is at the stage of initial description and interpretation of hydrothermal and associated phenomena at seafloor spreading centers. The papers are grouped by subject into eight sections each with an introductory paper that places the papers in that section in context of the subject as a whole. The subjects treated comprise geologic settings of hydrothermal activity at slow-, intermediate-, and fast-spreading oceanic ridges (section 1); the nature of hydrothermal circulation inferred from crustal sections exposed on land and modeled at seafloor spreading centers (section 2); the exchange of elements between basalt and seawater at elevated temperatures and pressures determined experimentally in the laboratory and the application of these findings to interpretation of natural hydrothermal systems and basalt alteration at oceanic ridges (section 3); comparison of chemical and physical processes at submarine oceanic ridges with those on Iceland, a subaerial portion of an oceanic ridge (section

4); the chemistry of submarine hot springs determined where primary hydrothermal solutions were first sampled at an oceanic ridge, and the role of such solutions in geochemical mass balances and budgets, as well as in the maturation of hydrocarbons (section 5); a reconsideration of ferromanganese deposits including manganese nodules in light of the new knowledge of seafloor hydrothermal processes (section 6); hydrothermal ore-forming processes directly observed at an oceanic ridge and inferred from mineral deposits in certain slices of oceanic crust exposed on land (section 7); the adaptation of biologic communities to the environment of active hydrothermal vents at oceanic ridges (section 8). A chronological list of landmarks in studies of hydrothermal processes at seafloor spreading centers presented in the Appendix helps to place the papers in perspective of the development of the entire field. The papers consolidate previous work and delineate a research frontier that is advancing rapidly.

The interdisciplinary overview reveals the immense potential of the study of hydrothermal processes at seafloor spreading centers for continued progress on both basic and applied levels. On a basic level the field is at an early stage in developing understanding of permeability and magmatic heat sources as controls of hydrothermal convection in oceanic crust; mass and heat transfer by high-temperature hydrothermal convection, low-temperature hydrothermal convection, and other submarine processes; the contribution of each of these processes to local and global geochemical mass balances and budgets and their roles in geochemical cycles; and the dynamics of biologic adaptation at active submarine hydrothermal vents. The development of understanding of hydrothermal processes at seafloor spreading centers has brought the field to the threshold of deciphering the role these processes in the evolution of the hydrosphere, lithosphere and biosphere.

On an applied level, basalt-seawater exchange in hydrothermal systems is a component to be considered in evaluation of the capacity of the oceans to assimilate toxic substances. The concentration of polymetallic mineral deposits by hydrothermal processes at seafloor spreading centers constitutes a resource for the future, and provides an analog to model ore-forming processes that apply to the understanding of and exploration for economic hydrothermal mineral deposits on land. The study of bacterial chemosynthesis at hydrothermal vents may lead to developments in mariculture, metal fixation, and medicine.

It is a pleasure to acknowledge those who made the NATO ARI and this volume possible. The NATO Special Program Panel on Marine Sciences under the chairmanship of J. D. Burton of Southampton University offered the opportunity to organize the ARI. The NATO Scientific Affairs Division provided funding with supplemental support donated by the Centre National Pour l'Exploitation des Oceans (CNEXO) of France, and the National Oceanic and Atmospheric Administration (NOAA) of the United States. F. Webster gave guidance and support in his dual capacities as member of the NATO Panel and as Assistant Administrator for NOAA Research and Development. N. A. Ostenso provided encouragement as Deputy Assistant Administrator for NOAA Research and Development. E. R. Oxburgh and M. I. Johnston, in their respective positions as Chairman and Administrator of the Department of Earth Sciences at Cambridge University, made available an ideal place to convene the ARI and facilitated local arrangements. P. M. Vann and L. Schmidt in their respective positions as Associate Editor and Assistant to the Managing Editor of the Plenum Publishing Corporation ably handled the production of this volume. Finally, acknowledgment is extended to the authors who presented the latest results of their investigations at the ARI, to other participants whose discussion contributed to vigorous interaction at the sessions, and to reviewers whose critical comments have benefited the papers in this volume.

Peter A. Rona
Kurt Boström
Lucien Laubier
Kenneth L. Smith, Jr.

CONTENTS

GEOLOGIC SETTING

HYDROTHERMAL CONVECTION

BASALT-SEAWATER INTERACTION

ICELAND AND OCEANIC RIDGES

MASS BALANCES AND CYCLES

FERROMANGANESE DEPOSITS

HYDROTHERMAL MINERALIZATION

BIOLOGY OF HYDROTHERMAL VENTS

APPENDIX 1

PHOTOGRAPHS

THE FOUR DIMENSIONS OF THE SPREADING AXIS

Tjeerd H. van Andel

Geology Department, Stanford University

Stanford, California 94305 U.S.A.

Even though Matthew Fontaine Maury had revealed the existence of the Mid-Atlantic Ridge a hundred years before, it was well past the mid-point of the present century before the reality and great extent of the global mid-ocean ridge system became impressed upon the collective consciousness of marine geoscientists, largely due to the efforts of Bruce C. Heezen (e.g., Ewing and Heezen, 1956), and later H.W. Menard (1958). Not until the late 1950's did the globe-girdling ridge evolve into a major oceanographic issue. Its image, however, remained rather two-dimensional, based as it was upon widely spaced ocean traverses; a broad, rather gentle rise, usually cleft longitudinally and deeply along its crest.

Long before the true function of the ridge as a divergent plate boundary had been recognized, its size and the peculiar crestal relief invited attention. In the early 1960's B.D. Loncarevic set out from the Bedford Institute to fill this gap with a detailed survey of the Mid-Atlantic Ridge crest at 45° N, and Woods Hole shortly followed suit some 20 degrees farther south. Both carried out what were then highly detailed bathymetric, geophysical, and geological studies. At a time when we worry about track mismatches as small as 20-50 m, a working scale of 16 inches to the degree no longer seems impressive, but for its day this involved some tricky navigation, and led to very pleasing results. Those studies (e.g., Loncarevic *et al.*, 1966; van Andel *et al.*, 1965; van Andel and Bowin, 1968) considerably increased our knowledge of the rift morphology even though, as is inevitable with wide-beam echosounders, the valley continued to resemble an irregular V with a crumpled base. Extensive, precisely controlled dredging (Aumento, 1970) laid the foundation for our current understanding of the complexity of the oceanic crust (e.g., Melson et al., 1968; Aumento et al., 1971) and removed the notion (Engel and Engel, 1966) that it consisted almost wholly of tholeiitic basalt.

The sophistication of our technology has now advanced far beyond this level, and the volume of our knowledge has increased exponentially with it, even though we have explored not even 1 percent of the axis of the global ridge system and far less of its flanks. It is amazing how much we have learned about the volcanic, tectonic, and hydrothermal processes on the ridge crest since those first "detailed" studies. It is equally amazing, as MacDonald (1982) has noted in his informative review of the state of our art, how many important issues remain unresolved. In that overview MacDonald, although by no means obscuring the open questions, takes the optimistic view and dwells largely on what we know. It seems therefore proper that, in the interest of symmetry, I deemphasize the known in the following pages and stress instead the questions that are challenging us as MacDonald has also done in this volume.

In less than two decades we have learned to apply navigation methods with a precision of better than 10 m, deep-towed geophysical and photographic equipment that has in some cases increased our resolution to centimeters, and multibeam high resolution echosounding that has produced the first deep-sea charts comparable in detail to a topographic map on land. Finally, nearly ten years ago already, we descended to the rift valley floor for a first hand look at the geology and its lava flows, fissures, and fault scarps, and to do precisely controlled sampling and geophysical experiments.

It is not surprising that this technological avalanche has changed not only our methods of study and our data sets, but also the very nature of the questions we pose or emphasize. Initially, our attention was focused to a large extent on the problem of the rift valley itself, its origin, its structure, the reasons for its absence at very fast and very slow spreading ridges (Menard, 1967; van Andel and Bowin, 1968), while the fact that there must be a volcanic beginning to all this was, of course, accepted but rarely further explored. Since then we have discovered that rift valleys are not necessarily always absent on fast spreading ridges, although they are shallow there or possibly concealed by large volumes of lava. The most striking exception to the rift valley image remains the horst on the slow spreading Reykjanes Ridge, presumably connected in some unknown fashion with the Icelandic volcanic plateau, itself as yet not fully explained. Rare in the Cenozoic ocean basins, such volcanic plateaus appear to have been far more common on crust generated more than about 80 m.y. ago, as even the briefest look at the morphology of the Pacific Ocean will show.

Twenty years of ridge study have generated a broad range of hypotheses to explain the origin of the rift valley itself and the apparent correlation of its characteristics with spreading rate, but little progress has been made past the very general hydraulic head-loss models of Sleep (1969), Osmaston (1971), and Lachenbruch (1973, 1976). Other mechanisms have been suggested (e.g., Anderson and Noltimier, 1973; Tapponier and Francheteau, 1978; Nelson, 1981) but none is currently favored strongly. There are good reasons for this because solid evidence for all models is still sparse and combinations of various suggested processes are very likely. Our lack of knowledge of the rheology of the deeper crust and mantle remains a major obstacle.

Only slightly better understood are the processes that convert slivers of the new crust on the rift floor into the terraces of the rift walls, and eventually into the crestal ranges and upper ridge flanks on mature lithosphere. Ramberg and van Andel (1977), and MacDonald (1977; MacDonald and Luyendyk, 1977; MacDonald and Atwater, 1978) have postulated an intermittent process whereby entire volcanic bodies (and sometimes volcano halves: MacDonald _et al_., 1980) are transported sideways and raised along faults evolved from tension fissures, to form first the terraces of the rift valley walls, and subsequently the bordering crestal ranges. As a result, the rift valley at any instant exists in one of two extreme tectonic states; wide with diffuse volcanic activity just before uplift and incorporation of a block into the walls, or narrow with more precisely aligned volcanic edifices after a period of uplift. The process of uplift itself, as well as its causes remain poorly defined and we are not certain why there should be a depressed rift valley floor or what drives the uplift of the new crust as it ages and moves away.

Once raised to the level of the crest, the blocks that make up the walls and terraces of the rift are further modified to the normal relief of the mature ridge. In the case of the rugged crest and flank topography of slow-spreading ridges, the amount of further deformation required is quite substantial. Here data fail us, because only rarely have the detailed studies that must generate the information extended so far out. On the other hand, we do not lack potential models as MacDonald's summary (1982) indicates. One thing, however, is quite clear. The linearity of the ridge structure increases very much with increasing spreading rate, from the short _en echelon_ fault scarps of the Mid-Atlantic Ridge to the tens to hundreds of kilometers-long faults that can be traced on the Galapagos Ridge (Laughton and Searle, 1979; Allmendinger and Riis, 1979). Once again, one can only guess at what this means, but that it is a fundamental phenomenon is made clear by an analogous inverse relationship between spreading rate and the spacing of transform faults, another as yet not carefully examined problem (MacDonald, this volume).

One may reasonably regard the rift valley floor as belonging to neither plate (Ballard and van Andel, 1977), but even beyond its inner walls the maturation of oceanic lithosphere is far from complete. This transitional zone is much in need of study with the advanced techniques that have rewarded us so highly in the rift valley. Because of the age, size, and sediment blanket of the upper ridge flanks, such study is likely to be tedious and at times frustrating, but it addresses fundamental problems, one of them having to do with hydrothermal processes, the subject of this volume. Thus far, virtually all active hydrothermal phenomena have been observed on the volcanic axis itself, although dead springs have been found at some small distance to either side, and hydrothermal deposits are commonly intercalated between basalt and pelagic sediments in DSDP holes. One of the exceptions is the Galapagos mounds field, (Williams et al., 1979) located some 25 km south of the rift axis on crust between 0.5 and 0.9 m.y. old. This field demonstrates, as everyone expected and theory suggests, that hydrothermal activity does not cease with the death of the volcanic activity at the rift, but continues, albeit in a different mode, for an unknown time and distance afterwards (Green et al., 1981). We do not know in what manner the hydrothermal circulation relates to the tectonic processes that give the lithosphere its final shape, nor are we even sure that the Galapagos mounds field is not the surface expression of some anomalous hot area of flank volcanism underneath. We will not be able to view the hydrothermal processes of the spreading plate edge in true perspective until we have a much better idea of what happens off the axis. Thus, the enthusiasm for spending time and money on more axial black smokers, tubeworms, and sulphide deposits ought to be at least partly diverted into less immediately spectacular but ultimately more rewarding directions. The same applies to the repeated observations and claims of hydrothermal activity in transform and other fault zones. It is necessary to acquire much better documentation of these events before we can develop, on the one hand, some confidence in their importance, and on the other hand assess their significance in the ridge hydrothermal cooling system.

To return to the rift valley, the extensive and rewarding field work of the last decade has raised a host of questions regarding the volcanic processes that generate new crust, questions that have overshadowed those related to the genesis of the rift valley itself and to the tectonic phase discussed above. One is concerned these days with the nature of the magma chamber, whether it be permanent of intermittent or extant at all, with the pathways the magma takes to the surface and with what happens to it on the way. Those issues have recently been ably summarized (e.g., MacDonald, 1982; East Pacific Rise Study Group, 1981; Rhodes and Dungan, 1979; Sleep _et al._, this volume). Further discussion here would be redundant. When the magma reaches the surface, volcanic edifices are constructed with which we have lately been so concerned, in part because they had been much neglected prior to the work of the FAMOUS project, and partly because the data on surface geology we now so conveniently obtain are best suited for this purpose. How does the volcanic layer of the upper crust form? What is its composition in terms of intrusive dike and extrusive lava complexes? What do lava flow forms tell us about the eruptive process? How do the constructive volcanic processes vary in space and time and what is their episodicity? It is with such questions that a growing literature deals and they loom large in one of the following papers (Ballard and Francheteau, this volume). However, if the data summaries and the hypotheses erected upon them might suggest that our knowledge is fairly complete, we need only remember that the discovery that sheet lava flows are a major part of the the volcanic edifice, a fact now an essential part of all models, only dates from about 5 years back (Lonsdale, 1977; van Andel and Ballard, 1979). Other such surprises may be in store, especially given the persistent difficulty we are experiencing in fitting the results of DSDP crustal drilling into models derived from surface geology and geophysical data (e.g., Hall and Robinson, 1979).

What happens at the rift as new crust is generated and transformed into oceanic lithosphere can be considered in four distinct, yet complexly interrelated dimensions. _First_, there is a _dimension perpendicular to the spreading axis_, the ridge cross section that attracted our attention from the very beginning. Its information content bears primarily on the tectonic phase, the tension and, at greater depth, shear stresses to which the crust is subjected as soon as it forms. This transverse dimension tells us mainly about the conversion of new crust into wall blocks and crestal ranges, and about the final modification, collapse as some models have it, that produces the typical abyssal hill terrain. The active volcanic zone, of course, also has a dimension perpendicular to the eruptive fissures. However, the events are so highly constrained in this direction by previous and contemporaneous tectonism that what we learn says little about the volcanic process except that it happens in a very narrow zone. Especially studies of magnetic anomaly transitions have brought out clearly that the generation of new crust, viewed from above, is an almost one-dimensional process.

This cannot be taken to imply a very narrow magma chamber underneath, however, as both geophysical data (MacDonald, 1982; this volume) and studies on ancient ophiolites (e.g., Coleman and Hopson, 1981) demonstrate.

While most investigators still carry some inclination to view the problems of the divergent plate edge in terms of a cross section perpendicular to its strike, the data acquisition necessary to assess what happens at right angles, along the rift axis, is just reaching the stage of marginal adequacy. In this dimension lie tectonic variations as great as and volcanic ones much greater than those observed across the strike (Ballard *et al.*, 1982; Ballard and Francheteau this volume and references therein), though bound at both ends by transform faults. Where the transforms are widely spaced, an astonishing linearity of volcanic and structural elements is evident and longitudinal variation is small (MacDonald, this volume). In the Atlantic on the other hand, at slow spreading rate, transforms are close together, active and recently active volcanoes far apart, and the terrain varies drastically on scales from a few hundreds of meters to a few kilometers. The data covering this dimension have been very sparse but now appear to be arriving almost faster than they can be digested.

With admitted oversimplification, the basic issue dividing the proposers of various models for longitudinal development of the spreading axis, (e.g., Ballard *et al.*, 1982; Ballard and Francheteau, this volume; MacDonald, 1982; Allmendinger and Riis, 1979; Crane, 1979) appears to be that of the chicken and the egg. Is segmentation by transform faults a primary event that controls both the extent of the magma chambers and the arrangement of volcanoes along the strike? Or is it longitudinal variation within the magma chamber that not only sets the pattern of volcanic eruptions, but also controls the presence, spacing, and perhaps sense of offset of at least the smaller transform faults? The data at this time leave room for both views and others as well, and underline sharply how much we have neglected the study of real fracture zones and the processes that make them, having perhaps become overconfident in the flush of Tuzo Wilson's elegant solution of the geometric dilemma. A whole range of transforms appears to exist from the highly ephemeral ones that Ballard and van Andel (1977) assumed to exist when a volcanic unit is transferred whole to one side or the other, to the very large ones in the eastern Pacific and central Atlantic whose roots go far back in time, and which seem invulnerable to the modifying influence of any magma chamber. In the intermediate category, where offsets range from a few to a few tens of kilometers, control of spacing and sense by crustal generation processes seems possible, but personally I am reluctant to accept that the sense of motion could be altered as easily and quickly as Ballard and Francheteau (this volume) imply. Clearly, there is much room for further study.

A closely related question is whether tectonic and volcanic change along the axis may be interpreted in temporal terms. Tectonic processes are likely to be more or less continuous, but material balance calculations suggest that, while at fast spreading rates major eruptive cycles should be spaced from 50 to 500 years apart, the spacing at a slow spreading ridge may be as much as 10,000 years. Do we, as we view along the spreading axis, see successive or randomly interspersed snapshots of major volcanic cycles and of the tectonic processes that accompany plate genesis? Do we find here the volcanic constructive phase in full swing, there quiescence, marked only by tension, faulting, and uplift? Ballard *et al.* (1982) have taken this view for a 30 km segment of the Galapagos Rift and have generated a temporal sequence from various segments of the rift valley. Various models for the tectonic evolution of the Mid-Atlantic rift similarly start from the assumption that at different points the rift valley exists in different stages of development. On the other hand, Ballard and Francheteau in this volume emphasize the view that variation along strike represents a spatial change in the behavior of the magma chamber and the extrusive phenomena above it, rather than an evolutionary volcanic history.

It is obvious that these controversies bear directly on one of the central themes of this symposium, because in different models hydrothermal activity is to be expected at different times and in different places. Questions such as where do springs arise, how are they timed relative to the volcanic cycle, are they episodic, how long do they last, are intricately dependent upon models of the volcanic and tectonic evolution of the rift.

The *third major dimension* of the spreading axis is vertical and inevitably closely tied to the fourth, the time dimension, in a much more direct and obvious fashion than for the two horizontal dimensions. In this context the issues (Sleep *et al.*, this volume) are the thickness of the crust above the magma chamber, the width, height, and shape of the chamber, the processes of differentiation, cumulate deposition, and roof underplating that may exist, and the nature of the intrusions that reach upward to form cupolas and dike complexes. Most of all, the burning question is whether there is a permanent magma chamber at all, or only an ephemeral one and, if the latter should prove to be true, how long its lifespan may be and how that affects the genesis of the crust. Here extreme views range from the fairly good geophysical evidence that magma chambers, presumably at least long-lived if not permanent, exist under ridges with moderate to rapid spreading rate (MacDonald, 1982), to the calculations of Lister (this volume) which suggest to him that a magma chamber of any but the shortest duration is impossible at any rate of spreading. The petrological evidence is really compatible only with a magma chamber existing long enough to produce extensive differentiation, but under the Mid-Atlantic Ridge this is at variance with the complete absence of any indication at all that a magma chamber exists or even could exist except briefly. At the moment, with crustal production rates in the Atlantic demanding a recurrence rate of volcanism of about 10,000 years, an intermittent magma chamber seems likely, while for the much shorter recurrence times in the Pacific permanence is indicated. There the matter rests for the time being, until either the geophysical data or theory evolve to the point where they can turn the tide of battle (Sleep et al., this volume).

Inevitably, with theory not altogether helpful, geophysical data yet so sparse, and drilling in the main still too shallow, one turns to ophiolite sections on land for further insight. This, of course, involves the assumption that at least some ophiolites represent ancient oceanic crust, and some discomfort is involved in using this key of the past to open the door of the present. Besides being in conflict with the more traditional reasoning from present to past, such an approach carries with it the risk of forgetting which is key and which is lock. Various papers in this volume address the implications of ophiolite sections for an understanding of the hydrothermal system on active ridges and as a source of information they are too rich to avoid, but caution is indicated.

The vertical dimension includes the nature of the volcanic complex, its contact, as yet not clearly identified, with the underlying sheeted dike zone which has but rarely been observed on an active mid-ocean ridge (DSDP site 504B), and the makeup of the volcanic body in terms of sheet and pillow flows. The models proposed most recently all assign a much smaller role to pillow flows than was customary in the past, and utilize observed lateral and longitudinal variation in the rift valley and what limited access is afforded by fault scarps as their main data base. Deep-sea drill cores so far have been less useful than they will be, in part because many were described before the common occurrence of sheet flows was recognized and in part because the distinction between sheet and pillow flows is not easy nor unambiguous in cores, or in outcrop for that matter.

Which, if any, of these models will turn out to be correct in terms of ratios of lava flow types, sequencing of flow processes, and in modeling lateral and along-strike variations, remains to be seen, but it is not evident that conclusive evidence will come from more field work identical to what has already been accomplished. Nevertheless, for individual eruptive cycles more information can surely be extracted from the fine details of lava flow types visible on bottom photographs (Holcomb and van Andel, in press), but little has been done so far and misinterpretations are not uncommon.

The fourth and last dimension, that of time, has appeared several times already and needs little further discussion. It is clearly part and parcel of the vertical aspects of the spreading axis and almost unfortunately intertwined with it. In the absence of solid stratigraphic data we continue to be perched on the edge of circular reasoning here. As far as tectonic evolution over time is concerned, much may still be learned by extending detailed structural studies beyond the rift valley walls and onto the upper ridge flanks. Better dating of tectonic events there is also necessary, but fortunately the time scale involved ranges from tens to hundreds of thousands of years and the required resolution in the main does not exceed that of conventional isotopic and stratigraphic methods. For the volcanic evolution, where more can still be expected from studies in the rift valley, the dating problem is as pressing but much less hopeful. He who comes up with a suite of techniques that will date lavas ranging in age from a few to thousands of years and spaced apart mere years or decades, will have made a major contribution to our understanding of crustal genesis. Without such a breakthrough in dating one is inclined to suggest that further survey work of the kind so common in recent years is rapidly approaching the point of diminishing returns.

There is little doubt that crustal genesis occurs intermittently on all except perhaps the fastest spreading ridges. It is also evident that each volcanic episode is, in itself, a complex event involving several, perhaps even numerous separate eruptive phases before another period of quiescence sets in. Aside from the importance of these second, and perhaps even third order cycles in terms of understanding crustal genesis, their impact on hydrothermal processes, recurrence rates, and spatial distribution is certain to be large but only very little understood. Lava flow types coupled with high resolution dating techniques constitute one possible approach that has not yet been tried.

One may end this introduction, then, as it began, by marveling at how much we learned in so short a time, and by marveling equally at one more repetition of the old experience that what all this investment of time and resources has done for us is mainly to sharpen the questions, to raise better ones, and to indicate, vaguely as always, that we should think about doing new and different things rather than repeating the approaches that have served us so well in the past.

References

Allmendinger, R.W., and Riis, F., 1979, The Galapagos Rift at 86° W; 1. regional morphological and structural analysis; Jour. Geophys. Res., v. 84, p. 5379-5389.

Anderson, R.N., Honnorez, J., Becker, K., Adamson, A.C., Alt, J.C., Emmermann, R., Kempton, P.D., Kinoshita, H., Laverne, C., Mottl, M.J., and Newmark, R.L., 1982, DSDP Hole 504B, the first reference section over 1 km deep through Layer 2 of the oceanic crust; Nature, v. 300, p. 589-594.

Anderson, R.N., and Noltimier, H.C., 1973, A model for the horst and graben structure of the mid-ocean ridge crest based upon spreading velocity and basalt delivery to the oceanic crust; Geophys. Jour. Roy. Astron. Soc., v. 34, p. 137-147.

Aumento, F., 1970, Improved positioning of dredges on the sea floor; Can. Jour. Earth Sci., v. 7, p. 534-539.

Aumento, F., Loncarevic, B.D., and Ross, D.I., 1971, Hudson Geotraverse: geology of the Mid-Atlantic Ridge at 45° N; Phil. Trans, Roy. Soc. London, ser. A., v. 268, p. 623-650.

Ballard, R.D., and Francheteau, J., this volume, Geologic processes of the mid-ocean ridge and their relationship to sulfide deposition; p.

Ballard, R.D., Holcomb, R.T., and van Andel, T.H., 1979, The Galapagos rift at 86° W: 3, sheet flows, collapse pits, and lava lakes of the rift valley Jour. Geophys. Res., v. 84, p. 5407-5422.

Ballard, R.D., and van Andel, Tj.H., 1977, Morphology and tectonics of the inner rift valley at lat. 36° 50' N on the Mid-Atlantic Ridge; Bull. Geol. Soc. Amer., v. 88, p. 507-530.

Ballard, R.D., van Andel, TjH., and Holcomb, R.T., 1982, the Galapagos rift at 86° W; 5, variations in volcanism, structure, and hydrothermal activity along a 30-km segment of the rift valley; Jour. Geophys. Res., v. 87, p. 1149-1161.

Coleman, R.G., and Hopson, C.A., 1981, (eds), Oman Ophiolite Special Issue; Jour. Geophys. Res., v. 86, p. 2495-2782.

Crane, K., 1979, The Galapagos rift at 86° W; morphological wave forms, evidence of propagating rift; Jour. Geophys. Res., v. 84, p. 6011-6018.

East Pacific Rise Study Group, 1980, Crustal processes of the mid-ocean ridge; Science, v. 213, p. 31-40.

Engel, A.E.J., and Engel, C.G., 1966, Composition of the oceanic crust and the underlying mantle; 2nd Int. Ocean. Congress, Moscow, Abstracts, p. 112.

Ewing, M., and Heezen, B.C., 1956, Some problems of Antarctic sumbarine geology; Antarctica in the I.G.Y., Geophys. Monogr. 1, p. 75-81.

Green, K.E., Von Herzen, R.P., and Williams, D.L., 1981, The Galapagos Spreading Center at 86°W: A detailed geothermal field study; Jour. Geophys. Res., v. 81, p. 979-986.

Hall, J.M. and Robinson, P.T., 1979, Deep crustal drilling in the North Atlantic Ocean; Science, v. 204, p. 573-586.

Holcomb, R.T., and van Andel, Tj.H., in press, Subaerial and submarine lava flows: an illustrated classification; Stanford University Press.

Lachenbruch, A.H., 1973, A simple mechanical model for oceanic spreading centers; Jour. Geophys. Res., v. 78, p. 3395-3417.

Lachenbruch, A.H., 1976, Dynamics of a passive spreading center; Jour. Geophys. Res., v. 81, p. 1883-1902.

Laughton, A.D., and Searle, R.C., 1979, Tectonic processes on slow-spreading ridges; Amer. Geophys. Union, Maurice Ewing Series v. 2, p. 15-32.

Loncarevic, B.D., Mason, C.S., and Matthews, D.H., 1966, The Mid-Atlantic Ridge near 45° N., I. The Median Valley; Can. Jour. Earth Sci., v. 3, p. 327-349.

Lonsdale, P., 1977, Abyssal pahoehoe with lava coils at the Galapagos rift; Geology, v. 5, p. 147-152.

MacDonald, K.C., 1977, Near-bottom magnetic anomalies, asymmetric spreading, oblique spreading, and tectonics of the Mid-Atlantic Ridge near 37° N; Bull. Geol. Soc. Amer., v. 88, p. 541-555.

MacDonald, K.C., this volume, A geophysical comparison between fast and slow spreading centers: constraints on magma chamber formation and hydrothermal activity; p.

MacDonald, K.C., and Luyendyk, B.P., 1977, Deep-tow studies of the Mid-Atlantic Ridge crest near lat. 37° N; Bull. Geol. Soc. Amer., v. 88, p. 621-636.

MacDonald, K.C., and Atwater, T., 1978, Evolution of rifted ridges; Earth Planet. Sci. Lett., v. 39, p. 319-327.

MacDonald, K.C., Becker, K., Spiess, F.N., and Ballard, R.D., 1980, Hydrothermal heat flux of the black smoker vents on the East Pacific Rise; Earth Planet. Sci. Lett., v. 48, p. 1-7.

MacDonald, K.C., 1982, Mid-ocean ridges: fine-scale tectonic, volcanic, and hydrothermal processes within the plate boundary zone; Ann. Rev. Earth Planet. Sci., v. 10, p. 155-190.

Melson, W.G., Thompson, G., and van Andel, Tj.H., 1968, Volcanism and metamorphism in the Mid-Atlantic Ridge, 22° North latitude; Jour. Geophys. Res., v. 73, p. 5925-5941.

Menard, H.W., 1958, Development of median elevations in ocean basins; Bull. Geol. Soc. Amer., v. 69, p. 1179-1186.

Menard, H.W., 1967, Seafloor spreading, topography, and the second layer; Science, v. 157, p. 923-924.

Nelson, K.D., 1981, A simple thermal-mechanical model for mid-ocean ridge topographic variation; Geophys. Jour. Roy. Astron. Soc., v. 65, p. 19-30.

Osmaston, M., 1971, Genesis of ocean ridge median valleys and continental rift valleys; Tectonophysics, v. 11, p. 387-405.

Ramberg, I., and van Andel, Tj.H., 1977, Morphology and tectonic evolution of the rift valley at lat. 36° 30' N, Mid-Atlantic Ridge; Bull. Geol. Soc. Amer., v. 88, p. 577-586.

Rhodes, J.M., and Dungan, M.A., 1979, The evolution of ocean floor basaltic magmas; Amer. Geophys. Union, Maurice Ewing Series, v. 2, p. 262-272.

Sleep, N.H., 1969, Sensitivity of heat flow and gravity to the mechanism of seafloor spreading; Jour. Geophys. Res., v. 74, p. 542-549.

Sleep, N.H., Morton, J.L., and Burns, L.E., this volume, Geophysical constraints on hydrothermal flow at ridge axes; p.

Tapponnier, P., and Francheteau, J., 1978, Necking of the lithosphere and the mechanics of slowly accreting plate boundaries; Jour. Geophys. Res., v. 83, p. 3955-3970.

van Andel, Tj.H., Bowen, V.T., Sachs, P.L., and Siever, R., 1965, Morphology and sediments of a portion of the Mid-Atlantic Ridge; Science, v. 148, p. 1214-1216.

van Andel, Tj.H., and Bowin, C.O., 1968, Mid-Atlantic Ridge between 22° and 23° North latitude and the tectonics of mid-ocean ridges; Jour. Geophys. Res., v. 73, p. 1279-1298.

Williams, D.L., Green, K., van Andel, Tj.H., Von Herzen, R.P., Dymond, J., and Crane, K., 1979, The hydrothermal mounds of the Galapagos rift: observations with the DSRV ALVIN and detailed heatflow studies; Jour. Geophys. Res., v. 84, p. 7467-7484.

GEOLOGIC PROCESSES OF THE MID-OCEAN RIDGE AND THEIR RELATION TO SULFIDE DEPOSITION

Robert D. Ballard and Jean Francheteau

Woods Hole Oceanographic Institution
Woods Hole, Massachusetts 02543 USA

Centre Oceanologique de Bretagne, BP 337
29273 Brest Cedex, France

ABSTRACT

Enough detailed data has been obtained along the axis of the Mid-Ocean Ridge to construct a kinematic model which attempts to describe observed variations in time, space, and spreading rate of volcanic, tectonic, and hydrothermal processes and their relationship to the deposition of massive sulfide deposits. Each accretionary segment or accretionary cell bounded by two transform faults is underlain by a unique magma reservoir. Because of cooling at the tranform fault edges, the magma reservoir pinches out and is most developed at a location away from the transform faults. Where the reservoir reaches its fullest development, there is a higher heat flux into the overlying crustal lid. The shallow region where the magma reservoir is most developed and where the crustal lid is thinnest should correspond to the most vigorous hydrothermal activity because of the higher energy content in the system at shallow depths. Because the crustal lid, at the high, is at its minimum thickness, rifting of the lid should result in lava flows having the most direct and shortest path from the magma reservoir to the young seafloor along the rifting axis. At least in the case of moderate to fast spreading segments, the flows nearest the topographic high should be copious surface-fed fluid lavas with the ratio of fluid to pillow flows decreasing down the topographic gradient above a domain of lid thickening. The farther from the topographic high, the more distal and channelized the flows would be resulting in more tube-fed pillow flows. We propose that areas of shallow sea floor at the axis of the Mid-Ocean Ridge are used as a prospecting tool in the search for active sulfide deposition along any given accretion segment. The model draws upon submersible and towed camera mapping programs in the FAMOUS area of the Mid-Atlantic Ridge where the spreading rate is approximately 2 cm/yr. In addition, recent efforts along various segments of the East Pacific Rise and Galapagos Rift have the most profound impact on the model as they involve spreading rates ranging from 6.0 to 10.2 cm/yr where hydrothermal circulation and associated massive sulfide deposition is well developed.

INTRODUCTION

Since their discovery in 1978 and 1979 at 21°N (Figure 1) on the East Pacific Rise (EPR), massive sulfides have received a considerable amount of attention (CYAMEX, 1979; RISE Project Group, 1980). The existing data about these deposits, however, is limited and is being severely stretched to answer the many

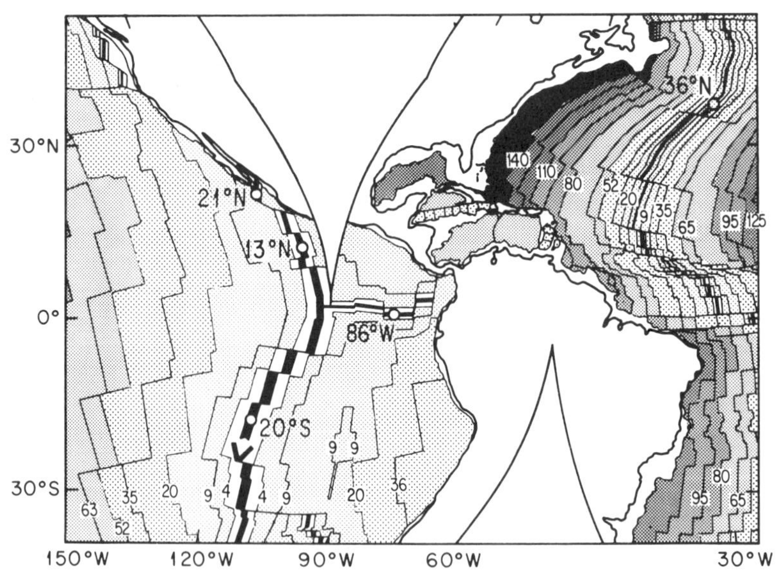

Fig. 1 The location of the study areas mentioned in the text. These include the FAMOUS area at 36°N on the Mid-Atlantic Ridge (2 cm/yr); the Galapagos Rift at 86°W (7.0 cm/yr); and the East Pacific Rise at 21°N (6 cm/yr), 13°N (10.2 cm/yr), and 20°S (16-18 cm/yr). This last site is discussed in detail in Francheteau and Ballard (in press). The source for the figure is from Sclater et al., 1981.

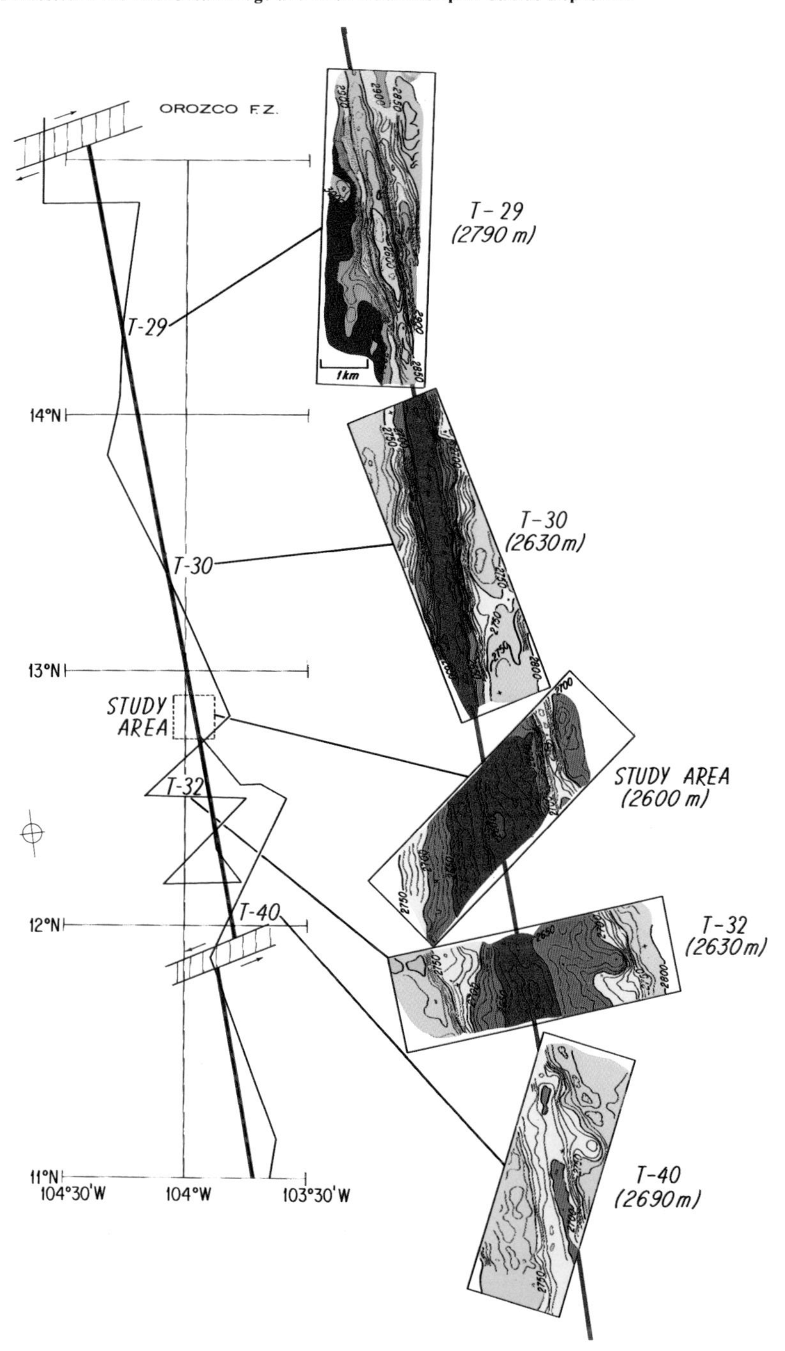

Fig. 4. SEABEAM coverage during SEARISE I cruise by the R/V JEAN CHARCOT. The black line is the ship track while the red line gives the location of the EPR axis between 11° and 15°N based upon the SEABEAM coverage. T-# is line designation beginning with the onset of the cruise. The lines shown in the figure go from T-28 to T-42, in addition to the detailed study area which received total coverage using acoustic transponder navigation. Figures on right are selected crossings of the EPR axis. The number in parentheses is the minimum depth of the axis during that crossing, in corrected meters. This depth decreases away from the Orozco Fracture Zone to the Study Area and then increases again.

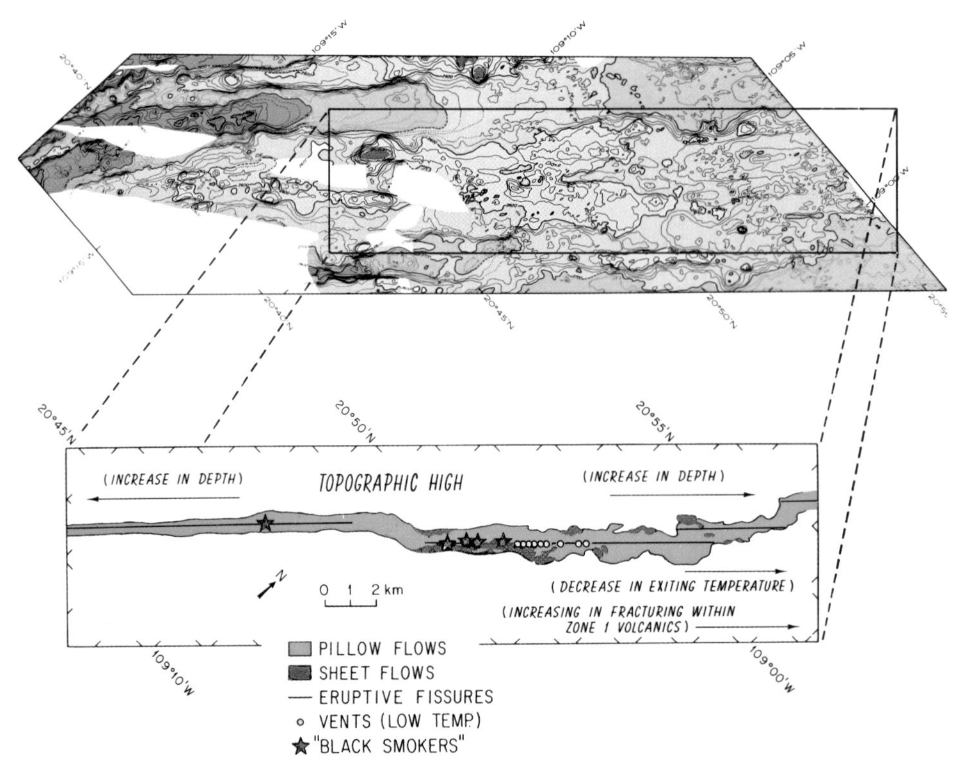

Fig. 5. (A) Bathymetric map of the EPR near 21°N based upon acoustically navigated SEABEAM coverage by the JEAN CHARCOT during 1980 SEARISE I cruise. Depth is in corrected meters. The black rectangle outlines the region represented by Figure 5B. (B) Simplified geology of Zone 1 volcanic belt showing dominance of pillow flows away from topographic high. The hydrothermal spring temperatures are 350°C ("black smokers") near topographic high, decreasing where observed to the northeast away from high. Similar changes in fracture density are also noted.

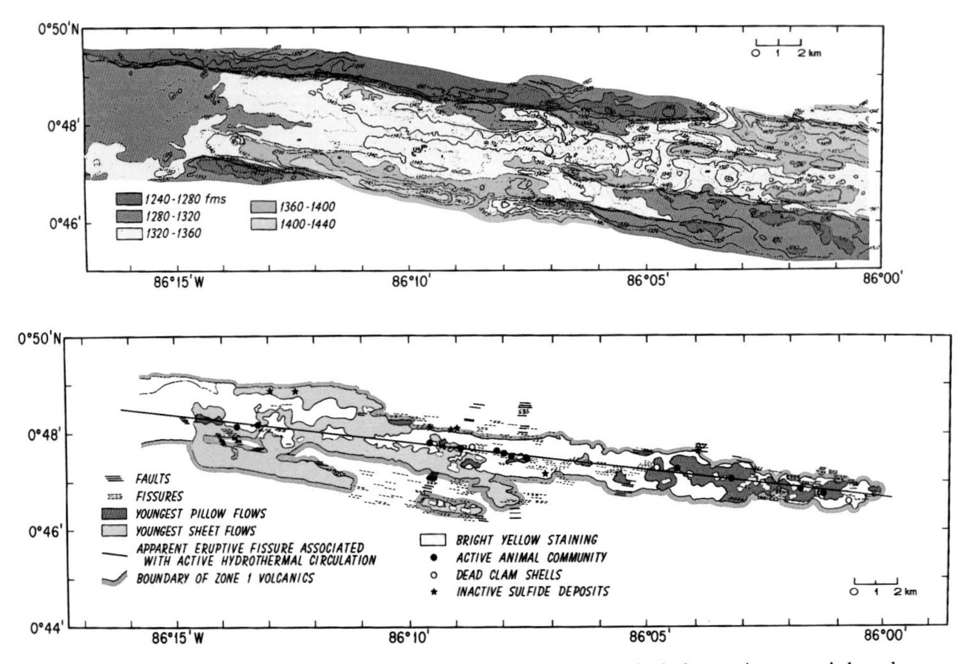

Fig. 6 (A) Bathymetric map of the Galapagos Rift at 86°W contoured in fathoms since map is based upon multi-narrow beam survey conducted by the Navy using U.S.N.S. DUTTON. More detailed revision found in Ballard et al. (1982). (B) Simplified geologic map of Zone 1 volcanic belt also found in Ballard et al. (1982). As in Figure 5, sheet/pillow flow ratio increases toward the topographic high, which in this case is to the west. Similar variations in exiting temperatures and fracture density also occur.

Fig. 5A. Bottom photograph of an active hydrothermal site located in the central graben of the Rise Crest near 12°50′N on E.P.R. at a depth of 2624 m (area "A") and visited by the submersible CYANA during dive CY 82-25 (Figs. 2, 3) (Hekinian et al., in press). Several small (about 1 m in height) chimneys are found to occur at the foot of a tall sulfide edifice. On 19 February 1982, the chimney was degraded and a measuring device made of stainless steel chain with markers every 10 cm was put on the top of the orifice, from which black fluids were exiting. CYANA's mechanical arm and the temperature probe are seen in the picture. Temperature measurements made on the chimney (with the chain) have a value of 232-270°C.

Fig. 5B. Bottom photograph of the "chain site" as revisited on 24 February 1982 during dive CY 82-30 (Hekinian et al., 1983). The marker chain is imbedded in an overgrowth (40 cm) of the chimney. On the far left of the picture there is another sulfide edifice topped with a dark-grey cone of powdery material built since the previous visit (dive CY 82-25) at the site. Diffused dark grey fluid was seen to exit from this cone.

Fig. 8A. Bottom photograph taken by the submersible CYANA during dive CY 82-31 at the "Bryce Canyon" site located near 12°47′N (area "B", Figs. 2, 3) at a depth of 2629 m. Notice the submersible mechanical arm reaching an active chimney (about 2 m in height) set at the foot of a tall edifice (about 4 m in diameter, right side of picture). This chimney was broken into two pieces, the top part was numbered CY 82-31-08 and the middle section represents sample CY 82-31-09 (Table 2).

Fig. 8B. Bottom photograph taken by CYANA during dive CY 82-31 at the "Bryce Canyon" site located near 12°47′N (area "B" in Fig. 2) on the E.P.R. at a depth of 2630 m. As many as four active chimneys (< 1 m in height) set on the top of a mound show discharge of black hydrothermal fluid at a temperature of 327°C. The mechanical arm of CYANA is reaching to sample the active chimney seen at the left front of the picture (sample CY 82-31-10). This sulfide edifice occurs less than five meters away from that where sample CY 82-31-09 was taken (Fig. 7A, 4).

J. Frederick Grassle
Introduction to the Biology of Hydrothermal Vents

Fig. 1. Dense mats of bacterial filaments belonging to the genus *Beggiatoa* growing on sediments in the Guaymas Basin.

Holger W. Jannasch
Microbial Processes at Deep Sea Hydrothermal Vents

Fig. 1. "Milky-bluish" water of ca. 15°C (as measured about 0.5 m above the bottom) emitted from a vent surrounded by various invertebrates (Musselbed, Galapagos Rift; Dive 887) (Photo by J. F. Grassle).

Fig. 1. Rose Garden, showing the flats adjacent to the large, central vestimentiferan/mussel thicket (background). Subsidiary vent fissures are clogged with mussels, which are replaced by clams (foreground) as the fissure dwindles. Translucent anemones dominte the surface of the flats (Photo by K. Crane)

Fig. 2. Rose Garden. Closeup of central vestimentiferan/mussel thicket. Mussels are numerous, particularly lower down (lower right); old byssal attachments are common on the vestimentiferan tubes. A few detritus-feeding limpets are seen on the mussels. *Bythograea* and *Alvinocaris* are both abundant. (Photo by R. Hessler)

Fig. 3. Rose Garden. Basalt flats a few meters away from the central vestimentiferan/mussel thicket. Irregular fissures with clams and mussels run through the field of translucent anemones with scattered mussels. A whelk is located front center. (Photo by R. Hessler)

Fig. 4. Rose Garden. Basalt flats with translucent anemones. Temperature 2.36°C. (Photo by R. Hessler)

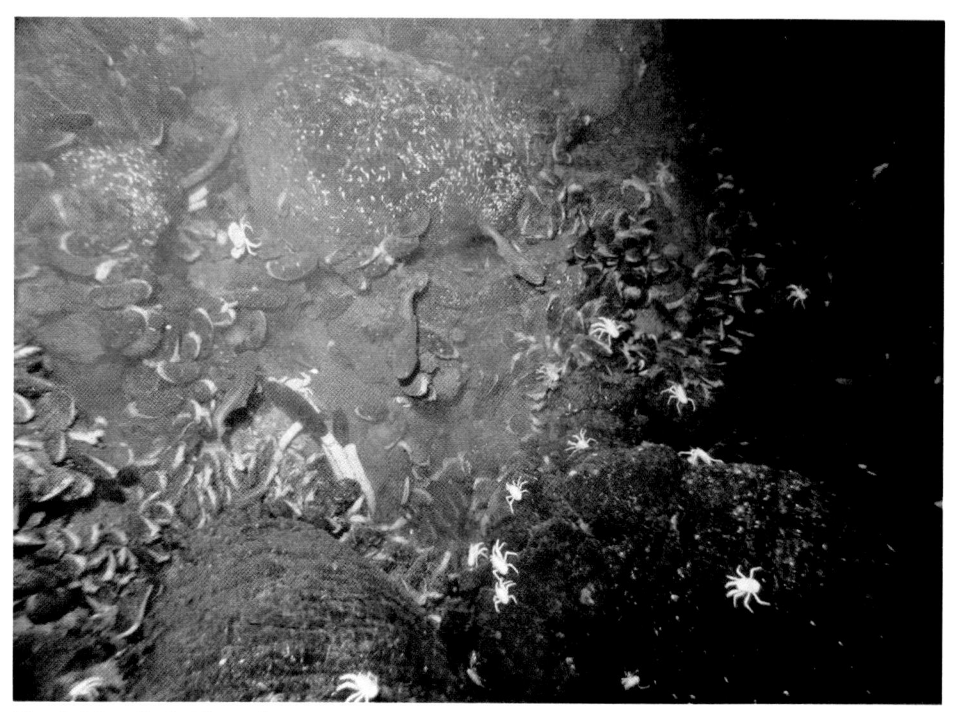

Fig. 5. Musselbed. The main, nearly vertical fissure occluded with mussels. The smoky cast is from venting water. Serpulids (top left and center) and galatheids (lower right) are abundant on adjacent pillow basalt.

Fig. 6. Musselbed. Detail of the main, mussel clogged vent. Mobile, deposit feeding limpets and *Alvinocaris* are abundant, along with one *Bythograea*. The mussel valves are covered with old byssal attachments. (Photo by R. Hessler)

questions being asked of it by land-based ore specialists as well as various private and governmental groups interested in the potential commercial significance of these deposits.

To answer many of these questions requires a broad understanding of the geologic processes occurring along the axis of the Mid-Ocean Ridge (MOR), processes we are only now beginning to understand. In addition, the majority of the literature treats the MOR as a steady-state system with crustal accretion occurring along a line of zero width in a perfectly symmetrical and two-dimensional manner. Although such a treatment has helped us understand the importance of age in the evolution of the lithosphere, it is a gross simplification.

As detailed mapping programs have continued in various segments of the MOR using manned submersibles and towed camera systems, significant variations in time, space, and spreading rate have been observed. Initially, these studies dealt with the processes of volcanism and tectonics, stressing their episodic nature in contrast to the steady-state model (ARCYANA, 1975; Moore et al., 1974; Ballard and van Andel, 1977; Lonsdale, 1977). Later studies began to deal with observed hydrothermal activity and its relationship to the volcanic cycles (van Andel and Ballard, 1979; Ballard et al., 1982; CYAMEX, 1981; Ballard et al., 1981).

Most recently, Francheteau and Ballard (in press; Ballard and Francheteau, 1982) have reviewed the results of these previous efforts as well as additional field programs just completed, presenting a kinematic model which seeks to explain the axial processes and their relationship to massive sulfide deposition. The purpose of this paper is to review those results.

Proposed Model

Figure 2 illustrates the model being proposed. This model deals with individual spreading segments isolated from others by offsetting transform faults. According to Sleep and Biehler (1970), the intersection of the spreading axis with an adjoining transform fault is characterized by a topographic depression. This depression is the result of increased cooling with the offset of crust placing colder and older material next to the spreading axis of zero age. The result is a loss in viscous head in the upwelling mantle material, causing the floor of the rift to rest at a greater depth. This cooling phenomenon at both ends of any given spreading segment results in a thermal gradient along the strike of the spreading cell, forming a single, thermally elevated topographic high (Figure 2). At this high, the vertical thermal gradient between seawater and the underlying magma chamber is the greatest. In other words, the rigid lid overlying the magma chamber at the topographic high should be at its minimum

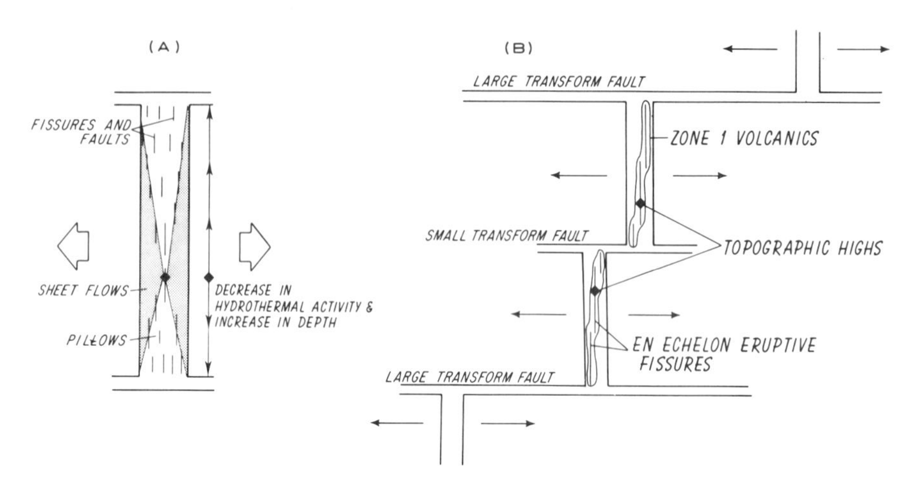

Fig. 2 Diagrams of the proposed model by Francheteau and Ballard (in press). (A) The variations observed in the volcanic, tectonic, and hydrothermal processes occurring along the strike of each individual spreading segment within the Zone 1 volcanic region of crustal accretion. The sheet/pillow flow ratio decreases away from the topographic high toward the adjacent transform fault intersection while the density of brittle fracturing increases. The degree of hydrothermal activity also decreases towards the intersections with massive sulfide deposition most probable at the topographic high. (B) Position of topographic high relative to series of offsetting transform faults of varying size. Lava flows within Zone 1 volcanic belt of zero-age associated with series of eruptive fissures arranged in en echelon pattern. Typically, but not always, the polarity of this pattern is the same as the adjacent transform faults.

thickness. As a result, rifting of the lid at or near the topographic high should result in lava flows having the most direct and shortest path from the chamber to the eruptive fissures along the rifting axis. In the case of moderate to fast spreading segments, the flows nearest the topographic high should be copious sheet flows with the ratio of sheet flows to pillow flows decreasing down the topographic gradient towards the transform intersections (Figure 2). The farther from the topographic high, the more distal and channelized the flows would be, resulting in more tube-fed pillow flows. The thinness of the overlying lid at the topographic high would also mean that tensional stresses, in the form of open fissures, would be more prevalent away from the topographic high where the lid should be thicker. Rifting near the topographic high would have a greater tendency to tap the chamber. Open fissures would therefore have a relatively shorter resident time and greatly reduced density (Figure 2). If the maximum thermal gradient in the overlying lid occurs at the topographic high, this is where one would also expect to find hydrothermal activity at its maximum. This would also mean that the topographic high should be the best site to find active massive sulfide deposition along any given spreading segment.

Data Base for Proposed Model

The model previously presented is based upon a series of mapping programs using French and American submersibles and unmanned towed camera systems in a series of cruises conducted along segments of the MOR having a wide range of spreading rates (Figure 2). The principal sources of information are the FAMOUS area of the MAR (ARCYANA, 1975; Ballard and van Andel, 1977; spreading rate of 2.1 cm/yr), Galapagos Rift (van Andel and Ballard, 1979; Ballard et al., 1982; spreading rate of 7 cm/yr), EPR at 21°N (CYAMEX, 1981; Ballard et al., 1981; spreading rate of 10.3 cm/yr), and EPR at 13°N (Francheteau and Ballard, in press; spreading rate of 16 cm/yr).

Existence of Single Topographic High. This aspect of the model is based upon multi-narrow beam maps of the FAMOUS area by Phillips and Fleming (1978; Figure 3), a SEABEAM survey between 10° and 15°N on the EPR by the JEAN CHARCOT (Francheteau and Ballard, in press, Figure 4), partial mapping efforts of the EPR at 21°N (Francheteau and Ballard, in press, Figure 5), and the Galapagos Rift at 86°W (Ballard et al., 1982, Figure 6).

Figure 3 is a simplified version of the Phillips and Fleming (1978) data base showing the presence of a single high along the strike of four spreading segments (dotted pattern), while the diagonal pattern denotes the topographic lows situated at all spreading axis/transform fault intersections. Another aspect of the model is the suggestion that the topographic high of any spreading segment will be displaced towards the smaller of the two adjoining transform faults. This is the case in the AMAR rift, while the others are approximately equidistant.

The best example of an asymmetrical shift toward the smaller transform is evidenced in the 10-12°N region of the EPR (Figure 4). Here, SEABEAM crossings by the CHARCOT clearly demonstrate the axis of the ridge rising slowly from the Orozco fracture zone, reaching a topographic high near 12°50'N, and then plunging down toward the smaller transform at 12°N.

Figures 5 and 6 show the topography for only a portion of the spreading segments of the EPR at 21°N and Galapagos Rift at 86°W,

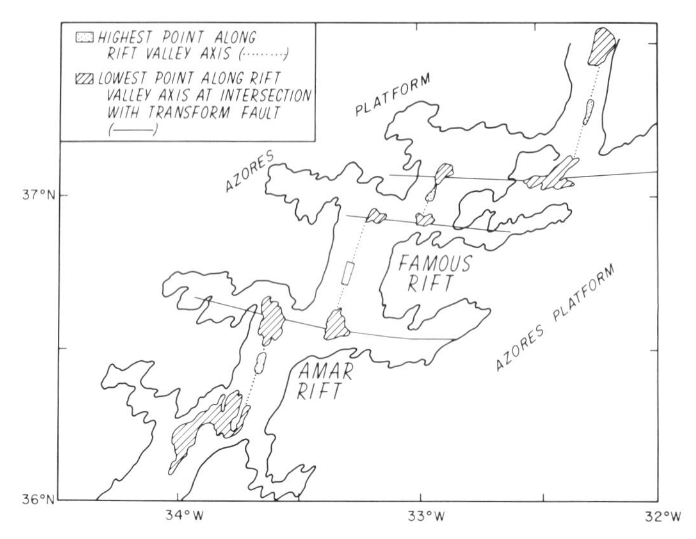

Fig. 3 Simplified spreading geometry of the FAMOUS area based upon multi-narrow beam data published by Phillips and Fleming (1978). The diagonal lines denote the topographic lows found at the intersections of transform faults (solid lines) and spreading segments (dotted lines). The dotted areas are the topographic highs.

Figures 4, 5, and 6 will be found in the color insert following page 18.

respectively. In the case of Figure 5, the topographic high is contained within the SEABEAM survey area, while in Figure 6 the SASS coverage shows a continuous gradient from east to west with the topographic high off to the west.

Variations Along Strike. Such variations are drawn primarily from our work in the Galapagos Rift (Ballard et al., 1982, Figure 6) and to a lesser degree at 21°N (Ballard et al., 1981, Figure 5). The mapping program at the Galapagos Rift by ANGUS and ALVIN began in 1977 following earlier work by Lonsdale (1974) and Crane (1978). The major difference between this spreading center and the slower spreading FAMOUS area described by Ballard and van Andel (1977) was the occurrence of copious sheet flows. An initial interpretation of the relationship between sheet flows and pillow flows was given by Ballard et al. (1979). Here, sheet flows are considered to be the submarine equivalent of surface-fed pahoehoe flows. Pillow flows are analogous to subaerial tube-fed pahoehoe flows. In that manuscript, emphasis was placed upon the degree of channelization within the flow with sheet flows expressing a more simplistic or direct distribution system commonly characterized by higher eruptive rates. In Ballard et al. (1981 and 1982, Figures 5 and 6), it was shown that these two basic flow types exhibit a systematic along-strike relationship keyed to the topographic gradient. Best seen in Figure 6, the sheet flow and pillow flows were expressed as a ratio with the sheet/pillow ratio increasing toward the topographic high. In the model by Francheteau and Ballard (in press), the magma chamber beneath the ridge is thought to have finite dimensions along strike with the topographic high being where its overlying lid is thinnest. Due to this thinness of the lid, the high should be a region where faulting should easily tap the chamber, leading to a short and direct distribution system; a distribution system which should favor the occurrence of sheet flows, resulting in a high sheet/pillow flow ratio. If the lid thickens toward the transform intersection, its brittle character should increase as well as resulting in a higher density of fractures and open gjar in that direction.

Variation in the exiting temperatures of the hydrothermal fluids along the strike of the rift was also observed in the Galapagos Rift (Ballard et al., 1982) and EPR at 21°N (Ballard et al., 1981). In the latter case, this difference was small, ranging from 8°-22°C from east to west, while at 21°N it varied from 8°-350°C along strike. In both cases, the temperature increased toward the topographic high as would be expected.

Additional Speculation

The model by Francheteau and Ballard (in press) reflects small wavelength variations along strike on the order of tens to hundreds of kilometers. Much less is known about larger wavelength anomaly patterns. Figure 7 presents the geometry of the EPR between 5°N and 25°S based upon GEBCO map sheets 5.07 (5th Edition, March 1982) and 5.11 (5th Edition, March 1980). Based upon an analysis of this data base, which has a varying contour interval between 100 and 500 meters, an effort has been made to resolve the spreading geometry and determine the minimum depth along each of 14 individual spreading segments. Although this is still preliminary and will be significantly modified by recent and on-going SEABEAM survey efforts, it does present an interesting picture. Beginning at the Galapagos Triple Junction near 2°N, the topographic highs of these spreading segments show a systematic increase in elevation up to 18°S, at which location the gradient reverses itself, stepping down to 22°S where it nears the complex area of the Easter Island microplate and "hotspot." Given this longer wavelength pattern of the EPR which contains many individual spreading segments, one can't help but think that the regional high at 17-18° might be a likely site for finding extensive sulfide deposits.

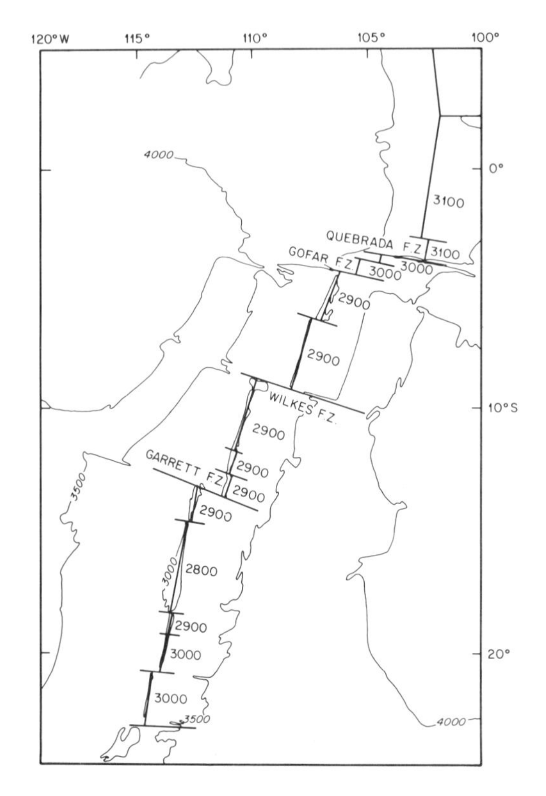

Fig. 7 Simplified bathymetric map of EPR between 5°N and 25°S based upon GEBCO sheets 5.07 and 5.11. Heavy lines are the inferred spreading segments offset by adjacent transform faults. Number associated with each spreading segment is the minimum depth of that segment along the EPR axis based upon GEBCO data base which has a contour interval which varies between 100 and 500 meters. For the axis depths, this interval was 100 meters.

REFERENCES

ARCYANA, 1975, Transform fault and rift valley from bathyscaph and diving saucer, Science, 190:108-116.

Ballard, R.D. and Francheteau, J., 1982, The relationship between active sulfide deposition and the axial processes of the Mid-Ocean Ridge, J. Mar. Tech. Soc., 16:8-22.

Ballard, R.D., Francheteau, J., Juteau, T., Rangan, C., and Normark, W., 1981, East Pacific Rise at 21°N: the volcanic, tectonic, and hydrothermal processes of the central axis, Earth and Plan. Sci. Letters, 55:1-10.

Ballard, R.D., Holccomb, R.T., and van Andel, Tj. H., 1979, The Galapagos Rift at 86°W: 3. Sheet flows, collapse pits, and lava lakes of the rift valley, J. Geophys. Res., 84:5407-5422.

Ballard, R.D., Morton, J., and Francheteau, J., 1981, Geology and high temperature hydrothermal circulation of ultra-fast spreading ridge: East Pacific Rise at 20°S, EOS, 62:912.

Ballard, R.D., van Andel, Tj. H., and Holcomb, R.T., 1982, The Galapagos Rift at 86°W: 5. Variations in volcanism, structure, and hydrothermal activity along a 30 km segment of the rift valley, J. Geophys. Res., 87:1149-1161.

Ballard, R.D., and van Andel, Tj. H., 1977, Project FAMOUS: Morphology and tectonics of the inner rift valley at 36°50'N on the Mid-Atlantic Ridge, Geol. Soc. Amer. Bull., 88:507-530.

Crane, K., 1978, Structure and tectonogenesis of the Galapagos inner rift, 86°10'W, J. Geol., 86:715-730.

CYAMEX Scientific Team, 1981, First manned submersible dives on the East Pacific Rise 21°N (Project RITA): General results, Mar. Geophys. Res., 4:345-379.

CYAMEX Scientific Team, 1979, Massive deep-sea sulphide ore deposits discovered on the East Pacific Rise, Nature, 277:523-528.

Francheteau, J. and Ballard, R.D., in press, The East Pacific Rise near 21°N, 13°N and 20°S: inferences for along-strike variability of axial processes of the Mid-Ocean Ridge, Earth and Plan. Sci. Letters.

Lonsdale, P., 1977, Structural geomorphology of a fast-spreading rise crest: The East Pacific Rise near 3°25'S, Mar. Geophys. Res., 3:251-294.

Lonsdale, P.F., 1974, Abyssal pahoehoe with lava whorls at the Galapagos Rift, Geology, 5:147-152.

Moore, J.G., Fleming, H.S., and Phillips, J.D., 1974, Preliminary model for the extrusion and rifting at the axis of the Mid-Atlantic Ridge, 36°48'North, Geology, 2:437-450.

Phillips, J.D., and Fleming, H.S., 1978, Multibeam sonar study of the Mid-Atlantic Ridge rift valley, 36-37°N, Geol. Soc. Am., Map Series, MC-19.

RISE Project Group, 1980, East Pacific Rise: Hot springs and geophysical experiments, Science, 207:1421-1433.

Sclater, J.G., Parsons, B., and Jaupart, C., 1981, Oceans and continents: similarities and differences in the mechanisms of heat loss, J. Geophys. Res., 86:11535-11552.

Sleep, N., and Biehler, S., 1970, Topography and tectonics at the intersections of fracture zones with central rifts, J. Geophys. Res., 75:2748-2752.

van Andel, Tj. H., and Ballard, R.D., 1979, The Galapagos Rift at 86°W: 2. Volcanism, structure, and evolution of the rift valley, J. Geophys. Res., 84:5390-5406.

A GEOPHYSICAL COMPARISON BETWEEN FAST AND SLOW SPREADING CENTERS: CONSTRAINTS ON MAGMA CHAMBER FORMATION AND HYDROTHERMAL ACTIVITY

Ken C. Macdonald

Department of Geological Sciences
Marine Science Institute
University of California
Santa Barbara, California 93106

ABSTRACT

While there are many similarities in the geologic structure of various spreading centers, there are some important differences which appear to be related to spreading rate. Taking recent studies of the fast-spreading East Pacific Rise and slow-spreading Mid-Atlantic Ridge, I have compiled a list of properties (summarized in Table 1) which distinguish the two spreading centers. These studies include seismic reflection and refraction, microearthquake studies, gravity and magnetic measurements, electromagnetic sounding, thermal models, observations of hydrothermal activity and geomorphic/tectonic studies. For each of the contrasting properties listed in Table 1, I briefly explain in the text the origin of the observation and emphasize possible limitations and areas of disagreement in the associated interpretation. Many of the important contrasts between fast and slow spreading centers are related to the increased thermal budget at fast spreading rates which allows for the maintenance of a steady state chamber along most of the length of the rise. At slow spreading rates, the axial magma chamber may persist at distances greater than 15-20 km from transform fault intersections, but given the finite width and spacing of transform faults on slow spreading centers, the axial magma chamber may be transient along most of the ridge's length. The mechanics and deformation of the lithosphere are also affected by the thermal budget as manifested by along strike topographic continuity, transform fault spacing and style of deformation along spreading center offsets. High temperature vents (350°C) are common at intermediate to fast spreading rates, but may be rare occurrences at slow speading rates (with the possible exception of the Reykjanes ridge and other hot-spot influenced ridges). At most slow-spreading ridges, the frequency and duration of such hydrothermal events may not be adequate to sustain the chemosynthetic benthic faunal communities which thrive on faster spreading centers.

INTRODUCTION

Exploration of the mid-ocean ridge (MOR) system in the last decade has revealed significant variations in tectonics, volcanism, hydrothermal activity and structure which may be related to differences in spreading rate. As early as 1967, Menard pointed out significant topographic differences between the Mid-Atlantic Ridge (MAR) and East Pacific Rise (EPR) which might be related to differences in spreading rate and thermal budget (Menard, 1967). Sleep (1969) pointed out that the dynamics of MOR's might be spreading rate dependent as well, whereby a deep rift valley is created at slow spreading rates by a viscous head loss mechanism, while a smooth, non-rifted rise is characteristic of fast spreading rates. In general, the axial rift valley is pronounced at slow total spreading rates of 1-5 cm/yr, very small (< 200 m deep) at intermediate spreading rates of 5-9 cm/yr, and disappears at fast spreading rates of 9-18 cm/yr, where it is replaced by a volcanic high with a shield volcano-like cross section.

While some of the fine-scale development of the axial neovolcanic zone and flanking tectonic zones is remarkably similar regardless of spreading rate (e.g., van Andel and Ballard 1979; Macdonald, 1982), marked differences are revealed by measurements of microearthquake activity, seismic refraction and reflection, magnetic anomalies, gravity, heat flow and geomorphology. In this paper I shall focus on and review the differences which appear to be spreading-rate related. Relative to fast spreading rates, oceanic crust generated at slow spreading rates is very heterogeneous in its physical properties as well as irregular and non-linear in its structure and geomorphology. These fundamental differences may be related to an axial magma chamber which is steady state throughout most of its length at fast spreading centers, but transient along most of its length at slow spreading centers. As a result crustal generation processes will be highly episodic and discontinuous at slow spreading rates compared to fast. These differences are compounded at slower spreading rates by the greater amplitude of tectonic deformation due to rift valley formation. The longevity and along strike continuity of the axial magma chamber may also influence the character of hydrothermal activity in the axial zone so that high temperature hydrothermal activity on the seafloor may be common at fast rates but rare at slow rates. I shall now briefly review recent studies involving seismic refraction and reflection, microearthquakes, gravity, magnetic anomalies, heat flow, electromagnetic sounding, thermal modeling, geomorphology, and spreading center offsets, in that order. In each case I shall mention the flaws and ambiguities in the data sets and interpretations. The results are simplistically summarized in Table 1.

Seismic Refraction and Reflection

A key result from seismic refraction studies on intermediate to fast spreading centers is the detection of a low velocity zone (LVZ) beneath the spreading axis. The first clear seismic evidence for a LVZ beneath a spreading center came from ocean bottom seismometer refraction studies on the axis of the EPR near 9°N. The primary evidence here is the presence of a significant shadow zone beginning at ranges of 15 km in which wave amplitudes are significantly attenuated and travel times are delayed (Orcutt et al. 1976; Rosendahl et al. 1976). Such data require a LVZ which is interpreted as a shallow crustal magma chamber beneath the spreading axis. Subsequent seismic studies have indicated axial magma chambers beneath the EPR near 22°N (McClain and Lewis, 1980), 21° N (Reid et al., 1977), 15°N (Trehu, 1981), 13°N (Orcutt et al., in press), 12°N (Lewis and Garmany, in press), 10°S (Bibee, 1979) and the Galapagos spreading center near 86° W (Bibee 1979). The one study area where evidence for an axial magma chamber is equivocal is the Juan de Fuca Ridge near 45°N, which is spreading at 2.9 cm/yr (Jung and Lewis, 1982; Morton et al., 1982). The original refraction results at EPR 9°N are substantiated by multi-channel seismic reflection work (Herron et al., 1980). A strong reflector appears at 2 km sub-basement corresponding to the 2 km-thick lid overlying the magma chamber found by Orcutt and others. In addition, reprocessing of the reflection data suggests that the reflection off the magma chamber is phase-shifted 180° (Hale et al., 1982). If so, this is strong evidence for a magma chamber containing partial melt at 2 km depth. Thus seismic evidence demonstrates the presence of a LVZ at eight out of nine carefully studied fast- and intermediate-rate spreading centers, which is interpreted as evidence for a shallow axial magma chamber.

In contrast numerous seismic refraction studies over the slow spreading MAR have not provided clear evidence for an axial magma chamber (Keen and Tramontini, 1970; Poehls, 1974; Whitmarsh, 1975;

Table 1. Comparison Summary: Fast vs. Slow Ridges (see text for explanation).

	Fast, Intermediate	Slow
1. Axial low velocity zone	Yes	No
2. Seismic reflection off magma chamber roof	Yes	No measurement
3. Axial low Q zone	Yes	No
4. Microearthquake focal depths	Max. 3 km along axis	Max. 10 km along axis
5. Harmonic tremors	Observed beneath axial zone	Not observed yet
6. Axial gravity anomaly	Magma chamber interpretation	No magma chamber interpretation, mechanically strong lithosphere in axial zone
7. Magnetic anomaly quality	Clear, two dimensional	Unclear, variable, very limited two-dimensionality
8. Magnetic polarity transition widths	Narrow, sharp	Highly variable
9. Inferred eruption rates	~50-500 yrs.	~1000-10,000 yrs.
10. High temp. (300-350°C) hydrothermal activity	Commonly observed on seafloor	Rare, not observed yet (may occur at depth within crust)
11. E-M sounding, resistivity	Pillow layer conductivity 10^3 greater than lower crust, vigorous shallow convection inferred	No deep measurements
12. Thermal models	Most models support a steady state magma chamber	Most models suggest a transient magma chamber

	Fast, Intermediate	Slow
13. Depth of axial zone	Smoothly varying	Highly variable, rough
14. Axial neovolcanic zone	High degree of continuity along strike (en echelon for intermediate rates)	Highly discontinuous
15. Topographic amplitude in axial zone	Small, 50-100 m.	Large, 100-2000 m
16. Transform fault spacing	90 km (most are OSC's)	50 km
offset > 30 km	500 km	170 km
17. Maintenance of transform fault pattern	Unstable for small offsets overlapping spreading centers common	Stable
18. Propagating rifts	Common	Rare
19. Petrologic data	Steady state magma chamber inferred, in agreement with most geophysical data	Steady state magma chamber inferred, in conflict with most geophysical data

Fowler, 1976; Nisbett and Fowler, 1978; Bunch and Kennett, 1980; Purdy et al., 1982). These experiments include four attempts in the FAMOUS area at 37°N, two at 45°N, one at 23°N, and one on the Reykjanes ridge.

Problems: On the slow-spreading MAR severe limitations are imposed by the closely spaced offsets of the spreading center transform faults (~50 km). It is difficult to find a long enough spreading segment to resolve deep crustal structure. (The experiment by Purdy and others (1982), however, was performed on a long ridge segment at 23°N and the oceanic crust appears normal, i.e. no magma chamber, throughout the length of the rift valley.) In addition, rift valley topography is large in amplitude and makes travel time corrections difficult. It is also possible for a magma chamber to exist at slow spreading rates which would be seismically undetectable if it were narrower than a seismic wavelength (approximately 1 km).

For both fast and slow-spreading centers, most of the data sets are not dense enough to escape the assumption of lateral homogeneity in the inversion modeling. Furthermore, since a LVZ cannot be established by travel time inversion alone, ray tracing and synthetic seismogram modeling is used in a forward modeling sense. Thus real "proof," and careful error and credibility bounds are difficult to assign to the LVZ determinations. A very thorough experiment has been completed recently on the EPR near 13°N in which data density is high enough to avoid many of the earlier assumptions (Orcutt et al., in press). Preliminary analysis of this recent experiment indicates an axial magma chamber which is approximately 12 km wide and whose roof is approximately 3 km deep, substantiating the results of previous studies.

Existing studies have not constrained the dimensions and shape of the chamber very well. The magma chamber is determined to be less than 10-20 km wide in most studies, but how much narrower it might be is not known (except in the EPR 13°N study). Its shape, whether it be triangular, funnel-shaped or a "blob" is not known and this would have profound effects on the relation between the magma chamber and the stratigraphy of the crust (Pallister and Hopson, 1981; Dewey and Kidd, 1977; Macdonald, 1982). There is some controversy as to whether the magma chamber is only 2 km subbottom or up to 6 km deep (Orcutt et al., 1976; Bibee, 1979). However, there is little controversy as to the existence of an axial magma chamber beneath eight out of nine intermediate to fast spreading centers which have been carefully studied. The Juan de Fuca Ridge is the one case where evidence for a magma chamber is highly equivocal. Its morphology and spreading rate of 2.9 cm/yr places it near the transitional realm between slow and intermediate rate spreading centers. In addition it is very close to the Cobb propagating rift which may disturb the local thermal structure and continuity of the axial magma chamber (Johnson et al., in press).

Microearthquake Seismicity

Microearthquake studies reveal at least three important pieces of information about axial volcanic, tectonic and hydrothermal activity: 1) delineation of zones of low Q where S waves are severely attenuated, perhaps by an axial magma chamber, 2) limits on focal depths which place implicit bounds for the depth of hydrothermal circulation and thickness of the brittle crustal lid overlying the magma chamber, and 3) unusual seismic events such as harmonic tremors, which may be caused by magmatic and/or hydrothermal activity.

During a microearthquake experiment on the EPR near the Rivera fracture zone, several earthquakes were detected on the rise axis at a range of 50 km. For seismic ray paths along the axis of the EPR, shear waves were severely attenuated. For a path from the earthquake epicenter to an ocean bottom seismometer only 5 km off-axis, the shear wave arrival was prominent (Reid et al., 1977). The small azimuthal extent of the array relative to the events makes radiation pattern an unlikely explanation. It appears that a narrow shear wave attenuation zone underlies the rise axis in which Q_b is less than 50, which suggests a narrow zone of partial melt in a sub-axial magma chamber. Shear wave attenuation of this type has not been observed on the slow-spreading MAR (Nisbett and Fowler, 1978; Reid and Macdonald, 1973; Francis et al., 1977).

What few measurements exist also suggest a contrast in earthquake focal depths. In a microearthquake study of the "black smoker" hydrothermal field (EPR 21°N), the deepest hypocenters were less than 3 km deep, while most were less than 2 km deep (Riedesel et al., 1982). This is in keeping with the earlier cited refraction results which indicate a LVZ at 2-3 km depth. In contrast, microearthquakes on the MAR at 45°N occur up to 10 km deep directly beneath the inner floor of the rift valley (Lilwall et al., 1978). In a very detailed seismicity survey of the MAR near 23°N, earthquakes were accurately located up to 8 $\pm$ 1 km beneath the inner floor of the rift valley (Toomey et al., 1982). Significantly, these "deep" earthquakes occurred tens of kilometers away from the nearest transform fault intersection. These studies preclude steady state crustal magma chambers at these two sites on the MAR. They also suggest that hydrothermal activity may extend to depths of 10 km on the axes of slow spreading ridges. Unfortunately, the resolution of teleseismic hypocentral depth determinations is not adequate to distinguish conclusively between the 0-5 km and 5-10 km depth ranges on the seafloor, so little can be added from that data set.

A third piece of information comes from the type of events recorded. Enigmatic seismic events resembling harmonic tremors have been recorded on the EPR at 21°N in 1974 and 1980, as well as on the Galapagos spreading center in 1972 (Riedesel et al., 1982; Prothero and Reid, 1982; Macdonald and Mudie, 1974). At 21°N, these events are characterized by emergent arrivals with no clear P or S waves, and by a peak in the power spectrum in the 2-10 Hz range (Riedesel et al., 1982). Such tremors have been associated with the recent eruptions of Mt. St. Helens and Kilauea volcano and are indicative of recent or impending volcanic eruption. Records from the EPR at 21°N are similar to the St. Helens events in wave form and frequency, and suggest that this area may be either ending or entering an active volcanic phase.

Problems: The greatest limitation in this data set is the paucity of microseismicity surveys on spreading centers in which control is good enough to yield hypocentral determinations. So far we only have focal depth determinations from the EPR at 21°N, the MAR at 45° N, and the MAR at 23°N. In addition, hypocentral resolution is severely limited by the small number of seismometers in the arrays. The geometry of these small arrays provides high resolution depth control for only very small areas within the array. The recent seismicity study on the MAR at 23°N, however, involved 10 instruments deployed for 3 weeks and had excellent velocity control from seismic refraction, so the hypocenters are well constrained. The deep hypocenters occur far away (> 20 km) from the nearest transform faults, a place where petrologists are the most adamant about an axial magma chamber persisting in the steady state (e.g. Bryan et al., 1979; Stakes et al., in press). The lack of extremely primitive or highly differentiated basalt samples from spreading center segments well removed from transform fault intersections is one of several petrologic arguments for a steady state magma chamber in these regions. Very shallow earthquakes (< 3 km) on the EPR are consistent with this petrologic argument, while deep earthquakes (up to 10 km) along the axis of the MAR directly conflict with the petrologic arguments.

Gravity Measurements

Long wavelength gravity anomalies over mid-ocean ridges ($\lambda > 800$ km) are small and suggest that on a large scale, mid-ocean ridges are isostatically compensated (Talwani et al., 1965; Watts, 1982). At intermediate wavelengths (~ 50 km) slow spreading rift valleys are marked by a negative gravity anomaly of up to -75 mgals (Woodside, 1972; Watts, 1982). This large anomaly may be caused by a dynamic hydraulic head loss mechanism which depresses the inner floor of the rift valley relative to the uplifted rift mountains (Sleep, 1969; Lachenbruch, 1973, 1976). An alternative explanation is that the lithosphere over slow spreading centers has significant mechanical strength, and the anomaly is associated with continuous necking of the lithosphere as the plates diverge (Cochran, 1979; Tapponier and Francheteau, 1978). The resolution of the data and the large topographic signal preclude obtaining any information regarding a magma chamber.

Recent studies on the EPR near 9°N to 13°N reveal a distinct 10 to 20 mgal positive free air anomaly over the spreading axis which is 20 to 30 km wide (Lewis, 1982). The anomaly is tentatively interpreted as being caused by a magma chamber whose density is actually slightly greater than the surrounding pervasively fissured crust. In this case, some forceful injection of the magma is implied and the region within 50 km of the rise axis may not be compensated locally (Lewis, 1982).

Addressing this problem on a much finer scale, on-bottom gravity measurements have been taken from ALVIN on a 7 km traverse across the EPR at 21°N (RISE, 1980). These measurements have a resolution of 0.2 mgal (an order of magnitude greater than surface ship measurements) and show a pronounced gravity minimum of -1.5 mgal. The anomaly is 3 km wide and centered over the axial neovolcanic zone. This indicates a region of low density which may be associated with crustal fissuring or shallow penetration of a magma chamber. Geological observations show a maximum of fissuring outside the axial neovolcanic zone and flanking it, while the anomaly is centered over it. This suggests that the anomaly is associated with upward penetration of the magma chamber as a shallow magma "cupola" beneath the present neovolcanic zone. This is consistent with geological observations in the area which constrain the axial neovolcanic zone to be very narrow, at most 2 km and generally less than 1 km wide (Cyamex, 1980; Ballard et al., 1981).

Problems: Interpretation of the negative gravity anomaly measured from ALVIN at EPR 21°N is speculative. Interpretation of the anomaly as a shallow penetration of the axial magma chamber is consistent with geologic observations and seismic refraction data which indicate a low velocity zone as shallow as 2 km sub-bottom. However, extensive crustal fissuring may also explain the anomaly. Furthermore, there are no comparable studies so far from other places on the EPR or the MAR.

More troublesome is the positive 10 to 20 mgal anomaly measured over the EPR at 9°N-13°N on three separate profiles (Lewis, 1982). The gravity data have been interpreted in terms of an axial magma chamber subject to the conditions that the chamber is deep in the crust or sub-crustal, that the magma is significantly denser (0.25 gm/cm^3) than the surrounding crust, and that the axial zone is not in isostatic equilibrium (Lewis, 1982). One of the gravity profiles at 9°N is in the same area as some of the best refraction and reflection data indicating a crustal magma chamber at only 2 km depth (Orcutt et al., 1976; Herron et al., 1980). Sleep et al. (this volume) have an alternative explanation which may reconcile the gravity observations with the seismic data. They suggest that a deep "root" of partialy molten material may buoyantly support the axial topography which produces the positive gravity anomaly.

Magnetic Studies

The study of marine magnetic anomalies on mid-ocean ridges has gone beyond the calculation of spreading rates to the study of oceanic crustal emplacement processes and fine scale spreading behavior. For example it is possible to estimate the width and stability of the crustal accretion zone by studying the magnetic polarity transition zones between Vine-Matthews lineations (Matthews and Bath, 1967; Harrison, 1968). The finite width of a polarity transition is a measure of the distance over which axial lava flows overlap, the width of the dike intrusion zone at the axis, and the overlap of magnetized gabbroic plutons at depth in the crust (Atwater and Mudie, 1973; Cande and Kent, 1976; Kidd, 1977). Deep tow magnetic studies of transition widths at intermediate- to fast-spreading centers suggest that the crustal accretion zone has remained narrow (1-4 km) and localized for millions of years (Klitgord et al., 1975; Macdonald et al., 1980a). At slow spreading rates, the zone appears to vary between 1 and at least 8 km in width (Macdonald, 1977). Thus, for all spreading rates, the crustal accretion zone is remarkably narrow compared to the lateral dimensions of the plates, but at slow spreading rates there may be a greater tendency for the crustal accretion zone to wander laterally or periodically to widen to 5 or more km within the rift valley floor.

There is also a significant contrast in the overall clarity and sharpness of magnetic anomalies. At fast spreading rates, magnetic lineations are generally sharply defined and quite two-dimensional while at slow spreading rates anomaly identification is often difficult. The contrast in clarity is underscored by recent DSDP and ALVIN-based studies. On the EPR near 21°N, a magnetic gradiometer was mounted on ALVIN to conduct a detailed study of magnetizations of volcanic outcrops across a Vine-Matthews transition. The magnetic lineations in this study were nearly perfectly homogeneous in polarity (Macdonald et al., in press). In contrast, DSDP results from the slow spreading Atlantic reveal an extremely heterogeneous vertical section of magnetic properties including multiple reversals and significantly non-dipolar inclinations (Hall, 1976). This contrast can be explained in part by the greater spatial separation of anomalies at faster spreading rates. In addition, the greater clarity of anomalies at fast rates may be achieved by a narrower and more stable zone of crustal accretion (discussed above) as well as by a more frequent rate of volcanic eruption. Magnetic and geomorphic studies suggest that major eruptions may occur every ~1000-10,000 years at slow spreading centers compared to 50-500 years at intermediate to fast rates (Bryan and Moore, 1977; Atwater, 1979; Hall, 1976; Lonsdale, 1977; Macdonald, 1982). The contrast in inferred eruption rates, accretion zone width, and stability suggests a higher degree of continuity in time and space for crustal accretion at fast spreading rates versus greater episodicity at slow rates. The presence or absence of a steady state axial magma chamber which is continuous along most of its length is one way of explaining these differences.

Problems: There are numerous assumptions in deriving crustal magnetizations from magnetic field data (e.g., Klitgord et al., 1975; Parker and Huestis, 1974; Macdonald, 1977). In addition, it is not entirely satisfactory to compare ALVIN results which bear on horizontal heterogeneity, with DSDP results which address vertical variations. However, both measurements give some idea of magnetic polarity heterogeneity, and at present data do not exist which permit direct comparison of Pacific vs. Atlantic DSDP magnetic results or Pacific vs. Atlantic ALVIN magnetic studies.

Although it does not affect the discussion here directly, there is also the lurking doubt of still not knowing which stratigraphic levels of the oceanic crust carry the bulk of the magnetic signal (e.g., Harrison, 1981; Blakely and Lynn, 1977; Cande and Kent, 1976). It is likely to be shared between the layer 2 extrusives and the layer 3 plutonic gabbros depending on crustal age and other factors.

Heat Flow

At both fast and slow spreading centers there is a significant discrepancy between measured conductive heat flow and theoretical cooling plate models which suggest that at least 40% of the heat loss at mid-ocean ridges and 20% of Earth's total heat loss is accomplished by hydrothermal circulation at mid-ocean ridges (Lister, 1972; Williams and Von Herzen, 1974; Anderson et al., 1977). In addition, conductive heat flow measurements are extremely variable over short distances near spreading centers, again suggestive of hydrothermal heat loss (Elder, 1965; Lister, 1972).

There are however, two differences in heat flow patterns at fast and slow spreading centers. First, on both the Galapagos spreading center as well as at the EPR at 21°N, conductive heat flow on the flanks of the spreading center shows a distinct oscillatory pattern which can be modeled by cellular convection of seawater in the oceanic crust (Williams et al., 1974; Green et al., 1981; Becker and Von Herzen, 1982). Whether the alternating highs and lows in heat flow represent topographic forcing of the convection is unclear, but they do suggest zones of hydrothermal discharge and recharge within the crust. Such regular patterns have not been clearly observed in the slow spreading Atlantic, but this may be a sampling problem due to the extreme topography and patchy sediment cover on the MAR.

A second difference is the direct observation and measurement of high temperature hydrothermal discharge at intermediate to fast spreading centers, still lacking at slow spreading centers (the Red Sea is an exception). The discovery of the RISE hydrothermal field in 1979 allowed the first direct estimate to be made of axial hydrothermal heat flux (Macdonald et al., 1980b). Here hydrothermal fluids flow from discrete chimneys in many cases, rather than diffusing around pillows. Water blackened by sulfide precipitates jet out at rates of 1-3 m/sec at temperatures of up to 350°C. These are by far the hottest water temperatures and highest flow rates ever measured at a seafloor hydrothermal vent. These fluids are thought to be pristine and undiluted by subsurface mixing (Edmond, 1980). The entire RISE hydrothermal field is 6.2 km long with at least half a dozen separate vent clusters of chimneys and mineralized mounds. The field lies entirely within the neovolcanic zone in a band less than 500 m wide (RISE Team, 1980). The vents here are also characterized by shallow microearthquakes and possible harmonic tremors of the type associated with volcanic eruptions (Riedesel et al., 1982). The resulting heat flux is approximately $(6 \pm 2) \times 10^7$ cal/sec for a vent cluster of four black smoker chimneys.* This is between three and six times the total theoretical heat flow along a 1 km ridge segment out to a distance of 30 km on each side (crustal age 1 m.y.). Several vent clusters were found in a narrow linear zone less than 1 km long; obviously the total heat flux from the vents is so great that it is highly unlikely that these vents are steady state. Macdonald et al. (1980b) estimate that individual vent clusters have a lifetime on the order of ten years This is in keeping with the time estimated for the sulfide mound edifices to be built by precipitation (Haymon and Kastner, 1981; Finkel et al.,

* A heat flux of $(6 \pm 2) \times 10^7$ cal/sec was originally the estimate for the heat flux of a single chimney (Macdonald et al., 1980b). However, an error by a factor of two was made in the chimney diameter, owing to misidentification of the collected hand sample used for the diameter measurement with the chimney in the videotapes which was used for the flow rate and heat flux calculations. This results in a factor of four error in the calculated heat flux. Thus the calculated heat flux is appropriate for a cluster of four chimneys. It is observed, in fact, that a single mound edifice usually consists of 3-9 individual chimneys (RISE, 1980) so our original estimate is reasonable for a single vent cluster rather than a single chimney. Our original flow rate estimates compare very well with those actually measured two years later (Converse et al., 1982).

1980), and with the age distribution of clams near the vents (Killingley et al., 1980; Turekian and Cochran, 1981).

Subsequent discoveries have revealed evidence of high temperature hydrothermal discharge on the EPR near 12°N, 11°N, 20°S, the Guaymas Basin, the Galapagos spreading center, and the Juan de Fuca Ridge, all of which are intermediate to fast spreading centers (Francheteau and Ballard, in press; Simoneit and Lonsdale, 1982; Normark et al., 1982; Malahoff, 1982). Extensive surveys in the Atlantic have not revealed such activity (e.g., Fehn et al., 1977). There is abundant evidence for low temperature activity on the seafloor, particularly at the TAG area at 26°N (Rona, 1980). There is also evidence that high temperature activity occurs at depth within the crust on the MAR (Rona et al., 1980).

Problems: The apparent lack of high temperature seafloor venting at slow spreading centers may be a sampling problem. Until recently, detection of active hydrothermal discharge at spreading centers has been a rather "hit and miss" operation. Lack of sediment cover hampers conductive measurements and rapid entrainment of ambient water by hydrothermal plumes makes direct thermal detection of vent water difficult. Photographic detection of exotic benthic faunal communities has been the most reliable indirect indicator to date (Corliss et al., 1979).

If further exploration in the Atlantic shows that the apparent contrast is real then we may hypothesize that high temperature hydrothermal discharge on the seafloor is charcteristic of intermediate and fast spreading centers, but is rare at slow ridges. This may be due to deeper and more extensive cracking at slow spreading centers coupled with a smaller thermal budget related to spreading rate. Cooling of the newly formed crust may be efficient enough to restrict high temperature activity to some depth in the crust. While high temperature activity on the seafloor may occur at times on slow spreading centers, the frequency, duration and spacing of such activity may not be favorable for the support of chemosynthetic biological communities. At faster spreading rates, high temperature activity may be sustained on the seafloor as well as at depth in the crust, providing for seafloor massive sulfide deposition and the energy source for the vent faunal communities.

Electromagnetic Sounding

A new technique for probing the oceanic crust is electromagnetic active source sounding. So far this technique has only been attempted on the flanks of the EPR at 21°N (Young and Cox, 1981). Their measurements show that there is no indication of a magma chamber only 7 km away from the spreading axis, which restricts the total width of the axial magma chamber to less than

14 km. In addition they suggest that the high permeability zone, in which hydrothermal circulation occurs, is only 1.4 km thick.

Comparable measurements at other spreading centers are lacking but there are two other experiments to be noted in this context. Resistivity measurements in DSDP hole 504B show an increase of three orders of magnitude in the resistivity going from the volcanic layer into the intrusive dike sequence (Becker et al., 1982). This dramatic change in resistivity and inferred porosity occurs between 550 and 700 m sub-basement. These measurements support Young and Cox's results, and suggest that the most vigorous hydrothermal circulation occurs in the thin volcanic portion of the oceanic crust. These measurements were made on the Costa Rica Rift which spreads at essentially the same intermediate rate as the EPR at 21°N.

Problems: Active E-M sounding and borehole resistivity measurements need to be made at other spreading centers for comparison. The data from this first active E-M experiment justified use of only a half-space conductivity model, so that the results are only approximate. It is likely that some degree of hydrothermal circulation penetrates to at least the base of the crust. Further work must be done to test if voluminous hydrothermal circulation extends deeper along the axes of slow spreading centers than fast.

Thermal Models

Numerous thermal models have been proposed for spreading centers (e.g., Sclater and Francheteau, 1970; Parker and Oldenburg, 1973; Sleep, 1975). Sleep's model addresses the fine scale thermal structure beneath the spreading center and predicts a steady state magma chamber at intermediate to fast spreading rates, but a small, transient magma chamber at slow spreading rates. More recent models which account for the importance of hydrothermal heat loss still allow for maintenance of a steady state magma chamber at intermediate to fast spreading rates, but not at slow rates (Sleep and Rosendahl, 1979; Sleep, this volume). Kuznir (1980), however, allows for hydrothermal cooling and still obtains a steady state magma chamber 3 km wide for a slow half-spreading rate of 1 cm/yr. At the other extreme, Lister (1977, and this volume), maintains that hydrothermal circulation may preclude the maintenance of a steady state magma chamber at even the fastest known spreading rates.

Problems: The conflict in thermal models stems from poorly constrained estimates for the magnitude and areal distribution of hydrothermal heat loss at spreading centers and the depth and rate of crustal fissuring. Two important and hotly debated unknowns concern the rate at which cracks propagate downward in the axial zone versus the importance of clogging of the through-going cracks, as well as by hydrothermal precipitation. Sleep (this volume) suggests that the rapid crustal cooling effect of deep cracking fronts is mitigated by hydrothermal precipitation, which clogs the cracks, crustal heterogeneities, and extremely dense, high salinity seawater at depth. Lister (this volume) maintains that the cracking front proceeds downward at rates of meters per year, rates far faster than any known seafloor spreading rates, so that a steady state magma chamber does not occur at any known time-averaged spreading rate. Catastrophic spreading at rates of meters per year is called upon to initiate short-lived magma chambers. However, seismic and petrologic data both support a steady stage magma chamber at faster spreading rates. Most petrologic studies suggest that a small steady state magma chamber is maintained even at slow spreading rates (e.g. Bryan and Moore, 1977).

Many of these models also assume two-dimensionality. Clearly transform fault or propagating rift offsets perturb the thermal structure. For example, an intermediate rate spreading center near a transform fault may resemble a slow spreading center in terms of thermal budget and structure.

Geomorphology

At fast spreading rates, the spreading center is marked by an axial shield volcano which is remarkably continuous, and is only interrupted by transform faults and places where the spreading center experiences en echelon and overlapping offsets (Macdonald and Fox, 1983). Recent Seabeam bathymetric data show segments of the EPR axial shield volcano to be nearly perfectly continuous for up to 100 km, often with a narrow summit graben less than 500 m wide (Macdonald and Fox, 1983). The depth of the axial volcano also varies smoothly along strike. At intermediate spreading rates, along strike continuity is diminished by en echelon offsets of the axial neovolcanic zone.

In contrast, slow spreading ridges have a highly discontinuous string of volcanoes of varying height and morphology in the axial neovolcanic zone. Only the gross tectonic structure of the rift valley shows tranform-to-transform continuity.

This contrast in morphology does not require but is consistent with an underlying axial magma chamber which is relatively continuous between ridge offsets at fast spreading rates. Such a steady state magma chamber provides a persistent zone of weakness near the axis or a "thermal memory" which helps to maintain a narrow, linear, and continuous neovolcanic zone. At slow spreading rates, a non-steady state magma chamber, or one which is steady state for only a short distance along strike, will result in episodic and discontinuous crustal accretion, and irregular topography.

Problems: The data are excellent with considerable Seabeam data from the Pacific and SASS and GLORIA data from the North Atlantic. Control is very poor in the South Atlantic.

The Role of Transform Faults and Axial Offsets

Transform faults and other axial offsets such as propagating rifts disturb the thermal structure of mid-ocean ridge segments by juxtaposition of a cold, rigid boundary with the spreading axis (Sleep and Biehler, 1970). The degree of the thermal disturbance and resulting lateral heat flow depends upon the crustal age contrast across the transform fault. Closely spaced offsets exacerbate the problem of maintaining a magma chamber at any spreading rate. In addition, oceanic crust accreted near a transform fault intersection may be thinner due to a reduced volcanic budget (Fox and Stroup, 1981). The crust is also deformed by shear stresses which are transmitted from the transform fault into the spreading center domain (Macdonald et al., 1979).

Based on the recent GEBCO bathymetric charts, the average spacing for transform faults with offsets greater than 30 km is 170 km in the North Atlantic and 500 km in the South Pacific. For the North Atlantic, where the GEBCO charts are reliable, the average spacing for all transform faults is 50 km. In the Pacific where the charts are less reliable for small offset transform faults, continuous coverage Seabeam data reveal that the average spacing for all transform faults and other offsets of the spreading axis is 90 km (Macdonald and Fox, 1983; Lonsdale, 1982). Thus the disrupting influences of transform faults upon spreading centers is greater at slow spreading centers than fast due to an apparent spreading rate dependence on their spacing. In fact, nearly every segment of the MAR is significantly influenced by transform fault edge effects along its entire length due to the effects of viscous head-loss, lateral heat flow, and propagation of shear stresses into the spreading center domain (Sleep and Biehler, 1970; Macdonald, 1979; Fox and Stroup, 1981).

Not only is the spacing of transform offsets spreading-rate dependent, but to some extent, so is their style of deformation. Two types of axial offsets occur commonly at fast spreading rates but rarely at slow rates; propagating rifts and overlapping spreading centers. Propagating rifts are spreading centers which propagate along strike though plates, creating a V-shaped wake in their path (see Hey et al., 1980 for details). Large-scale along strike propagation of fast spreading centers can account for many of the oblique offset magnetic patterns and plate boundary reorganizations in the Pacific. So far there is considerable evidence for propagating rifts at intermediate to fast-spreading centers, but little evidence in the slow spreading Atlantic. In contrast, seafloor spreading in the Atlantic seems to be characterized by remarkably stable "cells" which maintain their integrity for tens of millions of years (Schouten and Klitgord, 1982). These cells persist for transform offsets of only 20 km and even for periods when asymmetric spreading reduces the transform offset to near zero (Schouten et al., 1982).

An interesting hypothesis to test then is, "Are propagating rifts and their mechanism spreading rate dependent?" At fast spreading rates, the corresponding isotherms are shallower than at slow rates. Thus, for a given transform offset, the rigid lithosphere will be thinner and more easily fractured across a fast-spreading rise/transform intersection. At slow spreading rates, the rigid lithosphere is sufficiently thick, even at small offset transform faults, so that rift propagation is impeded. Along-strike growth of one spreading center at the expense of another may occur on a scale of a few km, for example at FZB (Ramberg et al., 1977; Macdonald and Luyendyk, 1977), however large scale rift propgation appears to be rare.

The second phenomenon which is common at fast spreading centers, but not at slow is the occurrence of overlapping spreading centers (OSC's). In a recent seabeam study of the EPR a new kind of volcano-tectonic geometry has been discovered. At several locations along the rise axis the neovolcanic zone is discontinuous, and is laterally offset a short distance (1-15 km). In contrast to a classic ridge-transform-ridge plate boundary, however, the offset ridge terminations overlap each other by a distance approximately equal to or greater than the offset. They curve sharply towards each other and often merge into one another along strike. Separating the overlapping spreading centers (OSCs) is a closed contour depression up to several hundred meters deep which is sub-parallel to the trend of the OSCs. Apparently the shear strength of the thin lithosphere is less than the resistive stresses across the transform fault, and it is not strong enough to sustain a classic transform rigid-plate boundary. In these cases, the offset rises overlap and curve toward each other, separated by a complex shear zone with no obvious transform parallel structures. Over 40 OSCs have been charted on the EPR.

Problems: The contrast in transform fault spacing is well documented in the North Atlantic by GEBCO charts and in the South Pacific by continuous Seabeam coverage. It is not clear if OSCs should be considered a type of transform fault, but in any case they do represent along strike disturbances of the spreading center. The hypothesis that propagating rifts and OSCs are peculiar to intermediate-to-fast-rate spreading needs further testing.

Conclusions

The principal purpose of this paper is to provide a short summary of geopysical differences between fast and slow spreading centers. The emphasis here is on shallow crustal structure in the axial zone. Some of the differences are well documented while others are quite speculative and require further observations and measurements. In each case I have tried to emphasize the limitations of the observations and interpretations. Clearly much work remains to be done.

Taken collectively, the observations summarized here point to a distinct contrast in the structure between fast and slow spreading centers. Between transform faults and other lateral offsets, fast spreading centers have a remarkable degree of along strike continuity of faults and volcanic structures. This, coupled with geophysical measurements to date, suggest that fast spreading centers are underlain by shallow magma chambers which are steady state along most of their length. This is in keeping with the now common observation of high temperature vents on the seafloor at fast spreading centers but not at slow. Intermediate rate spreading centers show less along strike continuity but share more properties with fast spreading centers than with slow. In contrast it is most unlikely that large, strike-continuous, steady state magma chambers are sustained at slow spreading rates, and this appears to be reflected in the lack of along strike continuity in topography and the nature of the hydrothermal activity. Magma chambers, volcanic eruptions, and high temperature hydrothermal activity obviously occur at times along even slow spreading centers. However, as yet we do not know if such activity is frequent enough in time and in space at slow spreading centers to provide an adequate sustained energy source for the chemosynthetic faunal communities observed commonly at intermediate to fast spreading centers.

It should be noted that there is still considerable disagreement on several issues. On the one hand Lister (this volume) argues that axial magma chambers are short-lived regardless of spreading rate, and that they occur sporadically when spreading rates accelerate to rates of meters per year for short bursts (Lister, 1982). On the other hand, the very narrow range of petrologic composition of basalt samples from the MAR suggest that axial magma chambers are steady state even at spreading half-rates of 1 cm/yr (Bryan et al., 1979; Stakes et al., in press). If magma chambers were non-steady state, very primitive and highly fractionated basalts should occur during the formative and waning stages of the magma chamber respectively. While this appears to conflict with most of the geophysical inferences, seismic and other data cannot exclude a steady state chamber which is less than a seismic wavelength in width (1-2 km) and which is confined to the shallower mid-sections of slow spreading rift valleys (i.e. > 15-20 km from transform intersections). However, even if a small magma chamber such as this persists in the steady state at slow spreading rates, the lack of significant along strike continuity of the magma chamber could explain many of the spreading rate-related geophysical differences summarized in Table 1.

Acknowledgments

This work was supported by the U.S. National Science Foundation and the Office of Naval Research (Code 425GG). The manuscript was improved by the reviews of P. J. Fox and T. H. Van Andel.

References

Anderson, R. N., Langseth, M. G., and Sclater, J. G., 1977, The mechanisms of heat transfer through the floor of the Indian Ocean, J. Geophys. Res., 82:3391-3409.

Atwater, T. M., 1979, Constraints from the Famous area concerning the structure of the oceanic section. Deep drilling results in the Atlantic Ocean: Ocean crust. Eds. M. Talwani, C. G. Harrison, and D. E. Hayes, 2:33-42.

Atwater, T. M., and Mudie, J. D., 1973, Detailed near-bottom geophysical study of the Gorda Rise, J. Geophys. Res., 78:8665-8686.

Ballard, R. D., Francheteau, J., Juteau, T., Rangan, C. and Normark, W., 1981, East Pacific Rise at 21°N: The volcanic, tectonic and hydrothermal processes of the central axis, Earth Planet. Sci. Lett., 55:1-10.

Becker, K., and Von Herzen, R. P., 1982, Heat transfer through the sediments of the mounds hydrothermal area, Galapagos Spreading Center at 86 W, EOS, 63:529-536.

Becker, K., Von Herzen, R. P., Francis, R. J. G., Anderson, R. N., Honnorez, J., Adamson, A. C., Alt, J. C., Emmerman, R., Kempton, P. D., Kinoshita, H., Laverne, C., Mohl, M. J., and Newmark, R. L., In situ electrical resistivity and bulk porosity of the ocean crust; Costa Rica Rift, Nature, in press, 1982.

Bibee, L. D., 1979, Crustal structure in areas of active crustal accretion, Ph. D. Thesis, University of California, San Diego, pp. 155.

Blakely, R. J., and Lynn, W. S., 1977, Reversal transition widths and fast spreading centers, Earth Planet. Sci. Lett., 33:321-330.

Bryan, W. B., and Moore, J. G., 1977, Compositional variations of young basalts in the Mid-Atlantic Ridge rift valley near lat. 36°49'N, Geol. Soc. Amer. Bull., 88:556-570.

Bryan, W. B., G. Thompson, and P. J. Michael, 1979, Compositional variation in a steady-state zoned magma chamber: Mid-Atlantic Ridge at 36° 50 N, Tectonophysics, 55:63-85.

Cande, S. C., and Kent, D. V., 1976, Constraints imposed by the shape of marine magnetic anomalies on the magnetic source, J. Geophys. Res., 81:4157-4162.

Cochran, J. R., 1979, An analysis of isostacy in the world's oceans, part 2, mid ocean ridge crests, J. Geophys. Res., 84:4713-4729.

Corliss, J. B., Dymond, J., Gordon, L. I., Edmond, J. M., Von Herzen, R. P., Ballard, R. D., Green, K., Williams, D., Bainbridge, A., Crane, K., and Van Andel, Tj. H., 1979, Submarine thermal springs on the Galapagos Rift, Science 203:1073-1083.

Converse, D. R., Holland, H. D., and Edmond, J. M., 1982, Hydrothermal flow rates at 21° N, EOS, 63:472.

CYAMEX Team, 1980, First manned submersible dives on the East Pacific Rise at 21° N (Project RITA): General Results, Mar. Geophys. Res., 4:345-379.

Dewey, J. F., Kidd, W. S. F., 1977, Geometry of plate accretion, Geol. Soc. Amer. Bull., 88:960-968.

Edmond, J. M., 1980, The chemistry of the 350° C hot springs at 21° N on the East Pacific Rise, EOS 61:992.

Elder, J. W., 1965, Physical processes in geothermal areas. In: Terrestrial Heat Flow Am. Geophys. Union Monogr., 8:211-239.

Fehn, U., Siegel, M. D., Robinson, G. R., Holland, H. D., Williams, D. L., Erickson, A. J., Green, K. E., 1977, Deep-water temperatures in the Famous area, Geol. Soc. Amer. Bull. 88:488-494.

Finkel, R. C., MacDougall, J. D., Chung, Y. C., 1980, Sulfide precipitates at 21° N on the East Pacific Rise: ^{226}Ra, ^{210}Pb and ^{210}Po, Geophys. Res. Lett., 7:685-688.

Fowler, C. M. R., 1976, Crustal structure of the Mid-Atlantic Ridge crest at 37° N, Geophys. J. Roy. Astr. Soc., 47:459-491.

Fox, P. J., and Stroup, J. B., 1981, The plutonic foundation of the oceanic crust, The Sea, 7:119-218.

Francheteau, J., and Ballard, R. D., 1982, The East Pacific Rise near 21° N, 13°N and 20°S: Inferences for along-strike variability of axial processes at the Mid-Ocean Ridge, Earth Planet. Sci. Lett., in press.

Francis, T. J. G., and Porter, I. T., 1973, Median valley seismology: The Mid-Atlantic Ridge near 45° N, Geophys. J. Roy. Astr. Soc., 34:279-311.

Francis, T. J. G., Porter, I. T., and McGrath, J. R., 1977, Ocean bottom seismograph observations on the Mid-Atlantic Ridge near 37° N, Geol. Soc. Am. Bull., 88:664-677.

Green, K. E., 1980, Geothermal processes at the Galapagos Spreading Center, Ph.D. Dissertation, WHOI-80-33.

Green, K. E., Von Herzen, R. P., Williams, D. L., 1981, The Galapagos Spreading Center at 86° W: A detailed geothermal field study, J. Geopys. Res., 86:979-986.

Hale, L. D., Morton, C. J., and Sleep, N. H., 1982, Reinterpretation of seismic reflection data over the East Pacific Rise, J. Geophys. Res., 87:7707-7719.

Hall, J. M., 1976, Major problems regarding the magnetization of oceanic crustal layer 2, J. Geopys. Res., 81:4223-4230.

Harrison, C. G. A., 1968, Formation of magnetic anomaly patterns by dyke injection, J. Geophys. Res., 73:2137-2142.

Harrison, C. G. A., 1981, Magnetism of the oceanic crust, The Sea, 7:219-237.

Haymon, R., and Kastner, M., 1981, Hot spring deposits on the East Pacific Rise at 21°N: preliminary description of mineralogy and genesis, Earth Planet. Sci. Lett., 53:363-381.

Herron, T. J., Stoffa, P. L., and Buhl, P., 1980, Magma chamber and mantle reflections - East Pacific Rise, Geophys. Res. Lett., 7:989-992.

Hey, R., Duennebier, F. K., and Morgan, W. J., 1980, Propagating rifts on mid-ocean ridges, J. Geophys. Res., 85:3647-3658.

Johnson, H. P., Karsten, J. L., Delaney, J. R., Davis, E. E., Currie, R. G., and Chase, R. L., A detailed study of the Cobb offset of the Juan de Fuca Ridge: Evolution of a propagating rift, J. Geophys. Res., in press.

Jung, H., and Lewis, B. T. R., 1982, Seismic refraction results from the southern Juan de Fuca Ridge, EOS, 63:1153.

Keen, C. E., and Tramontini, C., 1970, A seismic refraction survey on the Mid-Atlantic Ridge, Geophys. J. Roy. Astr. Soc., 20:473-491.

Kidd, R. G. W., 1977, A model for the process of formation of upper oceanic crust, Geophys. J. Roy. Astr. Soc., 50:149-183.

Killingley, J. S., Berger, W. H., Macdonald, K. C., and Newman, W. A., 1980, $^{18}O/^{16}O$ variations in deep sea carbonate shells from the RISE hydrothermal field, Nature, 288:218, 221.

Klitgord, K. D., Huestis, S. P., Parker, R. L., and Mudie, J. D., 1975, An analysis of near-bottom magnetic anomalies: Sea floor spreading, the magnetized layer, and the geomagnetic time scale, Geophys. J. Roy. Astr. Soc., 43:387-424.

Lachenbruch, A. H., 1973, A simple mechanical model for oceanic spreading centers, J. Geophys. Res., 78:3395-3417.

Lewis, B. T. R., 1982, Constraints on the structure of the East Pacific Rise from gravity, J. Geophys. Res., 87:8491-8500.

Lilwall, R. C., Francis, T. J. G., and Porter, I. T., 1978, Ocean bottom seismograph observations on the Mid-Atlantic Ridge near 45°N - further results, Geophys. J. Roy. Astr. Soc., 55:255-262.

Lister, C. R. B., 1972, On the thermal balance of a mid-ocean ridge, Geophys. J. Roy. Astr. Soc., 26:515-535.

Lister, C. R. B., 1977, Qualitative models of spreading center processes, including hydrothermal penetration, Tectonophysics, 37:203-218.

Lister, C. R. B., 1982, On the intermittency of seafloor spreading, EOS, 63:1153.

Lonsdale, P., 1977, Structural geomorphology of a fast-spreading rise crest: The East Pacific Rise near 3°25'S, Mar. Geophys. Res., 3:251-293.
Lonsdale, P., 1982, Small offsets of the Pacific-Nazca and Pacific-Cocos spreading axes, EOS, 63:1108.
Macdonald, K. C., 1977, Near-bottom magnetic anomalies, asymmetric spreading, oblique spreading and tectonics of the Mid-Atlantic Ridge near 37°N, Geol. Soc. Am. Bull., 88:541-555.
Macdonald, K. C., 1982, Mid-ocean ridges: Fine-scale tectonic, volcanic and hydrothermal processes within the plate boundary zone, Ann. Rev. Earth Planet. Sci., 10:155-190.
Macdonald, K. C., Miller, S. P., Huestis, S. P., and Spiess, F. N., 1980a, Three-dimensional modeling of a magnetic reversal boundary from inversion of deep-tow measurements, J. Geophys. Res., 85:3670-3680.
Macdonald, K. C., Becker, F., Spiess, F. N., and Ballard, R. D., 1980b, Hydrothermal heat flux of the "black smoker" vents on the East Pacific Rise, Earth Planet. Sci. Lett. 48:1-7.
Macdonald, Ken C., and Fox, P. J., 1983, Overlapping spreading centers: new accretion geometry on the East Pacific Rise, Nature, 302:55-58.
Macdonald, K. C., Kastens, K., Spiess, F. N., and Miller, S. P., 1979, Deep tow studies in the Tamayo Transform Fault, Marine Geophys. Res., 4:37-70.
Macdonald, K. C., and Luyendyk, B. P., 1977, Deep-tow studies of the structure of the Mid-Atlantic Ridge crest near lat. 37°N, Geol. Soc. Amer. Bull., 88, 621-636, 1977.
Macdonald, K. C., Luyendyk, B. P., Mudie, J. D., and Spiess, F. N., 1975, Near-bottom geophysical study of the Mid-Atlantic Ridge median valley near lat. 37°N: Preliminary observations, Geology, 3:211-215.
Macdonald, K. C., Miller, S. P., Luyendyk, B. P., Atwater, T. M., and Shure, L., Investigation of an Vine-Matthews magnetic lineation from a submersible: The source and character of marine magnetic anomalies, J. Geophys. Res., in press.
Macdonald, K. C., and Mudie, J. D., 1974, Microearthquakes on the Galapagos Spreading Center and the seismicity of fast-spreading ridges, Geophys. J. R. Astr. Soc., 36:245-257.
Malahoff, A., 1982, Massive enriched polymetallic sulfides of the ocean floor - a new commercial source of strategic minerals? OTC 4293.
Matthews, D. H., and Bath, J. 1967. Formation of magnetic anomaly patterns on the Mid-Atlantic Ridge, Geophys. J. Roy. Astr. Soc., 13:349-357.
Mammerickx, J., and Smith, S. M., 1978, Bathymetry of the Southeast Pacific, Geol. Soc. Amer., Map and Chart Series MC-26.
McClain, J. S., and Lewis, B. T. R., 1980, A seismic experiment of the axis of the East Pacific Rise, Marine Geol., 35:147-169.

Menard, H. W., 1967, Seafloor spreading, topography and second layer, Science, 157:923-924.
Morton, J. L., Tompkins, D. H., Normark, W. R., and Sleep, N. H., 1982, Structure of the southern Juan de Fuca ridge from multi-channel seismic reflection records, EOS, 63:1153.
Nisbet, E. G., ad Fowler, C. M. R., 1978, The Mid-Atlantic Ridge at 37°and 45°N, some geophysical and petrologic constraints, Geophys. J. Roy. Astr. Soc., 54:631-660.
Normark, W. R., Morton, J. L., and Delaney, J. R., 1982, Geologic setting of massive sulfide deposits and hydrothermal vents along the southern Juan de Fuca Ridge, USGS, open-file report 82-200A.
Orcutt, J. A., Kennett, B. L. N., and Dorman, L. M., 1976, Structure of the East Pacific Rise from an ocean bottom seismometer array, Geophys. J. Roy. Astr. Soc., 45:305-320.
Orcutt, J. A., McClain, J. S., Burnett, M., Seismic constraints on the generation, evolution and structure of the ocean crust, Geol. Soc. of London Spec. Pub., in press.
Pallister, J. S., and Hopson, C. A., 1981, Semail ophiolite plutonic suite: Field relations, phase variations, cryptic variation and layering, and a model of a spreading ridge magma chamber, J. Geopys. Res., 79:1587-1593.
Parker, R. L., and Huestis, S. P., 1974, The inversion of magnetic anomalies in the presence of topography, J. Geopys. Res., 79:1587-1593.
Parker, R. L., and Oldenburg, D. W., 1973, Thermal model of ocean ridges, Nature Phys. Sci., 242:137-139.
Poehls, K., 1974, Seismic refraction on the Mid-Atlantic Ridge at 37°N, J. Geophys. Res., 79:3370-3373.
Prothero, W., and Reid, I., 1982, Microearthquake results from the East Pacific Rise, J. Geophys. Res., 87:8509-8518.
Purdy, G. M., Detrick, R. S., and Cormier, M., 1982, Seismic constraints on the crustal structure at a ridge-fracture zone intersection, EOS, 63:1100.
Ramberg, I. B., Gray, D. F., and Raynolds, R. G. H., 1977, Tectonic evolution of the FAMOUS area of the Mid-Atlantic Ridge, lat. 35 50' to 37 20'N, Geol. Soc. Amer. Bull., 88:609-620.
Riedesel, M., Orcutt, J. A., Macdonald, K. C., and McClain, J. S., 1982, Microearthquakes in the Black Smoker Hydrothermal Field, East Pacific Rise at 21°N, 87:10613-10624.
Reid, I. D., and Macdonald, K. C., 1973, Microearthquake study of the Mid-Atlantic Ridge near 37°N using sonobuoys, Nature, 246:88-90.
Reid, I. D., Orcutt, J. A., and Prothero, W. A., 1977, Seismic evidence for a narrow zone of partial melting underlying the East Pacific Rise at 21°N, Geol. Soc. Amer. Bull., 88:678-682.

RISE Team: Spiess, F. N., Macdonald, K. C., Atwater, T., Ballard, R., Carranza, A., Cordoba, D., Cox, C., Diaz Garcia, V. M., Francheteau, J., Guerrero, J., Hawkins, J., Haymon, R., Hessler, R., Juteau, T., Kastner, M., Larson, R., Luyendyk, B., MacDougall, J. D., Miller, S., Normark, W., Orcutt, J., and Rangin, C., 1980, East Pacific Rise: Hot springs and geophysical experiments, Science, 207:1421-1433.
Rona, P. A., 1980, TAG hydrothermal field: Mid-Atlantic Ridge Crest at latitude 26°N, J. Geol. Soc. London, 137:385-402.
Rona, P. A., Bostrom, K., and Epstein, S., 1980, Hydrothermal quartz vug from the Mid-Atlantic Ridge, Geology, 8:569-572.
Rosendahl, B. R., Raitt, R. W., Dorman, L. M., Bibee, L. D., Hussong, D. M., and Sutton, G. H., 1976, Evolution of oceanic crust, 1. A physical model of the East Pacific Rise crest derived from seismic refraction data, J. Geophys. Res., 81:5294-5305.
Schouten, H., and Klitgord, K. D., 1982, The memory of the accreting plate boundary and the continuity of fracture zones, Earth Planet Sci. Lett., 59:255-266.
Schouten, H., Denham, C., and Smith, W., 1982, On the quality of marine magnetic anomaly sources and sea-floor topography, Geophys. J. Roy. Astr. Soc., 70:245-260.
Sclater, J. G., and Francheteau, J., 1970, The implications of terrestrial heat flow observations on current tectonic and geochemical models of the crust and upper mantle of the Earth, Geophys. J. Roy. Astr. Soc., 20:509-542.
Simoneit, B. R. T., and Lonsdale, P. F., 1982, Hydrothermal petroleum on mineralized mounds at the seabed of Guaymas Basin, Nature, 295:198-202.
Sleep, N. H., 1969, Sensitivity of heat flow and gravity to the mechanism of seafloor spreading, J. Geophys. Res., 74:542-549.
Sleep, N. H., 1975, Formation of ocean crust: Some thermal constraints, J. Geophys. Res., 80:4037-4042.
Sleep, N. H., Hydrothermal convection at ridge axes, P. Rona et al., eds., Hydrothermal Processes at Spreading Centers, Plenum, in press.
Sleep, N. H., and Biehler, S., 1970, Topography and tectonics at the intersections of fracture zones with central rifts, J. Geophys. Res., 75:2748-2752.
Sleep, N. H., Morton, J. L., Burns, L. E., Geophysical constraints on the volume of hydrothermal flow at ridge axes, P. Rona et al., eds., Hydrothermal Processes at Spreading Centers, Plenum, in press.
Sleep, N. H., and Rosendahl, B. R., 1979, Topography and tectonics of mid-ocean ridge axes, J. Geophys. Res., 70:341-352.

Stakes, D., Shervais, J. W., and Hopson, C. A., The volcano-tectonic cycle of the FAMOUS and AMAR valleys, Mid-Atlantic Ridge (36° 47'N): Evidence from basalt glass and basalt phenocryst compositional variations for a steady-state magma chamber beneath the valley midsections, J. Geophys. Res., in press.

Talwani, M., Le Pichon, X., and Ewing, M., 1965, Crustal structure of the mid-ocean ridges, J. Geophys. Res., 70:341-352.

Tapponnier, P., and Francheteau, J., 1978, Necking of the lithosphere and the mechanics of slowly accreting plate boundaries, J. Geophys. Res., 83:3955-3970.

Toomey, D. R., Murray, M. H., Purdy, G. M., Murray, M. H., 1982, Microearthquakes on the Mid-Atlantic Ridge near 23°N: new observations with a large network, EOS, 63:1103.

Trehu, A. M., and Solomon, S. C., 1981, Microearthquakes in the Orozco Fracture Zone: a closer look at the results from project ROSE, Trans. Amer. Geophys. Un., 62:325.

Turekian, K. K., and Cochran, J. K., Growth rate determination of a visicomyid clam from the Galapagos Spreading Center hydrothermal field using natural radionuclides, Earth Planet. Sci. Lett., in press.

Van Andel, T. H., and Ballard, R. D., 1979, The Galapagos Rift at 86°W, 2, volcanism, structure and evolution of the rift valley, J. Geophys. Res., 84:5390-5406.

Watts, A. B., 1982, Gravity anamolies over oceanic rifts, in: "Continental and Oceanic Rifts," G. Palmason, ed., Geodynamics Series 8, American Geophysical Union, 309 pp.

Whitmarsh, R. B., 1975, Axial intrusion zone beneath the median valley of the Mid-Atlantic Ridge at 37°N detected by explosion seismology, Geophys. J. Roy. Astr. Soc. 42:189-215.

Williams, D., and Von Herzen, R. P., 1974, Heat loss from the earth: new estimate, Geology 2:327-328.

Williams, D. L., Von Herzen, R. P., Sclater, J. G., and Anderson, R. H., 1974, The Galapagos Spreading Center: lithospheric cooling and hydrothermal circulation, Geophys. J. Roy. Astr. Soc. 38:587-608.

Woodside, J. M., 1972, The Mid-Atlantic Ridge near 45°N, the gravity field, Can. J. Earth Sci. 9:942-959.

Young, P. D., and Cox, C. S., 1981, Electromagnetic active source sounding near the East Pacific Rise, Geophys. Res. Lett. 8:1043-1046. slow rates. I shall now briefly review recent studies involving seismic refraction and reflection, microearthquakes, gravity,

GEOPHYSICAL CONSTRAINTS ON THE VOLUME OF HYDROTHERMAL FLOW AT RIDGE AXES

Norman H. Sleep, Janet L. Morton, and Laurel E. Burns

Dept. of Geophysics
Stanford University
Stanford CA 94305

and

Thomas J. Wolery

Earth Sciences Division
Lawrence Livermore Laboratory
Livermore CA 94550

ABSTRACT

Hydrothermal circulation at the ridge axis removes heat from the oceanic crust more rapidly than would conduction alone. The top of the axial magma chamber is thus deeper and possibly wider than the theoretical shape computed from conductive thermal models. At 9° N on the East Pacific Rise seismic reflection indicates that the roof of the magma chamber is relatively flat, 2 km deep, and extends 4 km from the axis. This is about a kilometer deeper than predicted by a purely conductive model.

We believe that the magma chamber is mostly filled with mush at ridges with both fast and slow spreading rates. At fast rates the mush is formed by crystallization at the top of a magma chamber that is wide and flat topped. At slow rates a narrow magma chamber is probably an anastomosing complex of partially molten dikes and associated cumulate layers. Thermal modeling indicates that the hydrothermal heat flux is between 0.7×10^{8} and 1.5×10^{8} cal/cm^2, or less than 1/10 of the total missing heat flux (the difference between obsereved and theoretical heat flow) at the ridge axis. By using the observation that Mg is totally depleted from exiting axial fluids, we find that the minimum amount of crust which reacts with axial hydrothermal flow is equivalent to a 80 m thick section of crust. A minimum thickness of 200 m is obtained from K which is leached from the basalt into the hydrothermal fluid. These estimates indicate that there is no requirement that the bulk of the oceanic crust react strongly with the axial hydrothermal fluid.

INTRODUCTION

Even before seafloor spreading was well understood, it was noted that the heat flow on mid-oceanic ridges was lower than would be expected from cooling of hot lithospheric material derived from the interior of the earth. The "missing heat flow" was immediately attributed to the unmeasured escape of heat through hydrothermal vents (Hess, 1965, page 133). Conductive models calibrated to the heat flow on the flanks of the ridge, where hydrothermal heat flow is minor, and to the topography of the ridge confirmed this discrepancy (Sclater and Francheteau, 1970). The total heat loss from hydrothermal circulation can be estimated from the difference between observed and theoretical heat flow (Anderson and Hobart, 1976; Anderson et al., 1977; Wolery and Sleep, 1976; Sleep and Wolery, 1978; Sclater et al., 1980). The results as compiled by Sleep and Wolery(1978) are given in table 1.

There was little interest in the thermal state in the immediate few kilometers of the ridge axis because the sediment cover necessary for conventional heat flow measurements do not exist there. Conductive thermal models which are valid in the axial region were published by Parker and Oldenburg (1973), Oldenburg (1975), and Sleep (1974, 1975). After Sleep (1975) quantified the inferences of Cann (1974) on the formation of oceanic crust and the shape of the magma chamber, thermal constraints were explicit or at least implicit in later petrological studies.

Actual high temperature vents have been sampled on several intermediate and fast spreading ridges. The temperature of exiting water is about 350°C . The heat loss from an individual vent was estimated as 6×10^7 cal/sec (Macdonald et al., 1980) and from a cluster of vents as 4 to 5×10^7 cal/sec (Converse et al., 1982). The heat loss from these vents is too large to be steady state. For example, all the missing heat flow on a 60 mm/yr full-rate ridge is equivalent to 3.4×10^7 cal/km-sec. Unless most of the hydrothermal heat flow exits from axial vents, axial vents must be intermittent and short-lived, as is indicated by studies of the vents themselves (Macdonald, 1982; Converse et al., 1982).

The long-term average heat flux from high temperature axial vents can be determined by balancing geochemical cycles or by determining the shape of the magma chamber from geophysics and modelling the heat and mass transfer. We use the latter approach in this paper, although a local average over time, rather than a global average, is obtained. The chemical fluxes then can be computed by extrapolation to the rest of the

Table 1. Comparison of inferred hydrothermal heat loss with available heat in the oceanic lithosphere.

Heat source	Heat per area of sea floor 10^8 cal/ cm^2	global heat flux 10^{18} cal/yr
Heat budget of ridge		
Total heat loss from seafloor spreading	39	117
Missing heat flow at ridge	17	51
Missing heat on crust less than 1 m.y.	4.6	14
Theoretical conductive heat within 1 km of axis[a]	1.5	4.5
Heat in oceanic lithosphere		
Latent heat of 5 km basalt	1.5	5
Total heat in 1 km basalt	1.4	4
Linear cooling to 8 km depth[b]	4.6	14
Linear cooling to 30 km depth[b]	17	50
Thermal model results[a]		
Axial heat sinks	0.7	2.1
Theoretical conductive heat within 2 km of axis	0.8	2.4
Heat removed by axial vents, upper bound	1.5	4.5

[a] Half spreading rate is 6 cm/yr.
[b] Does not include latent heat.

world. The size of the crustal chamber is fairly sensitive to the amount of hydrothermal heat loss as the total heat loss (that is cooling to the seafloor temperature) from one kilometer of oceanic crust is almost 1/10 of the hydrothermal heat loss (Table 1).

STRUCTURE OF THE MAGMA CHAMBER

Various kinematic models have been proposed for the magma chamber at mid-oceanic ridges. Minimum criteria for an acceptable model are that the stratification of oceanic lithosphere can be continually created and that the thermal structure is realizable (See Fox and Stroup, 1981). Some models have begun with this thermal structure, others have begun with the crustal structure inferred from ophiolitic complexes, and still others have begun with structure of the magma chamber inferred from geophysical data. It is thus not surprising that recently proposed models have differed in the extent and geometry of hydrothermal circulation. Four models are illustrated in Fig. 1, 2, and 3.

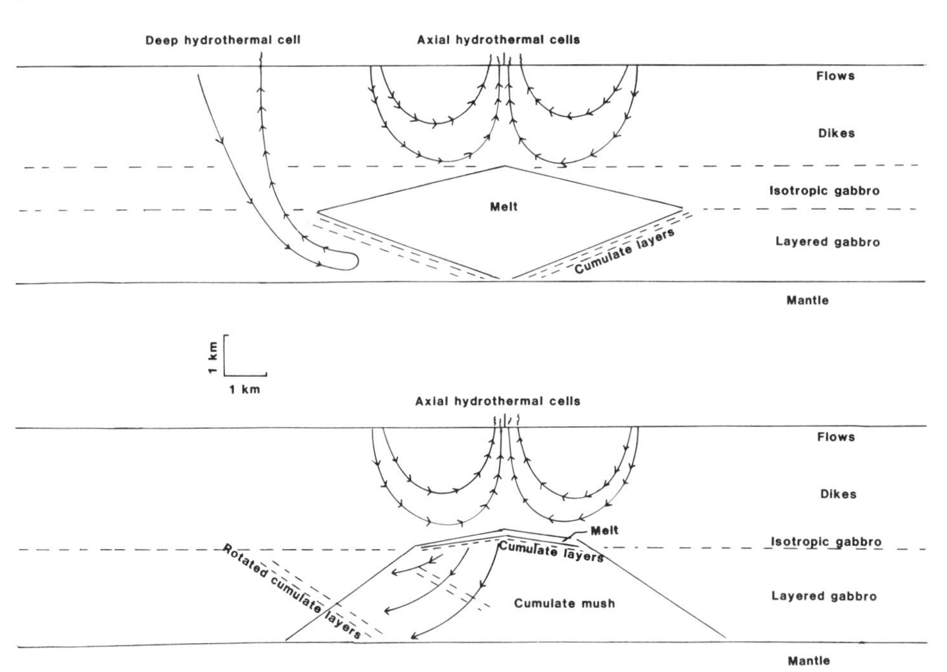

Fig. 1. Schematic drawing showing the Hopson (above) and Dewey-Kidd-Sleep magma chambers on the same scale. In the Hopson model the magma chamber is all melt and is cooled from above and below by hydrothermal circulation. The dip of the cumulate layering is primary. In the Dewey-Kidd-Sleep model the cumulate layers are formed at the top of a mush-filled chamber and rotate into a dipping orientation. The hydrothermal circulation is a minor perturbation on the shape of the chamber.

chamber, as obtained from seismic reflection, is used to constrain the amount of hydrothermal heat loss through axial vents.

A third (Lewis) model is based on the analysis of gravity and topography of intermediately spreading parts of the East Pacific Rise (Fig. 2). Extensive hydrothermal circulation on the flanks of the chamber causes the chamber to be narrow, flat-topped, and steep-walled. Extensive cracking related to the hydrothermal circulation causes the crust on the flanks of the chamber to be less dense than the molten material in the chamber and causes a free air gravity anomaly over the axis which is higher than expected from local isostatic compensation of thermally expanded topography. As noted by Lewis (1981, 1982) the topography alone is consistent with a simple conductive cooling model.

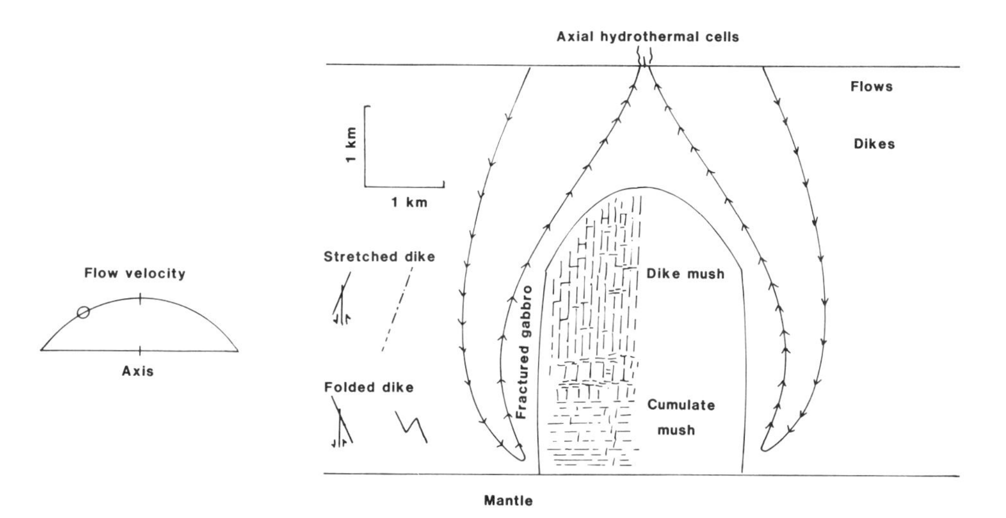

Fig. 2. Schematic diagram showing the intense hydrothermal circulation around the Lewis model magma chamber. The rock outside the chamber is highly fractured low density gabbro. The inside of the chamber is filled with mush originating as anastomosing dikes and cumulate sills. There is a greater portion of cumulates lower in the chamber. The entire mush complex itself moves upward as a slow viscous dike. The flow velocity across the chamber is parabolic. The direction of shear at a point on the left side of the chamber (circle) is shown below the scale. The dikes can be folded or stretched depending on their orientation. The kinematic model for the interior of the chamber is also applicable to slow ridges.

Lewis did not propose a kinematic model for the formation of layered mafic rocks in the magma chamber. Therefore, we propose a model which is likely to be applicable to slow ridges and also applicable to fast ridges if the magma chamber is as narrow as envisioned by Lewis (Fig. 2). The magma chamber is considered to be a network of anastomosing, subvertical dikes with some horizontal sills. Magma moves upward from one level of sill to the next and fractionally crystallizes when temporarily stagnant at some level. More ultra-mafic cumulates are thus formed at the deeper levels of the complex and mafic cumulates are formed at the higher levels. At the top of the complex and to some extent on the side, contamination with hydrated oceanic crust occurs and plagiogranite is produced.

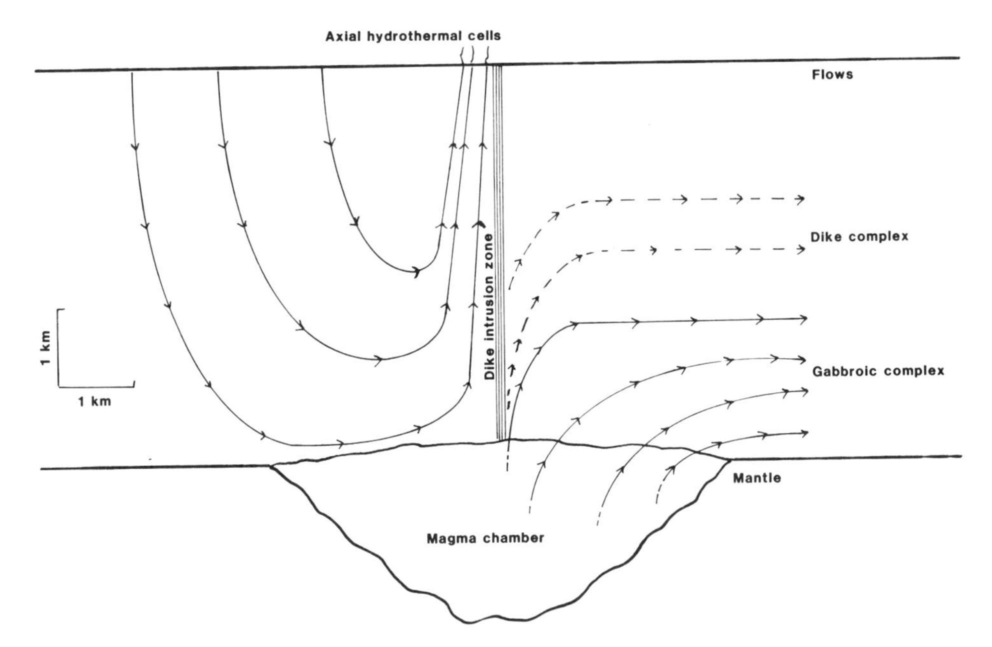

Fig. 3. A schematic diagram of the Bibbee model magma chamber has such intense hydrothermal circulation that the magma chamber is located below the level of the crust. A complex flow pattern is needed to get the dike complex and the gabbroic complex to their proper crustal levels.

Unlike the Dewey-Kidd-Sleep model, there is a systematic upward and outward movement of mush formed at deeper levels. (A wide magma chamber, with kinematics intermediate between the Dewey-Kidd-Sleep model and this model seems possible, but harder to illustrate.) Earlier cumulates are transposed into the transport direction (or plane of shear) which is subvertical and subparallel to the dikes. Earlier dikes are also intruded by later ones as material moves away from the axis of the magma chamber. There is also some tendency for bodies of mush to become mobilized and intruded as highly flow-banded dikes. All but the latest bodies are flow banded to some extent. When a more mafic dike intrudes less mafic gabbro, it will freeze and later become flow folded and intruded as the less mafic material continues to flow. As the dikes are nearly parallel to the plane of shear (the entire partially molten dike complex is considered to flow upward as a wide slow dike), either folds or boudins can develop depending on the precise orientation of the dike. Both features may develop as the plane of shear varies with respect to the plane of the dike during flow.

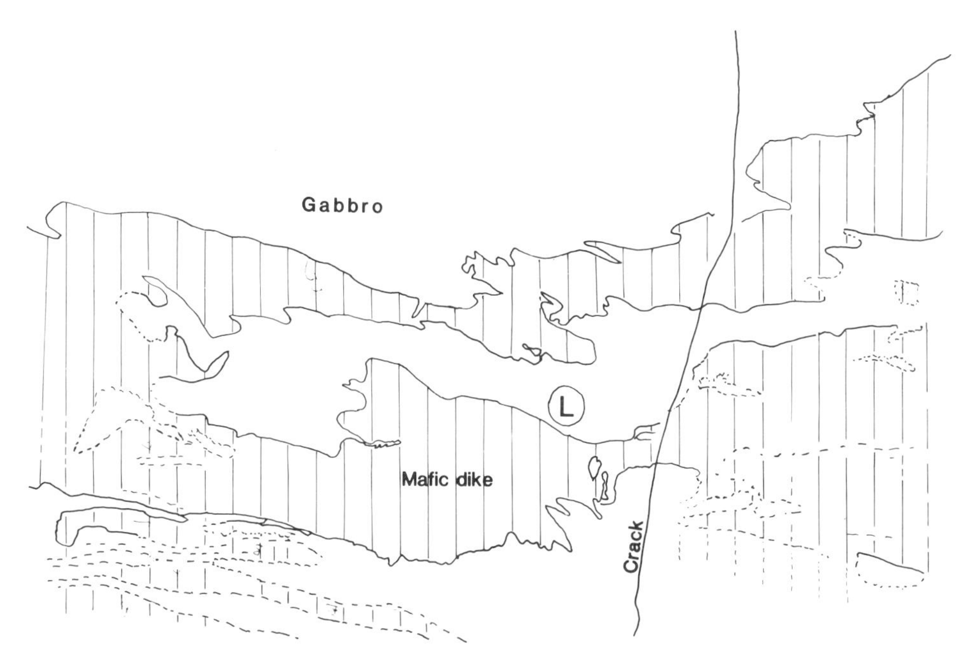

Fig. 4. Tracing of a photograph shows deformed mafic dikes and less mafic rock from Nelchina Glacier, Alaska. Lens cap (L) from 35 mm camera gives scale. The dike was both stretched and folded in a complex manner. The gabbro intrudes the mafic dikes and the dikes are partially assimilated in the lower part of the figure. Many of the dikes of this set can be traced for 10's of meters with varying degrees of deformation.

Field work by Burns in a gabbroic complex near Nelchina Glacier, Alaska has recognized these textures and fabrics (Fig. 4). Although this complex was probably formed beneath an arc rather than a ridge, we feel that these features may occur in ophiolites particularly those formed at slow ridges. Our experience with this gabbro indicates that a deeper level dike complex is hard to recognize as such and that excellent exposures are needed.

It is also conceivable that the small magma chamber is filled with melt rather than mush and that the cumulates plate on to the sides and walls of the magma chamber such that a section of cumulates as thick as the chamber is created (Casey et al., 1981). Field studies of the Bay of Islands ophiolite support this variation of the Lewis model. As previously noted, the amount of hydrothermal circulation to cool a chamber does not depend on whether the chamber is filled with mush or melt.

A fourth (Bibbee et al., 1983; Elthon et al., 1982) model assumes that the magma chamber occurs below the level of the moho (Fig. 3). This model is based on reinterpretation of the seismic refraction records which were used by Rosendahl (1976) and Rosendahl et al. (1976) to infer a magma chamber similar to that expected from a thermal model (Bibbee et al., 1983) and on the chemistry and structure of the basal cumulates at North Arm Mountain, Newfoundland (Elthon et al., 1982). A kinematic scheme, which would keep the bulk of the magma chamber below the level of the moho, is not evident. The large amount of low-temperature deformation required to get the gabbro from the magma chamber into the crust would render the material into a tectonite, rather than the mildly deformed material often exposed in ophiolitic complexes. At North Arm mountain, the cumulate tectonite is confined to the deepest levels of the complex (Elthon et al., 1982). The Elthon et al. (1982) variation of this model, therefore, has a significant amount of the magma chamber above the moho.

Gravity anomalies

We use the East Pacific Rise near 9°N to demonstrate the amount of axial hydrothermal circulation because gravity and seismic data are available and because this is one of the ridges used by Lewis (1981, 1982) to construct his crustal model. The seismic profile indicates that the magma chamber is about two kilometers beneath the sea floor (Herron et al., 1978; Hale et al., 1982). The reflector is one-sided probably because the profile crosses a small unmapped transform fault (Fig. 5). The polarity and moveout of the reflection are consistent with the reflector being the top of the low viscosity magma chamber. Seismic refraction data also indicate a low velocity region beneath the ridge axis (Orcutt et al., 1976). The base of the chamber was not clearly detected and the data is consistent with either a melt-filled Hopson chamber or a mush-filled Dewey-Kidd-Sleep chamber (Fig. 1).

The gravity highs near 9° N on the East Pacific rise axis are about 10 to 20 mgals and at least 5 to 10 mgal higher than that expected from simple isostasy (Lewis, 1981; 1982). Following the method of Sleep and Rosendahl (1979) we construct a fluid dynamic model of the ridge axis. In the numerical model, the topography is computed from the vertical pressure at the free surface. Masses at depth and the movement of the plates away from the ridge axis produce the flow. The calculation scheme was modified to give better resolution of topography and to compute gravity at the seafloor.

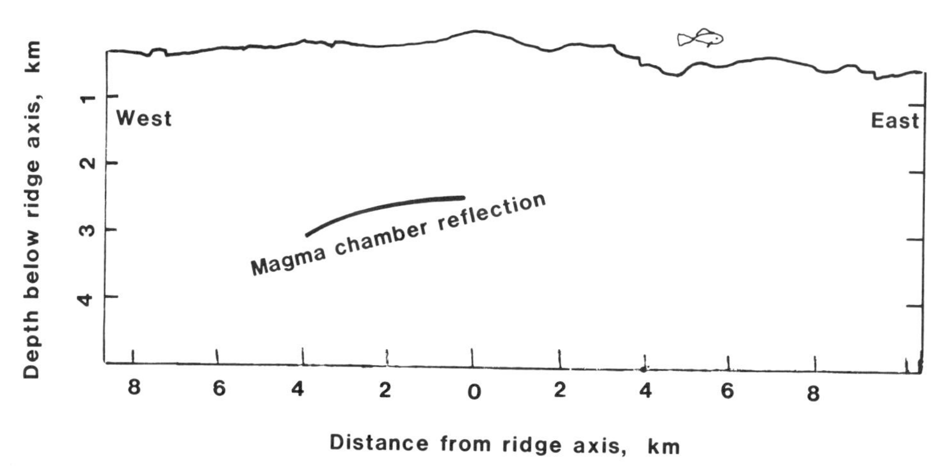

Fig. 5. The location of the magma chamber at 9°N as determined by seismic reflection is shown with the topography of the axis. The magma chamber was detected only on one side of the axis. The break in slope 3 km east of the axis may be a small transform fault which was cross at a highly oblique angle.

The models of fast ridges given by Sleep and Rosendahl (1979) are nearly isostatic and therefore will not explain the high gravity anomaly at the ridge axis (Fig. 6). A likely cause for the excess elevation at the ridge axis and hence the positive gravity anomaly is that as a steady-state condition some excess melt is retained beneath the ridge axis. We consider the melt as distributed, but a 10 km deep root of the magma chamber which is filled with cumulate mush as proposed by Elthon et al. (1982), would have a similar effect. By isostasy this excess melt causes a general uplift of the magma chamber and hence the uplift of the surface. Because the root of this uplift is deep, a positive gravity anomaly results.

In Fig. 6, the general amplitude of the observed topography and gravity is obtained once some melt is retained below the ridge axis. The inclusion of retained melt increases gravity over the magma chamber by about 5 mgal without modifying the other features of the computed profile. The density change of 0.02 gm/cm^3 corresponds to 3 to 6 percent melt retained in the mantle at the axis (Fig. 6). This amount does not seem to be excessive and we can conclude that steady-state retention of melt beneath the ridge axis is a plausible explanation for the high gravity anomalies at fast ridges. A numerical model that includes the two-phase flow involved with the segregation and migration of melt would better appraise the hypothesis.

THERMAL MODEL OF THE RIDGE AXIS WITH HYDROTHERMAL CIRCULATION

A theoretical model for the thermal structure at mid-oceanic ridges, which allows for distributed heat sources near the axis, has been developed. The formulation is similar to that of Sleep (1975), with a heat source term added to the right side of the steady-state heat flow equation. The heat sources can represent latent heat (positive) or heat removed by hydrothermal circulation (heat sinks, negative).

The seismic reflection profile across the East Pacific Rise near 9°N indicates that the roof of the magma chamber is about two kilometers deep and nearly flat-lying (Herron et al., 1978; Hale et al., 1982). Purely conductive thermal models predict that the magma chamber would be only one kilometer deep and that the roof would dip steeply away from the axis. By varying the amount of hydrothermal circulation and the placement of latent heat in the new model, a theoretical temperature structure was obtained that is compatible with the seismic constraints (Fig. 7).

In both the Hopson and the Dewey-Kidd-Sleep magma chamber models, most of latent heat of crystallization is released at the top of the chamber. (We do not imply that crystals necessarily settle from the roof to the floor of the chamber. The complex process by which latent heat is convected from the site of crystallization to the roof of the chamber is not considered in this paper.) In the model shown in Fig. 7, 70 percent of

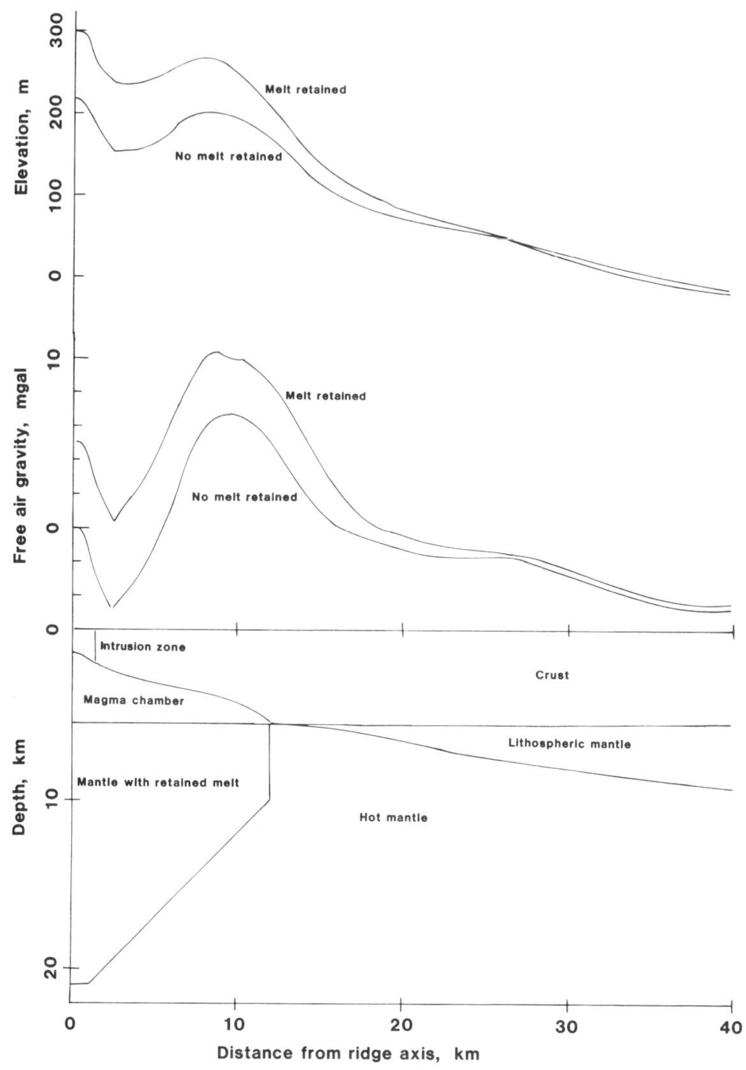

Fig. 6. The computed gravity and elevation is shown in the top of the figure and the rheological and density regions in the lower part. The thermal model and thermal expansion coefficient are compatible with the square root of age dependence of ridge elevation. The magma chamber has a relative density contrast of -0.1 gm/cm^3 and a viscosity between 10^{17} and 10^{18} poise. The upwelling hot mantle has a viscosity of 5×10^{19} poise and the intrusion zone has a viscosity of 10^{17} poise. The viscosity of cold crust and lithospheric mantle is 10^{23} poise. The mantle differs only in that the region below the magma chamber has a relative density contrast of -0.02 gm/cm^3 in the model with retained melt. This increases the gravity anomaly near the axis by 5 mgal but does not change the shape of the anomaly. The zero for the gravity and topography curves is arbitrary.

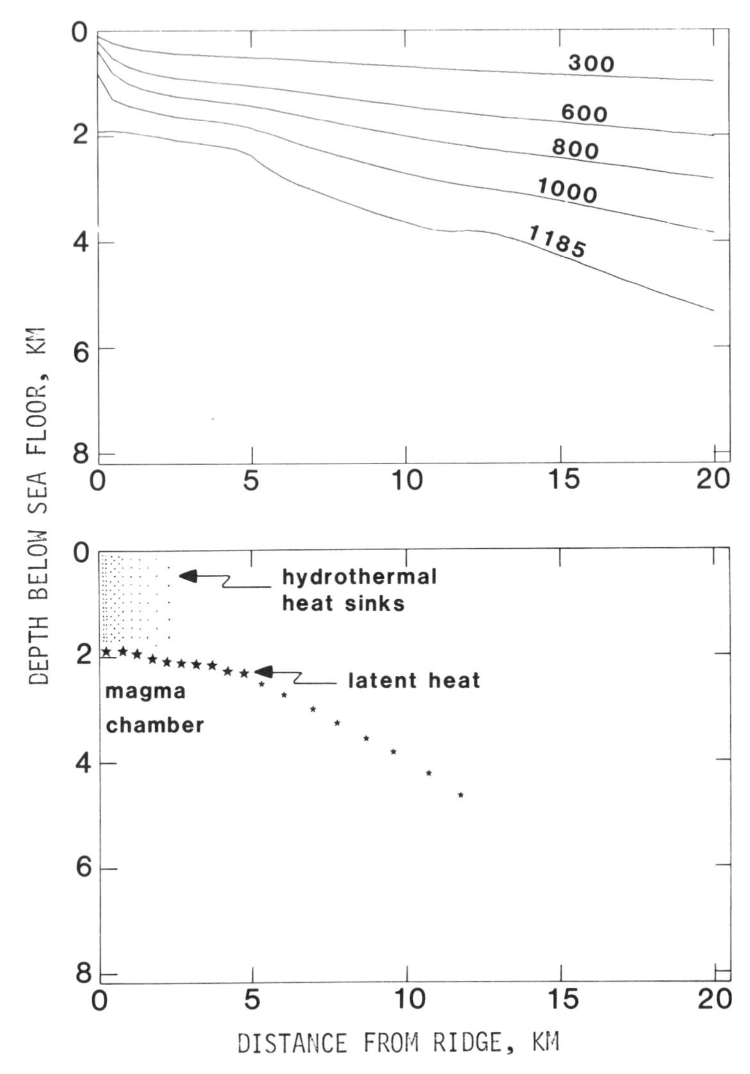

Fig. 7. Temperature in degrees Celsius computed for the East Pacific Rise at 9°N (top) and the distribution of hydrothermal heat sinks and latent heat used in the model (bottom). Above 5 kilometers depth, the computed magma chamber is indicated by the 1185°C isotherm. Seventy percent of the latent heat was placed along the top of the magma chamber (large stars). The two kinks in the 1185°C isotherm are due to the step in the input latent heat at 5 kilometers from the axis and the higher temperature in the mantle below 5 kilometers, respectively. Constants used in the calculation are: half spreading rate, 6.1 cm/yr; latent heat 245.7 cal/cm^3; specific heat 0.91 cal/cm^3-°C; conductivity, 0.006 cal/cm-sec-°C; and temperature at 100 kilometers depth, 1400°C.

the latent heat was placed along the top of the magma chamber, indicated by the 1185 °C isotherm; the remainder was placed along the side of the chamber. This produces a somewhat wider and flatter magma chamber than models with all of the latent heat at the axis. Heat sinks, representing heat removed by hydrothermal circulation, were then added in the upper two kilometers to produce a thicker lid over the magma chamber. For the model shown, the heat sinks were placed within two kilometers of the axis and total 14 cal/cm-sec. At a half spreading rate of 6.1 cm/yr, this is equivalent to 0.7×10^8 cal/cm^2. The resulting temperature distribution gives an integrated conductive heat flow to two kilometers from the ridge axis of 15 cal/cm-sec, or 0.8×10^8 cal/cm^2. In the model, the heat sinks were evenly distributed between the sea floor and a depth of two kilometers. It is likely that the actual hydrothermal circulation is greater in the porous, uppermost few hundred meters of the crust. Increasing the very shallow heat sinks would decrease the conductive heat flow, but would have little affect on the size and depth of the magma chamber. Hence, we feel that the actual axial hydrothermal circulation is greater than the total heat sinks, in this model 0.7×10^8 cal/cm^2, but less than the sum of the heat sinks and the conductive heat flow, in this model 1.5×10^8 cal/cm^2. This upper bound is approximately 1/10 of the total missing heat at the ridge axis as calculated by Sleep and Wolery (1978). The other 9/10's of the missing heat is presumably removed by off-axial hydrothermal flow. If we assume that comparable amounts of hydrothermal circulation occur at all mid-oceanic ridges, we can extrapolate our results to the global heat budget of the oceanic lithosphere. If 3 km^2 of new oceanic lithosphere is created each year, then the global heat flux of axial hydrothermal vents is between 2.1×10^{18} cal/yr and 4.5×10^{18} cal/yr (table 1).

The hydrothermal heat loss determined from the thermal modelling represents the long-term, average heat flux of the axial vents. The actual hydrothermal circulation is probably episodic, often occurring at short-lived, high temperature vents, such as those observed at 21 °N on the East Pacific Rise (Macdonald, 1982; Converse et al., 1982). For example, Converse et al. (1982) estimate that the heat loss from a cluster of high temperature vents is 4 to 5×10^7 cal/sec. The theoretical heat loss in the model is equivalent to such vent clusters occurring approximately once in every 14 to 36 km of ridge crest, or to more closely spaced but episodic vents. If the vent clusters are spaced at 1 km intervals along the ridge, they would only need to be active 1 part in 14 to 36. These short-lived vents will only cool the crust for a short distance around the actual flow, thus the temperature of the fluid will be lower than the average temperature of the crust at a given depth. We modeled the hydrothermal circulation has a heat sink, rather than convective flow, for this reason.

The theoretical temperature structure at the ridge axis shown in Fig. 7 is consistent with the Dewey-Kidd-Sleep magma chamber model. It is also consistent with the shape of the top of the Hopson magma chamber, but not with the bottom edge of the Hopson model. In the theoretical model, heat sinks could be added deeper in the crust to simulate hydrothermal circulation occurring beneath the magma chamber, as shown in Fig. 1. In the future this will be done to see if it is possible to obtain theoretical isotherms that are consistent with the Hopson magma chamber model.

DISCUSSION AND CHEMICAL IMPLICATIONS

The information we obtained from the thermal model gives an average of the hydrothermal flux over the time it takes the ridge to spread about the width of the magma chamber or 10^5 yr. We extrapolate our findings to the rest of the world although without explicitly accounting for the variation of hydrothermal flux with spreading rate. Our results are compatible with the mostly conductive Dewey-Kidd-Sleep model for the ridge axis and to some extent with the Hopson model. They are not compatible with the Lewis model or the Bibbee model where the size of the magma chamber is drastically reduced by hydrothermal circulation. The thermal models are insensitive to whether the magma chamber is mostly filled with melt or mostly filled with mush. At slow spreading ridges the magma chamber is narrow and it may be filled with mostly frozen dikes or with melt.

We use Mg and K as examples to discuss the chemical implications of our results because the importance of seawater-rock interaction to the cycles of these elements has been recognized for some time (Hart, 1973; Mottl, 1976; Wolery and Sleep, 1976). This section is based on Wolery (1978) and Wolery and Sleep (1983); the reader is referred to those papers for more detail. Because nearly all magnesium is removed from axial hydrothermal waters, the magnesium flux into the oceanic crust from this process is simply the concentration of magnesium in seawater multiplied by the hydrothermal flux. The magnitude of this sink can be appraised by comparing it to the river input of 5.5×10^{12} moles/yr of which 2.7×10^{12} moles/yr is not accounted for by known sinks (other than reaction with the oceanic crust) assuming the ocean is in steady state (Drever, 1974). If the total missing heat at the ridge is 51×10^{18} cal/yr and the vent temperature is 350°C the flow rate is 1.4×10^{17} gm/yr. As the Mg in seawater is 0.0544 molal, a potential Mg flux of 7.6×10^{12} moles/yr is obtained (Edmond et al., 1979). This would deplete the total exogenic mass of Mg of 2.7×10^{21} moles in 350 m.y. The time is sufficiently short that some explanation is needed. Either the Mg abundance in seawater was lower throughout much of geological time so that the potential amount of reaction was much less or most of the missing heat flow at the ridge is not related to axial hydrothermal flow. Our thermal modeling favors the latter possibility, as we found that less than 1/10 of the missing heat is explained by axial hydrothermal flow. (The off-axial hydrothermal flow to explain the other 9/10 of the heat does not react as extensively which the rock as the axial hydrothermal flow and thus is not as depleted in Mg.) This gives an axial Mg flux of 0.76×10^{12} moles/yr which is less than the missing sink for river flux.

A minimum amount of reaction of basalt with axial hydrothermal flow can be computed by noting that 0.06 gm of basalt is needed to remove the magnesium from 1 gm of seawater. Expressing our axial heat flux as 1.5×10^8 cal/cm^2 and again using 350°C as the exit temperature, we find that 2.5×10^4 gm(rock)/cm^2(seafloor) react with the axial fluid, which is equivalent to a section of 80 m of oceanic crust. This is a minimum estimate because once the Mg is removed from seawater, further reaction with the circulating water with rock cannot be detected using that element.

The amount of rock which reacts with axial hydrothermal flow can be also constrained by considering elements which are leached from the basalt, such as K (Edmond et al., 1979; Wolery and Sleep, 1983). The main difficulty here is that the composition of the basalt is not as well constrained as that of seawater. We use an abundance of K appropriate for fractionated basalt of 0.24 percent (60×10^{-6} moles/gm) and assume that all this K is leached. The ratio of K to heat in the Galapagos vents is 25×10^{-9} moles/cal (Edmond et al., 1979). Using our axial heat flux we find that 3.75 moles of K is leached from each square centimeter of seafloor. This flux requires a minimum of 6×10^4 gm/cm^2 or a 200 m section. This is probably a better minimum estimate than obtained from Mg because K is added rather than removed from the seawater. Both estimates indicate that the bulk of the oceanic crust does not necessarily react strongly with the axial hydrothermal flow.

Acknowledgements. We thank Rob Haar, Rick Schult, Grant Lichtman, Jerry van Andel, and an anonymous reviewer for their helpful comments. The research was supported in part by the National Science Foundation grants #EAR77-14479, EAR80-01076, and EAR-81-15522. Burns was supported in part by Alaska Division of Geological and Geophysical Surveys.

REFERENCES

Anderson, R. N., and Hobart, M. A., 1976, The relation between heat flow, sediment thickness and age in the Eastern Pacific, *J. Geophys. Res.*, *81*: 2968-2989.

Anderson, R. N., Langseth, M. G., and Sclater, J. G., 1977, The mechanism of heat transfer through the floor of the Indian Ocean, *J. Geophys. Res.*, *82*: 3391-3409.

Bibbee, L. D., Dorman, L. M., Johnson, S. H., and Orcutt, J. A., 1983, Crustal structure of the East Pacific Rise at 10°S, *J. Geophys. Res.*, *87*: (in press), (unseen).

Cann, J. R., 1974, A model for oceanic crustal structure developed, *Geophys. J. R. astron. Soc.*, *39*: 169-187.

Casey, J. F., Dewey, J. F., Fox, P. J., and Karson, J. A., 1981, Heterogeneous

nature of oceanic crust and upper mantle: A perspective from the Bay of Islans ophiolite complex, *in*: "The Oceanic Lithosphere, The Sea, Vol. 7," C. Emiliani, ed., John Wiley, New York, 305-338.
Converse, D. R., Holland, H. D., and Edmond, J. M., 1982, Hydrothermal flow rates at 21°N, *EOS, Trans. Amer. Geophys. Union*, *63*: abstract V551-4, 472.
Dewey, J. F., and Kidd, W. S. F., 1977, Geometry of plate accretion, *Geol. Soc. Amer. Bull.*, *88*: 960-968.
Drever, J. I., 1974, The magnesium problem, *in*: "Marine Chemistry, The Sea, Vol. 5," E. D. Goldberg, ed., Wiley-Interscience, New York, 337-357.
Edmond, J. M., Measures, C., McDuff, R. E., Chan, L. H., Collier, R., Grant, B., Gordon, L. I., and Corliss, J. B., 1979, Ridge crest hydrothermal activity and the balances of the major and minor elements in the ocean: The Galapagos data, *Earth Planet. Sci. Lett.*, *46*: 1-18.
Elthon, D., Casey, J., and Komor, S., 1982, Mineral chemistry of ultramafic cumulates from the North Arm Mountain massif of the Bay of Islands ophiolite: evidence for high-pressure crystal fractionation of oceanic basalts, *J. Geophys. Res.*, *87*: 8717-8734.
Fox, P. J., and Stroup, J. B., 1981, Geological and geophysical properties of the lower oceanic crust, *in*: "The Oceanic Lithosphere, The Sea, Vol. 7," C. Emiliani, ed., John Wiley, New York, 119-218.
Gregory, R. T., and Taylor, H. P., 1981, An oxygen isotope profile in a section of Cretaceous oceanic crust, Samail ophiolite, Oman: Evidence for $\delta^{18}O$ buffering of the oceans by deep (>5 km) seawater-hydrothermal circulation at mid-ocean ridges, *J. Geophys. Res.*, *86*: 2737-2755.
Hale, L. D., Morton, C. J., and Sleep, N. H., 1982, Reinterpretation of seismic reflection data over the East Pacific Rise, *J. Geophys. Res.*, *87*: 7707-7718.
Hart, R. A., 1973, A model for chemical exchange in the basalt-sea water system of oceanic layer II, *Can. J. Earth Sci.*, *10*: 801-816.
Herron, T. J., Ludwig, W. J., Stoffa, P. L., Kan, T. K., and Buhl, P., 1978, Structure of the East Paccific Rise crest from multichannel seismic data, *J. Geophys. Res.*, *83*: 798-804.
Hess, H. H., 1965, Mid-oceanic ridges and the tectonics of the sea-floor, *in*: "Submarine Geology and Geophysics," W. F. Whittard and R. Bradshaw, eds., Butterworths, London, 317-333.
Lewis, B. T. R., 1981, Isostasy, Magma chambers, and plate driving forces on the East Pacific Rise, *J. Geophys. Res.*, *86*: 4868-4880.
Lewis, B. T. R., 1982, Constraints on the structure of the East Pacific Rise from gravity, *J. Geophys. Res.*, *87*: 8491-8500.
Macdonald, K. C., 1982, Mid-ocean ridges: Fine scale tectonic, volcanic and hydrothermal processes within the plate boundary zone, Ann. Rev. Earth Planet Sci., 10: 155-190.
Macdonald, K. C., Becker, K., Spiess, F. N., and Ballard, R. D., 1980, Hydrothermal heat flux of the "black smoker" vents on the East Pacific Rise, *Earth Planet. Sci. Lett.*, *48*: 1-7.
Mottl, M. J., 1976, Chemical exchange between sea water and basalt during

hydrothermal alteration of the oceanic crust, Ph. D. thesis, Harvard University, Cambridge, Mass.

Oldenburg, D. W., 1975, A physical model for creation of the lithosphere, *Geophys. J. R. astron. Soc.*, *43*: 425-451.

Orcutt, J. A., Kennett, B. L. M., and Dorman, L. M., 1976, Structure of the East Pacific Rise from an ocean bottom seismometer survey, *Geophys. J. R. astron. Soc.*, *45*: 305-320.

Pallister, J. S., and Hopson, C. A., 1981, Samail ophiolite plutonic suite: Field relations, phase variation, cryptic variation and layering, and a model of a spreading ridge magma chamber, *J. Geophys. Res.*, *86*: 2593-2644.

Parker, R. L., and Oldenburg, D. W., 1973, Thermal model of ocean ridges, *Nature*, *242*: 137-139.

Rosendahl, B. R., 1976, Evolution of oceanic crust 2.. Constraints, implications, and inferences, *J. Geophys. Res.*, *81*: 5305-5313.

Rosendahl, B. R., Raitt, R. W., Dorman, L. M., Bibbee, L. O., Hussong, D. M., and Sutton, G. H., 1976, Evolution of oceanic crust, 1, A physical model of the East Pacific Rise crest derived from seismic refraction data, *J. Geophys. Res.*, *81*: 5294-5305.

Sclater, J. G., and Francheteau, J., 1970, The implications of terrestrial heat flow observations on current tectonic and geothermal models of the crust and upper mantle of the earth, *Geophys. J. R. astron. Soc.*, *20*: 509-542.

Sclater, J. G., Jaupart, C., and Galson, D., 1980, The heat flow through oceanic and continental crust and the heat loss of the earth, *Rev. Geophys. Space Phys.*, *18*: 269-312.

Sleep, N. H., 1974, Segregation of magma from a mostly crystalline mush, *Geol. Soc. Amer. Bull.*, *85*: 1225-1232.

Sleep, N. H., 1975, Formation of the oceanic crust: some thermal constraints, *J. Geophys. Res.*, *80*: 4037-4042.

Sleep, N. H., 1978, Thermal structure of mid-oceanic ridge axes, some implications to basaltic volcanism, *Geophys. Res. Lett.*, *5*: 426-428.

Sleep, N. H., and Rosendahl, B. R., 1979, Topography and tectonics of mid-ocean ridge axes, *J. Geophys. Res.*, *75*: 6831-6839.

Sleep, N. H., and Wolery, T. J., 1978, Thermal and chemical constraints on venting of hydrothermal fluids at mid-ocean ridges, *J. Geophys. Res.*, *83*: 5913-5922.

Wolery, T. J., 1978, Some chemical aspects of hydrothermal processes at mid-oceanic ridges -- A theoretical study. I. Basalt-sea water reaction and chemical cycling between the oceanic crust and the oceans. II. Calculation of chemical equilibrium between aqueous solutions and minerals. Ph. D. thesis, Northwestern University, Evanston, Ill., 263 pp.

Wolery, T. J., and Sleep, N. H., 1976, Hydrothermal circulation and geochemical flux at mid-ocean ridges, *J. Geol.*, *84*: 249-275.

Wolery, T. J., and Sleep, N. H., 1983, Interactions between exogenic cycles and the mantle, *in*: "Chemical Cycles and the Evolution of the Earth," R. M. Garrels, C. B. Gregor, and F. T. Mackenzie, eds., (in press).

HYDROTHERMAL CONVECTION AT RIDGE AXES

Norman H. Sleep

Dept. of Geophysics
Stanford University
Stanford CA 94305

INTRODUCTION

Our knowledge of hydrothermal circulation at the ridge axis is based largely on sampling of the hydrothermal fluid, indirect geophysical measurements of the oceanic crust, and studies of rocks which are believed to have undergone hydrothermal alteration at the ridge axis. Basic questions still remain to be answered: how does the hydrothermal seawater initially enter deep levels of the oceanic crust; how effective is hydrothermal circulation at cooling the crust; what is the geometry of hydrothermal circulation; what is the relationship between the hydrothermal circulation and the magma chamber; how does the oceanic crust react with the seawater; and how can the hydrothermal fluid which altered a rock sample be identified. In this presentation, the relationships between these questions are discussed. In this volume, Lister (1983a) is concerned with mainly the first three questions and Taylor (1983) with the last three. Lister's (1983a) approach is mainly theoretical, while Taylor (1983) has extensively studied the chemistry of rocks affected by hydrothermal fluids. Therefore, I discuss most extensively the topics not treated in those papers. That is chemical approaches to the questions addressed by Lister (1983a) and physical approaches to the questions discussed by Taylor (1983).

The conclusions reached by Lister (1983a) and Taylor (1983) are not in close agreement on one key issue. Lister (1983a) finds that hydrothermal circulation should cool the magma chamber so efficiently that steady-state magma chambers are precluded at all spreading rates. Taylor (1983), however, finds that the magma chamber of the Oman ophiolite was at least long lived and that moderate amounts of water entered the chamber at high temperatures but did not quickly cool the chamber. In this presentation, apparent conflicts between Taylor's (1983) and Lister's (1983a) results are discussed. An alternative geometry for axial hydrothermal circulation is proposed in an attempt to explain both the chemical and the physical constraints on hydrothermal systems. Before discussing physical approaches, I quickly review the size of the hydrothermal flow at axial vents.

Any theory for the origin of axial vents must explain the high temperature, 350°C, and the rapid flow rate of 100 kg/sec per vent cluster. The heat flux from the vent is the product of these numbers with the specific heat of water, 1 cal/gm-°C, or 4×10^7 cal/sec. The magnitude of this flux can be better visualized by computing the mass of water which escapes during the approximate 10 year life of the vent cluster. The mass of water is 3×10^{10} kg, which is equivalent to a 300-m cube of water or all the water in a cubic kilometer of crust at 3% porosity.

CRACK FRONT

New cracks must form in the oceanic crust at a rate that can supply at least the large amounts of hot water of the axial vents. Lister (1983a) obtains that a crack front should propagate downward into the magma chamber at rates of meters per year. The magma body is inferred to cool from magmatic to low temperatures during the passage of such a crack front. Basically, the process involves thermal contraction cracks associated with the initiation of cooling. These cracks permit further infiltration of water, further cooling, and more cracking. On a small scale thermal contraction cracking is responsible for columnar jointing in dikes and cracking in submarine lava flows.

The process examined by Lister (1983a) is an active "mining" of heat from the rock. Once the the crack front is done penetrating, a much longer period of passive "harvesting" of heat would follow (Lister, 1982). A general caveat in inferring the conditions of hydrothermal circulation from studies of rocks is that the circulation which produces the bulk of the metamorphism may not be the circulation which carries the bulk of the heat. In particular, the heat carrying circulation may be restricted in space and its metamorphism overprinted by later passive circulation.

The general implication of the crack front hypothesis is that crustal magma chambers should be rare and transient at all spreading centers. The rate of propagation of the crack front is inferred to be so much faster than plate tectonic rates that only a rapid, but episodic process could create a magma body. It is conceivable, however, that the oceanic crust is cooled by crack fronts but the rate is much slower than proposed by Lister (1983a). In addition, if a region becomes reheated after it has been cooled, a second crack front may propagate though the region (Lister, 1981). Evidence for rapid quenching thus must be examined separately from evidence for the crack front itself.

Observations Relevant to the Crack Front

As envisioned by Lister (1983a), the cracking front should pass through a particular region in the magma chamber in around 50 years. High temperature alteration by hydrothermal water should be rare ahead of the crack front and the material behind the crack front should be subjected to a long period of moderate-temperature alteration. Direct evidence for this prediction is obtained by Taylor's (1983) studies of $\delta^{18}O$ in plutons and ophiolites. Prolonged periods of very high temperature hydrothermal alteration occurred in several bodies, including the deep gabbros of the Oman ophiolite. Therefore, at least some hydrothermal cracks can form without catastrophic cooling in an igneous body, or the high temperature alteration occurred during reheating after the first crack front passed. For example, the deeper parts of the crack front could become detached from the upper part of the hydrothermal system owing to clogging at higher levels and be reheated (Lister, 1981).

Another implication of the crack-front hypothesis is that the magma chamber should continually be driven back at the expense of the crack front. Inclusion of hydrothermally altered material in the magma chamber thus might be expected to be rare, especially in isolated plutons such as the Skaergaard. However, Taylor (1983) has found that hydrothermally altered crust frequently contaminates the upper levels of magma chambers including that of the Skaergaard. The most probable explanation for this phenomena is that there are times when the roof of the magma chamber stopes upward faster than hydrothermal circulation can remove the heat. That is, the rate of crack front propagation even at the highest levels of the magma chamber is relatively slow or the rate of growth of the magma chamber is much faster than the average rate of sea-floor spreading.

The latter possibility has been discussed extensively by Lister (1983b). Basically, episodic periods of rapid sea-floor spreading over less than a thousand years result in an axial magma chamber. For the process to work, the rapid spreading must occur simultaneously along a significant portion of the mid-oceanic ridge system implying movement on ridge-ridge transform faults. Such rapid movement could conceivably be detected from detailed studies of the sea floor and of the geomorphology of transform faults on land, such as the San Andreas system.

Limitations of Crack Front Hypothesis

Are there any processes which might make hydrothermal circulation at the ridge axis so inefficient that a steady-state magma chamber can exist? It may be difficult for water to initially enter the crack front, the water may not efficiently cycle through freshly formed cracks, or the large scale hydrothermal system may be unable to transport heat rapidly from the crack front to the surface. As the nature of large scale convection in the oceanic crust is considered later in this presentation only the topics directly related to cracking are considered in this section.

The applicability of the crack-front theory to the heterogeneous rocks of the oceanic crust may be questioned because the theory was developed using data on more homogeneous substances, a shortcoming discussed by Lister (1983a). The main difference between a heterogeneous polycrystalline substance and an ideal one is that thermal contraction cracking must occur within grains to remove the stresses associated with the mismatch of grains with different thermal elastic properties. The cracking within grains can relieve the more regional stresses at the same time. The fine-scale network of cracks thus created would be much less permeable than more widely spaced, through-going cracks. However, the tendency in a cooling material is still to form some through-going cracks. With even half as many cracks as proposed by Lister (1983a), the crack front would still efficiently remove heat.

The immediate egress of water from actively forming cracks is most likely the rate limiting process near the crack front. Quartz and other silicates usually become less soluble with decreasing temperature, especially below 400°C (Walther and Helgeson, 1978). In Lister's (1983a) model, the rapid removal of heat after the first cracks formed is essential toward keeping the production of cracks self-exciting. Clogging occurring away from the crack front would tend to divide that convection system into a hot, deep system and a cool, shallow system rather than precluding hydrothermal flow altogether (Lister, 1981). Relatively minor amounts of hydrothermal minerals, such as epidote, could restrict any exit cracks. These minerals would be unlikely to redissolve upon further cooling of the rock mass. The total amount of heat which could be brought through a crack in a region with a large thermal gradient would be limited by the heat to precipitate ratio in the hydrothermal fluid. The effect of clogging on the pattern of the gross hydrothermal system is discussed in a later section of this presentation. Other conceivable clogging schemes related to the crack front are discussed below; all tend to restrict circulation at temperatures greater than about 400°C. None of the mechanisms have been examined in detail; it is unclear whether any of the clogging mechanisms are a basis of objection to the crack front hypothesis.

Clogging of cracks by hydrothermally precipitated minerals may restrict the flow of water near the crack front. This process is not likely to be important in the highest temperature cracks, because the common alteration products of basalt, with the exception of hornblende, are stable at somewhat lower temperatures. Taylor (1983) has shown that the highest temperature alteration does not result in significant precipitation of vein minerals. Minerals which precipitate upon an increase in temperature and thus block the tip of forming cracks are not likely at near magmatic temperatures. It is likely that quartz becomes less soluble with increasing temperature above 400°C (Walther and Helgeson, 1978; see also Fournier and Potter, 1982; and Fournier et al., 1982). Anhydrite, which becomes less soluble with increasing temperature, should be completely gone before the seawater reaches the highest temperature cracks.

Another possibility is that the crack tips may become filled with more dense, saline waters, although the salinity increase needed for stable stratification is much greater at high temperatures than in the ocean (Vanko and Bishop, 1983). Extensive reaction of seawater with rock to form hydrous minerals would increase the chlorine concentration in the remaining water. The higher temperature alteration of gabbros in the Cayman trough apparently is believed to have been caused by such chlorine-rich water (Ito, 1979). Another possibility is that seawater disproportionates into a chlorine rich-fluid and a chlorine-poor fluid above about 400°C beneath ridge axes (Delaney, 1982). The scapolite from St. Paul's rocks is believed to have formed by a hypersaline fluid created by a combination of these processes (Vanko and Bishop,1983). The former possibility can retard flow only in those parts of the hydrothermal system that are so stagnant that the water gets significantly used up by reaction with wall rocks. The later possibility would retard convection in a finely porous material but would have less effect in a through going crack because separate convection systems in each fluid might become established.

OBSERVED PATTERN OF HYDROTHERMAL ALTERATION

Oxygen and hydrogen isotopic variations delineate the extent of hydrothermal alteration in several igneous bodies. The data on the Skaergaard intrusion indicate that the extent of hydrothermal alteration within the intrusive is controlled by the properties of the host rock (Taylor, 1983). The intrusion is extensively altered where the host rock is permeable lava flows and much less altered where the host rock is impermeable gneiss. The hydrothermal circulation thus appears to have been largely restricted to the lava flows and the part of the intrusion which was above the basalt-gneiss unconformity. This data can be interpreted to indicate that the hydrothermal circulation did not create an extensive fracture network in the previously unfractured rocks. The Skaergaard intrusion is also believed to have remained above 700°C for the initial 150,000 years. The inefficacy of cracks penetrating the pluton can be appreciated, as this time is comparable to the cooling time of a crustal magma chamber obtained from purely conductive models.

In the Oman ophiolite, a stratified pattern of alteration is well developed. An upper system of hydrothermal alteration affected the pillow basalts and the sheet dike complex. The chemistry of the altered rocks indicates that the water/rock ratio was quite high (around 10/1 by mass). The lower system affected the gabbroic complex. This alteration occurred at a much higher temperature and lower water/rock ratio than the upper system. This pattern is similar to that observed within the Skaergaard intrusion in that only moderate amounts of water penetrated into previously uncracked regions.

The method of entry of water into the lower system is also of interest. The flow pattern outlined by Taylor (1983) consists of flow beneath a wing-shaped magma chamber (Fig. 1). The cumulate gabbros on the floor of this chamber are cooled from below by the hydrothermal circulation. An alternative is that the circulation through the gabbroic complex occurs off-axis and the cumulate complex is cooled from the top down. The adcumulous texture of the cumulate gabbro is interpreted to indicate freezing before the layers became deeply buried by subsequent cumulates (Pallister and Hopson, 1981). If instead the adcumulous texture is interpreted to indicate slow cooling and exchange between melt and crystals (Morse, 1969, page 85), the second alternative seems more likely.

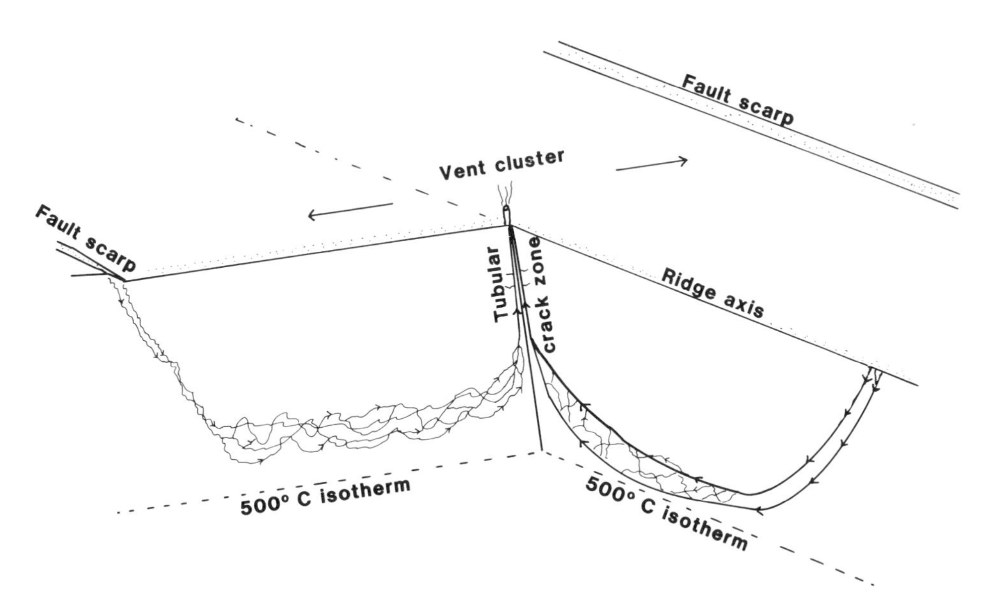

Fig. 1: A schematic diagram show the geometry of hydrothermal circulation proposed by Taylor (1983) (left) and in this presentation (right). Modest volumes of water flow underneath the chamber in Taylor's (1983) model cooling the cumulate layers from the bottom. In the model presented here the magma chamber is somewhat wider and a component of off-axis circulation penetrates into the deep crust. The structure of the axial circulation is the same in both models.

NATURE OF THE HYDROTHERMAL FLUID

Taylor (1983) presents strong evidence that the hydrothermal fluid in the intrusions which he studied was either meteoric water or seawater depending on the location of the body. The earlier view that hydrothermal fluids were mysterious emanations, which came from some great distance away (especially if they formed an economic deposit), is completely refuted. Taylor (1983) is even able to establish the climatic conditions during the alteration by meteoritic water of some of the bodies he studied. Taylor (1983) discusses alteration in subaerial environments by meteoritic waters and alteration by seawater beneath the ocean. The former process yields potassic granophyres while the later process yields sodium-rich plagiogranites. The difference among types of late stage intrusions formed by remelting of hydrothermally altered wall rocks is probably the most powerful method of determining the tectonic setting of older mafic complexes and perhaps even determining whether ridges were subaerial during Precambrian time.

Another environment, not discussed by Taylor (1983), is the hypersaline seas which form in narrow ocean basins soon after continental break-up. Hypersaline water would presumably produce scapolitic alteration, such as is observed in the Humbolt Lopolith in Nevada (Vanko and Bishop, 1983). It is not known whether plagiogranite formed by melting of scapolitized basalt would differ from those produced by melting of albitized basalt. A fourth environment, which might have both sodic and potassic granophyres is a sediment dominated ridge axis, such a the northern part of the Gulf of California. At present, no hypersaline or sediment dominated ophiolite complex has been recognized.

PHYSICS OF LARGE SCALE CONVECTION

A theory of convection at the ridge axis must be compatible with the high temperature of vented water, the short lifetime of individual vents, and localized points of egress of the fluid. Convection in the oceanic crust has been modeled as either flow through discrete, through-going crack zones or as flow through widely distributed porosity. Lister (1983a) considers the flow near the cracking front to be through closely spaced cracks and the flow above the cracking front to be through a porous medium. I will not repeat his results in detail.

Flow through porous media can be usefully modeled by using two dimensionless numbers. These numbers are also relevant to flow through discrete crack zones, but formulation of the physics is more complicated. The Nusselt number is the ratio of the actual heat transferred to the heat which would be transferred by conduction alone. It is by definition one if no convection is occurring, slightly over one in a weakly convecting system, and much greater than one in a vigorously convecting system. The Rayleigh number, which depends on the physical state of the system, determines the vigor of convection. Below a critical value of the Rayleigh number no convection occurs; at greater Raleigh numbers the convection becomes more vigorous. A quantitative theory of convection relates the Nusselt number to the Raleigh number.

For a porous material, the Raleigh number is

$$R = D\alpha_w g h \rho_w c_w \Delta T / k_r \nu_w \tag{1}$$

where D is the permeability, α_w is the thermal expansion coefficient of water, g is the acceleration of gravity, h is the thickness of the convecting region, $\rho_w c_w$ is the volume specific heat of water, ΔT is the temperature difference across the convecting region, k_r is the thermal conductivity of the rock, and ν_w is the kinematic viscosity of water. Around 350°C and at pressures corresponding to the upper oceanic crust, the volume heat capacity and the thermal expansion coefficient of water greatly increase with temperature and the viscosity of water decreases with increasing temperature. It is thus not particularly surprising that the exiting waters are approximately 350°C. The permeability is by far the most uncertain property in the Raleigh number for hydrothermal convection at the ridge axis.

An adequate theory for relating Nusselt number to Raleigh number has not been developed for porous flow. For feeble convection just above the critical Rayleigh number the amount of heat carried by convection is proportional to the Rayleigh number. For more vigorous convection, there is a weak power law dependence between Rayleigh number and Nusselt number. The theory of porous convection is discussed further by Lister (1983a)

As noted by Lister (1983a), a significant difference between convection in a porous material and the more familiar case of a convecting fluid are that only a small fraction of the material actually moves in a porous medium. Flow in a porous medium tends to change slowly with time because hot upwelling water must heat up a larger mass of rock before it can vent on the surface. Similarly, a hot region of upwelling water would remain hot for a significant time after the fluid flow from below was cut off. The episodic nature of axial hydrothermal flow is thus hard to explain if the hot water upwells over a broad region. For example, if the upwelling fluid came through a zone 200 m square by 500 m deep it would take several years to heat the zone to the temperature of the water.

Another property of flow through porous media is that the flow is dispersive. That is, a parcel of fluid becomes spread out laterally and along the direction of flow by the vagaries of movement through a complex network of passages. A new batch of fluid moving into a zone therefore mixes with the fluid already in the pore space. However, the highest temperature axial vents show little evidence of having mixed with lower temperature waters.

Convection through Crack Zones

An alternative theory to wide spread porous convection is convection through discrete crack zones. The stable mode of long lasting convention in through-going crack zones is for the flow to be confined in the plane of the crack zone (Strens and Cann, 1982). A two-dimensional numerical model perpendicular to the crack zone would not include this mode of convection. The pressures associated with the flow of water through cracks do not significantly modify the shape of the crack.

The main interest here is to model the high temperature venting at the ridge axis. The critical flow rate necessary for hot fluid to vent can be obtained from a dimensional approximation comparing the conductive heat loss from the upwelling fluid to the total amount of heat carried by the fluid following Lowell (1975) and Sleep and Wolery (1978). Hot water can be expected to vent if the heat loss from conduction is small compared with the heat in the upwelling fluid. Here, the axial vent at the mid-oceanic ridge may be modeled as a vertical tubular crack zone. For transient flow the amount of conductive heat loss is proportional to the volume of material which becomes heated while the flow is active. If the size of the crack zone is comparable to the size of the heated region, then the distance of heating away from the crack zone is proportional to $(\kappa t)^{1/2}$ where κ is thermal diffusivity and t is time. The heated area is thus

$$Area = \pi \kappa t \tag{2}$$

and the total heat loss is

$$Q = \pi \kappa t T h \rho c = \pi k T h t \tag{3}$$

where T is the temperature in the upwelling water, ρc volume specific heat, k is thermal conductivity, and h is the vertical extent of the crack zone. The heat loss per time is

$$F \equiv \frac{Q}{t} = kTh \tag{4}$$

where the π has been dropped to obtain a dimensional result. The heat loss, F, is independent of time. The critical flow rate for venting water is the flux, F divided by the amount of heat per volume of water or

$$W \equiv \frac{F}{\rho_w c_w T} = \kappa_{rw} h \tag{5}$$

which is independent of the temperature. (An exact solution, which can not be conveniently evaluated, is given Carslaw and Jaeger, 1959, page 336). For the axial vents, k = 0.006 cal/cm-sec-°C, T = 350° C, and h = 1 km seem appropriate. This gives a critical heat flux of 2×10^5 cal/sec compared with the observed flux of 4×10^7 cal/sec from a vent cluster. For $\rho_w c_w$ = 1 cal/cm³ °C, the critical flow rate is 0.6 l/sec compared with the observed 100 l/sec. The computed values of the critical conductive heat loss and the critical flow rate are so small compared with the actual values that we can conclude that the upwelling water may have lost very little heat to conduction. Another implication is that vents with much smaller water and heat fluxes could still vent hot water for a period of time.

At flow rates comparable to the critical flow rate, the exiting fluid loses a significant proportion of its heat so that warm water escapes. In this situation, vents would quickly clog with minerals precipitated by the cooling of the water. Clogging is most efficient at moderately high temperatures, 350°C, where there is much silica dissolved in the fluid and where the kinetics of precipitation are fast. For example, the concentration of silica would drop from 1000 ppm to 500 ppm upon cooling from 350°C to 250°C. This is about 0.2 cm^3 per liter of seawater. For a flow rate of 1 l/sec, the silica would clog a 100 m square area of a 1 cm wide through-going crack in one year.

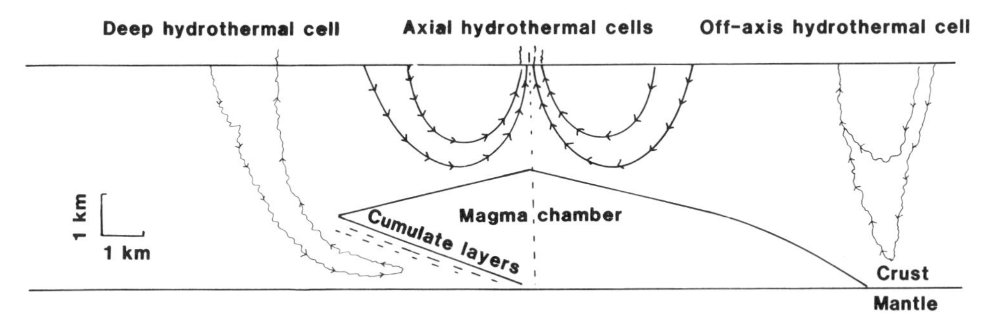

Fig. 2: A schematic cut-away diagram shows the hydrothermal system for an active vent cluster. The vent is fed by a narrow tubular crack zone. Hot water already present in cracks is tapped by the vent. Through-going recharge occurs on planar crack zones at the ridge axis. More widely distributed recharge occurs along fault zones near the axis. The bulk of the flow does not penetrate the 500°C isotherm. Clogging tends to self-seal leakage from the crack zone directly under the vent.

At significantly slower flow rates, the water cools on the way up and the precipitation of minerals is spread out over a larger region of the flow system. Weak discharges of water, therefore may persist for some time after a vent system has clogged. The majority of the veins in dredged samples from the mid-Atlantic ridge were probably formed over time by these weak discharge (Scott et al., 1982). Clogging may also explain why long-lasting porous convection does not quickly cool the oceanic crust and why isolated vents occur. Basically, any slowly upwelling water will cool and also mix with cool waters before it reaches the surface. On a small scale this would self-seal through-going crack zones from leaks. On a larger scale, regions with widespread upwelling would become clogged by gradual precipitation of minerals. Clogging would be unimportant in the downgoing parts of the system (except for perhaps anhydrite) and in relatively isothermal parts of the system. Once hot water in a crack zone reached the surface, the bulk of the dissolved material would escape into the ocean rather than clog cracks at depth. The fluid which escaped from the vent could be recharged by porous flow over a large area and tap water which was already hot from a broad zone around the vent. The geometry of this flow scheme is shown in figure 2.

Acknowledgements. The research was supported in part by the National Science Foundation grants #EAR80-01076 and EAR81-15522.

REFERENCES

Carslaw, H. S., and Jaeger, J. C., 1959, "Conduction of Heat in Solids, 2nd ed.," Oxford University Press, Oxford, 510pp.

Delaney, J. R., 1982, Generation of high Salinity fluids from seawater by two-phase separation (abst.), *EOS Trans. Amer. Geophys. Union, 63*: 1135-1136.

Fournier, R. O., and Potter, R. W. II, 1982, An equation correlating the solubility of quartz in water from 25°C to 900°C at pressures up to 10,000 bars, *Geochim. Cosmochim. Acta, 46*: 1969-1973.

Fournier, R. O., Rosenbauer, R. J., and Bischoff, J. L., 1982, The solubility of quartz in aqueous sodium chloride solution at 350°C and 180 to 500 bars, *Geochim. Cosmochim. Acta, 46*: 1975-1978.

Gregory, R. T., and Taylor, H. P., 1981, An oxygen isotope profile in a section of Cretaceous oceanic crust, Samail ophiolite, Oman: Evidence for $\delta^{18}O$ buffering of the oceans by deep (>5 km) seawater-hydrothermal circulation at mid-oceanic ridges, *J Geophys. Res., 86*: 2737-2755.

Ito, E., 1979, Oxygen isotope ratios of minerals from intrusive rocks of the Mid-Cayman ridge (abst.), *EOS Am. Geophys. Un. Trans., 60*: 410.

Lister, C. R. B., 1981, Rock and water histories during sub-oceanic hydrothermal events, *Oceanologica Acta, SP, Proceedings 26th International Geological Congress*: Geology of Oceans Symposium, 41-46.

Lister, C. R. B., 1982, "Active" and "Passive" hydrothermal systems in the oceanic crust: predicted physical conditions, *in*: "The dynamic Environment of the Ocean Floor", K. A. Fanning, and F. T. Manheim, eds., D. C. Heath, Lexington, Massachusetts, 441-470.

Lister, C. R. B., 1983a, The basic physics of water penetration into hot rock: (this volume).

Lister, C. R. B., 1983b, On the intermittency and crystallization mechanisms of sub-sea-floor magma chambers, *Geophys. J. Roy. Astron. Soc.*: (in press).

Lowell, R. P., 1975, Circulation in fractures, hot springs, and convective heat transport on mid-ocean ridge crests, *Geophys. J. Roy. Astron. Soc.*, *40*: 351-365.

Morse, S. A., 1969, The Kiglapait layered intrusion, Labrador, *Geol. Soc. Amer. Memoir*, *112*: 204 pp.

Palister, J. S., and Hopson, C. A., 1981, Samail ophiolite plutonic suite: Field relations, phase variation, cryptic variation and layering, and a model of a spreading ridge magma chamber, *J. Geophys. Res.*, *86*: 2593-2644.

Scott, R. B., Temple, D. G., and Peron, P. R., The nature of hydrothermal exchange between oceanic crust and seawater at 26°N latitude, Mid-Atlantic ridge, *in*: "The dynamic Environment of the Ocean Floor," K. A. Fanning and F. T. Manheim, eds., D. C. Heath, Lexington, Massachusetts, 441-470.

Sleep, N. H., and Wolery, T. J., 1978, Thermal and chemical constraints on venting of hydrothermal fluids at mid-ocean ridges, *J. Geophys. Res.*, *83*: 5913-5922.

Strens, M. R., and Cann, J. R., 1982, A model of hydrothermal circulation in fault zones at mid-ocean ridge crests, *Geophys. J. R. astron. Soc.*, *71*: 225-240.

Taylor, H. P. Jr., 1983, Oxygen and hydrogen isotope studies of hydrothermal interactions at submarine and subaerial spreading centers: (this volume).

Vanko, D. A. and Bishop, F. C., 1983, Occurrence and origin of marialitic scapolite in the Humboldt lopolith, N. W. Nevada, *Contribut. Mineral. Petrol.*: (in press).

Walther, J. V., and Helgeson, H. C., 1978, Calculation of the thermodynamic properties of aqueous silica and the solubility of quartz and its polymorphs at high pressures and temperatures, *Amer. J. Sci.*, *277*: 1315-1351.

OXYGEN AND HYDROGEN ISOTOPE STUDIES OF HYDROTHERMAL INTERACTIONS AT SUBMARINE AND SUBAERIAL SPREADING CENTERS

Hugh P. Taylor, Jr.

Division of Geological and Planetary Sciences
California Institute of Technology
Pasadena, California 91125

ABSTRACT

Oxygen and hydrogen isotope studies of three subaerial spreading centers, East Greenland (50-55 m.y.), Red Sea (22 m.y.) and Iceland (<12,000 y.) demonstrate that deep circulation and interaction of surface waters with layered gabbro intrusions and their associated basaltic country rocks is very common in such environments. Overall, water/rock ratios are very high (commonly >1), and much of the alteration and exchange takes place at very high temperatures (500°-1000° C) at depths of a least 5-10 km. The sheeted dike complexes themselves produce large-scale convective circulation of hydrothermal fluids that overlaps the hydrothermal systems produced by the gabbro plutons. In all these areas, assimilation and/or partial melting of hydrothermally altered country rocks in the vicinity of the magma chambers is an important petrological process. Detailed analysis of the differentiated Skaergaard layered gabbro body in East Greenland proves that only very minor amounts of H_2O diffused directly into the liquid magma, even though the hydrothermal system was initiated immediately after intrusion and operated for the entire 130,000-year period of crystallization. However, blocks of low -^{18}O roof rocks fell into the Skaergaard magma; these had undergone hydrothermal depletion in ^{18}O above the magma chamber prior to their assimilation. Most of the ^{18}O depletion observed in these environments took place after crystallization, with plagioclase becoming much more strongly depleted in ^{18}O than coexisting clinopyroxene. Essentially all of the above processes have also been documented in the submarine equivalents of such spreading centers, namely the ophiolite complexes. In particular, at the Samail ophiolite in Oman, the data of Gregory and Taylor (1981) indicate that pervasive subsolidus hydrothermal exchange with seawater occurred throughout the upper 75% of this

8-km-thick oceanic crustal section; locally, the H_2O even penetrated down below the Moho into the tectonized peridotite. Pillow lavas ($\delta^{18}O$ = 10.7 to 12.7) and sheeted dikes (4.9 to 11.3) are typically enriched in ^{18}O, and the gabbros (3.7 to 5.9) are depleted in ^{18}O. Integrating $\delta^{18}O$ as a function of depth for the entire ophiolite establishes (within geologic and analytical error) that the average $\delta^{18}O$ (5.7 ± 0.2) of the oceanic crust did not change as a result of all these hydrothermal interactions with seawater. Therefore the net change in $\delta^{18}O$ of seawater was also zero, indicating that seawater is buffered by MOR hydrothermal circulation. Under steady-state conditions the overall bulk ^{18}O fractionation (Δ) between the oceans and primary mid-ocean ridge basalt magmas is calculated to be +6.1 ± 0.3, implying that seawater has had a constant $\delta^{18}O \approx -0.4$ (in the absence of transient effects such as continental glaciation). The $\delta^{18}O$ data and the geometry of the mid-ocean ridge (MOR) magma chamber require that two decoupled hydrothermal systems must be present during much of the early spreading history of the oceanic crust (approximately the first 10^6 years); one system is centered over the ridge axis and probably involves several convective cells that circulate downward to the roof of the magma chamber, while the other system operates along the sides of the chamber in the layered gabbros. Upward discharge of ^{18}O-shifted water into the altered dikes from the lower system, just beyond the distal edge of the magma chamber, combined with the effects of continued low-T hydrothermal activity, produces the ^{18}O enrichments in the dike complex and pillow lavas. Whereas the dilute meteoric fluids have had relatively little influence on the chemical compositions of rocks and magmas in the subaerial spreading centers, this is not the case in the ophiolite complexes, where more saline, NaCl-rich fluids were involved. Oceanic plagiogranites are the equivalents of the low-^{18}O potassic granophyres in subaerial spreading centers; the chemical differences between the two types of rocks are due to subsolidus hydrothermal exchange and to the fact that the hydrothermally altered precursors in the oceanic environments were spilites with very high Na/K, Na/Ca, and Na/Rb ratios.

INTRODUCTION

The purpose of this paper is to review the oxygen and hydrogen isotope data recently obtained on subaerial and submarine spreading centers, and to examine some of the implications of such data with respect to contrasting rock types and mineral assemblages in these two environments. Oxygen and hydrogen isotope studies have clearly demonstrated that deep circulation and interaction between surface waters and igneous intrusions is a very common phenomenon (Taylor and Epstein, 1963; Taylor, 1968, 1971, 1974a, 1974b, 1977; Taylor and Forester, 1971; Sheppard et al., 1969; 1971). Convective movement of ground waters is particularly important around intrusions emplaced into young volcanic terranes, and in such areas one can expect

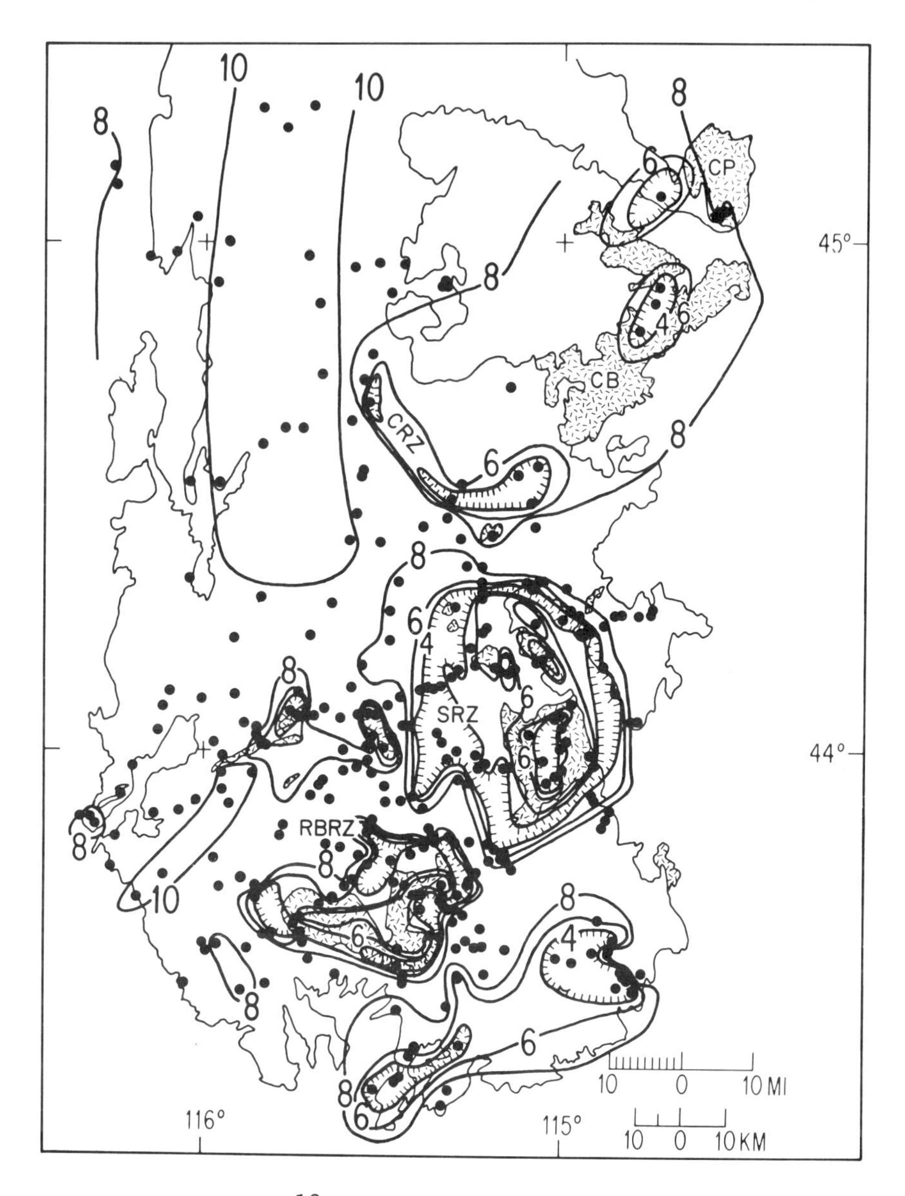

Fig. 1. Contour map of $\delta^{18}O$ values of feldspar (dots; including a few whole-rock analyses) from granitic rocks of the southern part of the Idaho batholith (after Criss and Taylor, 1982). Most of this composite batholith is made up of Jurassic and Cretaceous plutons. Some younger Eocene plutons are shown in stippled pattern. Contours are drawn at $\delta^{18}O$ = +10, +8, +6, and +4. Within the hatchured +4 contour the values go down as low as -6. The figure includes the earlier $\delta^{18}O$ data of Taylor and Magaritz (1978).

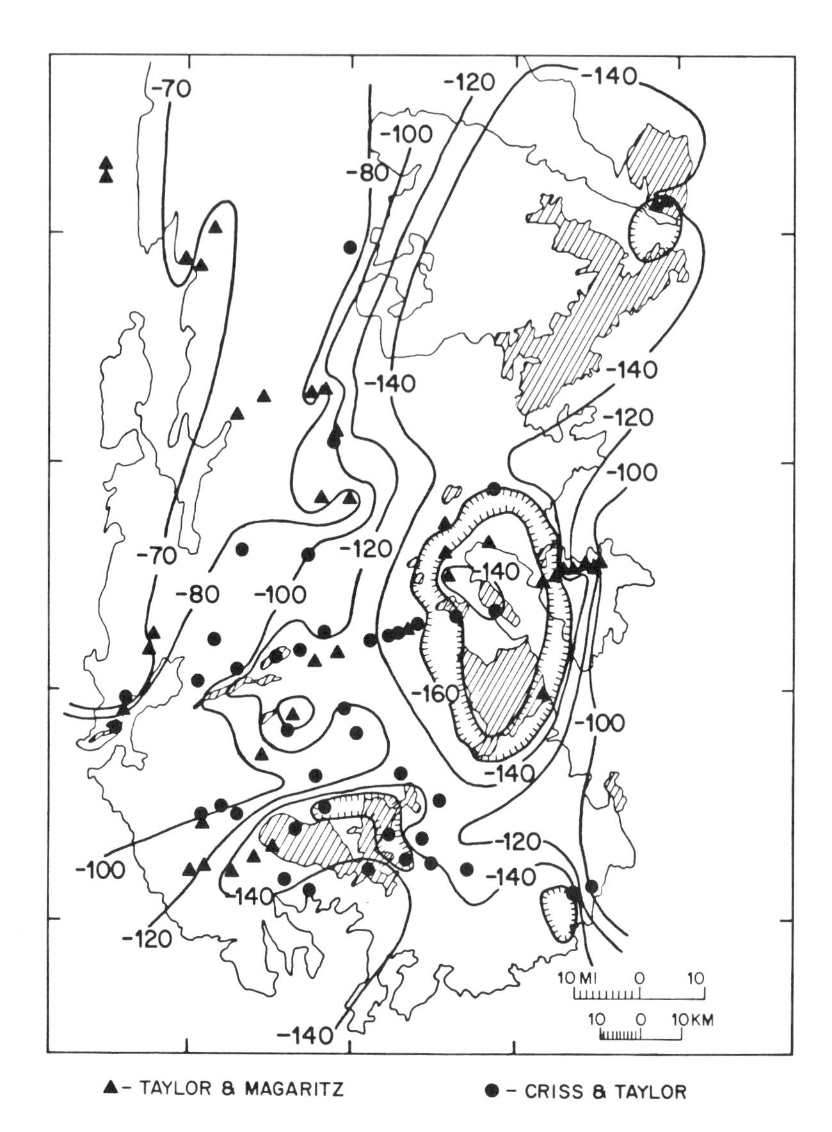

Fig. 2. Contours of δD values at 20 per mil intervals in biotite (±chlorite) in the southern part of the Idaho batholith (after Taylor and Magaritz, 1978, and Criss and Taylor, 1982). The -160 contour is hatchured, outlining a low-deuterium zone that corresponds very well with the extremely low $\delta^{18}O$ zones shown in Fig. 1. Normal δD values of about -70 per mil are found only along the western edge of the batholith. Pronounced D/H alteration effects (δD < -140) are found in the vicinity of all the Eocene plutons (diagonal lined pattern).

extremely large-scale, deep-seated, high-temperature hydrothermal effects.

To illustrate the kinds of large-scale isotopic changes that can occur in igneous rocks, Figs. 1 and 2 show the variations in $\delta^{18}O$ and δD observed in granitic rocks of the Idaho batholith, based on work by Taylor and Magaritz (1978) and Criss and Taylor (1982). The original $\delta^{18}O$ and δD values in these granitic plutons (largely Mesozoic in age) were approximately +9 to +11 and -65 to -80, respectively. Note that over an area of about 15,000 km^2 in southern Idaho, essentially all the δD values of the rocks and most of the $\delta^{18}O$ values of the feldspars were markedly changed (to values as low as $\delta^{18}O < -5$ and $\delta D < -160$). These effects were produced by widespread and deep (5 to 10 km depth) convective circulation of low-^{18}O, low-D meteoric ground waters during Eocene time, about 40 to 45 m.y. ago (Criss, Lanphere and Taylor, 1982). The hydrothermal systems were associated with batholith-sized Eocene granite plutons, which provided the heat necessary to drive the fluid movement. Hydrothermal effects were localized at the contacts of these plutons and in the vicinity of high-permeability ring-fracture zones (calderas). The calderas were formed by collapse of the volcanic centers following voluminous eruption of rhyolitic ash-flow tuffs that covered much of southern Idaho in the Eocene. The isotopic and hydrothermal effects now observed in these "fossil" hydrothermal systems in the Idaho batholith are thought by Criss and Taylor (1982) to be similar to what we might observe today in the Yellowstone volcanic field, Wyoming, if we could examine those rocks by drilling to 3 to 5 km depth (Fig. 3). The similarities are striking: in both areas the most intense hydrothermal effects are associated with the caldera ring fractures or with the central resurgent domes of granitic magma.

In such hot, water-rich environments as those described above, a variety of petrologic phenomena conceivably could take place, including (1) Hydrothermal alteration of the volcanic country rocks. (2) Melting of the hydrothermally altered country rocks above the magma chambers. (3) Stoping of the hydrothermally altered roof rocks and exchange or assimilation between these stoped blocks and the magmas. (4) Continued fractional crystallization and differentiation of such contaminated magmas, with consequent modification of the late-stage magmatic differentiates. (5) Possible direct influx of H_2O into the magma by diffusion. (6) Sub-solidus hydrothermal exchange between the crystalline magma body and the aqueous fluids; this can begin to occur at temperatures as high as 700° C, or at whatever temperatures below the solidus that through-going fractures first begin to form in the rocks. (7) Continued sub-solidus hydrothermal exchange down to temperatures as low as 50°-100° C during cooling of the volcanic-intrusive terrane.

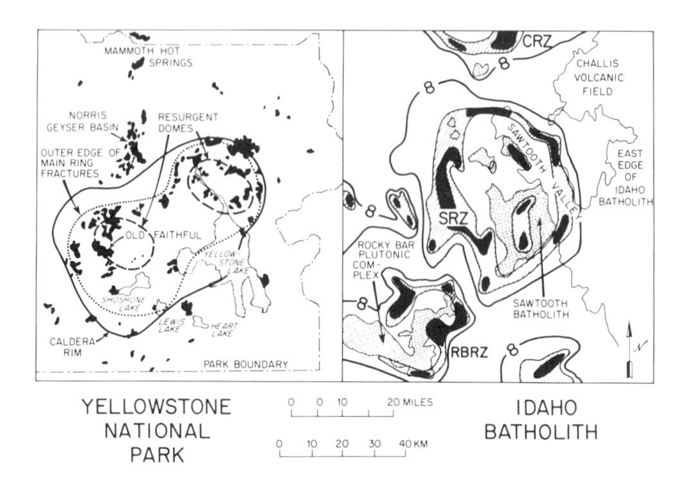

Fig. 3. Comparison of the low-^{18}O areas in the Idaho batholith (Criss and Taylor, 1982) with the Yellowstone rhyolite volcanic field, Wyoming (same scale). In the Yellowstone area, multiple ash-flow tuff eruptions during the last 2 million years produced major calderas and associated ring fractures, followed by resurgent doming (Smith and Christiansen, 1980; Eaton et al., 1975). The main areas of present-day hydrothermal activity are shown in solid black. In Idaho, the positions of the Eocene plutons (resurgent domes?) and the annular low-^{18}O zones ($\delta^{18}O$ feld $< +4$) are repeated from Fig. 1; the solid black areas within the low-^{18}O ring zones represent areas with $\delta^{18}O < +2$. Note the general similarities between the features on the two maps, particularly the association of hydrothermal alteration with either (1) ring fracture zones, or (2) resurgent domes.

The importance of such varied phenomena can be investigated by means of oxygen and hydrogen isotope studies, because major differences of $\delta^{18}O$ and δD are found among the various kinds of surface-derived waters throughout the world. Sea water, for example, is very uniform at $\delta^{18}O \approx 0$ and $\delta D \approx 0$, whereas meteoric waters show pronounced latitudinal and topographic variations in $\delta^{18}O$ and δD, from values near zero in tropical oceanic islands to values as low as -20 to -25 and -150 to -250, respectively, in high-latitude continental environments (Friedman et al., 1964; Craig, 1961; Epstein et al., 1965). If dilute meteoric waters are involved in the hydrothermal alteration, no major-element or trace-element chemical changes will accompany the isotopic exchange. However, if sea water is involved, there should be some dramatic chemical changes as well, because sea water has a much higher salinity, and also will have other obvious chemical signatures (such as high NaCl contents and high Na/K and Na/Ca ratios).

Below we shall discuss the isotopic effects produced by deep hydrothermal circulation in three subaerial spreading centers that formed at different times in different latitudes, and which can now be examined at different erosional levels. The isotopic contrasts in these three subaerial spreading centers (East Greenland, Iceland, Red Sea) will then be compared with analogous effects observed in ophiolite complexes formed in submarine environments, which are thought to be analogs of presentday mid-ocean ridges. We shall then examine some of the chemical differences between rocks formed in the meteoric-hydrothermal and marine-hydrothermal environments. If the kinds of petrologic processes outlined above are important, the only major differences between the various subaerial environments should be $\delta^{18}O$ and δD. Also, the submarine environments should all be substantially identical to one another, because ocean water is isotopically and chemically very uniform. On the other hand, the subaerial environments ought to differ markedly from the submarine environments in both isotopic characteristics and chemical characteristics.

EAST GREENLAND DIKE SWARM AND SKAERGAARD INTRUSION

Geological Relationships

The geologic setting of the Skaergaard intrusion and East Greenland dike swarm has been described by Wager and Brown (1967). The regional geology is indicated in Fig. 4. The $\delta^{18}O$ and δD values of the rocks in this area have been measured by Taylor and Forester (1979).

The Skaergaard intrusion is oval-shaped in plan, with dimensions of about 10 km by 7 km. It was intruded about 50-55 m.y. ago across a flatlying unconformity between an 8-10 km-thick section of tholeiitic plateau basalt lavas (with subordinate volcanic breccia and tuff) and an underlying Precambrian basement gneiss complex. Immediately after intrusion, the region was tilted 10°-20° to the SE, as a result of a monoclinal flexure that trends subparallel to the present Greenland coast. The lowest 1-2 km of the basaltic lava section was apparently deposited in a shallow marine basin (many of the flows exhibit pillow structures at their base, and

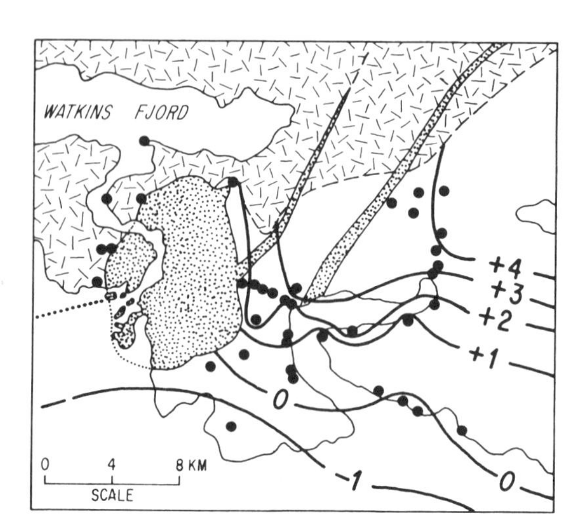

Fig. 4. Generalized geologic map of the coast of Greenland in the vicinity of the Skaergaard intrusion and East Greenland dike swarm, showing wholerock $\delta^{18}O$ contours in the plateau lavas and dikes (blank pattern, after Taylor and Forester, 1979). The dikes trend E-W, and are most abundant (~100% of the outcrop) in the vicinity of the 0 and -1 contours. The dense stippled pattern indicates the Skaergaard intrusion and two gabbro macrodikes of similar age. The random-dash pattern indicates the Precambrian basement gneiss, which underlies the basalts along an originally flat-lying unconformity (dashed line) that now dips 10°-15° to the SE.

are intercalated with units containing marine fossils); the rate of eruption then exceeded that of subsidence and the upper 8 km of the basalt pile were formed in dominantly subaerial conditions over a relatively short time interval, probably only a couple of million years just prior to emplacement of the Skaergaard intrusion (Soper et al., 1976).

Just prior to emplacement of the Skaergaard intrusion, numerous closely spaced basaltic dikes were emplaced along an E to NE-trending lineament during opening of the North Atlantic Ocean (Brooks, 1973). In the vicinity of the Skaergaard intrusion, this dike swarm is most intense right at the coast, where the rocks are typically 100% dikes that strike E-W (Fig. 4). Dikes having this orientation rapidly decrease in number to the north, and they are practically absent from the northern and central parts of the Skaergaard intrusion and the adjacent country rocks.

Isotopic Systematics

The $\delta^{18}O$ variations in the plateau lavas and the East Green-

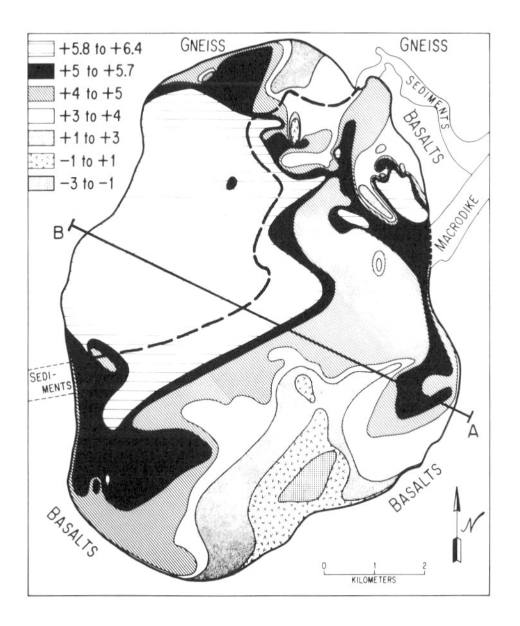

Fig. 5. Generalized distribution of $\delta^{18}O$ values of plagioclase within the Skaergaard intrusion (after Taylor and Forester, 1979). The "normal" $\delta^{18}O$ values (horizontal lined pattern, +5.8 to +6.4) are largely confined to the NW portion of the pluton, below the heavy dashed line; the latter represents a projected horizon defined by the base of a 250 m-thick section of sedimentary rocks that lies between the basalts and the gneiss.

land Dike Swarm are indicated by $\delta^{18}O$ contours on Fig. 4. The lowest $\delta^{18}O$ values in the basalts occur near the margins of the upper part of the Skaergaard intrusion (-0.7 to -2.3); however, on a regional scale the lowest values occur in the vicinity of the dike swarm along the Atlantic Coast, where dikes form almost 100% of the outcrops and $\delta^{18}O$ values are typically 0 to -1.9. The $\delta^{18}O$ contours are sub-parallel to the strike of the dike swarm (Fig. 4), and it is clear that the regional pattern is independent of the Skaergaard intrusion.

The East Greenland Dike Swarm is the subaerial equivalent of the sheeted dike complexes of ophiolites, and the $\delta^{18}O$ data (Fig. 4) indicate that the dikes themselves provided enough heat-energy to set up their own meteoric-hydrothermal system; this partially overlapped the Skaergaard system both in space and in time. The basalts are intensely jointed and fractured, and in the vicinity of the Skaergaard intrusion and the dike swarm they are heavily altered to chlorite, epidote, actinolite, calcite, prehnite, and secondary quartz. However, primary igneous textures and structures are well-preserved, as the alteration was produced without any accompanying penetrative deformation.

The Skaergaard intrusion exhibits a very complex pattern of $\delta^{18}O$ (Fig. 5), but in general the values decrease stratigraphically upward in the cumulate sequence; the lowest $\delta^{18}O$ plagioclase is found in the Upper Border Group (-2.4) and the highest value is found in the Lower Zone (+6.4). As shown in the generalized cross-section in Fig. 6a, the regional unconformity between the plateau basalts and the underlying gneiss clearly played an important role in the hydrothermal circulation pattern. Vigorous circulation in the permeable basalts allowed penetration of considerable amounts of H_2O into the gabbro _above_ the projected level of the unconformity (heavy dashed line on Fig. 5). Below this horizon very little H_2O penetrated into the gabbro, apparently because the gneiss was so much less permeable than the basalts. The cumulate gabbros beneath the projected unconformity horizon typically have normal $\delta^{18}O$ plagioclase (+6.0 to +6.4), as well as normal $\delta^{18}O$ pyroxene (+5.2 to +5.5). These values are typical of fresh, unaltered tholeiitic gabbros elsewhere in the world.

It is now well known that plagioclase-pyroxene $^{18}O/^{16}O$ systematics follow a characteristic pattern in layered gabbros that have been hydrothermally altered at high temperatures; the plagioclases are _much_ more susceptible to ^{18}O exchange with hot aqueous fluids and thus more ^{18}O-depleted than are the coexisting pyroxenes. The plagioclase-pyroxene pairs in such cases display steep, non-equilibrium, $^{18}O/^{16}O$ trajectories that are diagnostic of subsolidus exchange (see Fig. 21 below). For example, most of the Skaergaard plagioclase-pyroxene pairs lie close to a least-squares line given

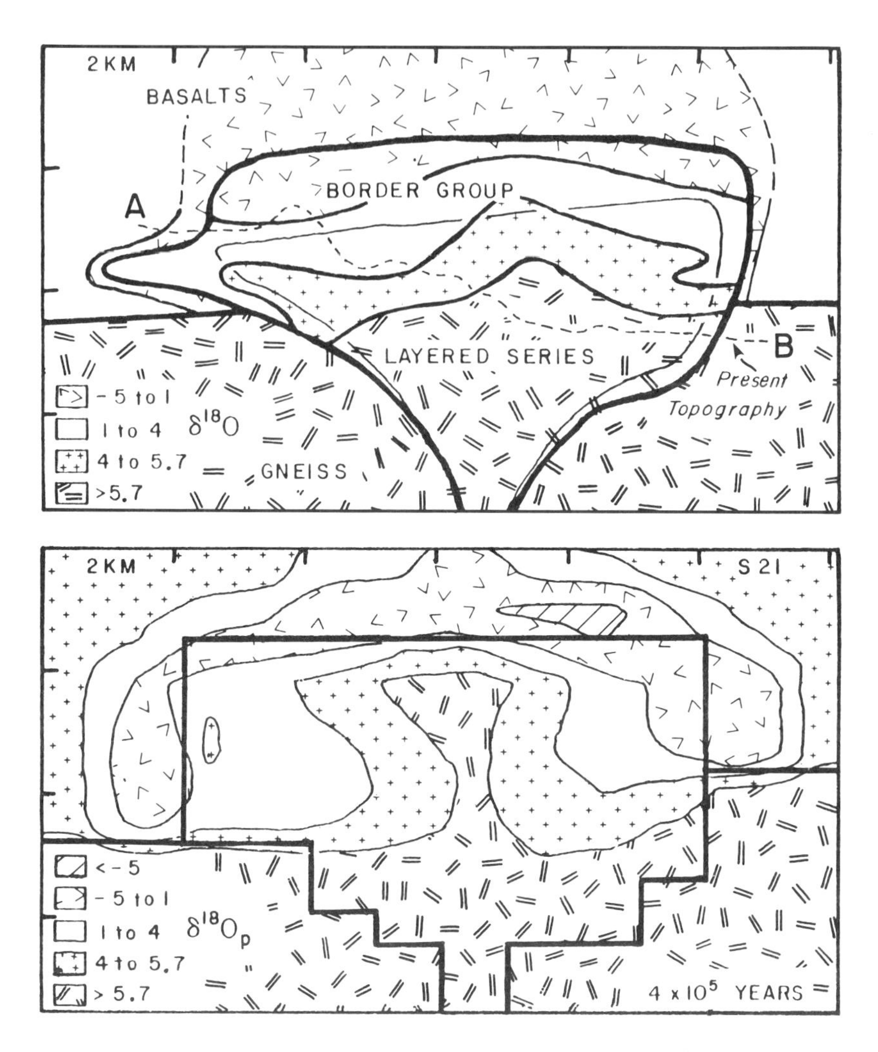

Fig. 6a. Composite geological cross-section of the Skaergaard intrusion, showing interpolated $\delta^{18}O$-values in plagioclase with respect to major rock types and topography along the restored geological section line A-B shown in Fig. 5 (after data by Taylor & Forester, 1979).

Fig. 6b. Calculated $\delta^{18}O$ values of plagioclase in an idealized cross-section of the Skaergaard intrusion at 400,000 years (see Fig. 8), based on the numerical modeling study of this hydrothermal system by Norton and Taylor (1979). The good match between Figs. 6a and 6b indicates that the bulk permeabilities assumed for the various rock types (gneiss = 10^{-16} cm^2, gabbro = 10^{-13} cm^2, basalts = 10^{-11} cm^2) are probably quite accurate.

by the equation:

$$\delta^{18}O \text{ plagioclase} \approx 4.7\ \delta^{18}O \text{ clinopyroxene} -20.0$$

Thus the dike swarm and the Skaergaard intrusion both set up meteoric-hydrothermal systems, whose intensities drop off sharply to the north or downward in the pluton, respectively, as the dikes become very sparse or as one passes down into the impermeable Precambrian basement gneiss. None of the analyzed basalts seem to have escaped ^{18}O exchange. The highest $\delta^{18}O$ values (+4.0 to +4.1) occur 12 to 13 km east of the Skaergaard intrusion, but these are still lower than in typical primary basaltic magmas. Sheppard et at. (1977) analyzed somewhat fresher, but still slightly chloritized basalts 100-120 km ENE of the Skaergaard intrusion; these have $\delta^{18}O$ = 4.1 to 4.9. Thus even the freshest, least altered East Greenland plateau basalts apparently underwent exchange with heated water and became somewhat depleted in ^{18}O relative to the "normal" value of +5.7, which is characteristic of most fresh tholeiitic basalts throughout the world.

One disclaimer has to be attached to the above statement, because Muehlenbachs et al. (1974) found that many fresh, post-glacial, tholeiitic basalt lavas on Iceland were erupted with abnormally low $\delta^{18}O$ values, many with $\delta^{18}O$ = +3.5 to +4.5 (see below). Such low -^{18}O basaltic magmas are very rare elsewhere in the world, but because the present-day geologic setting of Iceland (e.g. outpouring of plateau basalts in a subaerial, rift environment) bears many similarities to the early Tertiary situation in East Greenland, we cannot rule out the possibility that the original magmatic $\delta^{18}O$ values of some of the East Greenland plateau basalts were as low as +4.0. However, it was clearly demonstrated by Taylor and Forester (1979) that the original Skaergaard magma had a "normal" $\delta^{18}O \approx +5.7$, and it seems most likely that this was the probable $\delta^{18}O$ value of the bulk of the East Greenland plateau basalt magmas as well.

D/H ratios have been determined on only a few minerals from the Skaergaard intrusion and its country rocks (Fig. 7). One reason for this is that OH-bearing minerals are exceedingly rare in the intrusion itself. Except perhaps for tiny amounts of interstitial biotite in some of the rocks of the lowest part of the layered series, no primary igneous OH-bearing minerals occur in this body. Chlorite, actinolite, and stilpnomelane are, however, found as rare, late-stage alteration products, along fractures, and in miarolitic cavities.

As shown in Fig. 7, all of the rocks in the vicinity of the Skaergaard intrusion (the gabbros themselves, the basement gneiss, and the plateau basalts) have very low δD values indicative of meteoric-hydrothermal exchange. It thus appears that at least small amounts of meteoric water also circulated through the Precambrian basement gneiss. This is undoubtedly because of the large numbers of igneous intrusions found throughout the area, each of which was probably associated with some hydrothermal activity (Fig. 4). The overall water/rock ratios at such depths had to be less than about 0.05 (in weight units), however, because there is a general lack of any discernible ^{18}O depletion throughout most of the basement gneiss.

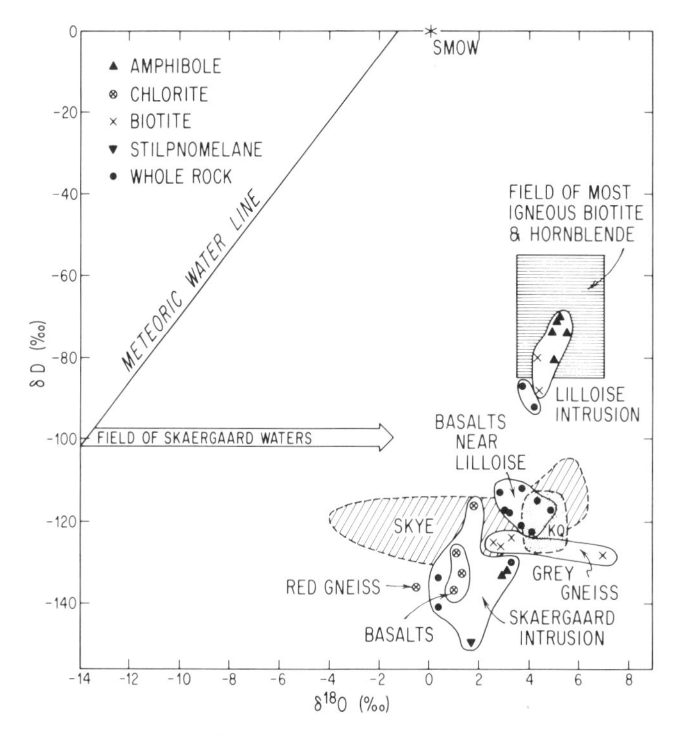

Fig. 7. Plot of δD vs. $\delta^{18}O$, showing the position of the meteoric water line and SMOW (standard mean ocean water, defined as equal to zero for both $\delta^{18}O$ and δD). The positions of the Skaergaard and Lilloise plutons and the East Greenland basalts and gneiss are indicated, as well as the general field of the Kangerdlugssuaq intrusion (KQ) and most primary igneous biotites and hornblendes (see text).

Any rocks whose $\delta^{18}O$ values have been measureably lowered by meteoric-hydrothermal interaction must have encountered high enough water-rock ratios to completely overwhelm the D/H of magmatic or metamorphic hydroxyl minerals originally present (see Fig. 8 in Taylor, 1977). The only localities where measurable ^{18}O depletions are observed in the gneiss are at the western contact with the Skaergaard intrusion, and along a north-trending fracture zone west of the intrusion where the feldspar in the gneiss has been transformed to a brick-red, turbid appearance and the mafic minerals have been totally chloritized (red gneiss $\delta D = -136$, Fig. 7). These data are important because they indicate that at least locally, along major fracture conduits, there was significant meteoric water circulation through the basement gneiss.

The δD values of OH-bearing minerals that have exchanged with a hydrothermal fluid can be used to calculate the δD of the fluid, utilizing the experimental calibration curves of Suzuoki and Epstein (1976). These calibration curves are sensitive to Fe/Mg ratio, explaining why the Fe-rich stilpnomelane has a lower δD value (-149) than the more Mg-rich chlorites and actinolites in Fig. 7. Assuming an average δD value of about -135 for the chlorites and actinolites, and temperatures of about 250°-400° C, the calculated δD_{H_2O} of the Skaergaard hydrothermal fluids is -100 ±5. Although much higher than the present δD values of melt water from the Greenland ice sheet, this would be a very reasonable value for meteoric surface waters if the Eocene climate was warmer and more temperate than the present one (which it certainly was based on faunal and floral evidence). This implies that the initial $\delta^{18}O$ values of the surface meteoric waters in southeast Greenland during the early Tertiary were about -14, because the isotopic compositions of such waters must have originally plotted on the meteoric water line of Craig (1961). These relationships are shown schematically in Fig. 7. The latitudinal dependence of δD and $\delta^{18}O$ in meteoric waters is well known (Friedman et al., 1964; Epstein and Mayeda, 1953). Geographically, the Skaergaard intrusion lay well to the north of the Scottish Hebrides in the early Tertiary, compatible with the 10-15 per mil difference in the calculated δD values of the meteoric waters in these two regions; there was significant plutonic and volcanic igneous activity in both areas during the Eocene, and the meteoric waters had $\delta D \approx -75$ to -90 at Skye and Mull (Forester and Taylor, 1977).

Taylor and Forester (1979) showed that the meteoric-hydrothermal system associated with the Skaergaard intrusion was established directly after the magma body was emplaced. Blocks of basaltic roof rock and of the early-crystallized Upper Border Group were depleted in ^{18}O by the hydrothermal activity before they fell into the magma. In one case, a 6 m-wide basalt xenolith in the Middle Zone of the layered series was found to have $\delta^{18}O = -4.0$. However, in spite of the fact that the hydrothermal system was operating for the entire 130,000-year crystallization history of the pluton (Norton and Taylor, 1979), there was no measurable depletion of ^{18}O in the liquid magma. Small amounts of H_2O may have diffused into the magma from the outside hydrothermal system, (e.g. in the trough-band localities, see Taylor and Forester, 1979), but the amounts were not sufficient to measureably deplete the silicate liquid in ^{18}O. This is proved by the fact that the pyroxenes throughout the layered series have relatively normal $\delta^{18}O$ values, commonly higher than the more easily exchanged plagioclase. Thus, most of the observed ^{18}O depletion in the Skaergaard intrusion was confined to the plagioclase, and this occurred after crystallization under sub-solidus conditions.

Numerical Modeling Study

Norton and Taylor (1979) carried out a computer simulation of the Skaergaard magma-hydrothermal system, producing detailed maps of the temperature, pressure, fluid velocity, integrated fluid flux, and $\delta^{18}O$ values in rock and fluid as a function of time for a two-dimensional cross-section through the pluton. An excellent match was made between calculated $\delta^{18}O$-values and the measured $\delta^{18}O$ values in the three principal rock units, basalt, gabbro, and gneiss, as well as in xenoliths of roof rocks that are now embedded in the gabbro cumulates (compare Figs. 6a and 6b). The best match was realized for a system in which the bulk rock permeabilities were 10^{-13} cm^2 (i.e. 0.01 millidarcy) for the intrusion, 10^{-11} cm^2 for basalt, and 10^{-16} cm^2 for gneiss; reaction domain sizes were 0·2 cm in the intrusion and gneiss and 0·01 cm in the basalts, and activation energy for the isotope exchange reaction between fluid and plagioclase was 30 kcal/mole (Norton and Taylor, 1979).

The thermal history calculated for the Skaergaard system by Norton and Taylor (1979) is characterized by extensive fluid circulation that is largely restricted to the permeable basalts and to regions of the pluton stratigraphically above the basalt-gneiss unconformity. Although fluids circulated all around the crystallizing magma, fluid flow paths were deflected around the impermeable magma sheet during the initial 130,000 years. At that time, crystallization of the final sheet of magma and fracture of the rock shifted the circirculation system toward the center of the intrusion (Fig. 8).

Transport rates of thermal energy out of the intrusion and of low-^{18}O fluids into the intrusion controlled the overall isotope exchange process. During the initial 150,000 years, the average temperature of the intrusion was high (>700° C) and reaction rates were fast; thus, fluids flowing into the intrusion quickly equilibrated with plagioclase. By 400,000 years, the pluton had cooled to approximately ambient temperatures, and the final $\delta^{18}O$ values were 'frozen in'. Reactions between hydrothermal fluid and the intrusion occurred over

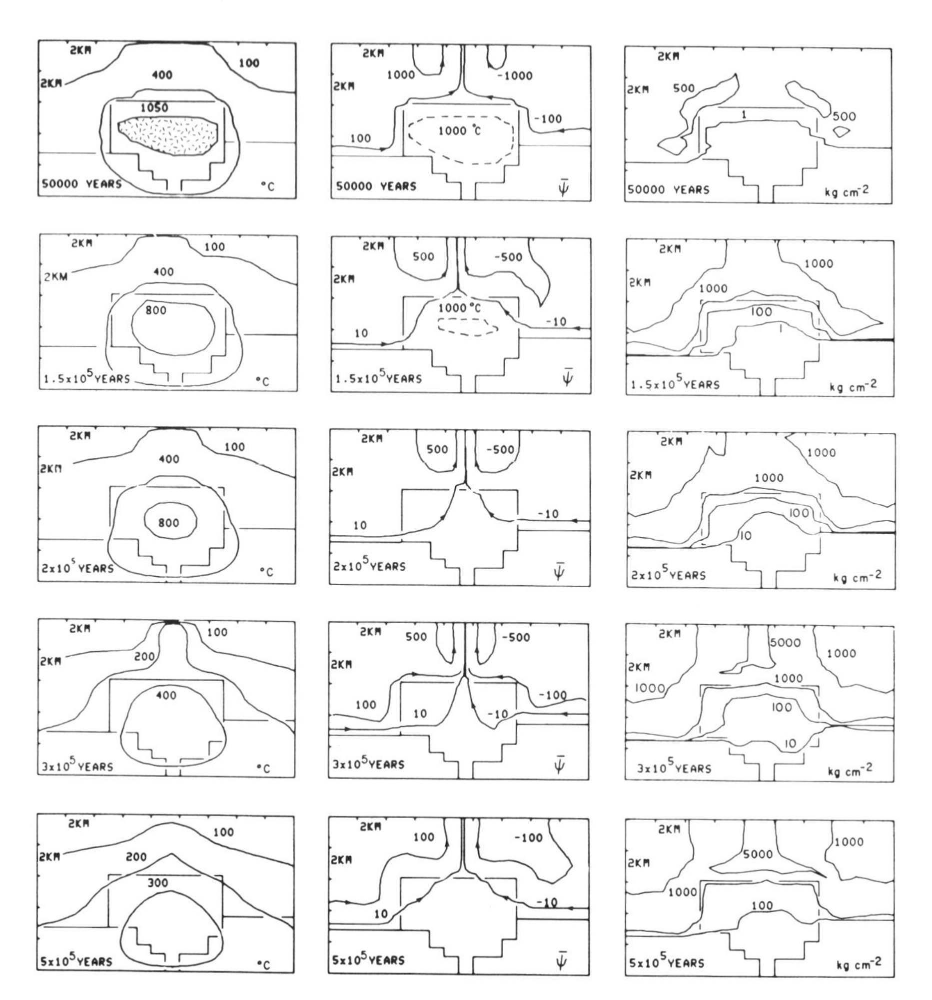

Fig. 8. Temperature (°C), streamlines (ψ), and integrated fluid flux (km cm^{-2}) determined as a function of time by Norton and Taylor (1979) in their numerical modeling study of the Skaergaard intrusion. In the streamline plots, note that at any instant, the largest amounts of H_2O are flowing through zones where the streamlines are closest together. Stippled pattern region represents magma and dashed contour in streamline plot shows region where intrusion is assumed to be impermeable, e.g. where T>1000°C. After 400,000 years, this history of groundwater flow leads to the calculated $\delta^{18}O$ pattern shown in Fig. 6b. During the first 500,000 years after emplacement, the bottom row of figures shows that in the upper part of the pluton: (1) The temperature drops from

a broad range in temperature, 1000°-200° C, but 75 per cent of the fluid circulated through the intrusion while its average temperature was >480° C (Norton and Taylor, 1979). The relative quantities of water to rock integrated over the entire cooling history were 0•52 for the upper part of intrusion, 0•88 for the basalt, and 0•003 for the gneiss (weight units). Almost all of the fluid flowed into the sides of the intrusion from the basalt host rocks, and only tiny amounts flowed inward from the gneiss. Convection transferred about 20 per cent of the total heat contained in the gabbro upward into the overlying basalts; the remaining 80 per cent of the heat was transferred by conduction.

One of the important conclusions of the study by Norton and Taylor (1979) is the demonstration that most of the sub-solidus hydrothermal exchange in the Skaergaard pluton took place at very high temperatures (400°-800° C); this is compatible with the general absence of hydrous alteration products in the mineral assemblages. Of course, outside in the country-rock basalts the average temperature of hydrothermal alteration was much lower (200°-400° C), and there was extensive development of hydrous minerals such as chlorite, epidote, and prehnite. These high temperatures of hydrothermal ^{18}O exchange seem to be a general characteristic of essentially anhydrous gabbro cumulates emplaced into permeable country rocks (see discussion of Jabal at Tirf and the Oman ophiolite, below). The alteration temperatures in these gabbros are much higher than in the typical plutonic granite, such as described earlier from the Idaho batholith.

Granophyres

Late-stage dike and sill-like bodies of granitic magma were sporadically emplaced into the Skaergaard intrusion just after crystallization. The most prominent of these is a 40 m-thick sill in the Upper Border Group, but smaller intrusive bodies are ubiquitous throughout the intrusion, particularly in the Middle Zone and lower part of the Upper Zone (Wager and Brown, 1967). These granitic "dif-

---

about 1100° C to about 250°-350° C; (2) A cumulative amount of 100 to 1000 kg of H_2O has flowed through any given square centimeter of cross-sectional area of gabbro, corresponding to an overall water/rock mass ratio of 0.52 (Norton and Taylor, 1979); and (3) Assuming that the volume of the intrusion above the level of the basalt-gneiss unconformity is 150 km^3, with a mass of 4.5 x 10^{14} kg (Wager and Deer, 1939), this means that a total of 2.3 x 10^{14} kg of H_2O (i.e. 230 km^3 of original ground water) has flowed into the upper part of the intrusion (principally through the sides) and then flowed back out into the country rocks (principally through the roof).

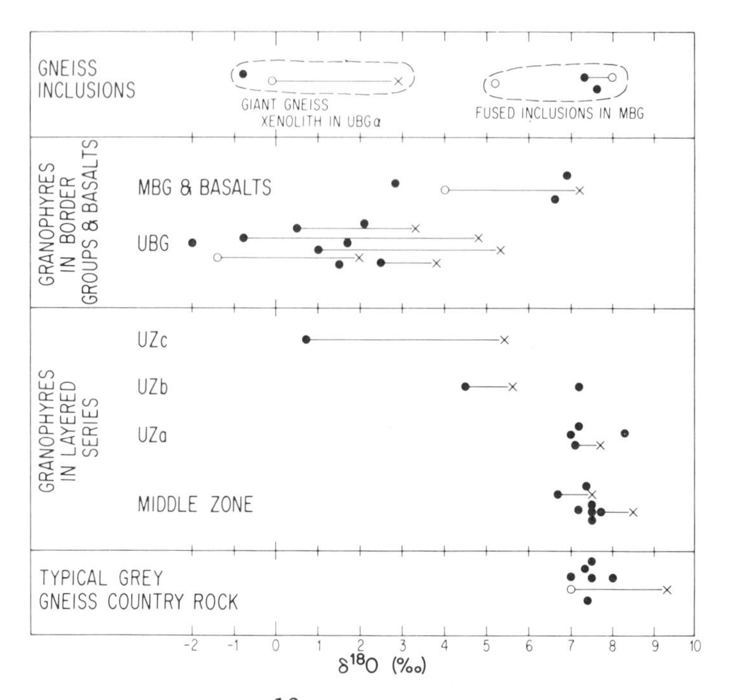

Fig. 9. Graph showing the $\delta^{18}O$ values of granophyres, gneiss inclusions, and country rock gneiss from the Skaergaard intrusion and vicinity (Taylor and Forester, 1979). It is important to analyze quartz (crosses) in such rocks because quartz is much more resistant to ^{18}O exchange with hydrothermal solutions than is feldspar (open circles). Filled circles = whole rock.

ferentiates" invariably lack chilled margins and show well-developed micropegmatite textures; some contain acicular quartz indicative of inversion from tridymite. They commonly have turbid alkali feldspar, are extremely leucocratic, and some contain miarolitic cavities. The rare mafic minerals are most commonly chlorite and opaques, although in some of the more melanocratic varieties Fe-rich clinopyroxene is common. The $\delta^{18}O$ values of some of these granophyres are shown in Fig. 9.

Gneiss inclusions, in various stages of fusion, are fairly common in the Marginal Border Group of the Skaergaard intrusion. These typically have $\delta^{18}O$ values very similar to the country-rock gneiss ($\delta^{18}O$ = 7.3 to 7.6). Only in one case, a giant 400 m-wide gneiss xenolith in the Upper Border Group (UBG), was any marked ^{18}O depletion observed in the gneiss xenoliths (Fig. 9). It is likely that the the ^{18}O depletion of this giant xenolith was principally produced after its incorporation into the pluton, because the adjacent (UBG) rocks have also all undergone extreme ^{18}O depletion.

The whole-rock $\delta^{18}O$ values of the granophyres are very uniform at about +7.5 in the lower, unexchanged portion of the intrusion where the gabbro plagioclase has $\delta^{18}O \geq +5.7$. These localities have only been incipiently affected by meteoric-hydrothermal fluids, so the measured $\delta^{18}O$ values are likely to be essentially identical to those of the primary granophyre magmas. However, above the gneiss-basalt unconformity, the granophyres are typically strongly depleted in ^{18}O, often having even lower $\delta^{18}O$ values than the rocks into which they are intruded (Fig. 9).

The whole-rock $\delta^{18}O$ values of the transgressive granophyres are a little too high to be the result of direct differentiation from the Skaergaard gabbro magma. They are, however, identical to the whole-rock $\delta^{18}O$ values of the typical samples of grey gneiss country rock and to the fused gneiss inclusions that have suffered only minor meteoric-hydrothermal alteration (Fig. 9). Therefore, the $^{18}O/^{16}O$ data strongly support the hypothesis (Wager and Brown, 1967) that the transgressive granophyres in the lower part of the Skaergaard intrusion were formed by partial melting of blocks of grey gneiss that fell into the Skaergaard magma chamber. Such a conclusion is also supported by a variety of other geochemical data, including $^{87}Sr/^{86}Sr$ measurements (Leeman and Dasch, 1978).

Higher up in the intrusion, the granophyres show extreme variability in ^{18}O, making any detailed explanations difficult. These rocks are very fine-grained and are thus quite susceptible to subsolidus hydrothermal alteration, so it is now virtually impossible to determine the exact $\delta^{18}O$ values that the granophyre magmas had when they were intruded. However, Taylor and Forester (1979) attempted to set limits on these magma $\delta^{18}O$ values by analyzing quartz separates (Fig. 9). In all cases, the analyzed quartz is much richer in ^{18}O than the whole-rock granophyre (indicating marked disequilibrium quartz-feldspar $\Delta^{18}O$ >3), but much lower in ^{18}O than the quartz in the grey gneiss. This contrasts sharply with the "normal" $\delta^{18}O$ quartz and $\Delta\ ^{18}O$ quartz-feldspar values found in the Middle Zone granophyres. These data indicate that although significant meteoric-hydrothermal ^{18}O depletion occurred after the high-level granophyres were emplaced, some of the granophyre magmas intruded into the Upper Border Group had initial $\delta^{18}O$ values of +3.0 to +5.0 and thus were very likely derived by melting of hydrothermally altered country rocks.

Alkalic Intrusions, East Greenland

Shortly after emplacement of the Skaergaard intrusion and the East Greenland Dike Swarm, several alkalic intrusions were emplaced in this area. Like the Skaergaard, both the Kraemers I. syenite just to the west of the area shown in Fig. 4 (Taylor and Forester, 1979), and the Kangerdlugssuaq intrusion (25 km to the northwest, Pankhurst et al., 1977; Sheppard et al., 1977) are depleted in ^{18}O and deuterium. The isotopic effects are much more complex than is indicated on Fig. 7, but it suffices to state that the above authors all conclude that some type of meteoric-water exchange is necessary to explain the isotopic data. However, at the present level of exposure these two intrusions everywhere abut against the basement gneiss complex. Why then are the $\delta^{18}O$ values so low? Pankhurst et al (1976) concluded that the meteoric water exchange had to occur at the magmatic stage, and they favored direct influx of H_2O into the magma. The above discussion of Skaergaard data makes this hypothesis unlikely, however; we therefore propose a different mechanism, because in any case ground water circulation through the basement gneiss must have been very limited.

The roof of the Kangerdlugssuaq intrusion definitely projected well up into the plateau basalt section. Many xenoliths of basalt are found throughout both this pluton and the Kraemers I. syenite, whereas only two were observed in the Skaergaard intrusion; these abundant xenoliths were very likely all low in ^{18}O, based on analyses from Skaergaard and the Kraemers I. syenite (Taylor and Forester, 1979). Therefore, we propose that exchange with the much greater proportion of low -^{18}O basalt xenoliths is the reason for the low ^{18}O of both of these syenite magmas. Thus material from the permeable section of roof basalts would be controlling the $\delta^{18}O$ of the syenite intrusions throughout their crystallization history, even though the present outcrops are totally surrounded by basement gneiss. Perhaps this occurred because no Upper Border Group developed above these magma bodies, so the magmas had direct access to their roof rocks.

The Lilloise alkali-hornblende gabbro 120 km ENE of the Skaergaard intrusion (Sheppard et al., 1977) provides another type of problem, because it contains amphibole and biotite with much higher δD values than found in the Kangerdlugssuaq-Skaergaard area, or in the outlying basalts well away from the Lilloise intrusion (Fig. 7). Sheppard et al. (1977) interpreted these data as indicating expulsion of magmatic water (under lithostatic pressure) from the H_2O-rich Lilloise magma body. There was also much less circulation of meteoric water and less depletion of the surrounding basalts in ^{18}O, principally because the dike swarm is very weakly developed at this locality, and also because the Lilloise intrusion was emplaced after the dike swarm and after the joints in the basalts had been largely sealed by alteration and mineralization.

These data from Lilloise demonstrate that deep-seated magmatic water in East Greenland 50 m.y. ago was isotopically totally different from the kinds of waters that affected the Skaergaard intrusion and other nearby plutons. No evidence for this type of primary magmatic water was found anywhere in the East Greenland Dike Swarm or in the Skaergaard intrusion or its associated country rocks. If present, the evidence has been totally obscured by the dominant role played by influx of meteoric ground waters. The later-stage, meteoric-hydrothermal effects associated with the alkalic plutons (Fig. 7) involved waters with slightly higher δD values than those that exchanged with the Skaergaard intrusion. These intrusions are known to post-date the emplacement of the Skaergaard and the major episode of rifting and dike intrusion by about 2-3 m.y. (Soper et al., 1976; Pankhurst et al., 1976). Thus the climatic regime perhaps changed from an interior continental type to a more maritime climate as the North Atlantic opened more widely. This would have been accompanied by an increase in δD of the meteoric waters, compatible with the timeshift and small δD shift indicated on Fig. 7.

JABAL AT TIRF COMPLEX, SAUDI ARABIA

Geological Relationships

The Jabal at Tirf complex on the east side of the Red Sea is of Miocene age, ~ 22 m.y. old, and is a remnant of the earliest stage of opening of the Red Sea rift zone. This subalkaline tholeiite complex includes layered gabbros, granophyres, diabase dike swarms, and rhyolitic dikes; their emplacement was followed by a long period of evaporite deposition in the Red Sea depression. Finally, nepheline-bearing alkali basalts and basanites were erupted on the tilted rift zone during the Pliocene and Pleistocene (Coleman et al., 1975).

The Jabal at Tirf complex is located just E of the Red Sea coastal plain at 17°N, 43°E, and from west to east it is made up of (1) a large mass of granophyre, 3 km wide and 6 km long, which lies stratigraphically above (2) a layered gabbro pluton 2 km wide and 8 km long, in which the layering now dips 40°-60° SW (toward the Red Sea), with a stratigraphic thickness of 1800 m. On its eastern side the lowest part of the gabbro is intruded by (3) a diabase dike swarm that ranges in width from 5 km to 1 km. The plutons and the dike swarm all trend NW parallel to the Red Sea axial trough. The older dikes dip 50°-60° to the NE while the younger dikes are practically vertical. The dike swarm has an exposed length of 60 km in the Jabal at

Tirf area, but it is readily inferred to extend at least as far north as Jiddah, and probably in the subsurface to the Gulf of Aqaba. On its E side the dike swarm gradually dies out as it invades Precambrian basement overlain by the Cambrian Wajid sandstone; the latter typically dips 30°-70° SW toward the Red Sea. In some areas, vesicular lavas with a similar dip and with poorly developed pillow structures are also invaded by the dikes (Coleman et al., 1975). Hydrothermal mineral assemblages in the dike swarm are similar to those listed above for the East Greenland dike swarm, and the mineralogical and petrological characteristics of the layered gabbro and granophyre are similar to analogous rocks in the Skaergaard intrusion. The major difference is in the much larger relative amount of granophyre at Jabal at Tirf, where the volume of granophyre is similar to that of the adjacent gabbro.

Isotopic Systematics

Oxygen isotope data have been obtained for all three of the major units of the Jabal at Tirf Complex (Taylor and Coleman, 1977).

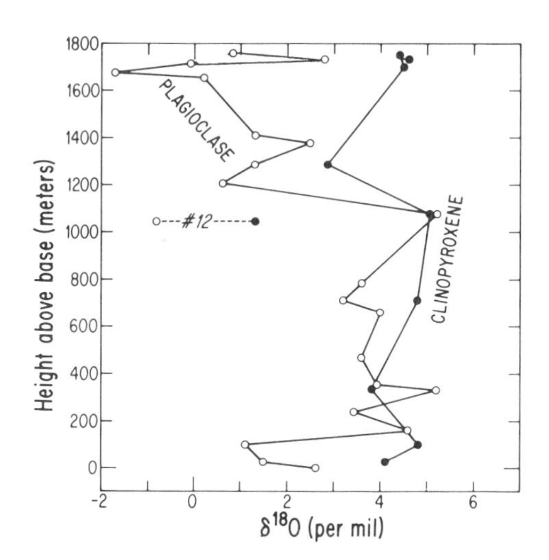

Fig. 10. Plot of $\delta^{18}O$ of plagioclase and clinopyroxene from Jabal at Tirf layered gabbro as a function of height above the base of the gabbro (after Taylor and Coleman, 1977, and Coleman et al., 1975). The samples are numbered consecutively from bottom to top; sample #12 has somewhat anomalous $\delta^{18}O$ values (see text). Lake quartz, clinopyroxene is much more resistant to $\delta^{18}O$ exchange with hydrothermal fluids than is feldspar.

The general $^{18}O/^{16}O$ variations (Figs. 10, 11, and 12) are practically identical to those described above for the Skaergaard intrusion and East Greenland dike swarm. Throughout the gabbro the plagioclase is much more strongly depleted in ^{18}O than the coexisting clinopyroxene (Fig. 10; also see Fig. 21 below). In addition the ^{18}O depletions in plagioclase become more pronounced upward, just as is the case in the Skaergaard intrusion. These features leave no doubt that deep convective circulation of meteoric-hydrothermal fluids also affected the entire Jabal at Tirf complex. At the time of opening of the Red Sea, this area was in the interior of the joint Arabian-African continent, explaining why meteoric ground waters were involved instead of sea water. Note that although the amount of variation of ^{18}O in the dike swarm, the granophyre, and the layered gabbro from Jabal at Tirf is practically identical to that in the analogous rock types from ophiolites (see Fig. 12), the average $\delta^{18}O$ value is about 5 to 7 per mil higher in the ophiolites. This difference is readily attributed to the fact that the ophiolites were altered by sea water, which has a significantly higher $\delta^{18}O$ value (based on analysis of a present-day hotspring, the $\delta^{18}O$ value of local meteoric water in the Jabal at Tirf area is -6 today; White, 1974).

Only a few differences need to be noted between the isotopic effects associated with the East Greenland and Red Sea spreading centers. First, the lowest 100 m of the Jabal at Tirf layered gabbro are also strongly depleted in ^{18}O (Fig. 10); this is undoubtedly due to hydrothermal activity associated with proximity to the younger (at least in large part) diabase dike swarm. Second, the diabase dike swarm is more ^{18}O-depleted near the gabbro-granophyre complex than

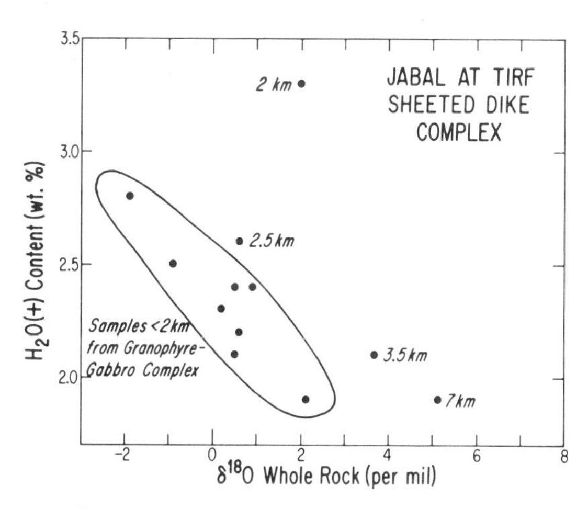

Fig. 11. Plot of $\delta^{18}O$ of whole rock samples of the Jabal at Tirf sheeted dike complex as a function of H_2O content (after Taylor and Coleman, 1977, and Coleman et al., 1975).

it is further away (Fig. 11). This probably indicates higher temperatures of hydrothermal alteration, due either to a greater depth of erosion, or to the added effects of the meteoric-hydrothermal system established by the gabbro-granophyre complex. The diabase dikes closest to the gabbro show a pretty good inverse correlation between $\delta^{18}O$ and H_2O content; the lowest $\delta^{18}O$ values are associated with the largest amounts of OH-bearing alteration minerals (Fig. 11).

Third, one sample of the layered gabbro (#12 on Fig. 10) is anomalous in a variety of ways: (a) The pyroxene is much more ^{18}O-depleted than any of the other analyzed pyroxenes. (b) For its height in the intrusion (1030 m), the plagioclase is also anomalously low in ^{18}O. (c) The plagioclase in this sample has a much higher Na_2O content (4.76 wt. %) and Sr content (500 ppm), and a lower K_2O (0.17 wt. %) than the other nearby samples in the layered gabbro (Coleman et al., 1975). This wholerock sample (JT-51 of Coleman et al., 1975) also has very high Fe_2O_3, high total Fe, low MgO, high Na_2O, high TiO_2, high Ba, low Cr, Low Cu, low Ni, and high V. It has all the appearances of having crystallized from a much more differentiated magma than the other parts of the layered gabbro complex, and the low $\delta^{18}O$ of the clinopyroxene indicates a likelihood that it was formed from a low-^{18}O magma. Unlike the Skaergaard intrusion, the Jabal at Tirf gabbro pluton did not form from a single pulse of basaltic magma. Overall, the plagioclase decreases only slightly in An content from bottom to top (from An_{72} to An_{62}) and there are a number of reversals in the trend. Coleman et al. (1975) concluded that the gabbro was emplaced in an open system and that separate batches of partially fractionated tholeiitic liquid invaded the volcanic rift zone and formed successive zones showing minor interlayer fractionation. Whereas most of the plagioclase-pyroxene pairs from Jabal at Tirf follow a characteristic non-equilibrium trajectory indicative of sub-solidus alteration (see Fig. 21 below), the pyroxenes in three samples are anomalously low in ^{18}O. These magma batches (sample #12 at 1040m, and possibly the samples at 340m and 1270m as well, see Fig. 10), thus probably crystallized from low-^{18}O magmas formed by partial melting, exchange, or massive assimilation of hydrothermally altered basaltic rocks at a lower level in the complex. The production of low-^{18}O magmas in this fashion is very likely in such an open-system environment that is undergoing active rifting.

The differentiated tholeiite that formed sample #12 could have formed from a magma with a $\delta^{18}O$ as low as +2.0, but probably no higher than +3.0. Judging by the clinopyroxene $^{18}O/^{16}O$ systematics (Fig. 21), at least seven out of the ten samples of layered gabbro crystallized from magmas with MORB-type $\delta^{18}O$ values of about +5.7. Some of these batches of partially fractionated tholeiitic liquids were also emplaced as diabase dikes and some possibly vented to form surface flows (Coleman et al., 1975). However, because of the subsolidus hydrothermal alteration that has affected all these rocks, it is not possible to state how much, if any, of the ^{18}O-depletion of the diabase dikes (Fig. 11) took place in the magmatic state prior to intrusion.

Granophyres

The Jabal at Tirf granophyre, which is chilled against and extends apophyses into the upper part of the gabbro, is far too large a mass to have simply differentiated from the exposed layered gabbro, because the two bodies are nearly equal in volume (Coleman et al., 1975). The granophyres range from 62 to 70 wt % SiO_2, 4.0 to 4.6% Na_2O, and 1.8 to 3.0% K_2O, with 16 to 26% normative quartz; on both AFM and Harker diagrams, the granophyres follow trends similar to the Skaergaard granophyres, and they probably have a similar origin (Coleman et al., 1975). Like the uppermost Skaergaard granophyres, the Jabal at Tirf granophyres are all strongly depleted in ^{18}O (compare Figs. 9 and 12). In view of the fact that the Jabal at Tirf

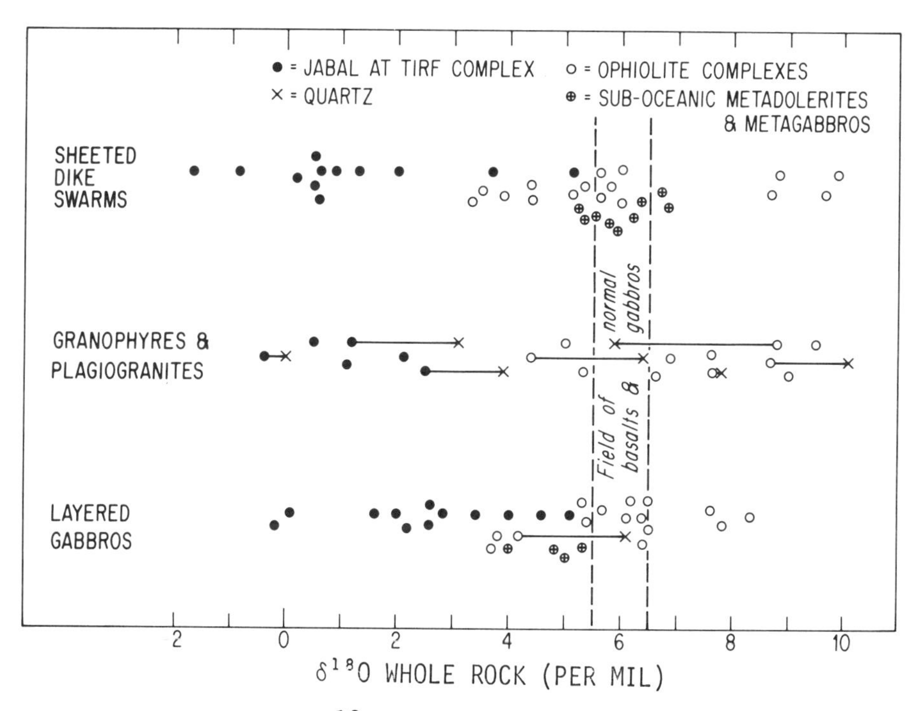

Fig. 12. Comparison of $\delta^{18}O$ values of quartz and whole-rock samples from the Jabal at Tirf complex (Taylor and Coleman, 1977) with analogous rocks from various ophiolites and submarine dredge hauls (data after Muehlenbachs and Clayton, 1971; Magaritz and Taylor, 1974; 1976; Heaton and Sheppard, 1977; Javoy, 1970; Spooner et al., 1974).

complex was emplaced into hydrothermally altered volcanic rocks as well as Precambrian granitic basement rocks, it is plausible that partial melting and/or assimilation of such country rocks contributed to the production of an abnormally large amount of late granophyric differentiate. The $^{87}Sr/^{86}Sr$ ratios of .7041 to .7057 support this hypothesis because they are considerably more radiogenic than in the layered gabbro or the diabase dikes, all of which are richer in Sr but lower in $^{87}Sr/^{86}Sr$ (.7031-.7036, Coleman et al., 1975). The low $\delta^{18}O$ values of the quartz in the Jabal at Tirf granophyres (Fig. 12) suggest that these bodies probably crystallized from low -^{18}O granitic melts, and therefore that they were derived in part from hydrothermally altered country rocks in the deeper and hotter parts of a vast meteoric-hydrothermal convective circulation system established along this 22 m.y. old axis of rifting and magmatic intrusion.

ICELAND

General Statement

The preceding discussion indicates that deep circulation of low -^{18}O meteoric ground waters is a normal phenomenon in subaerial spreading centers, based on examination of 22 m.y. and 50 m.y. old volcanic-plutonic complexes. Based upon isotopic analyses of minerals that are very resistant to sub-solidus ^{18}O exchange with hydrothermal fluids, namely quartz and pyroxene, it was also shown that low -^{18}O magmas were developed in such regions, most commonly in the late-stage granophyric "differentiates," but at least in one instance in a fractionated tholeiitic gabbro magma. Similar development of low -^{18}O basaltic and granitic magmas was previously demonstrated in the Eocene plutonic igneous complex at Skye in the Scottish Hebrides (see Fig. 21 below), which was also formed in a subaerial environment just prior to the opening of the North Atlantic Ocean, but on the edge of the European continental block rather than the Greenland block (Forester and Taylor, 1977).

It is relatively easy to demonstrate the importance of sub-solidus meteoric hydrothermal exchange in such ancient, deeply-eroded spreading centers, but because of these pervasive hydrothermal effects, considerable effort is required to prove the existence of low -^{18}O magmas in such environments. In other words, it is difficult to "see through" such effects and find out how much of the ^{18}O depletion was an original magmatic phenomenon. On the other hand, if one were to examine igneous rocks associated with a present-day spreading center, the reverse would be true. There would be no problem in determining whether or not low -^{18}O magmas exist, because one could analyze fresh, unaltered lava flows that have not been buried and therefore could not have suffered any sub-solidus hydrothermal alteration. However, because of the lack of erosion, without drilling there would be no way to directly determine the extent of deep hydrothermal alteration and exchange. If we can use the Jabal at Tirf, East Greenland, and Scottish Hebrides as models for the deeper parts of a typical subaerial spreading center, study of a modern example could provide evidence as to the near-surface characteristics of such hydrothermal systems. Even more important, this could provide direct and definitive evidence concerning the petrological character, relative abundance, and development of such low $-^{18}O$ magmas, together with all that this implies regarding the igneous history of a typical spreading center. Iceland is just such an example; it is the only major region of the Mid-Atlantic rift system exposed above sea level, and it represents a continuing stage of the opening of the North Atlantic begun 50 m.y. ago in the Eocene.

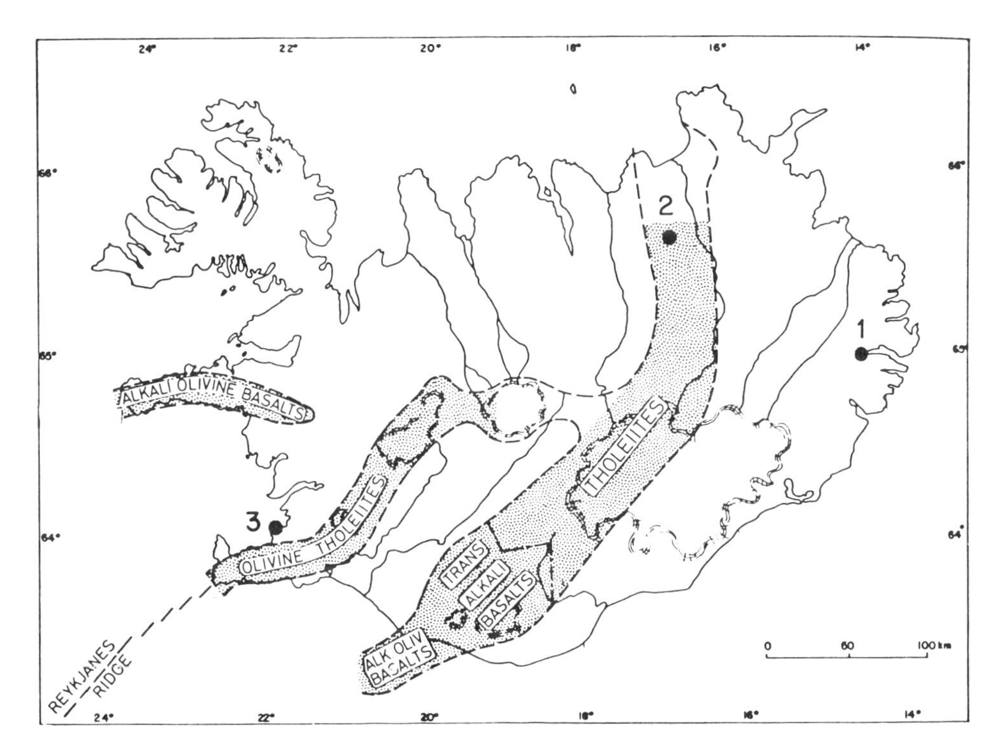

Fig. 13. Map of Iceland, showing areas of post-glacial volcanism and rifting, delineating the positions of the 4 major types of basalts (from Jacobsson, 1972). Also shown are the locations of the 3 drill sites (1 = Reydarfjordur; 2 = Krafla; 3 = Reykjavik) where Hattori and Muehlenbachs (1982) obtained samples for $^{18}O/^{16}O$ analysis (Fig. 17.)

Geological Relationships

Iceland is composed dominantly of flat-lying tholeiitic basalts with subordinate alkali olivine basalt and rhyolite, and there is abundant, near-surface meteoric-hydrothermal activity going on at the present time. The volcanism associated with this spreading center extends well back into the Tertiary on the western and eastern sides of the island, but post-glacial volcanic activity is confined to two major NE-trending rift zones and a minor E-W zone transverse to the others (Fig. 13). The western-most major rift zone occupies Reykjanes peninsula and extends northward through the southern half of Iceland; it joins up with the Reykjanes Ridge to the south. The eastern zone transects the entire island and is thought to be younger than the western zone (Saemundsson, 1974). Up to about 4 m.y. ago, however, the western zone apparently extended all the way across Iceland, joining up with the Kolbeinsey Ridge to the north. This was the major axis of spreading prior to 4 m.y. ago, but at that time the main axis of rifting shifted eastward and the eastern zone became active over its entire length, rupturing rocks that had formed earlier in the western zone and spread eastward (Ward, 1971; Saemundsson, 1974).

Thus, at the present time, rifting in southern Iceland is taking place along two NE-trending zones, but in northern Iceland it is confined to the eastern zone. The petrological characteristics of the post-glacial volcanic activity change systematically along these zones (Fig. 13). Alkali olivine basalts are confined to the southern periphery of the eastern rift zone and to the transverse zone along Snaefellnes Peninsula on the west edge of Iceland (Jakobsson, 1972).

Isotopic Systematics

Muehlenbachs et al. (1974) and Muehlenbachs (1973) have measured the $\delta^{18}O$ values of a number of fresh, post-glacial basalts and rhyolites from Iceland (Figs. 14, 15, and 16). Remarkably, practically all of these Holocene basaltic lavas on Iceland were found to be markedly depleted in ^{18}O, with very heterogeneous $\delta^{18}O$ values ranging from + 1.8 to + 5.7. The rhyolites and obsidians were found to be even more depleted in ^{18}O (+ 0.8 to + 4.8, Fig. 10). A series of SiO_2-rich xenoliths from the basaltic lavas show still greater heterogeneity and much more marked depletions in ^{18}O; quartz-tridymite-cordierite-feldspar-glass inclusions in the 1875 Askya pumice have $\delta^{18}O$ = + 2.5, much lower than the $\delta^{18}O$ of the plagioclase phenocrysts in the host andesite (+4.8). Basalts currently erupting at the large Krafla volcanic center are also low in ^{18}O (+4.6 to +4.8), and similar low-^{18}O magmas were erupted in Tertiary time as well (Hattori and Muehlenbachs, 1982; Muehlenbachs and Sigurdsson, 1977).

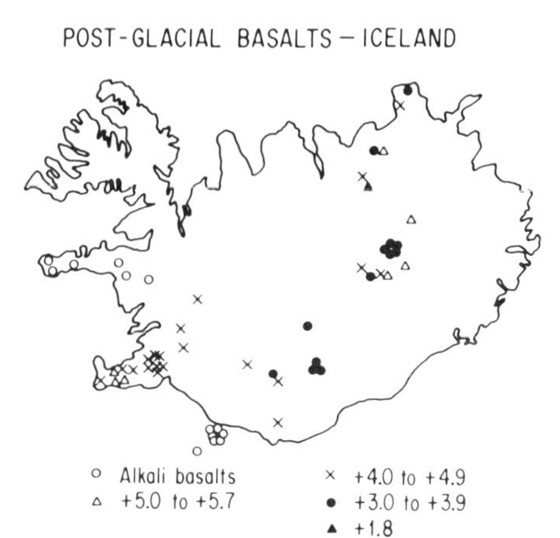

Fig. 14. Map of Iceland, showing locations and the range of $\delta^{18}O$ values determined by Muehlenbachs et al. (1974) for post-glacial tholeiitic and transitional alkali basalts. The locations of the analyzed alkali basalts ($\delta^{18}O$ > +5.2) are also shown.

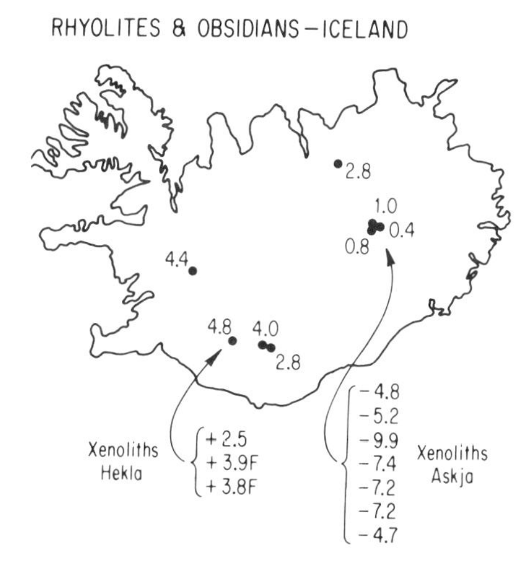

Fig. 15. Map of Iceland, showing the range of $\delta^{18}O$ values of rhyolites, obsidians, and siliceous xenoliths determined by Muehlenbachs et al. (1974) and Muehlenbachs (1973).

Hattori and Muehlenbachs (1982) report $\delta^{18}O$ values of drill-core samples from 3 different sites in Iceland, to depths of up to 3 km (Fig. 13). All of these samples are depleted in ^{18}O relative to normal tholeiitic magmas elsewhere in the world, and the deep samples of hydrothermally altered basalt at Krafla have the lowest $\delta^{18}O$ values yet measured (-10.3 to -10.5, see Fig. 17). Virtually all of these $\delta^{18}O$ values must be the result of subsolidus hydrothermal

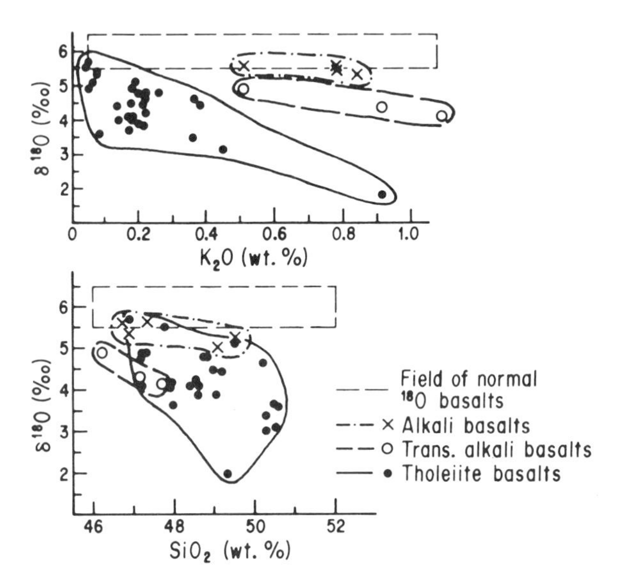

Fig. 16. Plot of $\delta^{18}O$ vs. K_2O and $\delta^{18}O$ vs. SiO_2 for post-glacial basalts from Iceland (after Muehlenbachs et al., 1974).

alteration (Hattori and Muehlenbachs, 1982). This is particularly clear in the case of Krafla where the volcanic rocks are all very young and the isotopic compositions of the meteoric ground waters are known ($\delta^{18}O$ = -10 to 13). The marked ^{18}O depletion associated with the major central volcano at Krafla is readily explained by the large amount of hydrothermal activity that would be expected around such a large energy source ("heat engine") in the crust.

The extreme variability in ^{18}O of the above-described rocks, together with the marked ^{18}O depletions that typically become more intense in the sequence basalt → rhyolite → xenoliths, all are

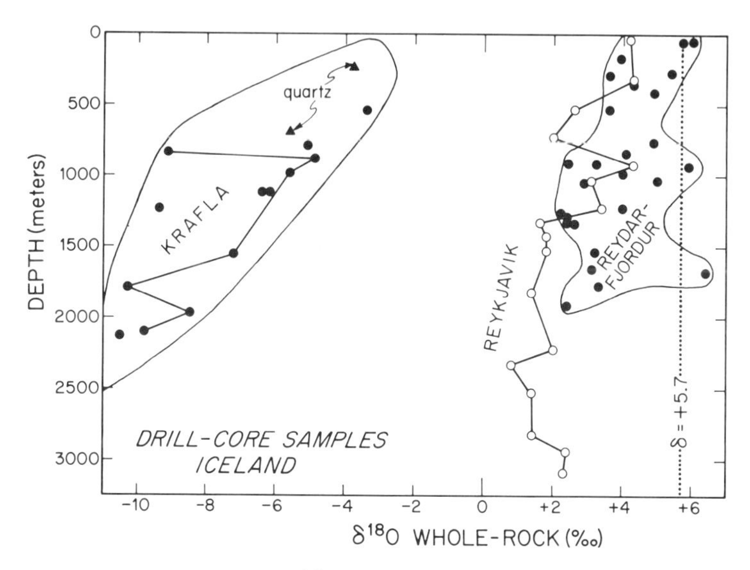

Fig. 17. Plot of depth vs. $\delta^{18}O$ (modified after Hattori and Muehlenbachs, 1982) for samples of hydrothermally altered basalt from 3 drill holes in Iceland (see Fig. 13 for locations). The vertical dotted line indicates the average $\delta^{18}O$ value of most terrestrial tholeiitic basalts (including MOR samples) and lunar basalts. This figure demonstrates that essentially all of the volcanic rocks in Iceland are depleted in ^{18}O down to at least 3 km, with the ^{18}O depletions being more pronounced with depth and with proximity to a central volcano (e.g. Krafla). The solid lines connect samples from a single drill core.

exactly what one would expect during development of low $-^{18}O$ magmas in the deeper parts of a volcanic-plutonic terrane undergoing meteoric-hydrothermal alteration (see above). Muehlenbachs et al. (1974) and Muehlenbachs (1973) also considered this to be a plausible explanation of their results. However, they tentatively rejected this hypothesis for a variety of reasons that do not now seem valid. For example, they cite the general lack of evidence for H_2O-rich magmas on Iceland and the lack of primary hydroxyl-bearing minerals. However, at depths of only a few kilometers, magmas are under too low a pressure to dissolve much H_2O, and in any case it has now been dramatically demonstrated in a variety of such terranes that the lowest -^{18}O rocks produced by hydrothermal alteration are typically also the "driest". Subsolidus hydrothermal alteration at temperatures of 500° -900°C at the Skaergaard, Jabal at Tirf, the Cuillin gabbro mass at Skye, and in the Samail ophiolite (see below) have all produced large volumes of very low -^{18}O, anhydrous gabbro, whose assimilation or re-melting would not produce extremely H_2O-rich magmas. Hydrothermal alteration at relatively shallow depths only produces H_2O-rich mineral assemblages if the temperatures are about 350°C or lower.

Low -^{18}O Magma Problem

Throughout the Earth and the Moon fresh basalts typically have very uniform $\delta^{18}O$ values of about +5.7 and this characteristic value also represents a well defined upper limit for the basaltic lavas from Iceland. However, Muehlenbachs (1973) has appealed to a possible low -^{18}O source region in the mantle to explain the low -^{18}O Icelandic magmas, citing the evidence of Garlick et al. (1971) for some low -^{18}O eclogites from a kimberlite pipe in South Africa. Certainly, some rare samples with low $\delta^{18}O$ values do come from the mantle, most plausibly materials like eclogites that have been hydrothermally altered near the Earth's surface, subducted, and then brought back up to the surface by some process such as explosive vulcanism. However, these are extremely rare, and in any case none are known with $\delta^{18}O$ values lower than +1.0.

To this author, it cannot be a coincidence that abundant, highly variable, ^{18}O-depleted basalt and rhyolite magmas are developed in the single geologic environment which is most favorable for extremely deep circulation and interaction with low -^{18}O meteoric ground waters. There must be a genetic connection between these phenomena. The major problem, and the one of most concern to Muelhlenbachs et al. (1974), is why such ^{18}O-depleted basaltic magmas should be so abundant on Iceland and so rare elsewhere. Probably the best answer is simply that this is a very active subaerial spreading center involving rocks that are extremely permeable to ground waters, and because of the northerly location the groundwaters are quite low in ^{18}O. In fact, a number of rhyolitic and dacitic ash-flow tuff samples have also been identified as ^{18}O-depleted magmas (Friedman et al., 1974; Lipman and Friedman, 1975), and these too are derived from highly fractured, permeable volcanic terranes that have undergone intensive meteoric-hydrothermal alteration. However, the answer to the above question also probably lies in some specific features of the geology of Iceland, as outlined below:

(1) The alkali olivine basalts (Fig. 16) have distinctly higher and more uniform $\delta^{18}O$ values than any of the other Icelandic basalts, +5.3 to +5.7; such values are similar to those found in most other basalts throughout the world. These alkali-olivine basalts are

found only on the periphery of the island, at the edges of any possible low -^{18}O meteoric-hydrothermal circulation systems. (2) The transitional alkali basalts occupy an intermediate geographic position in the eastern rift zone (Fig. 13) and also have intermediate $\delta^{18}O$ values, +3.9 to +4.9. (3) The olivine tholeiites from the the western rift zone (Fig. 13) also have intermediate $\delta^{18}O$ values, +4.0 to +5.7, with the highest values closest to the coast on the SW part of Reykjanes peninsula. (4) Among the Icelandic basalts, the lowest and most heterogeneous $\delta^{18}O$ values, +1.8 to +5.4, are all confined to the quartz tholeiites of the eastern rift zone. (5) There are some extremely crude correlations between $\delta^{18}O$ and chemical composition (Fig. 16); increasing SiO_2, K_2O, and TiO_2 tend to be accompanied by decreasing $\delta^{18}O$ for each petrologic class of basalts. In particular, the tholeiite with by far the most extreme $\delta^{18}O$ value (+1.8) also has by far the highest K_2O content. (6) Those rhyolites and obsidians with the lowest and most variable $\delta^{18}O$ values also all come from the eastern rift zone (Fig. 15).

The isotopic relationships described above strongly support the idea that assimilation and/or partial melting of hydrothermally altered country rocks in the deeper parts of the rift zone is the most likely mode of formation of the low -^{18}O magmas from Iceland. This explanation was previously proposed for the rhyolites, on the basis of other evidence, by G. E. Sigvaldason; Hattori and Muehlenbachs (1982) now also subscribe to this view. It is also compatible with the Pb isotope data of Welke et al. (1968) and the $^{87}Sr/^{86}Sr$ and rare-earth data of O'Nions and Gronvold (1973). Friedman and Smith (1951) measured a δD = -150 in an Icelandic obsidian (0.17 wt %, H_2O), clearly indicative that all of the H_2O in this sample is meteoric in origin ($\delta^{18}O$ = +4.4 on Fig. 15). Rhyolitic rocks are overall quite abundant on Iceland (~ 5% to 10% of the exposed rocks, Walker, 1966), and because of the lack of any evidence for remnant continental crust in this area, that type of source material is probably ruled out (Welke et al., 1968). Because of the abundance of such rhyolitic magmas, another plausible mechanism for the formation of the low-^{18}O basaltic magmas on Iceland is mixing or hydridization between primary basaltic magma and extremely ^{18}O-depleted rhyolite magma formed by partial melting of earlier, hydrothermally altered crust. This would explain the type of correlations shown in Fig. 16, for example.

It is probably significant that all of the extremely low -^{18}O basaltic and rhyolitic magmas are confined to the eastern rift zone, which has only been active during the past 3-4 m.y. (Saemundsson, 1974). The magmas in this rift zone are penetrating upward through a thick section of volcanic and plutonic rocks that has undergone considerable subsidence and lateral movement since being intensively hydrothermally altered in an earlier episode of magmatic activity at the time of original formation in the western rift zone. Thus, the magmas coming up through the eastern rift zone would be interacting with country rocks that had already suffered heterogeneous ^{18}O depletions and been strongly heated simply as a result of isostatic subsidence through a very steep geothermal gradient. Superimposed upon such earlier effects would be the normal deep circulation of meteoric ground waters that would always be expected to accompany magmatic activity in a very young rift system containing relatively open joints and fractures that had not yet been sealed by mineral deposition. This two-stage process may explain why both the basaltic and rhyolitic magmas in the eastern rift zone are so much more strongly depleted in ^{18}O compared to the post-glacial lavas elsewhere on Iceland. It would also call into question whether one should use the petrological and isotopic characteristics of the eastern-zone basalts as being representative of all the tholeiitic magmas poured out along this spreading center during the past 10-20 m.y.

Because very low-temperature (<150° C) hydration reactions that accompany simple burial tend to increase the $\delta^{18}O$ values of the rocks, it is not easy to determine the original magmatic $\delta^{18}O$ values of basalts that are Pleistocene or older. Nonetheless, working on some carefully selected samples, Muehlenbachs and Sigurdsson (1977) have determined a range of $\delta^{18}O = +4.5$ to $+5.1$ for some relatively unaltered 10-16 m.y. tholeiites from NW Iceland. This indicates some ^{18}O depletion, but nowhere near as much as typically observed in the post-glacial basalts from the eastern rift zone.

The low-^{18}O quartz tholeiites (hypersthene normative) of the eastern rift zone are more strongly differentiated than the higher-^{18}O olivine tholeiites of the western rift zone (the former are richer in Fe, Si, Ti, Na, and K and lower in Mg and Ca; Jacobsson, 1972). The olivine tholeiites of the western rift zone have much more "normal" $\delta^{18}O$ values, and they are also chemically more similar to the MOR tholeiites of the adjacent Reykjanes Ridge. The chemical differences between these two types of rift tholeiite can be attributed to combined processes of assimilation and partial melting of roof rocks in the eastern rift zone, which in turn must produce an enhanced degree of fractional crystallization in order to provide the appropriate heat balance (Taylor, 1980b). Mixing with either low-^{18}O rhyolitic melt or with stoped blocks of low-^{18}O basalt could account for the observed chemical and isotopic effects. Note that even if the stoped xenoliths are not thoroughly dissolved by the magma, ^{18}O exchange apparently occurs quite easily under these conditions (Shieh and Taylor, 1969). Sigvaldason (1969) in fact originally proposed that the postglacial lavas in Iceland had undergone marked chemical differentiation in the form of Fe enrichment. Note that many of the chemical distinctions outlined above in the eastern-rift tholeiites are very similar to those inferred for the anomalous low-^{18}O gabbro magma from Jabal at Tirf (sample #12, Fig. 10). If these oxygen isotopic effects were due to some primary feature or some process operating in the mantle source regions of these basaltic magmas, it would be reasonable to expect much more systematic correlations between $\delta^{18}O$ and chemical composition, not the extremely crude and poorly defined correlations such as shown in Fig. 16, or in the TiO_2-^{18}O plots of Muehlenbachs et al. (1974).

If the above-described model is correct, it has important implications, because it implies that virtually all the tholeiites from the eastern rift zone of Iceland have been contaminated in some manner with meteoric H_2O or hydrothermally altered roof rocks. It was the scale of this process that most bothered Muehlenbachs et al. (1974) when they rejected the meteoric-hydrothermal explanation. Something on the order of 200 km^3 of low -^{18}O tholeiite has been erupted in the eastern rift rift zone in the last 12,000 years (Jacobsson, 1972). However, this is only about 1/2 the amount of basaltic magma originally emplaced in the single magma chamber occupied by the Skaergaard intrusion (see above). Also, this constitutes only about 2 km^3 of erupted magma per 100 km^2 of area (Jacobsson, 1972), or a total of only 0.17 km^3/yr over the entire length of the eastern rift zone. Compared to the time scale of meteoric-hydrothermal processes, these amounts of exchanged basaltic magma are well within reason.

The low -^{18}O basaltic and rhyolitic magmas were probably formed well down into layer 3, which under Iceland begins at a depth of 2-3 km below sea level and extends down to 9 km depth (Palmasson, 1971). In Iceland, the depth to layer 3 is markedly greater in the areas where the more normal -^{18}O alkali basalts have been erupted, and the total thickness of crust is much larger there, as well. For comparison, note that meteoric-hydrothermal circulation in the 50 m.y. old East Greenland spreading center penetrated downward through at least 5-10 km of plateau basalt lavas (Taylor and Forester, 1979), and that Hattori and Muehlenbachs (1982) conclude that low -^{18}O altered basalts may exist at great depths (>25 km) in Iceland because known subaerial basalts have been shown to have isostatically subsided at a rate of 1 km per million years.

OPHIOLITE COMPLEXES

General Statement

D/H and $^{18}O/^{16}O$ studies of ophiolite complexes have been carried out by Javoy (1970), Magaritz and Taylor (1974, 1976), Spooner et al. (1974), Heaton and Sheppard (1977), and, most recently, by Gregory and Taylor (1981). The $\delta^{18}O$ data from these studies are shown in Figs. 12 and 18. The δD data are given in Fig. 19.

The $\delta^{18}O$ values of some samples of metadolerites and metagabbros from oceanic dredge hauls are also shown in Fig. 12, and these

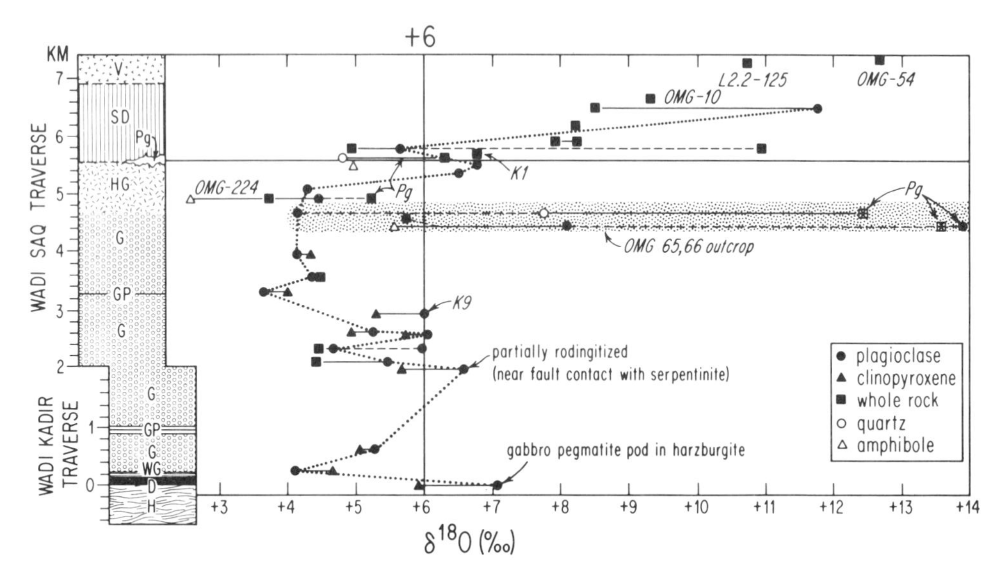

Fig. 18. Oxygen isotope data obtained by Gregory and Taylor (1981) for the combined Wadi Saq and Wadi Kadir sections from the Ibra area, Oman (see Fig. 20). OMG 10, OMG 54, K 1, K 9, and OMG 224 are samples from the Ibra area analyzed by McCulloch et al. (1981) and shown in Fig. 23. Sample L 2.2-125 is from Wadi Jizzi. Shaded area represents OMG 65 and OMG 66 outcrops, which are located at the same structural horizon in Wadi Saq. Dotted line connects analyses of plagioclase separates. Horizontal solid lines connect coexisting minerals from the same hand speciman. Horizontal dashed lines connect data points from the same outcrop but different hand specimens. Pg = plagiogranite; V = pillow lavas; SD = sheeted dikes; HG = high-level gabbro; G = cumulate gabbro; D = dunite; GP = gabbro pyroxenite; WG = wehrlite gabbro; H= harzburgite tectonite.

display a range of $\delta^{18}O$ identical to analogous rock types from ophiolite complexes (Muehlenbachs and Clayton, 1972). Excluding epidote, D/H values of OH-bearing alteration minerals such as chlorite and amphibole in ophiolite complexes typically range from $\delta D = -65$ to -35 (Magaritz and Taylor, 1974; 1976; Heaton and Sheppard, 1977). This is identical to the range of δD values in sub-oceanic samples (Sheppard and Epstein, 1970; Wenner and Taylor, 1973; Jehl et al., 1977; Stakes and O'Neil, 1981), and these data are all compatible with formation from a water with δD similar to ocean water at temperatures in the range 150°-600° C. Note that the average δD of ocean water has been relatively constant at -10 to 0 during the past 200 million years, principally fluctuating only in response to the size of the polar ice sheets.

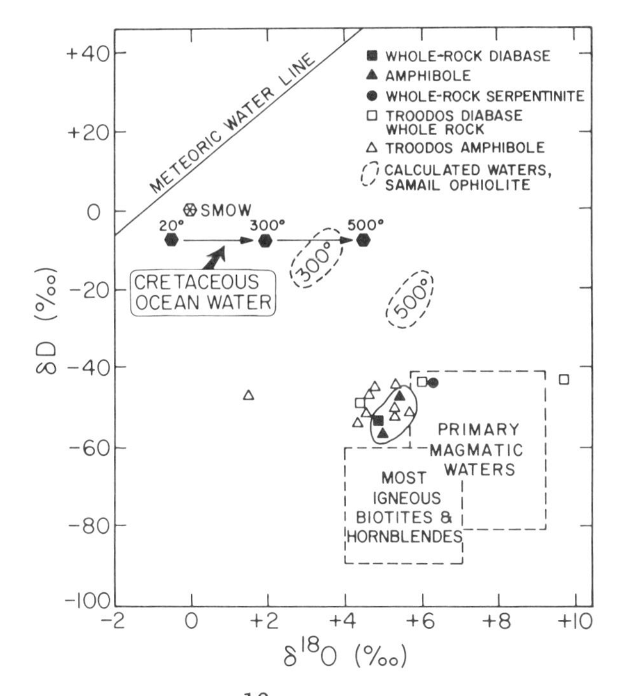

Fig. 19. Plot of δD versus $\delta^{18}O$ for rocks analyzed from the Oman and Troodos ophiolites, showing calculated isotopic compositions of ^{18}O-shifted Cretaceous ocean water and of waters that would be in equilibrium with the Oman samples at T = 300° C and 500° C (after Gregory and Taylor, 1981).

Ophiolite complexes are now generally recognized to represent fragments of ancient oceanic crust and mantle (Coleman, 1977), and many are thought to have formed at mid-ocean ridge spreading centers (e.g. the Cretaceous Samail ophiolite in Oman, Fig. 20). All described ophiolite complexes have suffered the same general types of mineralogical and petrological changes due to hydrothermal metamorphism that were outlined above for the subaerial spreading centers. The only major differences appear to be the more extensive development of hornblende (higher P_{H_2O}?) and the existence of much more saline, NaCl-rich waters in the case of the ophiolites. Sodium metasomatism and albitization, for example, are very prevalent in the upper parts of ophiolite complexes, which are commonly spilitized. The grade of hydrothermal metamorphism increases downward in the ophiolite complexes from zeolite facies → greenschist facies → amphibolite facies (see Coleman, 1977, p. 108-114), so in general, the hydrothermal temperatures increase in the sequence: pillow lavas → sheeted dike complex → plagiogranite → gabbro.

The $\delta^{18}O$ values of hydrothermal alteration products are strongly dependent upon both temperature and water/rock ratio, among other factors, but other things being equal the higher the temperature the lower the $\delta^{18}O$. This is why the whole-rock $\delta^{18}O$ values of the various units of the ophiolite complex also tend to decrease downward. The pillow lavas typically have $\delta^{18}O$ = +10 to +16 whereas the lower units in the complexes have $\delta^{18}O$ = +3 to +10 (Figs. 12, 18).

Samail Ophiolite, Oman

The lowest $\delta^{18}O$ values in an ophiolite complex ought to be produced in the deepest, hottest portions of the hydrothermal systems, assuming that the water/rock ratios are sufficiently high. Underneath this deepest zone of significant convective sea-water circulation, there should be a "reversal" in $\delta^{18}O$, below which the values ought to climb back up to the range typical of primary unaltered oceanic gabbros. This "reversal" has been beautifully demonstrated by Gregory and Taylor (1981) within the cumulate gabbro sequence in the Samail ophiolite, Oman. The latter is the largest, best exposed ophiolite complex in the world, with an intact stratigraphic section on the order of 20 to 30 km thick and a strike length of about 500 km (Fig. 20).

Moving upward from the ultramafic rocks at the base of the Samail ophiolite complex, one starts in rocks with normal magmatic $\delta^{18}O$ values (Fig. 18). Somewhere within the layered gabbro section the effective base of the pervasive marine-hydrothermal circulation system is encountered, and the plagioclase $\delta^{18}O$ values drop to values as low as +3.5, with reversed plagioclase-pyroxene $\Delta^{18}O$ fractionations indicative of subsolidus hydrothermal exchange (Gregory and Taylor, 1981). These cumulates are indistinguishable in hand specimen and in thin section from the normal -^{18}O lower cumulates. Mineralogical evidence for alteration is practically nonexistent in all these lower cumulates, even those that are markedly depleted in ^{18}O (Fig. 18). Thin reaction rims of talc + magnetite around olivine grains occur locally in the low-^{18}O rocks but are not universal. Plagioclase generally appears to be completely unaltered. Pyroxenes, although generally unaltered, may contain minor (<10%) replacement brown amphibole along grain cleavages and fractures. As at Skaergaard, these data indicate that the hydrothermal alteration of most of the gabbro samples occurred at extremely high temperatures and that subsequently there was only local influx and exchange with low-T waters.

As the gradational contact between the cumulate gabbro and high-level gabbro is approached, OH-bearing alteration minerals become more abundant. Brown amphibole is followed by green amphibole, then chlorite appears, and finally in some rocks, rare epidote occurs in the

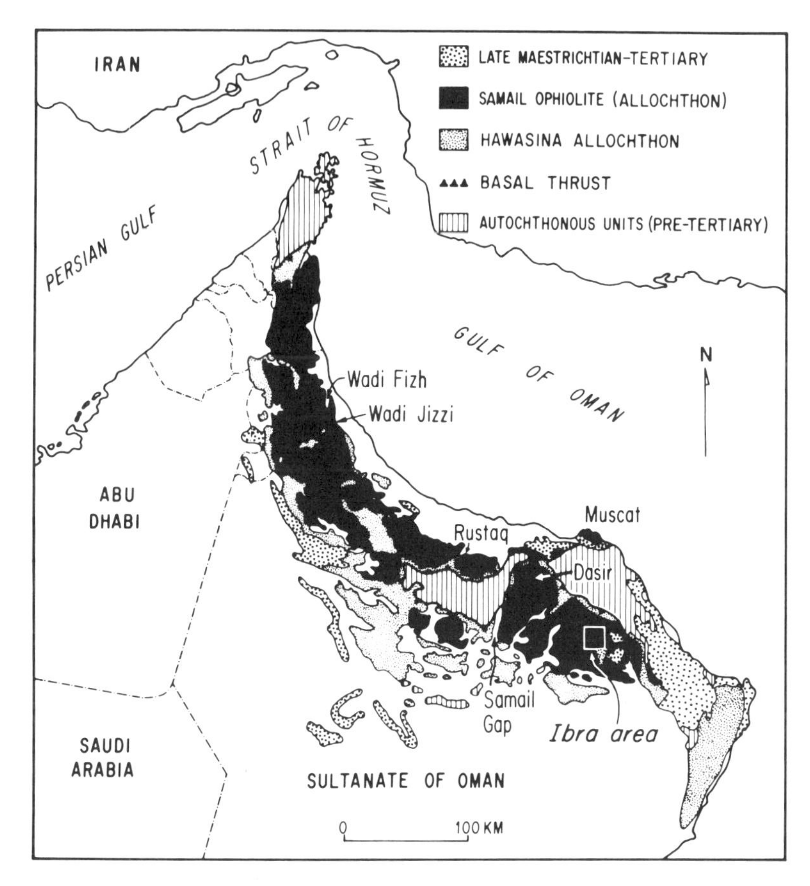

Fig. 20. Generalized geologic map of the Samail ophiolite, Oman, showing an inset of the Ibra area (modified after Glennie et al., 1974 and Gregory and Taylor, 1981).

high-level gabbros. The presence of talc + magnetite after olivine (and the lack of serpentine or chlorite) in some high-level gabbros again demonstrates that most of the hydrothermal fluids were flushed through at temperatures exceeding 400° C. In this part of the ophiolite section the $\delta^{18}O$ values overlap the normal magmatic range of $\delta^{18}O$ (Fig. 18). However, this similarity in $\delta^{18}O$ values is only a coincidence because all of these rocks have been hydrothermally altered; the average temperatures and water/rock ratios, integrated over the lifetime of the circulation system at this level in the ophiolite complex, just happen to be appropriate to produce exchanged minerals or new metamorphic minerals with about the same $\delta^{18}O$ values as the primary igneous minerals. Above the stratigraphic horizon where the $\delta^{18}O$ values are coincidentally identical to the primary magmatic values (typically within the high-level gabbros or the dike complex), the $\delta^{18}O$ values continue to systematically increase upward, becoming as high as +13 to +15 in the pillow lavas (Gregory and Taylor, 1981).

Neither Wadi Kadir nor Wadi Saq represent complete sections through the Samail ophiolite, but where they overlap, the agreement between geology, petrography, alteration mineralogy, and the $\delta^{18}O$ profile is very good (Fig. 18). The consistency of geology and $\delta^{18}O$ profiles in these two sections in the Ibra area separated by a lateral distance of over 15 km, as well as the basically similar geologic sections exposed over several hundred kilometers of the entire Oman mountain belt (Fig. 20), both suggest that we are looking at a representative cross section through Cretaceous oceanic crust. It also appears that the physical and chemical processes involved in formation of this oceanic crust had essentially reached steady state conditions in the 5- to 10-m.y. interval after igneous crystallization and before detachment. This steady state crustal section contains significant volumes of both low-^{18}O (<+6) plutonic rocks and high-^{18}O(>+6) hypabyssal and pillowed volcanic rocks.

When a $\delta^{18}O$ integration (material balance calculation) is done for the entire Samail ophiolite, it is seen that the contribution of ^{18}O-depleted rocks in the lower sequence almost exactly cancels out the contribution of the ^{18}O-enriched rocks in the upper sequence (Figs. 18 and 23). The net change in ^{18}O content of this section of oceanic crust produced by a long history of hydrothermal alteration appears to be essentially zero. The calculated average $\delta^{18}O$ = 5.8 ± 0.3 (Gregory and Taylor, 1981) is, within experimental and geologic error, identical to the average $\delta^{18}O$ of MOR basalts (+5.7), which is accepted to be the primary magmatic $^{18}O/^{16}O$ ratio of the oceanic crust.

If there has been no net change in $\delta^{18}O$ of the oceanic crust, there obviously cannot have been any net ^{18}O flux either into or out of ocean water as a result of interactions at MOR spreading centers. If we extrapolate the relationship discovered in the Samail ophiolite to all oceanic spreading centers, it is clear that seawater must have been in steady state isotopic balance, "buffered" with a $\delta^{18}O$ close to zero during the Cretaceous.

In addition to the above discussion and the D/H data (Fig. 19), evidence for deep circulation of seawater-derived hydrothermal fluids comes mainly from data on coexisting plagioclase and pyroxene. Throughout the cumulate gabbro section this mineral pair exhibits oxygen isotopic disequilibrium; based on analogous results from other localities (Forester and Taylor, 1977; Taylor and Forester, 1979), it it is clear that this feature is a result of subsolidus exchange.

Plagioclase-pyroxene pairs analyzed from rapidly quenched gabbroic magmas, such as terrestrial basalts and lunar microgabbros, all have $\Delta^{18}O$ plagioclase-pyroxene $\equiv \delta^{18}O$ plagioclase - $\delta^{18}O$ pyroxene ≈ 0.5, close to the equilibrium fractionation (Taylor, 1968; Anderson et al., 1971; Taylor and Epstein, 1970; Onuma et al., 1970). Therefore in normal midocean ridge tholeiites that have whole-rock $\delta^{18}O = 5.7$, the corresponding primary magmatic $\delta^{18}O$ plagioclase should be +6.0 and $\delta^{18}O$ pyroxene should be +5.5. On a plot such as Fig. 21, all primary magmatic plagioclase-pyroxene pairs from normal tholeiite basalts should therefore cluster around a spot having the coordinates (5.5, 6.0). Closed-system ^{18}O exchange and slow cooling of a plutonic gabbro down to 600° - 700° C could result in a change to (5.3, 6.4) and an increase to $\Delta^{18}O = 1.1$; note above that many Skaergaard samples underneath the regional gneiss-basalt unconformity have such $^{18}O/^{16}O$ values.

However, many of the cumulate plagioclase-pyroxene pairs from the Samail ophiolite plot well below the $\Delta = 0.5$ and $\Delta = 1.1$ lines, in the field forbidden under equilibrium conditions. In addition, the plagioclase-pyroxene pairs do not plot on an equilibrium trend with slope unity but instead plot in the vicinity of a least squares line (Fig. 21) given by the equation:

$$\delta^{18}O \text{ plagioclase} \approx 2.3\ \delta^{18}O \text{ clinopyroxene} - 6.6$$

The $\Delta^{18}O$ values associated with the line of slope 2.3 include some samples that have normal igneous fractionations, and in fact, this line passes through the primary magmatic spot. As plagioclase $\delta^{18}O$ values decrease, the $\Delta^{18}O$ values decrease to zero and then become reversed. Inasmuch as all equilibrium plagioclase-pyroxene $\Delta^{18}O$ values are known to be positive in the temperature range that is geologically reasonable ($T < 1300°$ C), the trend of data points in Fig. 21 constitutes clearcut evidence for open-system (hydrothermal) subsolidus exchange.

Trend lines analogous to those observed in Oman were described above for the Skaergaard intrusion and the Jabal at Tirf complex; they have also been found in the Cuillin gabbro of the Isle of Skye (Fig. 21). The slightly different position and slope of the plagioclase-pyroxene trajectory in the Oman rocks is either a consequence of the drastically different chemical and isotopic compositions of the hydrothermal fluids, or perhaps a result of the incipient development of hornblende in the Oman clinopyroxenes, thus opening up the structures and making them more susceptible to ^{18}O exchange.

The range of whole-rock $\delta^{18}O$ values of most diabase dikes from the Wadi Saq and Wadi Kadir sections is +6.8 to +10.9. However, the range of whole-rock $\delta^{18}O$ values in diabasic xenoliths (now horn-

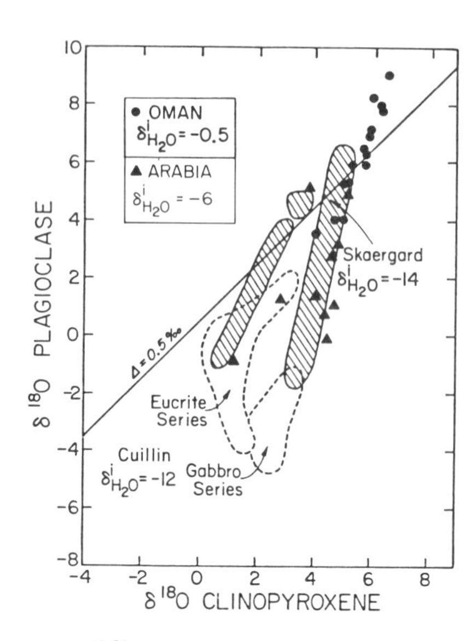

Fig. 21. Comparison of $\delta^{18}O$ data on clinopyroxene-plagioclase pairs from Oman (Gregory and Taylor, 1981) with analogous data from the Jabal at Tirf complex, Arabia, the Skaergaard intrusion (diagonal lined pattern), and the Cuillin gabbros of Skye (dashed envelopes); the latter 3 localities were altered by meteoric-hydrothermal fluids (Taylor and Forester, 1979; Forester and Taylor, 1977; Taylor and Coleman, 1977). Seven out of the ten data-points at Jabal at Tirf correspond very well with the data from Oman, the main Layered Series rocks at Skaergaard, and the Cuillin Gabbro Series. The other 3 samples from Jabal at Tirf contain unusually low-^{18}O pyroxenes and, like the eastern rift-zone basalts from Iceland (Muehlenbachs et al, 1974) and the Eucrite Series from the Cuillins, they probably crystallized from low -^{18}O magmas (particularly sample #12). However, note that the somewhat similar low -^{18}O clinopyroxene trend from Skaergaard (pattern to the left of the main trend) represents late-stage Fe-rich magmas that crystallized at Upper Zone C time. These were not low-^{18}O magmas; the pyroxenes in these samples actually crystallized as β ferrowollastonites and became depleted in ^{18}O during inversion to fine-grained aggregates of ferrohedenbergite grains during subsolidus meteoric-hydrothermal alteration.

fels) associated with plagiogranites from the Dasir area (about 70 km to the NW, see Fig. 20) is +3.7 to +6.3 (Gregory and Taylor, 1979). The Dasir xenoliths represent fragments of diabase dikes from the roof of the gabbro magma chamber that became incorporated into the magma by piecemeal stoping and foundering of the roof.

The stoped blocks at Dasir record a different, earlier stage of hydrothermal alteration than that shown in Fig. 18. The Dasir xenoliths in part preserved the low $\delta^{18}O$ values that they had attained just prior to their being stoped into the underlying magma near the ridge axis. In contrast, although the lower parts of the section shown in Fig. 18 probably also underwent a similar high-T, ^{18}O depletion, they did not preserve these low $\delta^{18}O$ values when they were later subjected to interaction with more ^{18}O-shifted, lower-temperature hydrothermal fluids in the vicinity of the distal portions of the magma chamber as the oceanic crust migrated away from the ridge axis (see Fig. 22). The difference in ^{18}O contents can be attributed to the different circulation systems, temperatures, and types of fluid that the two suites of diabases experienced.

The xenolith suite was altered directly above and virtually in contact with the magma chamber in a region where the hydrothermal circulation system was dominated by high-temperature fluids that circulated to depths of only 2-3 km and interacted only with basalt and diabase. Because of the high water/rock ratios (McCulloch et al., 1981), this fluid would not have suffered any large ^{18}O shifts and thus would be isotopically similar to seawater; the fluid path lines would not take it through the lower portions of the oceanic crust. However, as the crust continues to spread, the diabase section moves beyond the distal edge of the magma chamber (Fig. 22) into into a regime where the fluids are dominantly moving upward from deep in the cumulate gabbro section. Such fluids may still be at relatively high temperatures, but because of overall lower water/rock ratios in the less permeable, deeper parts of the ocean crust, they will have suffered much more dramatic ^{18}O shifts because of interactions with the lower cumulates. During continued spreading, the temperatures also continue to decrease, which also leads to ^{18}O enrichment of the altered rocks. When the high-^{18}O and/or low-T fluids interact with the sheeted diabase complex, it becomes enriched in ^{18}O, thus partly masking the earlier low-^{18}O exchange history. Another factor probably involved in the preservation of the low $\delta^{18}O$ values of the Dasir xenoliths is the fact that they were carried down to deeper levels; they now lie approximately 200 m below the diabase-gabbro contact, where they are associated with large masses of plagiogranite.

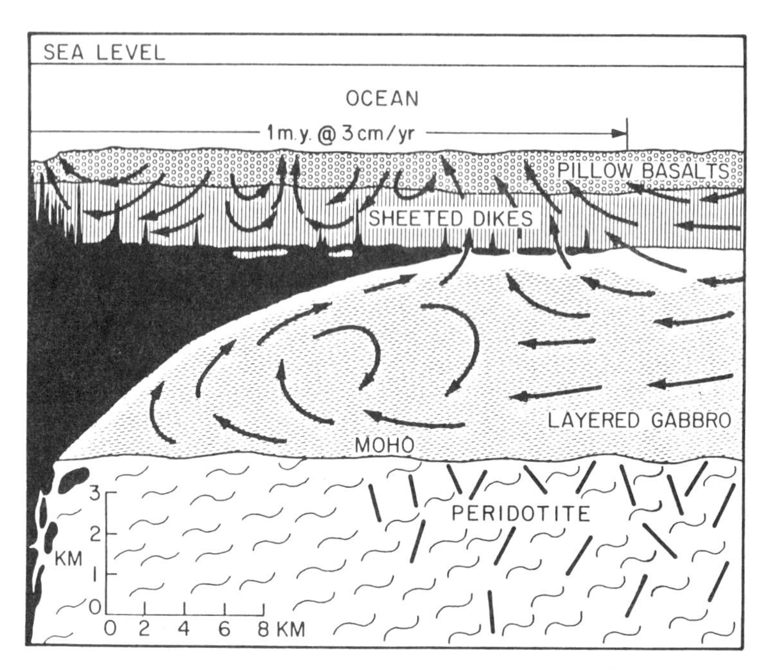

Fig. 22. Cartoon sketch, after Gregory and Taylor (1981), showing probable seawater circulation patterns in a cross section of the Samail ophiolite at the time of its formation. The solid black indicates magma. Note the existence of two essentially isolated circulation regimes, separated by a thin sheet of magma that is essentially impermeable to the fluids in the hydrostatic convective system. One system (the upper system) extends outward from the ridge axis and is located directly over the magma chamber, and the other (the lower system) lies underneath the wings of the magma chamber. The two systems interact with one another only at the distal edges of the magma chamber. Also shown is the lateral distance that would be traversed by this oceanic crustal section in 1 m.y., assuming a half spreading rate of 3 cm/yr. The wavy line pattern in the peridotite represents the tectonite fabric, and the short heavy lines schematically indicate gabbro dikes filling conjugate fractures that postdate the tectonite fabric; locally, the hydrothermal circulation penetrated these conduits. Note that only one side of the MOR spreading system is illustrated; a mirror image of the above diagram should exist to the left side of the figure.

This type of isotopic "aging" described above can also be seen on the scale of an individual outcrop. Samples from localities OMG 65 and OMG 66 (Fig. 18) in Wadi Saq exhibit the same general type of time-$\delta^{18}O$ trends. Samples from the host rock cumulate gabbros have $\delta^{18}O$ plagioclase = 4.2 to 5.8, whereas the later hornblende-gabbro segregations have $\delta^{18}O$ plagioclase = 8.1, and the still later plagiogranite dikes and veins have $\delta^{18}O$ whole rock = 12.4 and 13.6. The high-^{18}O plagiogranites occupy a prominent fracture system and have sharp nonchilled contacts against the host gabbros. In these fracture systems the fluids continued to circulate down to relatively low temperatures (as recorded by plagioclase turbidity and the presence of prehnite, epidote, and thullite). The coarser-grained host gabbros have partially preserved their earlier-formed, lower $\delta^{18}O$ values, and they also contain only amphibole as a new alteration phase (they have no epidote). Thus in the single OMG 65, 66 outcrop Gregory and Taylor (1981) were able to observe practically the entire range of $\delta^{18}O$ values in the Samail ophiolite, simply by sampling the host rocks as well as the later-stage rock types that occupy the vein and fracture systems.

The style of H_2O circulation proposed by Gregory and Taylor (1981) for the Oman ophiolite is illustrated in cartoon fashion in Fig. 22. On the basis of field mapping (Hopson et al., 1981; Pallister and Hopson, 1981), the Samail gabbro magma chamber appears to have had a shape similar to that indicated in Fig. 22. This is the basic shape proposed for the Troodos (Greenbaum, 1972) and Pt. Sal ophiolites (Hopson and Frano, 1977), and it is similar to the shape of many continental layered gabbro complexes formed in rift environments (e.g., the Muskox, Irvine and Smith, 1967, and Great Dyke, Worst, 1960). The floor of the magma chamber thus somewhat resembles the bottom of a wide, very long ship, with the poorly defined feeder dike system representing the keel; in this analogy the roof of the chamber is the deck of the ship. This shape probably closely approximates the geometry of the magma chamber at a fast spreading ridge. At a slow spreading ridge the 'wings' would be much smaller or nonexistent.

According to Gregory and Taylor (1981), the result of this particular geometry is that two decoupled regimes of hydrothermal circulation probably exist during most of the history of alteration. The first occurs directly over the magma chamber and is continuous across the ridge axis (this is termed the 'upper system'). The upper seawater-hydrothermal circulation system lies exclusively within the pillow lavas and the sheeted diabase-dike complex, and the H_2O penetrates downward only as far as the flat roof of the magma chamber ($<$ 3 km into the oceanic crust). The water cannot penetrate any more deeply than the joint and fracture system will allow, and thus (in geologically reasonable times) the water cannot cross the diabase-

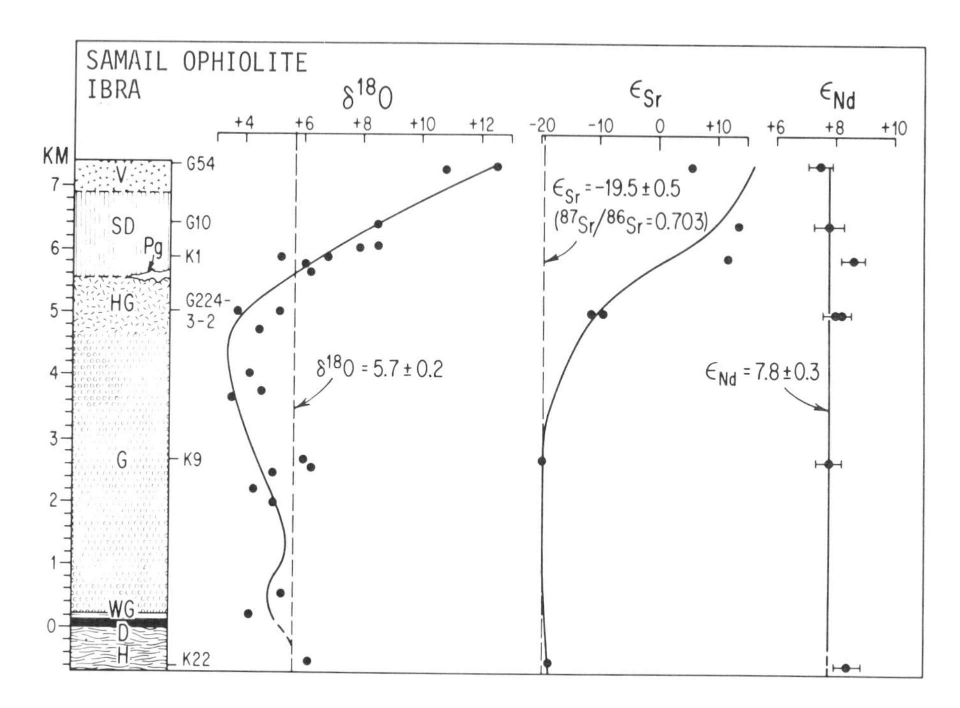

Fig. 23. The $\delta^{18}O$, ε_{Sr}, and ε_{Nd} values from a composite profile through the Samail ophiolite, Ibra section, after McCulloch et al. (1981). It can be seen that the ε_{Nd} values are, in general, within error of the primary magmatic value of ε_{Nd} = 7.8. In contrast, the $\delta^{18}O$ (from Gregory and Taylor, 1981) and ε_{Sr} values show large deviations from their primary magmatic values of $\delta^{18}O$ = 5.7 and ε_{Sr} = -19.5. These large variations in $\delta^{18}O$ and ε_{Sr} are a consequence of hydrothermal interaction with seawater, and imply a wide range of water/rock ratios, from 0.3 to about 50. The Sm/Nd systematics are not affected by this process because of the low concentration of rare earth elements in seawater (a water/rock ratio of 10^5 or more would be required).

magma chamber contact (Taylor and Forester, 1979; Norton and Taylor, 1979). In fact, some H_2O is undoubtedly added to the magma from the upper system, but this must come about through dehydration of stoped blocks of hydrothermally altered roof rocks that should be abundant in such a tectonically active, rifting environment (Taylor, 1977; Taylor and Forester, 1979; Gregory and Taylor, 1979; Taylor, 1980a). Because the stratigraphic thickness of rocks above the roof of the magma chamber is very small compared to the width of the chamber, the upper hydrothermal system probably involves a large number of separate convection cells, dominated by the ridge-axis system (Corliss et al., 1979). The upper system therefore must involve very large overall water/rock ratios (>10, based on the strontium isotope data of McCulloch et al., 1981), and the circulating seawater will be only slightly ^{18}O shifted away from its initial value ($\delta^{18}O = +1.6$, Craig, 1981).

The isotopic and alteration effects produced by the upper regime are in part destroyed or masked by later alteration effects ("aging") that come about when fluids from the 'lower system' finally are able to penetrate upward into the sheeted diabase complex at the distal edges of the magma chamber. This will happen as soon as the rocks (high-level gabbros and plagiogranites) are consolidated enough to fracture. In the cartoon (Fig. 22) this is indicated to occur discontinuously as small pockets of late-stage magma become isolated from one another due to the vagaries of the crystallization process in such a spreading environment. The seawater circulation system within the layered gabbro cumulates underneath the wings of the magma chamber involves very high temperatures (>400° C) and low water/rock ratios (closed system) of the order of 0.3-1.0 (weight units).

The fluid involved in the lower system is seawater that has moved laterally inward from well beyond the distal ends of the magma chamber. Because this water cannot move upward in any significant quantities directly through the liquid magma, it must either cycle (hence the closed system characteristic) or escape upward at the distal edge of the chamber when conduits in fractured, solidified rock become available at the gabbro-diabase contact (Fig. 22). Both processes undoubtedly occur. Therefore large quantities of this heated and strongly ^{18}O-shifted water ($\delta^{18}O_{H2O} = +4$ to $+8$) will be focused upward along the sloping base of the magma chamber, imposing a final alteration event upon the sheeted dike complex. Note that if any water does diffuse directly into the magma from the country rocks, it must be from the lower system. It is plausible that tiny amounts of the very low density, high-T H_2O in fractures below the 'wings' of the magma chamber could diffuse upward into the overlying magma, in the manner proposed for the trough bands of the Skaergaard intrusion (Taylor and Forester, 1979). The shape of the magma chamber shown in Fig. 22 is, in fact, ideal for such a process to operate. The strongly ^{18}O-shifted water of the lower system also locally penetrates down into the ultramafic rocks along conjugate fractures that postdate the peridotite tectonite fabric.

The style of hydrothermal circulation and the geometry of the magma chamber shown in Fig. 22 imply that there should be a dramatic shift or abrupt discontinuity in the isotopic record at the gabbro-diabase contact (i.e., at the fossil roof of the magma chamber). This is in fact just what is observed, as both the $\delta^{18}O$ (Fig. 18) and the $^{87}Sr/^{86}Sr$ values (Fig. 23) change very rapidly at this boundary, which must represent a discontinuity in both the average temperature of hydrothermal alteration and in the average, integrated water/rock ratio. Part of the explanation for higher water/rock ratios in the sheeted complex is the long history of hydrothermal exchange that these rocks undergo prior to crystallization of the high-level gabbro. However, in addition, there is almost certainly an abrupt permeability change across the diabase-gabbro contact as well. The highly jointed dike complex ought to be much more permeable than the gabbros (perhaps by a factor of 10, see Norton and Taylor, 1979). The combination of finer grain size and higher permeability thus also contributes to the much higher effective water/rock ratios in the sheeted complex relative to the gabbros. The variations in these parameters also help to explain the preservation of the high-temperature alteration assemblages and lack of low-temperature ^{18}O exchange in the gabbros. The finer-grained, more permeable diabase dikes and pillow lavas will undergo ^{18}O exchange down to much lower temperatures than the coarser-grained, less permeable gabbros. The latter rocks exhibit such effects only along fractures and veins.

Summing up, Gregory and Taylor (1981) showed that deep hydrothermal circulation of seawater has affected most of the feldspar-bearing rocks of the Samail ophiolite, including a large part of the section that is equivalent to oceanic layer 3. The deeper portions are depleted in ^{18}O relative to primary MOR basalts ($\delta^{18}O = +5.7$), whereas the shallower parts are enriched in ^{18}O. However, the final $\delta^{18}O$ profile in the ophiolite is the cumulative result of a long history of hydrothermal alteration ('isotopic aging'), beginning with high-temperature interactions with newly formed crust at the ridge axis and continuing for at least several hundred thousand years during spreading away from the ridge axis. The earlier high-temperature stages tend to produce ^{18}O depletions whereas the later low-temperature stages produce ^{18}O enrichments.

Integrating $\delta^{18}O$ as a function of depth for the entire ophiolite establishes (within geologic and analytical error) that the average $\delta^{18}O$ (5.7 ± 0.2) of the oceanic crust did not change as a result of all these hydrothermal interactions with seawater. Therefore the net change in $\delta^{18}O$ of seawater was also zero, indicating that seawater is buffered by MOR hydrothermal circulation, thus-providing strong support for a hypothesis originally put forth by

Muehlenbachs and Clayton (1976). Under steady state conditions the overall bulk ^{18}O fractionation (Δ) between the oceans and primary mid-ocean ridge basalt magmas was calculated to be +6.1 ± 0.3 by Gregory and Taylor (1981), implying that seawater has had a constant $\delta^{18}O$ ≈ -0.4 (in the absence of transient effects such as continental glaciation). Utilizing these new data on the depth of interaction of seawater with the oceanic crust, numerical modeling of the hydrothermal exchange shows that as long as worldwide spreading rates are greater than 1 km^2/yr, the oceans will be completely cycled through the oceanic crust on a time scale of a few tens of millions of years, and ^{18}O buffering of seawater will occur (Gregory and Taylor, 1981). Thus ocean water has probably had a constant $\delta^{18}O$ value of about -1.0 to +1.0 during almost all of Earth's history.

ORIGIN OF PLAGIOGRANITES

Plagiogranites, or oceanic trondhjemites, occupy essentially the same structural and petrological position in ophiolite complexes as do granophyres in subaerial spreading centers or continental layered gabbros. They typically form less than 5% of the total volume of the ophiolite, and are commonly sandwiched between the gabbro and the sheeted dike complex. They have granophyric or myrmekitic textures and small miarolitic cavities, and are composed of quartz, albite, chloritized hornblende, epidote, ilmenite, sphene, and rare clinopyroxene (Wilson, 1959); in fact, they have all of the mineralogical and textural features of typical continental granophyres except that K feldspar is absent.

Above, we have presented arguments that the granophyric and rhyolitic rocks at Skye, Jabal at Tirf, Iceland, and Skaergaard are dominantly formed by melting of hydrothermally altered roof rocks and/or assimilation of such roof rocks by the fractionally crystallizing gabbro magma, thus providing an abnormally large proportion of SiO_2-rich late differentiates. We here propose exactly the same origin for the oceanic or ophiolitic plagiogranites, the only difference being that in this case the hydrothermally altered roof rocks that are being melted or assimilated would all have previously been spilitized and exchanged with sea water. Thus, inasmuch as these hydrothermally altered roof rocks have much higher Na/K and Na/Ca ratios than the analogous rocks in continental environments, partial melting of such rocks will lead to very high-Na, low-K, SiO_2-rich magmas. Another factor to be considered is the possible incorporation of K-rich sialic basement rocks into the continental granophyres; such source materials are ruled out for ophiolites, and this could contribute to the differences between the two types of rocks.

If a major assimilated or melted component in many continental granophyres is meteoric-hydrothermally altered basalt, then there will be little or no $^{87}Sr/^{86}Sr$ enrichment, particularly if the roof rocks are about the same age as the granophyre. If the hydrothermally altered roof rocks are somewhat older, there could be time for a small build-up of radiogenic Sr in the rocks; this apparently happened locally in the Icelandic rhyolites, producing very slight enrichments in $^{87}Sr/^{86}Sr$ (O'Nions and Gronvold, 1973). If there is incorporation of continental basement, the $^{87}Sr/^{86}Sr$ values can get much higher, as at the Skaergaard and Jabal at Tirf. On the other hand, in terms of the above model, the plagiogranites should always show at least some $^{87}Sr/^{86}Sr$ enrichment, because of the high $^{87}Sr/^{86}Sr$ and Sr contents of ocean waters. However, the plagiogranites would also exhibit a well-defined upper limit of $^{87}Sr/^{86}Sr = 0.707$ to 0.709, the range of values in sea water during the past 150 million years. The $^{87}Sr/^{86}Sr$ of sea water has varied with time, and therefore the upper limit will vary depending on the age of the ophiolite.

All of the $^{87}Sr/^{86}Sr$ data so far obtained on continental granophyres, plagiogranites, and the Icelandic rhyolites confirm the above expectations. The $^{87}Sr/^{86}Sr$ ratios in plagiogranites are typically 0.704-0.7065, similar to values in the other hydrothermally altered portions of ophiolites (Fig. 23; also see Coleman and Peterman, 1975; Spooner, 1977). On a normative Ab-An-Or ternary diagram, the positions of plagiogranites are completely separated from the positions of the continental granophyres, with the plagiogranites all lying along or very near the Ab-An edge of the diagram (Coleman and Peterman, 1975). Similarly, on plots of K_2O vs. SiO_2, (Fig. 24) and Rb vs. Sr (Fig. 25), the plagiogranite field is totally separated from the continental granophyre field. These effects are all explicable in terms of the sea-water hydrothermal model; the K and Rb contents of ocean water are exceedingly low.

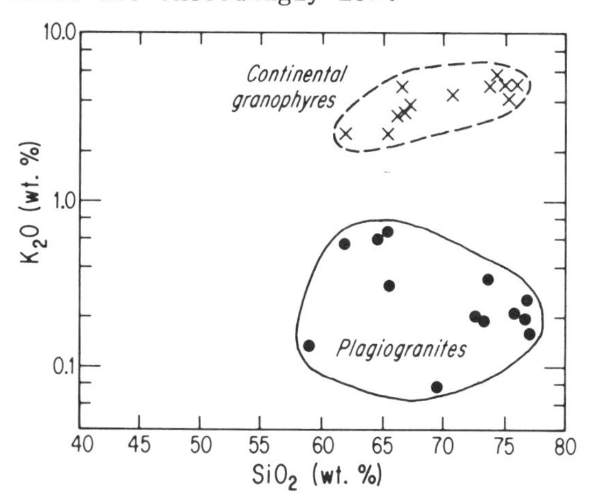

Fig. 24. Plot of K_2O vs SiO_2 comparing the field of continental granophyre (including Iceland rhyolites) with the field of plagiogranites (after Coleman and Peterman, 1975).

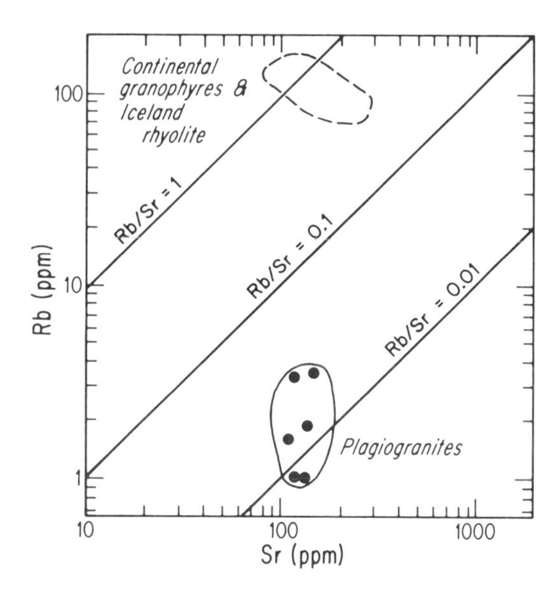

Fig. 25. Plot of Rb vs. Sr comparing the field of continental granophyres with that of plagiogranites (after Coleman and Peterman, 1975).

Gregory and Taylor (1979) and Gregory et al. (1980) have made detailed oxygen isotope studies of plagiogranites from Oman and Canyon Mtn., Oregon, and have basically confirmed the above model. They showed that xenoliths of hydrothermally altered roof rock were particularly abundant in the vicinity of plagiogranite bodies, and concluded that the above process was responsible for the build-up of a sufficiently high H_2O content to form magmatic hornblende in the high-level gabbro magmas; in areas where this build-up did not occur there was little or no development of plagiogranite. The typical lack of such magmatic hornblende in the subaerial environments is probably due to the lower P_{H_2O} values that result from the absence of a 3-4 km-thick column of ocean water above the top of the volcanic-rock section.

Although the model described above explains essentially all the geochemical and petrological contrasts between plagiogranites and continental granophyres, it is possible that many or most of these features in plagiogranites (e.g. the high Na/K ratios) are all subsolidus effects superimposed upon the rocks by the marine-hydrothermal solutions. Gregory and Taylor (1979) have shown that essentially all plagiogranites exhibit disturbed quartz-feldspar $\Delta^{18}O$ fractionations. Thus, because of the pervasive hydrothermal alteration that all these rocks have suffered after formation, it is very difficult to "see through" these effects and decipher the primary characteristics of the plagiogranite magmas. The same, of course, is true of all the low -^{18}O granophyres from the subaerial spreading centers, but in those cases we are dealing with dilute, meteoric waters which we know have not drastically affected the chemical compositions of the granophyres; the low $\delta^{18}O$ values (Figs. 9, 12, and 15) are the only clear-cut signatures of the hydrothermal process. We also have before us the examples of actual, low - ^{18}O, potassic rhyolite magmas (obsidians) from Iceland which have not suffered any post-crystallization alteration (Fig. 15).

ACKNOWLEDGEMENTS

I am indebted to Robert G. Coleman, Robert T. Gregory, Robert E. Criss, and Denis Norton for valuable discussion of this work. The research was supported by the National Science Foundation, Grants No. GA-30997X, DES 71-00510, EAR 76-21310, and EAR 78-16874, and by Department of Energy Grant No. EY-76-G-03-1305. The U.S. Geological Survey supported the field work and sampling in the Jabal at Tirf complex. This manuscript is an expanded, revised, and updated version of a paper (Taylor, 1980a) originally published as part of the 1977 Grenoble Symposium on the "Orogenic Mafic-Ultramafic Association" organized by Claude Allegre.

REFERENCES

Anderson, A. T., Clayton, R. N., and Mayeda, T. K., 1971, Oxygen isotope geothermometry of mafic igneous rocks, J. Geol., 79:715-729.

Brooks, C. K., 1973, The Tertiary of Greenland: a volcanic and plutonic record of continental break-up, Mem. Am. Assoc. Petrol. Geologists, 19:150-160.

Coleman, R. G., 1977, "Ophiolites. Ancient Oceanic Lithosphere?," Springer-Verlag monograph 12, Series Minerals and Rocks, 1-220.

Coleman, R. G., and Peterman, Z. E., 1975, Oceanic plagiogranite, J. Geophys. Res., 80:1099-1108.

Coleman, R. G., Fleck, R. J., Hedge, C. E., and Ghent, E. D., 1975, The volcanic rocks of southwest Saudi Arabia and the opening of the Red Sea, U.S. Geological Survey Saudi Arabian Project Report, 194:1-60.

Corliss, J. B., Dymond, J., Gordon, L. I., Edmond, J. M., von Herzen, R. P., Ballard, R. D., Green, K., Williams, D., Bainbridge, K., Crane, K., and Van Andel, T. H., 1979, Submarine thermal springs on the Galapagos Rift, Science, 203:1073-1083.

Craig, H. (1981) Personal communication.

Craig, H. (1961) Isotopic variation in meteoric waters, Science, 133: 1702-1703.

Criss, R. E., Lanphere, M. A., Taylor, H. P., 1982, Effects of regional uplift, deformation, and meteoric-hydrothermal metamorphism on K-Ar ages of biotites in the southern half of the Idaho batholith, J. Geophys. Res., 87:7029-7046.

Criss, R. E., and Taylor, H. P., 1982, An $^{18}O/^{16}O$ and D/H study of Tertiary hydrothermal systems in the southern half of the Idaho batholith, Bull. Geol. Soc. of America, in press.

Eaton, G. P., Christiansen, R. L., Iyer, H. M., Pitt, A. M., Mabey, D. R., Blank, H. R., Zeitz, I. and Gettings, M. E., 1975, Magma beneath Yellowstone National Park, Science, 188:787-796.

Epstein, S., Sharp, R. P., and Gow, A. J., 1965, Six year record of oxygen and hydrogen isotope variation in South Pole firn, J. Geophys. Res., 70:1809-1819.

Epstein, S., and Mayeda, T. K., 1953, Variation of ^{18}O content of waters from natural sources, Geochim. Cosmochim. Acta, 4: 213-224.

Forester, R. W., and Taylor, H. P., 1977, $^{18}O/^{16}O$, D/H, and $^{13}C/^{12}C$ studies of the Tertiary igneous complex of Skye, Scotland, Amer. J. Sci., 277:136-77.

Friedman, I., and Smith, R. L., 1958, The deuterium content of water in some volcanic glasses, Geochim. Cosmochim. Acta, 15: 218-228.

Friedman, I., Lipman, P. W., Obradovich, J. D., and Christiansen, R. L., 1974, Meteoric water in magmas, Science, 184:1069-1072.

Friedman, I., Redfield, A. C., Schoen, D., and Harris, J., 1964, The variation in the deuterium content of natural waters in the hydrologic cycle, Geophys. Rev., 2:177-224.

Garlick, G. E., McGregor, I. D., and Vogel, D. E., 1971, Oxygen isotope ratios in eclogites from kimberlites, Science, 172: 1025-1027.

Glennie, K. W., Bouef, M. G. A., Hughes Clark, M. W., Moody-Stuart, M., Pilaar, W. F. H., and Reinhardt, B. M., 1974, "Geology of the Oman Mountains", Trans. Roy. Dutch Geol. Mining Soc., 31: 1-423.

Greenbaum, D., 1972, Magmatic processes at ocean ridges: Evidence from the Troodos Massif, Cyprus, Nature, 238:18-21.

Gregory, R. T., and Taylor, H. P., 1981, An oxygen isotope profile in a section of Cretaceous oceanic crust, Samail ophiolite, Oman: Evidence for $\delta^{18}O$ buffering of the oceans by deep (>5km) seawater-hydrothermal circulation at mid-ocean ridges, J. Geophys. Res. 86:2737-2755.

Gregory, R. T., and Taylor, H. P., 1979, Oxygen isotope and field studies applied to the origin of oceanic plagiogranites, International Ophiolite Symposium, Geol. Surv. Dep. of Cyprus, Nicosia (abstract), 117-118.

Gregory, R. T., Taylor, H. P., and Coleman, R. G., 1980, The origin of plagiogranite by partial melting of hydrothermally altered stoped blocks at the roof of a Cretaceous mid-ocean ridge magma chamber, the Samail ophiolite, Oman, Geol. Soc. Amer.

Abst. with Prog., 12:437.
Hattori, K., and Muehlenbachs, K., 1982, Oxygen isotope ratios of the Icelandic crust, J. Geophys. Res., 87:6559-6565.
Heaton, T. H. E., and Sheppard, S. M. F., 1977, Hydrogen and oxygen isotope evidence for sea-water hydrothermal alteration and ore deposition, Troodos complex, Cyprus, in: "Volcanic Processes in Ore Genesis", Spec. Paper No. 7, Geol. Soc. Lond., 42-57.
Hopson, C. A., and Frano, C. J., 1977, Igneous history of the Point Sal ophiolite, southern California, in: "North American Ophiolites", Oreg. Dep. Geol. Miner. Resour. Bull., 95:161-183.
Hopson, C. A., Coleman, R. G., Gregory, R. T., Pallister, J. S., and Bailey, E. H., 1981, Geologic section through the Samail ophiolite and associated rocks along a Muscat-Ibra transect, southeastern Oman Mountains, J. Geophys. Res., 86:2527-2544.
Irvine, T. N., and Smith, C. H., 1967, The ultramafic rocks of the Muskox intrusion, in: "Ultramafic and Related Rocks", ed. P. J. Wyllie, John Wiley, New York, 38-49.
Jakobsson, S. P., 1972, Chemistry and distribution pattern of recent basaltic rocks in Iceland, Lithos, 5:365-386.
Javoy, M., 1970, "Utilisation des isotopes de l'oxygene en magmatologie", These de Doctorat e'Etat des Sciences Physiques, Faculte des Sciences de Paris, 1-196.
Jehl, V., Poty, B., Weisbrod, A., and Sheppard, S. M. F., 1977, Unpublished data, quoted in Heaton and Sheppard (1977).
Leeman, W. P., and Dasch, E. J., 1978, Strontium, lead, and oxygen isotopic investigation of the Skaergaard intrusion, East Greenland, Earth Planet. Sci. Lett., 41:47-59.
Lipman, P. W., and Friedman, I., 1975, Interaction of meteoric water with magma: An oxygen isotope study of ash flow sheets from southern Nevada, Bull. Geol. Soc. Amer., 86:695-702.
Magaritz, M., and Taylor, H. P., 1974, Oxygen and hydrogen isotope studies of serpentinization in the Troodos ophiolite complex, Cyprus, Earth Planet. Sci. Lett., 23:8-14.
Magaritz, M., and Taylor, H. P., 1976, Oxygen, hydrogen, and carbon isotope studies of the Franciscan formation, Coast Ranges, California, Geochim. Cosmochim. Acta, 40:215-234.
McCulloch, M. T., Gregory, R. T., Wasserburg, G. J., and Taylor, H. P., 1981, Sm-Nd, Rb-Sr, and $^{18}O/^{16}O$ isotopic systematins in an oceanic crustal section: Evidence from the Samail ophiolite, J. Geophys. Res., 86:2721-2735.
Muehlenbachs, K., 1973, The oxygen isotope geochemistry of siliceous volcanic rocks from Iceland, Carnegie Inst. Washington Yr. Bk., 72:593-597.
Muehlenbachs, K., Anderson, A. T., and Sigvaldason, G. E., 1974, Low $-^{18}O$ basalts from Iceland, Geochim. Cosmochim. Acta, 38: 577-588.
Muehlenbachs, K., and Clayton, R. N., 1972, Oxygen isotope geochemistry of submarine greenstones, Can. J. Earth Sci., 9:471-478.
Muehlenbachs, K., and Clayton, R. N., 1965, Oxygen isotope composition

of the oceanic crust and its bearing on seawater, J. Geophys. Res., 81:4365-4369.

Muehlenbachs, K., and Sigurdsson, H., 1977, Tertiary low -^{18}O basalts from N.W. Iceland, EOS, Trans. Amer. Geophys. Union, 58:529.

Norton, D., and Taylor, H. P., 1979, Quantitative simulation of the thermal history of crystallizing magmas on the basis of oxygen isotope data and transport theory: An analysis of the hydrothermal system associated with the Skaergaard Intrusion, J. Petrology, 20:421-486.

O'Nions, R. K., and Gronvold, K., 1973, Petrogenetic relationship of acid and basic rocks in Iceland: Sr isotopes and rare earth elements in late and post-glacial volcanics, Earth Planet. Sci. Lett., 19:397-409.

Pallister, J. S., and Hopson, C. A., 1981, Samail ophiolite plutonic suite: Field relations, phase variation, cryptic variation and layering, and a model of a spreading ridge magma chamber, J. Geophys. Res., 86:2593-2644.

Palmasson, G., 1971, Crustal structure of Iceland from explosion seismology, Soc. Sci. Islandica, 40:1-187.

Pankhurst, R. J., Beckinsale, R. D., and Brooks, C. K., 1976, Strontium and oxygen isotope evidence relating to the petrogenesis of the Kangerdlugssuaq alkaline intrusion, East Greenland, Contrib. Mineral. Petrol., 54:17-42.

Saemundsson, K., 1974, Evolution of the axial rifting zone in northern Iceland and the Tjornes fracture zone, Bull. Geol. Soc. Amer., 85:495-504.

Sheppard, S. M. F., Brown, P. E., and Chambers, A. D., 1977, The Lilloise intrusion, East Greenland: Hydrogen isotope evidence for the efflux of magmatic water into the contact metamorphic aureole, Contrib. Mineral. Petrol., 68:129-147.

Sheppard, S. M. F., and Epstein, S., 1970, D/H and ^{18}O/^{16}O ratios of minerals of possible mantle or lower crustal origin, Earth Planet. Sci. Lett., 9:232-239.

Sheppard, S. M. F., Nielsen, R. N., and Taylor, H. P., 1969, Oxygen and hydrogen isotope ratios of clay minerals from porphyry copper deposists, Econ. Geol., 64:755-777.

Sheppard, S. M. F., Nielsen, R. N., and Taylor, H. P., 1971, Hydrogen and oxygen isotope ratios in minerals from porphyry copper deposits, Econ. Geol., 66:515-542.

Shieh, Y. N., and Taylor, H. P., 1969, Oxygen and carbon isotope studies of contact metamorphism of carbonate rocks, J. Petrol., 10:307-331.

Sigvaldason, G. E., 1969, Chemistry of basalts from the Icelandic rift zone, Contr. Mineral. Petrol., 20:357-370.

Smith, R. B., and Christiansen, R. L., 1980, Yellowstone Park as a window on the earth's interior, Scientific American, 242: 104-117.

Soper, N. J., Higgins, A. C., Downie, C., Matthews, D. W., and Brown, P. E., 1976, Late Cretaceous-early Tertiary stratigraphy of the

Kangerdlugssuaq area, East Greenland, and the age of opening of the north-east Atlantic, J. Geol. Soc. Lond., 132:85-104.

Spooner, E. T. C., 1977, Hydrodynamic model for the origin of the ophiolitic cupriferous pyrite ore deposits of Cyprus, in: "Volcanic Processes in Ore Genesis", Spec. Paper No. 7, Geol. Soc. Lond., 58-71.

Spooner, E. T. C., Beckinsale, R. D., Fyfe, W. S., and Smewing, J. D., 1974, ^{18}O enriched ophiolitic metabasic rocks from E. Liquria (Italy), Pindos (Greece), and Troodos (Cyprus), Contrib. Mineral. Petrol., 47:41-62.

Stakes, D., and O'Neil, J. R., 1981, Mineralogy and stable isotope geochemistry of hydrothermally altered oceanic rocks, Earth Planet. Sci. Lett., 57:285-304.

Suzuoki, T., and Epstein, S., 1976, Hydrogen isotope fractionation between OH-bearing silicate minerals and water, Geochim. Cosmochim. Acta, 40:1229-1240.

Taylor, H. P., Jr., 1968, The oxygen isotope geochemistry of igneous rocks, Contrib. Mineral. Petrol., 19:1-71.

Taylor, H. P., Jr., 1971, Oxygen isotope evidence for large-scale interaction between meteoric ground waters and Tertiary granodiorite intrusions, western Cascade Range, Oregon, J. Geophys. Res., 76:7855-74.

Taylor, H. P., Jr., 1974a, Oxygen and hydrogen isotope evidence for large-scale circulation and interaction between ground waters and igneous intrusions, with particular reference to the San Juan volcanic field, Colorado, in: "Geochemical Transport and Kinetics", ed. by Hofmann, A. W., Giletti, B. J., Yoder, H. S., and Yund, R. W., Carnegie Institution of Washington Publ., 634:299-324.

Taylor, H. P., Jr., 1974b, The application of oxygen and hydrogen isotope studies to problems of hydrothermal alteration and ore deposition, Econ. Geol., 69:843-883.

Taylor, H. P., Jr., 1977, Water/rock interactions and the origin of H_2O in granitic batholiths, J. Geol. Soc. Lond., 133:509-558.

Taylor, H. P., Jr., 1980a, Stable isotope studies of spreading centers and their bearing on the origin of granophyres and plagiogranites, in: "Proceedings of the International Meeting of Mafic-Ultramafic Association in Orogenic Belts", edited by C. Allegre, Centre National de Recherche Scientifique, Grenoble, France, 149-165.

Taylor, H. P., Jr., 1980b, The effects of assimilation of country rocks by magmas on $^{18}O/^{16}O$ and $^{87}Sr/^{86}Sr$ systematics in igneous rocks, Earth Planet. Sci. Lett., 47:243-254.

Taylor, H. P., Jr., and Coleman, R. G., 1977, Oxygen isotopic evidence for meteoric-hydrothermal alteration of the Jabal at Tirf complex, Saudi Arabia, EOS, Trans. Amer. Geophys. Union, 58:516.

Taylor, H. P., Jr., and Epstein, S., 1963, $^{18}O/^{16}O$ ratios in rocks and coexisting minerals of the Skaergaard intrusion, J. Petrology, 4:51-74.

Taylor, H. P., Jr., and Epstein S., 1970, $^{18}O/^{16}O$ ratios of Apollo 11 lunar rocks and minerals, Geochim. Cosmochim. Acta. Suppl. 1, 2:1613-1626.

Taylor, H. P., Jr., and Forester, R. W., 1971, Low $-^{18}O$ igneous rocks from the intrusive complexes of Skye, Mull, and Ardnamurchan, western Scotland, J. Petrol., 12:465-497.

Taylor, H. P., Jr., and Forester, R. W., 1979, An oxygen and hydrogen isotope study of the Skaergaard intrusion and its country rocks: A description of a 55 m.y. old fossil hydrothermal system, J. Petrol., 20:355-419.

Taylor, H. P., Jr., and Magaritz M., 1978, Oxygen and hydrogen isotope studies of the Cordilleran batholiths of western North America, in: "Stable Isotopes in the Earth Sciences", ed. B.W. Robinson, New Zealand Dept. of Sci. and Ind. Research Bull., 220:151-174.

Wager, L. R., 1947, Geological investigations in East Greenland, Pt. 4. The stratigraphy and tectonics of Knud Rasmussen Land and the Kangerdlugssuaq region, Medd. om Gronland, 134:1-62.

Wager, L. R., and Brown, G. M., 1967, "Layered Igneous Rocks", W. H. Freeman & Co., San Francisco, 588 pp.

Wager, L. R., and Deer, W. A., 1939, Geological investigations in East Greenland, Pt. 3. The petrology of the Skaergaard intrusion, Kangerdlugssuaq, East Greenland, Medd. om Gronland, 105:1-352.

Walker, G. P. L., 1966, The acid volcanic rocks of Iceland, Bull. Volcanologique, 29:375-406.

Wenner, D. B., and Taylor, H. P., Jr., 1973, Oxygen and hydrogen isotope studies of serpentinization of ultramafic rocks in oceanic environments and continental ophiolite complexes, Amer. J. Sci., 273:207-239.

Ward, P. L., 1971, New interpretation of the geology of Iceland, Geol. Soc. Amer. Bull., 82:2991-3012.

Welke, H., Moorbath, S., Cumming G. L., and Sigurdsson, H., 1968, Lead isotope studies on igneous rocks from Iceland, Earth Planet. Sci. Lett., 4:221-231.

White, D. L., 1974, Diverse origins of hydrothermal ore fluids, Econ. Geol., 69:954-973.

Wilson, R. A. M., 1959, The geology of the Xeros-Troodos area, Mem. Geol. Surv. Cyprus, 1:1-135.

Worst, B. G., 1960, The great dyke of Southern Rhodesia, S. Rhodesia Geol. Surv. Bull., 47:1-234.

THE BASIC PHYSICS OF WATER PENETRATION INTO HOT ROCK

C.R.B. Lister

School of Oceanography, WB-10
University of Washington
Seattle, Washington 98195

ABSTRACT

There are a number of rival ideas about how heat is transferred to the hot fluids of major geothermal areas. One of these is the theory of water penetration into hot rock, where thermal shrinkage of the rocks permits them to crack on a relatively small scale and then transport heat away from the hot boundary thus formed. The theory has the advantage of requiring a relatively small area of contact between water and hot rock to produce the high thermal outputs of large geothermal areas: of the order of 1 km^2. This is because the thermal boundary layer is established by the advance of a cracking front into the rock itself, and is a function of front velocity rather than pre-existing geological structure. A critical review of the physics of the process shows that the weakest areas of the theory are in the understanding of the mechanics of the cracking process and in the structure of the porous-medium convection that discharges the heat. The problems in both these areas stem from a basic lack of knowledge of the physics, and not from weaknesses in the theory itself. The one-dimensionality of the treatment is not a serious limitation because the predicted crack spacing, of the order of a meter, is much smaller than the kilometer scale of three-dimensionality in the geometry.

The theory predicts correctly the order-of-magnitude of the hot water temperature even when data not appropriate to mafic crustal rocks has to be used. Examination of exposed regions of ophiolite suites that should have undergone the cracking process is at too primitive a stage to confirm or disprove the theory. A partially-controlled experiment where large volumes of water were pumped onto an advancing lava flow confirms that the cracking process does take place, and the numbers are in general agreement with the theory in spite of conditions being substantially different from those treated in the calculations.

INTRODUCTION

As is usual in geophysics, the nature of a geothermal system can only be explored from the surface manifestations downwards. There is little dispute that fumaroles, steaming ground, geysers or hot springs indicate the presence of a major circulation system in the earth's crust. In a few places, such as at Wairakei, New Zealand, the flow system has been characterized by exploratory drilling prior to power development. There, the upwelling plume was well-defined, with a lateral scale of about 1 km, and a flux level of about 400 MW. The permeability of the surface rocks was only moderate, leading to a mushroom-shaped plume like those observed in convection experiments at moderate Rayleigh numbers (Elder, 1965). It is not surprising that many subsequent theoretical studies considered convection in a moderately permeable layer, heated from below to an extent that would drive convection at up to a few times the critical Rayleigh number (e.g., Ribando et al., 1976; Norton and Knight, 1977; Lowell, 1980). The approach has the advantage that the observed water temperatures can be matched simply by making the permeability low enough. A cartoon of the essential features of these models is shown in Figure 1.

There are two pieces of evidence that suggest a lack of physical reality to this approach. The latest, and most obvious, is the discovery of hot smokers on the ocean floor, where water of about the same temperature and flow rate as at Wairakei appears at a few nearby vents (Spiess et al., 1980). The equivalent Rayleigh number of such a concentrated plume is enormously high, and its very existence belies the assumption of moderate permeability. The other evidence is hidden in the large power output of the plume and the heat flow values typical of the terrain in which geothermal areas are found. Such a heat flow may be 2 HFU (μcal $cm^{-2}s^{-1}$) near major land geothermal areas, so that a plume of 400 MW needs to draw on about 5000 km^2 of the earth's surface. This is most unlikely, since a saturated permeable layer overlying such a large area would break down into a number of separate convection cells, unless the layer were much thicker than anyone expects (e.g., Hartline and Lister, 1981). So a special, concentrated source of heat is needed: if the permeable layer were 5 km thick, the <u>maximum</u> possible diameter of a source supplying a single plume would be 10 km, implying a heat flow of 127 HFU and a barrier between 400°C water and magma about 300 m thick. Remember, this is a <u>maximum</u> thickness, and must be maintained over the entire 10 km diameter.

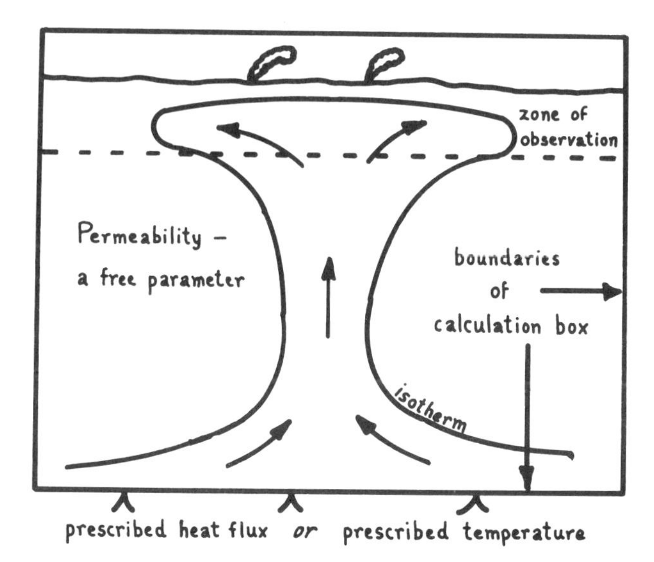

Fig. 1. Cartoon of a geothermal area as seen by computer modellers. The side boundaries are treated in different ways depending on the goal of the model and the power of the program.

Further evidence that the geothermal heat source must be concentrated can be found in the distribution of geothermal areas along the quaternary volcanic province of New Zealand (Bolton, 1975). There are many of them, and several are as large, or larger, than that at Wairakei; they are much too close together to draw their heat from a large area of normal heat flow. If the magma chamber with the thin cap begins to seem a little unreasonable as an alternative, consider that the model in Figure 1 begs the question of how the calculation box became permeable in the first place, let alone how it achieved the right value to match observed hot water temperatures.

There is a simple solution to all these problems: the idea that the water itself penetrates into hot rock, cooling it and cracking it by thermal contraction as it goes. The boundary temperature gradient is no longer limited by a plausible distance to circulating magma, but only by the rate at which water can penetrate. The germ of this idea appeared when Bodvarsson and Lowell (1972) showed that a single thin crack of kilometer extent could circulate enough water to cool a large volume of oceanic crust, and itself could be produced by a minute amount of thermal contraction. In a slightly different form, it was factored into the calculations of Harlow and Pracht (1972), who assumed that, below a temperature where thermal contraction compensated for overburden pressure, cracks nominally 5 cm apart appeared and opened up by further contraction. They used a variant of the Kozeny formula for flow through tortuous channels in a porous medium (see Scheidegger, 1974), with an assumed 'hydraulic radius', corresponding, in effect, to an assumed crack spacing.

At one extreme a single crack was considered; at the other, isotropic permeability from a fixed, small crack spacing. The real physics has still not been entered into the problem, because no actual mechanism of cracking was proposed, nor how the process might be controlled. The first attempt to include all the relevant physics is the theory of water penetration into hot rock, presented by Lister (1974). The treatment is one-dimensional, and considers the advance of a horizontal cracking front separating a permeable medium above and hot rock below. Although this extreme simplification is necessary to make the problem tractable, it is not as severe a limitation as might at first appear. Local regions in a geothermal system are close enough to planar if the crack spacing is small compared to the overall scale, and the order of magnitude of convective heat transport is not sensitive to the angle of the heated surface.

The picture of a geothermal area is now modified from Figure 1 to Figure 2; a body of hot rock replaces the hot boundary as the source of the heat. Note that the surface manifestations do not change, as they are dependent on the properties of the surface layer and the chemistry of the hot fluids. The theory applies strictly only to the base of the permeable region that is eating into the hot rock; the value of the results lies mainly in the order of magnitude found for the heat extraction per unit area and the hot water temperature. The remainder of this paper will discuss the key points in the physics of the process in a non-mathematical but critical manner; for more details the reader can look at Lister: 1974, 1975, 1981, 1982.

THE THERMAL BOUNDARY LAYER, CONTRACTION AND CRACKING

The breakdown of the problem into manageable parts is based on the concept of a macroscopically distinct cracking front that separates the permeable and impermeable regions (Fig. 3). The alternative is a set of essentially independent macrofractures that propagate by causing local contraction (Bodvarsson, 1975). The choice between these philosophically very different approaches depends only on whether the natural crack spacing is less than the characteristic thickness of the conductively cooled and thermally stressed layer, or is larger. If the crack spacing is small, the macroscopic thermal and stress fields provide a natural

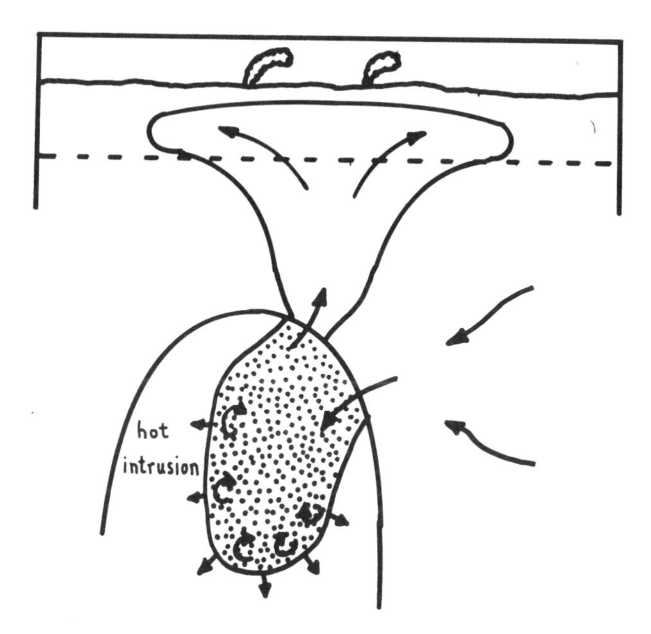

Fig. 2. The same geothermal area as in Figure 1 with a heat source derived from the idea of water penetration into hot rock. The mode of entry of the water into a cooling intrusion is far from being established, as is the mode of entry of the intrusion into the country rock, unless it is associated with general rifting (as in New Zealand).

stabilization, and ensure that an individual crack cannot move ahead of the general front. Hence the logical procedure is to postulate the concept, proceed with the analysis, and see if the results remain consistent with the original postulate. This test was not in fact consciously applied to the results of the original theory, but they do survive it (Lister, 1974, Table 1); the revised theory (Lister, 1982, Table 24-1) is more marginal, but, as will be seen, contains an implied boundary condition for the cracking that is inconsistent with the results.

If the rock is only moderately hot, the temperature of cracking is determined by the point at which thermal contraction from the original state, applied to the appropriate elastic modulus, brings it into horizontal tension (e.g., Harlow and Pracht, 1972). If it starts out very hot, like a recently crystallized pluton, then rock creep limits the stress achievable at the highest temperatures, and determines when tension can build. In either case the temperature of cracking is a slowly varying function of depth, and can be treated as locally constant. The thermal problem is thus the response of a medium of constant initial temperature flowing toward a boundary held at another constant temperature. This is a standard problem and the temperature distribution is exponential, with a characteristic scale of u/κ, where u is the flow (or front advance) velocity, and κ is the diffusivity of the medium (Fig. 3).

In the case of the infinite horizontal cracking front, the appropriate elastic modulus is simply the planar modulus, and the stresses can be converted into equivalent vertical uniaxial stress for easy comparison to the overburden pressure (Lister, 1974). In the more likely case of a three-dimensional cracked zone (Lister,

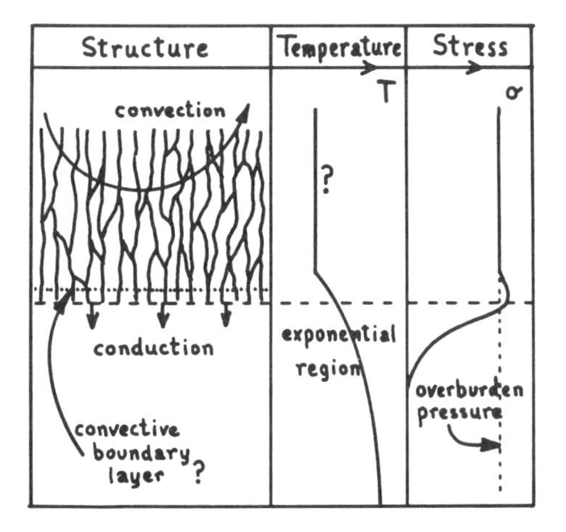

Fig. 3. A sketch of the essential features of water penetration theory. On the left, a series of sub-vertical cracks, forming irregular polygons in plan view, is propagating downward at some defined velocity. The cracking temperature sets a boundary condition for solution of the conductive region, while the heat is carried away by fluid flow. The resulting temperature distribution is shown in the center panel, and the vertical equivalent stress due to horizontal thermal contraction is shown on the right. A small excess horizontal tension is sufficient to propagate the cracks.

1975) a stressed wall has to be developed that is equivalent to the wall of a pressure case, and the stresses are larger than the overburden pressure. This is because the hydrothermal system, connected by channels to the surface, can only be at the hydrostatic pressure due to its depth, but plastic hot rock must be at the lithostatic pressure caused by the column of rock above it (both being increased by the hydrostatic pressure of any overlying ocean). In either case, the elastic modulus of igneous rock is so high that only a few degrees of cooling are needed to nullify the overburden pressure at a depth of several kilometers. The limiting factors are the creep of the rock and its strength, particularly its static fatigue strength for applied stresses of long duration. Neither of these is as easily measured as expansion coefficient or elastic modulus, particularly as the rock starts in an annealed pristine condition not readily duplicated in the laboratory. The overall amount of creep possible in response to thermal contraction is small, and therefore the creep remains in the region called transient creep, where stress increases with strain even at constant temperature and strain rate. In fact the temperature is changing, and so only an educated guess can be made at the appropriate cracking temperature for a given depth of overburden. Such an estimate was made in Lister (1974): 800°K at 5 km depth, but the problem can be circumvented by treating the cracking temperature as an independent variable, and preparing tables of the results versus cracking temperature, as was done also. This last procedure has the advantage that the theory can be applied to cases where the rock is not a fresh hot pluton, but rock with a prior cooling history, perhaps including a previous phase of hydrothermal penetration, followed by healing of the cracks due to rock alteration (Lister, 1981). The remaining correction, due to differing initial temperatures, is not large.

To propagate cracks into any material requires an excess tensile stress over any overburden or other pre-existing stresses that would tend to close them. The question is how much net tension is required for a given crack geometry and material. The geometric problem is the easiest to discuss, since the excess tensile stress must decline to zero behind the cracking front as the rock becomes simply a set of polygonal columns supporting the overburden (Fig. 4). It was shown by Irwin (1958) that a crack tip stress intensity factor (CTSIF) is responsible for crack propagation, and that this is a product of the stress and the square root of the stressed length of crack. Above a certain level, crack propagation is catastrophic, such as in the sudden failure of brittle materials under tension, but below this critical level, crack propagation is not zero, but merely slow and dependent on other factors such as temperature and water vapor pressure. A random polygonal geometry cannot be solved exactly, but a fair approximation can be made by assuming that the excess tensile stress declines linearly to zero in a distance equal to

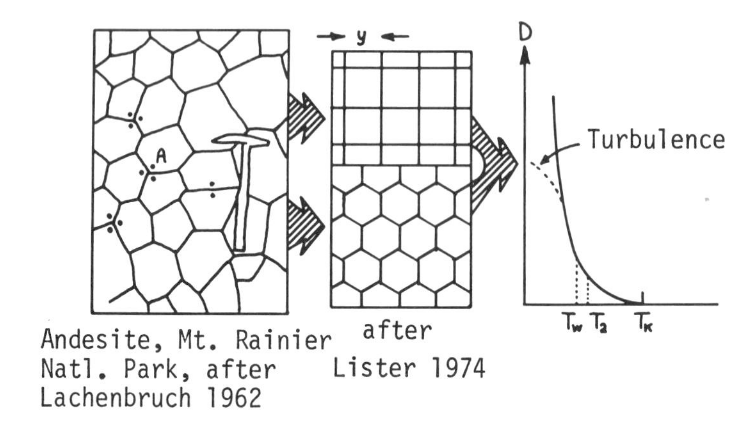

Fig. 4. The permeability derivation follows the arrows in this diagram. A real irregular polygonal pattern on the left is abstracted by considering two regular arrays easily amenable to calculation. The permeability of these is the same for the same crack spacing and shrinkage, and allows that of the irregular pattern to be based solely on the mean crack spacing. A plot of permeability against temperature is drawn on the right; it is proportional to the cube of the drop below the cracking temperature. A hot water temperature and effective permeable boundary temperature are sketched in, as is a potential deviation from linear permeability if flow through the cracks should become turbulent.

about half a mean column diameter. This allows use of a solution by Lachenbruch (1961), strictly only applicable to an infinite plane crack, to arrive at $\sqrt{y}/3$ for the geometric factor, where y is the mean crack spacing.

The material problem is a severe one because I have been able to find only one report of experiments on crack propagation in a silicate. This is the work of Martin (1972) on the stable propagation of cracks in single-crystal quartz, a material that is available in large flaw-free blocks and transparent enough for the propagation to be observed and measured. It is not the mafic material typical of geothermal areas, and it is not a multi-grain aggregate like a rock. Grain boundaries could be effective in stopping the propagation of a crack in a grain; on the other hand, a crack in a ceramic material is a separation at grain boundaries, with breakage only of those grains geometrically unable to separate cleanly. It is not obvious whether a mafic rock will be stronger or weaker than quartz, and the uncertainty is serious enough to warrant inclusion in all the calculations of a fudge factor ϕ. An approximate solution of Martin's (1972) geometry produces numerical values for the excess tension needed to propagate cracks at spacing y(m) and velocity u(m/yr), expressed as equivalent vertical uniaxial tension: $\sigma_c = \phi y^{-1/2}$ (0.003 + 0.00041 ℓn u) kb (T = 600°K). For reasonable values of y in the range of 0.1 to 10 m, and velocities of 1-50 m/yr, the excess tension is very small compared to the overburden pressure of 5 km of rock, which is about 1 kb or 0.1 GPa above the hydrostatic pressure at that depth. However, it is comparable to the overburden pressure in a surface basalt flow that is in the process of turning into a columnar basalt, an unusually regular exemplar of hydrothermal penetration according to this theory. The stress decreases with increasing temperature according to an Arrhenius relation with an activation energy of 26 kcal/mole (Quartz: Martin, 1972).

PERMEABILITY AND CONVECTION

The cracking stress analysis shows that the system favors an increased crack spacing as long as this remains compatible with the stressed-layer concept. Before deciding what the limiting phenomena are, it is necessary to consider the problem of just what the convection should be like at a high-flux geothermal boundary. The permeability of the cracked matrix is the dominant parameter controlling the convection, so this should be looked at first.

We postulate an irregular structure of sub-vertical cracks forming a polygonal pattern in the horizontal plane, and characterized by the mean crack spacing, or polygon diameter, y (Fig. 4). If rock creep has effectively ceased, there is a temperature at which the cracks would be just closed, but the columns support the overburden pressure: it should be close to the temperature at which the stress curve in Figure 3 first crosses the overburden stress line, and is close to the actual cracking temperature (in the case of large overburden pressure) because the excess cracking tension is so small. At any temperature below this, the columns shrink and the cracks open up. Flow through a parallel-sided slot under laminar conditions is proportional to slot width cubed: $d^3/12$; whether the flow really is laminar or not under the actual conditions remains a problem. The permeability is a strong function of temperature, especially near the cracking temperature where the phenomena of most interest occur.

The permeability of an irregular polygonal crack structure can be estimated by calculating the permeability of two regular crack patterns: squares and hexagons. In both of the regular patterns, the horizontal permeability is the same and isotropic, while the vertical permeability is exactly twice the horizontal. Convection theory and experiments are still learning about the phenomena in isotropic porous media, but this level of anisotropy is unlikely to have much effect on the structure of the convection and is dismissed by taking as an equivalent isotropic permeability the harmonic mean of the two: $\sqrt{2}\, d^3/12y$. Since d is, before clogging by chemical processes, simply the shrinkage of the rock below the cracking temperature (T_K), $d \approx \alpha t(T_K - T)$, the final estimate for permeability is $D = (\sqrt{2}/12)\alpha^3 y^2 (T_K - T)^3$.

Here it is implicitly assumed that the creep of the rock effectively ceases once the stress stabilizes at the overburden pressure. The problem was analyzed carefully in Lister (1974) and it was found that the creep rate at constant stress is such a rapid function of temperature that the residual creep after cracking should be small enough to be neglected.

The boundary of the porous medium is diffuse in the permeability sense, but, again, this is unlikely to influence the structure of the convection, and it is merely necessary to choose, as the boundary temperature, one where the permeability is some reasonable fraction of that in the main body of hot water. The fraction actually chosen was about 2/3 (due to an error in Lister, 1974); given the cubic variation with temperature (Fig. 4) this does not seem unreasonable. Below the general hot water temperature T_w, the permeability continues to rise, subject to changes due to the onset of turbulent flow in the cracks, but the viscosity of the water rises and its expansion coefficient declines. While the convection is not quite like that in an experiment, or theoretical model, where all the properties are constant, the compensation due to the fluid properties may prevent the convection from becoming radically different in form. Not so, however, for fluids whose chemistry enables them to precipitate walls that prevent advection of cooler fluid toward the hot plume, as may be the case for the ocean floor smokers.

It would be possible to proceed logically toward a definition of the thermal regime near the cracking front, if convection in porous media were either simple or well-understood. About the only thing that has been established unequivocally is the critical Rayleigh number for a porous slab bounded by horizontal perfect conductors. In this case the Rayleigh number is defined easily as $\mathcal{R} = D\alpha_w gh\, \Delta T/\kappa' \nu_w$, where α_w and ν_w are the volumetric expansion coefficient and kinematic viscosity of the fluid, g is gravity, h is slab height, ΔT is the applied temperature difference and κ' is a thermal diffusivity based on the conductivity of the matrix as a whole divided by the heat capacity of the fluid alone. The theoretical value for the onset of convective instability is $4\pi^2$ (Lapwood, 1948) and this has been verified convincingly in a Hele-Shaw cell (Hartline and Lister, 1977).

The next fundamental question is how the heat transport varies with Rayleigh number once convection has been established. A dimensionless number, the Nusselt number, is used, and is defined as the convective heat transport divided by the heat that would be transported by conduction in the absence of motion. Just above the critical Rayleigh number, both experiment and theory agree that the Nusselt number should be linearly proportional to Rayleigh number $\mathcal{N} \simeq \mathcal{A}/\mathcal{A}_c$ (Busse and Joseph, 1972; Combarnous and Bories, 1975). However, at some Rayleigh number less than 1000, all the experiments show a breakover to a lower power dependence, perhaps as low as 1/3 (Fig. 5). They do not agree on the level of the breakover, and no other parameter has been extracted from the published data that explains the variation. The problem is serious because the heat transport influences the advance of the cracking front directly by removing the heat of the cooling rock, and also indirectly by affecting the crack spacing through the thickness of the convective boundary layer. Due to the scale of a geothermal area and the vigor of the convection, the Rayleigh number is extremely high: from greater than 10^5 in Lister (1974) to 10^9 in Lister (1982). The main reason for this difference is that a linear relationship was used for the former, and the 1/3 power law shown in Figure 5 for the latter. The convective system can maintain the heat transport near the same order of magnitude (kw/m^2) because the permeability is such a strong function of crack spacing and water temperature. The fundamental problem can be summarized as follows. In the case of free convection, the structure at high Rayleigh number is well-known: very thin boundary layers eject random plumes into a well-mixed core of fluid that is nearly isothermal on average (e.g., Katsaros et al., 1977). This means that the real heat transport is independent of cell height and the Nusselt number is linearly proportional to height; as the free-convection Rayleigh number is proportional to height cubed, the Nusselt number must depend on the 1/3 power of Rayleigh number if it is to be a single-valued function. The experimental results bear this out (although there is also a small dependence on the Prandtl number, it would remain constant as the cell height is varied). If the structure of convection in a porous medium were similar, Nusselt number would have to be linearly proportional to Rayleigh number, since the latter depends only linearly on height. The departure of the experimental results from this

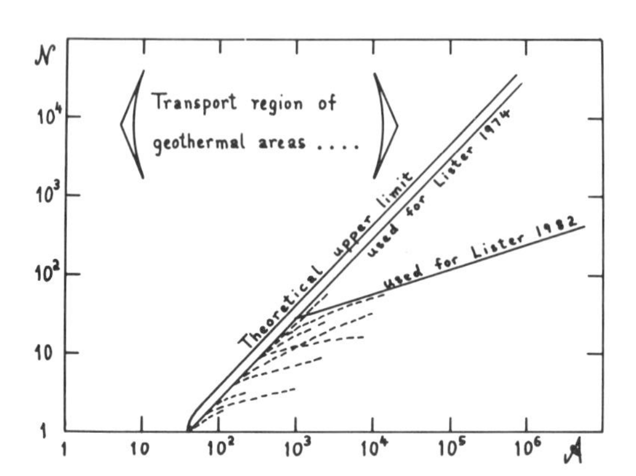

Fig. 5. A logarithmic plot of the heat transport, or Nusselt, number against Rayleigh number for porous medium convection. Dashed lines are the results of experiments; solid lines are the theoretical upper limit of Busse and Joseph (1972), and the nominal relations used in the two prior publications on water penetration theory. The experimental data are from Schneider (1963), Elder (1965), Elder (1967), Bories (1970), Combarnous (1970), and Buretta and Berman (1976), all replotted in such a way as to demonstrate the same initial slope: according to unpublished work by Booker and Hartline.

relationship would be viewed as the breakdown of the porous medium approximation, when the pore size in the real medium became larger than the thickness of the convective boundary layer.

The calculation of a boundary layer thickness is easy if the Nusselt number is known; by definition, it is half the cell height divided by that number. This is because heat transport into the convective boundary layer is by conduction; as the conductivity is the same as that of the whole porous layer, the thermal gradient must be steeper by the same factor as the increase in heat transport. Nusselt numbers of 10 were routinely achieved (see the caption to Fig. 5), so the two boundary layers would be each 1/20 of the cell height in thickness. The cells were usually filled with beads, and sometimes the bead diameter was as large as 1/8 of the cell height (to attain an adequate permeability). Thus the convection might be viewed as "free convection in the presence of obstacles", and slope breakovers to something like the 1/3 power law of free convection would be expected. This aspect of the problem has been discussed by Elder (1967).

The argument so far is plausible; the problem remains as to whether a structure similar to that in free convection actually develops in a porous medium. The keys are the unsteadiness of the convective plumes, and their ability to mix into the bulk core fluid. If the porosity of the porous medium is nearly 100%, such as in an experiment I have conducted recently, a plume can part from the boundary, displace cooler fluid, and retain its thermal identity. If the porosity of the medium is low, such as in cracked rock, the plume would be cooled instantly upon moving into cool matrix; thermally driven flow can only develop over time as the matrix is heated up. Boundary-layer heat is not stored in fluid able to physically peel away from the boundary, but mostly in the stationary matrix. So the random variability of the free convection cannot develop; instead one could imagine very small-scale steady plumes forming, with recharge in between.

For boundary-layer phenomena the high porosity porous medium is similar to free fluid, while the low porosity medium is not. In the core that appears isothermal when horizontally averaged, the two porous media are fundamentally different from the free fluid. There, the scale on which friction operates is larger away from the boundary, and a small temperature difference can propel a large glob of fluid at a high enough velocity to transport the convective heat. The isothermal core is not only isothermal when the temperature is averaged horizontally, but is nearly so even when measured point by point. In the porous medium, motions on a scale larger than the inside of the pores meet with resistance from the matrix, whatever the size of the glob of fluid or plume. This means that the rising velocity of a blob of hot fluid is dependent only on its temperature excess, or buoyancy, and not on its size. Now, in the case of free convection, the net convective transport need only go up as the one-third power of temperature difference, so that the bulk fluid becomes relatively more and more isothermal as the Rayleigh number rises. In the porous medium, the convective heat transport should go up linearly with Rayleigh number, and therefore with temperature difference. As the velocity of rising or falling fluid is only proportional to its temperature offset from the mean, the temperature of the plumes must remain at the same fraction of the boundary temperature as the Rayleigh number rises. Yet the thickness of boundary layer that can be heated on one circuit of the fluid decreases as the velocities are higher and the contact time is less. Here is the fundamental inconsistency; height invariance requires a linear dependence of Nusselt number on Rayleigh number, but this can only be achieved if well-defined plumes of fluid nearly at the boundary temperature persist. The apparent difference between high porosity and low porosity media disappears again, since the existence of well-defined plumes precludes a statistically-determined structure with strong temporal variability.

The results of this discussion are summarized in Figure 6. At Rayleigh numbers just above critical, the form of the circulation cells in both porous medium and free convection are similar. As the velocities increase, and boundary layer residence times decrease, a true boundary layer structure develops in free convection. In the porous medium, the cell aspect ratios at first decline to retain a plume structure consistent with a thinner boundary layer. The tendency for this was noted in the Hele-Shaw cell experiments of Hartline (1978), but was deliberately suppressed in the strictly two-dimensional apparatus by carefully initiating the convection near the critical Rayleigh number. Since only an integral number of circulation cells can exist in an apparatus of finite length, the number tended to remain constant unless the experiment was disturbed. No such restrictions apply in the more complex pattern of three-dimensional convection, and it is likely that the convection seeks an aspect ratio dependent on Rayleigh number.

At very high Rayleigh numbers, the aspect ratio would become vanishingly small, and this seems unrealistic. The alternative is a breakdown into multi-tier circulation, where small cells contribute to heating a thick boundary layer that then forms a large-scale plume. Such a structure would be consistent with the geophysical observations: a single large hot plume for each geothermal area. However, the small cells no longer see the full temperature difference across the porous slab, and so they cannot transport the quantity of heat required by the theoretical upper bound. On a diagram like Figure 5, where the logarithm of Nusselt number is plotted against the logarithm of Rayleigh number, the true curve must depart from the linear upper bound, and follow some unknown power law. It is not likely to be lower than one third because a boundary layer analysis of flow in a porous medium at constant aspect ratio produces such a law (J. R. Booker, unpublished work, in which revisions of the calculations can also produce other (higher) power laws), and multi-tier convection must be more efficient than that. The slopes used by Lister (1974) and Lister (1982) are therefore upper and lower bounds if the breakover point is not radically wrong, and it will be seen that the main effect of the different laws is on the crack spacing and Rayleigh number of the convection, rather than power extraction or

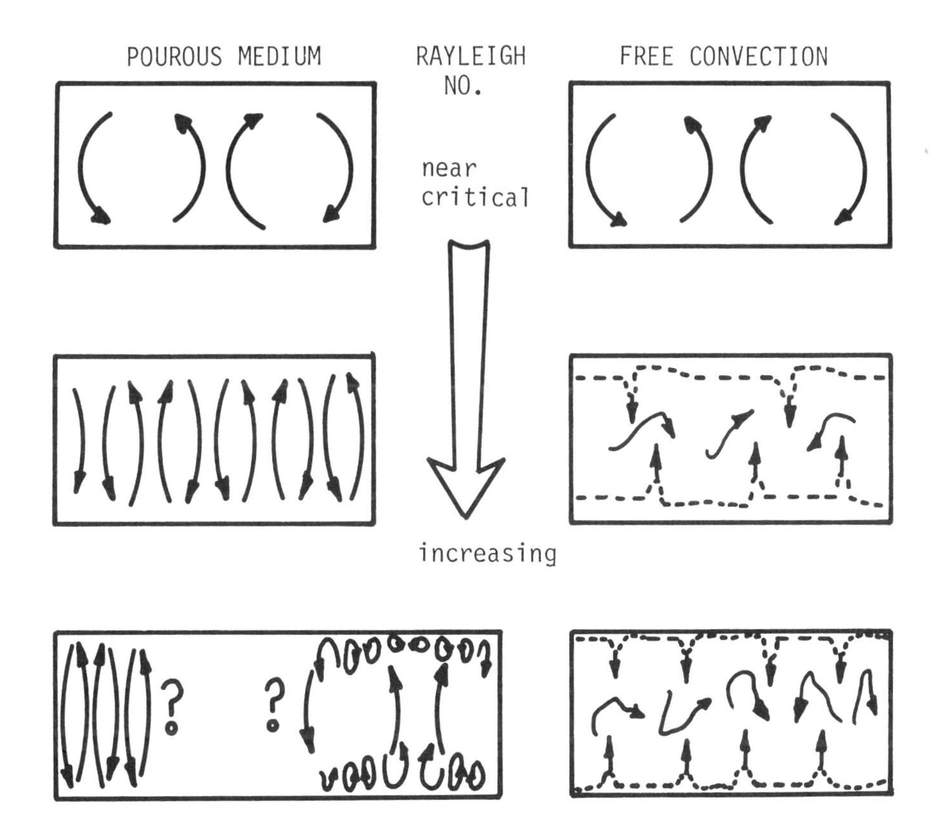

Fig. 6. Cartoons of the forms of convection in a porous slab and a layer of free fluid, with Rayleigh number increasing downward. The inertially controlled, time-varying structure of free convection cannot be duplicated in a porous medium, especially if the porosity is low; heat is stored in the matrix and not in fluid able to move as a body.

water temperature. The porosity of the medium is likely to have some effect on the level, if not the slope, of the heat transport curve; the advection of small sub-cells toward the main plume is easier in the high porosity medium than in the low one.

CRACK SPACING

(This section is not as non-mathematical as the rest of the paper, since it contains some needed clarification of the presentations in prior work.)

The foregoing discussion of convection is necessary to establish whether the form of the circulation near the cracking front is likely to have any significant influence on the mechanism that controls crack spacing. A propagating crack tends to relieve tension over a lateral distance comparable to its length (Lachenbruch, 1961), and so when a body of mud dries, the spacing of the active cracks increases as the desiccated layer thickens. On the other hand, in a columnar lava, proposed as a specially regular case of thermal crack penetration into hot rock, the columns are small in diameter compared to their length. Even when the jointing extends inward from a convex surface, the columns are not pie-shaped, but seem to maintain their characteristic scale all the way in toward the center of curvature. Some process limits the extent of the natural tendency to increase scale, and it is related to the generation of hoop tension around a column cooled radially as well as axially. A brittle column cooled suddenly from its surface would break up due to the development of radial cracks. This is the process of subdivision if the column becomes too large; in the case treated here, the more finely-spaced cracks already exist, and so the process is really one of preventing the opposite, obdivision, from happening.

The geometry can perhaps be visualized by referring back to Figure 4, where a triple corner has been highlighted and labelled 'A'. If this corner fails to propagate downward, it leaves a nine-sided polygon defined by the circumferentially radiating cracks. If three of these cracks also cease to propagate downward, by the loss of two more triple corners and one single crack, the remaining column is again hexagonal, but larger. In a natural system of propagating cracks, some triple corners will tend to lag behind the main cracking front, but any radial cooling of the larger column would help that corner from being completely left behind. The key to the problem is therefore the amount of cooling that occurs directly to the fluid in the cracks, instead of by vertical conduction into the convective boundary layer.

If the convective boundary layer is very thick compared to the crack spacing, advective cooling due to convective motions does not become significant until far enough away from the cracking front not to influence the propagation of the cracks. The only fluid advection that occurs is due to the propagation and widening of the cracks themselves, and is purely vertical for a plane horizontal cracking front. This is one extreme case, a lower bound for radial cooling and an upper bound for permissible crack spacing. The opposite bound is the situation where the convective boundary layer is very thin compared to the crack spacing, and essentially all of the heat is extracted from the cracks within one or two column diameters from the front. In a real geothermal system, there may be some advective activity close to the boundary (e.g., Fig. 6), but it is likely to be small compared to this extreme case. Final results can be tested for self-consistency by using the Nusselt number to calculate the effective boundary-layer thickness, and comparing this to the crack spacing.

The real problem of determining how much radial cooling is needed to stabilize the propagating polygonal crack spacing is clearly very difficult. What is needed here is the zeroth-order approximation that will allow a solution to proceed with order-of-magnitude accuracy. To do this, we derive two temperature quantities associated with the process, one ΔT_R, the radial temperature difference associated with advective cooling, and ΔT_κ, the temperature difference that causes the excess horizontal tension and drives the main cracking process. The applicable ratio of these should be of order unity, and any difference from equality can be lumped into the fudge factor ϕ that already contains any cracking differences between mafic rocks and quartz, as well as the geometric differences between a polygonal crack system and the plane parallel cracks treatable by analytic theory.

Only order-of-magnitude accuracy is needed, so the radial temperature difference is calculated for the steady cooling of a cylinder of diameter y by either the vertical advection of fluid into the widening cracks, or complete cooling at a rate equal to the vertical temperature gradient in the conductive boundary layer, both scaled by the crack-advance velocity u:-

$$\Delta T_R = \frac{u^3 y^3 \alpha \rho_w c_w}{8\kappa^3 \rho c} \left(T_1 - T_\kappa\right)^2 \quad \text{or} \quad \frac{u^2 y^2}{16\kappa^2}\left(T_1 - T_\kappa\right)$$

(Lister, 1982; 1974).

These represent the lower and upper bounds for ΔT_R, where $\rho_w c_w$ is the heat capacity of the fluid and ρc that of the rock. T_1 is the initial temperature. Transient effects and the real geometry have been neglected deliberately to produce the simplest possible formulae for further use.

To calculate an approximate value for ΔT_κ, we need the Martin (1972)-derived formula for the stress needed to propagate cracks at velocity u:-

$$\sigma_c = 0.00041\, \phi y^{-1/2} \left(\ell n\, u - 14.6 + \frac{25.8}{RT}\right)$$

where u is expressed in m/yr (for this formula only), y is in m (as opposed to the cm of Lister, 1974), and the stress is assumed to be in the logarithmic range for computational simplicity. This is a good assumption for reasonable crack advance rates of order 30 m/yr and temperatures of 800°K or less. In addition, we need to know how the creep stress of the thermally shrinking rock is increasing with decreasing temperature, since

$$\Delta T_K = - \sigma_c \left| \frac{d\sigma}{dT} \right|_{\dot{\varepsilon}}^{-1} ,$$

where σ is the vertical-equivalent creep stress and $\dot{\varepsilon}$ the effective strain rate, approximately constant through the small change of temperature ΔT_K. In Lister (1974) some creep data for wet Lherzolite were analyzed by means of the impurity-site theory of Ree, Ree and Eyring (1960) to produce a formula for transient creep of an annealed rock that extrapolates to low temperatures and strain rates in a physically reasonable manner. The formula may or may not give answers of the right order of magnitude in a region far away from the data, and the rocks at the base of a geothermal area may or may not behave like wet Lherzolite. They will not be as wet, but, since the strain rates are very low, they may be wet enough, either with water or dissolved carbon dioxide; if they are not as mafic as a Lherzolite, the activation energies may be low enough without the presence of much in the way of volatiles. Here is a second major materials problem to be lumped into the fudge factor ϕ, the point being that the important property of the rock is the ratio of its deformability to its crackability; the absolute value of the former appears in the choice of a cracking temperature, and not in the spacing determinant ϕ.

The logarithmic form of the creep equation (50) in Lister (1974) can be readily differentiated with respect to temperature:-

$$- \frac{d\sigma}{dT} = 0.74 \frac{40}{RT^2}$$

independent of $\dot{\varepsilon}$ and therefore of u. Since u now only appears as a logarithm in the equation for σ_c, and changes modify a number of about 11, a plausible value can be entered to make the final equation for y non-transcendental. The interesting result is that

$$\Delta T_K = \phi y^{-1/2} [0.00036 \; T_K - 3 \times 10^{-7} T_K^2]$$

and is nearly invariant over most of the plausible range of cracking temperatures:-

T_K	500°	600°	700°	800°	900°	1000°K
[ ]	0.104	0.107	0.103	0.094	0.078	0.056

Bearing in mind what has already gone into ϕ, one can write

$$\Delta T_\kappa = 0.1\ \phi y^{-1/2}$$

(the only difference from prior work being the retention of a non-dimensional ϕ while converting y to meters; the equation has an implied dimensionality which should be attached to ϕ in checking derivative equations for correct dimensions). The remarkable simplicity of this result allows a solution for y in terms of u by equating ΔT_κ and ΔT_R :-

$$y = \left| \frac{0.8\phi\ \kappa^3\ \rho c}{\alpha u^3 \rho_w c_w} (T_1 - T_\kappa)^{-2} \right|^{2/7} \quad \text{or} \quad \left| \frac{1.6\phi\kappa^2}{u^2} (T_1 - T_\kappa)^{-1} \right|^{2/5}$$

where the first is the upper bound, and the second the lower bound. Note that the fudge factor ϕ appears at a fairly low power in either result, so that it would have to be large to have a significant effect on the crack spacing. As will be seen later, there is another parameter that acts to stabilize the overall system by varying the permeability, the hot water temperature, so that the power extraction capability of a geothermal system is in fact quite well determined.

In sum, fairly exhaustive discursion into cracking theory and the creep of rocks has produced the result that the crack spacing is almost linearly related to the conductive boundary-layer thickness κ/u. I expected that to be the case when I began the analysis ten years ago, just on general physical principles, but I could not see what the ratio should be, or why. The mechanism, radial cooling of the columns stabilizing them against obdivision, is not one that lends itself to easy analysis, either from the properties of materials aspect or from the geometry. The main advance since the first investigation is the realization that the scale of the convective boundary layer influences the amount of cooling through the cracks themselves, and therefore the permissible size of the columns. It is useful to evaluate the two bounds for y with fairly typical values of the parameters: $u = 34$ m/yr; $\kappa = 9 \times 10^{-7} m^2/s$; $\rho c = 3.2 \times 10^6 j/m^3 °K$; $\rho_w c_w = 4.2 \times 10^6 j/m^3 °K$; $\alpha = 1.5 \times 10^{-5}/°K$(linear); $T_1 - T_k = 700°K$; $\phi = 1$; whereupon 7.6 cm < u < 42 cm. The range is not inconsiderable, particularly when one remembers that the permeability is proportional to y^2. It remains to finish the solution of the problem to see if the results will permit the range to be narrowed.

SOLUTIONS AND RESULTS

The problem can be closed quite simply by noting that the

heat transported away from the hot boundary by the convective system must equal the heat supplied by the cooling of the rock from the initial temperature to the temperature of the water in the hot region. If the effective boundary temperature (discussed previously: see Fig. 4) is written T_2 and the cold water source temperature T_0:-

$$Q = \mathcal{N} \kappa \rho c (T_2 - T_0)/h = u \rho c (T_1 - T_w)$$

where T_1 is the initial temperature and T_w the hot water temperature. The Nusselt number $\mathcal{N}$ is determined primarily by the Rayleigh number, and that depends on the permeability, the temperature difference, and the thermal properties of the fluid and matrix. The permeability depends on y^2, and, since y has been found as a function of u, the equation can be re-arranged by simple algebra to give u. The uncertainty in $\mathcal{N}$ can be dealt with in the same way as that for y, by doing separate calculations with formulae that represent the upper and lower bounds for the heat transport. The result is a rather complex expression, where the basic parameters of fluid and matrix are raised to various powers, as is the convective height, and the hot water temperature appears in a polynomial with the other important temperatures. For example, if the variation of water properties with temperature is given the simple form $\alpha_w/\nu_w = 0.126\ (T_w - 273)^2\ s/m^2\ {}^\circ K$ (Elder, 1965), the case C treated by Lister (1974) generates the polynomial

$$\frac{(T_w - 273)^2\ (T_\kappa - T_w)^3\ (T_2 - T_0)^2}{(T_1 - T_w)\ (T_1 - T_\kappa)^{0.8}}$$

where T_w appears in all the brackets except one, because $T_2 \simeq (1/9)\ (T_\kappa + 8\ T_w)$ relates the effective boundary permeability to that in the bulk hot water. Here the symbol T_κ replaces the less printable one in the prior work that indicated the temperature at which the overburden pressure was just cancelled, and before the excess cracking tension was generated. This is justifiable because their difference, ΔT_κ, is less than a degree for reasonable crack spacing, and so they are interchangeable except when that difference is required.

The value of T_w is found by maximizing the polynomial expression. The physical system should adjust itself to maximize the heat transfer by maximizing the mobility of the hot water in the matrix. If the boundary region is too hot, cooler water can penetrate the hot region by virtue of its greater mobility and heat transfer capability; if it is too cool, local regions of hotter fluid can grow by transporting more heat away from the boundary than the average. The maximization should include the water properties as it is the total fluid mobility and boundary coolability that matters; the difference between case C here and the calculation in Lister (1974) is that the water properties were left out of the maximization. The results of completing the calculation are given in Table 1 for the most likely cracking temperature, in the middle of the plausible range, and the four possible cases created by having two bounds each for the crack spacing and the heat transport relation.

Table 1

Parameter Case	A	B	C	D	
Initial temperature	1,500	1,500	1,500	1,500	°K
Cracking temperature	800	800	800	800	°K
Hot water temperature	633	633	574	574	°K
Front velocity	3.2	8.2	75	254	m/yr
Thermal flux	0.28	0.73	7.1	24	kw/m^2
Nusselt number	1,300	3,360	38,000	130,000	
Convective boundary layer	1.9	0.74	0.065	0.019	m
Conductive boundary layer	8.8	3.4	0.37	0.11	m
Crack spacing	0.5	1.4	0.040	0.075	m
Height dependence of u	-10/23	-14/33	none	none	(power)

Cases: A = lower bound for y, lower bound for Nusselt number
B = upper bound for y, lower bound for $\mathcal{N}$ (Lister, 1982)
C = lower bound for y, upper bound for $\mathcal{N}$ (Lister, 1974)
D = upper bound for y, upper bound for $\mathcal{N}$

At first glance, it appears as if the inclusion of all four possible cases in the table has sustantially widened the range of possible front velocities from those given in prior work. The real effect of the dramatic change in heat transport relation is about a factor of thirty on the thermal output per unit area. This is not small, but it is much less than the change in Nusselt number for a given high Rayleigh number--not fully shown in Figure 5 so that the large scale could indicate the experimental data. The bounds on the heat transport are gross bounds, and, in spite of a probable influence by the porosity, there should be a fairly definite relationship against Rayleigh number for low-porosity large-scale systems. It is probably closer to the lower bound than to the upper bound at very high Rayleigh numbers.

The reasonableness of the convective picture can be tested by contemplating the effective boundary layer thicknesses. These are nominal, for boundary layers with a linear temperature slope from the boundary to the mean (or core) temperature. In complex, multi-tier convection (Fig. 6) the boundary layer thermal gradient should decline steadily away from the boundary, so that the physical boundary layer could be substantially thicker, and produce a more reasonable plume for a circulation with a vertical scale of several kilometers. The layers for cases C and D are nevertheless clearly ludicrous for a large-scale geothermal area, being mere centimeters in thickness.

A more interesting aspect of the boundary layer thicknesses is their relation to the crack spacing, and the assumptions behind the derivation of the two bounds for y. The lower bound for y assumes that convection is active on a scale smaller than the crack spacing, thus cooling the cracks where this can still influence the cracking front. In case A, and somewhat less so in C, the convective boundary layer is thicker than the crack spacing, invalidating the basic assumption. Conversely, where no local convection was assumed, in cases B and D, the convective boundary layer is thinner than the crack spacing, invalidating that assumption also. In other words, the crack spacing must settle somewhere between the bounds and is influenced directly by the form of the convection near the boundary. This is unfortunate from the analytic standpoint, since a firm relation between u and y cannot be established, and a key parameter is influenced by the one aspect of the physics about which the least is known: the form of high Rayleigh number convection in a porous medium. It will not stop the geothermal area from convecting, however, and an assessment of the somewhat nebulous relation between the heat transport bounds and reality suggests case B as the closest to real conditions. It is reassuring to note that the crack spacing is in all cases less than the thickness of the conductive boundary layer (though marginally in the ludicrous case D), validating the original concept of a cracking front.

TESTING THE THEORY

As pointed out in Figures 1 and 2, the surface manifestations of a geothermal area offer little clue to the processes beneath that generate the hot water. Even if a drill hole were projected into the depths of the plume, it would be more luck than management if it intersected an actively penetrating region of the source (Fig. 2). Once there, the convective boundary layer is only a meter thick and is advancing about ten meters per year (Table 1), so that the time during which it could be observed in any one spot is comparable to the time needed for a normal drilling disturbance to subside. Nevertheless, there are three things that suggest themselves as indirect tests: geothermometry of the waters, examination of rocks that should have undergone water penetration, and experiments where water is poured onto hot rock under controlled conditions.

The silica content of the waters has long suggested that the hot region of a geothermal area is at about 600°K (e.g., Ellis and Mahon, 1977) even if the surface discharge is limited by boiling or dilution to 373°K or less. The conclusion has been dramatically confirmed by the discovery of black smokers on the ocean floor, where a heavily mineralized solution at 600-700°K discharges directly into the deep ocean at localized vents (e.g., Spiess et al., 1980). The correspondence between these temperatures and those of Table 1 is remarkable, especially as the cracking temperature of about 800°K was estimated from creep data before the answer to the overall calculation was known (Lister, 1974). It should be pointed out, though, that the estimate is a very crude one, and the nature of the real rocks could be different from those that generated the experimental data. However, it is difficult to imagine cracking much outside the range of 700-900°K at moderate overburden pressures, and, once a cracking temperature is chosen, the water temperature is the most stable of all the estimates (Table 1). Even a gross change in the convective heat transport relation alters it only 60°K, and an alteration to the crack spacing mechanism not at all, since only (T_1-T_K) enters into that calculation. The physical reason for this is simple and is clearly embodied in the temperature polynomial given in the last section: decreasing water temperature increases the permeability but decreases the effective hot boundary temperature that is available to drive the convection. Between these two trends is a maximum heat transport that stabilizes the convection so well that even the exclusion of strongly temperature-dependent water properties has relatively little effect: about 45°K for case B (Lister, 1982, p. 464). One can conclude that water penetration theory predicts hot water temperatures in the right general region, and provides a reason for the relative constancy of this parameter among widely differing geothermal areas around the world.

The tectonic stability of most land areas, and therefore the unusual nature of the conditions that give rise to a land geothermal area, make it difficult to find currently exposed outcrops that can be unequivocally associated with former high-temperature geothermal systems. On the ocean floor, however, all crustal rocks have been emplaced at high temperature directly beneath the cold ocean water, and the expected hot springs have been found. This means that an ophiolite suite now exposed should have once had within it a high-temperature circulation; in fact the primary cooling mechanism of the crustal rocks appears to be hydrothermal (e.g., Lister, 1980). Some difficulty arises in the interpretation of what is found, in that the rocks have been translated from the ocean floor to a position high in a continental section to be exposed. The emplacement process is not now thought to be as violent as once was supposed, and it seems reasonable to assert that rocks thermally cracked on a scale of meters are unlikely to suffer much further mechanical degradation.

The most difficult problem is that field geologists have their own biases, and these differ from those of the author of a water penetration theory: would they recognize the right kinds of crack even if they saw them? I have visited small parts of only two ophiolites, and feel that I saw the right kinds of crack even in rocks from deep in the section that would have been declared as untouched hydrothermally by any field geologist at that time. They expect obvious signs of hydrothermal alteration, such as are found high in the section: veins filled with minerals and bordered by alteration halos. These signs are reasonable for rocks exposed to hydrothermal waters in the ascending plume for considerable periods of time, and also from a zone where the waters are cooling and depositing minerals. Where the water is being heated, it is <u>extracting</u> minerals, and the chemistry of sea water is complex enough that the hot surfaces could even become coated with a protective precipitate layer, like a boiler scale, but one that could later be redissolved by cooler circulating water (e.g., Mottl and Seyfried, 1980). The time for which an individual block of rock would be bathed by hot water is short; it remains in the hot convective boundary layer only for about a month, and the water temperature is decreasing continuously (Table 1). How far alteration should penetrate into a massive rock during that time has not been established, since rock/sea water interaction studies have so far used only finely powdered basalt.

A major unresolved question in the physics remains as to what the temperature should be beyond the boundary layer, in the relatively stagnant core of the convective circulation. At moderate Rayleigh number and between two impermeable conducting boundaries, the core settles to the mean temperature; at high Rayleigh number and with the upper boundary effectively open, the core temperature could be much lower. On the other hand, the permeability pattern in the crustal rocks could easily be such as to restrict cold water access to the pluton being attacked (e.g., Fig. 2) in which case the core water could be relatively hot. The problem is worth some thought because it may be the answer to an apparent paradox: the water from the black smokers bears signs of heavy extractive interaction with crustal rocks, yet the majority of rocks in ophiolite suites do not show any sign of massive alteration. Postulating that all the sulphides come from the small volume of rock in the highly altered trace of the hot upwelling plume is probably an unwarranted dismissal of a problem that may lead to valuable discoveries. I think it is fair to conclude that there is, so far, no strong confirmatory evidence of the theory in ophiolitic rocks, but at the same time no evidence against it either. The mere fact that the water of the smokers is so heavily laden with extracted minerals is a good argument for a theory that proposes fine-scale cracking of the rocks, exposing the necessary large surface area for the extraction.

The third obvious test of the theory, pouring cold water onto some hot rock, might appear wildly impractical. Due to other, overriding imperatives, however, the experiment has actually been done on an adequate scale to prove the principle. During the 1973 eruption on Heimaey (a part of Iceland), a lava flow threatened the town, and a pumping system was set up to spray water onto it so that early solidification might form a rock dam and divert the flow. Subsequent excavation by bulldozer confirmed that the rock under the spray was heavily cracked on a scale much smaller than nearby lava that had cooled naturally. The crack advance rate was about 330 m/yr, yielding cracks 10-20 cm apart, and the thermal output was about 40 kw/m^2 (Bjornsson et al., 1982). These figures are in basic agreement with the theory, even though the cooling was by direct boiling of the water and there was no significant overburden pressure.

CONCLUSIONS

The development of a theory that determines how water may penetrate into hot rock is potentially an important step toward understanding where the heat of a major geothermal area comes from. Although this is of limited practical interest to those who develop the surface regions of existing geothermal areas for the production of power, it is of critical importance in unravelling the geochemistry of the systems. The recent discoveries of massive sulphide bodies, freshly emplaced on the ocean floor (Malahoff, 1982), has focused attention on the process of hydrothermal mineral extraction. Not only is the exchange between seawater and crustal rocks important for the chemical balance of the oceans, but hydrothermal processes are probably responsible for much of the minable mineral resources of the world. The extant theory might be called a zeroth order approximation: there are great gaps in our knowledge of the applicable physics and mechanics, but gross estimates of these produce results that appear to be of the right order of magnitude. The difficulties with the physics do not allow the theory to offer clear pointers to what specific conditions may apply to the chemical reactions, and an understanding of these is most likely to develop by careful comparison between the results of laboratory chemical experiments and the field information. The largest current gap in the latter is an enlightened investigation of a well-preserved and well exposed ophiolite suite to look at what the waters leave behind. Combined with rate experiments on relatively massive blocks of rock, the chemistry may provide the key to unravelling the physics of a process that cannot be modelled in the laboratory.

ACKNOWLEDGEMENTS

I wish to thank Peter Rona for stimulating this work by organizing the Cambridge Conference, and then making completion possible by stretching the submission deadline. I also wish to thank John Elder for inadvertently suggesting the best way of introducing the subject. The typing was paid for by National Science Foundation grant No. EAR-OCE-81-11413, thanks to the courtesy of John Delaney. This is Contribution No. 1316 of the School of Oceanography, University of Washington.

REFERENCES

Björnsson, H., Björnsson, S., and Sigurgeirsson, Th., 1982, Penetration of water into hot rock boundaries of magma at Grimsvötn, Nature, 295:580-581

Bodvarsson, G., and Lowell, R. P., 1972, Ocean floor heat flow and the circulation of interstitial waters, J. Geophys. Res., 77:4472-4475.

Bodvarsson, G., 1975, Thermoelastic phenomena in geothermal systems, Proc. 2nd United Nations Symposium on Development and Use of Geothermal Resources, 2:903-907.

Bolton, R. S., 1975, Recent developments and future prospects for geothermal energy in New Zealand, Proc. 2nd United Nations Symposium on Development and Use of Geothermal Resources, 1:37-42.

Bories, S., 1970, Sur les mécanismes fondamenteaux de la convection naturelle en milieu poreux, Rev. Gen. Therm., 108:1377-1401.

Buretta, R. J., and Berman, A. S., 1976, Convective heat transfer in a liquid-saturated porous layer, J. Applied Mech., 43:249-253.

Busse, F., and Joseph, D. D., 1972, Bounds for heat transport in a porous layer, J. Fluid Mech., 54:521-543.

Combarnous, M., 1970, Convection naturelle et convection mixte dans une couche poreuse horizontale, Rev. Gen. Therm., 108:1355-1375.

Combarnous, M., and Bories, S., 1975, Hydrothermal convection in saturated porous media, Adv. Hydrosci., 10:231-307.

Elder, J. W., 1965, Physical processes in geothermal areas, Am. Geophys. Union Monogr., 8:211-239.

Elder, J. W., 1967, Steady free convection in a porous medium heated from below, J. Fluid Mech., 71:379-389.

Ellis, A. J., and Mahon, W. A. J., 1977, "Chemistry and Geothermal Systems," Academic Press, N.Y., 392 pp., Chapter 4.

Harlow, F. H., and Pracht, W. E., 1972, A theoretical study of geothermal energy extraction, J. Geophys. Res., 77:7038-7048.

Hartline, B. K., 1978, Topographic forcing of thermal convection in a Hele-Shaw cell model of a porous medium, Ph.D. Thesis, University of Washington, Seattle, WA.

Hartline, B. K., and Lister, C. R. B., 1977, Thermal convection in a Hele-Shaw cell, J. Fluid Mech., 79:379-389.

Hartline, B. K., and Lister, C. R. B., 1981, Topographic forcing of supercritical convection in a porous medium such as the oceanic crust, Earth Planet. Sci. Lett., 55:75-86.

Irwin, G. R., 1958, Fracture, in: "Handbuch der Physik", S. Flugge, ed., 6:551-590, Springer, Berlin.

Katsaros, K. B., Liu, W. T., Businger, J. A., and Tillman, J. E., 1977, Heat transport and thermal structure in the interfacial bounday layer measured in an open tank of water in turbulent free convection, J. Fluid Mech., 83:311-335.

Lachenbruch, A. H., 1961, Depth and spacing of tension cracks, J. Geophys. Res., 66:4273-4292.

Lachenbruch, A. H., 1962, Mechanics of thermal contraction cracks and ice-wedge polygons in permafrost, Geol. Soc. Amer. Spec. Papers, 70:69 pp.

Lapwood, E. R., 1948, Convection of a fluid in a porous medium, Proc. Cambridge Phil. Soc., 44:508-521.

Lister, C. R. B., 1974, On the penetration of water into hot rock, Geophys. J. R. astr. Soc., 39:465-509.

Lister, C. R. B., 1975, Qualitative theory on the deep end of geothermal systems, Proc. 2nd United Nations Symposium on Development and Use of Geothermal Resources, 1:459-463.

Lister, C. R. B., 1980, Heat flow and hydrothermal circulation, Ann. Rev. Earth Planet. Sci., 8:95-117.

Lister, C. R. B., 1981, Rock and water histories during sub-oceanic hydrothermal events, Oceanologica Acta, SP, Proc. 26th International Geological Congress, Geology of Oceans Symposium, pp. 41-46.

Lister, C. R. B., 1982, "Active" and "passive" hydrothermal systems in the oceanic crust: prediced physical conditions, in: "The Dynamic Environment of the Ocean Floor," K. A. Fanning and F. T. Manheim, eds., D. C. Heath, Lexington, Mass., pp. 441-470.

Lowell, R. P., 1980, Topographically driven subcritical hydrothermal convection in the oceanic crust, Earth Planet. Sci. Lett., 49:21-28.

Malahoff, A., 1982, Polymetallic sulphides from the oceans to the continents, Sea Tech., 23(1):51-55.

Martin, R. J., 1972, Time-dependent crack growth in quartz and its application to the creep of rocks, J. Geophys. Res., 77:1406-1419.

Mottl, M. J., and Seyfried, W. E., 1980, Sub-seafloor hydrothermal systems: rock versus seawater dominated, in: "Seafloor Spreading Centers: Hydrothermal Systems," P. A. Rona and R. P. Lowell, eds., "Benchmark Papers in Geology," 56:66-82, Dowden, Hutchinson and Ross, Stroudsburg, Penn.

Norton, D., and Knight, J., 1977, Transport phenomena in hydrothermal systems: cooling plutons, Amer. J. Sci., 277:937-981.

Ree, F. H., Ree, T., and Eyring, H., 1960, Relaxation theory of creep of metals, Amer. Soc. Civ. Eng., Eng. Mech. Div. J., 86, EM-1:41-59.

Ribando, R. J., Torrance, K. E., and Turcotte, D. L., 1976, Numerical models for hydrothermal circulation in the oceanic crust, J. Geophys. Res., 81:3007-3012.

Scheidegger, A. E., 1974, "The Physics of Flow Through Porous Media," Third Edition, Univ. of Toronto Press, Toronto, pp. 135-144.

Schneider, K. J., 1963, Investigation of the influence of free thermal convection on heat transfer through granular material, Proc. 11th Int. Cong. Refrig., Munich, Paper 11-4:247-254.

Spiess, F. N., Macdonald, K. C., Atwater, T., Ballard, R., Carranza, A., Cordoba, D., Cox, C., Diaz Garcia, V. M., Francheteau, J., Guerrero, J., Hawkins, J., Hayman, R., Hessler, R., Juteau, T., Kastner, M., Larson, R., Luyendyk, B., Macdougall, J. D., Miller, S., Normark, W., Orcutt, J., and Rangin, C., 1980, East Pacific Rise: hot springs and geophysical experiments, Science, 207:1421-1433.

BASALT - SEAWATER EXCHANGE: A PERSPECTIVE FROM AN EXPERIMENTAL VIEWPOINT

Jose Honnorez

Rosenstiel School of Marine and Atmospheric Science
University of Miami
4600 Rickenbacker Causeway, Miami, Florida 33149

The hydrothermal circulation of seawater through the ocean crust near spreading centers is not only responsible for dissipating about 30% of the heat generated by the emplacement of new crustal material (Sclater *et al.*, 1981), but also for substantial chemical exchanges between the crust and the ocean. It has even been suggested that basalt-seawater interactions "buffer" the composition of the ocean with respect to certain elements. Submarine hot springs appear to discharge into the ocean quantities of manganese, rubidium and lithium equivalent to three, seven, and ten times the river fluxes of these three elements (see Table 1; also G. Thompson, this volume). Hydrothermal inputs in calcium and silica amount to 1/3 and 1/2, respectively, of the river fluxes whereas those in barium and potassium are of about the same order of magnitude as the river fluxes (see Table 1). On the other hand, quantities of magnesium of about the same order of magnitude as the river input seem to be taken up by the altered crust. The actual amounts of most of the various elements remobilized from crustal rocks by hydrothermal circulation are probably higher than those measured in the hot springs debouching on the sea floor because substantial quantities are left behind in the crust when secondary minerals precipitate in the cracks to form veins and cement of breccias. This is particularly true for calcium and silica which form the abundant veins of calcite, quartz, prehnite and various zeolites found in dredged and cored samples.

Significant quantities of transition metals, mainly Fe, Zn and Cu along with traces of Pb and Ag, are also extracted from crustal rocks by the hydrothermal solutions, and reprecipitated within or onto the sea floor as sulfides forming the so-called "massive sulfide" chimneys observed at various locations along the Pacific spreading centers (Haymon and Kastner, 1981; Oudin, 1981; Picot and Février, 1980; Styrt et al., 1981) or the disseminated mineralizations found in cored (Honnorez-Guerstein et al., in press) or dredged crustal rocks (Bonatti et al., 1976). Substantial amounts of these sulfides are injected into the bottom water column as suspended particle matter by the "black smokers" emitted from the chimneys, but do not appear to settle either quickly or close to their sources. Such an extremely fine suspension in the bottom waters could be one of the original sources of the metalliferous sediments sometimes found close to the base of sedimentary sequences generated near the spreading centers. These sediments are particularly rich in Mn and Fe, but also in such traces as B, Ba, V, As, Hg, Cu, Zn, Co, Ni, Pb and Ag confirming their assumed hydrothermal origin (Skornyakova, 1964; Bonatti and Joensuu, 1966; Boström and Peterson, 1966).

Altered rocks provide records of the evolution of former hydrothermal systems whereas submarine hot springs give us an instantaneous picture of present day hydrothermal systems. Therefore, petrologists can ascertain the relict chemical and mineralogical effects of extinct hydrothermal systems by studying dredged and cored crustal rocks; however, they are generally unable to assess the effects of active hydrothermal systems because of the low crystallization rate of most secondary minerals. On the other hand, water geochemists can study ongoing reactions when sampling the hydrothermal solutions debouching on the seafloor but do not know what is left behind in the altered rocks. They can only guess the chemical composition of old solutions. Our direct key to this question is the study of fluid inclusions representing tiny "fossils" of hydrothermal solutions trapped in secondary minerals (Jehl, 1975, Le Bel and Oudin, in press; Delaney et al., subm.). Further information about pre-existing hydrothermal solutions can be inferred from strontium and stable isotope chemistry of the solutions and/or secondary minerals (Albarede et al., 1981; Muehlenbachs and Clayton, 1982; Stakes and O'Neil, 1980; Vidal and Clauer, 1981). Therefore petrologists and water geochemists do not actually complement each other, as is usually thought, in studying the solid and liquid phases resulting from one and the same hydrothermal process. Both specialists should rely upon the results of hydrothermal experiments during which basaltic material reacts with seawater under controlled conditions. Could hydrothermal experiments duplicate in the laboratory the effects of natural hydrothermal activity on oceanic crust and seawater?

Experiments have been carried out during which oceanic basalt powders reacted with seawater under hydrothermal conditions (Bischoff and Dickson, 1975; Bischoff and Seyfried, 1978; Hajash,

1975; Mottl and Holland, 1978; Seyfried and Bishoff 1977: Seyfried *et al*., 1978; Seyfried and Bischoff, 1981; Seyfried and Mottl, 1982; Rosenbauer and Bischoff, this volume). Water:rock mass ratio usually varied from 1:1 to 125:1, pressure from 500 to 1000 bars, and temperature from 70 to 500°C, but most of the experiments have been carried out at temperatures of 150°C or higher with only one run at 70°C. Such conditions correspond to a seawater-dominated "ocean-floor metamorphism" (Miyashiro, 1973) under pressure-temperature conditions of the zeolite to greenschist facies of the regional metamorphism. The experiments demonstrated that heated seawater can extract the transition metals and other elements such as the alkalis, transition metals, Ca, B, Ba etc. from the basalts and transport them in the hydrothermal solutions, whereas seawater Mg^{2+} and $SO_4{}^{2-}$ were lost by the solutions through precipitation of smectite and anhydrite. The latter mineral can also use up basaltic sulfur previously released by oxidation of primary sulfide during low temperature alteration ("weathering") of the oceanic basalts, (Alt *et al*., in press). However, to date, anhydrite has rarely been found in oceanic basement rocks. It has only been observed as a minor component of hydrothermal veins from DSDP Hole 504B (Anderson *et al*, 1982; Alt *et al*., in press) even though the present-day temperatures at this site are favorable to the preservation of anhydrite. This discrepancy between the mineralogies of laboratory and natural hydrothermal systems can be explained either by the high solubility of anhydrite in cold seawater (calcite pseudomorphs after anhydrite have been observed in 504B rocks indicating that the sulfate has been dissolved before the site was reaheated to its present temperature (Alt, pers. comm.), or below $a_{SO_4^=}$ due to reduction of seawater sulfate to sulfide (McDuff and Edmond, 1982).

The hydrothermal experiments did not and probably cannot exactly duplicate nature because they are ipso facto simplified versions of the natural systems. For instance, the crustal lavas are replaced by fine powders of oceanic basalts with extremely large reactional surface areas and, as a consequence, the hydrothermal reactions were generally complete in the laboratory whereas ocean floor samples are rarely completely altered for many reasons. The actual proportion of the ocean crust affected by alteration is still unknown. Dredged and drilled samples of hydrothermally altered rocks generally contain igeous mineral relics, and secondary mineral parageneses indicate that several alteration processes occurred successively but that they rarely reached completion or equilibrium. The amount of altered rock is a function of kinetics rock-seawater reactions, and therefore it depends not only on the reaction temperature and duration, but also on reactional surface area and permeability, i.e., on how the flow rate of the solution through the rock compares to the reaction rates between rock and solution. The experimental systems are closed whereas the oceanic crust is essentially an open system. Another difference between laboratory and natural hydrothermal alterations results from the fact that, during the experiment, a given mass of seawater and rock completely reacted with each other, and no solution with constantly changing composition flowed through the rock more or less "armored" by secondary minerals as occurs in nature. For similar kinetic reasons, only smectite could form during these experiments, even at temperatures as high as several 100°C. In contrast, chlorite is commonly found in oceanic rocks whose parageneses indicate that they have undergone hydrothermal metamorphism at these temperatures.

Finally, the water:rock ratios assumed to occur in active hydrothermal systems of the East Pacific Rise range from 0.7 to 4.2 (Edmond _et al._, 1979; Edmond, 1980; Craig _et al._, 1980), i.e. these systems are rock-dominated contrary to the experiments. It is important at this point to realize that most of the laboratory experiments had been carried out before the first oceanic hydrothermal springs had been discovered and their waters analyzed.

The very concept of the water:rock ratio is, to use Mottl's own words, "necessarily a somewhat ambiguous" concept (Mottl, 1976; Mottl, this volume). Its definition varies with authors (see discussion in Hajash and Chandler, 1981) and terms such as "integrated" versus "instantaneous" ratios (Spooner _et al._, 1977), "effective ratio" (Ohmoto and Rye 1974) or "static" ratio (Ellis, 1967, 1970), have been used with different significances. Mottl (1976) and Seyfried (1977) defined the water:rock ratio for a hydrothermal system as "the total mass of water which has passed through the system during its lifetime divided by the mass of rock within the system which has been altered" (Seyfried and Mottl, 1982). This definition coincides with the experimental water:rock ratio since the finely powdered samples completely reacted with all of the seawater available. Defining the water:rock ratio of natural systems is even more difficult because it probably changes through time depending on the evolution of the "plumbing network" geometry of the hydrothermal system. If seawater indeed reacts with basalts already partly weathered in the recharge portion, then an evolved solution reacts further with crustal material at high temperature. In other words, what constitutes "altered rocks" and the "solutions" varies through time and space in any natural system under consideration.

TABLE 1

Elemental Inputs to Ocean by Hydrothermal Solutions at Spreading Centers and Rivers

	Hydrothermal Input to Ocean (moles/yr)	River Input to Ocean (moles/yr)	Ref.
Ba	6.1×10^9	10×10^9	a
K	1.25×10^{12}	1.9×10^{12}	a
Mn	16.5×10^9	4.5×10^9	b
Li	160×10^9	13.5×10^9	a
Rb	2.7×10^9	$.37 \times 10^9$	a
Ca	4.3×10^{12}	12×10^{12}	a
Si	3.1×10^{12}	6.4×10^{12}	a
Mg	-7.7×10^{12}	5.3×10^{12}	a

a - Edmond *et al.*, 1979

b - M. Lyle, 1976

REFERENCES

Albarede, F., A. Michard, J.F. Minster and G. Michard. 1981. $^{87}Sr/^{86}Sr$ ratios in hydrothermal waters and deposits from the East Pacific Rise at 21°N. Earth Planet. Sci. Lett. 55: 229-236.

Alt, J.C., J. Honnorez, H. Hubberten and E. Saltzman. In press. Occurrence and origin of anhydrite from DSDP Leg 70, Hole 504 B, Costa Rica Rift. DSDP Vol.: 69-70.

Anderson, R.N., J. Honnorez, K. Becker, A.C. Adamson, J.C. Alt, R. Emmermann, P.D. Kempton, H. Kinoshita, C. Laverne, M.J.

Mottl and R.L. Newmark. 1982. DSDP Hole 504B, the first reference section over 1 km through Layer 2 of the oceanic crust. Nature 300: 589-594.

Bischoff, J.L. and F.W. Dickson. 1975. Seawater-basalt interaction at 200°C and 500 bars: implications for origin of sea-floor heavy-metal deposits and regulation of sea-water chemistry. Earth Planet. Sci. Lett. 25: 385-397.

Bischoff, J.L. and W.E. Seyfried. 1978. Hydrothermal chemistry of seawater from 25° to 350°C. Am. Jour. Sci. 278: 838-860.

Bonatti, E. and O. Joensuu. 1966. Deep-sea iron deposits from the South Pacific. Science 154: 385-402.

Bonatti, E., B.M. Honnorez-Guerstein and J. Honnorez. 1976. Copper iron sulfide mineralizatons from the equatorial Mid-Atlantic Ridge. Economic Geology 71: 1515-1525.

Boström, K. 1973. The origin and fate of ferromanganoan active ridge sediments. Acta Univ. Stockholm, Stockholm Contr. Geol. 27: 149-243.

Boström, K. and M.N.A. Peterson. 1966. Precipitates from hydrothermal exhalations on the East Pacific Rise. Econ. Geol. 61: 1258-1265.

Craig, H., J.A. Welhan, K. Kim, R. Poreda and J.E. Lupton. 1980. Geochemical studies at 21°N EPR hydrothermal fluids, Abscract. EOS 61: 992.

Delaney, J.R., D.W. Mogk and M.J. Mottl. Quartz-cemented, sulfide-bearing greenstone breccias from the Mid-Atlantic Ridge---samples of a high-temperature hydrothermal upflow zone. Subm. to Science.

Edmond, J.M. 1980. The chemistry of the 350° hot springs at 21°N on the East Pacific Rise, Abstract. EOS, 61: 992.

Edmond, J.M., C. Measures, R.E. McDuff, L.J. Chan, R. Collier, B. Grant, L.I. Gordon and J.B. Corliss. 1979. Ridge Crest hydrothermal activity and the balances of the major and minor elements in the ocean: the Galapagos data. Earth Planet. Sci. Lett. 46: 1-18.

Ellis, A.J. 1967. The chemistry of some explored geothermal systems, pp. 465-514. In: Barnes, H.L., ed., Geochemistry of Hydrothermal Ore Deposits. Holt, Rinehart and Winston, Inc., N.Y.

Ellis, A.J. 1970. Quantitative interpretation of chemical characteristics of hydrothermal systems. Geothermics, Sept. Issue 2: 516-528.

Hajash, A. 1975. Hydrothermal processes along Mid-Ocean Ridges: an experimental investigation. Contrib. Mineral. Petrol. 53: 205-226.

Hajash, A. and G.W. Chandler. 1981. An experimental investigation of high-temperature interactions between seawater and rhyolite, andesite, basalt and peridotite. Contr. Mineral. Petrol. 78: 240-254.

Haymon, R., and M. Kastner., 1981. Hot spring deposits on the East Pacific Rise at 21°N: preliminary description of mineralogy and genesis. Earth Planet. Sci. Lett., 53: 363-381.

Honnorez-Guerstein, B.M., J. Alt, J. Honnorez and D. Laverne. In press. Zn, Cu, Pb, Fe-sulfide mineralizations in DSDP Hole 504 B: a buried equivalent of the black smokers? DSDP Vol. 83.

Jehl, V. 1975. Le metamorphisme et les fluides associés des roches océaniques de l'Atlantique nord. Ph.D. Thesis, Université de Nancy I, 242pp.

Le Bel, L. and E. Oudin. 1982. Fluid inclusion studies of deep-sea hydrothermal sulphide deposits on the East Pacific Rise near 21°N. Chem. Geol., 37: 129-136.

Lyle, M. 1976. Estimation of hydrothermal manganese input to the oceans. Geology, 4: 733-736.

McDuff, R.E. and J.M. Edmond. 1982. On the fate of sulfate during hydrothermal circulation at mid-ocean ridges. Earth and Planet. Sci. Lett. 57: 117-132.

Miyashiro, A. 1973. Metamorphism and metamorphic belts, p. 432. John Wiley and Sons, New York.

Mottl, M.J. 1976. Chemical exchange between seawater and basalt during hydrothermal alteration of the oceanic crust. 107. pp. Ph.D. Thesis, Harvard Univ., Cambridge, Mass.

Mottl, M.J. and H.D. Holland. 1978. Chemical exchange during hydrothermal alteration of basalt by seawater - 1. Geochim. Cosmochim. Acta, 42: 1103-1115.

Mottl, M.J. and W.E. Seyfried. 1980. Sub-seafloor hydrothermal systems, rock - vs. seawater-dominated. In Rona, P.A. and R.P. Lowell, Eds., Seafloor Spreading Centers: Hydrothermal Systems. Dowden, Hutchinson & Ross Inc., Stroudsburg, Pa., 66-82.

Muehlenbachs, K.and R.N. Clayton. 1972. Oxygen isotope studies of fresh and weathered submarine basalts. Can. Jour. Earth Sci., 9: 172-184.

Ohmoto, H. and R.O. Rye. 1974. Hydrogen and oxygen isotopic compositions of fluid inclusions in the Kuroko deposits, Japan. Econ. Geol., 69: 947-953.

Oudin, E., C. Fouillac, and L. Le Bel. 1981. Etudes Minéralogique et géochimique des dépôts sulfurés sous-marins actuels de la ride Est-Pacifique (21°N). Campagne Rise. BRGM Publ. No. 25: 241 pp.

Picot, P. and M. Février. 1980. Etude minéralogique d' échantillons du Golfe de Californie (campagne CYAMEX). BRGM Publ. No. 20: 50 pp.

Sclater, J.G., B. Parsons and O. Jaupart. 1981. Oceans and continents: similiarities and differences in the mechanisms of heat loss. Jour. Geophys. Res., 86: 11535-11552.

Seyfried, W.E. Jr. 1977. Seawater-basalt interaction from 25° - 300°C and 1-500 bars: implications for the origin of submarine metal-bearing hydrothermal solutions and regulation of ocean chemistry. Ph.D. Thesis, Univ. Southerm California, Los Angeles, 216 pp.

Seyfried, W.E. Jr. and M.J. Mottl. 1977. Origin of submarine metal-rich hydrothermal solutions: experimental basalt-seawater interaction in a seawater-dominated system at 300°C, 500 bars, 173-180. In Pacquet, H., and Y. Tardy, eds., Proc. Second Internat. Symp. on Water rock Interaction, Strasbourg, France.

Seyfried, W.E. Jr. and M.J. Mottl. 1982. Hydrothermal alteration of basalt by seawater under seawater-dominated conditions. Geochem. Cosmochem. Acta., 46: 985-1002.

Seyfried, W. E. Jr., M. J. Mottl and J. L. Bischoff. 1978. Seawater/basalt ratio effects on the chemistry and mineralogy of spilites from the ocean floor. Nature, 275: 211-213.

Skornyakova I.S., 1964. Dispersed iron and manganese in Pacific Ocean sediments. Int. Geol. Rev., 7(12): 2161-2174.

Spooner, E.T.C., H.J. Chapman and J.D. Smewing. 1977. Strontium isotopic contamination and oxidation during ocean floor hydrothermal metamorphism of the ophiolitic rocks of the Troodos massif, Cyprus. Geochem. Cosmochim. Acta, 41: 891-912.

Stakes, D.S. and J.R. O'Neil. 1982. Mineralogy and stable isotope geochemistry of hydrothermally altered oceanic rocks. Earth and Planet. Sci. Lett. 57: 285-304.

Styrt, M.M., A.J. Brackmann, H.D. Holland, B.C. Clark, U. Pisutha-Arnond, C.S. Elridge and H. Ohmoto. 1981. The mineralogy and the isotopic composition of sulfur in hydrothermal sulfide/sulfate deposits on the East Pacific Rise, 21°N latitude. Earth and Planet. Sci. Lett. 53: 382-390.

Vidal, P., and N. Clauer. 1981. Pb and Sr isotopic systematics of some basalts and sulfides from the East Pacific Rise at 21°N (project Rita). Earth and Planet. Sci. Lett. 55: 237-246.

UPTAKE AND TRANSPORT OF HEAVY METALS BY HEATED SEAWATER:

A SUMMARY OF THE EXPERIMENTAL RESULTS

Robert J. Rosenbauer and James L. Bischoff

U.S. Geological Survey

Menlo Park, CA

ABSTRACT

In general, the chemistry of seawater experimentally reacted with basalt is in accord with the observed chemistry of the 350°C vent waters from 21°N on the East Pacific Rise. Experiments at 350°C, 500 bar, and a water/rock ratio of $\leq$10 reproduce most of the major components in the vent waters, in particular the low Mg and SO_4 and high SiO_2, Ca, and K. In comparison with the vent waters however the experimental fluids are deficient in H^+ and heavy metals. Experiments with natural and evolved seawater (Mg and SO_4 depleted) at $\geq$400 C and 500 bars or 375°C and ~375 bars under rock-dominated conditions more closely reproduce the vent-water chemistry.

INTRODUCTION

The hydrothermal cycle that gives rise to sea-floor metal deposits can be separated into three distinct processes: (1) leaching of heavy metals from basaltic rocks by heated seawater at the maximum temperature of the cycle, (2) transport of the metals in solution from the zone of leaching to the sea floor, and (3) deposition of the metals on the sea floor. Each of these processes is poorly understood at present.

Observations from manned submersibles of the discharging hydrothermal fluids and their resulting deposits in such areas as 21°N on the East Pacific Rise relate directly to the last stage of this cycle. The mechanisms of initial metal uptake by the fluid and subsequent metal transport cannot be directly observed but are more readily approached by experimental study. The purpose of this report is to summarize these experimental results and to relate them to observations of the vent fluids themselves.

Experiments have indicated that the major variables which determine the extent of metal mobilization are temperature, pressure, and water/rock ratio, where the water/rock ratio is the proportion of seawater that chemically alters a given amount of rock. This definition is distinct from that used in heat-transfer considerations which relates the volume of water to the amount of rock from which the heat is extracted, wherein heat exchange can occur without chemical reaction.

Previous Work

Results of the initial experimental basalt-seawater studies were reported by Bischoff and Dickson (1975), Hajash (1975), and Mottl and Holland (1978). These experiments defined the general direction and magnitude of change for the major components of seawater, and found that seawater changes from a slightly basic Na-Mg-Cl-SO_4 solution to a slightly acidic Na-Ca-Cl solution during basalt alteration. The role of acidity and related Mg uptake were found to be important in the solution of metals. These workers also pointed out the apparent inconsistency between the absence of anhydrite in submarine rocks and the ubiquitous appearance of anhydrite in all the seawater experiments. The behavior of heated seawater itself was investigated by Bischoff and Seyfried (1975), who first discovered the precipitation of a magnesium-hydroxy-sulfate-hydrate (MHSH) and determined its role in the generation of acidity. Bischoff and Rosenbauer (1983) and Janecky and Seyfried (in press) have extended this study to higher temperatures and helped define the stoichiometry of the MHSH. This material which has recently been identified in the mounds at 21°N on the East Pacific Rise (Haymon and Kastner, 1981), apparently has the status of a new mineral.

The importance of the water/rock ratio was first noted by Seyfried and Bischoff (1977) and Seyfried et al. (1978), who found that metals are drastically leached and effectively held in solution during basaltic alteration under seawater-dominated conditions. They also attributed certain mineralogic and chemical heterogeneities in sea-floor basalt to varying water/rock ratios. Mottl and Seyfried (1980) and Mottl (in press) followed up on the role of excess Mg in defining the parameters of rock-versus seawater-dominated systems and defined the transition ratio as 50, below which seawater is rock dominated and does not transport metals. This transition ratio seems to hold for temperatures as high as 350°C; however, higher temperatures seem to override this water/rock-ratio effect. The data of Mottl et al. (1979) and Pohl and Dickson (1979; in press) over the temperature range 400°-500°C indicate that significant amounts of metals can also be transported under rock-dominated conditions.

Table 1. Concentrations (in ppm) of normal deep-ocean seawater and of the 350°C vent water exiting from hydrothermal vents at 21°N on the East Pacific Rise (Edmond, 1981).

	pH	Mg	Ca	K	SO_4	H_2S	SiO_2	Fe	Mn	Zn	Cu
Seawater	7.9	1315	420	393	2755	0	10	0	0	0	0
Vent water	3.6	0	862	978	0	200	1291	100	34	7.8	1

Conditions of Hydrothermal Circulation

The vent sampled at 21°N is an acidic metal-rich fluid depleted in Mg and SO_4 and enriched in Ca, K, and SiO_2 relative to normal seawater (Edmond, in press, Table 1). The concentrations of Ba, B, Li, and K in the vent water indicate that it was derived from basalt-seawater interaction at a water/rock ratio of < 3, which points to a strongly rock dominated system. Above 200°C, Ba, B, Li, and K are considered soluble elements because they are strongly partitioned into the fluid phase (Mottl and Holland, 1978). The concentrations of these elements are thus supply limited and at 21°N suggest a rock-dominated system.

What might be the controls and limits on the maximum subsurface pressure and temperature of this fluid? Generalizations regarding the geometry of land-based geothermal systems were provided by Henley (1973). Temperature and pressure relations for the cycle of geothermal fluids have been described by Bischoff (1981, Figure 1). In the large area of the recharge zone of a hydrothermal cycle, flow rates are presumably low, so heating is gradual.

Attempts to assess the limits of pressure and temperature within the hydrothermal cycle (Bischoff, 1981) suggest that at the point of maximum penetration, the pressure likely exceeds the critical pressure of seawater (Sourirajan and Kennedy, 1962). Maximum temperature, therefore, might be determined by buoyancy rather than boiling. As a supercritical fluid is heated, it expands, and its rate of expansion is related to the pressure. At some point, the degree of expansion overcomes the driving force of the system, and the fluid rises quickly, possibly adiabatically. Thus, the first step in delimiting the point of maximum conditions is to define the range of pressures that controls the rate of fluid expansion. Geophysical evidence at 21°N on the East Pacific Rise (McDonald and Luyendyk, 1981), and sections of exposed ophiolite in Oman (Gregory and Taylor, 1981), provide information about the depth of hydrothermal circulation. A range of pressures for the depths of circulation can be

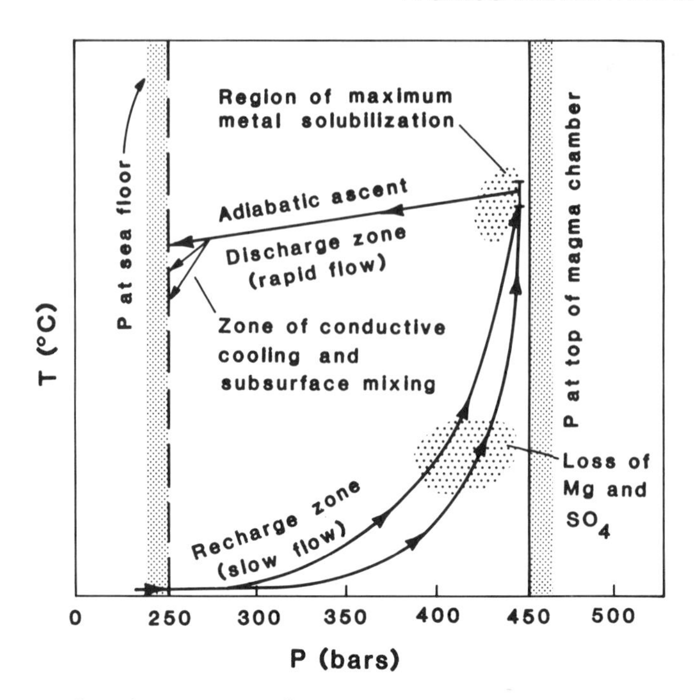

Fig. 1. Schematic temperature versus pressure relations suggested for cycle of geothermal fluid at 21°N on the East Pacific Rise (from Bischoff, 1981, Figure 8).

defined from 250-290 bars at the sea floor, through ~350 bars at the base of the pillow lavas, to ~450 bars at the base of the sheeted dikes on top of the magma chamber. The depth of penetration is initially constrained by the top of the melt but as the magma solidifies, hydrothermal circulation may extend into the layered gabbro to a depth of 5 km at a pressure of 700 bars. During the lifetime of the magma chamber, therefore, the circulating fluid would be exposed to a pressure range of 250-700 bars.

The second step in delimiting the maximum conditions of basalt-seawater interaction is to observe the effects of pressure on the fluid with respect to temperature. What constraints does pressure place on an expanding fluid as it is heated? In the absence of experimental pressure-volume-temperature (PVT) data for seawater or equivalent NaCl in this pressure-temperature (PT) region we can only use the data of pure water as an analogy. At 250 bars there is a large inflection in the specific volume of pure water at 375°C. At any temperature above this point, the fluid greatly expands, is more gas-like, and possesses a high degree of buoyant force. At 450 bars the specific volume slope begins to rise at 400°C. But at 700 bars the specific volume remains relatively uniform to about 500°C. Therefore, if pressure is limited to 450 bars, the temperature of a circulating pure-water fluid is limited to ~420 °C by buoyancy. Chen (1981) has estimated that the presence of NaCl would increase this temperature slightly. Upon complete crystallization of the magma, temperatures of 500 °C are possible by penetration to depths that result in 700 bars pressure.

Generalized Experimental Results

We report here typical generalized results compiled from the various studies that have application to an understanding of the observed vent water at 21°N. The systems studied include: seawater alone, basalt-seawater at low water/rock ratio (≤ 360°C), basalt-seawater at high water/rock ratio (≤360° C), basalt-seawater at low water/rock above 360°C, and basalt-evolved seawater above 350° C.

Chemistry of Heated Seawater

Progressively heated seawater by itself undergoes changes that serve as a background to an understanding of basalt-seawater interactions (Figure 2). As seawater is heated, its chemistry remains unchanged until about 150°C at which temperature Ca and SO_4 concentrations begin to decrease in equivalent proportions due to anhydrite precipitation. Above 250°C, Mg concentration begins to decrease, and H^+ concentration increases. MHSH precipitates at this temperature and above with a structure accommodating both $MgSO_4$ and $Mg(OH)_2$ components. Varying MHSH stoichiometries are apparently possible for this phase, depending on the pressure, temperature, and Mg concentration (Bischoff and Rosenbauer, 1983; Janecky and Seyfried, in press). MHSH continues to precipitate up to 500°C and extracts enough SO_4 from solution that anhydrite partially redissolves. The incorporation of a brucite like component into the MHSH is responsible for the increase in H^+ concentration to as much as 11.5 mmol, which produces a highly buffered solution. The basic reaction involving Mg in the production of H^+ is the formation of this brucite like component:

(1) $Mg^{+2} + 2H_2O = Mg(OH)_2 + 2H^+$

in MHSH:

(2) $4/3Mg^{+2} + SO_4^{-2} + H_2O = MHSH + 2/3H^+$

where the brucite "molecule" occurs as part of the MHSH structure. The acidity of heated seawater and related Mg depletion are important in metal mobilization during basalt-seawater interaction at elevated temperatures.

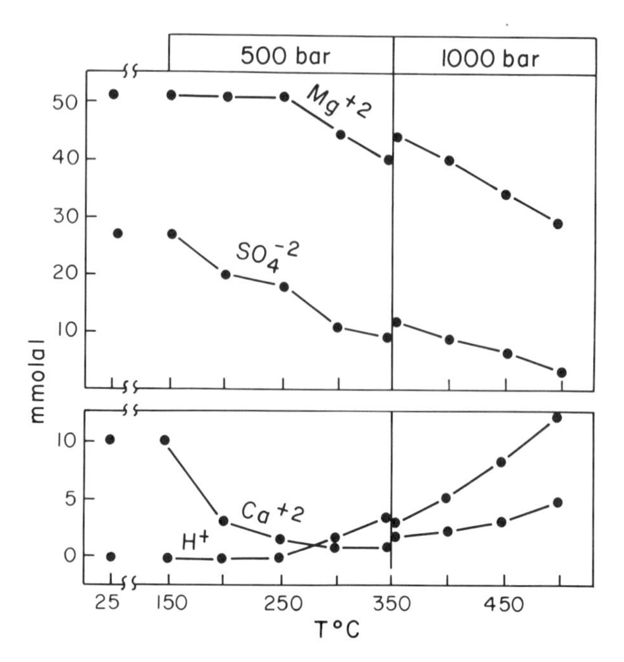

Fig. 2. Concentrations of Mg, Ca, SO_4, and H^+ in seawater alone at various temperatures and pressures. Data from Bischoff and Rosenbauer (1983).

Basalt-Seawater Interaction at <360°C and Low Water/Rock Ratio

The composition of seawater undergoes extensive changes on reaction with basalt at elevated temperature. Figure 3 plots the typical pattern of rock-dominated systems; this experiment was carried out at 360°C and 700 bars using the procedures discussed by Bischoff and Seyfried (1978) and Seyfried and Bischoff (1979, 1981). The types of changes illustrated in Figure 3 are representative in degree and direction of all low-water-rock-ratio systems between 150° and 360°C. At the lower temperatures within this range, reaction rates are slower, but the pattern of change is similar.

Important generalizations characteristic of a rock-dominated system are as follows: (1) pH decreases rapidly with a parallel depletion of Mg. (2) pH then rises and levels off after Mg has been completely removed. (3) Metals are released early to solution and reach a maximum concentration corresponding to the minimum pH. (4) Metal concentrations afterward decrease as pH rises, and fall to levels well below those observed in the discharging fluids at

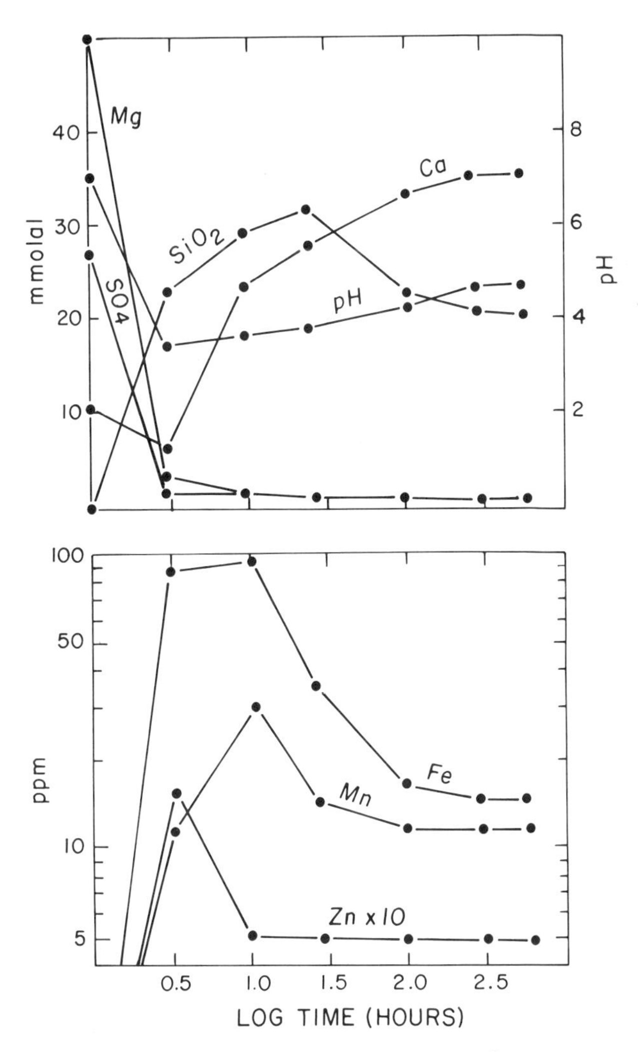

Fig. 3. Concentrations of Ca, Mg, SiO_2, SO_4, Fe, Mn and Zn and pH (25° C) in seawater during reaction with basaltic glass at 360° C, 700 bars and an initial water/rock ratio of 3.

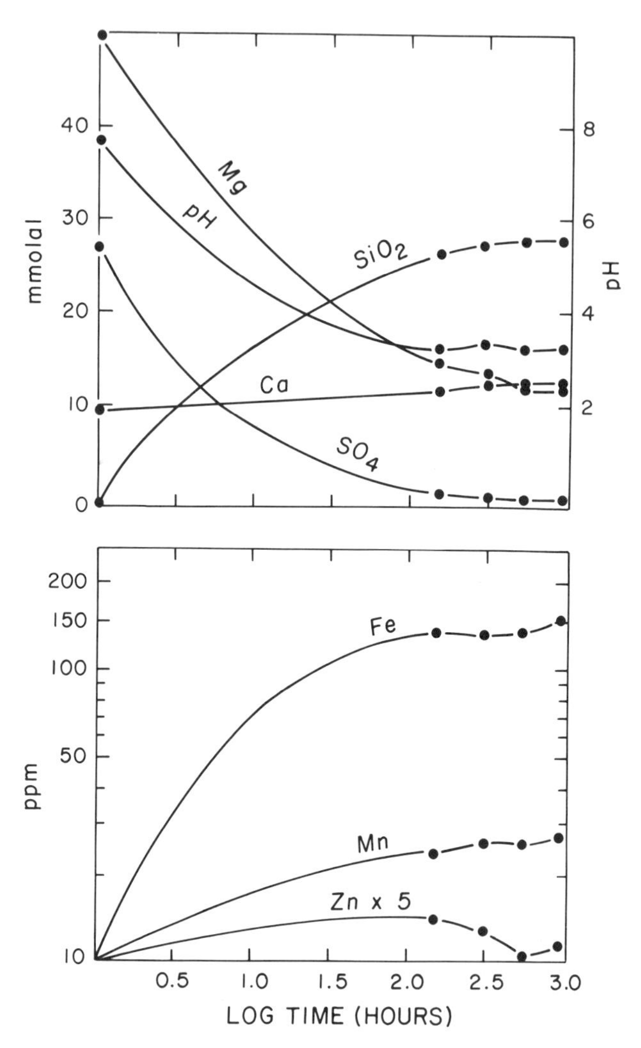

Fig. 4. Concentrations of Ca, Mg, SiO_2, SO_4, Fe, Mn, and Zn and pH (25°C) in seawater during reaction with basaltic glass at 300°C, 500 bars and an initial water/rock ratio of 62. Data from Seyfried and Mottl (1982).

21°N. (5) SO_4 is quantitatively removed from solution as anhydrite; some SO_4 may be reduced to sulfide and mix with sulfide leached from the basalt. (6) Ca is leached from the rock and is the main component in ultimately balancing the Mg removal. About half the Ca leached from the rock is precipitated with seawater SO_4 as anhydrite; the remaining Ca concentration depends on the water/rock ratio because of the SO_4 reservoir in seawater. (7) K is uniformly and almost quantitatively leached from the rock, and thereafter is maintained at a constant concentration in solution; the increase of K in solution is inversely proportional to the water/rock ratio. (8) The changes in Na concentration are generally less than the degree of analytical precision; Na removal from solution is believed to occur at water/rock ratios of < 5 and Na enrichment in the fluid occurs at higher ratios (Mottl, in press). (9) SiO_2 is released from the rock into solution at concentrations near quartz saturation (Kennedy, 1950). (10) The dominant alteration products are smectite, smectite/chlorite, anhydrite, and minor quartz, hematite, wairakite, truscottite, and pyrite.

Mg removal appears to be responsible for the pH decrease, in this case as a brucite interlayer of a mixed-layer chlorite-smectite. An important generalization for rock-dominated systems is that Mg is quantitatively removed from solution; thus H^+ production is limited, and excess H^+ is later consumed in silicate-hydrolysis reactions. The subsequent rise in pH is apparently responsible for the decrease in metal concentration. The generalized form of the reaction involving Mg in the production of H^+ in the presence of basaltic rock is

$$(3) \quad 5Mg^{+2} + \text{anorthite} + 8H_2O + SiO_2 = \text{clinochore} + Ca^{+2} + 8H^+$$

where clinochore represents the smectite/chlorite alteration phase.

In rock-dominated systems, the basaltic rock has excess capacity to take up Mg, and so reaction 3 proceeds until all the seawater Mg is depleted, and all the excess H^+ generated by this reaction is then consumed in silicate hydrolysis. The resulting fluid at steady state contains virtually no Mg or SO_4.

Basalt-Seawater Interaction at <360°C and High Water/Rock Ratio

Figure 4 plots data from Seyfried and Mottl (1982) to represent the pattern of chemical change in seawater-dominated systems. These results are typical of all seawater-dominated systems over the temperature range 150° - 350°C. Important generalizations are as follows: (1) pH decreases rapidly and remains low thereafter. (2) Mg initially decreases, paralleling the decrease in pH, but then levels off to a constant concentration. (3) Metals are leached from the rock and maintained in solution at high concentrations, equivalent to or exceeding those reported for the discharging fluids at 21°N. (4) SO_4 is quantitatively removed from solution as anhydrite.

(5) Ca is removed from the rock, a portion of which precipitates with SO_4 as anhydrite; the final concentration of Ca is lower in comparison with that in the rock-dominated system because of additional available SO_4. (6) K is leached from the rock but is supply limited because apparently all the available K is removed from the rock. (7) SiO_2 concentration approaches quartz saturation at steady state. (8) Smectite, smectite/chlorite, anhydrite and quartz are the dominant reaction products.

The initial direction of chemical change observed in the seawater-dominated system resembles that seen in the rock-dominated system. The extent of change is less for some species, such as Ca, which reacts with a larger reservoir of SO_4 to precipitate as anhydrite and for the supply limited elements such as K, which are effectively diluted by the larger volume of seawater. The extent of Mg depletion is limited by the capacity of the rock to take up Mg. In the presence of excess Mg, reaction 3 proceeds until all the rock has been altered; because the rock has been completely altered, the reactions that consume H^+ have also gone to completion. The effect of excess Mg then is to generate excess H^+, which maintains metals in solution. The reversals in pH and metal concentrations upon complete removal of Mg that characterize a rock-dominated system are not evident in a seawater-dominated system. Thus, in a water-dominated system, metals remain relatively concentrated in solution, and the pH remains low. Experiments by Mottl and Seyfried (1980) indicate that the transition from low to high water/rock ratio occurs sharply at about 50/1.

Although a fluid resulting from high-water/rock-ratio basalt-seawater interaction can transport metals, it cannot account for the vent waters at 21°N, which are rich in metals but contain no Mg. Thus, it is unclear how basalt-seawater interaction at a low water/rock ratio can generate a metal-containing fluid at $\sim 350\,°C$. One hypothesis which we tested experimentally, is that flow rates are rapid in the system at 21°N and that Mg is removed quickly, but that the fluid reaches the seafloor before the metals can precipitate. This possibility was tested by samples taken at short intervals during the first 24 hours from a low-water/rock-ratio basalt-seawater experiment at 360°C. The results indicate that the metals actually precipitate before Mg is completely removed (Figure 5); therefore, we discount this hypothesis. This conclusion suggests that the rate of H^+ removal during hydrolysis is more rapid than the rate of H^+ production during Mg removal.

Thus, experimental results indicate that under rock-dominated conditions and a maximum temperature of 360 °C seawater-basalt interaction cannot create the chemistry reported for the vent water at 21°N by Edmond et al. (in press). The answer may lie with differences in the reaction products between the experimental and natural systems, or it may be that the subsurface temperatures are much

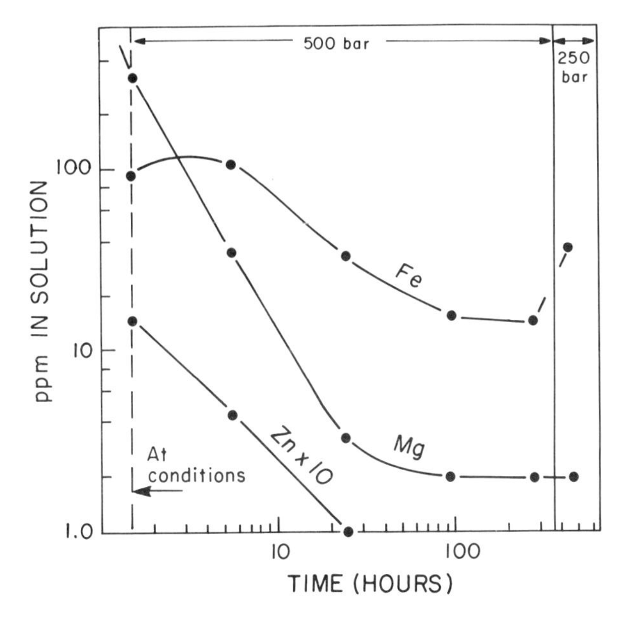

Fig. 5. Concentrations of Fe, Zn, and Mg during reaction of seawater with basaltic glass at 360 °C and an initial water/rock ratio of 3. Note the early sample points and the relative rates of decrease between Fe and Zn and Mg.

higher than the exit temperatures. Results of Mottl et al. (1979) indicate that large amounts of metals can be mobilized above 400° C.

Basalt-Seawater Interaction at ≥400°C and Low Water/Rock Ratio

Figure 6 illustrates the changes in the composition of seawater reacted with basalt at ≥400°C, according to the procedures described in Bischoff and Seyfried (1978) and Seyfried and Bischoff (1979, 1981). Although the pattern of chemical change resembles that in a rock-dominated system at lower temperatures, some characteristics of a water-dominated system are also present. Important generalization are as follows: (1) pH sharply decreases, then slowly increases to a value slightly above the minimum pH. (2) Mg removal from solution is rapid and complete. (3) The pattern of metal solubilization resembles that at 350°C, in which concentrations initially rise to some maximum and then fall to a lower, steady-state value. These steady-state concentrations are high in terms of effective metal transport; maximum metal concentrations correspond to minimum pH. (4) SO_4 is quantitatively removed from solution as anhydrite; some

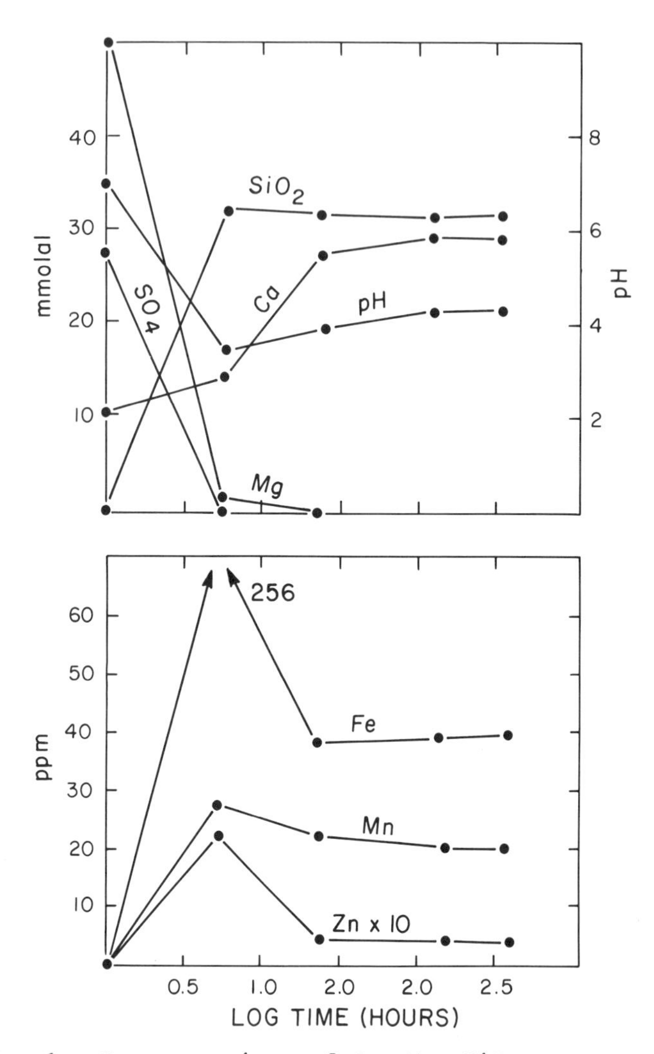

Fig. 6. Concentrations of Ca, Mg, SiO_2, SO_4, Fe, Mn, and Zn and pH (25°C) in seawater during reaction with basaltic glass at 400°C, 1000 bars and an initial water/rock ratio of 10.

SO_4 initially precipitated as anhydrite probably is reduced to H_2S. (5) Ca is simultaneously removed from the rock in exchange for Mg and H^+, and removed from seawater until all the SO_4 is removed as anhydrite. The Ca concentration continuously increases in solution but its rate of increase is greater immediately after SO_4 is removed; Ca concentration then levels off and finally decreases slightly, possibly in response to the formation of a Ca silicate. (6) SiO_2 concentration rapidly increases to quartz saturation. (7) K is leached from the rock to similar extent to that at 200°C and 300°C for similar water/rock ratios.

The most important difference is comparison with temperatures below 400°C in a rock-dominated system is in the steady-state metal concentrations, which are generally sufficient to account for the reported metal concentrations in the vents at 21° N. The basic composition of the 350°C fluid exiting on the seafloor at 21°N is reproduced experimentally at 400°C. Both fluids have reacted with basalt at a low water/rock ratio and have the same characteristics of zero Mg and SO_4 concentrations, low pH and high metal concentrations, and intermediate Ca concentration.

Higher temperatures also make SO_4 reduction more likely. SO_4 reduction would be accompanied by anhydrite dissolution, evident by a gradual increase of Sr in solution which had originally co-precipitated in anhydrite. A transition apparently occurs between 350°C and 400° C within rock-dominated systems with respect to metal concentrations, which are negligible at 350°C but very high at 400°C. Metal concentrations in experiments above 400°C increase strikingly with increasing temperature (Mottl et al., 1979; Pohl and Dickson, 1979).

Table 2 summarizes the experimental results for rock-dominated systems between 350°and 500°C. The data at 350°C are from Mottl and Seyfried (in press), and the remaining experiments were recently carried out in our laboratory, generally following the procedures of Bischoff and Seyfried (1978) and Seyfried and Bischoff (1979, 1981). Important generalizations are as follows. (1) pH appears to decrease with increasing temperature at constant pressure, and also to decrease with decreasing pressure at constant temperature. (2) Metal concentrations appear to be directly related to acidity, and both increase rapidly with increasing temperature. Metal concentrations over this temperature range also seem to respond to expansion, because they increase with decreasing pressure. (3) The steady state Ca concentration is high but not so high as in experiments at 350°C. Ca fixation into solid phases at higher temperatures may be a possible source of H^+.

The fluid produced experimentally at 400° C and 600 bars most closely resembles the reported composition of the vent fluids at

Table 2. Steady-state concentrations of dissolved components (in ppm) of nine separate experiments (>350° C) reacting basaltic glass with seawater at the temperature, pressure, and water/rock indicated. pH is measured at 25 °C. NA, not analyzed. Data at 350°C from Mottl and Seyfried (in press).

Temperature °C	Pressure (bars)	W/R	pH	Ca	SiO_2	Fe	Mn	Zn	Cu
350	500	10	5.0	1358	1008	3	2	0.1	0.1
360	500	3	4.3	1405	NA	14	11	0.1	0.1
360	250	3	3.9	1420	1002	35	22	0.2	0.1
375	500	15	4.1	1015	1280	35	16	0.1	0.1
375	250	15	3.7	970	934	191	29	0.2	0.1
400	1000	10	4.2	1120	1941	48	15	0.1	0.1
400	600	3	3.3	1460	1578	117	44	0.6	0.1
450	1000	10	3.4	930	2073	236	44	0.4	0.1
500	1000	10	2.8	686	2075	781	74	0.4	0.1

21° N. An experimental fluid at 375° C, extrapolated between the results for 250 and 500 bars might also resemble the vent fluids; this fluid is our preferred analog for the vent water. On expansion to the 200-250 bar sea-floor pressure at 21° N such a fluid would cool somewhat, perhaps close to the 350°C exiting temperature observed at several vents on the East Pacific Rise. If this analog is correct, then the maximum penetration of the hydrothermal fluid would be limited to only about the top 1 km of oceanic crust.

Evolved-Seawater Experiments

It may be that seawater has been significantly altered before it reaches the point of maximum conditions, because of likely slow flow rates in the large recharge zone of the hydrothermal cycle (Bischoff, 1981). Most of the SO_4 will have been removed as anhydrite, and Mg has been removed in smectite in exchange for Ca, and so the fluid reacting with fresh glassy basalt at the highest temperature and pressure probably differs from seawater and is a dominantly Na - Ca - K - Cl solution. Therefore, we prepared a solution to resemble a fully reacted (evolved) seawater at 300°C (Seyfried and Bischoff, 1981) and allowed this fluid to react with fresh basalt (at temperatures from 350° to 500°C (Table 3)).

The results of our evolved seawater experiments resembled those of the natural-seawater experiments. (1) pH decreased with increasing temperature; pH values were only 0.2 to 0.4 higher than in the

Table 3. Steady-state concentrations of dissolved components (in ppm) of six separate experiments (>350°C) reacting basaltic glass with an evolved seawater, including the starting composition of the evolved seawater (see text). pH is measured at 25° C. Composition of the vent water at 21°N on the East Pacific Rise is listed for comparison.

Temperature °C	Pressure (bars)	pH	Ca	SiO_2	Fe	Mn	Zn	Sr
300 (start)	--	5.9	1057	0	0.1	<0.1	<0.1	<0.1
350	162	4.4	-	732	25	12	2.5	2.2
375	220	3.3	923	795	153	42	2.5	2.6
400	292	3.2	779	691	580	55	2.5	2.6
400	1000	4.4	845	1638	29	14	<0.5	3.9
450	1000	3.7	750	1889	224	16	<0.5	3.1
500	1000	3.2	444	1904	839	88	<0.5	3.6
Vent water								
350		3.6	862	1292	100	34	7.8	7.8

the corresponding experiments with natural seawater. (2) Significant amounts of heavy metals were leached and maintained at about the same concentrations as in the natural-seawater experiments. (3) Ca concentration decreased continuously from initially high values to steady state values slightly lower than the corresponding concentrations in the natural-seawater experiments. Because the evolved-seawater experiments were longer in duration than the natural seawater experiments (3 months versus 1 week, respectively) more crystallized Ca phases may have developed with time resulting in lower Ca concentrations. Higher Ca concentrations in the seawater experiments may also reflect anhydrite dissolution due to SO_4 reduction. (4) Steady-state SiO_2 concentrations were somewhat lower than in the corresponding experiments with natural seawater.

These results are somewhat surprising because the fluid became acidic in the absence of Mg, which is the apparent source of H^+ in the natural-seawater experiments. This observation raises the question as to the source of H^+ in the evolved-seawater experiments. The starting solution contained a Ca concentration about 3 times that of natural seawater that decreased to lower steady-state values during the course of the experiments. This Ca removal suggests that Ca fixation in some hydrous phase may be related to the H^+ production observed at high temperature in a reaction perhaps analogous to Mg fixation at lower temperatures. Ca-bearing epidote is commonly found in sea-floor meta-basalt of greenschist facies (Humphris and

Thompson, 1978). No Ca-containing phases were detected during X-ray diffraction scans of the bulk solids remaining at the end of our experiments. Such phases, however, may be poorly crystallized or present below the levels of detection by X-ray diffraction.

Petrographic studies by Humphris and Thompson (1978) suggest that during metamorphism, epidote is formed at the expense of plagioclase. Their observations, in combination with the experimental Ca depletions and H^+ production, suggest that a possible reaction generating epidote is:

$$\text{(4)} \quad \underset{\text{plagioclase}}{NaAlSi_3O_8 \cdot 3CaAl_2O_8} + Ca^{+2} + 2H_2O = \underset{\text{epidote(clinozoisite)}}{2Ca_2Al_3Si_3O_{12}(OH)} + \underset{\text{albite}}{NaAlSi_3O_8} + 2H^+$$

This reaction would effectively reduce the anorthitic component of the plagioclase and thus relatively increase the amount of albite, without a net uptake of Na from seawater. Albite is also a principal mineral of greenschist-facies basalt (Humphris and Thompson, 1978).

The increased metal concentrations at higher temperatures under rock-dominated conditions may result from this additional H^+ production due to Ca metasomatism or simply from steady-state mineral equilibria. The possible consumption of H^+ by reaction with such accessory metal oxides or sulfides as magnetite, hematite, pyrite, or chalcopyrite in the basalt may release metals to solution. In experiments at 400°and 450°C, the total amount of metals leached from the basalt is less than the amount of Ca removed from solution, and so excess H^+ is generated. At 500°C Ca removal from solution is balanced by Fe+Mn leached from the basalt on a molal basis.

An alternative explanation is that the increased metal concentrations may simply result from the increased solubility of metals with respect to the alteration phases at increasing temperatures. Metals originally precipitated in smectite/chlorite may be released at the higher temperatures; however, there is net H^+ production and buffering at these higher temperatures.

In summary, the metal concentrations produced in the evolved-seawater experiments at 400°C and 1000 bars compare favorably with the reported chemistry of the vent fluids 21°N on the East Pacific Rise.

Problems

From the earliest experiments (Bischoff and Dickson, 1975; Hajash, 1975; Seyfried, 1977; Mottl and Holland, 1978), it was recognized that certain mineralogic and chemical discrepancies exist between the experimental and natural systems. The discoveries of active hydrothermal systems on midoceanic ridges and Deep Sea Drilling Project (DSDP) results have helped to resolve some of these problems as well as to raise new ones.

Although our experiments have successfully predicted the major-component chemistry of the hydrothermal fluids, they have not reproduced the alteration mineralogy of dredged oceanic metabasalt. The main problems are the following. (1) The major discrepancy is the absence of chlorite in the experimental studies. Chlorite is ubiquitous in oceanic greenstone (Humphris and Thompson, 1978), but only smectite and mixed-layer smectite/chlorite were found in the experiments up to 500° C. Even in seawater-dominated systems containing excess Mg, a mixed-layer smectite persists apparently metastably over chlorite. (2) Albite, a common component of metabasalt, is generally absent in the alteration products of the experiments. The formation of albite may, however, depend on the water/rock ratio. Although small changes in dissolved Na are difficult to detect analytically, Mottl (in press) suggests that at water/rock ratios of $\lesssim$ 5-10, Na is taken up from seawater but at higher ratios is leached from the rock. At higher water/rock ratios, Na is leached from the rock in exchange for Mg when the supply of leachable Ca and K has been exhausted.

Albite has been formed in NaCl experiments (Seyfried and Bischoff, 1981), and the newly formed plagioclase in some 300° C seawater experiments may have been albite (Mottl and Holland, 1978). Our own experiments at a water/rock ratio of 3/1 between 350° -400° C did not produce detectable (by bulk X-ray diffraction) albite. Even basalt-NaCl brine experiments have not produced significant albite (Rosenbauer and Bischoff, unpublished data). The explanation for this discrepancy may be the H^+ as well as the Na requirements of albitization, which depend on the phase being albitized. (3) The anhydrite problem, recognized from the earliest experiments (Bischoff and Dickson, 1975; Seyfried, 1977; Mottl and Holland, 1978), continues to be discussed at present (for example, see McDuff and Edmond, 1982 and Shanks et al., 1982). Experiments predict that massive anhydrite should be found in the recharge zone of the submarine hydrothermal cycle; however, anhydrite has not been observed in any dredged metabasalt. Anhydrite has only recently been found in veins from DSDP Leg 53, hole 504B from 907-1076 m at 140° C (Anderson et al., 1982), but otherwise is absent. It may be that retrogressive low-temperature alteration dissolves earlier anhydrite; alternatively anhydrite may dissolve or never even form owing to reduction of seawater SO_4 by Fe^{+2} in the basalt. A $\delta^{34}S$ value of 2.1 for hydrothermal-vent sulfides (Arnold and Shepard, 1981) suggests that most of the sulfide is apparently leached from the basalt and only a small (perhaps 10%) component results from the reduction of seawater SO_4. Shanks et al. (1982) have pointed out some conditions of inorganic-sulfate reduction and demonstrated the sensitivity of the bulk sulfur-isotopic ratio to increasing amounts of seawater sulfate. (4) Truscottite, a disordered Ca silicate, and wairakite form in experiments, whereas epidote is found in nature. (5) An unresolved discrepancy exists between Zn concentration in the vent fluids and in the experimental fluids, even where the Fe and Mn concentrations agree. The vent fluids at 21° N are reported to contain 7 ppm Zn. Pohl and Dickson (1979) and Mottl et al. (1979) reported 2 ppm Zn in 400°C rock-dominated experiments. However, Zn contamination is difficult to avoid. The erratic variation of Zn concentration with time in the study by Pohl and Dickson suggests contamination, and the samples used by Mottl were experimental quenches and obtained with a stainless-steel needle. Our own experiments have also been plagued by Zn contamination. The sensitivity of Zn to contamination was shown during our control experiments, in which we traced contamination to several sources, including the reaction-cell system, sampling equipment, storage containers, and even the acid-bottle wash. After great effort to remove all sources of contamination we have found that Zn appears early in our experiments at 400°C at concentrations ~ 3 ppm and then decreases rapidly within 24 hours to below 0.5 ppm. Zn apparently coprecipitates in the alteration products and is strongly held even at higher temperatures. (6) Cl depletions have been observed for some vent fluids, something we have not observed in the experiments. Current experiments on the physico-chemical behavior of seawater along its isocompositional curve near the critical point may help explain these Cl depletions.

SUMMARY

(1) Experimental basalt-seawater interaction at 350° C, 500 bars and any water/rock ratio does not generate a fluid equivalent in chemical composition to that reported for the vent fluid at 21° N on the East Pacific Rise. (2) Significant amounts of metals (Fe, Mn) are mobilized during rock-dominated basalt-seawater interaction at temperatures of >350°C, but Zn concentration remains low even at 500°C. Excess H^+ is important in maintaining metals in solution. (3) On the basis of experimental evidence, maximum conditions of interaction at 21° N of ~400° C, and 600 bars and a water/rock ratio of 3 are necessary to produce the reported vent-fluid composition. A rock-dominated system at 375°C and 250 - 500 bars generates a fluid similar to a 400°C fluid at 600 - 1000 bars. (4) Evolved seawater (free of Mg and SO_4), on reaction with basalt, produces a fluid similar to that in a natural-seawater system. Ca may be responsible for H^+ production at >400°C. (5) The minerals produced in the experiments are apparently metastable with respect to normal greenschist facies basalt alteration in nature: smectite/chlorite is formed in place of chlorite, hydrated Ca silicates in place of epidote, and only minor albite and major anhydrite where we would expect major albite and minor anhydrite.

REFERENCES

Anderson, R. N., Honnorez, J. J., Adamson, A. C., Alt, J. C., Becker, K., Emmermann, R., Kempton, P. D., Kinoshita, H., Laverne, C., Mottl, M. J., and Newmark, R., 1982, Cruise summary, Leg 83, deep drilling on the Costa Rica Rift, p. 16 in JOIDES Journal, v. 8, no. 1, February.

Arnold, M. and Sheppard, S. M. F., 1981, East Pacific Rise at latitude 21°N: isotopic composition and origin of the hydrothermal sulfur, Earth Planet. Sci. Lett., v. 56, p. 148-156.

Bischoff, J. L., 1981, Geothermal system at 21° N, East Pacific Rise: Physical limits on geothermal fluid and role of adiabatic expansion, Science, v. 207, p. 1465-1469.

Bischoff, J. L. and Dickson, F. W., 1975, Seawater-basalt interaction at 200°C and 500 bars: Implication for origin of sea-floor heavy metal deposits and regulation of seawater chemistry, Earth Planet. Sci. Lett., v. 25, p. 385-397.

Bischoff, J. L. and Rosenbauer, R. J., 1983, A note on the chemistry of seawater in the range 350°-500° C, Geochim. Cosmochim. Acta, vol. 47, p. 139-144.

Bischoff, J. L. and Seyfried, W. E., 1978, Hydrothermal chemistry of seawater from 25°C to 350°C, Am. J. Sci., v. 278, p. 838-860.

Chen, C. A., 1981, Geothermal system at 21°N, Science, v. 211, p. 298.

Edmond, J. M., von Damm, K. L., McDuff, R. E., and Measures, C. I., The chemistry of the hot springs on the East-Pacific Rise and the dispersal of their effluent, (in press), Nature.

Gregory, R. T. and Taylor, H. P., 1981, An oxygen-isotope profile in a section of Cretaceous oceanic crust, Samail Ophiolite, Oman: Evidence for $\delta^{18}O$ buffering of the oceans by deep (>5 km) seawater-hydrothermal circulation at mid-ocean ridges, Jour. Geophys. Res., v. 86, no. B4, p. 2737-2755.

Hajash, A., 1975, Hydrothermal processes along mid-ocean ridges: An experimental investigation, Contrib. Mineral. Petrol., v. 75, p. 1-13.

Humphris, S. E. and Thompson, G., 1978, Hydrothermal alteration of oceanic basalts by seawater, Geochim. Cosmochim. Acta, v. 42, p. 107-125.

Janecky, D. R. and Seyfried, W. E., The solubility of magnesium hydroxide sulfate hydrate in seawater at elevated temperatures and pressures, (in press), A. J. Sci.

Kennedy, G. C., 1950, A portion of the system silica-water, Econ. Geol., v. 45, p. 629-653.

MacDonald, K. C. and Luyendyk, B. P., 1981, The crest of the East Pacific Rise, Sci. Am., May, p. 100-116.

McDuff, R. E. and Edmond, J. M., 1982, On the fate of sulfate during hydrothermal circulation at mid-ocean ridges, Earth Planet. Sci. Lett., v. 57, p. 117-132.

Mottl, M. J., Metabasalts, axial hot springs, and the structure of hydrothermal systems at mid-ocean ridges, (in press), Geol. Soc. Am. Bull.

Mottl, M. J. and Holland, H. D., 1978, Chemical exchange during hydrothermal alteration of basalt by seawater - I. Experimental results for major and minor components of seawater, Geochim. Cosmochim. Acta, v. 42, p. 1103-1115.

Mottl, M. J., Holland, H. D., and Corr, R. F., 1979, Chemical exchange during hydrothermal alteration of basalt by seawater - II. Experimental results for Fe, Mn, and sulfur species, Geochim. Cosmochim. Acta, v. 43, p. 869-884.

Mottl , M. J. and Seyfried, W. E., 1980, Sub-seafloor hydrothermal systems rock- vs. seawater-dominated, in: Seafloor Spreading Centers: Hydrothermal Systems, P. A. Rona and R. P. Lowell, eds., Dowden, Hutchinson and Ross, Inc., Stroudsburg, PA, 1980, p. 66-82.

Pohl, D. C. and Dickson, T. W., 1979, Basalt-seawater reaction at 400°and 500°C and 1 kb; Implications as to depositional processes at spreading centers, Eos, v. 60, no. 46, p. 973.

Seyfried, W. E., 1977, Seawater-basalt interaction from 25° -300° C and 1-500 bars: Implications for the origin of submarine metal-bearing hydrothermal solutions and regulation of ocean chemistry, (Doctoral dissertation), Univ. of So. Calif.

Seyfried, W. E. and Bischoff, J. L., 1977, Hydrothermal transport of heavy metals by seawater: The role of seawater/basalt ratio, Earth Planet. Sci. Lett., v. 34, p. 71-77.

Seyfried, W. E., Mottl, M. J., and Bischoff, J. L., 1978, Seawater/basalt ratio effects on the chemistry and mineralogy of spilites from the ocean floor, Nature, v. 275, no. 5677, p. 211-213.

Seyfried, W. E. and Bischoff, J. L., 1979, Low temperature basalt alteration by seawater: An experimental study at 70° C and 150° C, Geochim. Cosmochim. Acta, v. 43, p. 1937-1947.

Seyfried, W. D. and Bischoff, J. L., 1981, Experiment seawater - basalt interaction at 300° C, 500 bars. Chemical exchange secondary mineral formation and implications for the transport of heavy metals, Geochim. Cosmochim. Acta, v. 45, p. 135-147.

Seyfried, W. E. and Mottl, M., 1982, Hydrothermal alteration of basalt by seawater under seawater-dominated conditions, Geochim. Cosmochim. Acta, v. 46, p. 985-1002.

Shanks, W. C., Bischoff, J. L., and Rosenbauer, R. J., 1981, Seawater sulfate reduction and sulfur isotope fractionation in basaltic systems - Interaction of seawater with fayalite and magnetite at 200°- 350°C, Geochim. Cosmochim. Acta, v. 45, p. 1977-1995.

Sourirajan, S. and Kennedy, G. C., 1962, The system H_2O-NaCl at elevated temperatures and pressures, Am. J. Sci., v. 260, p. 115-141.

von Damm, K. and Edmond J., 1982, Chemistry of 21° N East Pacific Rise hydrothermal solutions, Eos 63, p. 1015.

HYDROTHERMAL PROCESSES AT SEAFLOOR SPREADING CENTERS: APPLICATION OF BASALT-SEAWATER EXPERIMENTAL RESULTS

Michael J. Mottl

Department of Chemistry
Woods Hole Oceanographic Institution
Woods Hole, MA 02543

ABSTRACT

Both the chemistry of seafloor hot springs and the chemical changes exhibited by basalts during alteration to greenschist facies assemblages have been accurately predicted by laboratory experiments reacting seawater with basalt. Although the experiments were run as an isothermal, closed-system, batch process, they largely succeeded in duplicating the products of the natural open-system, continuous flow process. For the solutions, this resulted mainly from rapid reaction rates at high temperature, relative to flow rates in the natural systems, so that equilibrium with the secondary mineral assemblage represented a significant control on solution composition both in the experiments and in nature. For the rocks, it resulted from a similar alteration history in which largely unreacted seawater reached greenschist facies temperatures before reacting with the basalts, and from element exchanges between rock and solution which were coupled via charge balance constraints so that the batch process in the experiments simulated the incremental process in nature.

The key concept in relating the batch process to the incremental process is the seawater/rock ratio, which because of the nature of the chemical exchanges involved can best be estimated from the uptake of seawater Mg by the altered rock. The experiments predict a systematic change in rock chemistry and mineralogy as alteration proceeds to higher seawater/rock ratios. The prediction is borne out for the fluxes of Mg and Ca, the flux directions of Na, Si, and Mn, and the mineral abundances of chlorite, quartz, and actinolite. It is not borne out for the Fe flux, the magnitude of the Na flux, and the abundances of albite and epidote, because the experiments failed as batch processes to allow for local redistribution of elements via diffusion. This latter process is important in altered rocks from the natural systems for Fe^{2+}, which diffuses into zones where chlorite is forming preferentially due to influx of seawater Mg, and for Na^{+}, which accumulates as albite in zones of lesser Mg influx, in exchange for Ca.

INTRODUCTION

Chemical processes in submarine hydrothermal systems at seafloor spreading axes play an important role in alteration of the oceanic crust, in transfer of elements between the crust and the oceans, and in formation of polymetallic sulfide deposits on and within the crust. These processes are inherently difficult to study because they occur in the subsurface beneath the ocean floor and are inaccessible to direct observation and sampling, except by drilling. As no drilling has yet been done into a high-temperature, axial sub-seafloor hydrothermal system, we have only three indirect ways of studying these processes.

The first and most direct is by sampling the effluent from these systems at hydrothermal vents along the mid-ocean ridge axis. This has been accomplished at four sites to date, all in the Pacific: the Galapagos Rift at 86°W, the East Pacific Rise at 21°N and 13°N, and the Guaymas Basin in the Gulf of California (Von Damm et al., this volume). An analogous subaerial hydrothermal system has been sampled on the Reykjanes Peninsula of Iceland, both from vents and by deep drilling (Stefansson, this volume).

The second way is by sampling the solid products of subsurface hydrothermal processes: the altered basalts and gabbros (Thompson, this volume). These are accessible along fault scarps of axial valley walls and transverse fracture zones. As well-developed axial valleys are a unique feature of slow-spreading ridges, our sample collection is heavily biased toward these ridges, particularly the Mid-Atlantic Ridge. An equally significant bias may result from the processes which uplift and expose hydrothermally altered rocks, especially the mettagabbros, along axial valley walls. These walls typically consist of a series of steep fault zones separated by fault-block terraces arranged like a giant staircase. The vertical relief within an individual fault zone rarely exceeds 200-300 m, and the throw on any one fault within a zone is even smaller. It is difficult to see how these scarps can expose deep levels of the crust, yet they often yield gabbros and hydrothermally altered rocks from both the greenschist and, less commonly, the amphibolite facies. Metabasalts and metagabbros are not always exposed along axial valley walls, however; many sections have yielded only unmetamorphosed basalts, in spite of intensive dredging. We do not know at present whether the distribution of exposed metabasalts reflects their distribution at depth within the crust, or some other factor of crustal constitution or tectonics which leads to their preferential exposure in some places and not in others. Thus it is difficult to evaluate their significance, or to know how far we can generalize the nature of their alteration to the mid-ocean ridge as a whole.

The third way to study chemical processes in sub-seafloor hydrothermal systems is to simulate subsurface reaction conditions of high temperature and pressure in the laboratory. A large number of experiments have now been performed reacting seawater with basalt over the range of conditions expected to prevail within the oceanic crust at mid-ocean ridges (Bischoff and Dickson, 1975; Hajash, 1975; Furnes, 1975; Seyfried and Bischoff, 1977, 1979, 1981; Mottl and Holland, 1978; Mottl et al., 1979; Menzies and Seyfried, 1979; Thomassin and Touray, 1979, 1982; Mottl and Seyfried, 1980; Hajash and Archer, 1980; Hajash and Chandler, 1981; Seyfried and Mottl, 1982; Crovisier et al., 1983; and Bischoff and Rosenbauer, this volume). The effects of five major variables have been evaluated in these experiments: temperature from 20-500°C, pressure from 1-1000 bars, basalt crystallinity from glass to holocrystalline (diabase), seawater/rock mass ratio from 1-125, and duration from 14-602 days. Besides basalt-seawater experiments, the behavior of seawater on heating has been studied up to 350°C (Bischoff and Seyfried, 1978) and some studies have used rocks other than basalt (e.g. Seyfried and Dibble, 1980; Bischoff et al., 1981; Hajash and Chandler, 1981; Shanks et al., 1981). Unlike the natural systems studied to date (except for Reykjanes), these experiments have allowed us to study the effects of alteration on both the solution and the solids under known physical conditions. They have thus proved useful in interpreting the conditions of alteration in the subsurface based on data from both hot spring solutions and metabasalts, and to infer the nature of alteration in one from data on the other (Mottl, 1983).

DESIGN AND OBJECTIVES OF THE EXPERIMENTS

Two principal objectives of the laboratory experiments have been to determine the composition of solutions formed during reaction of seawater with basalt under a variety of conditions and to identify the critical parameters which are likely to control the concentrations of the various chemical species in solution in the natural setting.

An obvious limitation of the experimental results is that reaction in the experiments has been treated operationally as a closed-system, isothermal, batch process. Reaction in the natural systems, by contrast, occurs over a range of temperatures in an open system between a fluid which is changing in temperature and composition as it travels through the crust, and rock which is in some places fresh and in other places already altered. Whether the experimental solutions will approximate those in nature depends largely on the extent to which solution and alteration minerals equilibrate with one another within a given temperature zone in a natural system. If water-rock reaction rates in natural systems are rapid relative to flow rates, then the experimental results are likely to be applicable, because the solution will tend to equilibrate with the alteration mineral assemblage forming at a given temperature before it moves on to a different temperature.

The reaction model used here is illustrated schematically in Figure 1. Two initial conditions generally prevail when solution of some composition enters a volume of fresh or partially altered rock at some elevated temperature and pressure: 1) the primary igneous minerals and glass are thermodynamically unstable in contact with an aqueous solution and 2) the solution is out of equilibrium with the solids

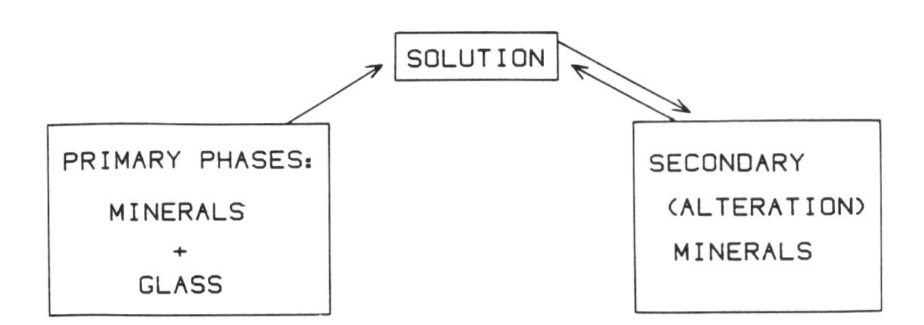

Fig. 1. Model of water-rock reaction applicable to the experiments and the natural hydrothermal systems.

present. Reaction ensues which involves congruent or incongruent dissolution of the primary phases and any unstable secondary phases, accompanied by precipitation of any secondary minerals with which the solution becomes supersaturated. The tendency throughout is for the solution to approach equilibrium with the secondary minerals which are forming. How closely this equilibrium is approached depends on 1) the reaction rate for a) dissolution of primary solid phases (minerals and glass) and b) dissolution and precipitation of secondary solid phases vs. 2) the flow rate of solution through the volume of rock.

The above model applies to those chemical species which participate in the secondary mineral-solution equilibria. Concentrations of these species in solution will either be controlled by equilibrium with the alteration minerals at the prevailing temperature and pressure or determined by a balance among the relative rates of primary phase dissolution, secondary phase formation, and flow of solution through the volume of rock being considered. Some species, however, will not participate in secondary mineral-solution equilibria except as trace elements, because their dissolved concentrations never become high enough to saturate the solution with a mineral bearing them as a major component. These include the so-called "soluble elements" (Ellis, 1970), which are characteristically present in low concentrations in the rock and are strongly partitioned into the solution rather than the solid phases. The concentrations of these elements in solution are limited ultimately by their depletion from the rock; in a closed system, the increase of their concentration in solution would be inversely proportional to the amount of water present divided by the amount of altered rock, i.e., the water/rock ratio.

Thus there are three principal mechanisms which can determine the concentration of a chemical species in solution: equilibrium, kinetics (reaction vs. flow rates), and depletion from the rock. For equilibrium the solution will carry no record of its previous alteration history, whereas for the other two mechanisms the dissolved concentrations will depend in part on the path the solution has followed through the system, including the pressure-temperature regimes in which it has reacted and the type of rock it has encountered, whether fresh or altered. In contrast with the solutions, the altered rocks as a whole will invariably retain a record of their alteration history.

Two approaches have been used in the experiments to study the reaction path of the solution vs. that of the rock. The first has been via duration, either by varying the length of time an experiment ran or by monitoring the change in solution chemistry with time using sampling apparatus (Seyfried et al., 1979). The second has been via the seawater/rock ratio. This parameter is defined operationally in the experiments as the mass of seawater initially present in the reaction cell divided by the mass of fresh rock. When finely powdered rock is used, the practical lower limit for this parameter in order to recover solution for analysis is 1:1. Its instantaneous value for a given volume within a natural hydrothermal system is much less than 1:1, considering the small pore volume of most rocks. In applying the experimental results to natural systems, therefore, a different definition is required. For natural submarine hydrothermal systems as a whole, the water/rock ratio may be defined as the total mass of seawater which has passed through the system, integrated through time, divided by the total mass of altered rock within the system. The same definition could be applied to a small volume within a hydrothermal system, provided that some correction is made for the case in which all of the water entering the volume is not unaltered seawater. The key words in the definition are "integrated through time" and "altered" rock. By running an experiment at a water/rock mass ratio of 10 or 50, the intent is to simulate via a batch process the end results of an incremental process which has run through a large number of increments. This concept and its applications have been discussed at length by Mottl and Holland (1978), Mottl and Seyfried (1980), Seyfried and Mottl (1982), and Mottl (1983). The overall success of the experiments in duplicating both the altered solutions and the altered rocks from the natural systems, detailed in the sections which follow, indicates that this approach has been valid in general. Some exceptions do exist, notably for the distribution of Fe and Na in the altered rocks. These exceptions can be readily attributed to the nature of the experiments as a closed-system, batch process.

REVIEW OF EXPERIMENTAL RESULTS

Basalt-seawater experiments under hydrothermal conditions have established a number of key points which have enhanced their applicability to the natural systems. First, reaction rates have been found to be rapid at temperatures of 150°C or greater (Figure 2). Secondly, solution composition has been found to approach a steady state for most chemical species well within the duration of the experiments. Thirdly, the direction of net transport between rock and solution for some species has been successfully reversed by using artificial starting solutions with enhanced concentrations of those species (Mottl and Holland, 1978). The latter two points both indicate that formation of the secondary alteration mineral assemblage has acted as a significant control on solution composition.

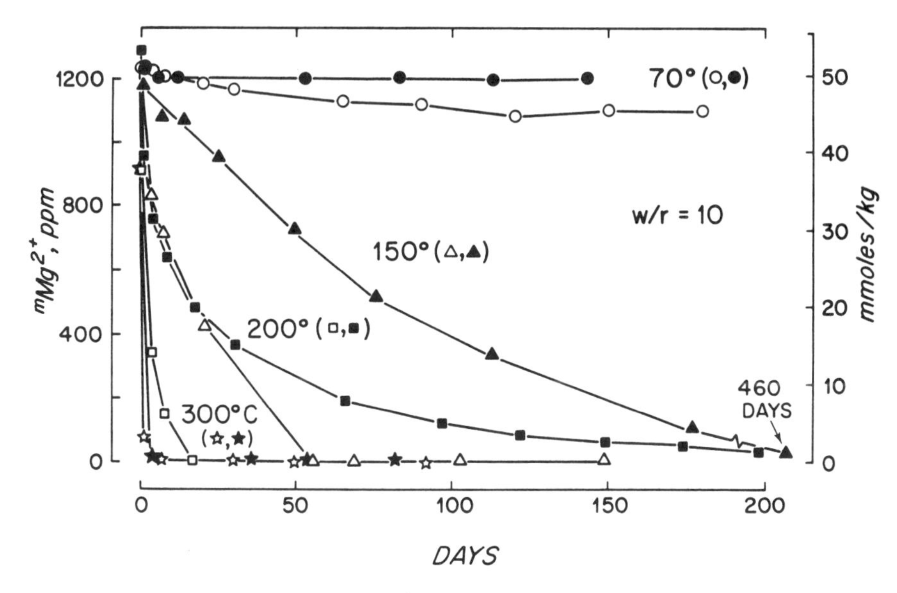

Fig. 2. Concentration of Mg^{2+} with time in some experiments reacting seawater with basalt glass (open symbols) or diabase (closed symbols) at 70°C, 1 atm, and 150-300°C, 500 bars, at a water/rock mass ratio of 10. Data are from Bischoff and Dickson (1975), Seyfried (1977), and Seyfried and Bischoff (1979, 1981).

The secondary mineral assemblage formed in the laboratory experiments resembles that found in nature, as does the order of susceptibility to alteration of the primary phases. There are some important discrepancies, however. The chief problems are the formation of smectite or mixed layer smectite-chlorite in the experiments at temperatures at which chlorite forms in nature, and the failure to form epidote. Minerals which did form include smectite, smectite-chlorite, analcime, wairakite, truscottite, albite, oligoclase, tremolite-actinolite, talc, sphene, quartz, prehnite (?), anhydrite, pyrite, pyrrhotite, chalcopyrite, hematite, and magnetite. There is a tendency for the first appearance of some minerals to occur at somewhat higher temperatures in the experiments than in the natural systems; for example, tremolite-actinolite formed in experiments at 350°C (Mottl and Seyfried, unpublished data) but not at 300°C whereas data from Iceland geothermal areas suggest that it probably first forms in nature at about 290°C (Kristmannsdottir, 1976).

Despite the discrepancies in alteration mineral assemblage, the experiments succeeded in closely duplicating the solution chemistry of the natural systems, as shown in Table 1 and Figure 3. This implies that reaction rates are generally rapid relative to flow rates in these systems. The experiments which most closely duplicated the natural solutions from the Galapagos Rift, the East Pacific Rise at 21°N, and the Reykjanes Peninsula were those performed at low water/rock ratios of 1:1.

The most characteristic feature of basalt-seawater interaction at elevated temperatures is the rapid removal of Mg^{2+} from seawater into secondary solids such as smectite, chlorite, tremolite-actinolite, or talc (Figure 2). This process occurs over a temperature range of at least 70-500°C. At 150°C and above, and possibly also at lower temperatures, almost complete removal of Mg^{2+} from seawater occurs at seawater/rock ratios up to 50, i.e., basalt undergoing alteration has the capacity to remove all Mg^{2+} from an amount of seawater up to 50 times its own mass. At higher seawater/rock ratios Mg^{2+} removal is necessarily less than complete (Mottl and Seyfried, 1980). Removal occurs rapidly at 150°C and above, in a matter of hours to months in the experiments (Figure 2).

Removal of Mg^{2+} from seawater occurs in the form of a $Mg(OH)_2$ component which is incorporated into secondary silicates, leaving H^+ behind in solution (Bischoff and Dickson, 1975). Because Mg^{2+} is removed rapidly, H^+ generation is also rapid and the solution pH falls dramatically. Even if removal of Mg^{2+} into secondary silicates were not sufficiently rapid this drop in pH would occur, in response to formation of a $Mg-OH-SO_4$ phase which precipitates on simple heating of seawater to 250-300°C and above (Bischoff and Seyfried, 1978). This compound would form in nature in the absence of high concentrations of dissolved silica and Ca^{2+} (Janecky and Seyfried, 1982). As long as the Mg^{2+} concentration is appreciable, as would have to be the case at water/rock ratios >50, the solution remains quite acid; when the Mg^{2+} concentration drops to a low value, H^+ is rapidly consumed by silicate hydrolysis reactions and pH rebounds to near neutrality (Seyfried and Bischoff, 1977). This process is illustrated schematically in Figure 4. Mottl and Seyfried (1980) have referred to these two stages as reaction under "seawater-dominated" vs. "rock-dominated" conditions. Because seawater-dominated conditions are characterized by a very low pH, only a few secondary minerals can form in equilibrium with the acid solution. These include smectite, mixed layer clay, or chlorite, and quartz, hematite, and anhydrite. Under rock-dominated conditions, by contrast, a complex assemblage of

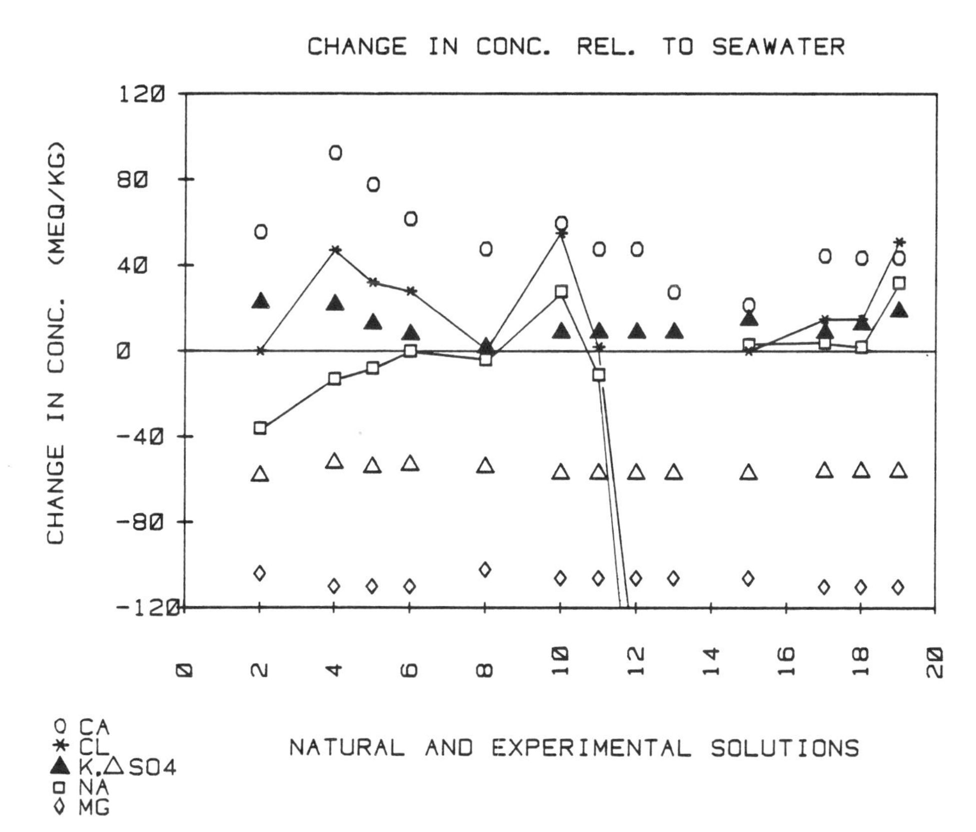

Fig. 3. Change in the concentrations of six major species in seawater which has reacted with basalt either in laboratory experiments or in natural hydrothermal systems. Key: 2: Reykjanes Peninsula, drill hole 8. 4-6: Experiments of Mottl and Holland (1978) at 300°C, 600 bars, and water/rock (w/r) mass ratios of 1, 2, and 3, respectively. 8: Experiment of Mottl and Seyfried (unpubl.) at 350°C, 500 bars and w/r = 10. 10-13: High-temperature (∿350°C) end-member at four vent areas on the Galapagos Rift at 86°W (Edmond et al., 1979): Clambake, Garden of Eden, Dandelions, Oyster Beds. 15: East Pacific Rise at 21°N: 350°C solution (Edmond et al., 1982). 17-19: Experiments of Mottl and Holland (1978) at 400°C; 700 bars, and w/r = 3, 2, 1, respectively.

Table 1. Natural vs. experimental hydrothermal solutions produced by reaction of seawater with basalt (ppm)

	T, °C	w/r[7]	pH at 25°C	SiO_2	K	Ca	Sr	Ba	Fe	Mn	Zn	H_2S	ΣCO_2
[1]Reykjanes No. 8	270	~1	6.6	600	1397	1590	9.1	10.3	0.5	2.0	0.07	45	1900
[2]Experiments	300	1	5.4-6.0		1250	1224-2178[8]	7.7	1.5-5	~2	2.2-7		66-211	169-207
[3]Experiment	350	10	5.0	1008	472	1358	2.1	0.8	3.1	2.5	<0.04	91	125
[4]Galapagos R.,86°W	350	~1		1296	735	986-1611	7.6	2.4-5.9		20-63			408-510
[5]E.Pacific R.,21°N	350	~1	~4.0	1292	978	862	7.9	4.8-13	101	34	7.2	221	252
[6]Experiments	400	1	3.6-4.1	1652	1146	1059-1304[8]	7.8	2.0-3.6	103	37-55	~2	227	500-680

[1]Bjornsson et al. (1972); Mottl et al. (1975); Arnorsson et al. (1978); Olafsson and Riley (1978).

[2]Mottl and Holland (1978); Mottl et al. (1979); range or average for 4 experiments using basalt of intermediate crystallinity (xt, x+g); pressure = 600 bars

[3]Mottl and Seyfried (in preparation); pressure = 500 bars

[4]Edmond et al. (1979, 1982)

[5]Edmond (1980); Craig et al. (1980); Edmond et al. (1982)

[6]Refs. as in note 2; range or average for 3-6 experiments; pressure = 700 bars

[7]Seawater/rock mass ratio

[8]Adjusted to zero SO_4 to correct for anhydrite dissolution on quench; Mg and SO_4 are near zero in all solutions

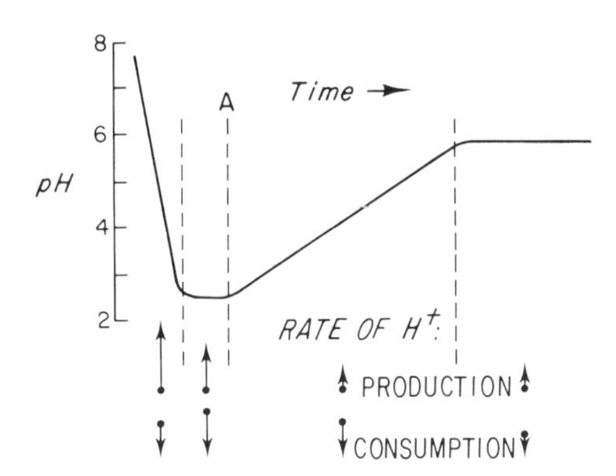

Fig. 4. Idealized profile of pH vs. time for rapidly heated seawater undergoing reaction with basalt. Length of arrows indicates the relative rates of H^+ production via Mg^{2+} removal and H^+ consumption via silicate hydrolysis reactions. The time period to the left of point A represents reaction under seawater-dominated conditions; to the right, rock-dominated. From Seyfried and Mottl (1982); see Mottl and Seyfried (1980) for actual profiles from experiments.

secondary silicates can form, including zeolites, albite, tremolite-actinolite, sphene, and epidote.

The heavy metals Fe, Mn, Zn, and Cu are leached from the rock into solution during basalt-seawater interaction. Their concentrations are strongly dependent on both temperature-pressure and pH, and thus they respond to the change from seawater- to rock-dominated conditions. Under seawater-dominated conditions Mn, Zn, and Cu can be almost completely leached from the rock into solution (Seyfried and Mottl, 1982).

Removal of Mg^{2+} from seawater is balanced with respect to electrical charge in solution ultimately by leaching mainly of Ca^{2+} from the rock. (Mottl and Holland, 1978; Seyfried and Bischoff, 1981). About half of the Ca^{2+} leached at 150°C and above combines initially with seawater sulfate to form anhydrite. The other half stays in solution. Ca in anhydrite may later be returned to solution either by dissolution at lower temperature or by dissolution accompanying reduction of sulfate to sulfide. Reduction of seawater sulfate is kinetically slow below 350°C (Shanks et al., 1981; Mottl et al., 1979). At water/rock ratios >5 it would play a minor role in the fate of Ca and SO_4 relative to that of anhydrite formation and dissolution, because the quantity of seawater SO_4 would greatly exceed the capacity of the rock to reduce it in such a system (Mottl et al., 1979).

Some Na^+ as well as Mg^{2+} is removed from seawater at seawater/rock ratios <5, into secondary albite or analcime (Hajash, 1975; Mottl and Holland, 1978). Na^+ removal is likewise balanced largely by leaching of Ca^{2+} from basalt. At a water/rock ratio of 10, Na^+ was removed from seawater at 350°C (Mottl and Seyfried, in prep.) but leached from the rock at 300°C (Seyfried and Bischoff, 1981). At ratios >10, Na^+ is leached from the rock along with Ca^{2+}. Both Na^+ and Ca^{2+} can be leached from the rock completely at ratios >50.

K^+ is leached from the rock at 150°C and above. At 300°C and above it is leached almost completely, even at low water/rock ratios. The effect of water/rock ratio on K can be seen in Figure 3. No K-rich phases formed in any of the experiments.

The concentration of dissolved SiO_2 equalled or exceeded quartz saturation in experiments at 150-200°C. At 300-500°C quartz saturation was also equalled or exceeded in most experiments, as evidenced by high SiO_2 concentrations, actual formation of quartz, or both. The Reykjanes, Iceland, geothermal system is saturated with quartz at 220-270°C, as are other Iceland geothermal systems at 180°C and above, and often at 130-180°C (Arnorsson, 1975).

Of the three mechanisms outlined in the previous section which determine the concentration of a species in solution, equilibrium with the secondary mineral assemblage at the prevailing temperature and pressure has been shown to control the concentrations of Mg, SiO_2, and SO_4 (Mottl and Holland, 1978). The main minerals involved are smectite or chlorite, quartz, and anhydrite, respectively. Reaction kinetics, specifically the rate of formation of Na-rich secondary phases such as albite, control the concentrations of Na and, via charge balance constraints, Ca under rock-dominated conditions. Concentrations of these two elements are thus a function of temperature, pressure, reaction rate vs. flow rate, and crystallinity (glass content) of the basalt being altered. Depletion from the rock, and hence water/rock ratio, determines the dissolved concentrations of K, Ba, B, Li, and Rb under rock-dominated conditions at temperatures above 150-200°C. Under seawater-dominated conditions depletion from the rock controls in addition the concentrations of Na, Ca, Mn, Zn, and Cu.

APPLICATION TO SOLUTIONS FROM MID-OCEAN RIDGE HYDROTHERMAL SYSTEMS

The vent solutions sampled to date from hydrothermal systems along the axis of the mid-ocean ridge range from a few degrees above ambient bottom water temperatures to 350°C and possibly higher (RISE, 1980; MacDonald et al., 1980). Their chemistry indicates, however, that the cooler temperatures generally result from shallow subsurface mixing of a high-temperature (∿350°C) end-member with cold groundwater (Corliss et al., 1979; Edmond et al., 1979). Allowing for adiabatic cooling, the temperature at the base of the upflow zone feeding the springs is probably in the range 350-375°C for typical axial hydrothermal systems. This is about in the middle of the range estimated for the greenschist facies (∿250-450°C) at pressures below 1 kbar and oxygen fugacities between the nickel-nickel oxide and hematite-magnetite buffers (Moody et al., 1983).

The seawater/rock ratio in the axial systems has been estimated both from oxygen isotopic data (Craig et al., 1980) and from the concentrations of soluble elements K, Ba, Li, and Rb in the solutions (Mottl et al., 1975; Edmond et al., 1979; Edmond, 1980). The estimates range from 0.7 to about 3 by mass for the Reykjanes Peninsula, Galapagos Rift, and East Pacific Rise (21°N) systems, with a value near one being most likely. As noted earlier, the high-temperature solutions from these systems have been closely duplicated in laboratory experiments at similar water/rock ratios (Table 1 and Figure 3), indicating that reaction rates are generally rapid relative to flow rates in these systems.

Note that the natural solutions have much higher Fe and Mn and a lower pH than does the experimental solution produced at 350°C and 500 bars, whereas they greatly resemble the experimental solutions produced at 400°C and 700 bars. Adiabatic cooling alone requires that the solutions from the East Pacific Rise at 21°N be somewhat hotter than 350°C in the subsurface, but whether they are as hot as 400°C and have cooled slightly by conduction on the way up is unknown. Additional experimental data bearing on this problem are presented by Bischoff and Rosenbauer (this volume).

One feature of the natural solutions which has not been duplicated in the experiments is the depletion in chloride ion shown by some springs. Two of the four springs from the Galapagos Rift shown in Figure 3 exhibit a large depletion in Cl accompanied nearly mole-for-mole by Na. While these depletions are based on long extrapolations (Edmond et al., 1979) and are therefore somewhat dubious, smaller depletions of up to 10 percent have been found in the high-temperature solutions from the East Pacific Rise at 21°N. The cause of these depletions is presently unknown.

APPLICATION TO HYDROTHERMALLY ALTERED ROCKS

Rock Chemistry

The two most detailed studies to date of the chemistry of hydrothermally altered basalts from the seafloor are those of Cann (1969) and Humphris and Thompson (1978). Both studies found a large difference in chemistry and mineralogy between the originally glassy portion of the basalts, typically pillow rims and interpillow haloclastites, and the more crystalline pillow interiors. Cann (1969, 1979) attributed these differences to a process of metamorphic differentiation resulting from the relative kinetics of diffusion vs. nucleation of a secondary phase, from primary minerals vs. glass. One problem with Cann's (1969) model is that it predicted that albite would form more rapidly by albitization of igneous plagioclase than by nucleation and growth from basalt glass. Mottl and Holland (1978) showed that in basalt-seawater laboratory experiments at 300-500°C the opposite was the case.

Seyfried et al. (1978) reinterpreted Cann's data in light of the experimental results as due to differing seawater/rock ratios during alteration of the pillow rims vs. cores. They reasoned that, because the spaces between pillows represent permeable pathways for the flow of hot water, the adjacent pillow rims and interpillow breccia would react with a larger volume of Mg-rich, unaltered seawater than would the relatively impermeable pillow interiors. Mottl (1983) used the data of Humphris and Thompson (1978) and additional data on greenstones from the Mid-Atlantic Ridge to show that the model of Seyfried et al. (1978) could be applied in general to metabasalts from the seafloor.

Because the altered rocks tend to reflect their alteration history to a greater extent than do the altered solutions, one would expect them to deviate more from the experimental results than do the solutions. In applying the experimental results to altered rocks, the design of the experiments requires that a rock sample be treated as if it were altered over a restricted temperature range by a batch process. A rock which has been altered in an open system by a continuous, flow-through process will resemble one altered by a batch process if, for the volume occupied by the sample, 1) all mass transport across the boundaries of the volume occurred via infiltration, i.e. direct transfer between rock and solution along a reactive surface combined with flow of the solution along the surface and out of the volume, 2) the effects of temperature-pressure variations in time and space were small, and 3) the exchanges of the various chemical species between rock and solution bore definite and constant relationships to one another during alteration. These relationships among species would result from chemical equilibria and/or coupled reaction rates and from charge balance constraints. This third requirement accounts for the general case in which the incoming solution is not unaltered seawater. The extent to which these conditions held for greenstones altered within the oceanic crust can be evaluated by comparing the actual chemical fluxes they have experienced with those predicted from the experiments.

Figure 5 shows the fluxes of Ca, Na, Fe^{2+}, and Mg calculated for Mid-Atlantic Ridge greenstones by Mottl (1983) assuming constant Al_2O_3 during alteration. The predictions from the experiments regarding element transfer are borne out for Mg vs. Ca and Na: Ca is generally leached from the rocks in exchange for Mg gained on a mole-for-mole basis, while Na is leached at water/rock ratios >10 to 20 but gained at lower ratios. The seawater/rock mass ratio has been estimated in Figure 5 from the MgO flux by assuming that the seawater initially contained 1300 ppm Mg which has all been transferred to the altered rock. This procedure is justified by the nearly complete removal of seawater Mg from both the natural and experimental solutions and by the fact that nearly all the greenstones analyzed, whether from pillow rim or core, show Mg gain and Ca loss. The calculated water/rock ratio for a given sample is a minimum value because the incoming solutions may not always have been unreacted seawater which then lost all its Mg within the volume of rock sampled. Nevertheless, because the net transfer of elements between rock and solution during basalt-seawater interaction occurs largely in response to uptake of Mg into alteration minerals and the accompanying generation of H^+, the water/rock ratio so calculated is an accurate indicator of the product (extent of reaction) x (flux) of seawater, where extent of reaction is defined in terms of the exchange of major ions between rock and solution.

Besides Ca, the data show that K, Si, and Mn are typically leached from greenstones, as predicted by the experiments. Fe^{2+} and total Fe, however, are gained as often as lost, as shown in Figure 5C. Both Fe and Mn fluxes, moreover, show a positive correlation with Mg flux and seawater/water ratio rather than the negative correlation predicted by the experiments. The positive correlation indicates that Fe and Mn are taken up along with Mg into chlorite during alteration.

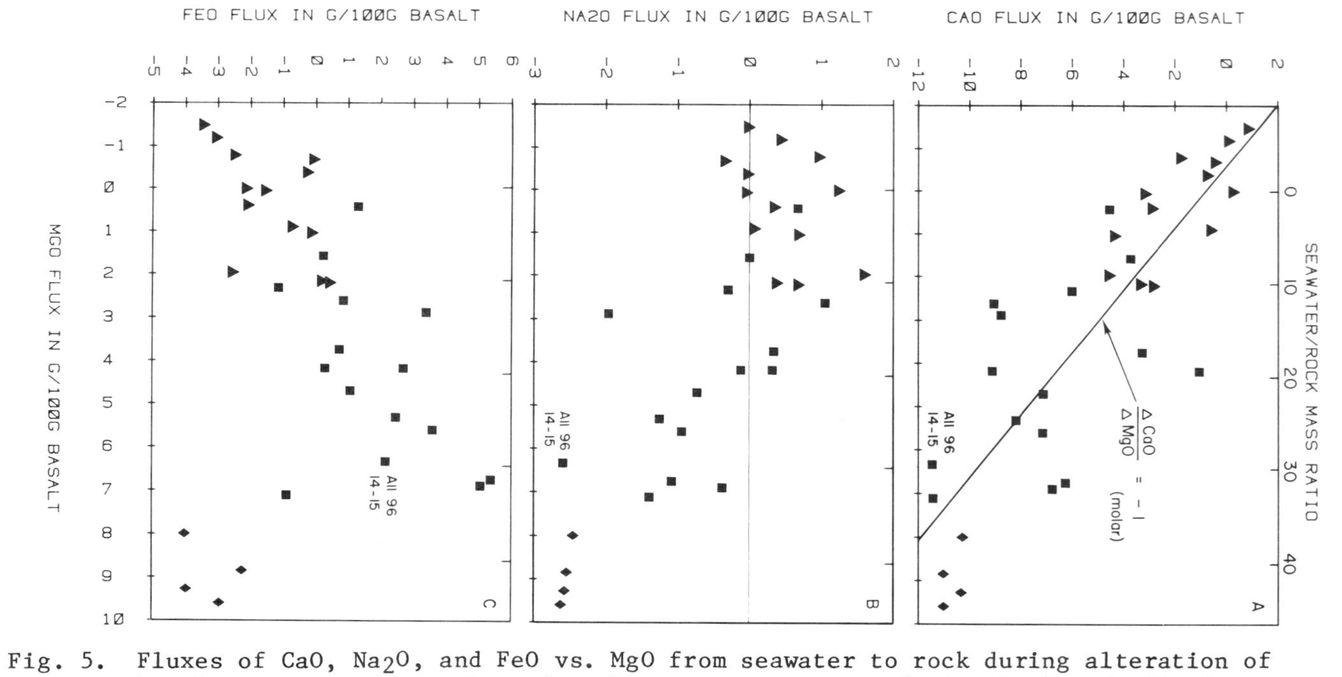

Fig. 5. Fluxes of CaO, Na_2O, and FeO vs. MgO from seawater to rock during alteration of basalt to greenstone. Data for chlorite-quartz-poor (triangles) and chlorite-quartz-rich rocks (squares) from the Mid-Atlantic Ridge at 22°S, 4°S, and 22°N are from Humphris and Thompson (1978) and at 24°N from Mottl (1983). Shown for comparison are data from Seyfried and Mottl (1982) for basalts altered experimentally (diamonds) under seawater-dominated conditions at 300°C and water/rock ratios of 50, 62 (x2), and 125, for which fluxes have been calculated by the same method used for the natural samples. Note that for FeO, the pattern for the experimentally altered samples is very different from that for the rocks altered in nature. The plot for total Fe as FeO is very similar to that above.

This mineralogical control superimposes a positive correlation on the net fluxes regardless of their direction. The net flux directions (Mg into the rock, Mn and Si out, Fe either in or out) indicate that these rocks were altered by relatively unreacted seawater still rich in Mg and poor in Mn, Si, and presumably Fe. The observed Fe enrichment in the most Mg-enriched rocks therefore probably derives from the Fe-depleted rocks, i.e., redistribution of Fe has occurred due to preferential growth of chlorite in the most Mg-enriched zones, those which reacted with the largest amount of through-flowing seawater. Diffusion must have transported Fe via the pore solutions into those zones where chlorite was forming most rapidly and the concentration of dissolved Fe was lowest.

This conclusion is reinforced by data from samples altered under seawater-dominated conditions. Four experiments performed under these conditions at seawater/rock ratios of 50-125 (Seyfried and Mottl, 1982) are shown as diamonds in Figure 5. These data may be contrasted with the square labelled "AII 96 14-15", which represents a basalt from the Mid-Atlantic Ridge altered at a similarly high seawater/rock ratio. This rock was originally a plagioclase-clinopyroxene phyric, vesicular basalt, but it has been replaced almost completely by chlorite and quartz while retaining its original texture. Its mineralogy is thus similar to that produced in the seawater-dominated experiments. A few relict Na-rich patches within former plagioclase phenocrysts which are now chloritized indicate that the rock was altered initially under rock-dominated conditions and that the seawater/rock ratio and corresponding Mg uptake increased gradually by increments into the seawater-dominated range. This alteration history contrasts with that in the experiments, in which seawater-dominated conditions were imposed initially and instantaneously on fresh basalt. The resulting difference can be seen in Figure 5C: the experimentally altered rock has of necessity lost Fe which was leached into the relatively large mass of solution, whereas the natural sample has actually gained Fe, presumbaly by diffusion from the adjoining rock (unsampled) where less chlorite was forming because the supply of Mg from the circulating seawater was lower. This Fe gain has occurred in spite of the nearly complete leaching of Ca, Na, Sr, Cu, K, Mn, Zn, and CO_2 from the rock (Mottl, 1983).

A second case in which the experiments failed to duplicate the chemistry of the altered rocks has likewise been attributed to differentiation by diffusion. Seyfried et al. (1978) noted that while the direction of Na flux was predicted correctly by experiments at low water/rock ratios, the magnitude of Na uptake was much larger for the naturally altered pillow basalt cores. They argued that Na diffused from the circulating pore fluds into the pillow cores where albite was forming due to the low water/rock ratio prevailing there and the resultant low Na/Ca ratio in solution. The case of Na differs somewhat from that of Fe, in that the source of Na enrichment could be either seawater or Na-depleted metabasalt, whereas Fe enrichment could derive only from Fe-depleted rock.

This process of local redistributon of elements via diffusion, known as metamorphic differentiation, could not occur in the experiments because they were conducted as a batch process, with all reactants available for reaction simultaneously, and because the rock was finely ground and suspended in the solution, so that transport of species via infiltration overwhelmed that via diffusion. Except for the large effects of this process on the distribution of Fe and Na, however, the experimental results are obviously useful for interpreting the chemical fluxes and alteration conditions experienced by seafloor metabasalts.

Mineralogy

The experiments predict a systematic change in secondary mineral assemblage and relative abundances as a result of reaction of the rock with increasingly large quantities of seawater (Figure 6). These changes result from two effects illustrated by the experiments: 1) the change in bulk chemistry of the altered solids as more seawater Mg is taken up in exchange for Ca and eventually Na, and 2) the transition from rock-dominated to seawater-dominated conditions at a seawater/rock ratio of 50:1, above which the solution remains highly acid and outside the stability fields of all but the most acid-resistant secondary minerals.

Figure 6 has been constructed from chemical data from the various experiments at seawater/rock ratios of 1-125, combined with data on mineral assemblages and mineral compositions from Mid-Atlantic Ridge greenstones. The secondary assemblages were calculated in the manner of an igneous rock norm. Details are given by Mottl (1983). The figure is constructed for 300°C but it probably applies approximately to the entire range over which the greenschist facies assemblage is stable.

The model in Figure 6 predicts the following assemblages with increasing seawater/rock ratio:

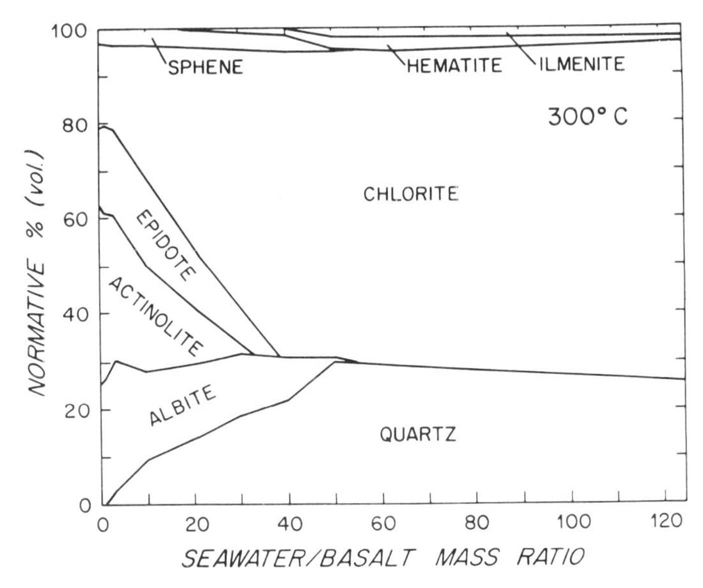

Fig. 6: Model predicting the mineral assemblages and proportions which are produced when basalt reacts with varying amounts of seawater within the greenschist facies. The model is based on chemical data from basalt-seawater experiments at 300°C and 500-600 bars. See text for method of construction.

0- 2: chl + ab + ep + act
2-35: chl + ab + ep + act + qtz
35-50: chl + ab + qtz
$>$ 50: chl + qtz

The model paradoxically predicts that the quartz content of metabasalts will increase with seawater/rock ratio in spite of an increasing proportion of Si leached from the rock. This irony results from an increase in chlorite content as more Mg is taken up from seawater. Because chlorite has the lowest Si content of any of these silicate minerals, formation of additional chlorite frees Si. The capacity of the solution to transport this Si away is, of course, limited by the solubility of quartz and the water/rock ratio, so that most of the freed Si precipitates in place as quartz.

The model based on the experimental data is compared with mineral abundances from Mid-Atlantic Ridge greenstones in Figure 7. The predicted positive correlation between chlorite and quartz holds, as does the negative correlation between chlorite and actinolite. The rocks contain more albite than predicted, however, and less epidote, due to the effects of metamorphic differentiation discussed in the previous section: the rocks have gained Na in exchange for Ca via diffusion, in excess of the amounts exchanged in the experiments via infiltration alone.

In summary, the data in Figures 5 and 7 show that the model in Figure 6 based on the laboratory experiments describes fairly well the chemical and mineralogical changes which occur progressively as basalt reacts with increasing quantities of seawater within the greenschist facies. The chief discrepancy results from the fact that the experiments were run as a batch process, and thus did not allow for localized transport of elements across the boundaries of a rock volume via diffusion rather than infiltration. The effects of temperature-pressure gradients in time and space are apparently small within the greenschist facies. Local redistribution of elements via diffusion ("metamorphic differentiation") in the natural samples is an important and probably ubiquitous process chiefly for Fe^{2+}, which diffuses into zones where chlorite is forming due to influx of seawater Mg, and Na^{+}, which accumulates as albite in zones of lesser Mg influx, in exchange for Ca.

RELATIONSHIP BETWEEN ALTERED ROCKS AND SOLUTIONS

The rocks in Figure 7 fall into two groups: those rich in chlorite and quartz and those poor in these two minerals. The latter group was apparently altered at low water/rock ratios: 10 judging from their Mg uptake as shown in Figure 5. These rocks are thus complementary to the hot spring solutions, which reflect reaction at similarly low water/rock ratios. The chlorite-quartz-rich rocks, by contrast, were altered at water/rock ratios of about 10–50, again judging from Figure 5. These rocks are not directly complementary to the venting solutions. Their large Mg uptake indicates that they were altered by relatively unreacted seawater which, despite having attained greenschist facies temperatures, still retained most of its Mg when it entered the volume occupied by these rocks. Given the rapid rate of Mg removal from seawater at elevated temperatures (Figure 2), this type of alteration almost certainly occurred within the downwelling limb of a convection

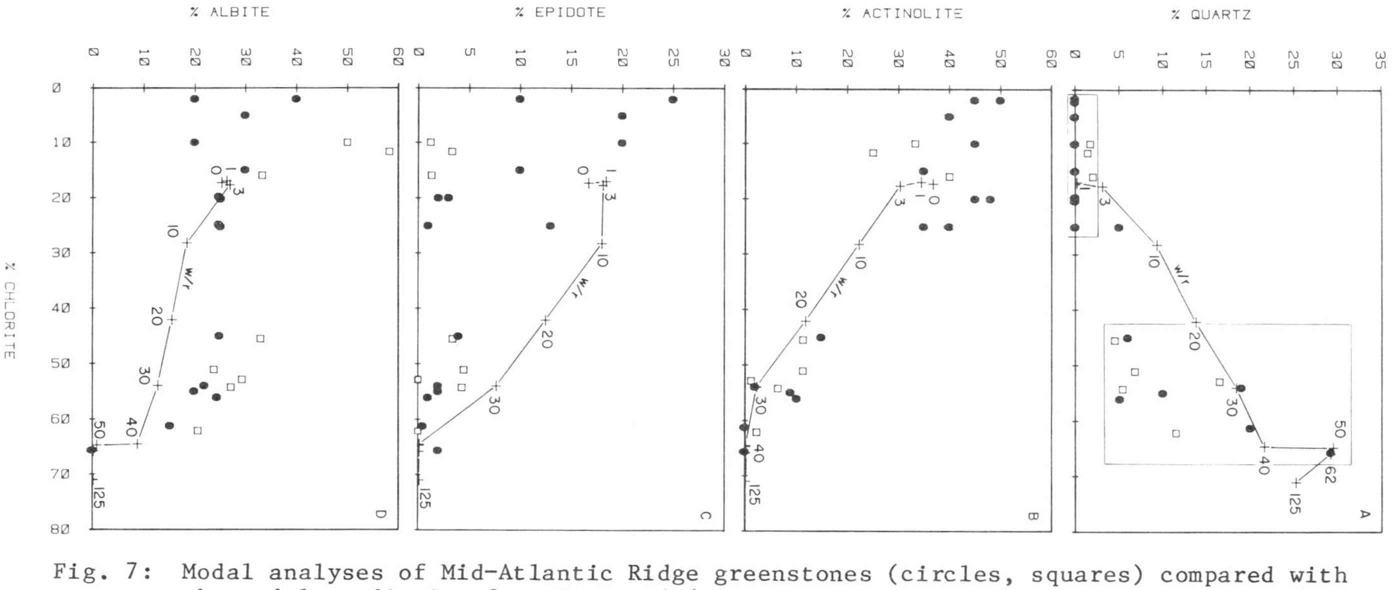

Fig. 7: Modal analyses of Mid-Atlantic Ridge greenstones (circles, squares) compared with the model prediction from Figure 6 (plusses connected by lines, with seawater/rock ratio noted). Boxes enclose chlorite-quartz-poor and chlorite-quartz-rich rocks. Closed circles are modes as measured; open squares are modes corrected for the presence of up to 40 percent unaltered groundmass. Data are from Mottl (1983) and from Humphris and Thompson (1978).

cell and, since many of the rocks are pillow lavas, within the upper 1-2 km of crust. Thus, downwelling seawater must have been heated rapidly to 250-450°C within shallow levels of the crust along the Mid-Atlantic Ridge. The fact that the bulk of Mg uptake occurs within the greenschist facies implies that seawater reached these temperatures with little or no prior reaction.

This circumstance is undoubtedly one of the reasons that the experimental results for both solutions and solids so closely resemble the natural products, as this is precisely the manner in which most of the experiments were conducted: unaltered seawater and basalt were heated together essentially instantaneously to some high temperature and allowed to react. If this circumstance holds generally for mid-ocean ridge axial hydrothermal systems, it would have important implications for the behavior of chemical species which react differently at low vs. high temperature. These include sulfate, which will precipitate as anhydrite at high temperature unless first reduced to sulfide at lower temperature (McDuff and Edmond, 1982); oxygen isotopes, which fractionate much more at low than high temperature during water-rock interaction; and the alkali elements K, Li,and Rb plus Ba and B, all of which are taken up into secondary minerals at low temperature but leached into solution at high temperature. Whether seawater circulating through axial hydrothermal systems can generally be inferred to reach high temperatures with little prior reaction depends on how typical and representative are the greenstones dredged from fault scarps along slow-spreading ridges, as discussed in the introduction.

CONCLUSION

Because the laboratory experiments reacting seawater with basalt have been relatively successful in duplicating the product solutions and solids from natural hydrothermal systems along the mid-ocean ridge axis, they have provided considerable insight into the interpretation of these natural products. A major limitation to these interpretations continues to be a lack of convincingly representative and complementary samples of both altered solutions and altered solids from a single system. All of the hot springs sampled to date are from ridges in the Pacific spreading at intermediate to fast rates. The vast majority of hydrothermally altered rocks have been obtained from slow-spreading ridges in the Atlantic, and even these are questionable as to what they represent. These limitations should eventually be overcome by deep crustal drilling on young crust, possibly into an active axial hydrothermal system, along segments of ridge spreading at different rates.

ACKNOWLEDGEMENTS

I wish to thank Dr. Peter Rona for inviting me to participate and the NATO ARI for paying my expenses to attend. This work has benefitted from helpful discussion with many persons, but especially my friend and colleague Dr. William E. Seyfried, Jr. Writing of this paper was supported by grants OCE 80-24930 and 81-10913 from the U. S. National Science Foundation. This is WHOI contribution number 5251.

REFERENCES

Arnorsson, S. (1975) Application of the silica geothermometer in low temperature hydrothermal areas in Iceland, Am. Jour. Science 275, 763-784.

Arnorsson, S., Gronvold, K., and Sigurdsson, S. (1978) Aquifer chemistry of four high-temperature geothermal systems in Iceland, Geochim. Cosmochim. Acta 42, 523-536.

Bischoff, J. L. and Dickson, F. W. (1975) Seawater-basalt interaction at 200°C and 500 bars: implications for origin of sea-floor heavy-metal deposits and regulation of seawater chemistry, Earth Planet. Sci. Lett. 25, 385-397.

Bischoff, J. L. and Seyfried, W. E. (1978) Hydrothermal chemistry of seawater from 25° to 350°C, Am. Jour. Science 278, 838-860.

Bischoff, J. L., Radtke, A. S. and Rosenbauer, R. J. (1981) Hydrothermal alteration of graywacke by brine and seawater: roles of alteration and chloride complexing on metal solubilization at 200°C and 350°C, Econ. Geol. 76, 659-676.

Bjornsson, S., Arnorsson, S. and Tomasson, J. (1972) Economic evaluation of Reykjanes thermal brine area, Iceland, Am. Assoc. Petrol. Geol. Bulletin, 56, 2380-2391.

Cann, J. R. (1969) Spilites from the Carlsberg Ridge, Indian Ocean, Jour. Petrol. 10, 1-19.

Cann, J. R. (1979) Metamorphism in the Ocean Crust: In Talwani, M., Harrison, C. G., and Hayes, D. E. (eds.), Deep Drilling Results in the Atlantic Ocean: Ocean Crust, 230-238: Am. Geophys. Union, Washington, D.C., 431 p.

Corliss, J. G., Dymond, J., Gordon, L. I., Edmond, J. M., von Herzen, R. P., Ballard, R. D., Green, K., Williams, D., Bainbridge, A., Crane, K. and van Andel, Tj. H. (1979) Submarine thermal springs on the Galapagos rift, Science 203, 1073-1083.

Craig, H., Welham, J. A., Kim, K., Poreda, R., and Lupton, J. E. (1980) Geochemical studies of the 21°N EPR hydrothermal fluids (abst.), EOS 61, 992.

Crovisier, J. L., Thomassin, J. H., Juteau, T., Eberhart, J. P., Touray, J. C., and Baillif, P. (1983) Experimental seawater-basaltic glass interaction at 50°C: study of early developed phases by electron microscopy and x-ray photoelectron spectrometry, Geochim. Cosmochim. Acta, 47, 377-387.

Edmond, J. R., Measures, C., McDuff, R. E., Chan, L. H., Collier, R., Grant, B., Gordon, L. I., and Corliss, J. B. (1979) Ridge crest hydrothermal activity and the balances of major and minor elements in the ocean: the Galapagos data, Earth Planet. Sci. Lett. 46, 1-18.

Edmond, J. R. (1980) The chemistry of the 350°C hot springs at 21°N on the East Pacific Rise (abst.), EOS 61, 992.

Edmond, J. R., Von Damm, K., and McDuff, R. E. (1982) Chemistry of hot springs on the East Pacific Rise and their effluent dispersal, Nature 297, 187-191.

Ellis, A. J. (1970) Quantitative interpretation of chemical characteristics of hydrothermal systems. Geothermics Spec. Issue 2, 516-528.

Furnes, H. (1975) Experimental palagonitization of basaltic glass of varied composition, Contrib. Mineral. Petrol. 50, 105-113.

Hajash, A. (1975) Hydrothermal processes along mid-ocean ridges: an experimental investigation, Contrib. Mineral. Petrol. 53, 205-226.

Hajash, A. and Archer, P. (1980) Experimental seawater/basalt interactions: effects of cooling, Contrib. Mineral. Petrol. 75, 1-13.

Hajash, A., and Chandler, G. W. (1981) An experimental investigation of high-termapture interactions between seawater and rhyolite, andesite, basalt and peridotite, Contrib. Mineral. Petrol. 78, 240-254.

Humphris, S. E. and Thompson, G. (1978) Hydrothermal alteration of oceanic basalts by seawater, Geochim. Cosmochim. Acta 42, 107-125.

Janecky, D. R., and Seyfried, W. E., Jr. (1982) The solubility of magnesium hydroxide sulfate hydrate in seawater at elevated temperatures and pressures, submitted to Am. Jour. Science.

Krismannsdottir, H. (1976) Hydrothermal alteration of basaltic rocks in Icelandic geothermal areas, In: Proceedings, Second U.N. Symposium on the Development and Use of Geothermal Resources, v. 1, U. S. Gov't Printing Off., Wash., DC, p. 441-445.

MacDonald, K. C., Becker, K., Spiess, F. N., and Ballard, R. D. (1980) Hydrothermal heat flux of the "black smoker" vents on the East Pacific Rise, Earth Planet. Science Lett. 48, 1-7.

McDuff, R. E. and Edmond, J. M. (1982) On the fate of sulfate during hydrothermal circulation at mid-ocean ridges, Earth Planet. Sci. Lett. 57, 117-132.

Menzies, M. and Seyfried, W. E., Jr. (1979) Basalt-seawater interaction: trace element and strontium isotopic variations in experimentally altered glassy basalt. Earth and Plan. Sci. Lett. 44, 463-472.

Moody, J. B., Meyer, D., and Jenkins, J. E. (1983) Experimental characterization of the greenschist/amphibolite boundary in mafic systems. Amer. J. Sci. 283, 48-92.

Mottl, M. J., Corr, R. F. and Holland, H. D. (1975) Trace element content of the Reykjanes and Svartsengi thermal brines, Iceland (abstract). Abstracts with Programs, Geol. Soc. Amer. Ann. Meetings, 7, 1206-1207.

Mottl, M. J. and Holland, H. D. (1978) Chemical exchange during hydrothermal alteration of basalt by seawater - I. Experimental results for major and minor components of seawater, Geochim. Cosmochim. Acta, 42, 1103-1115.

Mottl, M. J., Holland, H. D., and Corr, R. F. (1979) Chemical exchange during hydrothermal alteration of basalt by seawater - II. Experimental results for Fe, Mn and sulfur species, Geochim. Cosmochim. Acta, 43, 869-884.

Mottl, M. J. and Seyfried, W. E. (1980) Sub-sea-floor hydrothermal systems: rock vs. seawater-dominated, In: Rona, P. A., and Lowell, R. P. (eds.), Seafloor Spreading Centers: Hydrothermal Systems, p. 66-82: Dowden, Hutchinson and Ross, Inc., Stroudsburg, Pennsylvania, 424 p.

Mottl, M. J. (1983) Metabasalts, axial hot springs, and the structure of hydrothermal systesm at mid-ocean ridges. Bull. Geol. Soc. Amer., 94, 161-180.

Olafsson, J., and Riley, J. P. (1978) Geochemical studies on the thermal brine from Reykjanes (Iceland), Chem. Geol. 21, 219-237.

RISE (1980) East Pacific Rise: hot springs and geophysical experiments, Science 207, 1421-1433.

Seyfried, W. E., Jr. (1977) Seawater-basalt interaction from 25° - 300°C and 1 500 bars: implications for the origin of submarine metal-bearing hydrothermal solutions and regulation of ocean chemistry, Ph.D. thesis, University of Southern California, 242 p.

Seyfried, W. E., and Bischoff, J. L. (1977) Hydrothermal transport of heavy metals by seawater: the role of seawater/basalt ratio, Earth Planet. Sci. Lett. 34, 71-77.

Seyfried, W. E., Jr., and Bischoff, J. L. (1979) Low temperature basalt alteration by seawater: an experimental study at 70°C and 150°C, Geochim. Cosmochim. Acta 43, 1937-1947.

Seyfried, W. E., Jr., and Bischoff, J. L. (1981) Experimental seawater-basalt interaction at 300°C and 500 bars: chemical exchange, secondary mineral formation and implications for the transport of heavy metals, Geochim. et Cosmochim. Acta 45, 135-147.

Seyfried, W. E., Jr., Gordon, P. C., and Dickson, F. W. (1979) A new reaction cell for hydrothermal solution equipment, Amer. Mineralogist 64, 646-649.

Seyfried, W. E., Jr., and Dibble, W. E. Jr. (1980) Seawater-peridotite interaction at 300°C and 500 bars: implications for the origin of oceanic serpentinites, Geochim. Cosmochim. Acta 44, 309-321.

Seyfried, W. E., Jr. Mottl, M. J., and Bischoff, J. L. (1978) Seawater-basalt ratio effects on the chemistry and mineralogy of spilites from the ocean floor, Nature 275, 211-213.

Seyfried, W. E., Jr. and Mottl, M. J. (1982) Hydrothermal alteration of basalt by seawater under seawater-dominated conditions, Geochim. Cosmochim. Acta 46, 985-1002.

Shanks, W. C. III, Bischoff, J. L., and Rosenbauer, R. J. (1981) Seawater sulfate reduction and sulfur isotope fractionation in basaltic systems: interation of seawater with fayalite and magnetite at 200-350°C, Geochim. Cosmochim. Acta 45, 1977-1995.

Thomassin, J. H. and Touray, J. C. (1979) Etude des premiers stades de l'interaction eau de mer - verre basaltique: doneés de la spectrométrie de photoélectrons (XPS) et de la microscopie électronique à balayage, Bull. Minéral. 102, 594-599.

Thomassin, J. H. and Touray, J. C. (1982) L'hydrotalcite, un hydroxycarbonate transitoire précocément formé lors de l'interaction verre basaltique - eau de mer, Bull. Minéral. 105, 312-319.

BASALT - SEAWATER INTERACTION

Geoffrey Thompson

Woods Hole Oceanographic Institution
Woods Hole, MA 02543

ABSTRACT

The upper part of the igneous oceanic basement consists of basaltic lava. This basalt undergoes reaction with seawater over a range of temperatures, time and locale. This reaction is a major source and sink for various ions in seawater, and is a major process in buffering seawater composition and forming metalliferous ores in the marine environment.

The nature of the chemical reaction and the fluxes of ions exchanged between the oceans and the igneous basement are mostly dependent on the temperature of reaction and the relative proportion of the reactants. These vary in respect to the location of the water circulation and distance from the heat source. Four examples of seawater-basalt interactions are considered and the net exchange fluxes are calculated; these examples cover the range of temperature and water:rock ratios typically found in the ocean floor.

Low temperature, high water:rock ratio is typical of the exchange in the upper few meters of the oceanic basement. Only about 0.1% of new oceanic crust undergoes this reaction which extends over a time period of tens of millions of years. The annual fluxes produced are not very large. Low temperature, low water:rock ratio is typical of the reaction in the deeper parts of the oceanic basement. About 8% of newly formed oceanic crust can be expected to undergo this reaction over a period of a few million years. Reactions and fluxes on the flanks of spreading centers are at moderate temperatures and water:rock ratios. These reactions are relatively short lived, but the fluxes produced are quite high. High temperature reaction of seawater and basalt (in excess of 100°C) takes place at spreading centre axes. These reactions are fast but result in very high fluxes and formation of polymetallic sulfides or iron and manganese oxides. The products of this reaction and the direction of exchange for some ionic species are quite different compared to the lower temperature reactions.

The net effect of the basalt-seawater exchange is the sum of all the reacton fluxes over the full temperature range. This calculated net flux indicates that the basalt is a source for ions such as Si, Ca, Ba, Li, Fe, Mn, Cu, Ni, Zn and hydrogen ions. It also is the sink for ions such as Mg, K, B, Rb, H_2O, Cs and U. The annual fluxes calculated for some of these species is of the same order of magnitude as the annual river influxes into the ocean.

INTRODUCTION

One of the fundamental problems in Chemical Oceanography is to explain the composition of seawater. It has long been recognised that the major inputs to the ocean come from the rivers. However, the composition of river water is very different from that of seawater - see Table 1. Moreover, it is not just a question of difference in total dissolved matter, the proportions of the different anions and cations are markedly different. As can be seen in Table 2, if river water continued to flow into the oceans without any chemical changes some elements would reach present seawater levels in a very short time and we would have a highly alkaline, bicarbonate lake. However, seawater has a remarkably constant composition, (at least in the ratios of the major constituents Na^+, Mg^{2+}, Ca^{2+}, K^+, Cl^-, SO^{2-}), and has maintained that composition, or very close to it, over geological time represented by the last 500 million years. Obviously ions must be removed or added to seawater as fast as they are supplied by the rivers, i.e. there are some sort of buffering reactions which maintain the oceans at steady state composition.

The past two decades have seen chemical oceanographers address this problem in a variety of ways. Essentially, the concepts invoked include a chemical equilibria model, a steady state model or a kinetic model. Sillen (1961, 1967) has been most influential in addressing this problem. He suggested an equilibrium model in which the control on the composition of seawater came from chemical equilibria between seawater and components of

Table 1. Major constituents of river and seawater

Constituents	Seawater[1] ppm	River Water[2] ppm
Cl^-	19,350	8.1
Na^+	10,760	6.9
Mg^{2+}	1,294	3.9
SO_4^{2-}	2,712	10.6
K^+	399	2.1
Ca^{2+}	412	15.0
HCO_3^-	145	55.9
SiO_2	6	13.1
NO_3^-	--	1
Fe^{2+}	--	0.67
Al	0.001	0.01
Br^-	67	--
CO_3^{2-}	18	--
Sr^{2+}	7.9	--
B	4.6	--
F	1.3	--
Dissolved Org. C	0.5	9.6

[1]Data from Pytkowicz and Kester (1971) and Goldberg (1957) based on chlorinity ratio x 19.35.

[2]Data from Gibbs (1972) and Livingstone (1963).

Table 2. Mass of major dissolved constituents* delivered to oceans annually and times for these constituents to reach oceanic amounts (from Garrels and Mackenzie, 1971).

Constituent	Mass delivered by rivers to ocean annually (units of 10^{14} g)	Mass in ocean (units of 10^{20} g)	Storage time - time for river fluxes to attain oceanic amounts (units of 10^6 years)
Fe^{2+}	0.223	0.0000137	0.00006
Al	0.003	0.0000137	0.0046
SiO_2	4.26	0.08	0.02
HCO_3^-	19.02	1.9	0.1
Ca^{2+}	4.88	6	1.24
K^+	0.74	5	8
SO_4^{2-}	3.67	37	10.7
Mg^{2+}	1.33	19	15.4
Na^+	2.07	144	108
Cl^-	2.54	261	230
Diss. org. C	3.2	0.007	0.0002
H_2O	325,000	13,550	0.042

*corrected for atmospheric cycling.

oceanic sediments. Sillen's work spurred further evaluation and discussion by such workers as Holland (1965, 1972), Mackenzie and Garrels (1966), and Garrels and Mackenzie (1971) who all suggested a steady state model with supply and removal being balanced over geologic time. Like Sillen's model, their model invoked reactions between seawater and sedimentary components as the buffering reactions. In particular they pointed out the importance of reverse weathering, i.e. transformation of amorphous aluminium silicate to a clay mineral and carbon dioxide. Consumption of bicarbonate and alkali metals and production of hydrogen ions are the important requirement in the overall mass balance and are needed to counterbalance primary weathering on the continents. Broecker (1971) has shown that kinetic regulation by the rate of supply of individual components and interaction between biological cycles and mixing cycles is an important concept and accounts for some of the present distribution of chemical properties in the ocean.

In general the steady state model of the oceans has gained wide acceptance (Siever, 1968; Garrels and Mackenzie, 1971); the various controls on that steady state have been discussed by Stumm and Morgan (1981). For the time scale of 10^4 to 10^9 years this model and approach are best suited for consideration of global geochemical budgets and mass balances as suggested by Mackenzie and Garrels (1966). In recent years attempts to measure directly some of the chemical reactions suggested to be important by Sillen (1967) and Mackenzie and Garrels (1966) have not been successful (e.g. Russel, 1970; Drever, 1971).

The steady state model depends on identification of the major reservoirs, and fluxes between those reservoirs. In the past the main reservoirs considered have been the oceans, the continents, the atmosphere, and oceanic sediments. For certain components, particularly volatiles such as H_2O, CO_2, HCl, input from the earth's mantle needed to be invoked for mass balance calculations (Rubey, 1951). Even in this context it has not been possible to make mass balances for some of the components. In particular Drever (1974) showed a major imbalance for magnesium. In this instance Drever showed that, in considering the then known mechanisms for magnesium removal, only about half of the magnesium input could be accounted for - see Table 3. An imbalance such as this implies that: (1) the inputs of magnesium have changed drastically with time; or (2) seawater composition has changed. Given (1) and (2) do not seem likely, then we must infer that there is another reaction, or sink, controlling magnesium in the ocean that we have not recognised.

Table 3. Mg^{2+} removal from the oceans (from Drever, 1974)

Process	Estimated Mg^{2+} removal ($\times 10^{14}$ g)	% of River Flux
Carbonate Formation	0.075	6
Ion Exchange	0.097	8
Glauconite Formation	0.039	3
Mg-Fe Exchange	0.29	24
Burial of Interstitial Water	0.11	9
	0.61	50

The present day concept of plate tectonics had its roots in studies of the seafloor and the recognition of generation of new oceanic crust along the spreading or constructional margins of the plates. The surface expression of that new crust is basaltic lava. Studies of the present day oceanic spreading centers have shown that seawater and the basaltic lavas undergo reaction (e.g. Deffeyes, 1970; Hart, 1971; Corliss, 1971; Hart, 1973; Thompson, 1973; Spooner and Fyfe, 1973; Wolery and Sleep, 1976; and references therein). Indeed, the past decade has seen a growing interest and research in the idea that the basalt-seawater interaction is an important process and a potential source and sink in the chemical exchanges that control seawater composition. In this context, and the recognition that seafloor is subsequently consumed at the subduction zones, the earth's mantle becomes a major reservoir to be considered in global geochemistry.

In this paper we will examine the potential of the seawater-basalt reaction as a major source and sink for various ions in seawater. A variety of observations and measurements relevant to the extent, magnitude and directions of exchange during seawater-oceanic basement reaction will be reviewed. In the light of these observations and evidence, some of the major controls influencing the fluxes will be considered. It is important to realise that this process is not yet fully understood or the fluxes well known. In one sense it is a little premature to present a review of an active ongoing area of research in chemical oceanography. Nevertheless the implication of the seawater-oceanic basement reaction, and subsequent rates of exchange of ions, is far reaching and deserves close scrutiny. It is important to realize that basalt-seawater reaction takes place over a wide range of temperatures with reaction and exchange being very much dependent on this parameter. Calculation of fluxes must consider the full range of reactions and temperature before final budgets and mass balances can be put together.

DEDUCTIONS FROM OBSERVATIONS OF SEAFLOOR ROCKS

1. Low Temperature Alteration

Dredging and drilling of the oceanic crust has clearly shown the ubiquitous presence of basalts that have undergone some reaction with seawater. The majority of these rocks are generally referred to as 'weathered' basalts i.e. they have interacted with seawater at temperatures close to those of bottom waters. This is particularly true of the dredged basalts. They are generally characterized by only small changes in mineralogy from the original igneous precursor, with palagonitization of the glass, partial replacement of the igneous minerals by smectites, and oxidation of the ferrous oxides being the major effects. However, more complex alteration with formation of zeolites such as phillipsite and celadonite is often found, particularly in the drilled basalts (e.g. Bass et al., 1973). Honnorez (1981), in a review of low temperature alteration of oceanic basalts, suggests 'weathering' should include all interaction with seawater up to, but not including, the temperatures where unequivocally metamorphic minerals such as laumontite, wairakite or chlorite are formed. Cann (1979) has also suggested a range of temperatures of interaction might be applicable for so-called weathered basalts. He suggests the term brownstone facies for this range, with zeolite, greenstone and other facies resulting from increasing temperatures of basalt-seawater interactions - see figure 1. Thus, low temperature alteration should include surface weathering (0-3°C) at bottom water temperatures, through diagenesis, and possibly up to interactions that took place around 50-100°C. It is interesting to note in this context, that experimentally the results reacting basalt and seawater at 70°C differed markedly from reactions at 150°C (Seyfried and Bischoff, 1979). The former having results more in concordance with effects seen for low temperature altered basalts i.e. uptake of alkalis by the rock rather than leaching of the alkali metals.

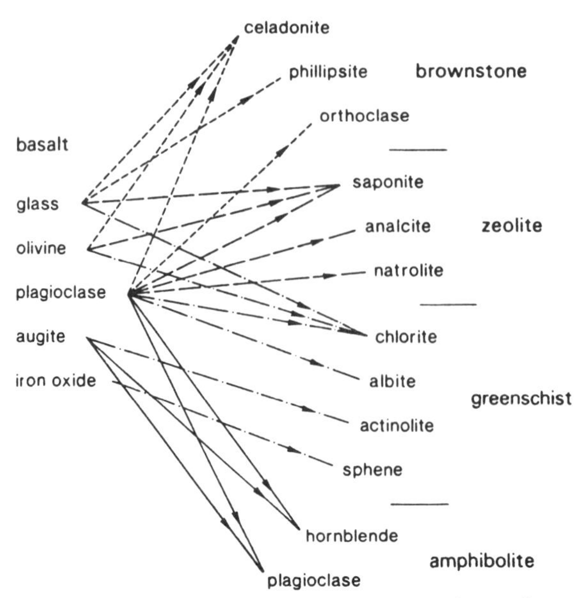

Fig. 1. Metamorphic facies and congruent mineral reactions for basalt-seawater reactions over a range of temperatures (from Cann, 1979).

Observations of low temperature alteration of oceanic basalts have been well covered in an extensive review by Honnorez (1981), and will not be covered in detail here. Most of the workers who have studied the alteration of oceanic basalts have shown that submarine glass alters differently and at different rates to the bulk holocrystalline rock. Honnorez (1981) based on the observations of such workers as Nicholls and Bowen (1961), Bonatti (1965, 1970), Moore (1966), Honnorez (1967, 1972), Hay and Ijiima (1968a,b), Jakobsson (1972, 1978), Thompson (1973), Melson (1973), Melson and Thompson (1973), Andrews (1978), concluded that submarine alteration of basaltic glass at low temperatures resulted in the formation of phillipsite, smectites and Fe-Mn oxides, and occurs in three stages. The initial stage is characterized by Na > K phillipsite, and K-Mg-rich smectite. The bulk glass is hydrated and oxidized and enriched in K, Na, and Mg but depleted in Ca. The second (mature) stage has similar mineralogy but the phillipsite is now K > Na, and Fe-Mg saponite or nontronite with high concentrations of K is common. The third (final) stage has Ca poor K > Na phillipsite, K-Mg-Fe smectite and Fe-Mn hydrous oxides. Uptake from seawater is K > Mn > Na, with Ca, Mg and Si released from the rock. Some workers, for example Hart (1969), Philpotts et al. (1969), Thompson (1973), Frey et al. (1974), Ludden and Thompson (1978, 1979), have shown that alkali metals such as Rb, Cs, Li, in addition to elements such as B and light rare earth elements, are also taken up by the palagonitized glass.

For dredged basalts, Wiseman (1937) first noted some marked chemical changes, notably oxidation of iron accompanying uptake of water. Other workers, e.g. Hart (1970, 1973), Hekinian (1971), Thompson (1973, and Stakes and Scheidegger (1981), suggested that progressive alteration of basalt was seen with distance from mid-ocean ridge, i.e. age. A number of workers have studied dredged basalts with a view to observing the chemical exchanges that have taken place e.g. Matthews (1971), Hart (1969, 1971), Paster (1968), Hart and Nalwalk (1970), Miyashiro et al. (1969), Miyashiro (1973), Thompson (1973), Hart (1970, 1973), Shido et al. (1974), Scott and Hajash (1976), Muehlenbachs and Clayton (1972, 1976), and Thompson and Humphris (1977). A detailed summary of many of these observations concerning chemical exchanges are tabulated in Honnorez (1981). In general these workers found oxidation and hydration of the basalt to be ubiquitous, uptake of K, Cs, Rb, B, Li and O^{18} is most common, and is accompanied by loss of Ca, Mg, and Si. Elements such as Fe, Mn, Na, Cu, Ba and Sr show variable effects, either gain or loss from the rock, and generally Al, Ti, Y, Zr and heavy REE show little or no effects.

More recently, studies of drilled basalts recovered during the DSDP Program from depths up to 500 m in the oceanic basement, have been reported on. Observations by such workers as Bass et al. (1973), Bass (1975, 1976), Andrews (1977, 1978), Robinson et al. (1977), Sommer and Ailin-Pyzik (1978), Donnelly et al. (1979a,b,c), Honnorez et al. (1978), Baragar et al. (1979), Böhlke et al. (1980), Hart and Staudigel (1978), Staudigel et al. (1979), and Mevel (1979), have shown that alteration deep in the crust can differ somewhat from that observed in dredged surface basalts. Most of these different effects are characterized by reactions under low water:rock ratios, and non-oxidative conditions. Although uptake of alkalis seems ubiquitous, changes in Ca and Mg appear variable. The upper few meters of the drilled holes are apparently altered under oxidative conditions and are generally similar to dredged basalts in their mineralogical and chemical changes. Honnorez (1981) has discussed in detail the results and observations of low temperature alteration for five deep sea drilling sites, ranging in age from 1.5 to 110 million years, and indicated the apparent diversity of the reactions from site to site, or within a given site.

In the context of determining the chemical exchanges taking place during low temperature reaction of basalt and seawater, and particularly the rates and the fluxes of elements, observations from drilled basalts are very important. Studies of deep sections of drill holes give information on the nature of circulation deep in the crust, and reactions at low water:rock ratios. Dredged basalts, and the upper few meters of the drill sites, give indication of reactions at high water:rock ratios. Calculations of fluxes are given in the more detailed case histories that follow this section.

Studies of low temperature altered basalts have shown that changes in isotopes such as $^{87}Sr/^{86}Sr$ (e.g. Chapman and Spooner, 1977; Spooner et al., 1977; McCulloch et al., 1980) or ^{18}O (e.g. Garlick and Dymond, 1970; Muehlenbachs and Clayton, 1972, 1976) are sensitive indicators of seawater-rock interaction. Drilled basalts show that such interactions have taken place to at least depths of 500 m in the oceanic crust. Hart and Staudigel (1978), Richardson et al. (1980), and Staudigel et al. (1981), in a study of the vein material in altered basalts from drill holes ranging in age from 3.5 to 110 million years in the Atlantic, have concluded that most reaction in the crust takes place in a fairly short time period (probably within 3 million years). This observation is based on the $^{87}Sr/^{86}Sr$ ratios in smectites, celadonites, analcites and carbonates in the vein materials, and the known variation of seawater $^{87}Sr/^{86}Sr$ with time. They have suggested that alteration proceeds in four stages: Stage I, formation of palagonites. Stage II, formation of smectites. These stages take place within the first 3 million years of crust formation and produce the major chemical fluxes. Stage III, formation of carbonates, probably has a lifespan of about 10 million years and involves principally contribution of Ca and minor Sr to circulating seawater with concomitant carbonate precipitation. Stage IV, compaction and dehydration of the crust, does not result in chemical changes but is principally a dehydration of very hydrous phases. These observations and conclusions have only been inferred for Atlantic-type crustal regimes i.e. plates with a low spreading rate.

2. High Temperature Alteration

In addition to the other geological and geophysical evidence that hydrothermal circulation takes place in the oceanic crust, the recovery of metamorphosed rocks, ranging from zeolite to amphibolite facies, is yet further confirmation that seawater-rock interaction is widespread. Such rocks were first recovered during the Challenger Expedition (Murray and Renard, 1891) and the first dredge hauls from the Mid-Atlantic Ridge (Shand, 1949; Quon and Ehrlers, 1963). Melson et al. (1966), Melson and Van Andel (1966), Melson et al. (1968), and Cann (1969) showed that these rocks had undergone major chemical, as well as mineralogical changes, compared to their precursors. Deffeyes (1970) suggested this might be evidence of an important reaction that affected seawater composition. Humphris and Thompson (1978a) have listed many of these observations of metamorphosed rocks (see Table 4), and noted that they have been recovered predominantly from the axial valley of the slower spreading mid-ocean ridges and the large fault scarps of transform faults. They are not abundant in DSDP drill holes (Cann, 1980) nor in the spreading centers of fast spreading plates. By far the most abundant of metamorphosed basalts recovered in dredge hauls are greenstones, i.e. metabasalts of the greenschist facies. Since the observations listed in Table 4, reports on metamorphosed gabbros, as well as some metabasalts, have been made by Miyashiro et al. (1979), Ito (1979), Tiezzi and Scott (1980), Scott et al. (1981), Stakes and O'Neil (1982). Studies of vein fillings in metabasalts have been made by Rona et al. (1980), and Delaney et al. (1980, 1981).

Humphris and Thompson (1978a, b) concluded that greenschist facies metamorphism of oceanic basalts resulted from hydrothermal circulation and reaction with seawater. Resulting mineralogical and chemical changes include:

(1) Plagioclase ⟶ albite ⟶ chlorite
⟶ albite + epidote
Plagioclase + pyroxene ⟶ chlorite and epidote
Olivine ⟶ chlorite (+ pyrite)
Pyroxene ⟶ actinolite
Glass ⟶ chlorite-actinolite intergrowth + chlorite
Vein minerals include chlorite, actinolite, epidote, quartz, pyrite and occasionally Cu-Fe-Zn sulphides.

(2) Mg is taken up by the basalt and Ca is leached in approximately a 1:1 molar basis. The Mg fluxes range from 1 to 10g MgO uptake per 100 cm^3 of rock. K, Si, B, and Li are generally leached from the rock but may be reprecipitated locally.

Table 4. Metabasalts dredged from active oceanic ridges and their reported mineralogies (from Humphris and Thompson, 1978a)

Metamorphic Facies	Reported Minerals	References
Zeolite	Analcite, stilbite, heulandite, natrolite-mesolite-scolectite series, mixed layer chlorite-smectite	1
Prehnite-pumpellyite	Prehnite, chlorite, calcite, epidote	2
Greenschist	Albite, actinolite, chlorite, epidote, quartz, sphene, hornblende, tremolite, talc, magnetite, nontronite	3
Amphibolite	Hornblende, plagioclase, actinolite, leucoxene, quartz, chlorite, apatite, biotite, epidote, magnetite, sphene	4

[1]Aumento et al. (1971); Miyashiro et al. (1971); Shido et al. (1974)

[2]Hekinian and Aumento (1973); Mevel (1981)

[3]Melson et al. (1968); Miyashiro et al. (1971); Aumento et al. (1971); Shido et al. (1974); Aumento and Loncarevic (1969); Melson et al. (1966); Melson and Thompson (1971); Bonatti et al. (1975); Thompson and Melson (1972); Cann (1971); Hekinian (1968); Chernyseva (1971)

[4]Cann and Funnell (1967); Cann (1971); Aumento et al. (1971); Bonatti et al. (1971); Bonatti et al. (1975); Bogdanov and Ploshko (1968); Rozanova and Baturin (1971).

(3) Gains or losses in other elements such as Na, K, Fe, Mn, Sr, Ba, Co, Cr and Ni varied according to the mineral assemblage, particularly the relative amounts of chlorite-epidote-pyrite. Although only samples from the median valley were considered in this study, some retrograde reactions (low temperature alteration) may have caused subsequent uptake of K, Na, B, Li and Rb.

Mottl (1982) has extended these studies of Humphris and Thompson and looked at a wider range of compositions represented by dredged greenstones. In this excellent and important paper, Mottl (1982) has shown that the mineralogy and composition of the greenstones can be predicted by laboratory experiments. Because of this Mottl has also been able to assess relevant information such as temperature and water:rock ratio of the hydrothermal system that caused the metamorphism. In the majority of cases, the chemical changes that occurred in the dredged metabasalts, and in agreement with the experimental data, include (1) uptake of Mg and loss of Ca on a mole for mole basis; (2) uptake of Na at water:rock ratios of less than 10, but loss of Na at higher water:rock ratios; (3) loss of K, Si, and Mn. Changes not predicted by the experiments, but seen in the rocks, include uptake or loss of Fe, with a positive correlation between Fe, Mn and Mg fluxes, (apparently due to uptake of Fe in chlorite), and uptake of Na in excess of that observed in the experiments. Mottl (1982) suggests these effects resulted from local redistribution of the elements via diffusion. Most of the changes are consistent with reaction of basalt and previously unreacted, heated seawater such as one might expect in a downwelling limb of a hydrothermal system.

Mottl (1982) also very elegantly combined the experimental data concerning the mineralogy of the solid products, with the observed mineralogy of dredged metabasalts. His model, see figure 2, suggests the following assemblages with increasing seawater:rock ratio:

Ratio	Assemblage
0-2:	chlorite + albite + epidote + actinolite
2-35:	chlorite + albite + epidote + actinolite + quartz
35-50:	chlorite + albite + quartz
>50:	chlorite + quartz

The model assumes that any anhydrite formed has subsequently been redissolved. Excess SiO_2 is not present at low water:rock ratios, but as the chlorite content increases, due to increasing Mg uptake, more SiO_2 is released to make quartz.

Mottl (1982) noted that, based on their mineralogy, dredged metabasalts fall into two groups: 1) chlorite-quartz-poor rocks altered at water rock ratios of < 10 (this group contains most of the metabasalts described by Humphris and Thompson who subdivided them into two groups depending on the epidote content). This group forms the bulk of the recovered rocks and can be re-

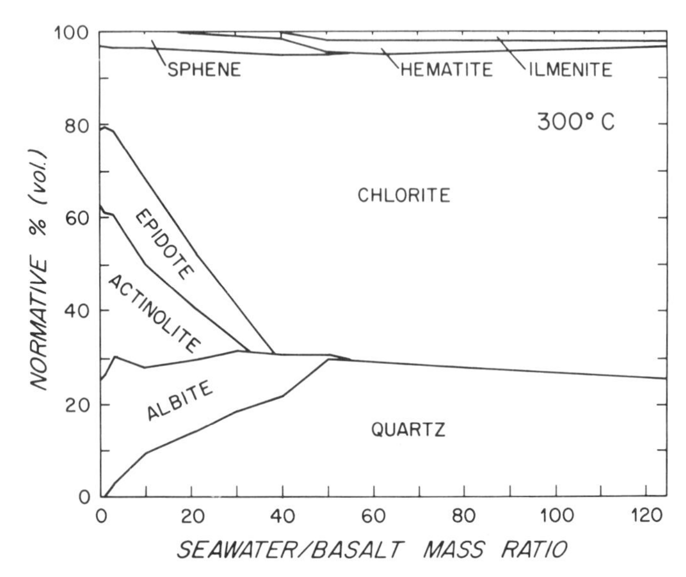

Fig. 2. Model predicting mineral assemblages and proportions produced when basalt reacts with seawater in different water/rock mass ratios (from Mottl, 1982). The model is based on experimental data but is close to actual observed assemblages in recovered greenschist facies metabasalts. See Mottl (1982) for discussion of details.

lated to the experimental reaction of heated seawater and basalt at low water:rock ratios. 2) The second group of metabasalts are chlorite-quartz-rich rocks altered at water:rock ratios of 10 to 50. Examples of these two groups can often be seen within a single pillow, as shown by Humphris and Thompson (1978a), where alteration presumably occurred at different seawater:rock ratios from rim to core. The distinction between chlorite-quartz-rich rim and chlorite-quartz poor core is identical to the distinction between hyalospilite and orthospilites seen by Cann (1969) and suggested by Seyfried et al. (1978) - see Table 5. Mottl further notes that within this second group there is a bimodal distribution of the composition of the chlorite. The chlorites are generally more Mg rich than chlorites from the chlorite-quartz-poor

Table 5. Gains (+) and losses (-) of chemical components in core (orthospilite) and rind (hyalospilite) of metabasalt pillow compared to the fresh precursor, and similar changes in basalt glass experimentally altered at water: rock ratios of 10 and 62 respectively. (from Cann, 1969, and Seyfried et al., 1978).

Component	Ortho-spilites	Basalt glass w/r = 10	Hyalo-spilites	Basalt glass w/r = 62
SiO_2	+7.02	-1.63	-7.77	-3.39
Al_2O_3	0	0	0	0
Fe_2O_3	+0.19	-0.10	+10.78	+0.46
MgO	+1.15	+1.97	+4.98	+9.46
CaO	-2.16	-2.92	-8.93	-9.32
Na_2O	+1.88	+0.13	-1.14	-1.61
K_2O	-0.06	-0.10	-0.08	-0.08
H_2O^+	+2.83	+6.18	+7.76	+11.38
Mineralogy:	Albite, chlorite, actinolite, pyrite, quartz, sphene.	Albite, saponite, chlorite-saponite, actinolite, anhydrite, quartz, pyrite	Chlorite	Chlorite smectite, anhydrite.

group, due to extensive uptake of Mg from seawater as predicted for higher water:rock ratios. Other quartz-chlorite rocks, however, contain Fe-rich chlorites - see figure 3. These Fe-chlorite-quartz-rich rocks also contain minor pyrite, chalcopyrite and pyrrhotite. Chemically they are enriched in Fe, Mn, Cu, Zn, CO_2 and depleted in Mg, Ti, P, V, Y, Zr compared to other basalts. They are believed to have formed as vein infillings, in an upflow zone of the hydrothermal system, from a highly reacted seawater solution rich in Fe, Mn, Cu, Zn, H_2S and SiO_2 and poor in Mg.

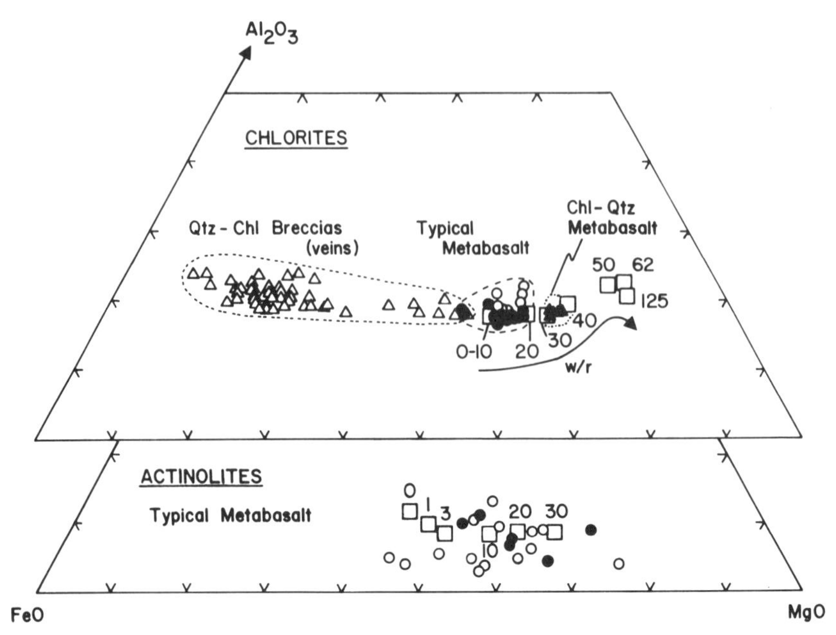

Fig. 3. Composition of chlorites and actinolites from seafloor greenstones compared with compositions inferred from a normative model based on experimental data (from Mottl, 1982). Most common metabasalts recovered are depicted by open and closed circles and closed triangles, and are chlorite-quartz poor or chlorite-quartz-rich varieties. Chlorites from quartz-chlorite vein filling breccias are more Fe-rich and are depicted as open triangles. The open squares represent the model predictions for various seawater/rock ratios. See Mottl (1982) for discussion of details.

Thus studies of rocks that have undergone reaction with heated seawater can provide data on the nature of that reaction. To date most studies have been done on metabasalts of the greenschist facies. Very few recoveries of rocks from other facies have been described. If this is typical of hydrothermal reactions within the oceanic crust it would suggest rapid heating of seawater to temperatures in the range 200-400°C in the upper part of the crust. The paucity of zeolite facies rocks suggests the temperature transition must be relatively sharp.

In Table 6 I have summarized the various mineral assemblages

Table 6. Metamorphic facies and reported minerals observed in each facies

Facies	Mineralogy
Halmyrolysis	Celadonite, phillipsite, palagonite, saponite, montmorillonite, nontronite, Fe-Mn hydroxide orthoclase
Zeolite	Analcite, stilbite, heulandite, natrolite-mesolite-scolectite, chlorite-smectite, saponite
Prehnite-pumpellyite	Prehnite, chlorite, calcite, laumontite, epidote
Greenschist	Albite, actinolite, chlorite, epidote, quartz, sphene, hornblende, tremolite, talc, magnetite, nontronite
Amphibolite	Hornblende, plagioclase, actinolite, leucoxene, quartz, chlorite, apatite, biotite, epidote, magnetite, sphene

resulting from seawater-basalt reaction at increasing temperatures. In Table 7 I have summarized the chemical exchanges that most commonly occur at low and high temperature reactions.

ESTIMATES OF HYDROTHERMAL FLUXES

In the previous sections evidence was presented that seawater-igneous oceanic basement reactions were widespread, and chemical exchange, heat extraction, and formation of new products occurred. The key question, from the viewpoint of chemical oceanography, is what is the magnitude and rate of the chemical exchange compared to other inputs and outputs that affect seawater composition. In considering the various lines of evidence of seawater-rock interactions, it was noted that the experimental observations often closely predicted those observed in the field. In the experiments it was clear that temperature and water:rock ratio were the principal factors influencing the net fluxes in the reactions.

Thus, in this section of the review I have tried to estimate the fluxes as a function of these two parameters. To do so I have chosen to describe four combinations of these parameters that I believe are critical in governing the direction and extent of exchange and, equally as important, have been observed in the oceans and may be typical of large regions of the oceanic lithosphere. The cases considered include (a) low temperature, high

Table 7. Summary of observed chemical gains and losses from basaltic rock during seawater reactions at different temperatures. Species in perentheses do not always show the loss or gain as indicated and are probably dependent on other parameters such as oxidation, rock-water ratio, presence of sulphides or other phases.

CHEMICAL EXCHANGE

Rock Gains		Rock Losses	
	Low Temperature (< 100°C)		
H_2O		Si	
K		Ca	
P		Mg	
Mn		(Na)	
(Fe)			
B	U	(Sr)	
Li	Cu		
Rb	Zn		
Cs	LREE		
	(Ba)		
	High Temperature (> 150°C)		
H_2O		Si	
Mg		Ca	
(S)		K	
		Mn	
		(Fe)	
		B	Sr
		Li	(Cu)
		Rb	(Ni)
		Cs	(Zn)
		Ba	(U)

water:rock - probably typical of the basement surface rocks over all the ocean floor; (b) low temperature, low water:rock - probably typical of the deeper oceanic basement; (c) medium temperature, medium water:rock - probably typical of flanks of mid-ocean ridges and possibly of basement topographic highs; (d) high temperature, low water:rock - probably typical of mid-ocean ridge axes.

Fluxes from Basement Surface Rocks - Low Temperature, High Water:Rock.

Basalts recovered from the oceanic basement surface by dredging or drilling show signs of having interacted with seawater at low temperature, in many cases presumably that of ambient bottom seawater. A number of workers have reported on the effects and chemical exchanges, (see Honnorez, 1981 for an extensive review), but only a few have calculated actual fluxes. Hart (1970), based on analyses of 112 basalts reported in the literature, showed changes in chemical composition of the basalts with

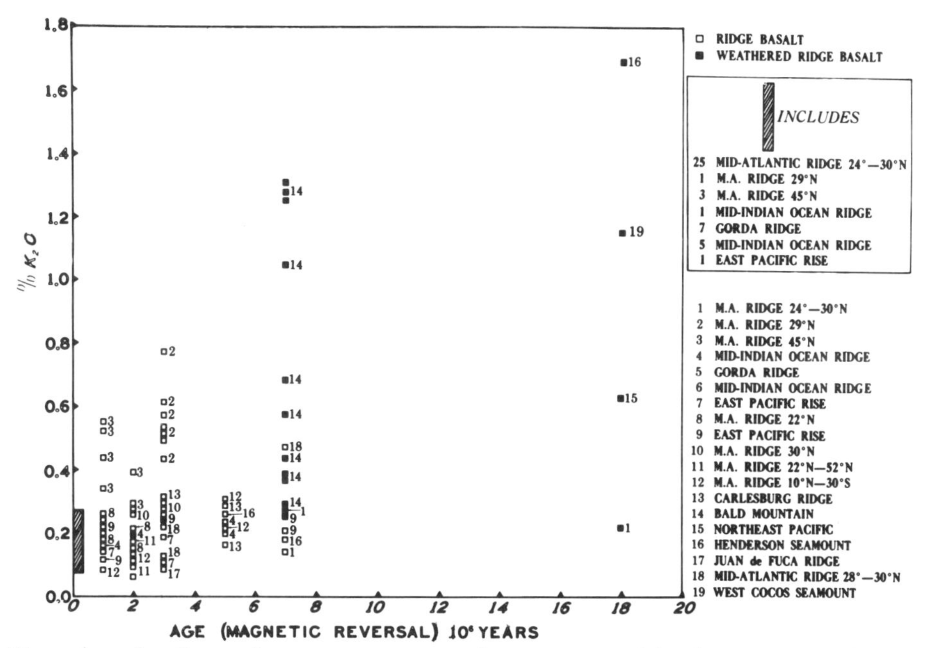

Fig. 4. A plot of percent potassium vs. residual magnetic field reversal age for 112 analyses of ridge basalts (from Hart, 1970).

distance from ridge spreading centers. Although the scatter in the data is large, and information is only available up to 18 million years, linear trends could be fitted, e.g. fig. 4. Assuming constant Al, and an average tholeiite composition as the starting material, estimated annual changes in mass were calculated by Hart - see column 8 in Table 8. Later, Hart (1973) refined his calculations by measuring the composition of 7 weathered basalts recovered by dredging or from the upper parts of drilled holes, and measured the density, and compressional wave velocity of each. The change in chemical composition was plotted as a function of compressional wave velocity. Further, Hart noted that compressional wave velocities of the upper part of the oceanic basement decreased as a function of age - see figure 5. Assuming this is the result of seawater-rock interaction affecting the overall density, then the rate of change of seismic velocities of layer II divided by the rate of change of the major oxides with seismic velocity equals the rate of change of the major oxides during seawater alteration of the upper part of the oceanic basement at low temperatures. The results of Hart's calculation for some of the major elements are shown in column 7 of Table 8.

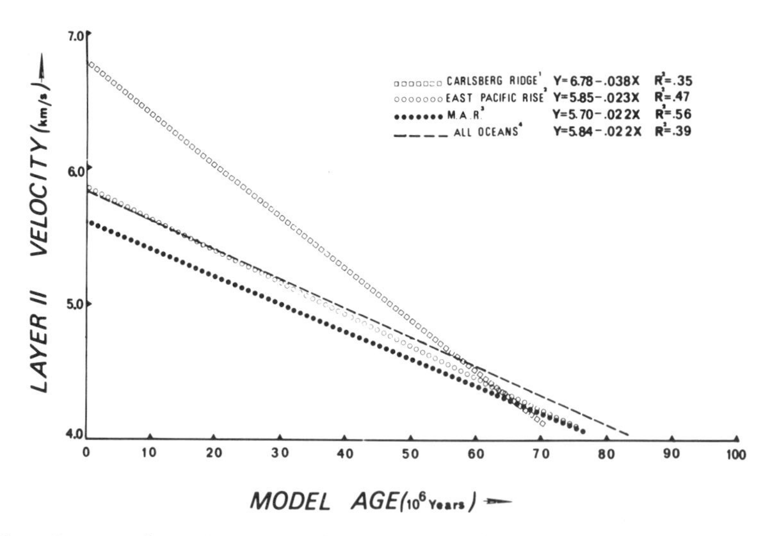

Fig. 5. Layer 2 seismic refraction velocities as a function of age for three mid-ocean ridge systems and an average for all three systems. For data sources and discussion, see Hart (1973).

Table 8

Comparison of different estimates of chemical exchange during basalt weathering in units of 10^{-9}gm/cm^3/year and the calculated elemental fluxes in units of 10^{14}g/yr.

Gained	1	2	3	4	5	6	7	8	Range	Average	Flux*(10^{14}g/yr)	
H_2O	+3.0	+2.6	+1.0	+1.0	+6.1	+3.0	+2.0	+3.7	+1.0–6.1	+2.8	H_2O	+0.066
K_2O	+0.8	+0.22	+0.6	+0.5	+1.2	+0.5	+0.5	+1.1	+0.2–1.2	+0.68	K	+0.013
Lost												
SiO_2	-7.3	-3.3	-6.3	-3.2	-11.5	-2.7	-2.2	-7.4	-2.2–11.5	-5.5	Si	-0.064
CaO	-4.0	-2.5	-2.1	-2.2	-2.7	-3.1	-2.0	-2.9	-2.0–4.6	-2.7	Ca	-0.045
MgO	-3.8	-1.2	-1.5	-0.4	-0.9	-1.1	-1.5	-3.8	-0.4–3.8	-1.8	Mg	-0.025
Na_2O	-0.7	-0.2	-0.1	-0.1	-0.4	-0.2	-0.3	+0.3	-0.1–0.7	-0.3	Na	-0.005

Col. 1. From Thompson et al. (in preparation) - based on analyses of 30 dredged samples from 0 to 57 million years old.
Col. 2 From Thompson et al. (in preparation) - based on detailed analysis of a 30 million year old pillow basalt.
Col. 3 As col. 2 for a 35 million year old pillow.
Col. 4 As col. 2 for a 57 million year old pillow.
Col. 5 From Thompson (1973) - DSDP hole 10 basalt, age 16 million years.
Col. 6 From Thompson (1973) - dredged basalt 80 million years old - based on data from Matthews (1971).
Col. 7 From Hart (1973) - based on analyses of 7 basalts and change in seismic velocity of oceanic crust and samples.
Col. 8 From Hart (1970) - based on analyses of 112 dredged basalts from 0 to 18 million years old.

* Assuming average annual exchange continues for 80 my and to a depth of 10 meters (see text for discussion).

Thompson (1973) presented analyses of a number of dredged samples and drilled basalt that had undergone alteration by seawater. The change in composition relative to the unaltered precursor was measured, and the annual mass exchange was calculated from the age of the sample. Column 5, Table 8 shows the exchange calculated for the drilled sample, and column 6 Table 8 the data for a dredged sample based on analytical data from Matthews (1971). Later, in an attempt to do a more definitive and rigorous study of the change of composition of seafloor basalts, a study was made of a series of dredged basalts ranging in age from zero to 57 million years (Thompson and Rivers, 1976; Henrichs and Thompson, 1976; Thompson and Humphris, 1977; Ludden and Thompson, 1978, 1979; and Thompson et al., in preparation). In this study a series of basalts, dredged from the ridge axis and at subsequent exposures out to over 750 km on the flank of the ridge, were studied. Detailed analyses were made of pillows from the margins through the interiors, as well as whole composite samples of sections through the individual pillows, so that mass balances could be computed. As figure 6 shows, the data are comparable whether the analytical data are normalized to constant volume, constant aluminum or constant titanium. Columns 2, 3 and 4 of Table 8 show the annual mass exchange for individual pillows of different ages, following the technique of Rivers (1976), but recalculated for the actual change in each pillow rather than against an average tholeiite. Thompson et al. (in preparation) also analyzed over 30 pillow basalts, including the outer 1 cm margins and the whole of the interiors. Based on change in composition as a function of age, they noted two important points. By 4 to 5 million years most margins had completely altered with respect to most major and trace components; but the interiors of the basalts continued to change in composition, becoming increasingly more altered up to 57 million years - see figure 7. Regression calculation of normalized composition (to constant volume corrected for change in density) as a function of age gives an annual mass exchange. Column 1 Table 8 shows the calculation for major elements, and Table 9 the data for trace elements. The mass exchange for the trace elements are shown for interiors (from the slope of normalized composition vs. time) and for margins (based on average composition of margins after 5 million years).

The remarkable thing about Table 8 is that all these calculations for annual mass exchange, whether based on individual pillows, or integrated data for many samples, or from analogy with change in physical properties of the upper oceanic crust, give values for the major components that are within an order of magnitude. The range and average for the annual mass exchange

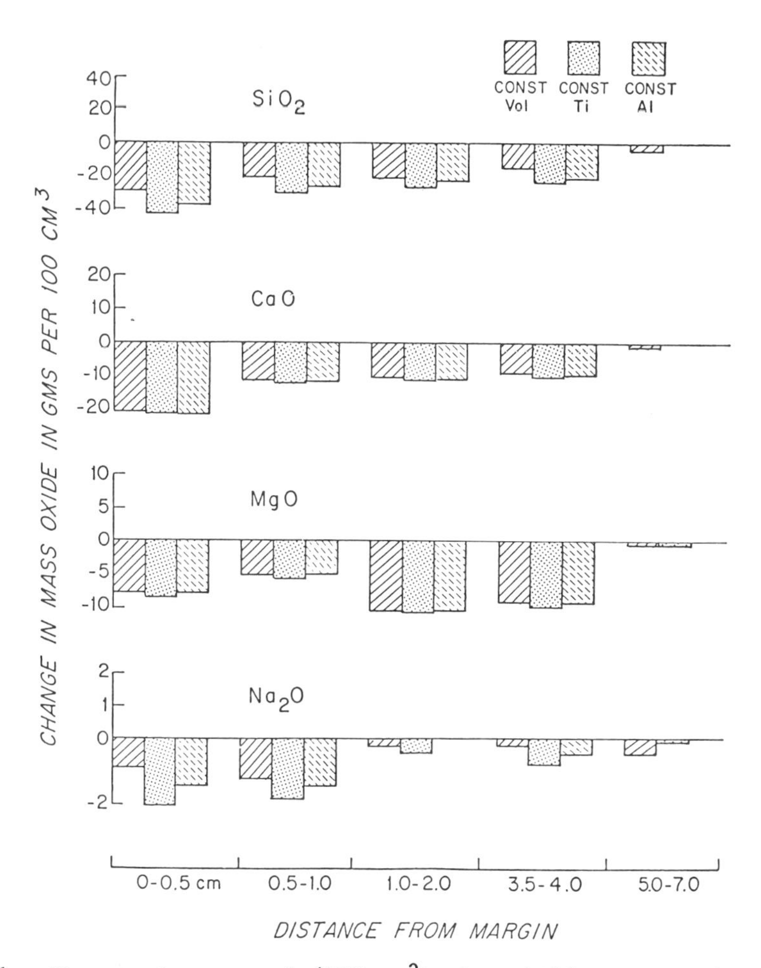

Fig. 6. Change in mass ($g/100\ cm^3$) for different oxides as a function of distance from the outer margin of a single pillow basalt. The data are shown normalized to constant volume, constant TiO_2, and constant Al_2O_3 (from Thompson et al., in preparation). The differences resulting from the different normalization procedures are relatively small.

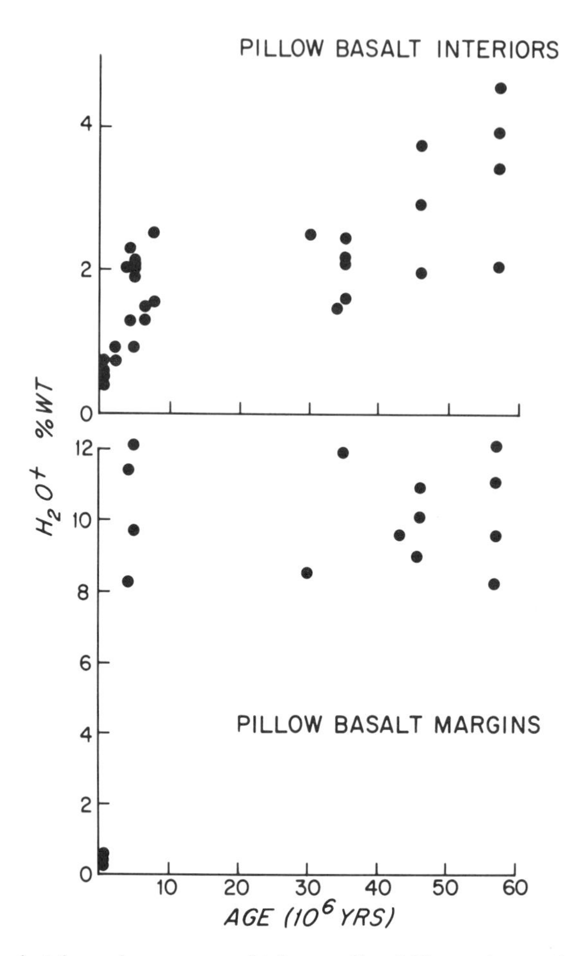

Fig. 7A. Variation in composition of pillow interiors and margins as a function of age (from Thompson et al., in preparation). (A) H_2O vs. age for interiors and margins.

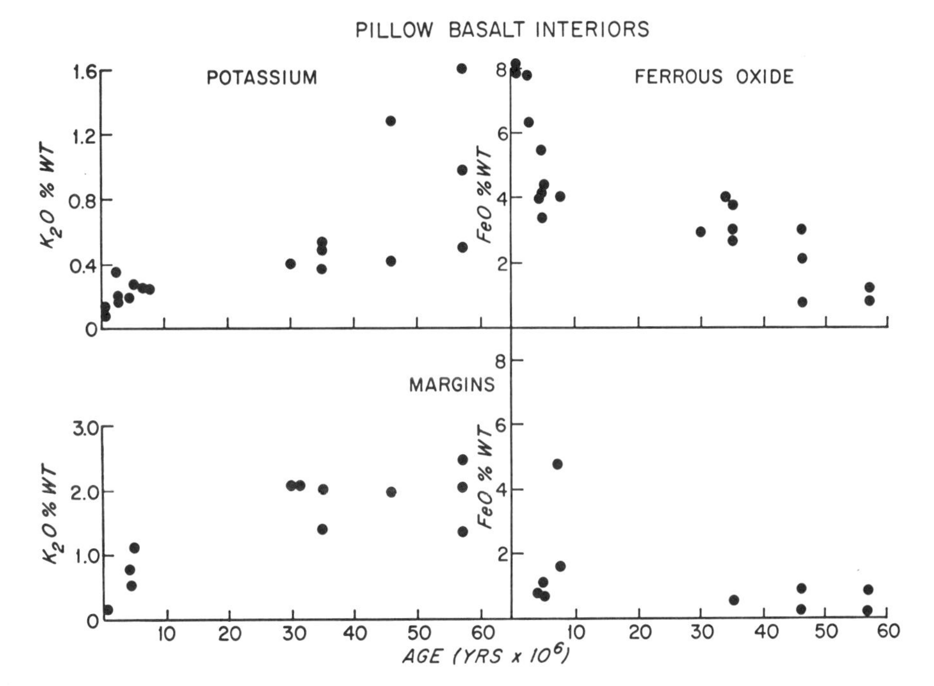

Fig. 7B. Variation in composition of pillow interiors and margins as a function of age (from Thompson et al., in preparation). (B) K_2O and FeO vs. age for interiors and margins. Note that the pillow interiors continue to show increases in H_2O and K_2O, and oxidation of ferrous iron for over 60×10^6 years; the pillow margins rapidly oxidize and hydrate in less than 10×10^6 years, and potassium uptake is very rapid also.

are given in Table 9 for certain components. Data for other components show small and variable changes with alteration. Al_2O_3 and TiO_2 show the least change; FeO and MnO vary showing both gains and losses possibly due to the addition of ferromanganese precipitates on the surface and along cracks in the rock; P 0 varies from gains of 0.25×10^{-9}g/cc/yr, to losses of 0.07×10^{-9}g/cc/yr with the preponderance of evidence favoring some small uptake with low temperature alteration. Ludden and Thompson (1978, 1979) showed some uptake of the light rare earth elements; for other trace elements Henrichs and Thompson (1976), and Thompson et al. (in preparation), show that alteration has very little effect on V, Cr, Co, Ni, Ga, Y and Zr

Table 9. Chemical exchange during surface basalt weathering in units of 10^{-12}gm/cm^3/yr and estimated fluxes in units of 10^{10} g/yr.

Element Gained	Annual Exchange for Pillow Interior[1]	Annual Exchange for Pillow Margin[2]	Fluxes[3]
B	0.9 ± 0.5	60 ± 40	+ 0.45
Li	1.3 ± 0.5	35 ± 30	+ 0.44
Rb	0.49 ± 0.08	9 ± 6	+ 0.14
Cu	1.4 ± 0.8	140 ± 130	+ 0.88
Zn	2.5 ± 0.6	30 ± 5	+ 0.71
Ba	1.3 ± 0.5	31 ± 27	+ 0.43

1 Exchange ± 1 standard deviation around least squares fit regression of 30 pillows ranging in age from 0 to 57 million years.

2 Exchange ± 1 standard deviation around the average for 20 margins from 0 to 57 million years old. Rate calculated assuming exchange completed in 5 million years.

3 Calculated assuming average annual exchange continues for 80 million years (interior) and 5 million years (margins) to a depth of 10 meters.

contents. Sr, however, is extremely variable showing large gains or losses.

The data presented in Tables 8 and 9 show estimated annual chemical exchanges. Fluxes can be calculated knowing the total volume of basalt that undergoes this reaction. Hart (1973), based on the observation of change in seismic velocity with age, suggested that the reaction continues up to 85 million years - see figure 5. The data of Thompson et al. suggest that it continues for at least up to 57 million years. Hart and Staudigel (1978) and Staudigel et al. (1981) have recently suggested that most interactions, including those at low tempera-

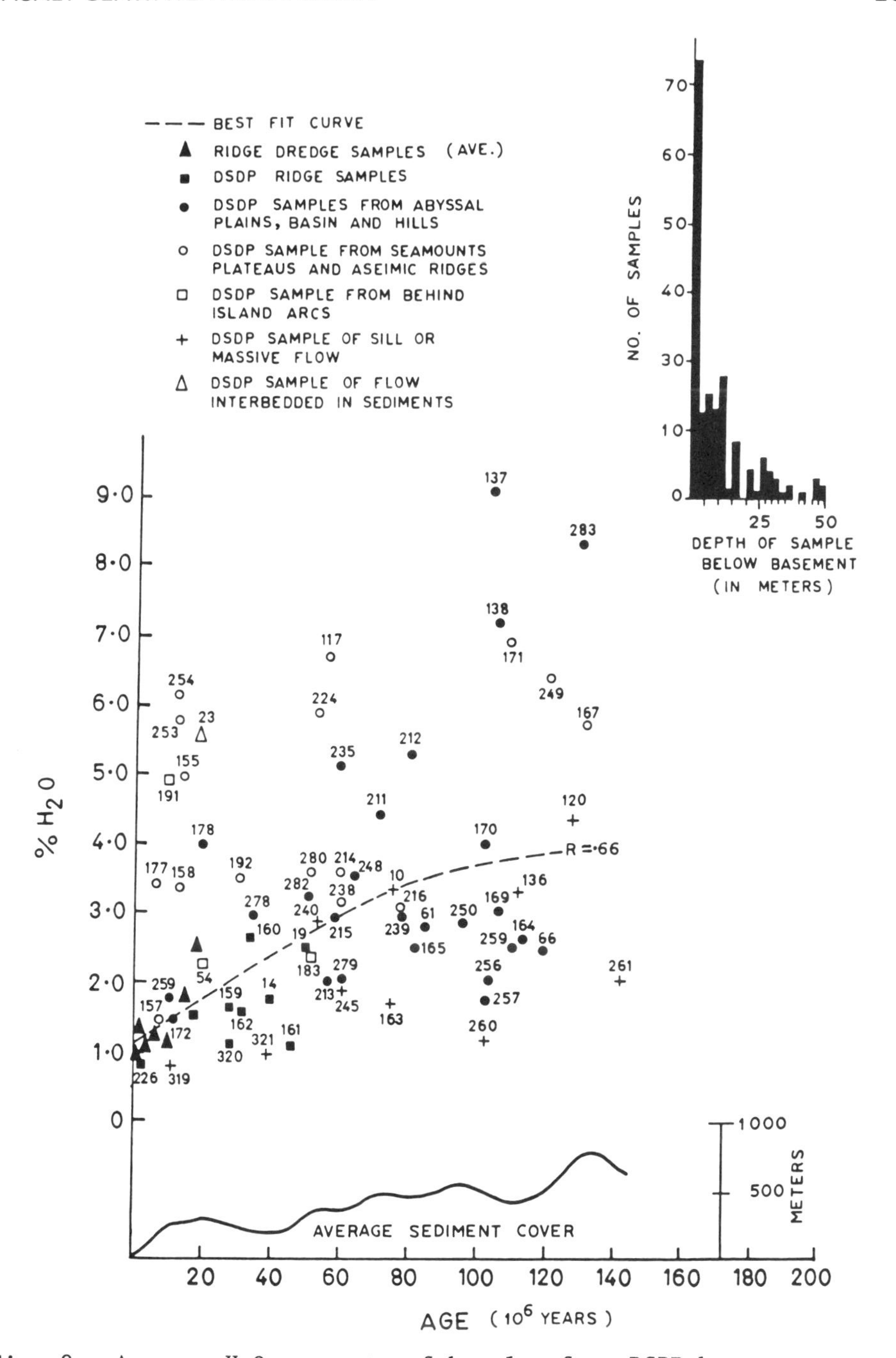

Fig. 8. Average H_2O contents of basalts from DSDP basement sites as a function of age (from Hart, 1976). Most of these sites are shallow (less than 10m penetration) and continue to show water for over 100 million years.

ture, are completed within 10 million years. However, their data are based on deep drill holes where probably low water:rock ratios apply, rather than the surface samples studied here where high water:rock ratio is more likely the case. Hart (1976) has shown for dredged samples, and the upper parts of deep drill holes, there is increasing water content with age out to the order of 100 million years - see figure 8. Thus, for calculation of fluxes, I suggest the data in Tables 8 and 9 can be reasonably considered as viable for the oceanic crust up to 80 million years and for a depth of 10 meters. The annual areal rate of new crust formation is 2.94km^2, depth is about 6 km. For the surface basement rocks I have taken the average annual exchange shown in Tables 8 and 9 and calculated the fluxes assuming the upper 10 m of oceanic crust is involved and that reaction occurs for 80 x 10^6 years i.e. only about 0.1% of the new oceanic crust undergoes this reaction.

Fluxes from Deeper Basement Rocks - Low Temperature, Low Water: Rock

Drilling in the ocean basins over the past ten years has given much insight into the nature and kinds of alteration that can occur in the igneous basement rocks with depth. Although maximum penetration has only been around 500 m and limited to a few holes, there have been many holes up to 100 m and some generalizations can be made. Unlike the upper few meters of the drilled basement, or dredged rocks, most of the deeper basement rocks are not as extensively altered. Original lithologies, glass and minerals are often well preserved even out to 180 million year old crust. Penetration by seawater has obviously taken place, even down to depths of 500 meters, but extensive alteration is restricted to narrow horizons often between massive flows or within the more porous pillow lava piles.

An extensive review of the kinds of alteration and effects in some of the deeper holes can be found in Honnorez (1981). It is clear that the deeper basement basalts have been altered, much the same as the dredged basalts, by interaction with seawater at low temperatures and similar exchanges of elements has occurred although not as extensively as in dredged basalts. Donnelly et al. (1979c) have done a detailed comparison of sites 332 (3.5 million years), 335 (13 million years), and sites 417D, 418A and 417A (110 million years) and which had penetration of basement from 100 m to nearly 600 m. With the exception of site 417A, which will be discussed later, they concluded there were not great differences in the alteration of these basalts. Apparently most of the chemical exchange took place relatively early in the crust formation. Aging of the crust resulted only in some reordering of the alteration minerals, and reactions involving the trapped water in the crust, such as calcite precipitation, but little exchange with overlying seawater (presumably because of sealing by the overlying sediments). As indicated earlier, Hart and Staudigel (1978), Staudigel et al. (1981), Richardson et al., (1980), from their dating of alteration phases in these same sites, had concluded most reactions were completed by about 3 million years.

Calculation of fluxes resulting from these deep basement rocks have not been done in detail. Hart and Staudigel (1982) have attempted an evaluation of the chemical fluxes at site 418A, although only for K, Rb, Cs and U. They chose site 418A because it is one of the deepest drilled (550 meters), core recovery was high (greater than 70%), and many detailed analyses are available (see part 2 of DSDP Initial Reports, leg 51-53). Furthermore, it is similar to other deep drill holes (Donnelly et al., 1979c; Andrews, 1977; Böhlke et al., 1980; Honnorez, 1981).

In Table 10 I show some of the analytical data from Site 418A. Column I represents an average of 24 bulk rock analyses taken at intervals down the core, and chosen to be representative of the fresher, least altered material. Column 2 is a similar compilation, but only obviously weathered material was selected. Following core descriptions of Donnelly et al. (1979c), Johnson (1979), and Hart and Staudigel (1982), the average for site 418A (column 3, Table 10) is calculated assuming the core is made up of 75% fresher rock and 25% of altered rock. It is apparent from the studies of the so-called 'fresher' rocks, that they have also been somewhat affected by seawater reactions. Based upon the analyses of fresh glass (Byerly and Sinton, 1979) and consideration of mineralogical and chemical compositions, a least altered sample was chosen as representative of the original composition - column 4, Table 10. Although different approaches and criteria were used to calculate average and least altered compositions, these averages compare reasonably well with those of Hart and Staudigel (1982), for K and Rb (K=0.48 and 0.05% wt., Rb=9.3 and 0.48 ppm for average and least altered rock respectively).

In Table 11 the chemical exchange, for those components which show significant differences, have been calculated assuming alteration was completed in 3 million years. In the final assessment of chemical fluxes I assume that site 418A is typical of the alteration of 500 m of oceanic crust, i.e. about 8% of the new crust formed.

Table 10. Chemical compositions used to calculate rates of chemical exchange in basalts from DSDP Site 418A.

wt.%	Whole Rock[1]	Weathered Rock[2]	Average[3]	Least Altered[4]
SiO_2	48.78	48.20	48.63	49.55
TiO_2	1.22	1.24	1.22	1.13
Al_2O_3	16.32	16.24	16.30	16.57
Fe_2O_3	4.04	4.90	4.26	3.82
FeO	5.35	4.29	5.08	5.63
MnO	0.16	0.13	0.15	0.16
MgO	6.81	6.69	6.78	7.55
CaO	12.81	11.89	12.58	12.41
Na_2O	2.14	2.56	2.24	1.80
K_2O	0.17	0.75	0.32	0.05
P_2O_5	0.14	0.14	0.14	0.13
H_2O	1.01	2.25	1.32	0.76
CO_2	0.97	0.91	0.96	0.45
ppm				
Y	32	27	31	29
Zr	79	61	74	73
Rb	2.04	9.1	3.81	0.4
B	17.5		21	14
Li	4		10	4
Sr	124	118	122	110
Ba	7.3		14	7.1

1. From Staudigel et al. (1979), average of 24 whole rock analyses selected as generally representative of drill core.
2. From Humphris et al. (1979), average of 22 whole rocks selected as representative of weathered basalts in drill core.
3. 75% whole rock, 25% weathered basalt. B, Li and Ba averages from Donnelly et al. (1979c).
4. Selected from Staudigel et al. (1979).

Table 11. Chemical exchange in basalts from DSDP Site 418A.

	Change in Mass[1] (g/100cc)	Annual Mass Exchange[2] (g/cc/yr x 10^{-9})	Elemental Fluxes[3] (10^{14}g/yr)	
SiO_2	-7.5	-25.1	Si	-0.517
MgO	-2.9	-9.7	Mg	-0.258
CaO	-0.8	-2.6	Ca	-0.819
Na_2O	+1.0	+3.5	Na	+0.115
K_2O	+0.75	+2.5	K	+0.092
P_2O_5	+0.015	+0.05	P	+0.001
H_2O	+1.5	+4.9	H_2O	+0.216
CO_2	+1.4	+4.6	CO_2	+0.202
		(g/cc/yr x 10^{-12})		(10^{10}g/yr)
Rb	+9.5 x 10^{-4}	+3.1		+1.37
Ba	+18.6 x 10^{-4}	+6.2		+2.73
B	+18.2 x 10^{-4}	+6.1		+2.69
Li	+16.4 x 10^{-4}	+5.5		+2.42

1. Assuming constant volume and density change of 2.9 (least altered) to 2.8 (average rock).

2. Assuming alteration completed in 3 x 10^6 years.

3. Assuming 500 m of crust involved in exchange for 3 million years.

Fluxes from Flanks of Mid-Ocean Ridges - Medium Temperature, Medium Water: Rock

Observations of warm water circulating in the oceanic basement at some distance from the axis of spreading have been reported. The mounds area at Galapagos Ridge shows warm waters in crust of 700,000 years. DSDP site 504 on the Costa Rica Rift shows basement waters with temperatures of about 80°C in 6 million years old crust (Mottl et al., 1982). The TAG area in the Mid-Atlantic Ridge has warm water exiting through the seafloor some distance off axis in crust less than 1 million years old (Rona, 1980). The work of Anderson et al. (1979) on heat flow measurements suggests continued circulation in the basement away from the spreading centers. Indeed, the curves of heat loss versus age, figure 9, show that although most of the heat loss is close to the axis, heat loss must continue for some distance away. This phenomenon has been discussed by Lister (1981) in the context of active and passive hydrothermal circulation; circulation on the flanks being typical of passive circulation. Accepting that this must occur, but then as the distance from the heat source at the spreading axis increases, the temperature of the circulating water will decrease. Thus, we should find reactions occurring between the seawater and the igneous basement rocks on the flanks of ridges but at lower temperatures than at the axis.

DSDP site 417A in the Atlantic in 110 million years old crust may be an excellent example of such reaction. Here the basalts of the upper 200 m of the drill site are extensively altered by reaction with seawater at temperatures probably around 30°C (Donnelly et al., 1979a, b, c; Lawrence, 1979). As might be expected from analogy with the experimental observations of Seyfried and Bischoff (1979), these reactions at low temperatures are different from the higher temperature reaction in the rates and exchanges of elements. For example, the lower temperature reactions lead to uptake of alkalis from seawater, similar to the effect observed in dredged and drilled rocks that have undergone weathering by cold seawater. Donnelly et al. (1979a,b) have shown that the site 417A basalts are extensively altered but in a similar direction to dredged basalts. Donnelly et al. (1979a,b) calculated a mass balance of the chemical exchange in a 200 meter column of basalt at site 417A, and indicated that the amount of exchange was much greater than seen in dredged basalts or other drilled basalts, presumably because of the higher temperature and the greater water:rock ratio. The warmer temperatures not only accelerated the reactions, but the heat presumably acted as the driving agent for greater circulation of water through the crust.

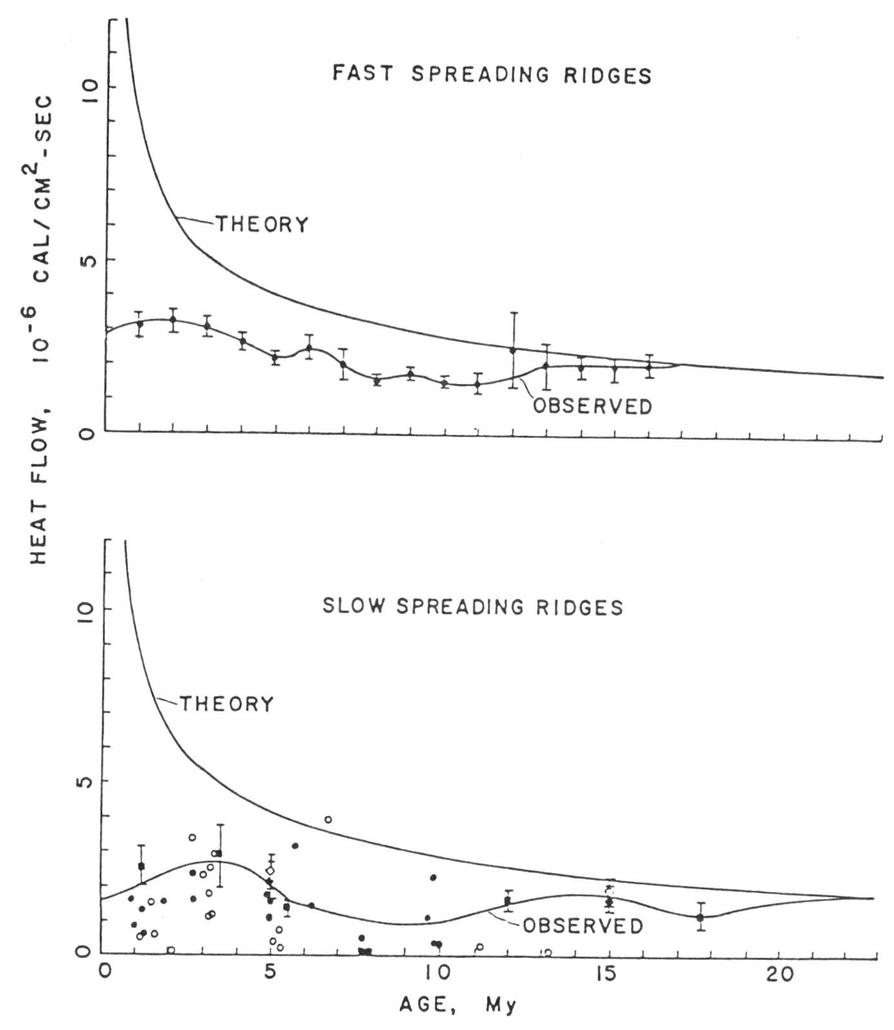

Fig. 9. Theoretical vs. observed heat flow profiles as a function of crustal age for fast spreading and slow spreading mid-ocean ridges (from Wolery and Sleep, 1976). Error bars indicate standard errors for points obtained by averaging a number of individual measurements. For data sources and discussion, see Wolery and Sleep (1976).

Table 12. Chemical compositions used to calculate rates of chemical exchange in basalts from DSDP Site 417A.

wt.%	Hyaloclastite[1]	Basalt[1]	Average[1]	Least altered[2]
SiO_2	53.6	49.6	49.9	49.4
TiO_2	1.13	1.53	1.50	1.5
Al_2O_3	18.1	17.9	17.9	17.0
Fe_2O_3	10.9	6.87	7.16	1.55
FeO	0.40	3.62	3.39	8.09
MnO	0.18	0.15	0.15	0.16
MgO	5.80	5.41	5.44	6.14
CaO	3.68	10.8	10.3	13.55
Na_2O	1.70	2.25	2.21	2.40
K_2O	4.36	1.59	1.79	0.12
P_2O_5	0.04	0.27	0.25	0.13
ppm				
Rb	61	26	29	2
Ba	12	80	75	5
B	100	44	48	15
Li	56	29	31	7

1. From Donnelly et al. (1979a), Table 3A, Site 417A upper 200 m.

2. From Donnelly et al. (1979a), Table 3A, site 417D average of freshest basalts.

In Table 12 I have attempted to calculate the fluxes resulting from the reactions at site 417A. Following Donnelly et al. (1979a,b), and consistent with the approach made for dredged basalts and site 418A drilled basalts, the composition of the highly altered and least altered components of site 417A are shown in columns 1 and 2 of Table 12 respectively. The average composition of the upper 200 m is shown in column 3, as calcu-

Table 13. Chemical exchange in basalts from DSDP 417A.

	Change in Mass[1] (g/100cc)	Annual Mass Exchange[2] (g/cc/yr x 10^{-9})	Elemental Fluxes[3] (10^{14} g/yr)	
SiO_2	-7.1	-23.7	Si	-0.22
MgO	-3.0	-10.0	Mg	-0.11
CaO	-11.2	-37.3	Ca	-0.47
Na_2O	-0.9	-3.0	Na	-0.04
K_2O	+4.5	+15.0	K	+0.22
P_2O_5	+0.3	+1.0	P	+0.008
		(g/cc/yr x 10^{-12})		(10^{12}g/yr)
Rb	+73 x 10^{-4}	+24		+4.23
Ba	+190 x 10^{-4}	+63		+1.10
B	+88 x 10^{-4}	+29		+5.12
Li	+64 x 10^{-4}	+21		+3.70

1. Assuming constant volume and density change of 2.9 (least altered) to 2.73 (average rock).

2. Assuming alteration completed in 3 x 10^6 years.

3. Assuming 200 m of crust involved in exchange for 3 million years.

lated by Donnelly et al., (1979a,b). A least altered rock, in this case from the nearby site 417D because no fresh rock is left in the upper 200 m of site 417A, is shown in column 4 and assumed to be similar in composition to the precursor for site 417A altered basalts. The calculated annual mass exchange is shown in Table 13, assuming that these reactions were completed in 3 million years following the data of Hart and Staudigel (1978), Staudigel et al. (1981), and Richardson et al. (1980). In making the overall assessment of fluxes from the seafloor I assume that site 417A is typical of reactions occurring on the flanks of the mid-ocean ridges and affects about 200 m of oceanic crust, i.e. about 3% of new crust formed.

Fluxes from Axes of Mid-Ocean Ridges - High Temperature, Low Water: Rock

Samples of solutions resulting from reaction of seawater and basaltic rocks of the oceanic crust at high temperatures have been taken from two locations on the sea floor - Galapagos spreading center and the East Pacific Rise at 21°N. Although the solutions from the Galapagos Ridge had undergone some mixing with bottom seawater, extrapolations from the analytical data suggested reactions at about 350°C had originally occurred and the composition of the end member hydrothermal solution was calculated (Edmond et al., 1979a, b) - see Table 14. For the solutions at 21°N not much data is available as yet. The preliminary sampling of the emitting solutions was not as good as in Galapagos because of the unexpected high temperature and complication resulting from the precipitation of solids on cooling and entrainment of bottom waters (Speiss et al., 1980; Edmond, 1981). Resampling of the vents under better conditions, with properly prepared equipment, has only just been successful. However the preliminary data for the 1979 dives, suggest similar findings to the Galapagos area. Extrapolation to the 350°C end member shows similar composition to the Galapagos 350°C solution (Edmond et al., 1981) - see Table 15.

Calculating actual fluxes from the hydrothermal solution data is not easy. Edmond et al. (1979a) have accepted the arguments of Jenkins et al. (1978) regarding the helium flux. From the observation of the ^3He anomaly in the Galapagos springs versus the temperature of the solution, Jenkins et al. calculated the ratio of transported heat to dissolved ^3He anomaly to be 7.6×10^{-3} cal/atom. The oceanic flux of ^3He is 6.5×10^{26} atoms/year from the global integration of the anomaly observed over mid-ocean ridges (Craig et al., 1975). Assuming all this ^3He is injected at the ridge axes in hydrothermal waters, then the heat flux is $6.5 \times 10^{26} \times 7.5 \times 10^{-7} = 4.9 \times 10^{19}$ cals/year, in good agreement with the geophysical estimates of 4×10^{19} cals/year (Wolery and Sleep, 1976). Edmond et al. (1979a) thus took the measured temperature

Table 14. Chemical data for the Galapagos hydrothermal vents (from Edmond et al., 1979a, b).

	Li[1]	K[2]	Rb[1]	Mg[2]	Ca[2]	Ba[2]	SO_4[2]	F[2]	Cl[2]	Si[2]	Mn[1]	Na
Concentration Range	28-59	10.08-10.37	1.32-1.88	52.7-51.1	10.31-11.0	0.15-1.1	28.55-27.39	0.074-0.072	34.41 34.74	0.165-0.775	1-33	
Gradients from Si regression equation (μmol/kcal)												
Clambake	3.2	25	0.0545	-154	86	0.122	-75	-165	156	61	3.8	76
Garden of Eden	3.2	25	0.0569	-154	69	0.049	-75	-216	6	61	1.3	-32
Dandelion	3.2	25	0.0459	-154	69	0.049	-75	-223	-419	61	1.6	-424
Oyster Bed	1.9	25	0.0347	-154	41	0.049	-75	-223	-629	61	1.2	-580
Extrapolated 350°C Solution Contents												
Clambake	1142	18.7	20.3	0	40.2	42.6	0	0	595	1296	1140	487
Garden of Eden	1142	18.8	21.2	0	34.3	17.2	0	0	543	1296	390	451
Dandelion	1142	18.8	17.3	0	34.3	17.2	0	0	395	1296	480	313
Oyster Bed	689	18.8	13.4	0	24.6	17.2	0	0	322	1296	360	259
Seawater Contents	28	10.1	1.32	52.7	10.3	0.145	28.6	70.8	541	0.160	0.002	461
Average tholeiite composition (mol/kg)	1.45	30	13		2.13	73		5.3	1.27		29	1.2

[1] μmol/l concentration

[2] mmol/l concentration

Table 15. Comparison of the estimated composition of a 350° C hydrothermal end member based on extrapolation of the Galapagos data with that observed at 21°N. The ranges in the Galapagos results derive from the different trends for composition versus heat observed between individual vent fields.

	Galapagos	21°N	Sea Water
Li μmol/kg	1142 - 689	820	28
K mmol/kg	18.8	25.0	10.1
Rb μmol/kg	20.3 - 13.4	26.0	1.32
Mg mmol/kg	0	0	52.7
Ca mmol/kg	40.2 - 24.6	21.5	10.3
Sr μmol/kg	87	90	87
Ba μmol/kg	42.6 - 17.2	95 - 35	0.145
Mn μmol/kg	1140 - 360	610	0.002
Fe μmol/kg	+	1800	-
Si mmol/kg	21.9	21.5	0.160
SO_4 mmol/kg	0	0	28.6
H_2S mmol/kg	+	6.5	0

\+ non-conservative to sub-surface mixing

\- sea water concentration not accurately known

dependence of the concentration anomalies for the other species (moles/cal), and combined them with the heat flux calculated by Jenkins et al. (1978), to give the hydrothermal flux for each species. Those fluxes are shown in Table 16 for each of the vents at Galapagos, and compared to the river flux. Although not yet calculated for 21°N hot springs, the fluxes are assumed to be reasonably similar. The ^3He data for 21°N are exactly comparable to the Galapagos fluxes (Speiss et al., 1980; Craig et al., 1980). Although many of the metal species were lost from the Galapagos springs in the zone of mixing in the crust (eg. Fe, S, Cu, Ni, Cr, U) this is not the case at 21°N. However, many of these metals obviously precipitate on entrainment with the cold bottom water. How much of these metals does finally enter the seawater in solution is not yet known. The data in Table 16 suggest that the fluxes from hydrothermal reactions at ridge axes are comparable to the river fluxes. For Mg and SO_4 the hydrothermal reaction more than takes care of all the river input. For Li and Rb the hydrothermal fluxes are much greater than the river fluxes and finding a sink for such inputs becomes a major problem. Similarly for K, Ba and Si the increased inputs into the ocean are half or more of the river input and compound the problem of finding appropriate sinks for such elements.

Edmond et al. (1979a,b,c), Edmond (1980), and particularly Corliss et al. (1979) have recognized the problems rising from this simplistic calculation of fluxes, and noted for some species it cannot be correct unless one assumes large and variable thicknesses of oceanic crust have been affected by the hot seawater reactions. Consideration of heat loss strongly suggests that not all the heat is removed at the axis but continues to be

Table 16. Hydrothermal fluxes for the Galapagos vents compared to river fluxes (from Edmond et al., 1979 a, b).

	Li	Rb	Ba	F	K	Mg	Ca	SO_4	Cl	Si
	x 10^9				x 10^{12}					
Flux (mol/year)										
Clambake	160	2.7	6.1	-8.25	1.25	-7.7	4.3	-3.75	7.8	3.1
Garden of Eden	160	2.8	2.45	-10.8	1.25	-7.7	3.5	-3.75	0.3	3.1
Dandelions	160	2.3	2.45	-11.5	1.25	-7.7	3.5	-3.75	-21.3	3.1
Oyster Beds	95	1.7	2.45	-11.5	1.25	-7.7	2.1	-3.75	-31.0	3.1
River Flux (mol/year)	13.5	0.37	10	165	1.9	5.3	12	3.7	6.9	6.4

removed for some distance away from the axis. The amount of crust available at the axis is not sufficient to provide the flux of some species even if all that element were removed from the rock by seawater. Wolery (1979), Hart and Staudigel (1982) and Mottl (1982) have suggested, from considerations of heat flow models, heat extraction, and mass balance of elements such as the alkalis, that the hydrothermal flux estimates of Edmond et al. (1979a,b,c) may be too high by a factor of 6 (Mottl, 1982), an order of magnitude (Wolery, 1979), slightly over an order of magnitude for K and Rb (Hart and Staudigel, 1982). These disagreements have not yet been resolved.

CONCLUSIONS

Much of the evidence presented here suggests that seawater-igneous oceanic basement reactions are widespread in space and time. At the axis of spreading centers where new crust is generated the reaction probably takes place at temperatures in excess of 100°C, possibly up to 400°C. Reactions continue for some distance away from the ridge axis, probably at decreasing temperatures and intensity, and possibly for a few million years.

The nature and distribution of the hydrothermal circulation is not well known at present. It is clearly linked to the volcanic cycle that forms new crust, and is probably dependent on such factors as the size, shape and duration of the magma chamber beneath the axis. There are probably distinct differences between fast and slow spreading ridges. Much work needs to be done yet in fully understanding the physics and chemistry of hydrothermal circulation. The two hot springs so far sampled are quite different in their plumbing characteristics and the nature of the solutions debouching on the ocean floor. The Galapagos solutions have already lost most of their heavy metal contents within the crust at some shallow depth. The 21°N solutions precipitate their metals directly on the sea floor. However, both solutions probably had similar chemical compositions at the initial site of reaction and temperatures of about 350°C. The nature of the circulation and the solutions percolating through oceanic crust on the flanks of ridges is poorly understood, although this is clearly an important process in advecting heat (and probably chemical species) from the crust.

The nature of the chemical reaction and the fluxes produced between the oceans and the igneous basement are mostly dependent on the temperature of reaction and the water:rock ratio. These will vary mostly in respect to the location of the circulation and distance from the heat source. In attempting to calculate the total flux involved in the seawater-rock reaction, I have considered four cases that may typify the principal reactions occurring. For each of the cases considered fluxes were calculated.

In Table 17 I have summarized the actual fluxes that occur for each of the cases considered. Only those species where relatively large chemical changes take place, or where there is sufficient information available, are considered. Many other species are involved, of course. The data for Table 17 have been calculated in the following manner:

The annual areal rate of new crust formation is $2.94 km^2$, depth is about 6 km. For the surface basement rocks I have taken the average annual exchange shown in Table 8 and 9 and calculated the fluxes assuming the upper 10 m of oceanic crust is involved and that reaction occurs for 80×10^6 years i.e. about 0.1% of the new oceanic crust undergoes this reaction. As can be seen in Table 17 this reaction does not lead to very large annual fluxes compared to other fluxes.

For the deeper parts of the oceanic crust I have taken the data in Table 11 and assumed this reaction is typical of the next 500 m of the oceanic crust i.e. about 8% of the new crust. This reaction takes place over about 3×10^6 years.

For the reactions occurring on the flanks of the spreading centers where temperatures remain moderate and circulation is more active than the colder upper parts of the crust, I have used the reaction typified by DSDP Site 417A in Table 13. I have assumed this reaction is typical for the next 200 m of oceanic crust, i.e. about 3% of the new crust formed annually. I have also assumed that this reaction takes place over 3 million years.

For the hydrothermal fluxes resulting from high temperature reaction of seawater and basalt at spreading center axes I have used the data of Edmond et. al (1979a, b) shown in Table 16. However, because of the uncertainty in these values as discussed, I have also assumed a second case where, for those species leached from the rock, the published fluxes are reduced by an order of magnitude. For species added to the rock from seawater, such as Mg, it is more difficult to calculate a flux. The data of Humphris and Thompson (1978a), based on observations of altered rock, suggested uptake of 1 to 10 g $MgO/100 cm^3$ of basalt. Assuming an average of 3 g $Mg/100 cm^3$ taken up by basalt at high temperature reaction and 1 km of the newly formed oceanic crust (i.e. about 16%) is altered, then the flux is about half

Table 17. Estimates of hydrothermal fluxes between oceanic basement and seawater

	x 10^{14} g/yr				x 10^{10} g/yr			
	Si	Ca	Mg	K	B	Li	Rb	Ba
Case A								
Surface[1]	-0.006	-0.045	-0.03	+0.013	+0.45	+0.44	+0.14	+0.43
Basins[2]	-0.52	-0.082	-0.26	+0.09	+2.69	+2.42	+1.37	+2.73
Flanks[3]	-0.2	-0.47	-0.11	+0.22	+5.12	+3.7	+4.23	+1.1
Axis[4]	-0.87	-1.3	+1.87	-0.49	(-)*0	-111	-20.5	-46
Total	-1.60	-1.90	+1.47	-0.17	+8.26	-104.54	-14.76	-41.74
River	-1.99	-4.88	-1.33	-0.74	-47.0	-9.4	-3.2	-137.3
Basement Flux as % of River Flux	80.4	38.9	110.5	23.0	17.6	1112	461	30.4
Case B								
Surface[1]	-0.006	-0.045	-0.03	+0.013	+0.45	+0.44	+0.14	+0.43
Basins[2]	-0.52	-0.082	-0.26	+0.09	+2.69	+2.42	+1.37	+2.73
Flanks[3]	-0.2	-0.47	-0.11	+0.22	+5.12	+3.7	+4.23	+1.1
Axis[4]	-0.087	-0.13	+1.0	-0.049	(-)*0	-11.1	-2.05	-4.6
Total	-0.82	-0.73	+0.6	+0.27	+8.26	-4.54	+3.69	-0.34
River	-1.99	-4.88	-1.33	-0.74	-47.0	-9.4	-3.2	-137.3
Basement Flux as % of River Flux	41.2	14.9	45.1	36.5	17.6	48.3	115.3	0.2

1. From Table 8 and 9
2. From Table 11
3. From Table 13
4. From Table 16 (reduced by order of magnitude in case B, except Mg. See text for discussion)
   + = Gained by rock, lost from seawater
   - = Lost from rock, gained by seawater

* Boron is not seen in the Galapagos vents but is at 21°N on the East Pacific Rise. Edmond (personal communication) believes the low temperature subsurface mixing at Galapagos results in B uptake from the hydrothermal solution. Calculation of the B flux for high temperature reaction will have to await the completion of analytical data for the 21°N vents.

of that calculated by Edmond et al. (1979a), i.e. 0.95 compared to 1.87 x 10^{14} g/yr. This is slightly less than that suggested by Mottl (1976) - 4.12 g Mg/100 cm^3, and Wolery and Sleep (1976): - 6 to 11 Mg/100 cm^3. Thus in Table 17, case B, I have taken the flux to be about 1 x 10^{14} g/yr.

The total flux calculated in Table 17 in both cases is the sum of the four reactions. The river flux to the oceans is also shown and the total seawater-rock reaction flux is shown as a percentage of the river flux. The important points to note are that the rock-water interaction process acts as a sink for species such as Mg, i.e. about 50% of the river flux in case B and close to the flux required to balance Drever's calculation of the Mg budget (see Table 3). The basalt-seawater interaction also acts ultimately as a sink for alkali metals and boron (in Case B) in spite of the fact that high temperature hydrothermal reactions leach those metals from the rock. Li is an apparent exception and sufficient is leached from the rock at high temperatures to counteract subsequent uptake at lower temperatures. Excess Li in marine sediments has been commented on by Holland (1978). In Case A the alkali metals are added to seawater in even greater fluxes than those from river water.

Table 18. Summary of net exchange between basalt and seawater considering the full range of temperature reactions. The basalt is considered as the source or sink for the various species shown. Actual fluxes have been calculated in Table 17 for the species in the upper part of the columns.

Source	Sink
Si	Mg
Ca	K
Ba	B
Li	Rb
———	———
Fe	H_2O
Mn	Cs
Cu	U
Ni	
Zn	

Si and Ca are additional fluxes to the ocean from seawater-rock reaction. However, both species are easily precipitated as quartz and calcite and probably the sediments are the ultimate sink for these species.

Data on other species are lacking, but as indicated by Hart and Staudigel (1982), the basalt-seawater reaction also ultimately acts as a sink for Cs and U. The latter is important in global heat balances and has important consequences for subsequent subduction of oceanic crust and hence return of heat producing species like U to the mantle. Volatiles such as CO_2, He and noble gases are added to the oceans. Water is added to the oceanic crust. Table 18 summarizes the role of the rock as a source or sink for various species.

Finally I should stress that all these estimates of hydrothermal fluxes are subject to revision. Many of the assumptions underlying the calculations are subjective. Much work is to be done in quantifying and understanding the seawater-igneous basement reaction. The main purpose of this review is twofold: (1) to indicate that it is potentially an important process in geochemical mass balances and the buffering reactions that maintain seawater at steady state composition; (2) to indicate that it is a complex process and takes place over a range of temperatures with different exchanges and fluxes. The actual flux must be the sum effect of all these reactions.

ACKNOWLEDGEMENTS

This work was supported by NSF Grants 24930 and 08120. I thank Mike Mottl, John Edmond, Stan Hart and Jose Honnorez for access to unpublished data. I also thank Margaret Sulanowska for editing and drafting the manuscript and Margaret Harvey for the typing. This is WHOI Contribution No. 5360.

REFERENCES

Anderson, R. N., M. A. Hobart, and M. G. Langseth, 1979: Geothermal convection through oceanic crust and sediments in the Indian Ocean. Science, 204, 828-832.

Andrews, A. J., 1977: Low temperature fluid alteration of oceanic layer 2 basalts, DSDP Leg 37. Can. J. Earth Sci., 14, 911-926.

Andrews, A. J., 1978: Petrology and geochemistry of alteration in layer 2 basalts, DSDP Leg 37. Ph. D. Thesis, Univ. of Western Ontario, London, Canada. 327 pp.

Aumento, F., and B. Loncarevic, 1969: The Mid-Atlantic Ridge near 45°N. III. Bald Mountain. Can. J. Earth Sci., 6, 11-23.

Aumento, F., B. Loncarevic, and D. I. Ross, 1971: Hudson geotraverse: Geology of the mid-Atlantic ridge at 45°N. Phil. Trans. Roy. Soc. London A, 268, 623-650.

Baragar, W. R. A., A. G. Plant, G. J. Pringle, and M. Schau, 1979: Diagenetic and postdiagenetic changes in the composition of an Archean pillow. Can. J. Earth Sci., 16 (II), 2102-2121.

Bass, M. N., R. Moberly, J. M. Rhodes, C. Y. Shih, and S. E. Church, 1973: Volcanic rocks cored in the central Pacific, Leg 17, Deep Sea Drilling Project. In: Initial Reports of Deep Sea Drilling Project, 17, 429-503. U. S. Gov't. Printing Office.

Bass, M. N., 1975: Secondary minerals in oceanic basalts. In: Ann. Rept. Dept. Terr. Magnetism-Carnegie Inst. Year Book, 74, 234-240.

Bass, M. N., 1976: Secondary minerals in oceanic basalt, with special reference to Leg 34, Deep Sea Drilling Project. In: Initial Reports of Deep Sea Drilling Project, 34, 393-432. U. S. Gov't. Printing Office.

Bogdanov, Y. A. and V. Ploshko, 1968: Igneous and metamorphic rocks from the abyssal Comanche Depression. Dokl. Akad. Nauk SSSR, 177. 173-176.

Böhlke, J. K., J. Honnorez, and B. H. Honnorez-Guerstein, 1980: Alteration of basalts from site 396B, DSDP: Petrographic and mineralogic studies. Contrib. Mineral. Petrol. 73, 341-364.

Bonatti, E., 1965: Palagonite, hyaloclastites and alteration of volcanic glass in the ocean. Bull. Volcanol., 28, 257-269.

Bonatti, E., 1970: Deep sea volcanism. Naturwiss., 57, 379-384.

Bonatti, E., J. Honnorez, and G. Ferrara, 1971: Peridotite-gabbro-basalt complex from the equatorial Mid-Atlantic Ridge. Phil. Trans. Roy. Soc. London A, 268, 385-402.

Bonatti, E., J. Honnorez, P. Kirst, and F. Radicati, 1975: Metagabbros from the Mid-Atlantic Ridge at 6°N: contact-hydrothermal-dynamic metamorphism beneath the axial valley. J. Geol., 83, 61-78.

Broecker, W. S., 1971: A kinetic model for the chemical composition of sea-water. Quat. Res., 1, 188-207.

Byerly, G. R. and J. M. Sinton, 1979: Compositional trends in natural basaltic glasses from Deep Sea Project Holes 417D and 418A. In: Initial Reports of the Deep Sea Drilling Project, 51-53, Part 2, 957-971. U. S. Gov't. Printing Office.

Cann, J. R., 1969: Spilites from the Carlsberg Ridge, Indian Ocean. J. Petrol., 10 (I), 1-19.
Cann, J. R., 1971: Petrology of basement rocks from the Palmer Ridge, NE Atlantic. Phil. Trans. Roy. Soc. London A, 268, 605-618.
Cann, J. R., 1979: Metamorphism in the ocean crust. In: Deep Drilling Results in the Atlantic Ocean: Ocean Crust, 230-238. Talwani, M., C. G. Harrison, and D. E. Hayes (Eds.) Amer. Geophys. Union, Maurice Ewing Series 2.
Cann, J. R., 1980: Availability of sulfide ores in the oceanic crust. J. Geol. Soc., 137, 381-384.
Cann, J. R., and B. Funnell, 1967: Palmer Ridge: a section through the upper part of the ocean crust. Nature, 213, 661-664.
Chapman, H. J., and E. T. C. Spooner, 1977: ^{87}Sr enrichment of ophiolite sulphide deposits in Cyprus confirms ore formation by circulating seawater. Earth Planet. Sci. Letters, 35, 71-78.
Chernyseva, V. I., 1971: Greenstone altered rocks of rift zones in median ridges of Indian Ocean. Int. Geol. Rev., 13, 903-913.
Corliss, J. B., 1971: The origin of metal-bearing submarine hydrothermal solutions. J. Geophys. Res., 76, 8128-8138.
Corliss, J. B., L. I. Gordon, and J. M. Edmond, 1979: Some implications of heat/mass ratios in Galapagos rift hydrothermal fluids for models of seawater-rock interaction and the formation of oceanic crust. In: Deep Drilling Results in the Atlantic Ocean: Ocean Crust, 391-402. Talwani, M., C. G. Harrison, and D. E. Hayes (Eds.). Amer. Geophys. Union, Maurice Ewing Series 2.
Craig, H., W. G. Clarke, and M. A. Beg, 1975: Excess ^{3}He in deep water on the East Pacific Rise. Earth Planet. Sci. Letters, 26, 125-132.
Craig, H., J. A. Welham, K. Kim, R. Poreda, and J. E. Lupton, 1980: Geochemical studies of 21°N EPR hydrothermal fluids. EOS, 61, 992.
Deffeyes, K. S., 1970: The axial valley: a steady state feature of the terrain. In: Megatectonics of Continents and Oceans, 194-222. Johnson, H., and B. L. Smith (Eds.) Rutgers Univ. Press.
Delaney, J. R., D. W. Mogk, and M. J. Mottl, 1980: High temperature, sulfide bearing hydrothermal system on the mid-Atlantic ridge at 23.6°N. Geol.-Soc. Amer. Ann. Mtg., 12, 411 (Abstract).

Delaney, J. R., D. W. Mogk, and M. J. Mottl, 1981: Quartz-cemented greenstone breccias - samples of high temperature hydrothermal fluid flow channels from the mid-Atlantic ridge. Abstr. Amer. Geophys. Union Chapman Conf., Airlie, VA.

Donnelly, T. W., G. Thompson, and M. H. Salisbury, 1979a: The chemistry of altered basalts at site 417, Deep Sea Drilling Project Leg 51. In: Initial Reports of the Deep Sea Drilling Project, 51-53, 1319-1330. U. S. Gov't Printing Office.

Donnelly, T. W., G. Thompson, and P. T. Robinson, 1979b: Very-low-temperature hydrothermal alteration of the oceanic crust and the problem of fluxes of potassium and magnesium. In: Deep Drilling Results in the Atlantic Ocean: Ocean Crust, 369-382. Talwani, M., C. G. Harrison, and D. E. Hayes (Eds.) Amer. Geophys. Union, Maurice Ewing Series 2.

Donnelly, T. W., R. A. Pritchard, R. Emmerman, and H. Puchelt, 1979c: The aging of oceanic crust: synthesis of the mineralogical and chemical results of Deep Sea Drilling Project, Legs 51 through 53. In: Initial Reports of the Deep Sea Drilling Project, 51-53, 1563-1577. U. S. Gov't Printing Office.

Drever, J. I., 1971: Early diagenesis of clay minerals, Rio Ameca Basin, Mexico. J. Sediment. Petrol. 41, 982-994.

Drever, J. I., 1974: The magnesium problem. In: The Sea, 5, 337-357, Goldberg, E. J., Ed. Wiley, N. Y.

Edmond, J. M., 1980: Ridge crest hot springs: the story so far. EOS, 61 (2), 129-131.

Edmond, J. M., 1981: Hydrothermal activity at mid-ocean ridge axes. Nature, 290 (5802), 87-88.

Edmond, J. M., C. Measures, R. E. McDuff, L. H. Chan, R. Collier, B. Grant, L. I. Gordon, and J. B. Corliss, 1979a: Ridge crest hydrothermal activity and the balances of the major and minor elements in the ocean: the Galapagos data. Earth Planet Sci. Letters, 46, 1-18.

Edmond, J. M., C. Measures, B. Mangum, B. Grant, F. R. Sclater, R. Collier, A. Hudson, L. I. Gordon, and J. B. Corliss, 1979b: On the formation of metal-rich deposits at ridge crests. Earth Planet. Sci. Letters, 46, 19-30.

Edmond, J. M., J. B. Corliss, and L. I. Gordon, 1979c: Ridge crest-hydrothermal metamorphism at the Galapagos spreading center and reverse weathering. In: Deep Drilling Results in the Atlantic Ocean: Ocean Crust, 383-390. Talwani, M., C. G. Harrison, and D. E. Hayes, (Eds.) Amer. Geophys. Union, Maurice Ewing Series 2.

Edmond, J. M., K. vonDamm, and R. E. McDuff, and C. I. Measures, 1981: Chemistry of hot springs on the East Pacific Rise and their effluent dispersal (a review). Nature, 297 (5863), 187-191.

Frey, F. A., W. B. Bryan, and G. Thompson, 1974: Atlantic Ocean floor: geochemistry and petrology of basalts from Legs 2 and 3 of the Deep Sea Drilling Project. J. Geophys. Res., 79, 5507-5527.

Garlick, G. D., and J. R. Dymond, 1970: Oxygen isotope exchange between volcanic materials and ocean water. Geol. Soc. Amer. Bull., 81, 2137-2142.

Garrels, R. M., and F. T. Mackenzie, 1971:. Evolution of Sedimentary Rocks, Norton, N. Y. 394 p.

Gibbs, R. J., 1972: Water chemistry of the Amazon River. Geochim. Cosmochim. Acta, 36, 1061-1066.

Goldberg, E. G., 1957: Biogeochemistry of trace metals. In: Treatise on Marine Ecology and Paleoecology. Ed. J. W. Hedgepath, Geol. Soc. Amer. Mem., 67, 345-357.

Hart, R. A., 1970: Chemical exchange between sea-water and deep ocean basalts. Earth Planet. Sci. Letters, 9, 269-279.

Hart, R. A., 1973: A model for chemical exchange in the basalt-seawater system of oceanic layer II. Can. J. Earth Sci., 10, 799-816.

Hart, R. A., 1976: Progressive alteration of the oceanic crust. In: Initial Reports of the Deep Sea Drilling Project, 34, 433-437. U. S. Gov't Printing Office.

Hart, S. R., 1969: K, Rb, Cs contents and K/Rb, K/Cs ratios of fresh and altered submarine basalts. Earth Planet. Sci. Letters, 6, 295-303.

Hart, S. R., 1971: K, Rb, Cs, Sr and Ba contents and Sr isotope ratios of ocean floor basalts. Phil. Trans. Roy. Soc. London A, 268, 573-587.

Hart, S. R., and A. M. Nalwalk, 1970: K, Rb, Cs and Sr relationships in submarine basalts from the Puerto Rico Trench. Geochim. Cosmochim. Acta, 34, 145-156.

Hart, S. R., and H. Staudigel, 1978: Oceanic crust: age of hydrothermal alteration. Geophys. Res. Letters, 5, 1009-1012.

Hart, S. R., and H. Staudigel, 1982: The control of alkalies and uranium in sea water by ocean crust alteration. Earth Planet. Sci. Letters 58 (2) 202-212.

Hay, R. L., and A. Iijima, 1968a: Petrology of palagonite tuffs of Koko Craters, Oahu, Hawaii. Contrib. Mineral. Petrol., 17, 141-154.

Hay, R. L., and A. Iijima, 1968b: Nature and origin of palagonite tuffs of the Honolulu Group on Oahu, Hawaii. Geol. Soc. Amer. Mem. 116, 331-376.

Hekinian, R., 1968: Rocks from the mid-oceanic ridge in the Indian Ocean. Deep-Sea Res., 15, 195-213.

Hekinian, R., 1971: Chemical and mineralogical differences between abyssal hill basalts and ridge tholeiites in the Eastern Pacific Ocean. Mar. Geol., 11, 77-91.

Hekinian, R., and F. Aumento, 1973: Rocks from the Gibbs Fracture Zone and the Minia Seamount near 53°N in the Atlantic Ocean. Mar. Geol., 14, 47-72.

Henrichs, S., and G. Thompson, 1976: The low temperature weathering of oceanic basalts by seawater: 2. trace element fluxes. Geol. Soc. Amer. Ann. Mtg. 6, 1098.

Holland, H. D., 1965: The history of ocean water and its effect on the chemistry of the atmosphere. Proc. Natl. Acad. Sci., 53, 1173-1183.

Holland, H. D., 1972: The geologic history of seawater - an attempt to solve the problem. Geochim. Cosmochim. Acta, 36, 637-657.

Holland, H. D., 1978: The Chemistry of the Atmosphere and Oceans. J. Wiley, N. Y., 351 pp.

Honnorez, J., 1967: La palagonitization: l'alteration sous-marine du verre volcanique basique de Palagonia (Sicile). Ph. D. Thesis, Univ. of Bruxelles.

Honnorez, J., 1972: La palagonitization: l'alteration sous-marine du verre volcanique basique de Palagonia (Sicile). Vulkaninstitut I. Friedl. Zurich, 9, 1-132.

Honnorez, J., 1981: The aging of the oceanic crust at low temperature. In: The Oceanic Lithosphere. Ed. Emiliani, E. The Sea, 7, 525-587. J. Wiley, N. Y.

Honnorez, J., J. K. Bohlke, B. M. Honnorez-Guerstein, 1978: Petrographical and geochemical study of the low temperature submarine alteration of basalt from Hole 396B, Leg 46. In: Initial Reports of the Deep Sea Drilling Program, 46, 299-329. U. S. Gov't. Printing Office.

Humphris, S. E. and G. Thompson, 1978a: Hydrothermal alteration of oceanic basalts by seawater. Geochim. et Cosmochim. Acta, 42, 107-125.

Humphris, S. E. and G. Thompson, 1978b: Trace element mobility during hydrothermal alteration of oceanic basalts. Geochim. et Cosmochim. Acta, 42, 127-136.

Humphris, S. E., R. N. Thompson, and G. F. Marriner, 1979: The mineralogy and geochemistry of basalt weathering, Holes 417A and 418A. In: Initial Reports of the Deep Sea Drilling Project, 51-53, 1201-1217. U. S. Gov't Printing Office.

Ito, E., 1979: High temperature metamorphism of plutonic rocks from the mid-Cayman Rise: a petrographic and oxygen isotopic study. Ph.D. Thesis, Univ. of Chicago, Ill.

Jakobsson, S. P., 1972: On the consolidation and palagonitization of the Tephra of Surtsey Volcanic Island, Iceland. Mus. Natl. Hist. Reykjavik Papers, 60, 1-8.

Jakobsson, S. P., 1978: Environmental factors controlling the palagonitization of the Surtsey Tephra, Iceland. Bull. Geol. Soc. Denmark, 27, 91-105.

Jenkins, W. J., J. M. Edmond, and J. B. Corliss, 1978: Excess ^{3}He and ^{4}He in Galapagos submarine hydrothermal waters. Nature, 272, 156-158.

Johnson, D. V., 1979: Crack distribution in the upper oceanic crust and its effects upon seismic velocity, structure, formation permeability and fluid circulation. In: Initial Reports of the Deep Sea Drilling Project, 51-53, 1479-1490. U. S. Gov't Printing Office.

Lawrence, J. R., 1979: Temperatures of formation of calcite veins in the basalts from DSDP holes 417A and 417D. In: Initial Reports of the Deep Sea Drilling Project, 51-53, 1183-1184. U. S. Gov't Printing Office.

Lister, C. R. B., 1981: Active and passive hydrothermal systems in the oceanic crust: predicted physical conditions. In: The Dynamic Environment of the Ocean Floor, 441-470. Fanning, K. A., and F. T. Manheim, (Eds.) Lexington Books.

Livingstone, D. A., 1963: Chemical composition of rivers and lakes. U. S. Geol. Surv. Prof. Paper, 440G.

Ludden, J. N., and G. Thompson, 1978: Behavior of rare earth elements during submarine weathering of tholeiitic basalt. Nature, 274, 147-148.

Ludden, J. N., and G. Thompson, 1979: An evaluation of the behavior of the rare earth elements during the weathering of sea-floor basalt. Earth Planet. Sci. Letters, 43, 85-92.

Mackenzie, F. T., and R. M. Garrels, 1966: Chemical mass balance between rivers and oceans. Amer. J. Sci., 264, 507-525.

Matthews, D. H., 1971: Altered basalts from Swallow Bank, an abyssal hill in the N. E. Atlantic, and from a nearly seamount. Phil. Trans. Roy. Soc. London A, 168, 551-571.

McCulloch, M. T., R. T. Gregory, G. J. Wasserburg, and H. P. Taylor, 1980: A neodymium, strontium, and oxygen isotopic study of Cretaceous Samail ophiolite, and implications for the petrogenesis and seawater-hydrothermal alteration of oceanic crust. Earth Planet. Sci. Letters, 46, 201-211.

Melson, W. G., 1973: Basaltic glasses from the Deep Sea Drilling Project - chemical characteristics, compositions of alteration products, and fission track "ages". EOS Trans. Am. Geophys. Union, 54 (11) 1011-1014.

Melson, W. G., and Tj. H. van Andel, 1966: Metamorphism in the mid-Atlantic ridge, 22°N latitude. Mar. Geol., 4, 165-186.

Melson, W. G., V. T. Bowen, Tj. H. Van Andel, and R. Siever, 1966: Greenstones from the central valley of the mid-Atlantic ridge. Nature, 209, 604-605.

Melson, W. G., G. Thompson, and Tj. H. van Andel, 1968: Volcanism and metamorphism in the mid-Atlantic Ridge, 22°N latitude. J. Geophys. Res., 73, 5925.

Melson, W. G., and G. Thompson, 1971: Petrology of a transform fault zone and adjacent ridge segments. Phil. Trans. Roy. Soc. London A, 268, 423-441.

Melson, W. G., and G. Thompson, 1973: Glassy abyssal basalts, Atlantic seafloor near St. Paul's Rocks: petrography and composition of secondary clay minerals. Geol. Soc. Amer. Bull., 84, 703-716.

Mevel, C., 1979: Mineralogy and chemistry of secondary phases in low temperature altered basalts from DSDP legs 51, 52, and 53. In: Initial Reports of the Deep Sea Drilling Project, 51-53, 1299-1312. U. S. Gov't Printing Office.

Mevel, C., 1981: Occurrence of pumpellyite in hydrothermally altered basalts from the Vema Fracture Zone. Cont. Min. and Petrol., 76, 386-393.

Miyashiro, A., 1973: The Troodos ophiolitic conplex was probably formed in an island arc. Earth Planet. Sci. Letters, 19, 218-224.

Miyashiro, A., F. Shido, and M. Ewing, 1969: Diversity and origin of abyssal tholeiite from the mid-Atlantic ridge near 24° and 30°N latitude. Contrib. Mineral. Petrol., 23, 38-52.

Miyashiro, A., F. Shido, and M. Ewing, 1971: Metamorphism in the mid-Atlantic Ridge near 24°N and 30°N. Phil. Trans. Roy. Soc. London A, 268, 589-603.

Miyashiro, A., F. Shido, and K. Kanehira, 1979: Metasomatic chloritization of gabbros in the Mid-Atlantic Ridge near 30°N. Mar. Geol., 31, M47-M52.

Moore, J. G., 1966: Rate of palagonitization of submarine basalt adjacent to Hawaii. U. S. Geol. Surv. Prof. Paper 550, D163-D171.

Mottl, M. J., 1976: Chemical exchange between seawater and basalt during hydrothermal alteration of the oceanic crust. Ph. D. Thesis, Harvard University.

Mottl, M. J., 1982: Metabasalts, axial hot springs and the structure of hydrothermal systems at mid-ocean ridges. Geol. Soc. Amer. Bull. (in press).

Mottl, M. J., R. M. Anderson, W. J. Jenkins, and J. R. Lawrence, 1982: Chemistry of waters sampled from basaltic basement in DSDP holes 501, 504B and 505B. In: Initial Reports of the Deep Sea Drilling Program, 68-70 (in press).

Muehlenbachs, K., and R. H. Clayton, 1972: Oxygen isotope geochemistry of submarine greenstones. Can. J. Earth Sci.,

9, 471-478.
Muehlenbachs, K., and R. H. Clayton, 1976: Oxygen isotope composition of the oceanic crust and its bearing on seawater. J. Geophys. Res., 81, 4365-4369.
Murray, J., and A. F. Renard, 1891: In: Deep Sea Deposits, Report on Sci. Results of Voyage of H. M. S. Challenger. Chapt. 5. HMSO, London.
Nicholls, G. D. and V. T. Bowen, 1961: Natural glass from beneath red clay on the floor of the Atlantic. Nature, 192, 156-157.
Paster, T. P., 1968: Petrologic variations within submarine basalt pillows of the South Pacific-Antarctic Ocean. Ph. D. Thesis, Florida State Univ.
Philpotts, J. A., C. C. Schnetzler, and S. R. Hart, 1969: Submarine basalts: some K, Rb, Sr, Ba, rare earths, H_2O and CO_2 data bearing on their alteration, modification by plagioclase and possible source materials. Earth Planet. Sci. Letters, 7, 293-299.
Pytkowicz, R. M., and D. R. Kester, 1971: The physical chemistry of seawater. In: Ann. Rev. Ocean. Mar. Biol., 9, 11-60.
Quon, S. H., and E. G. Ehrlers, 1963: Rocks of northern part of mid-Atlantic ridge. Geol. Soc. Amer. Bull., 74, 1-8.
Richardson, S. H., S. R. Hart, and H. Staudigel, 1980: Vein mineral ages of old oceanic crust. J. Geophys. Res., 85, 7195-7200.
Rivers, M., 1976: The chemical effects of low temperature alteration of seafloor basalt. M. Sc. Thesis, Geology Dept., Harvard Univ.
Robinson, P. T., M. F. J. Flowers, H. U. Schminke, and W. Ohnmacht, 1977: Low temperature alteration of oceanic basalts, DSDP Leg 37. In: Initial Reports of the Deep Sea Drilling Program, 37, 775-793. U. S. Gov't Printing Office.
Rona, P. A., 1980: TAG hydrothermal field: mid-Atlantic ridge crest at 26°N. J. Geol. Soc. London, 137, 385-402.
Rona, P. A., K. Bostrom, and S. Epstein, 1980: Hydrothermal quartz vug from the Mid-Atlantic ridge. Geology 8, 569-572.
Rozanova, T. V., and G. N. Baturin, 1971: Hydrothermal ore shows in the floor of the Indian Ocean. Oceanology, 11, 874-879.
Rubey, W. W., 1951: Geological history of seawater: an attempt to state the problem. Bull. Geol. Soc. Amer., 62, 1111-1147.
Russel, K. L., 1970: Geochemistry and halmyrolysis of clay minerals, Rio Ameca, Mexico. Geochim. Cosmochim. Acta, 34, 893-907.

Scott, R. B., and A. Hajash, 1976: Initial submarine alteration of basaltic pillow lavas: a microprobe study. Amer. J. Sci., 276, 480-501.

Scott, R. B., D. G. Temple, and P. R. Peron, 1981: The nature of hydrothermal exchange between oceanic crust and seawater at 26°N latitude, MAR. In: The dynamic environment of the sea floor. Fanning, K. S., and F. T. Manheim, (Eds.): Heath and Co., p. 381-416.

Seyfried, W. S., M. J. Mottl, and J. L. Bischoff, 1978: Seawater/basalt ratio effects on the chemistry and mineralogy of spilites from the ocean floor. Nature, 275, 211-213.

Seyfried, W. E., and J. L. Bischoff, 1979: Low temperature basalt alteration by seawater: an experimental study at 70°C and 150°C. Geochim. et Cosmochim. Acta, 43, 1937-1947.

Shand, S. J., 1949: Rocks of the mid-Atlantic ridge. J. Geol. 57, 89-92.

Shido, F., A. Miyashiro, and M. Ewing, 1974: Compositional variation in pillow lavas from the mid-Atlantic ridge. Mar. Geol. 16, 177-190.

Siever, R., 1968: Sedimentological consequences of a steady-state ocean-atmosphere. Sedimentology, 11, 5-29.

Sillen, L. G., 1961: The physical chemistry of seawater. In: Oceanography, p. 549-581. Sears, M., Ed.: Amer. Assoc. for Adv. Sci., Washington, D. C.

Sillen, L. G., 1967: Gibbs phase rule and marine sediments. In: Equilibrium Concepts in Natural Water Systems. Adv. in. Chem. series, 64, 56-69, Washington, D.C.

Sommer, S., and L. Ailin-Pyzik, 1978: Microscale spatial distribution of trace elements in altered portions of DSDP basalts. EOS, 59, 1111 (Abstract).

Speiss, F. N., K. C. MacDonald, T. Atwater, R. Ballard, A. Carranza, D. Cordoba, C. Cox, V. M. Garcia, J. Francheteau, J. Guerrero, J. Hawkins, R. Haymon, R. Hessler, T. Juteau, M. Kastner, R. Larson, B. Luyendyk, J. B. Macdougall, S. Miller, W. Normark, J. Orcutt, and C. Rangin, 1980: East Pacific Rise: hot springs and geophysical experiments. Science, 207, 1421-1933.

Spooner, E. T. C., and W. S. Fyfe, 1973: Sub-seafloor metamorphism, heat and mass transfer. Contr. Mineral. Petrol. 42, 287-304.

Spooner, E. T. C., H. J. Chapman, and J. D. Smewing, 1977: Strontium isotopic contamination and oxidation during ocean floor hydrothermal metamorphism of the ophiolitic rocks of the Troodos Massif, Cyprus. Geochim. Cosmochim Acta, 41, 873-890.

Stakes, D. S. and K. F. Scheidegger, 1981: Temporal variations in secondary minerals from Nazca plate basalts, diabases, and microgabbros. In: Nazca Plate: Curstal Formation and Andean Convergence. Geol. Soc. Am. Mem. 154, 109-130.

Stakes, D. S., and J. R. O'Neil, 1982: Mineralogy and stable isotope geochemistry of hydrothermally altered oceanic rocks. Earth Planet. Sci. Letters, 57, 285-304.

Staudigel, H., W. B. Bryan, and G. Thompson, 1979: Chemical variation in glass-whole rock pairs from individual cooling units in holes 417D and 418A. In: Initial Reports of the Deep Sea Drilling Project, 51-53, 977-986. U. S. Gov't. Printing Office.

Staudigel, H., S. R. Hart, and S. H. Richardson, 1981: Alteration of the oceanic crust: processes and timing. Earth Planet. Sci. Letters, 52, 311-327.

Stumm, W., and J. J. Morgan, 1981: Aquatic Chemistry. 2nd Edn. J. Wiley, N. Y., 780 pp.

Thompson, G., 1973: A geochemical study of the low temperature interaction of seawater and oceanic igneous rock. Trans. Amer. Geophys. Union, 54, 1015-1019.

Thompson, G., and W. G. Melson, 1972: The petrology of oceanic crust across fracture zones in the Atlantic ocean: evidence of a new kind of seafloor spreading. J. Geol., 80, 526-538.

Thompson, G., and M. Rivers, 1976: The low temperature weathering of oceanic basalts by seawater: 1. major element fluxes. Geol. Soc. Amer. Ann. Mtg., 8, 1137 (Abstract).

Thompson, G., and S. E. Humphris, 1977: Seawater-rock interaction in the oceanic basement. In: Proc. 2nd Internatl. Sym. Water-Rock Interact., v. III, 3. Eds. Pacquet, H., Tardy, Y.: Univ. L. Pasteur, Strasbourg.

Tiezzi, L. J., and R. B. Scott, 1980: Crystal fractionation in a cumulate gabbro, MAR 26°N. J. Geophys. Res., 85, 5438-5454.

Wiseman, J. D. H., 1937: Basalts from the Carlsberg Ridge, Indian Ocean. Geological and mineralogical investigations John Murray Expedition. Brit. Mus., Natl. Sci., 3, 1-28.

Wolery, T. J., 1979: Seawater-ocean crust hydrothermal chemistry: some theoretical considerations. EOS, 60, 863 (Abstract).

Wolery, T. J., and N. H. Sleep, 1976: Hydrothermal circulation and geochemical flux at mid-ocean ridges. J. Geol., 84, 249-276.

THE PHYSICS AND CHEMISTRY OF ICELAND VERSUS MID-OCEAN RIDGE HYDROTHERMAL SYSTEMS: CONTRASTING BOUNDARY CONDITIONS

Roger N. Anderson

Lamont-Doherty Geological Observatory
Columbia University
Palisades, N.Y. 10964

INTRODUCTION

This section contains three papers on the chemistry and physics of hydrothermal systems on Iceland and on mid-ocean ridges. These contributions detail the contrasting styles of the two systems in detail. But before examining the specifics of these hydrothermal circulation systems, it is appropriate to step back and examine the differences from a broad perspective. The physical and chemical processes of a subaereal and a deep subaqueous hot spring are similar, except that the boundary conditions are dramatically different. These changing boundary conditions produce very different surface expressions of hydrothermal convection, and it is important to identify cause and effect. Below two examples are given of differences caused by contrasting land versus deep sea hydrothermal activity that result from the same physical and chemical principles operating under atmospheric and greater than 200 atmospheres of pressure, respectively. First, the effects of boiling or the lack thereof in the two systems is quite different. Second, the pattern of fracturing which controls permeability is, itself, controlled by very different pore pressure and thus effective stress regimes on land versus in the deep sea.

CHEMICAL DIFFERENCES

A thorough review of the chemistry of hydrothermal systems operating on mid ocean ridges follows from the paper by Thompson (1983). Geochemical detail of the Iceland hydrothermal system is then presented in a paper by Kristmannsdottir (1983). One of the principal interests of the remarkable discovery of 'black smokers' spewing 350° mineral-laden water onto the deep sea floor is to examine metal-

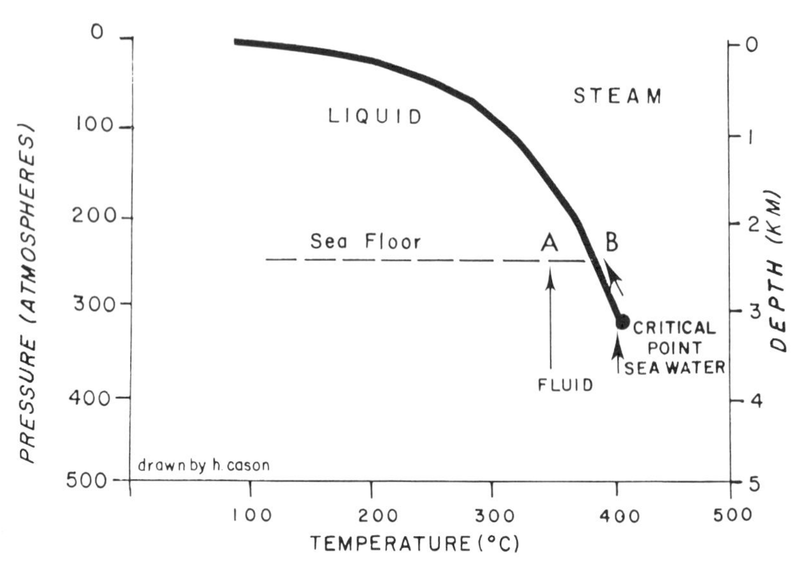

Fig. 1. Pressure-Temperature (P,T) curve for boiling sea water (solid line. If hydrothermal fluid passing up a chimney at a mid-ocean ridge axis intersects the sea floor before reaching the P,T boiling curve, no steam will occur (case A above). If, however, fluid is hot enough and/or the ridge crest is shallow enough, fluid will intersect the P,T boiling curve, BEFORE reaching the sea floor; it will then flash to steam, leaving behind highly saline fluid and taking gold and silver to higher levels of the chimney (case B). Figure adapted from Delaney and Cosens, 1982.

logenesis in progress. As Thompson (1983) points out, such deep-sea hydrothermal systems probably deposited a significant percentage of the minerals mined in the world today. Alternatively, such a system on Iceland produces superheated steam at the surface rather than superhot black fluid. The difference is caused by the effect of pressure on the flash point of water (Figure 1). As Stefanssen (1983) shows, most metal deposition occurs at the depth of rapid single-phase cooling of the hydrothermal fluid. On land this happens at great depth within the hot springs with fluid flashing to steam at shallower depths. At a deep sea ridge crest, deposition of metals caused by single phase cooling takes place much closer to the surface and flashing to steam hardy ever occurs.

On the East Pacific Rise, metal rich sulfides are deposited directly from fluid exiting at the sea floor. Although high metallic concentrations have been found in the Krafla Field (Stefansson, 1981), ore deposits are not common on Iceland. Presumably, metal deposition is occurring, but at great depth within the subaereal hydrothermal system. In the deep sea this process is occurring at much shallower depths. But an interesting separation occurs between metals when steam is formed. As Delaney and Cosens (1982) point out, steam carries precious metals such as gold and silver away with it. Two-phase separation of metals would be significant if boiling occurs on deep sea ridges. Such boiling would not be expected because of the very high temperature required to form steam at 200 atmospheres of pressure (Fig. 1). So it would be quite a surprise if evidence for flashing to steam were found in a deep sea hydrothermal system.

It is ironic that the oldest of marine geological techniques has presented just such a surprise from the mid-ocean ridges. To a large extent, high technology has been responsible for the recent series of discoveries that have remolded our concepts of how the sea floor cools at the volcanically active mid-ocean ridge. For example, submersible-carried bottom television, deep-sea drilling, and deeply towed remote sensing packages, have been the kinds of tools that were used first to discover, then to describe the 'black smoker' hot springs. It is also remarkable that these hot springs were found for the first time only in 1979, at just about the same time that the first pictures from Jupiter's moons were arriving back on Earth.

A dredge such as that used on H.M.S. CHALLENGER to return the first rock samples from the deep sea floor over 100 years ago, has delivered new scientific wonders. Delaney and Cosens (1982), report inclusions of super-high salinity water, in basalts dredged from the Mid-Atlantic Ridge. This leads them to conclude that boiling has occurred within the deep-sea hot spring system that generated the fluid inclusions. Evidence of boiling water has never before been found on a mid ocean ridge.

First, why would boiling be such a radical process under such pressures? Obviously, the higher the pressure, the higher the temperature required to transform liquid into gas (Figure 1), simply because energy is required to expand the molecules from fluid to gas, and the molecules have to push out harder (using more heat energy) under 200 atmospheres of pressure than that required under one atmosphere of pressure.

For boiling to occur on the Mid-Atlantic Ridge, hot spring water must move upward from its heat source, magma within a volcano, faster than it loses heat to the rock during its traverse to the surface. This requires a chimney, such as those observed by the diving submersible ALVIN on the East Pacific Rise crest (Rise Project Group, 1980). Hot hydrothermal liquid mixing with cold water will precipitate a thick coating of metals on the wall of a chimney. At some depth within a Mid-Atlantic Ridge chimney, boiling occurred in the past when the temperature of the hot water became high enough to intersect the liquid-vapor phase transition curve below the sea floor (Fig. 1). The steam that then formed left behind liquid of very high salinity. Some of that liquid ended up as fluid bubbles in the rock dredged, fortuitously, millions of years later.

We know so little about the deep-sea hydrothermal system, that before the Delaney and Cosens (1982) discovery, boiling was thought to be an absurd concept beneath the sea floor. Now, scientific tools such as those that might be lowered into a deep sea drilling borehole or placed down a chimney by a diving submersible might find a steam vent where we could observed for the first time the _in situ_ precipitation of _precious_ as well as the more common non-precious metals at shallow depths in the oceanic crust.

The major effect of the difference in pressure between the sub-aereal and submarine hydrothermal systems is that in the land case, metallogenesis occurs deep within the subsurface, whereas deposi-

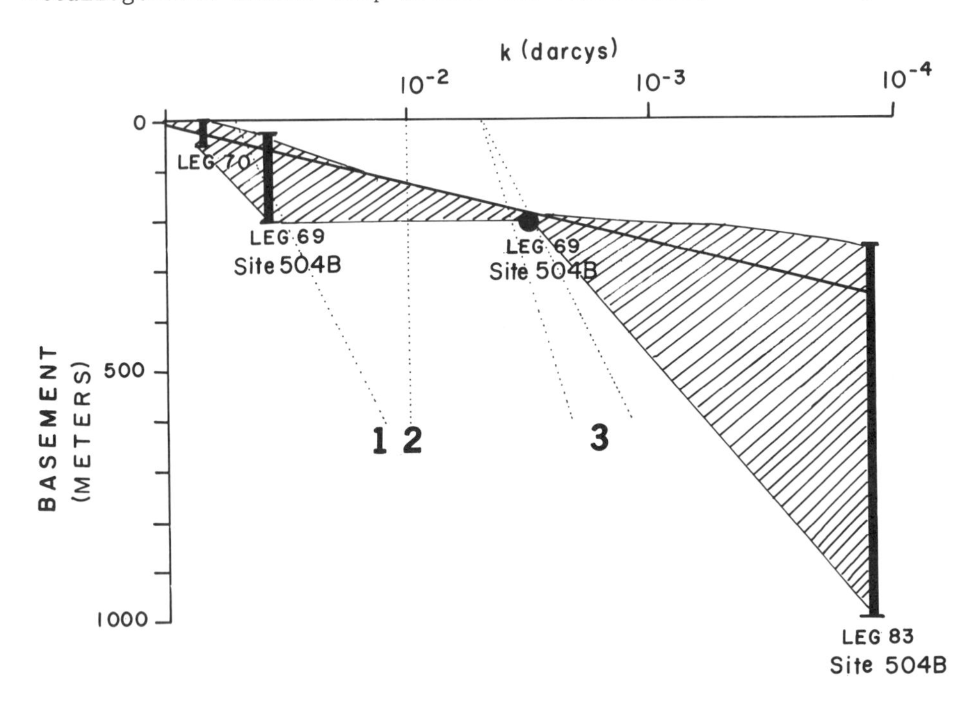

Fig. 2 Permeability measured _in situ_ in 504B, using inflatable packer and slug-type injection flow tests. 1,2,3 are theoretical models of ocean crustal permeability made before these tests (Anderson and Zoback, 1982, Anderson et al., 1983).

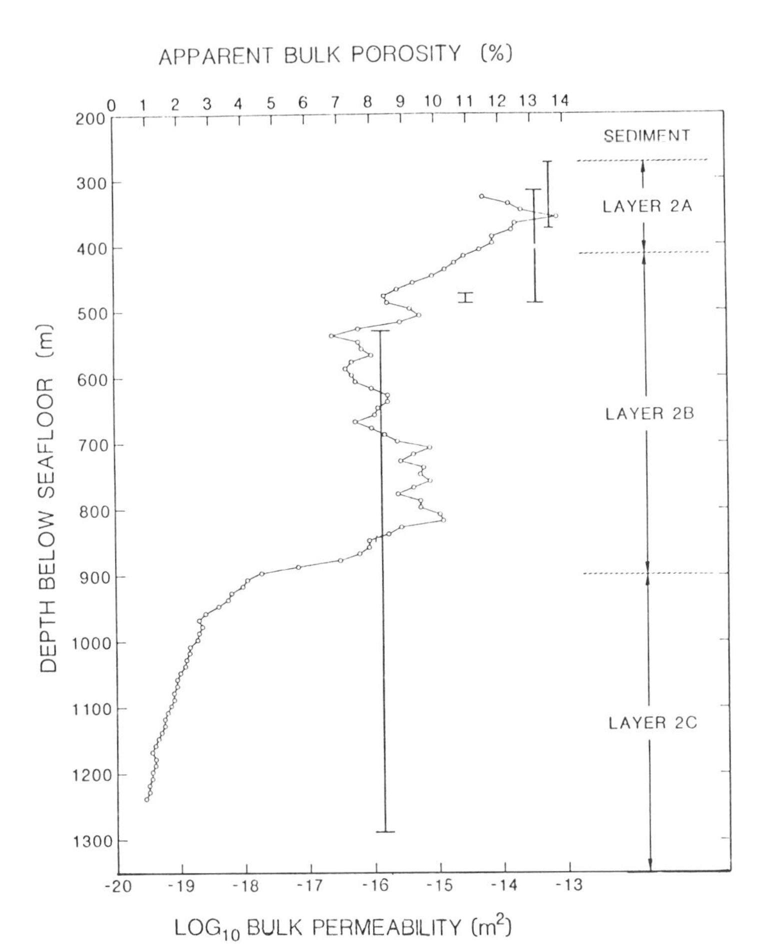

Fig. 3. (Open circles) apparent bulk porosity (scale at top) from electrical resistivities measured by Becker et al. (1982) in site 504B borehole (corrected for temperature using Archie's Law). Vertical bars are measured permeabilities from Anderson and Zoback (1982) and Anderson et al. (1983),(Scale at bottom defines meters squared in scale of permeability). No direct correlation between top and bottom scales is implied (after Becker et al., 1983).

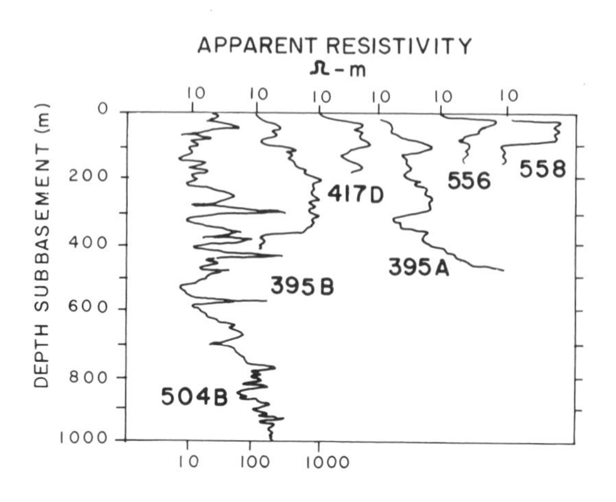

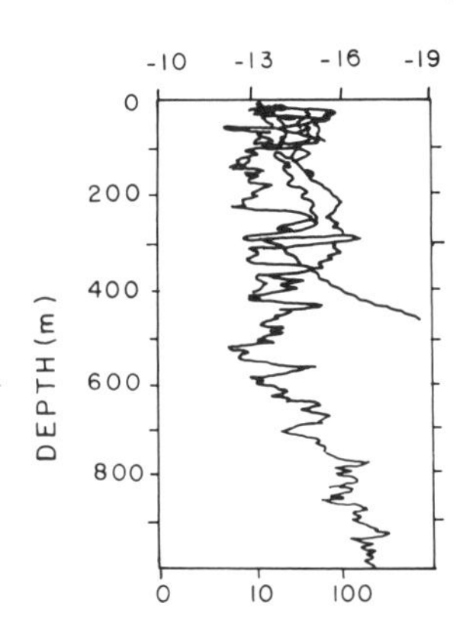

Fig. 4. Electrical Resistivities from dual laterlogs of several DSDP holes plotted against log from 504B to show similar porosity (below) for oceanic crust at these geographically variable sites. Ω-m is ohm-meters. None of these curves are corrected for downhole temperature increase. All but 504B (corrected in Fig. 5) are cold throughout the drilled section. Locations are all in the North Atlantic (396B: 23°05'N, 43°30'W; 417D; 25°04'N, 68°05'W; 395A: 22°45'N, 46°05'W; 556: 38°56'N, 34°41'W; 558: 37°46'N, 37°20'W) except for 504B; 1°15'N, 83°45'W (Ref. various DSDP Initial Reports volumes).

tion of metals can occur directly onto the deep sea floor. If steam could conceivably form in a deep sea environment, it would be 'metal-laden' and useless for, say, power generation purposes. The opposite is the case on land. Steam is abundant and comparatively clean and free of metals, because the metals have been deposited deep within the system. Therefore, the steam is useful for power generation.

PHYSICAL DIFFERENCES

The permeability of a porous medium undergoing convection is the single most important physical parameter controlling the form and magnitude of hydrothermal circulation. Again, the difference in pressure between the land and deep sea systems produces significantly different permeability distributions. Stefansson (1983) demonstrates that the topographic control on convection is the opposite on land than in the deep sea, but the change in permeability with depth is also profoundly different in the two environments. The sum total of our knowledge of permeability in the deep sea floor comes from *in situ* flow tests in two deep sea drilling boreholes, one in the Atlantic and one in the Pacific (Anderson and Zoback, 1982; Hickman et al., 1983; Anderson et al., 1983). In only one of these, at site 504B on the south flank of the Costa Rica Rift, Eastern Equatorial Pacific, is the variation with depth known. At site 504B an integrated suite of geophysical logging experiments has turned up a correlation between permeability and fracture porosity which promises to allow us to measure simply and cheaply the permeability versus depth function in all deep sea drill holes which have electrical resistivity logs run in them. Specifically, Anderson et al. (1983a) found that the permeability measured directly using flow tests, decreased approximately exponentially with depth at site 504B; decreasing from 70 millidarcies (MD) in highly fractured layer 2A to 2 MD in the fractured but altered pillow basalts of layer 2B, then to much lower values in the sheeted dikes of layer 2C (Fig. 2).

The form of this permeability decrease was similar to the decrease in apparent bulk porosity recorded by electrical resistivity logging of the hole at site 504B (Becker et al., 1982, Becker et al. 1983; Anderson et al., 1982) (Fig. 3). By examining electrical resistivity logs from other deep sea drilling boreholes published in the past, one can see that an exponential decreasing electrical resistivity is a general result (Fig. 4). The prediction is then that the surficial kilometer or so of the oceanic crust is up to six orders of magnitude more permeable than the deeper oceanic crust. Of course, no measurements have yet been made in the gabbros of oceanic layer 3, simply because we have not yet been able to drill deep enough. This is clearly one of the highest priority measurements to make in geophysics today.

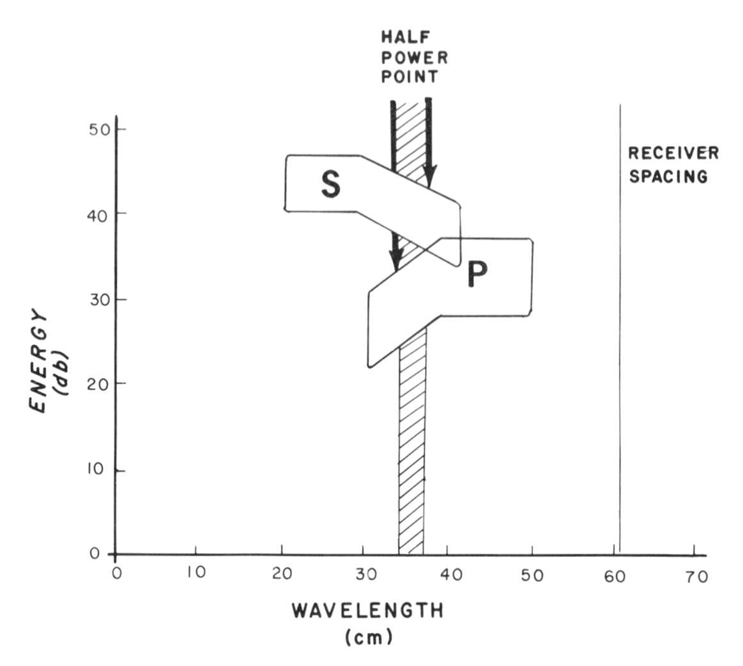

Fig. 5. Energy-wavelength cross-plot from cross-spectral analysis of multiple-receiver sonic log of dikes at bottom of hole at 504B, (Anderson et al., (1983). Note the drop in energy of sonic wavelengths between 30 and 40 cm long. This notch filtering characteristic of the dikes is thought to be caused by scattering from cyclical zones of intense horizontal fracturing in the dikes. The wavelength is complexly related to the spacing of the fracture zones. P and S are compressional and shear wave coda.

Continuing our contrast, Iceland is made up of a thick crust of horizontally stratified basaltic flow units. Stefansson (1983) clearly demonstrates that hydrothermal fluids penetrate high permeability rock to much greater depths on Iceland, than that given by the thickness of high permeability, low electrical resistivity pillow units in the oceanic crust. The difference is related to the way the two rock systems are fractured. Anderson et al. (1983b) examined the fracture distribution with depth in hole 504B and found distinct signs of fracture cyclicity in the dikes. They found that the dikes act like a notch-or band-reject filter to the propagation of sonic waves.

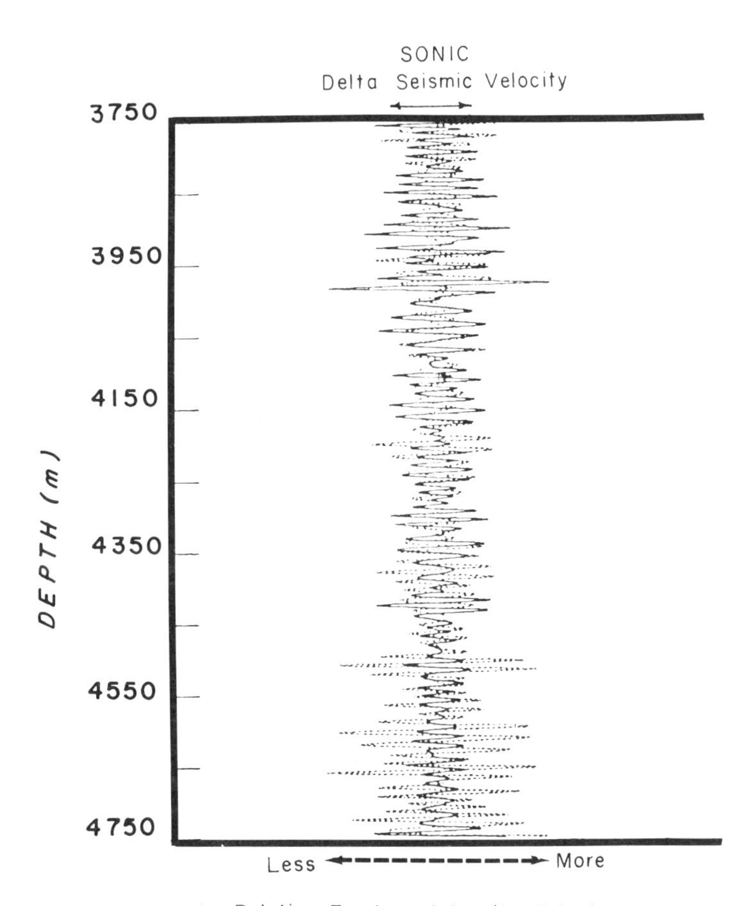

Fig. 6. The cyclical fracturing shows up throughout the 504B basaltic section when comparing borehole televiewer images with the sonic compressional wave velocity after band-pass filtering of all but (10-20 m) wavelengths. In-phase signals indicate correlation of zones of intense fracturing, low electrical resistivity and low sonic velocity. From Newmark et al.(1983).

That is, 30 to 40 cm wavelength sonic energy was being severely scattered (Figure 5). They interpreted this to be caused by a cyclical pattern to the fracturing of the dikes. Newmark et al. (1983) further identified a cyclical fracture pattern throughout the pillows as well as the dikes (Fig. 6). They found that zones of low electrical resistivity correlated with zones of intense fracturing seen on the borehole televiewer imagery of the wellbore, which in return correlated with zones of low seismic velocity (Fig. 6). The pillow basalts were more intensely fractured than the dikes, but the cyclicity continued into the dike sequence. In addition, Zoback and Anderson (1982) found that the upper and lower pillow layers (2A and 2B), though fractured intensely, were altered differently. The upper pillows contained fractures open to sea water, whereas the lower pillows contained fractures filled with abundant clay alteration products. All of these geophysical observations lead to a model for the permeability of the oceanic crust in which abundant fracturing occurs throughout the upper crust: more so in the pillows than in the dikes, however. Alteration products plug-up the plumbing from bottom to top in the crust as aging progresses. Therefore, a relatively shallow zone (c. 1 km) of rock with high permeability exits in young ocean as compared to Iceland (c. 5 km). Notice there is not yet a causal mechanism in this hypothesis.

A clue to why fracturing attenuates or decreases with depth more readily in the ocean crust versus Iceland comes from the ratio of volume of extrusives between the two settings. Fracturing is obviously going to be greater where quenching and subsequent shattering is favored. Extrusives fracture more easily than intrusives, but more fundamental physical processes are involved. Zoback and Anderson (1982) and Newmark et al. (1983) observed an additional remarkable fact in the borehole televiewer images of the pillows and dikes of the oceanic crust; most of the fractures were horizontal to sub-horizontal. Horizontal fracturing results when the principal horizontal stresses exceed that of the overburden vertical stress. In Iceland, this is the predominant condition at the time of extrusion since atmospheric pressure is often the only overburden pressure. But in the deep sea, even extrusives are subjected to 200 bars of pressure, and the dikes intrude into ~300 bars of vertical pressure. How then do they fracture horizontally? Hydrothermally controlled pore pressures must be larger than the overburden pressure at the time of fracturing in the ocean crust. Such high pore pressures probably limit the degree of fracturing in the intrusive layers of the oceanic crust. Conversely, the lack of high pore pressures on Iceland allows for more extensive fracturing to greater depth than in the ocean case.

SUMMARY

We have considered boiling point and permeability as two examples of the effects of differing boundary conditions acting on the physics and chemistry of a hydrothermal convection system. Additional examples are cited in the following papers. As one reads further, the overwhelming importance of hydrothermal circulation both on land and under the sea becomes more and more obvious. Diverse geological phenomena such as the chemical constancy of ocean water, metallogenesis, and the basic fracture pattern of volcanic rocks all are directly controlled by convection in the crust. While our knowledge of this system is still primitive, this section will convince one of the great strides currently being made in understanding this basic process, both on land and under the deep sea.

ACKNOWLEDGEMENTS

Sincere thanks go to Peter Rona for encouraging the author to complete this paper and to the organizers for the splendid NATO conference without which this book would not be possible. M.D. Zoback, S. Hickman, D. Moos, R. Newmark and D. Leslie contributed significantly to the concepts outlined in this paper. This work was completed under NSF grants OCE-79-27026 and OCE-81-10901 and ONR contract TO-0098, scope HH.

REFERENCES

Anderson, R.N., and Zoback, M.D., 1982, The permeability of the upper oceanic crust, J. Geophys. Res., 81: 2860-2868.

Anderson, R.N. et al., 1982, DSDP hole 504B, The first reference section over 1 km through layer 2 of the oceanic crust, Nature, 300: 589-594.

Anderson, R.N., Zoback, M.D., and Newmark, R.L., The Permeability versus depth function of the oceanic crust, J. Geophys. Res.: submitted, 1983a.

Anderson. R.N., Leslie, D., and Newmark, R.L., The Fracture Spectrum of the Oceanic Crust at Site 504B, Eastern Equatorial Pacific, J. Geophys. Res.: submitted, 1983b.

Becker, K., et al., 1982, In Situ electrical resistivity and bulk porosity of the oceanic crust, Costa Rica Rift, Nature, 300: 594-598.

Becker, K., Langseth, M.G., Von Herzen, R.P., and Anderson, R.N., Deep crustal geothermal measurements, Costa Rica Rift, J. Geophys. Res.: in press, 1983.

Delaney, J.R., and Cosens, B.A., Boiling and Metal Deposition in Submarine Hydrothermal Systems, Marine Tech. Journal, 16: 62-67, 1982.

Delaney, J.R., Mogk, D.W., and Mottl, M.J., Quartz-cemented, sulfide-bearing greenstone breccias from the Mid-Atlantic Ridge - samples of a high-temperature hydrothermal upflow zone, Science: in press, 1983.

Hickman, S., Langseth, M.G., and Svitek, J., 1983, Permeability and pore pressure measurements at site 395 on the Mid Atlantic Ridge, J. Geophys. Res.: submitted.

Kristmannsdóttir, H., 1983, Chemical evidences from Icelandic geothermal systems as compared to submerged geothermal systems, this volume.

Rise Project Group, 1980, East Pacific Rise; hot springs and geothermal experiments, Science, 207: 1421-1433.

Stefánsson, V., 1982, The Krafla Geothermal Field, Northeast Iceland, in "Geothermal Systems. Principals and Case Histories", L. Rybach and L.P.L. Muffler, eds., John Wiley, New York: 273-294.

Stefánsson, V., 1983, Physical Environment of hydrothermal systems in Iceland and on submerged Oceanic ridges: this volume.

Thompson, G., 1983, Basalt-Sea Water Interaction: this volume.

Zoback, M.D., and Anderson, R.N., 1982, Borehole Televiewer Imagery of the upper oceanic crust, DSDP site 504B, Nature, 295: 345-379.

CHEMICAL EVIDENCE FROM ICELANDIC GEOTHERMAL SYSTEMS AS COMPARED TO SUBMARINE GEOTHERMAL SYSTEMS

Hrefna Kristmannsdóttir

National Energy Authority
Department of Natural Heat, Geochemistry Section
Grensásvegur 9, 108 Reykjavík, Iceland

ABSTRACT

The Reykjanes geothermal system is the most interesting of the Icelandic geothermal areas for chemical comparison with submarine geothermal systems. The geothermal water at Reykjanes is seawater. Base temperatures in the Reykjanes geothermal system is considerably lower than the estimated temperatures of the submarine systems. In the Krafla geothermal system the base temperatures are almost as high as in the submarine geothermal systems but the geothermal water is of meteoric origin with low content of dissolved solids. Magmatic activity has strongly influenced the chemistry of the Krafla geothermal system.

In the Icelandic geothermal systems Mg and K are enriched at upper levels. Silica, Na, Ca, Fe and many of the trace elements have been mobilized, but a significant trend of depletion or increase has not been demonstrated. Mass fluxes by the rock alteration in the Reykjanes and Krafla area appear to be generally less than in hydrothermally altered metabasalts from the oceanic crust. This may partly be due to different sampling methods. The main reason for the difference is believed to be that all rock samples from the Icelandic areas are from the upflow zones in the geothermal system. A Mg enrichment and Ca depletion of the same magnitude as that demonstrated in submarine greenstones would rather be expected in the deep inflow zones of the Icelandic geothermal systems.

INTRODUCTION

Investigation of geothermal areas in Iceland been pursued for a long time. Surface mapping and some geophysical measurements have been done in most known high-temperature areas. Many areas have been drilled, the drillholes logged, and the rocks and fluids sampled (see Pálmason et al., 1979). In some areas steam drillholes have been sampled regularly for periods of 10-20 years. Variation in surface activity has been followed for tens of years in many areas.

In contrast the existence of the submarine geothermal systems has only been known for a short time (Wolery and Sleep, 1976), none has been drilled into and sampling of sea floor discharge is rather scanty (Corliss et al., 1979, Craig et al., 1980, Edmond 1980). Rocks probably altered by geothermal activity have been extensively dredged from the seafloor and recovered from drillholes.

Extensive analyses have been performed on samples from drilling and dredging on Mid-Ocean Ridges (see Humphris and Thompson, 1978, Mottl, 1982). In geothermal drillings in Iceland samples are mostly cuttings. Analyses of the samples are mostly limited to routine XRD runs. Chemical analyses of the rocks and minerals are only done in rare cases as they are considered to be of little direct use in the investigation of the geothermal field.

The data sets from geothermal areas in Iceland and on Mid-Ocean ridges are thus rather poorly compareable. From Iceland there exist ample chemical data on the fluids but scarce data on the altered rocks almost wholly limited to drill cuttings. From the Mid-Ocean ridges vast data exist of chemistry of altered rocks, but only sparse data on the fluids. The location of the analysed altered rocks within the geothermal systems is seldom known with any certainty. Chemical analyses of cuttings give an average composition of rocks over several meters of penetration (depending on drilling conditions) and short range variation is obscured.

The analyses of rocks from the Icelandic geothermal areas thus give the mass fluxes from one rock formation to another within a geothermal system. In contrast, analyses of altered rocks from Ocean Ridges give short range transport, but not in context of the geothermal systems as a whole.

In suboceanic geothermal systems the water is of seawater origin and highly saline. In most geothermal systems in Iceland the water is meteoric with low content of dissolved solids. However, geothermal areas on the Reykjanes peninsula (see Fig.1 for location) are at least partly fed by seawater (see Pálmason et al., 1979). In the Svartsengi area the deep fluid contains about 14000 ppm Cl

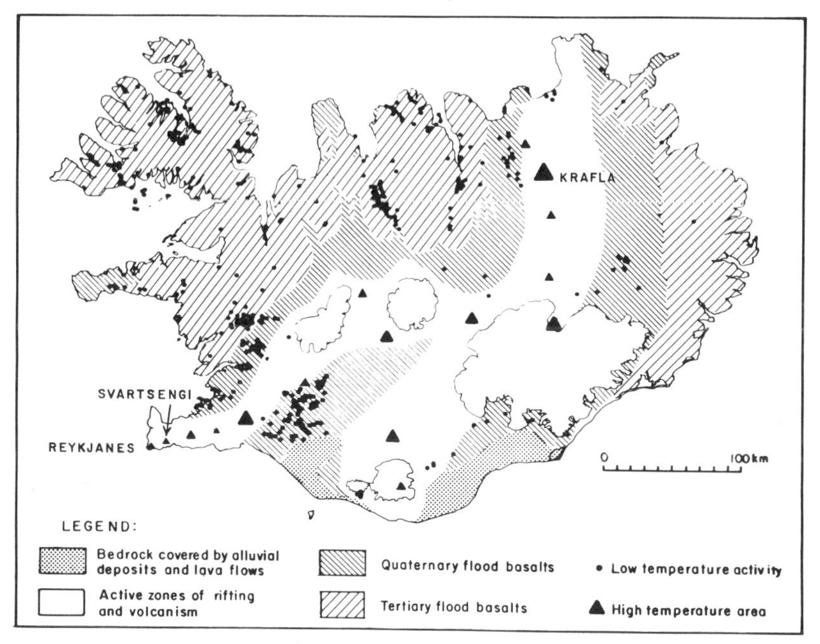

Fig. 1. A simpfied geological map of Iceland (see Sæmundsson, 1979) showing the sites of geothermal activity. The location of the Reykjanes, Svartsengi and Krafla high temperature geothermal areas is noted.

and that of the Reykjanes area contains about 19800 ppm which is similar to normal seawater.

The lithology of rocks is of great importance for the course and rate of hydrothermal alteration. The Icelandic geothermal areas are all within the zone of active volcanism and rifting. The upper part of the volcanic pile has been built up during glaciation. An upper formation of hyaloclastites (about 1000 m thick) succeeded by a pile of subaerial lavaflows is typical for the stratigraphic section in the high-temperature areas (see Kristmannsdóttir, 1975, Kristmannsdóttir, 1979). Intrusives become generally more frequent with increase in depth and in many fields they dominate below 2 km depth.

The basalts in the oceanic crust are probably fairly massive (Wolery and Sleep, 1976) although they may be brecciated in places. They are succeeded downwards by intrusives which probably are fairly impermeable.

Of the Icelandic high-temperature geothermal areas the Reykjanes area is most similar to submarine geothermal systems. As mentioned above certain differences exist. Several papers have dealt with the Reykjanes geothermal area (a.o. Björnsson et al., 1970, Björnsson et al., 1972, Kristmannsdóttir, 1976, Kristmannsdóttir, 1979, Kristmannsdóttir, 1981, Tómasson and Kristmannsdóttir, 1972, Arnórsson 1978, Ólafsson and Riley, 1978). Numerous data about chemistry of rocks and fluids are either unpublished or exist only as internal reports (Hauksson, 1981). This paper presents a synthesis of the present knowledge of the chemistry of the Reykjanes geothermal area. For certain features references are made to the Krafla geothermal area (Kristmannsdóttir, 1979, Kristmannsdóttir, 1981 and Kristmannsdóttir in prep.) which is one of the most densely drilled and investigated areas in Iceland. The Krafla area is also magmatically active at present, which influences the chemistry of the system.

THE REYKJANES GEOTHERMAL FIELD

General Description

The Reykjanes geothermal area lies on the southern tip of the Reykjanes peninsula. Surface alteration covers an area of 1 km^2, but geothermal circulation has much greater extent below 1000 m depth. A model of the geothermal system based on the present available data from the area is shown in Figure 2. Locations of existing drillholes are also shown.

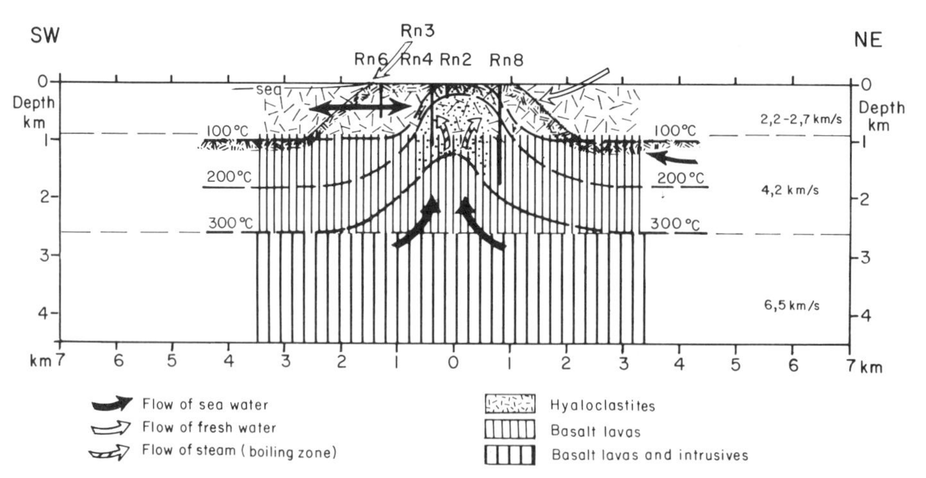

Fig. 2. A model of the Reykjanes geothermal area, slightly modified from Björnsson et al., 1970.

The subsurface rocks encountered in the drillholes are basaltic hyaloclastites, tuffaceous sediments and lavas down to approximately 1000 m. Below that depth basalt lavas are dominant. Intrusives are scarcely encountered in the drilled section. Based on geophysical survey and comparison with other areas intrusives probably become more common below approximately 2500 m depth. The rocks are assumed to have a fairly high permeability as the porosity of the hyaloclastites and sediments is 20-30%. No porosity data are available of the lavas, but based on their texture the porosity is estimated to be on average 2-5%. The main aquifers are found below 1000 m depth in the lava formation. The maximum measured temperature is 290°C.

Alteration of the underground rocks is extensive and up to 90% of the rocks are recrystallized. The three main alteration zones found in Icelandic geothermal areas are all represented (Kristmannsdóttir, 1979). Amphibole is found sporadically at lower levels, but is not found continously or commonly enough to define the fourth epidote - actinolite alteration zone as encountered in the Krafla geothermal field (Kristmannsdóttir, 1981 and Kristmannsdóttir in prep.). The degree of alteration thus corresponds with the zeolite to the lowest part of the greenschist metamorphic facies.

Alteration Mineralogy

The distribution of alteration minerals and a simplified stratigraphic section of drillhole Rn-8, the deepest drillhole in the geothermal area, are shown in Fig. 3. As shown in Fig. 2, this drillhole is outside the area of surface alteration and the zone of boiling above 600 m.

The most voluminous alteration minerals (Kristmannsdóttir, 1976) are the sheet-silicate sequence: Fe-rich saponites → mixed - layer chlorite/saponite, swelling chlorites → chlorites. Locally illite/chlorite mixed-layer minerals are found. Selected microprobe analyses of sheet silicates from drillhole Rn-8 are presented in Table 1. Mineral composition is calculated for 22(O,OH) as the silica composition is in all cases too high for any of them being normal chlorite. In samples 6-10 chlorite and swelling chlorite are encountered by XRD. In samples 1-3 smectite is dominant in the XRD analysis, but mixed-layer minerals dominate in samples 4 and 5.

Analysis no. 8 is of quite unusual composition. Minerals of such low aluminium, high silica and varying Fe/Mg contents were analysed in several samples at all depths in Rn-8, mainly as linings in cracks and pores. In some of the samples illite - chlorite mixed - layer minerals were analysed (by XRD) but not in all samples.

Table 1. Selected Micoprobe analyses of Sheet silicates in samples from drillhole Rn-8 on Reykjanes.

Sample no	SiO_2	TiO_2	Al_2O_3	FeO	MnO	MgO	CaO	Na_2O	K_2O	Total	Mineral Composition
1	44.56	1.47	10.70	11.07	0.21	17.99	3.05	0.85	0.16	90.06	$Si_{6.5}Ti_{0.2}Al_{1.8}Fe_{1.4}Mg_{3.9}Ca_{0.5}(O,OH)_{22}$
2	45.27	0.0	8.69	18.49	0.18	15.11	1.36	1.02	0.27	90.39	$Si_{6.8}Al_{1.5}Fe_{2.3}Mg_{3.4}Ca_{0.3}(O,OH)_{22}$
3	40.78	0.68	10.52	15.61	0.27	18.34	1.91	0.15	0.04	88.37	$Si_{6.2}Al_{1.9}Fe_{2.0}Mg_{4.2}Ca_{0.3}(O,OH)_{22}$
4	32.45	0.61	14.01	20.21	0.32	17.75	2.41	0.42	0.00	88.71	$Si_{5.2}Al_{2.6}Fe_{2.7}Mg_{4.2}Ca_{0.4}(O,OH)_{22}$
5	36.05	2.22	11.80	13.28	0.27	19.98	2.82	0.47	0.00	86.89	$Si_{5.6}Ti_{0.3}Al_{2.2}Fe_{1.7}Mg_{4.7}Ca_{0.5}(O,OH)_{22}$
6	37.16	0.06	12.41	18.85	0.25	18.54	0.51	0.62	0.00	88.42	$Si_{5.8}Al_{2.3}Fe_{2.5}Mg_{4.3}(O,OH)_{22}$
7	41.75	1.69	8.65	17.66	0.27	15.57	3.52	0.88	0.01	90.18	$Si_{6.4}Ti_{0.2}Al_{1.6}Fe_{2.3}Mg_{3.5}Ca_{0.6}(O,OH)_{22}$
8	56.11	0.03	2.14	10.67	0.06	24.58	0.14	0.25	0.20	94.23	$Si_{7.2}Al_{0.3}Fe_{1.1}Mg_{4.3}(O,OH)_{22}$
9	39.09	0.02	11.97	18.55	0.14	18.04	0.15	0.22	0.00	88.21	$Si_{6.0}Al_{2.2}Fe_{2.4}Mg_{4.2}Ca_{0.2}(O,OH)_{22}$
10	39.95	0.07	18.23	11.36	0.18	11.53	3.43	1.05	0.00	85.85	$Si_{6.1}Al_{3.3}Fe_{1.5}Mg_{2.6}Ca_{0.6}(O,OH)_{22}$

1 Greenish smectite from hyaloclastite at 154 m depth
2 Greenish mineral in altered tuffaceous sediment at 294 m depth
3 Green mineral in tuffaceous sediment at 508 m depth
4 Grey and greenish minerals in hyaloclastite breccia at 716 m depth
5 Brownish mineral in hyaloclastite breccia at 760 m depth
6 Green mineral in coarse sediment at 1158 m depth
7 Brownish green mineral in coarse sediment at 1158 m depth
8 Brownish green mineral in basalt at 1598 m depth
9 Greenish mineral in basalt at 1598 m depth
10 Greenish chlorite in basalt at 1696 m depth

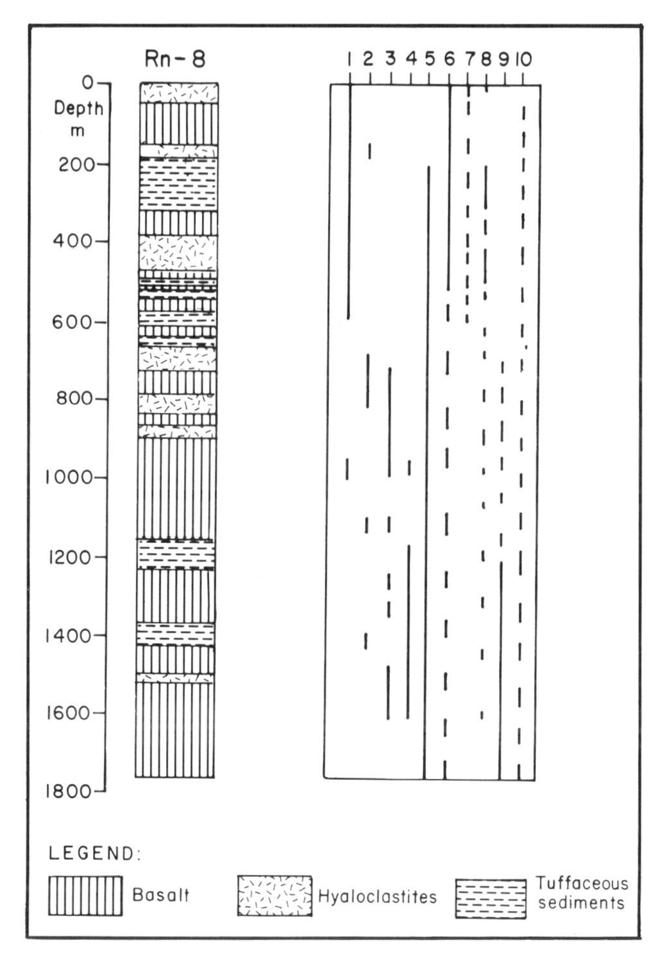

Fig. 3. A simplified stratigraphic section and distribution of main alteration minerals by depth in drillhole Rn-8 of the Reykjanes geothermal area. 1: smectite, 2: mixed-layer smectite/chlorite, 3: swelling chlorites, 4: chlorite, 5: quartz, 6: calcite, 7: zeolites, 8: prehnite, 9: epidote, 10: anhydrite.

Poorly crystalline, clay-like Mg/Fe silicates, were found to precipitate inside the casing of drillhola Rn-8 during two years closure (Kristmannsdóttir, 1981 b). Due to the unusual and varying chemical composition, the poor crystallinity of the substance and the undefinitive results of stability calculations relative to the brine the formation of this material was believed to result from precipitation of a Fe, Mg rich silica gel from the brine and later recrystallisation of the mineraloids. The range in d_{001} spacings of the mineraloids and the composition calculated on basis of $(O,OH)_{22}$ are shown in Table 2. The calculated log K against temperature for the substance $Mg_4Fe_2Si_8O_{20}(OH)_4$ and the respective activity products in the geothermal fluid from Rn-8 (mean) by step boiling from 300°C to 100°C are shown in Figure 4. The brine appears to be distinctly undersaturated with respect to the substance, but in light of the many approximations the fit is surprisingly good. Those findings of Al poor Fe/Mg silicates as precipitates from the fluid are quite important for the mechanics of clay-mineral formation and also for explanation of chemical fluxes in geothermal systems. As flashing seems to be necessary for supersaturation to occur such minerals are probably not formed at depth in submarine geothermal systems. The clay minerals in the geothermal system formed by recrystallization of both glass and by alteration of minerals in the basalts are found to be of the same kind at the same depth levels. Variations in composition are found, but not more than between minerals in similar lithologic units at somewhat different depth levels.

Quartz is a common precipitation mineral at all depths and so is calcite. In the uppermost 100 m cristobalite is recorded together with quartz. Zeolites are not very common and always in small quantities. The main types are mordenite, stilbite, heulandite, mesolite and analcime. Analcime is found dispersed deep below the main zeolite zone and also traces of wairakite. Anhydrite is dispersed in the section, but only in large quantities in a few narrow zones at high levels where influx of cold seawater is believed to have occurred. Pyrite is dispersed througout the section.

Table 2. XRD data and mineral composition of precipitates from drillhole Rn-8.

Sample	d 001 in Å			Calculated mineral composition
	Untreated	Glycol saturated	Heated	
1	-	-	9.7	$Si_{8.2}Al_{0.2}Fe_{1.2}Mg_{3.8}(O,OH)_{22}$
2	13.4	14.0	9.8	$Si_{7.8}Al_{0.6}Fe_{2.2}Mg_{2.7}(O,OH)_{22}$
3	14.0	16.5	9.6	$Si_{7.8}Al_{0.6}Fe_{2.2}Mg_{2.7}(O,OH)_{22}$

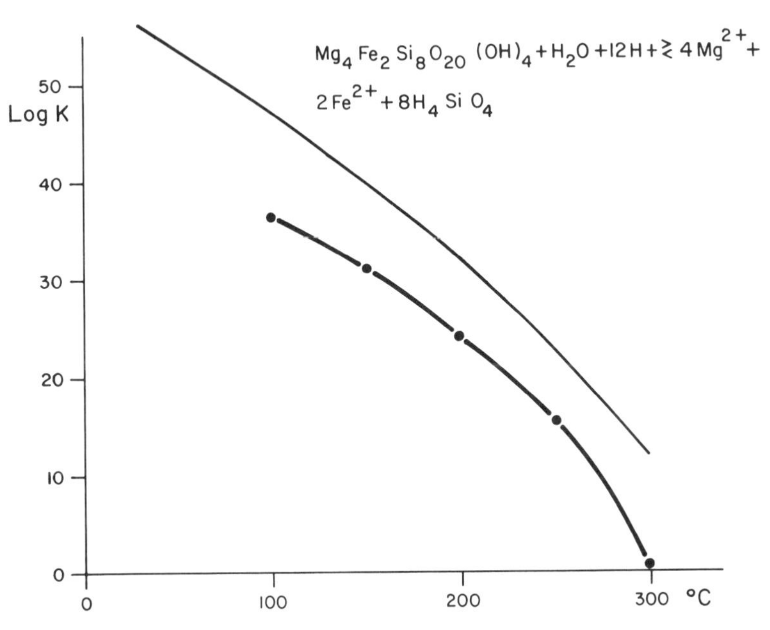

Fig. 4. The solid line shows Log K plotted against temperature for the reaction: $Mg_4Fe_2Si_8O_{20}(OH)_4 + 8H_2O + 12H^+ \rightleftharpoons 4Mg^{2+} + 2Fe^{2+} + 8H_4SiO_4$. The points connected by a line show the corresponding activity products in the brine calculated for step boiling of the brine from 300°C to 100°C.

Amorphous iron oxides are found in intimate intergrowths with the sheet silicates, especially above the chlorite-epidote alteration zone. Prehnite is locally rather common and epidote is continuously found at depths below 1100 m. Adularia is found locally but not in any big quantity. Plagioclase is albitized in some samples and very fresh in others. Microprobe analyses of some altered plagioclases in samples from drillhole Rn-8 are shown in Table 3. The high potassium content of plagioclase sample 1 from a depth of 154 m is remarkable. Precipitation of adularia is known from the Icelandic geothermal fields, but replacement of plagioclase by K-feldspar is rare. A strong increase of potassium has been demonstrated (Kristmannsdóttir, 1976) at upper levels in altered rocks from Icelandic geothermal areas, especially in the boiling zone. Mordenite, K-feldspar and clay-minerals were pointed out as the most likely sinks for the accumulated potassium.

Table 3. Microprobe analyses of "amphibole" from altered basalts of the Reykjanes geothermal field.

Sample no	SiO_2	TiO_2	Al_2O_3	FeO	MnO	MgO	CaO	NaO_2	K_2O	Total
1	50.42	1.80	13.80	10.26	0.31	6.30	12.42	2.31	0.17	98.02
2	49.43	0.11	6.35	13.68	0.51	9.02	18.33	1.17	0.00	98.60
3	49.04	0.17	15.23	7.10	0.21	8.27	13.97	1.68	0.13	95.80
4	46.28	0.66	11.98	6.63	0.16	9.40	15.15	1.46	0.08	91.75

	Mineral Composition	Samples
1	$Si_{7.1}Al_{2.3}Fe_{1.2}Mg_{1.3}\ Ca_{1.9}Na_{0.6}O_{22}(OH)_2$	from 294 m depth
2	$Si_{7.3}Al_{1.1}Fe_{1.7}Mg_{2.0}\ Ca_{2.9}Na_{0.3}O_{22}(OH)_2$	from 716 " depth
3	$Si_{7.0}Al_{2.6}Fe_{0.9}Mg_{1.8}\ Ca_{2.1}Na_{0.5}O_{22}(OH)_2$	from 1158 " depth
4	$Si_{7.0}Al_{2.1}Fe_{0.8}Mg_{2.1}\ Ca_{2.5}Na_{0.4}O_{22}(OH)_2$	from 1598 " depth

Analyses of alteration minerals tentatively identified as amphibole in thin section are shown in Table 4. The composition of the minerals is close to that of hornblende. The composition of some of the minerals suggests that pyroxene has been only partially replaced by the hornblende.

Table 4. Microprobe analysis of "albitized" plagioclase in samples from drillhole Rn-8 of the Reykjanes geothermal field.

Sample no	SiO_2	TiO_2	Al_2O_3	FeO	MnO	MgO	CaO	Na_2O	K_2O	Total
1	51.89	0.55	25.64	0.54	0.03	0.37	10.13	1.50	5.48	96.15
2	64.56	0.01	17.92	1.26	0.02	0.39	3.38	8.96	0.53	97.05
3	61.14	0.04	20.69	0.24	0.02	0.02	3.32	8.32	0.57	94.35
4	48.47	0.00	23.58	2.28	0.00	1.82	10.09	3.59	0.18	89.92

	Mineral Composition	Samples
1	$Si_{2.5}Al_{1.5}Ca_{0.52}Na_{0.14}K_{0.34}O_8$	from 154 m depth
2	$Si_{3.0}Al_{1.0}Ca_{0.17}Na_{0.8}\ K_{0.03}O_8$	from 508 " depth
3	$Si_{2.9}Al_{1.1}Ca_{0.17}Na_{0.75}K_{0.03}O_8$	from 760 " depth
4	$Si_{2.5}Al_{1.4}Ca_{0.55}Na_{0.35}K_{0.01}O_8$	from 1598 " depth

Chemical Composition of the Altered Rocks

XRF analyses of samples of altered rocks from the Reykjanes drillholes are shown in Table 5. The most prominent difference in the chemistry of the altered rocks as compared to the fresh rocks is hydration. Sulphur and carbonate have not been analysed in the rocks. A maximum content of 10% CO_2 in the boiling zone has been estimated from the calcite content. On average the CO_2 is assumed to be below 2%. Analyses of six samples from one of the drillholes (Gunnlaugsson, 1977) show contents of 830-5350 ppm.

Pure hydration and enrichment in sulphur and carbonate will lower the calculated weight percent of the main oxides, but not change their relative contents. The effects of pure dilution can be demonstrated by plotting the weight percent of the element against the contents of volatiles. In Figure 5 all the main oxydes are plotted against H_2O content. For comparison a line showing the effects of pure dilution has been drawn for mean composition of olivine-tholeiite from Reykjanes (Jakobsson, 1975).

A rather big scattering is observed for SiO_2, mostly showing local increases. On average MgO does not show a great deviation from the line of dilution. MgO increase is found in hyaloclastites in the smectite-zeolite alteration zone. A marked MgO decrease was observed at a very shallow level in a hyaloclastite. No extensive

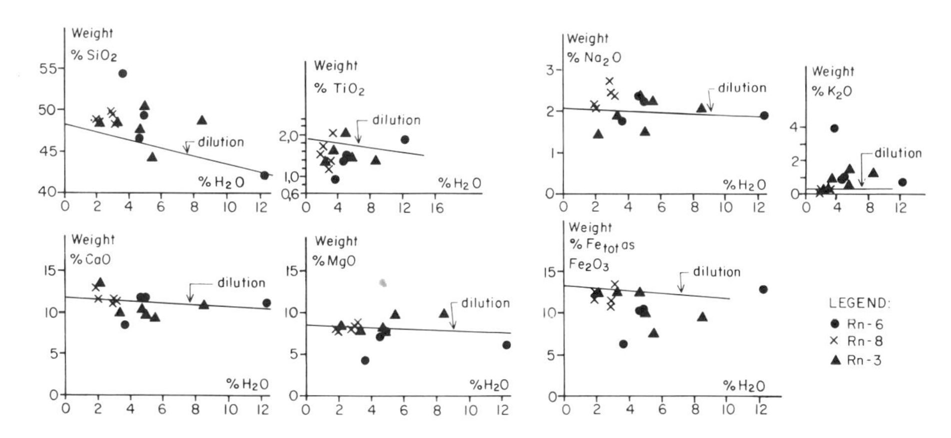

Fig. 5. The content of main oxides against water content in altered rocks from drillholes in Reykjanes. The dilution line is drawn for mean composition of olivine tholeiite from Reykjanes (Jakobsson, 1975).

Table 5. Chemical analysis (XRF) of hydrothermally altered rocks from the Reykjanes geothermal.

	Major elements in weight %												Trace elements in ppm				
Sample no	SiO_2	TiO_2	Al_2O	Tot.Fe as FeO	MnO	MgO	CaO	Na_2O	K_2O	P_2O_5	Igneous loss	Total	Sr	Rb	Ni	Zr	Y
1	42.04	1.83	10.03	14.30	0.19	6.13	11.19	1.89	0.62	0.19	10.78	99.19	209	14	65	105	35
2	54.33	0.82	10.87	6.98	0.25	4.35	8.40	1.74	3.98	0.08	5.08	96.88	52	61	75	97	36
3	49.12	1.54	11.47	11.62	0.25	7.59	11.76	2.20	0.87	0.17	3.25	99.84	172	16	76	103	25
4	46.67	1.39	12.39	11.37	0.22	7.29	11.79	2.35	0.75	0.13	5.28	99.63	223	25	107	108	25
5	48.39	1.40	12.26	10.40	0.22	9.92	10.86	2.04	1.12	0.14	6.40	103.15		23	-	99	16
6	44.02	1.45	13.13	8.26	0.14	9.60	9.15	2.22	1.36		9.43	99.76		22	-	120	29
7a	48.39	1.65	10.96	13.83	0.19	7.59	9.95	1.85	0.84	0.16	4.01	99.42	135	11	84	104	17
7b	43.71	1.48	10.59	11.88	0.18	7.44	8.36	2.22	1.13	0.13	10.89	98.01	44		77		
7c	46.79	2.50	11.18	15.55	0.20	8.35	9.76	1.86	0.34	0.26	2.91	99.70	154		99		
8	47.57	2.06	11.21	13.84	0.20	8.13	10.35	2.39	0.39	0.23	3.36	99.73	125	9	110	131	31
9	50.34	1.20	11.37	10.81	0.17	7.60	9.57	1.50	0.69	0.11	6.01	99.37	145	29	92	128	-
10	48.26	1.35	9.86	13.61	0.18	8.38	13.55	1.41	0.17	0.08	2.53		70	24	88	98	29
11	49.13	0.51	12.21	10.19	0.15	10.51	11.94	1.42	0.12	-	2.63		60	-	141	-	-
12	46.78	0.90	13.01	8.92	0.15	7.80	9.85	2.88	0.47	0.07	9.29		37	-	77	-	-
13a	48.24	2.03	10.93	14.79	0.21	8.77	11.18	2.36	0.22	0.20	2.08		207	14	119	124	29
13b	46.73	1.61	11.24	13.83	0.21	8.33	12.09	2.39	0.56	0.22	1.43		198		125		
14	49.37	1.40	11.66	12.71	0.20	8.31	11.50	2.31	0.13	0.13	1.77		135	23	68	87	34
15	48.91	1.55	11.09	12.87	0.20	8.06	12.85	2.05	0.11	0.16	1.65		159	29	85	104	29
16	49.72	1.16	11.83	12.00	0.27	8.11	11.04	2.74	0.11	0.10	2.65	99.91	156	35	93	86	25
17	49.29	1.75	11.24	13.85	0.20	7.71	11.46	2.07	0.17	0.18	1.84	99.76	124	14	78	124	27

1	Hyaloclastite from	Drillhole nr 6	58 m depth
2	Tuffaceous sediment from	Drillhole nr 6	346 m depth
3	Basalt lava from	Drillhole nr 6	512 m depth
4	Hyaloclastite from	Drillhole nr 6	570 m depth
5	Tuffaceous sediment from	Drillhole nr 3	302 m depth
6	Tuffaceous sediment from	Drillhole nr 3	392 m depth
7	Hyaloclastite breccia from	Drillhole nr 3	466 m depth
	a) is total sample		
	b) is from the matrix		
	c) is basalt fragment		
8	Hyaloclastite from	Drillhole nr 3	610 m depth
9	Hyaloclastite from	Drillhole nr 3	822 m depth
10	Basalt lava from	Drillhole nr 3	1128 m depth
11	Hyaloclastite from	Drillhole nr 8	140 m depth
12	Hyaloclastite from	Drillhole nr 8	306 m depth
13	Hyaloclastite breccia from	Drillhole nr 8	712 m depth
	a) is total sample		
	b) is basalt fragment		
14	Basalt lava from	Drillhole nr 8	814 m depth
15	Basalt lava from	Drillhole nr 8	1300 m depth
16	Basalt lava from	Drillhole nr 8	1494 m depth
17	Basalt lava from	Drillhole nr 8	1754 m depth

Mg increase comparable with that of the greenstones of the Mid-Atlantic Ridge is found (Humphris and Thompson, 1978). Total iron scatters significantly and irregularly. CaO shows some scattering, but no great changes in total contents are observed. Na_2O shows considerable variation. K_2O shows a wide scattering and a strong increase in the rocks within the boiling zone. The degree of oxidation appears to diminish by depth on the basis of the few analyses of Fe^{2+}.

In a rock suite a relation exists between most major elements and the content of Zr. Zr is considered to be relatively immobile during greenschist facies metamorphism (Pearce and Cann, 1973). The regular covarience of an element against Zr would thus be disturbed by mobilization of that element. Plots of major elements against Zr in altered rocks from Reykjanes are shown in Figure 6. The original rocks do not span a wide range in composition and the variation in Zr content is rather small (86-131 ppm). The alkalies, iron and silica show the greatest scattering, but considerable scattering is also found for calcium and magnesium.

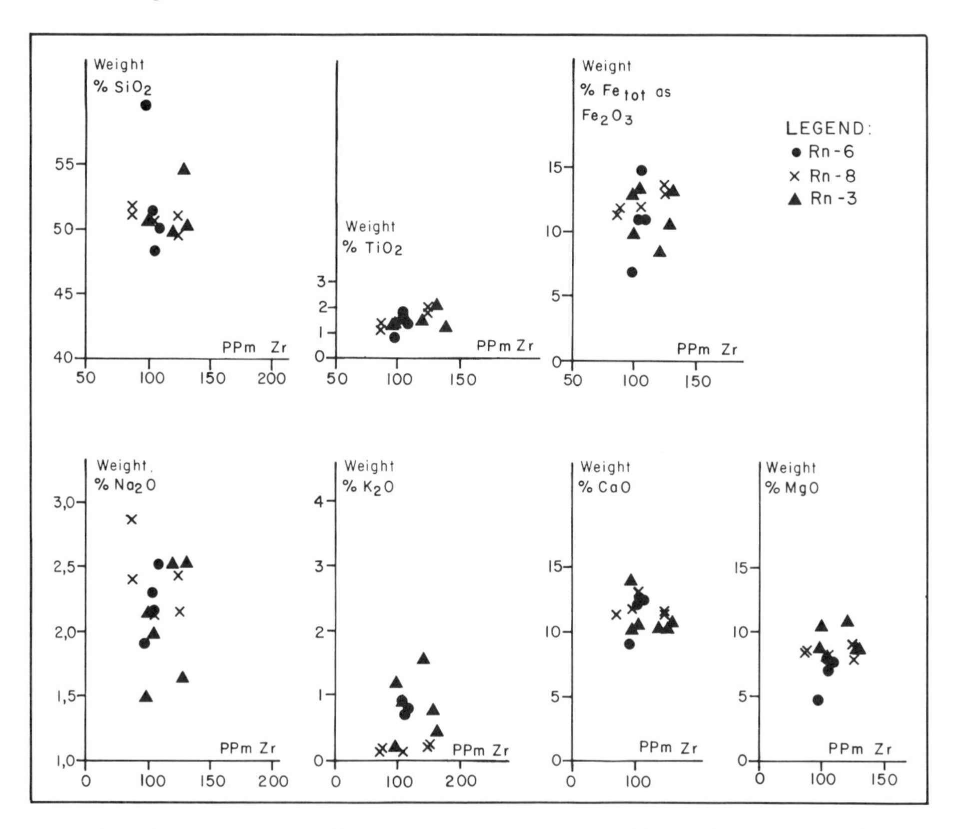

Fig. 6. The content of major oxides in samples of altered rocks from drillholes in Reykjanes against Zr content.

The contents of major oxides against depth in drillhole no 8 located on the edges of the geothermal field is shown in Figure 7. An enrichment of K_2O at upper levels is slight if any, but a trend of MgO increase is shown at upper levels as well as an increase in CaO with depth. The increase of MgO almost coincides with the smectite - zeolite alteration zone. The smectites are the most voluminous alteration minerals at this depth level. They are also the only group of secondary minerals present capable of incorporating magnesium in considerable quantities. The increased content of potassium is presumably bound mostly in zeolites (mordenite) and feldspar. Only a small amount can be incorporated in the types of clay minerals occurring in the altered rocks (Kristmannsdóttir, 1976).

At higher temperatures calcium is bound in secondary epidote and prehnite. Calcite is precipitated in the boiling zone. The calcium concentration is also linked with equilibrium reactions with anhydrite, zeolites and smectites. The Mg for Ca exchange observed in greenstones (Humphris and Thompson, 1978, Mottl, 1982) is not found in the Reykjanes rocks. This process would also be expected in the inflow zones of the geothermal system, but not in the zone of upflow and boiling.

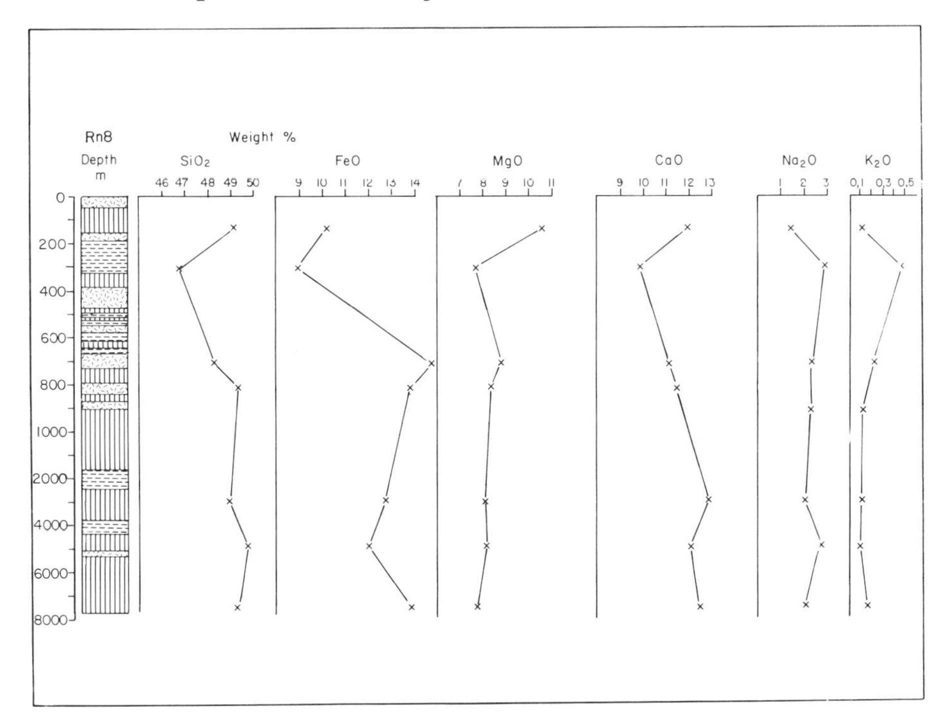

Fig. 7. The contents of some major elements in altered basaltic rocks against depth from drillhole Rn-8 in the Reykjanes geothermal area.

Chemistry of the Geothermal Fluid

Ground water in the Reykjanes peninsula is saline and believed to be of seawater origin. The tides can be observed in open fissure over 100 m offshore and also in shallow drillholes just outside the Reykjanes geothermal field. A thin (30-60 m) layer of fresh water floats on top of the saline water. The fresh water layer thickens inland and the salinity of the ground water decreases.

The main components (as calculated from mean analyses) of inflow into two drillholes and a hot spring in the Reykjanes geothermal field are shown in Table 6. For comparison the contents of dissolved solids in seawater of $34^{0}/_{00}$ salinity are also shown (Turekian, 1969). The composition of the deep geothermal brine as calculated from mean analysis of water and steam from drillhole Rn-8 is shown in more detail in Table 7. This drillhole has been sampled regularly over a period of 12 years (Fig. 8) and the samples analysed for main dissolved solids. Much older analysis exist of samples from the various hot springs in the area. The composition of hot spring samples depends on many poorly controllable factors such as boiling and degassing, and cooling and reaction at the surface. Sampling by separator at well head at known pressure

Table 6. Main components in geothermal water at Reykjanes (in ppm).

	"Hotspring 1918"	Rn-2	Rn-8	Sea-water
Temperature °C	100	225	270	
SiO_2 ppm	562	355	588	6.0
Na "	15900	10700	9520	10470
K "	2130	1400	1380	380
Ca "	2400	1790	1580	398
Mg "	24	1.1	1.43	1250
SO_4 "	180	75.6	40.8	2630
Cl "	28450	20500	19200	18800
F "	0.25	0.15	0.15	1.26
Total dissolved solids	49800	34800	33300	33900
CO_2	42.1	2110	1930	100

NEA data and Turekian, 1969

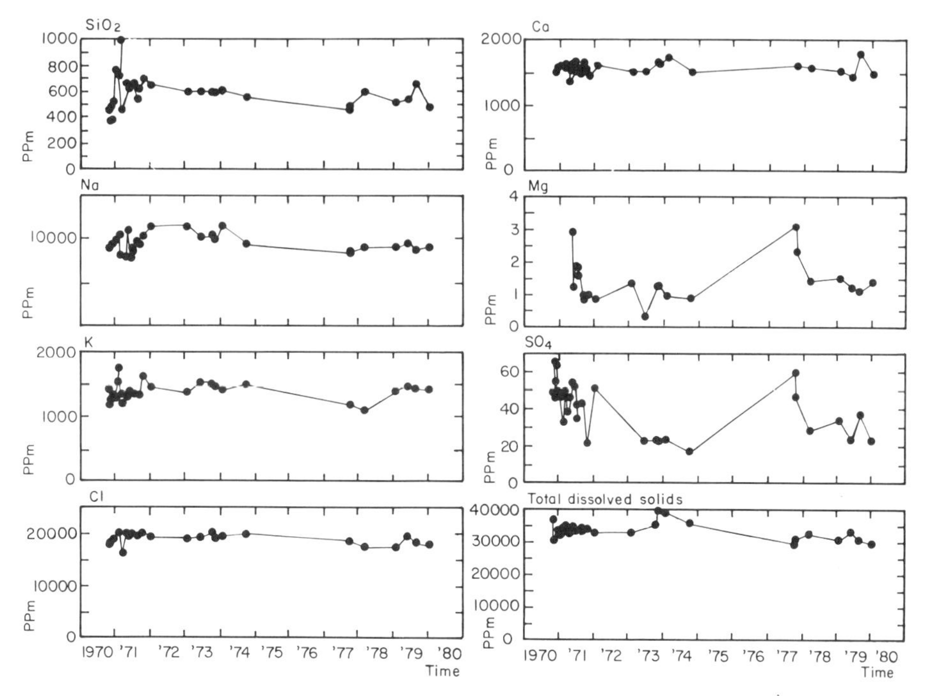

Fig. 8. The change in composition of the geothermal brine from drillhole Rn-8 over a 10 year period as compiled by Hauksson, 1981.

gives much more comparable data and enables recalculation to a one phase fluid. The geothermal fluid at depth has nearly the same chloride content as seawater. Relative to seawater it is depleted in Mg^{++} and SO_4^{--} and enriched in K^+, Ca^+ and SiO_2. The SiO_2 content is in equilibrium with quartz, as expected at temperatures exceeding 180°C. The quartz geothermometer gives 275°C as the temperature of the brine. The geothermal fluid is in equilibrium with calcite and anhydrite and pyrite. The Na/K geothermometers (see Ellis and Mahon, 1977) give a poor fit for the deep brine and so does direct relation to thermodynamic data from Helgeson (1969). The calculated Na/K deep temperatures are only 210-240°C. Supersaturation, with respect to albite, adularia (and microcline), pyrite and laumontite, occurs by boiling. As indicated in Figure 4 the layer silicate $Fe_2Mg_4O_{20}(OH)_4$ could be precipitated from the fluid by boiling. No thermodynamic data exist for the sheet -silicates formed by alteration of the basalt by the brine. However, that process involves both simple ionic exchange reactions and complex reactions involving Al, Si, Mg and Fe.

Table 7. The composition of the geothermal brine in the Reykjanes geothermal field (in ppm).

pH/°C	SiO_2	Na	K	Ca	Mg	SO_4	Cl	F	CO_2	H_2S	H_2	CH_4	N_2	Al	B	Fe	Total dissolved solids
4.98/270	588	9522	1378	1579	1.43	40.8	19200	0.15	1930	36.5	0.24	0.49	10.8	0.0005	11	0.0150	33282

Calculated from an average of 40 analyses of discharge from drillhole Rn-8 (NEA data), except for Al, Fe and B which are based on 1 - 4 analyses.

The fluid is calculated to be undersaturated with respect to montmorillonite (Na, Ca and Mg) and Mg-chlorite.

Data of other elements in the geothermal brine exist but mostly as analyses of a single sample (Hauksson, 1981). Those data are summarized below in Table 8. The Br/Cl ratio is similar in the geothermal brine to that in seawater. Li and Rb are about 40 times the concentration in seawater. The Sr content is similar to that of water, but Ba concentration can be as much as 5000 times that of seawater. B is doubled relative to seawater and Pb and Mn are considerably enriched. The same is true for As, Cu, Zn and Hg. As compared to the sampled emanations from the Pacific (Corliss et al., 1979, Craig et al., 1980 and Edmond, 1980), the Reykjanes brine is less acidic and contains less of the heavy metals. The reservoir temperature is inferred to be 50-70°C lower at Reykjanes both from measurements and from the calculated quartz temperature.

Data of stable isotope ratios in the geothermal brine are rather restricted (Árnason, 1976, Ólafsson and Riley, 1978, Sakai et al., 1980). The $\delta^{18}O$ values are close to zero (÷1.1 - ÷0.25 ‰) whereas the δD values are around ÷20 ‰ (÷23 - ÷11). The δD values were tentatively explained by boiling of the brine and mixing with local fresh water. However, this is not compatible with the geochemistry of the fluids. Gudmundsson et al., (1981) propose that the δD in the brine is controled by water-rock interaction due to an enchanced reaction rate in the highly ionized brine. The $\delta^{39}S$ in SO_4^{2-} in the fluid is around +20 ‰. Sakai et al., (1980) state that the sulfate and sulfide in the Reykjanes fluid are considerably out of isotopic equilibrium with each other. Further they state that from partial isotopic equilibrium the fluids are supposed to have a residence time shorter than 50 years in the system.

Table 8. Contents of some minor elements in the Reykjanes brine (in ppm)

Sampling place	Br	B	Li	Rb	Sr	Ba	Pb	Mn
"Hotspring 1918" (Arnórsson, 1969)		12.2					1.34	
Rn-2 (Arnórsson, 1969)		12.0					0.57	
"Hotspring 1918" (Ólafsson and Riley, 1978)	98.1	11.8	7.4	6.3	9.2	5.8		3.7
Rn-2 (Ólafsson & Riley,1978)	83.8	11.5	6.0	4.5	8.4	5.7		0.51
Rn-8 (Ólafsson & Riley,1978)	87.3	11.3	6.6	5.2	8.9	10.0		2.6
Sea-water (Turekian, 1969)	67.3	4.45	0.17	0.12	8.1	0.02	0.00003	0.01

Compiled by Hauksson, 1981

SELECTED CHEMICAL EVIDENCES FROM THE KRAFLA AREA

The main characteristic of the Krafla geothermal field is the present magmatic activity. This activity drastically disturbs the chemical balance in the geothermal system. The emanation of magmatic gases into the geothermal fluids has significantly changed the composition of fumarolic steam and discharge from drillholes. As an example, the pH of discharge from one of the wells decreased from about 9 to 1.8. Another well yielded superheated steam with varying contents (up to 100 ppm) of Cl in the form of free HCl. "Black water" discharge is common due to extensive iron sulphide precipitation by flashing. Extensive corrosion of drillhole casing and surface pipes has occurred and also scaling and blocking by unusual precipitates (Kristmannsdóttir and Svantesson, 1977, Kristmannsdóttir in prep.). Precipitation of iron sulphides, -oxides and - silicates together with calcite and cristobalite were found to block two of the production wells, one of them in only three weeks time. The fluids of the Krafla area contain normally about 0.02 ppm of iron, but contents up to 60 ppm have been measured after pH-drop due to magmatic events. The "black water" in Krafla is comparable to the "black smokers" of the submerged systems as both types of fluids are highly charged with iron sulphide precipitations.

The rock alteration is also shown to be affected by the volcanic activity (Kristmannsdóttir, 1979, Kristmannsdóttir, 1981). Changes in the general alteration pattern have been demonstrated after the outset of the activity. Skarn minerals as wollastonite, andradite, hedenbergite etc. are fairly common at dike contacts where hydrothermal precipitates have been baked and recrystallized. The underground rocks in Krafla area (Kristmannsdóttir, 1979) are tholeiite basaltic hyaloclastites and lavas in the upper 1000 - 1500 m. Dikes are common from about 400 m depth and the rocks are mostly rather fresh, but a few are intensively altered. A magma chamber has been located at 3-8 km depth (Björnsson et al., 1979) below the geothermal field by seismic methods. The intrusive rocks in the lower part of the profile are thus realistic counterparts to the sheeted dike complexes of ophiolites and young oceanic crust (Spooner and Fyfe, 1973, Wolery and Sleep, 1976).

The maximum temperatures in the Krafla convection system are much higher than in the Reykjanes system. Temperatures up to 360°C are indicated by measurements in drillholes. Oxygen isotope fractionation between quartz and epidote give an estimate of the alteration temperature of 350-400°C (Hattori and Muehlenbachs, 1982). The temperatures in the Krafla system are thus comparable to the convecting fluids in geothermal systems at Mid-Ocean Ridges.

A two phase inflow into the Krafla drillholes occurs at lower levels (below 1000-1500 m) and at temperatures exceeding 300°C. Many of the wells have yielded superheated steam by production. The Geothermal water is of meteoric origin and subsequently contains much less of dissolved solids (Table 9) than the fluids at Reykjanes. Many of the drillholes yield a mixture of water from an upper aquifer of about 200°C and a water/steam mixture from the lower 360°C hot aquifer (Ármannsson et al., 1982). The composition of drillhole discharge varies due to fluctuation in magmatic influences. The calculation of composition of deep fluids (Ármannsson et al., 1982) is rather complicated and uncertain. The distribution of alteration minerals by depth in one of the drillholes in Krafla, together with a simplified geologic section and estimated rock temperature are shown in Figure 9. The main difference in alteration mineralogy is the occurrence of skarn minerals, the common occurrence of amphibole, vermiculite and talc at great depths. Zeolites are more common and calcium silicates as gyrolite, truscottite and reyerite are found. Anhydrite is very rare. Pyrrhotite is rather common in Krafla, but is not found at Reykjanes. Leucoxene and sphene are observed to replace ilmenite in Krafla but were not observed in Reykjanes. A sequence of sheet - silicates occur in Krafla area similar to that in the Reykjanes field, but the zone of mixed - layer minerals and swelling chlorite is much narrower and normal ripidolite chlorites (Kristmannsdóttir, 1979, Kristmannsdóttir in prep.) are found in the chlorite - epidote zone. The chlorites tend to become more Mg-rich by depth. The

Table 9. Examples of chemical composition of geothermal water at depth in the Krafla area.

	Shallow aquifer	Deep aquifer
pH/°C	6,96/206	7.12/303
SiO_2	300	681
Na	151	143
K	11.8	20.2
Ca	2.1	1.1
Mg	0.03	0.02
SO_4	157.4	73.1
Cl	23.7	11.8
F	0.16	0.46
CO_2	271.7	10013.0
H_2S	63.4	92.4

composition of epidote in the Krafla area varies, but no compositional trend is observed. Very few data of epidote composition exist from the Reykjanes field. However, the epidote composition at Reykjanes appears to be rather similar to that from the Krafla area. The amphibole found in the Krafla area is both actinolite and hornblende (Kristmannsdóttir, 1979, and Kristmannsdóttir in prep.).

About sixty samples of cuttings and cores of basalt lavas, hyaloclastites and acid intrusives from drillholes in the Krafla area have been analysed chemically (Kristmannsdóttir, in prep.). The main difference in composition between altered and fresh rocks appears to be hydration, and increase in sulphur (Gunnlaugsson, 1977) and carbonate. However, the altered rocks show much more variation in composition than recent basalts in the area (Fig. 10). In Fig. 11 and 12 are shown plots of the main oxides against water content for the basaltic rocks and for all rock types against Zr content.

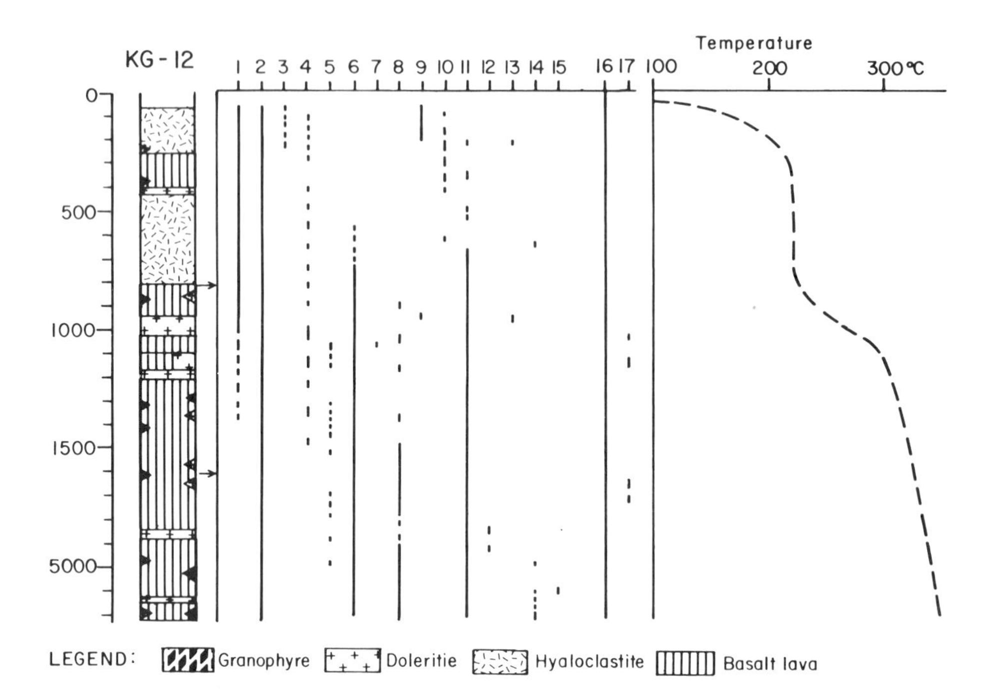

Fig. 9 A simplified stratigraphic section and distribution of alteration minerals by depth in drillhole KG-12 in the Krafla area. Also shown is estimated rock temperature. 1: calcite, 2: quartz, 3: heulandite, 4: wairakite, 5: wollastonite, 6: epidote, 7: garnet, 8: amphibole, 9: smectite, 10: mixed-layer minerals, smectite/chlorite, 11: chlorite, 12: vermiculite, 13: chlorite/vermiculite mixed layers, 14: illite and illite/smectite mixed-layer minerals, 15: talc, 16: pyrite, 17: pyrrhotite.

The alkalies, magnesium and iron show considerable scattering relative to the dilution line, but the scattering for the other elements is slight. Some scattering is observed for the "fresh" basalts, which are in most cases intrusives and probably belong to different magmatic events that built up the main pile of altered rocks. The correlation to Zr is not very good, but by no means completely destroyed. Silica, total iron, calcium and magnesium all show considerable variation but no consistent trend. The alkalies show a depletion trend and some scattering. In general the chemical changes by alteration in the Krafla rocks are similar as in the Reykjanes system. The deviations from the composition of fresh rocks at Krafla are even less than in the Reykjanes rocks.

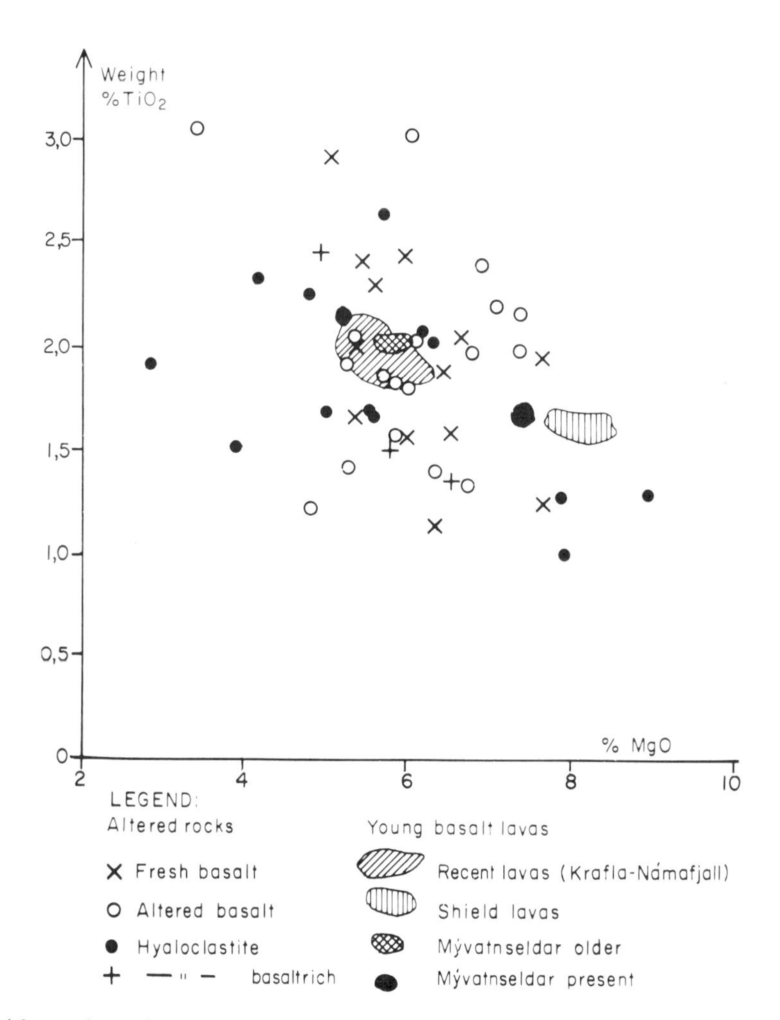

Fig. 10. The TiO_2 content plotted against MgO for altered rocks from drillholes in the Krafla area and from young basalts in the Krafla-Námafjall area (Grönvold, 1976).

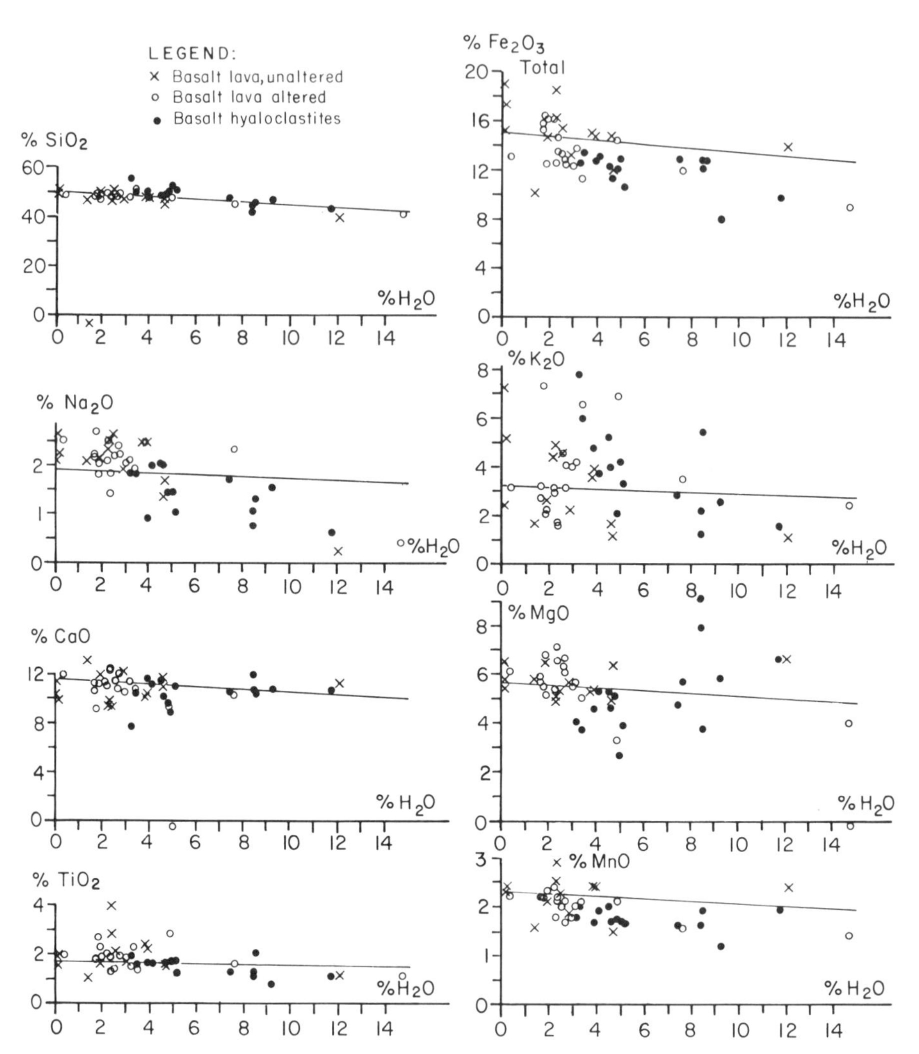

Fig. 11. Contents of main oxides plotted against water contents for altered and relatively fresh basaltic rocks from drillholes in the Krafla geothermal area.

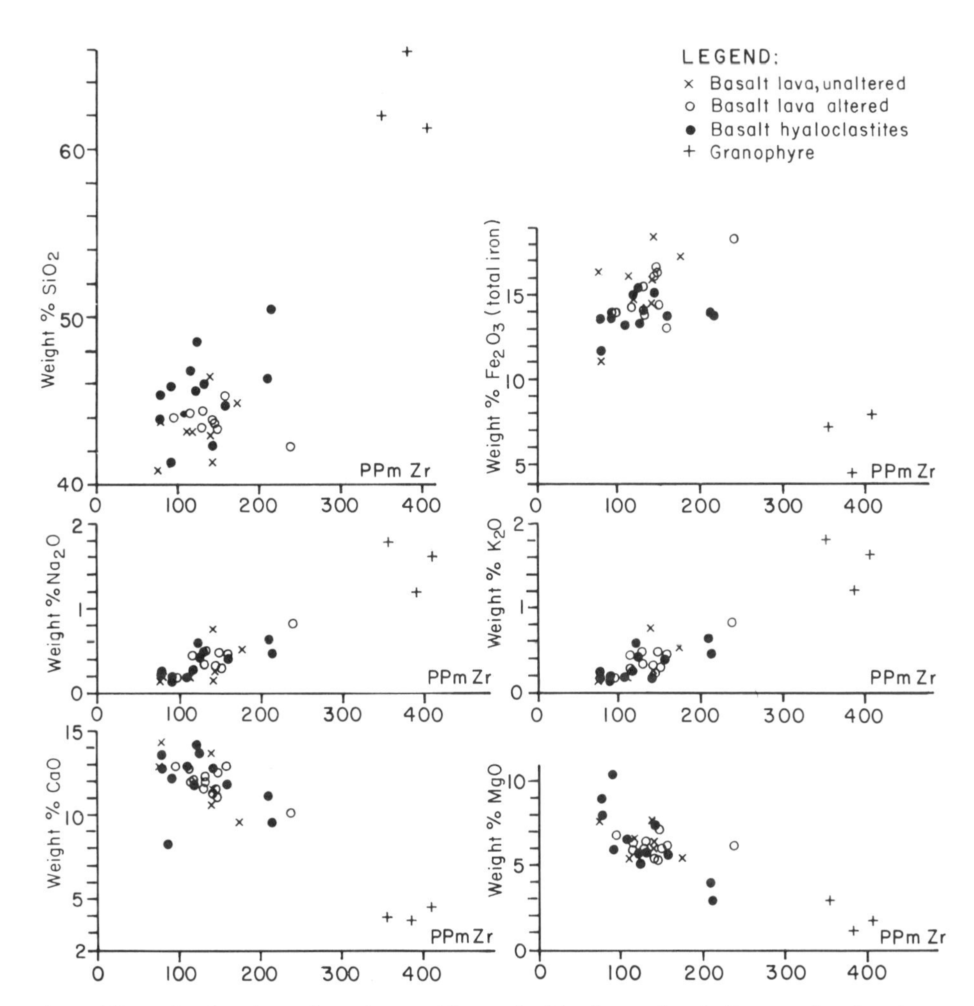

Fig. 12. Contents of main oxides plotted against Zr contents in rocks from drillholes in the Krafla geothermal area.

CHEMICAL CHARACTERISTICS OF ICELANDIC GEOTHERMAL SYSTEMS COMPARED WITH SUBMARINE SYSTEM

As pointed out in the Introduction, the data from Iceland and the submarine systems are not readily comparable. The most extensive data pertain to the chemistry of fluids in the Icelandic geothermal areas, and to the altered rocks from Mid-Ocean Ridges. The rock samples from the Icelandic areas are mostly cuttings and give the average composition of several meters of penetration. Their place in the geothermal system is known and is mostly in the zones of upflow. The highly altered rocks from the Mid-Ocean Ridges are dredged from the seafloor and their location in the oceanic crust is not exactly known. The alteration of the oceanic crust as encountered from drillcores (Mottl, 1982, Böhlke et al., 1981) is mostly low-temperature alteration, not in connection with submarine geothermal systems.

As pointed out by Spooner et al (1973) the main differences between submarine and subaerial geothermal systems are due to the fact that boiling does not occur in the underground rocks of submarine systems, which significantly affects the alteration process in the sub-aerial system. Many of the mineralogical and chemical characteristics of the Icelandic areas are the result of the boiling of the fluid in the upflow zones. The strong increase of potassium found in this zone is due to boiling and is not to be expected nor is it found in submarine systems. The same is true for the precipitation of Al-poor Mg/Fe sheet-silicates. A distinct Mg enrichment and a collateral depletion of Ca has been demonstrated in submarine greenstones (see Humphris and Thompson, 1978) and similar features are found generally in hydrothermally altered basalts from the ocean crust (Mottl, 1982). A gain of Fe is common in the chlorite-rich rocks (Humphris and Thompson, 1978) even though Mn, Cu, Ni and Zn have been leached out of the rocks. In rocks from the Icelandic geothermal fields a slight Mg increase is found in the smectite-zeolite alteration zone relative to the chlorite - epidote alteration zone. The iron content does not show a distinct increase, MnO is leached at those upper levels, Ni content varies and data of Cu and Zn do not exist. The samples from the Icelandic systems are mostly from the zones of upflow, where the ascending fluids are highly depleted in Mg. The main Mg-increase in the altered rocks would probably be in the deep inflow zones of the geothermal system. The slight Mg increase in the smectite-zeolite zone in the Icelandic areas might be connected with the precipitation of Fe/Mg silicates by boiling of the fluid is already mentioned.

The temperature at the base of the upflow zone in a typical axial hydrothermal system (see Mottl, 1982) is probably 350-370°C. In the Reykjanes geothermal area the maximum temperature in the convection system is estimated to be just below 300°C. This difference in alteration temperature must be reflected in both mineralogy and chemistry of rocks and fluids. The altered sequence in the Reykjanes geothermal area is mainly within the zeolite metamorphic facies whereas among the dredged rocks at Mid-Ocean Ridges zeolite facies rocks are rather rare as compared with greenschist facies rocks (Mottl, 1982). The subsurface temperature of the Krafla geothermal area is nearer to the submarine systems, but there the fluids are non-saline and consequently much less mineralized. The mass fluxes in the system would thus be expected to be less significant at Krafla than at Mid-Ocean Ridges. The magmatic effects occurring in the system in Krafla would be expected in many of the submarine geothermal systems.

The chemistry of seafloor hot springs and that of the Reykjanes brine have been closely predicted by laboratory experiments reacting seawater with basalt (Mottl and Seyfried, 1980, Mottl, 1982). The chemical changes due to alteration in greenschist facies metabasalts from the Mid-Atlantic Ridge are also closely predicted by the basalt seawater experiments and can be explained by a reaction with rapidly heated downwelling seawater. The experiments did not succeed in producing alteration mineralogy matching that found in nature, probably due to kinetic problems. The changes due to alteration in Icelandic geothermal systems do not compare as well with the experiments either chemically or mineralogically. Alteration of the Icelandic rocks appears to involve different mass fluxes than these found in metabasalts from the Mid-Atlantic Ridge. The alteration of the Icelandic rocks has occurred by reaction with seawater of already moderated composition due to heating and reaction in the downwelling zones. By drilling outside the geothermal area in the inflow zones the mass fluxes in the altered rocks would be more comparable to that of the experiments.

ACKNOWLEDGEMENTS

I am grateful to dr. K. Grönvold for valuable help in analysing the secondary minerals in drillcuttings from Reykjanes. Dr. M. Mottl was kind enough to give me a copy of his unpublished manuscript (submitted to GSA Bulletin) which was of great help in preparation of my paper.

REFERENCES

Ármannsson, H., G. Gíslason and T. Hauksson, 1982. Magmatic gases in well fluids aid the mapping of the flow pattern in a geothermal system. Geochim. Cosmochim. Acta 46: 167-177.
Árnason, B., 1976. Groundwater Systems in Iceland traced by deuterium. Soc. Sci. Islandica, Publ. 42: 236 p.
Arnórsson, S., 1969. A geochemical study of selected elements in thermal waters of Iceland. Ph.D. Thesis, University of London, 353 p.

Arnórsson, S., 1978. Major element chemistry of the geothermal seawater at Reykjanes and Svartsengi, Iceland. Miner. Mag. 42: 209-220.

Björnsson, S., S. Arnórsson and J. Tómasson, 1970. Exploration of the Reykjanes thermal Brine field. Geothermics, Spes. iss. 2: 1640-1650.

Björnsson, S., S. Arnórsson and J. Tómasson, 1972. Economic evaluation of Reykjanes thermal brine area, Iceland. Amer. Ass. Petrol Geol. Bull. 56: 2380-2391.

Björnsson, A., G. Johnsen, S. Sigurðsson, G. Thorbergsson and E. Tryggvason, 1979. Rifting of the Plate Boundary in North Iceland 1975-1978. J. Geophys. Res. 84: 3029-3038.

Böhlke, J.K., J. Honnorez, B.M. Honnorez - Guerstein, K. Muehlenbachs and N. Petersen, 1981. Heterogeneous alteration of the upper oceanic crust: Correlation of rock chemistry, magnetic properties and O isotope ratios with alteration patterns in basalts from site 396B, DSDP. J. Geophys. Research. 86: 7935-7950.

Corliss, J.G., J. Dymond, L.I. Gordon, J.M. Edmond, R.P. von Herzen, R.D. Ballard, K. Green, D. Williams, A. Bainbridge, K. Crane and Tj.H van Andel, 1979. Submarine thermal springs on the Galapagos rift. Science. 203: 1073-1083.

Craig, H., J.A. Welham, K. Kim, R. Poreda and J.E. Lupton, 1980. Geochemical studies of the 21°N EPR hydrothermal fluids (abst.). EOS. 61: 992.

Edmond, J.R., 1980. The chemistry of the 350°C hot springs at 21°N on the East Pacific Rise (abst.). EOS. 61: 992.

Ellis, A.J. and W.A.J. Mahon, 1977. Chemistry and Geothermal Systems. Academic Press, N.Y. 392 p.

Grönvold, K., 1976. Variation and origin of magma types in the Námafjall area, North Iceland. Bull. Soc. Géol. France. 18: 869-870.

Gudmundsson, J.S., T. Hauksson and J. Tomasson, 1981. The Reykjanes geothermal field in Iceland: Subsurface exploration and well discharge characteristics. Proc. 7. Workshop Geothermal Res. Eng. Stanford: 61-69.

Gunnlaugsson, E., 1977. The origin and distribution of sulfur in fresh and geothermally altered rocks in Iceland. Ph.D. Thesis, University of Leeds, 192 p.

Hattori, K. and K. Muehlenbachs, 1982. Oxygen isotope ratios of the Icelandic crust. J. Geophys. Res. (in press).

Hauksson, T., 1981. Reykjanes. "Contents of dissolved solids in the geothermal brine". Internal report (mimeographed) at NEA, Iceland. 53 p.

Helgeson, H.C., 1969. Thermodynamics of hydrothermal systems at elevated temperature and pressure. Amer. J. Sci.: 729-804.

Humphris, S.E. and G. Thompson, 1978. Hydrothermal alteration of oceanic basalts by seawater. Geochim. Et Cosmochim. Acta. 42: 107-126.

Jakobsson, S.P., 1975. Chemistry and distribution pattern of recent basaltic rocks in Iceland. Lithos. 5: 365-386.

Kristmannsdóttir, H., 1975. Hydrothermal alteration of basaltic rocks in Icelandic geothermal areas. Proceedings of the Second U.N. Symposium on Development and use of Geothermal Resources: 441-445.

Kristmannsdóttir, H., 1976. Types of clay minerals in hydrothermally altered basaltic rocks, Reykjanes, Iceland. Jökull. 26: 30-39.

Kristmannsdóttir, H. and J. Svantesson, 1977. "Scalings in drillholes in Krafla". A mimeographed report (in Icelandic) from the National Energy Authority. 31 p.

Kristmannsdóttir, H., 1979. Alteration of basaltic rocks by hydrothermal activity at 100-300°C. In International Clay Conference 1978 (ed. Mortland M.M. and V.C. Farmer). Elsevier Sci. Publ. Comp. Amsterdam: 359-367.

Kristmannsdóttir, H., 1981. Wollastonite from hydrothermally altered basaltic rocks in Iceland. Miner. Mag. 44: 95-99.

Kristmannsdóttir, H., 1981 b. Clay-like minerals formed in a geothermal brine in the Reykjanes high-temperature geothermal system, Iceland. Abstracts from the 7th Int. Clay Conf. Bologna and Pavia: 161-162.

Mottl, M. and W.E. Seyfried, 1980. Sub - sea - floor hydrothermal systems: rock vs. seawater-dominated. In Seafloor Spreading Centers: Hydrothermal Systems. (Ed. by Rona, P.A. and R.P. Lowell) Dowden Hutchinson and Ross, Inc. Stroudsburg Pensylvania: 66-82.

Mottl, M., 1982. Metabasalts, Axial Hot Springs and the Structure of hydrothermal systems at Mid-Ocean Ridges. G.S.A. Bull. (in press).

Ólafsson, J and J.P. Riley, 1978. Geochemical studies on the thermal brine from Reykjanes (Iceland). Chem. Geol. 21: 219-237.

Pálmason, G., S. Arnórsson, I.B. Friðleifsson, H. Kristmannsdóttir, K. Sæmundsson, V. Stefánsson, B. Steingrímsson, J. Tómasson and L. Kristjánsson, 1979. The Iceland Crust: Evidence from drillhole data on structure and processes, in Deep Drilling Results in Atlantic Ocean, Ocean Crust. (Ed. by Talvani, M., C.G. Harrison and D.E. Hayes) Maurice Ewing Series, 2, AGU, Washington, D.C.: 43-65.

Pearce, J.S. and J.R. Cann. 1973. Tectonic setting of basic volcanic rocks determined using trace element analysis. Earth Planet. Sci. Letters. 19: 290-300.

Sakai, H., E. Gunnlaugsson, J. Tómasson and J.E. Rouse, 1980. Sulfur isotope systematics in Icelandic geothermal systems and influence of seawater circulation at Reykjanes. Geochim. Cosmochim. Acta. 44: 1223-1231.

Spooner, E.T.C. and W.S. Fyfe, 1973. Sub-sea floor metamorphism, heat and mass transfer. Contr. Miner. Petrol. 42: 287-304.

Sæmundsson, K., 1979. Outline of the geology of Iceland. Jökull, 29: 7-29
Tómasson, J. and H. Kristmannsdóttir, 1972. High temperature alteration minerals and geothermal brine, Reykjanes, Iceland. Contr. Miner. Petrol. 36: 123-134.
Turekian, K.K., 1969. The Oceans, streams and athmosphere. Ch 10 in Handbook of Geochemistry. Vol 1, Wedepohl K.H. ed. Springer -Verlag, Berlin: 297-323.
Wolery, T.J. and N.S. Sleep, 1976. Hydrothermal circulation and geochemical flux at Mid-Ocean Ridges. J.Geology. 84: 249-275.

PHYSICAL ENVIRONMENT OF HYDROTHERMAL SYSTEMS IN ICELAND AND ON SUBMERGED OCEANIC RIDGES

Valgardur Stefánsson

ORKUSTOFNUN

Grensásvegur 9, 108 Reykjavik, Iceland

Contents:

ABSTRACT

The details of magmatic and hydrothermal activity are much better known in Iceland than at submerged rift zones. Therefore, it is of interest to compare hydrothermal system in Iceland and on other parts of mid-ocean ridges. The effect of topography on hydrothermal systems for submarine systems is opposite to that of systems discharging into air. The reason for this is the simple fact that these two kinds of hydrothermal systems discharge into different kind of fluids as compared to the hydrothermal fluid. The working pressure of submarine hydrothermal systems is usually much higher than for systems on land. This fact, together with the higher salinity of the submarine systems, is most likely the cause of the metallic depositions observed on the sea floor. Downward penetration of cold water into hot rock is the only known process which can explain the high heat flux density of some hydrothermal systems in Iceland and it seems natural to assume that this process is of major importance for submarine systems.

A controlled cooling experiment of hot lava during the Heimaey eruption of 1973 supports further that this process is a natural phenomenon. By comparing the heat flux through Iceland with the energy released on submarine spreading axes, it is found that neither the volcanic nor the hydrothermal activity in Iceland is exceptionally high as compared to estimates for submerged ridges. Extrapolation of knowledge of hydrothermal systems is therefore found to be realistic. In Iceland the cooling of the crust by hydrothermal and volcanic activity is concentrated to a certain location, the central volcanoes. The spacing of cooling spots in the Icelandic rift zones is approximately 12-15 km. This distance is approximately the same as the estimated thickness of the crust. It is proposed that the cooling of the submerged rift zones might proceed in a similar way and that the spacing of cooling spots is related to the spreading rate.

INTRODUCTION

Active high temperature hydrothermal systems are located at the plate boundaries of the Earth's crust. Figure 1 shows the location of major identified high temperature geothermal systems. A striking feature of this distribution is that the majority of known hydrothermal systems are located on land in association with destructive plate boundaries. The reason for this is simply the fact that most geological features are better known on land than on the floor of the oceans. The major portion of plate boundaries found on land are destructive, whereas the majority of constructive plate boundaries are situated on the sea-floor. The only constructive plate boundaries on dry land are the rift zones of Iceland and the East African rift, and high temperature geothermal fields are as frequent on these segments of constructive plate boundaries as on destructive plate boundaries. Ocean heat flow measurements have revealed that a large part of the heat flow on mid-ocean ridges is due to convection (Sclater and Francheteau 1970, Corliss 1971, Anderson 1972, Von Herzen and Anderson 1972, Lister 1974, Bottinga 1974, Sclater et al. 1974, Anderson and Hobart 1976, Anderson et al. 1977, Anderson 1979) and several powerful hydrothermal vents have been identified (Williams et al. 1974, Weiss et al. 1977, Corliss et al. 1979, Edmond et al. 1979a, Jenkins et al. 1979, Williams et al. 1979, Cyamex 1979, Rise 1980, Shank and Bischoff 1977, Rona 1976, Rona et al. 1976, Scott et al. 1974, Scott et al. 1976, Rona 1980). It seems, therefore, natural to assume that hydrothermal systems on the sea-floor are approximately as common as those on dry land, and that the apparent skewness in the distribution shown on Fig. 1 is simply due to the difference in knowledge of geological features on land and on the seafloor.

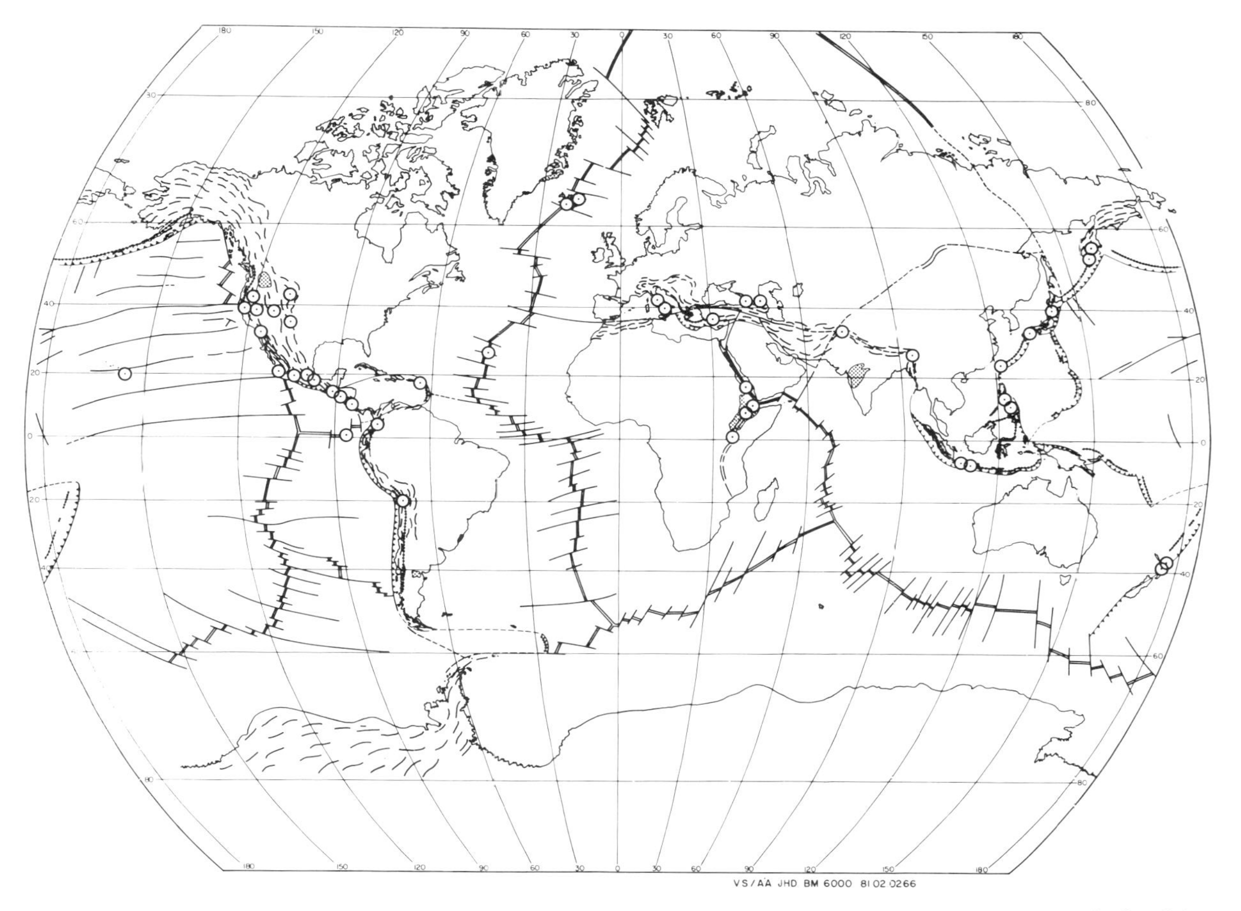

Fig. 1. Global distribution of major identified high temperature geothermal fields.

Hydrothermal systems on land are more readily accessible than those on the ocean floor, and details of hydrothermal systems on land are better known than the details of systems on the ocean floor. Drilling has proved to be essential to the understanding of hydrothermal systems on land, and should be equally critical in studies of suboceanic hydrothermal systems. At the present, thousands of wells have been drilled into hydrothermal systems on land, but no hole has so far been drilled into a submarine system. The situation for the submarine systems can in this respect be compared with the situation on land about one hundred years ago.

Iceland is a part of the Mid-Atlantic Ridge. The details of magmatic and hydrothermal activity are much better known in Iceland than at submerged rift zones. It is therefore of interest to compare the known properties of submarine hydrothermal systems to those of Iceland, and to investigate to what extent the observations made on Icelandic hydrothermal systems can be transferred to submarine systems.

The present paper discusses some of the differences in the physical environments of hydrothermal systems in Iceland and on the submerged ocean ridges. Furthermore, the heat transfer processes in high temperature geothermal systems are reviewed and a general cooling process at constructive plate boundaries is proposed.

THE DIFFERENCE IN PHYSICAL ENVIRONMENT

There are two major differences in the physical environments of the hydrothermal systems on land and on the ocean floor. Firstly, the systems discharge into different fluids. The fluid above hydrothermal systems on land is air, which has quite different physical properties from those of the hydrothermal fluid, in most cases water. On the other hand submarine hydrothermal systems discharge into sea water, which has almost the same physical properties as the hydrothermal fluid, hot sea water. Secondly, most of the submarine hydrothermal systems are located at such great oceanic depths that discharge at the ocean floor does not cause a phase transition (boiling), whereas boiling and condensation is of great importance for the dynamical behaviour of hydrothermal systems on land.

These differences are quite trivial, but some of the consequences are significant for the hydrothermal processes on land and on the sea floor.

Hydrothermal processes, both on land and on the sea floor, are primarily governed by buoyancy caused by the density difference between the hot fluid in the upflow zone and the surrounding colder fluid. Two-phase convection can occur in hydrothermal systems on land and at shallow oceanic depths. In such cases narrow chimneys are formed in the groundwater systems and the pressure difference between such upflow zones and the surroundings can be relatively large. High temperature geothermal areas on land are often situated on hills and mountains which shows that the buoyancy in these systems can overcome a substantial hydrostatic head. However, for single phase convection, the effect of topography is different on land than at the sea floor. This is due to the fact that hydrothermal systems on land and at the sea floor discharge into different fluids, as will be shown in the next section.

The hydrothermal vents so far identified on mid ocean ridges are at ocean depths of 2-3 km. The working pressure in such systems is of the order 20-60 MPa (megapascal = 10 bar) whereas the working pressure in hydrothermal systems found on land is usually less than 20-30 MPa. The two systems are thus working in different pressure ranges.

The temperatures in submarine hydrothermal systems are expected to be substantially higher than those found on land. It is therefore noteworthy that the highest temperature found in submarine vents, 350-400°C (Speiss et al. 1980, Edmond et al. 1979b, Bischoff 1980), is in the same range as the highest temperatures measured in geothermal systems on land (Muffler and White 1969, Castillo et al. 1981, Stefánsson 1981). Both theoretical (Strauss and Schubert 1977) and experimental studies (Dunn and Hardee 1981) have shown dramatic increases in the convective heat transfer near the critical temperature. Dunn and Hardee (1981) found the heat transfer rates in permeable media to increase by a factor of 70 in the vicinity of the critical point, and that this enhanced heat transfer might yield extraction rates of 12 kW/m^2. Bischoff (1980) has discussed physical limits on geothermal fluid. He points out that the great expansibility and increased buoyancy of water above the critical temperature at pressures lower than 45 MPa must act as a barrier to deeper penetration and greater heating of the fluid. If sea water is considered to be a pure NaCl solution the critical temperature would be 408°C and the critical pressure 30.4 MPa (Sourirajan and Kennedy 1962). So, in spite of different pressure ranges for submarine and landbased hydrothermal systems, the maximum working temperature for both kinds of systems might be of the order of 400°C.

In the temperature range 400-800°C the process of "chemical boiling" might be of greater importance for the heat transfer than adiabatic expansion of pure water. This process is assumed to be similar to two phase convection, where instead of the rising vapor there is a low salinity fluid with a density of 100-400 kg/m^3. Figure 2 shows an example of a density-temperature relation for the system $NaCl$-H_2O. Henley and McNabb (1978) have proposed a fluid

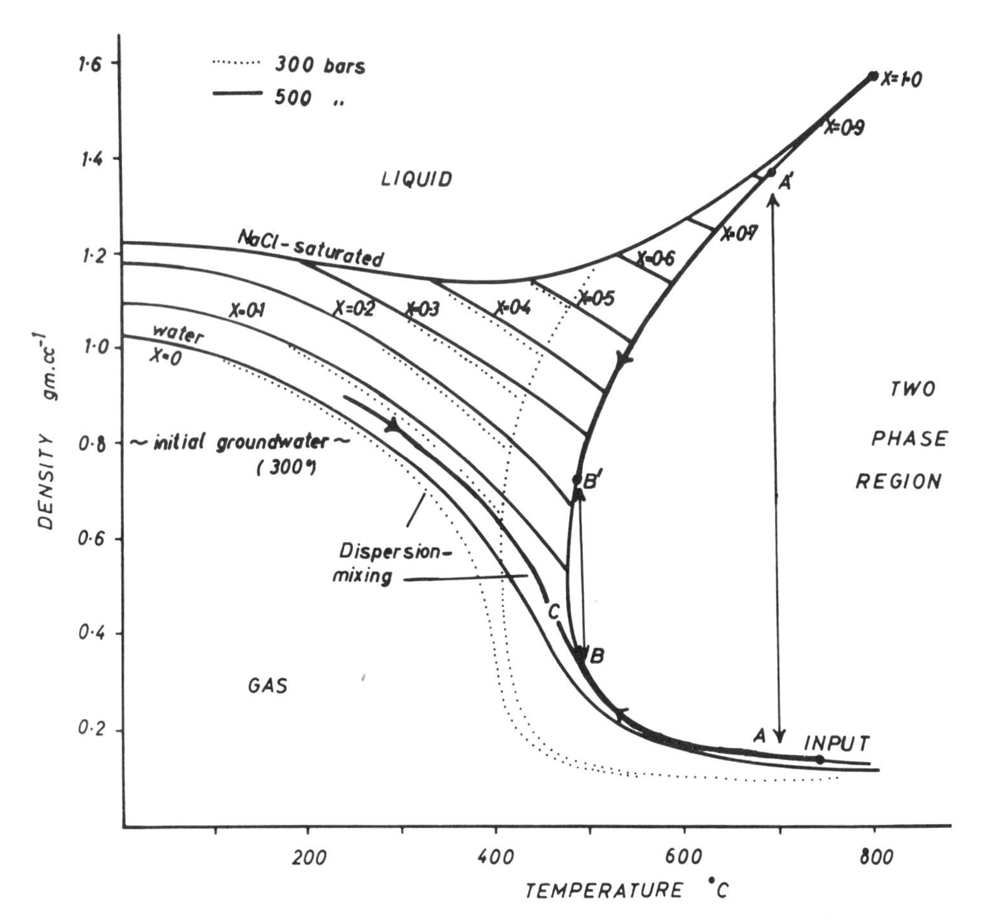

Fig. 2. Density-temperature relations in the system NaCl-H_2O at 500 bars and 300 bars fluid pressure. From Henley and McNabb (1978).

dynamic model for porphyritic copper emplacement which focused on the interaction of a buoyant low salinity magmatic "vapor" plume with surrounding ground water. As the low salinity fluid rises and cools, high salinity liquid condenses in a two-phase plume core.

In general, however, it can be stated that little is known about the details of hydrothermal processes above the critical temperature and pressure. Direct studies of such active systems on land require drilling down to 3-6 km depth, whereas such systems might be found at relatively shallow depths beneath the ocean floor.

One of the most marked differences between submarine and sub-aereal hydrothermal systems is the different chemical output of the systems. Metal rich deposits are the attributes of submarine systems, but similar deposits are not found near the surface in active geothermal systems on land. In fact, the enrichment of metals in sediments near mid ocean ridges was the first direct sign of hydrothermal convection at the sea floor (Skornyakova 1965 ; Bostrom and Peterson 1966) and the suggestion that iron and manganese are drived from local volcanics can be traced back to the last century (Bonatti and Nayudu 1965). In recent years numerous works have appeared in the literature (Bonatti and Joensuu 1966, Corliss 1971, Dymond et al. 1973, Bonatti 1975, Cronan 1976, Rona 1976, Seyfried and Mottl 1977, Seyfried and Bischoff 1977, Shanks and Bischoff 1977, Heath and Dymond 1977, Bonatti 1978, Corliss et al. 1978, Hoffert et al. 1978, Fyfe and Lonsdale 1978, Edmond et al. 1979b, Francheteau et al. 1979, Hekinian et al 1980, Lonsdale et al. 1980, Arnold and Sheppard 1981, Seyfried and Bischoff 1981, Styrt et al. 1981, Haymon and Kastner 1981, Oudin et al. 1981, Cronan et al. 1982, Reyss et al. 1982).

The reason for the difference in chemical output is believed to be twofold. First is the discharge of submarine vents at the sea floor causes a rapid single phase cooling of the fluid. The physical state of such systems can only be compared with the state of geothermal systems on land at several kilometers depth in the crust. Hydrothermal ore deposits are assumed to be formed at that depth.

Secondly, when sea water is heated by circulation in the crust, the redoxion of Mg from solution causes a low pH value of the fluid (Bischoff and Dickson 1975, Hajash 1975, Seyfried and Mottl 1977, Seyfried and Bischoff 1977, Mottl and Holland 1978, Bischoff and Seyfried 1978, Seyfried and Mottl 1982) and due to the low pH value, the metal solubility of heated sea water is considerably higher than that of heated meteoric water at the same temperature. The concentration of metals in the discharge of submarine vents is therefore in general much higher than in the discharge of geothermal systems on land. Exceptionally high metallic concentrations have, however, been found in the fluid at Salton Sea (Muffler and White 1969) and in the Krafla field in Iceland (Ármannsson et al. 1982, Stefánsson 1981). The Krafla case can be compared to submarine systems since injection of magmatic gases (SO_2, CO_2) causes temporarily low pH values of the hydrothermal fluid (pH values as low as 1.8 have been measured). Furthermore, such a fluid, contaminated by magmatic gases, has been found to cause rapid deposition of iron and silica in flowing wells. Concentrations of iron up to 10 ppm have been found in the hydrothermal fluid in Krafla, whereas a common concentration in other geothermal fields in Iceland is usually lower by a factor of 10^3-10^4. From an economic point of view it can therefore be stated that the chemical composition of submarine hydrothermal fluids prevents common utilization of these systems as geothermal energy resources.

THE EFFECT OF TOPOGRAPHY

Low temperature geothermal fields in Iceland are predominantly located in valleys and other topographic lows. Figure 3 shows the distribution of such fields in northern Iceland. Low temperature geothermal fields can be compared with submarine hydrothermal systems in the sense that both systems are pure single phase systems. On the other hand, high temperature geothermal systems often have two phase convection which makes the correlation with topography more diffuse.

At the sea floor, however, the high heat flow, caused by hydrothermal circulation, is strongly correlated with topographic highs (Lister 1972, Sclater et al. 1974, Anderson and al. 1977, Anderson and Hobart 1976, Crane and Normark 1977, Williams et al. 1974, Green et al. 1981).

Figure 4 shows a cross section of topography and heat flow measurements on the Explorer Ridge. Lister (1972) first pointed out that heat flow across the Explorer Ridge correlated with topography in the opposite manner to that of refraction of conductive heat flux. He suggested that the circulation is controlled topographically in such a way that topographic highs act as "chimneys" for the hot flow, whereas valleys are the natural recharge sites. Lowell (1980) studied hydrothermal convection in a model with wave-like oceanic crustal topography. He found that the fluid ascends under topographic highs and descends at topographic lows. Williams et al. (1974) and Sclater et al. (1974) present convincing evidence for this type of convection. In spite of that Sclater et al. (1974) state: "Though physically possible it is still somewhat difficult to accept that almost all the heat flux through the area occurs at a few outcropping basement highs".

Fig. 3. Distribution of low temperature geothermal areas in North-western Iceland.

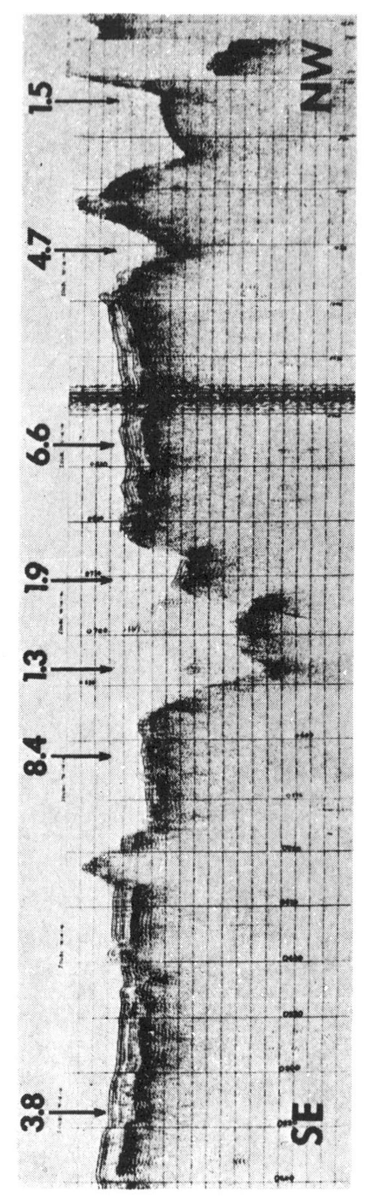

Fig. 4. Heat flow and seismic-reflection profile across the Explorer Ridge. From Lister (1972).

It seems, therefore, that the effect of topography on the hydrothermal circulation in submarine systems has appeared to be something strange. However, a very simplified consideration of the hydrostatic pressure differences involved shows that it is natural that discharge vents are predominately located at topographic highs at the ocean floor whereas the reverse is true for subaereal systems. Figure 5 shows the two cases for subaereal systems where the discharge is located downhill or upphill with respect to the recharge area. Taking the simplest case where the downflow is isothermal with a fluid density ρ_c, and the upflow is also isothermal with a fluid density of ρ_h, we can calculate the static pressure difference between P_A and P_B. In this case the hydrostatic head facilitates circulation where the recharge area is located uphill. On the other hand, in submarine systems (Figure 6) the reverse situation is true. In this case the chimney effect is dominant and discharge areas are predominantly located uphill with respect to the recharge area. The reason for this different effect of topography on subaereal and submarine hydrothermal circulation is simply the fact that these two systems discharge into different fluids. The density contrast between water and air is approximately 10^3 whereas the density difference between cold and hot sea water is less than 3. The pressure difference between discharge and recharge locations on land can therefore be neglected which is not the case for submarine systems.

In fact this chimney effect can be seen on land in hydrothermal systems where the circulating fluid has a density similar to that of air. These circumstances are often at hand in new scorious lavas, where steam is observed rising preferably from topographic highs. An example of this effect in nature is shown on Fig. 7. This picture was taken two years after the end of the volcanic eruption on Heimaey in 1973. The lava is still hot, but it is (successively) cooled by the rain. The rainwater percolates into the hot lava where it boils and the steam rises preferably through the chimneys of hills and other topographic highs.

A geothermal system has been identified on the Kolbeinsey-Ridge approximately 100 km north of the coast of Iceland. Located on the mid-ocean ridge, it is assumed to be potential high temperature area (Fig. 8). The Kolbeinsey area is of interest since the site is relatively accessable with the depth to the vent(s) of only 90 m. All other submarine hydrothermal vents found to date are at far greater depths.

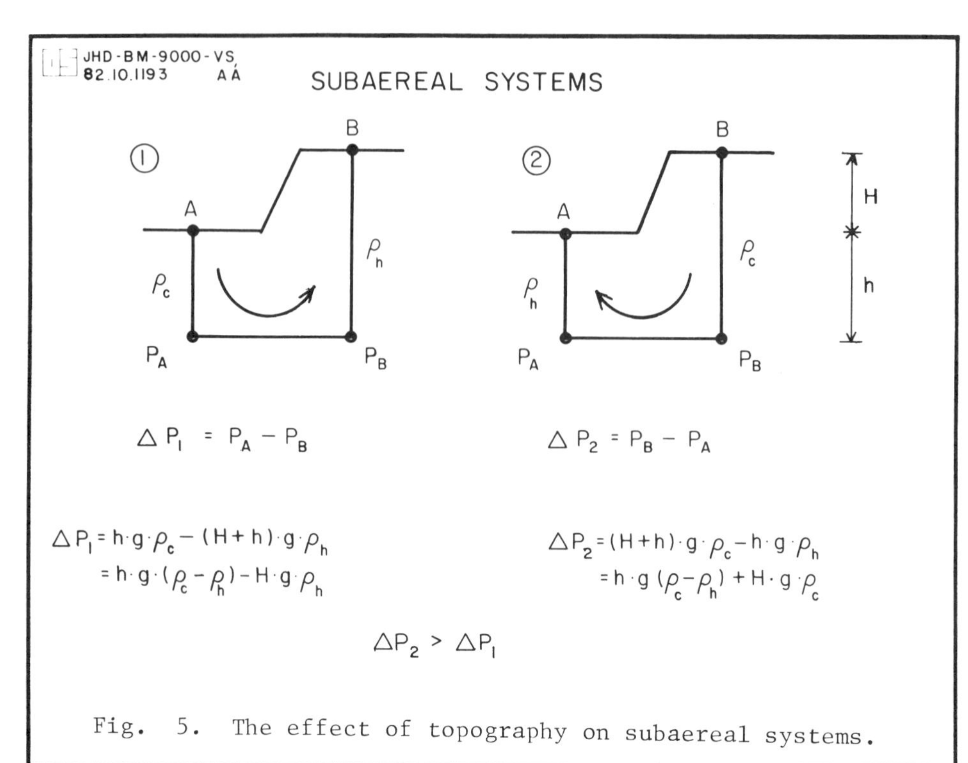

Fig. 5. The effect of topography on subaereal systems.

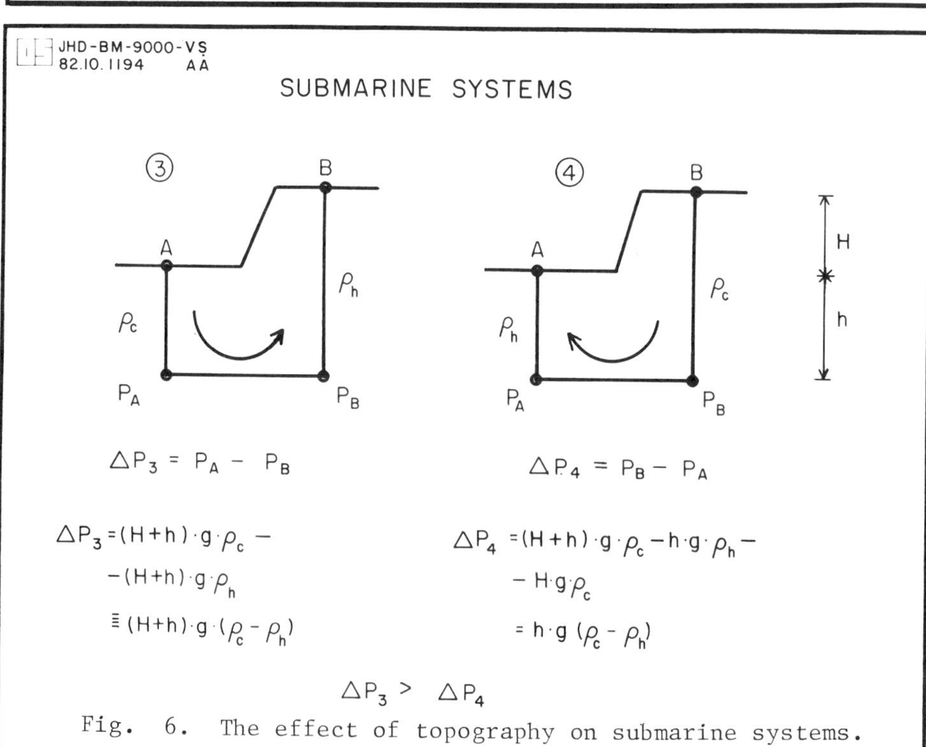

Fig. 6. The effect of topography on submarine systems.

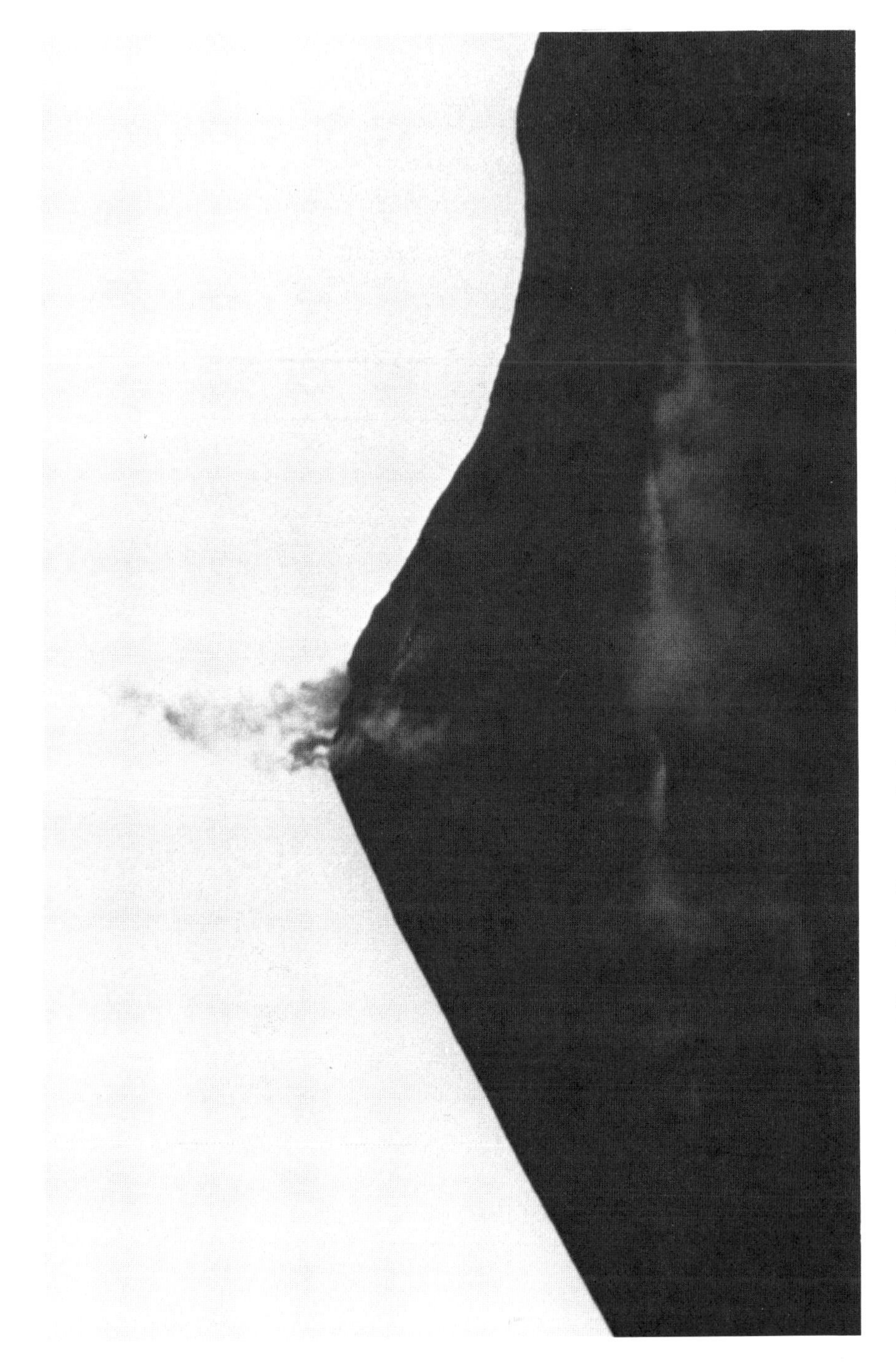

Fig. 7. Steam rising from Eldfell in Vestmannaeyjar.

DOWNWARD PENETRATION OF COLD WATER INTO HOT ROCK

The thermal energy released on mid-ocean ridges has been found to be a substantial part of the total heat loss of the Earth. Sclater et al. (1980) estimate the total heat loss of the Earth to be 42 TW and the heat loss by submarine hydrothermal circulation to be 10 TW, or 24% of the total heat loss of the Earth. In spite of the fact that submarine advection of magma seems to be included in the term hydrothermal circulation as used by Sclater et al. (1980) the heat flux from spreading axes must be considered exceptionally large as the active spreading axes cover only about 1% of the surface of the Earth. Macdonald et al. (1980) estimate a heat flow of 250 MW through a single vent of 30 cm diameter on the East Pacific Rise, and Björnsson (1974) estimates the heat flux from the Grímsvötn high temperature area in Iceland to be 5000 MW. The heat transfer processes at spreading axes are obviously quite effective.

Even though there has been agreement about molten magma or hot intrusions being the heat sources for high temperature hydrothermal systems, the heat transfer processes have been discussed for some decades (Bodvarsson 1951, 1961; White 1957, 1968; Banwell et al. 1957, Banwell 1963).

The relatively low heat conductivity of igneous rocks puts severe limitations on the possible rate of heat transfer from rock to water across a given surface area. It is difficult to explain the high heat flux through the surface of some geothermal fields unless the water actually penetrates into the hot rock, which is in the process of cooling. This process was first suggested by Bodvarsson in 1951 (Bodvarsson 1951, page 6), and has recently been elaborated further (Bodvarsson and Lowell, 1972; Bodvarsson 1979, 1982). Lister (1974, 1976, 1977a, 1977b, 1980, 1982) has presented a conceptual model of the downward penetration of water into hot rocks by the process of cooling and thermal cracking. The results obtained by Bodvarsson and Lister are similar, even though their approach is somewhat different. The process of downward penetration into hot rock appears to be an extraordinarily efficient form of heat transfer from the rock to the fluid, and it seems to be the only known process which can describe the high heat fluxes observed at some locations on spreading axes. It is, therefore, of interest to review some observations from Iceland which are relevant for this process.

The Grímsvötn high temperature geothermal area in Iceland is located under the icecap of Vatnajökull (Figure 9). Because of

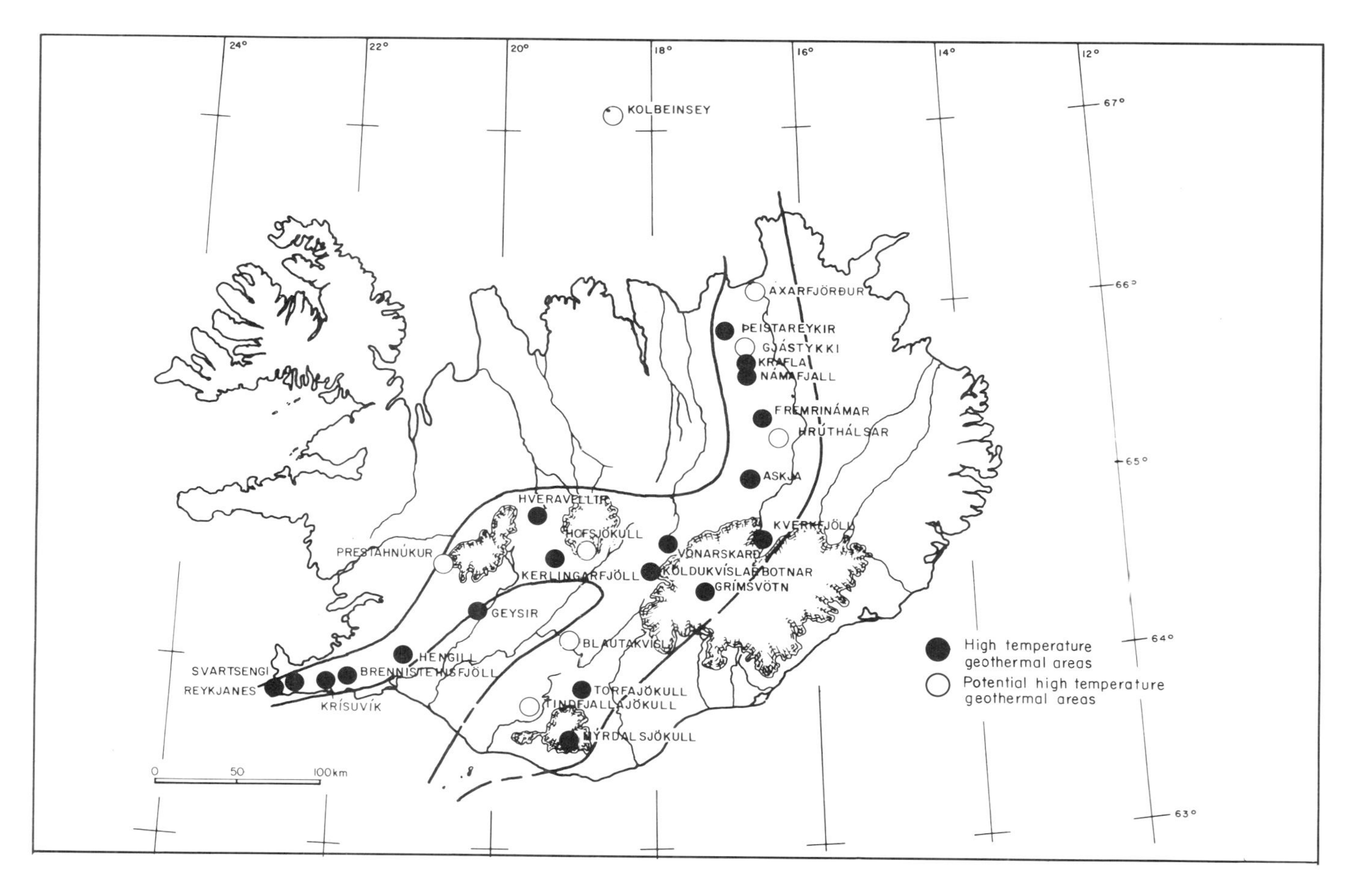

Fig. 8. Location of high temperature geothermal areas in Iceland.

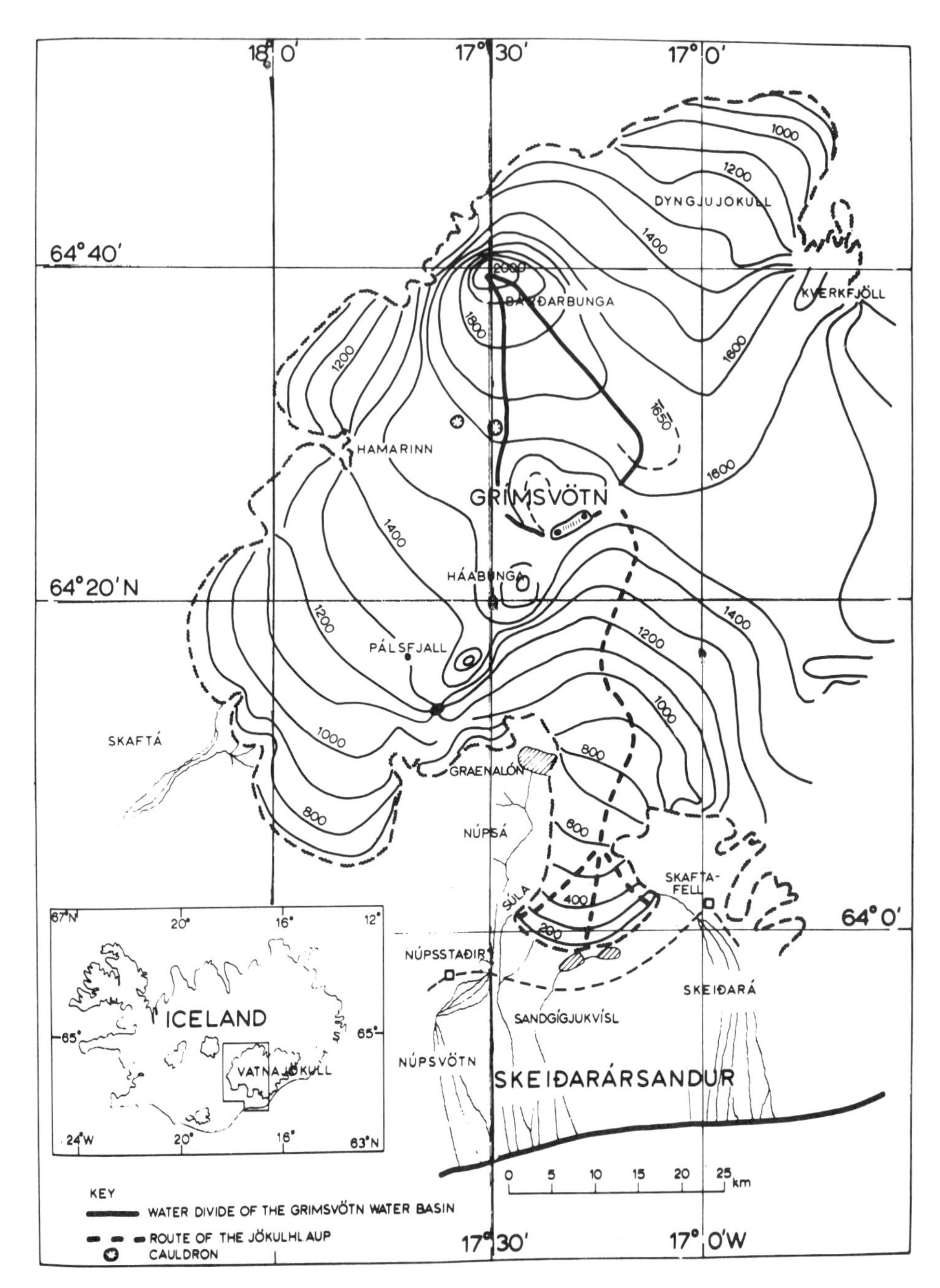

Fig. 9. The Grimsvotn high temperature geothermal area in Iceland. From Bjornsson (1974).

this location it has been possible to use calorimetric methods to estimate the heat flux of this geothermal system (Björnsson 1974). The value obtained is 5000 MW. The geothermal area in Grímsvötn is assumed to be no larger than 100 km^2 which gives a mean heat flow of 50 W/m^2. Periodic outbursts of meltwater (jökulhlaup) have occurred every decade for at least 350 years according to Icelandic annals and the first report on such a jökulhlaup dates back to AD 1332 (Björnsson et al. 1982). In order to sustain this flow of heat for 600 years 30 km^3 of magma need to be cooled down, which appears to be geologically possible. If this cooling of magma is taking place over the whole geothermal area, (100 km^2), a 300 m thick slab is cooled from 1200°C to 400°C in this time (600 yrs.). However, in order to sustain a conductive heat flow of 50 W/m^2 through a wall which separates magma at 1200°C from a fluid at 400°C, the wall must be not thicker than 30 m over the whole extent (100 km^2) of the geothermal area. This value is an order of magnitude lower than the thickness of the cooled rock. This indicates that the cooling front between magma and hydrothermal fluid can hardly be stationary, but is migrating downwards with a velocity of at least 0,5 m/year. If the contact area between magma and fluid is 10 km^2, the downward velocity of the cracking front would be 5 m/year. Lister (1982) estimated theoretically that the downward velocity of the cracking front might be of the order of 10 m/year (calculated range 1-20 m/year), and Bodvarsson (1982) gives the estimate 0,3 to 3 m/year.

The observation made in the Grímsvötn area in Iceland seems therefore to be in agreement with the theory of downward penetration of cold water into hot rock as described by Bodvarsson and Lister.

Similar heat flow values are obtained in a few other geothermal areas in Iceland. Friedman et al. (1972) estimated the heat flux of the Kverkfjöll geothermal area to be 1200-2300 MW and the areal extent to be 10-20 km^2. This gives a heat flow in the range 60-230 W/m^2. Bodvarsson (1961) estimated the heat flux from the Torfajökull area to be 2000 MW and the areal extent to be 100 km^2.

It is difficult to explain the high heat flux and heat flow from the three geothermal areas in Iceland (Grímsvötn, Kverkfjöll and Torfajökull) without assuming downward penetration of cold water into hot rock. But there exists a more direct observation from Iceland showing that downward penetration of cold water into hot rock actually takes place in nature.

The volcanic eruption on Heimaey in 1973 was literally within the fishing town of Vestmannaeyjar. In order to protect the town, an immense amount of cold water was pumped on the lava flow to divert it from the town. This method turned out to be very valuable, and much property was saved.

In one location 100 l/s were pumped on 7000 m^2 area (Jonson and Matthiasson 1974). After 14 days of continuous pumping, a hole was drilled into the lava. It was found that the water had cooled the molten lava down to a 12 m depth, and that the temperature increased from 100°C up to 1050°C within 1 m interval. Assuming that all the water has evaporated this experiment gives a heat transfer of 40 kW/m^2 and a downward penetration rate of 0.9 m/day (Björnsson et al. 1982). The temperature increase of approximately 1000°C within an interval of 1 metre is striking, but in agreement with the theory of downward penetration of cold water into hot rock.

Excavation of lava after the eruption showed a great difference in the structure of the lava which had been water-cooled as compared to places where no water had been applied during the solidification. The water-cooled rock was intensely fractured and broken into cubes commonly with an edge of 10-20 cm, whereas the non-water-cooled lava consisted of large blocks (Björnsson et al. 1982).

The experiment at Heimaey supports in many respects the theory of downward penetration of cold water into hot rock.

1. Rapid downward migration (0.9 m/day) of the cooling front is observed.

2. Large heat transfer (40 kW/m^2) from the rock to the fluid is observed.

3. The thickness of the cooling front is very thin (less than 1 mm).

4. Only the water-cooled lava is highly fractured.

The observations made in Heimaey, Grímsvötn, Kverkfjöll, and Torfajökull seem, therefore, to justify the statement that downward penetration of cold water into hot rock is a process which actually takes place in nature.

ICELAND IN RELATION TO OTHER PARTS OF MID OCEAN RIDGES

Iceland is a conspicuous part of the Mid-Atlantic Ridge. Several anomalous patterns are found, the most obvious one that the country is above sea level. The uniqueness of Iceland among other parts of mid ocean spreading axes has raised questions like: is the magmatic activity of Iceland greater than corresponds to the spreading rate?; is the thermal current through Iceland higher than at submarine spreading axes of similar spreading rate?; is the upper mantle under Iceland in some way anomalous?; and so on. Present knowledge prevents unambiguous answers to such questions.

Table 1. Energy currents and flows in Iceland, the oceans and of the Earth.

	ICELAND Area $10^5 km^2$ Rift axes 300 km						OCEANS Area $3 \cdot 10^8 km^2$ Rift axes $5 \cdot 10^4 km$			EARTH Area $5 \cdot 10^8 km^2$ Rift axes $5 \cdot 10^4 km$	
	Current GW		Flow mW/m^2		Current per km of ridge axes mW/km		Current GW	Flow MW/m^2	Current per km	Current GW	Flow MW/m^2
Reference	B82 A	P73 B	B82 C	P73 D	B82 E	P73 F	S80 G	Calc. H	Calc. I	S80 J	Calc. K
1. Total conduction	15	30	150	300			20300	68		31700	63
2. Conduction in excess of normal	9	24	90	240	30	78	3000	10	56	3000	6
3. Advection by lava	7	6	70	60	23	21					
4. Heat loss on spread-in axes due to advections of water and conduction	8.5	10.5	85	105	28	35	10100	34	191	10300	21
5. Total through rift axes	24.5	40.5	245	405	82	134	13100	44	247	13300	27
6. Total through area	30.5	46.5	305	465			30400	102		42000	84

B82: Bodvarsson 1982
P73: Palmason 1973
S80: Sclater et al. 1980

Calc.: Calculated from data in Sclater et al. 1980

However, it is necessary to try to find out to what extent the situation in Iceland can be compared to the situation on the more common submerged mid ocean spreading axes. We will in this section consider heat flow and the distribution of cooling spots along spreading axes.

Sclater et al. (1980) have made a detailed study of the heat flow through oceanic and continental crust and estimated the total heat loss of the Earth. The energy current through Iceland has been studied by Bodvarsson and by Palmason (Bodvarsson 1954, Palmason 1973, Bodvarsson 1982). The approach taken by these authors is quite different, and comparison of results can not be done in a straight forward manner. A comparable representation of the results is therefore appropriate. The representation of Bodvarsson (1982) is chosen for the comparison. The mechanical power factor of Bodvarsson (1982) is, however, omitted as this term is neither considered by Sclater et al. (1980) nor by Palmason (1973). The comparison is shown in Table 1. The values given by Bodvarsson (1982) are taken unchanged and enter rows A, C and E in Table 1.

The basic values estimated by Palmason (1973) are.

a) the heat transported laterally into the lithosperic plates, equal to 78 MW/km. This value enters F2 in Table 1.

b) the heat transported to the surface by volcanism, equal to 21 MW/km. This value enters F3 in Table 1.

c) the heat transported to the surface in the volcanic zone by advection of thermal water and by conduction. This sum is calculated to 35 MW/km. This value enters F4 in Table 1.

Other values in rows B and D in Table 1 are calculated from the above given values.

The values presented in Table 1 for the oceans and for the Earth are obtained from Sclater et al. (1980). Basic values are the total heat loss of the oceans, 30.4 TW, which enters G6 in Table 1. The heat lost due to hydrothermal circulation, 10.1 TW, which enters G3-4 of Table 1. It seems, that this figure contains also the contribution from submarine volcanism, which justify the placement in Table 1. The total energy current through the rift axes in the Oceans, 247 MW/km is obtained from Table 10 in Sclater et al. (1980), where the total heat loss through ocean floor of the age 0-20 Ma is estimated as $3.5 \cdot 10^{12}$ cal/s =14.6 TW. A normal heat flow of 1.5 TW is assumed for this area giving the 13.1 TW which enters G5 in Table 1. Other values for the oceans are calculated.

Heat flow values for the Earth are also based on the estimate by Sclater et al. (1980). The total heat loss of the Earth, 42 TW enters J6 in Table 1. Heat loss due to continental volcanism is given by Sclater et al. to be approximately 200 GW, and a value of 10.3 TW enters J3-4 in Table 1. Other values in rows J and K in Table 1 are calculated.

Some interesting points are obtained by comparing the figures in Table 1. Firstly, we see that the estimates for the energy current through a unit length of ridge axis is smaller by a factor of 2-3 for Iceland than the estimates for oceans.

Secondly we see that the sum of the estimates of magmatic and hydrothermal activity of Iceland is 51-56 MW/km whereas the corresponding value for the oceans is 191 MW/km. The estimates for the oceans are thus 3-4 times larger than the estimate for Iceland. These figures do not support the opinion that the volcanic or hydrothermal activity of Iceland is exceptionally great. On the contrary we would expect magmatic and hydrothermal activity on submerged ridges to be higher than in Iceland. Here we must point out, that in Iceland, the contributions from magmatic and hydrothermal cooling are known to be approximately equal, whereas the relative importance of these contributions is not known for the oceans. It is possible that the heat loss from the oceans of 191 MW/km is either dominated by submarine volcanism or by hydrothermal activity. If the heat loss of 191 MW/km is mostly due to submarine volcanism, the explanation of the elevation of Iceland above sea level becomes problematic. On the other hand if the heat loss of 191 MW/km is dominated by hydrothermal activity at the sea floor we would expect the hydrothermal activity on the sea floor to be approximately ten times more intensive than in Iceland.

Considering the above discussion we conclude that it is justified to extrapolate the knowledge of hydrothermal systems in Iceland to the submerged ridges, as we expect the submarine hydrothermal activity to be at least as intensive as in Iceland.

A very strong attribute of the high temperature geothermal areas in Iceland is their close association with central volcanoes. In fact, volcanism and high temperature geothermal systems in Iceland are so closely interrelated that the two phenomena can hardly be treated separately. The close association between central volcanoes and high temperature geothermal areas can be exemplified by the fact that all except four of the 19 known geothermal areas are within or connected to central volcanoes, and of all presently active central volcanoes, there are only three where high temperature systems have not been identified. These three central volcanoes (Öræfajökull, Hekla and Snæfellsjökull) are situated outside of the axial rift zone.

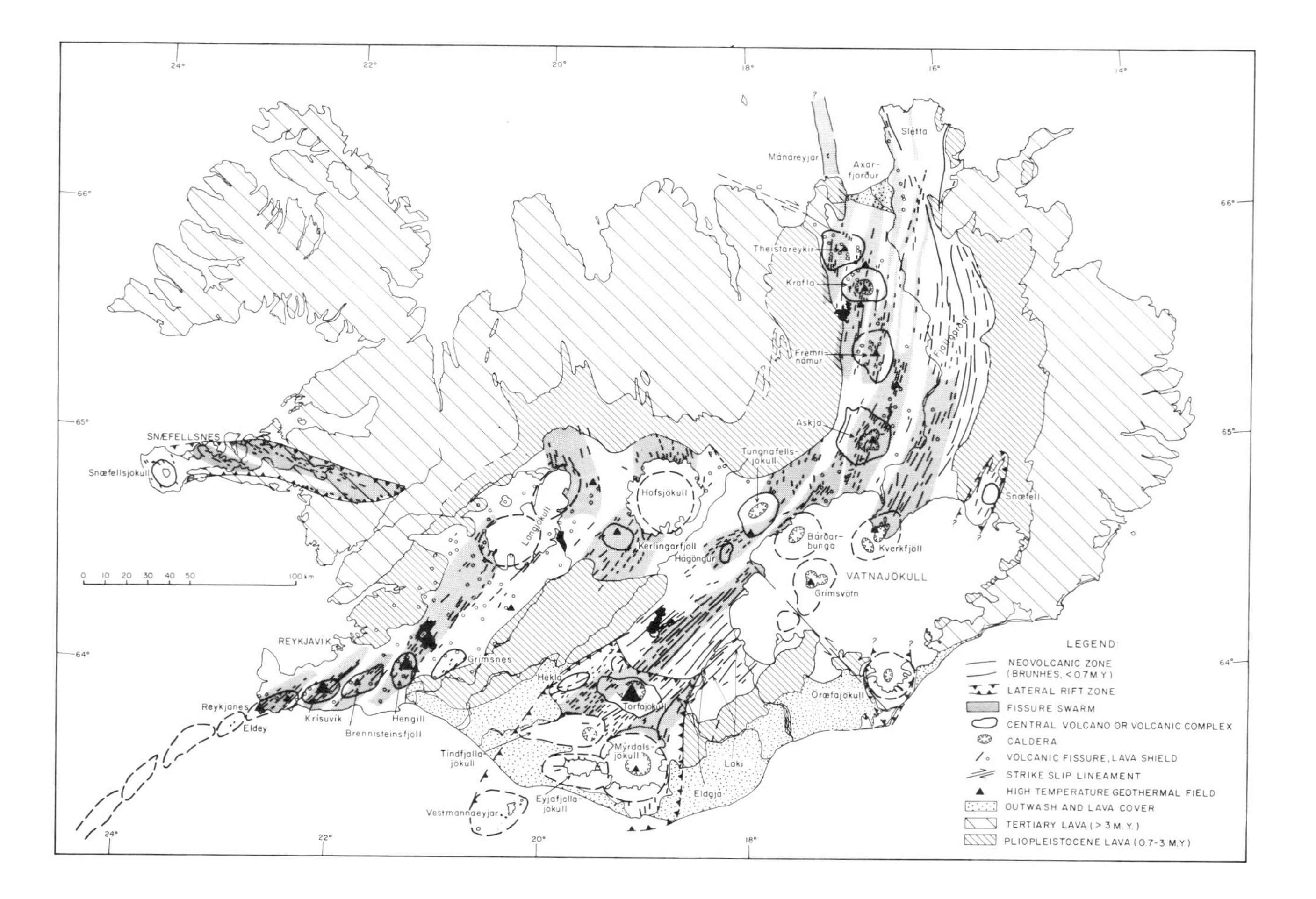

Fig. 10. Geological map of Iceland. Compiled by K. Saemundsson.

The distribution of high temperature geothermal areas in Iceland is shown in Figure 8. There are 19 identified areas and 8 potential areas. Figure 10 is a geological map of Iceland and shows the close association between high temperature areas and central volcanoes. There are 23 active central volcanoes in the axial rift zone.

The first striking feature of the distribution of magmatic and hydrothermal activity in the axial rift zones of Iceland is the fact that these activities are concentrated to 20-25 locations in the axial rift zone. These locations are the central volcanoes and the high temperature geothermal areas. As the heat flow is obviously much larger at these locations than elsewhere in the axial rift zone we prefer to refer to them as cooling spots. The total length of the axial rift zone in Iceland is approximately 300 km. The mean distance between the cooling spots is therefore 12-15 km. Adapting the energy estimated in Table 1 we might expect an average magmatic energy current of up to 275-345 MW and hydrothermal activity of the order 335-525 MW in each cooling spot.

These considerations immediately raise the question how the crust of the axial rift is cooled in the intervals between the cooling spots. There is strong evidence indicating that we can look upon this cooling as if it were concentrated to the cooling spots.

Starting with the magmatic activity, we have to mention some of the observations made on the Krafla central volcano during the last years. Figure 11 shows the general geological situation in NE-Iceland, central volcanoes and associated fissure swarms. A closer view of the Krafla area is shown in Figure 12 where the fissure swarm through the Krafla central volcano is shown in detail. Rifting and magmatic activity has been taking place there continuously since December 1975 (Björnsson et al. 1977, 1979). There is a high temperature geothermal field inside the volcano which is under exploitation (Stefánsson 1981). This magmatic-hydrothermal system is now probably the best known system of this kind in Iceland and can serve as an example for other parts of the rifting system. It has been found that a small magma chamber is situated at a depth of 3-7 km below the center of the Krafla caldera (Einarsson 1978). The high temperature geothermal field under exploitation is situated above the magma chamber (Stefánsson 1981). The magmatic activity has influenced the hydrothermal system (Ármannsson et al. 1981). Geothermal wells have been drilled into the hydrothermal systems, and the location of these wells compared to the magma chamber is shown in Figure 13. Some of the chemical impacts on the geothermal fluid during the initial phase of the magmatic activity are shown in Figure 14.

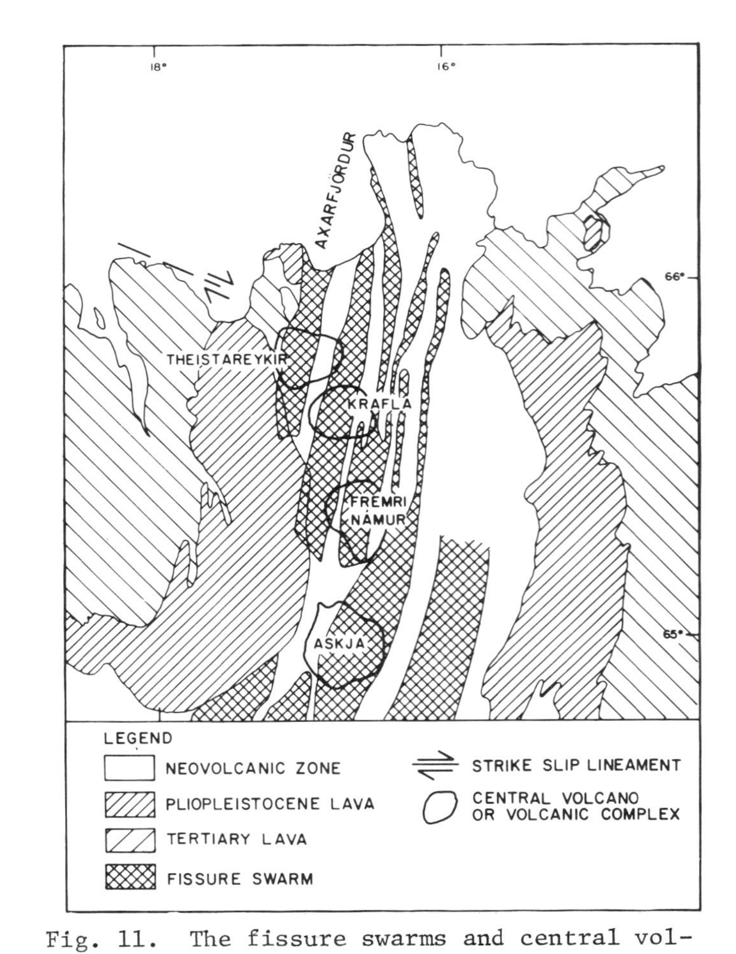

Fig. 11. The fissure swarms and central volcanoes in the Neovolcanic zone in Northeast Iceland. Compiled by K. Saemundsson.

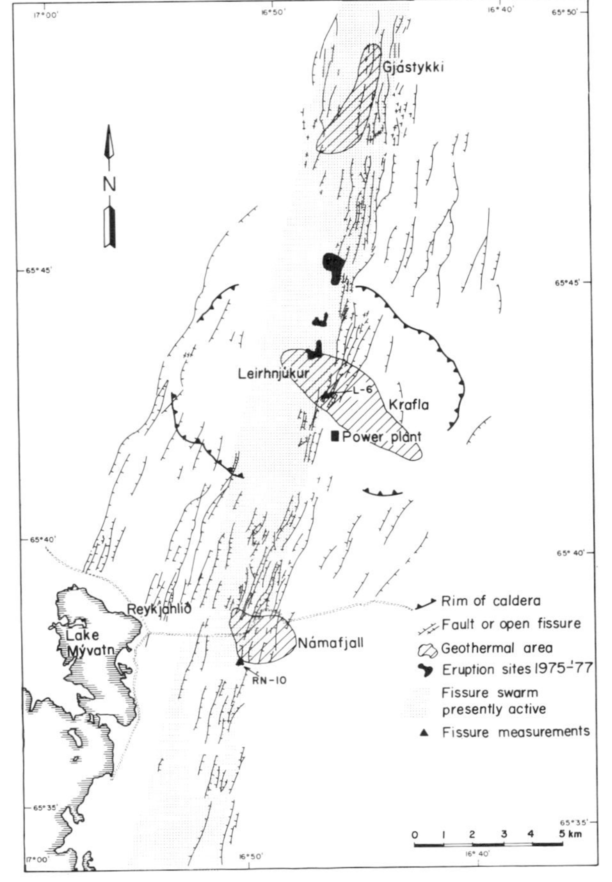

Fig. 12. The Krafla fissure swarm. Compiled by K. Saemundsson.

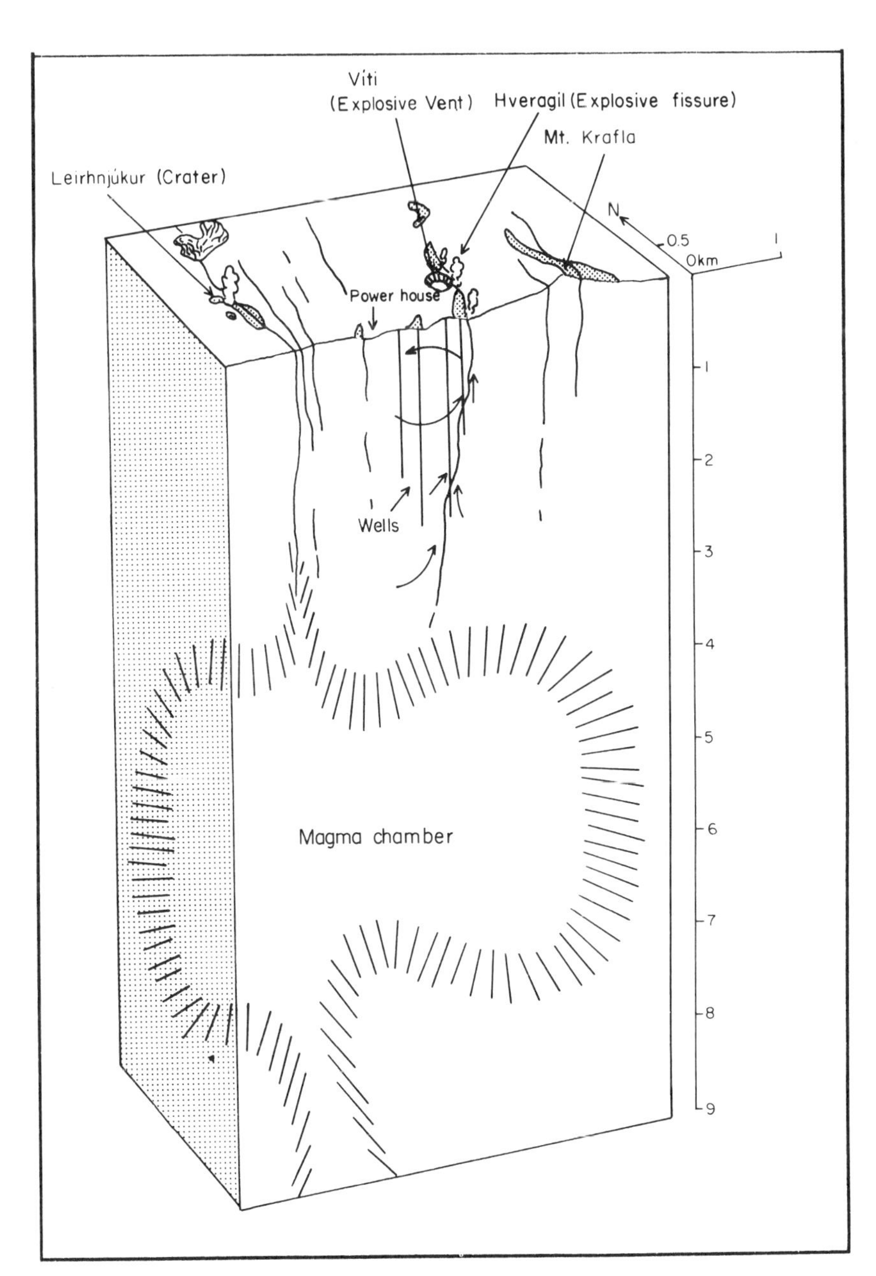

Fig. 13. The geothermal wells in Krafla in relation to the magma chamber.

Figure 15 shows a schematic model of the magmatic activity in the Krafla central volcano. The magma chamber below the volcano has been continuously supplied with magma from beneath during the last seven years. The surface is inflating most of the time, but at certain times the magma flows out of the magma chamber either to the surface or laterally into the fissure swarm. Figure 16 shows how the elevation of the surface has changed during this period and how the deflation events are associated with the magmatic activity either as volcanic eruptions or lateral propagation of magma into the fissure swarm. In many cases both effects have been observed. During one of these magmatic events, a small volcanic eruption occurred in Krafla but magma was also observed flowing southward through the fissure swarm to a different geothermal field in the Námafjall area, situated 10 km south of Krafla. About 1 m^3 of the magma erupted through a discharging drillhole in Námafjall. As a matter of curiosity it can be mentioned that this steam well in Námafjall did not change in character because of this "volcanic eruption" and the well is still in use for steam production.

An important feature of the behaviour of the Krafla volcano is that the bulk of the magma is rising through the cooling spot of the central volcano, but part of the material is flowing horizontally several tens of km into the fissure swarm. We can therefore look upon this cooling process as concentrated to the cooling spot of the central volcano, but where a larger part (20-40 km) of the rift zone absorbs the magmatic heat.

Lateral movement of magma along the rift axes in Iceland has so far only been observed and studied in the Krafla volcano (Einarsson and Brandsdóttir 1980). This process is, however, well known in Hawaii (Swanson et al. 1976) and there is growing evidence indicating that this process might be common in the volcanic activity of Iceland (Sigurdsson and Sparks 1978, Helgason and Zentilli 1982, Robinson et al. 1982). It can even be assumed that this process takes place on submarine rift axes, where the crustal stress and tectonic pattern is quite similar to that of the rift zones of Iceland.

Turning now to the hydrothermal activity in the rift zones of Iceland we note that there are no signs of thermal flow at the surface outside of the high temperature geothermal areas. On the other hand, the few estimates of the thermal output of geothermal areas carried out so far, indicate a typical heat flux of some hundreds of megawatts with a few exceptions up to several thousands of megawatts. This indicates clearly that the high temperature geothermal areas can be looked upon as the major cooling spots of hydrothermal advection in the axial rift zones of Iceland.

Thermal conditions between high temperature areas in the volcanic zones of Iceland are not very well known as there are few

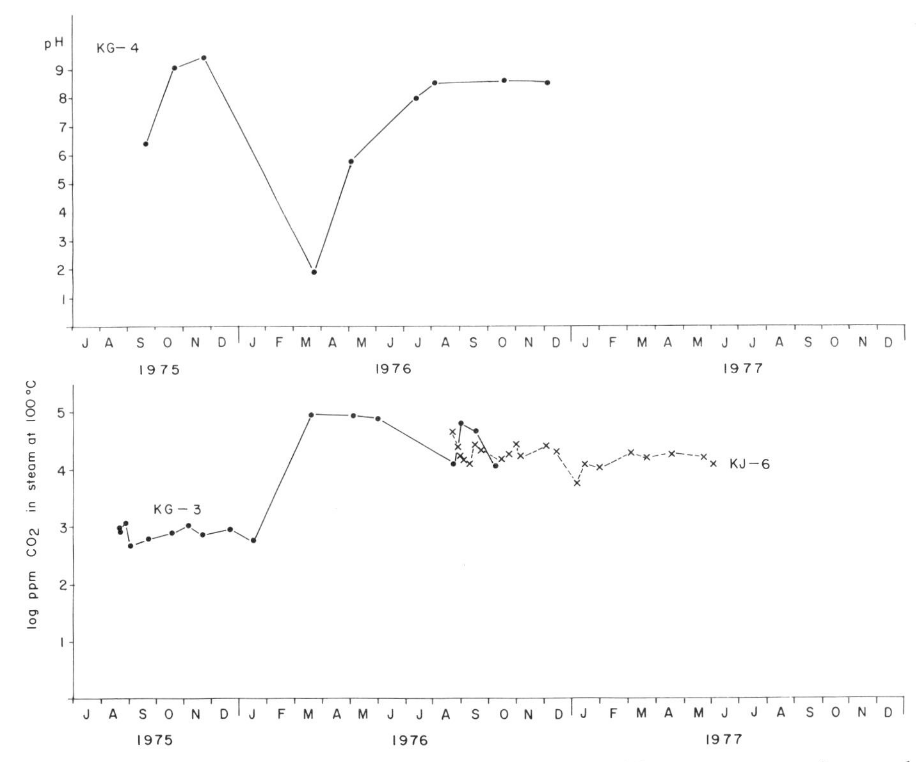

Fig. 14. The effect of magmatic gases on CO_2 concentration and pH value of well discharge in Krafla.

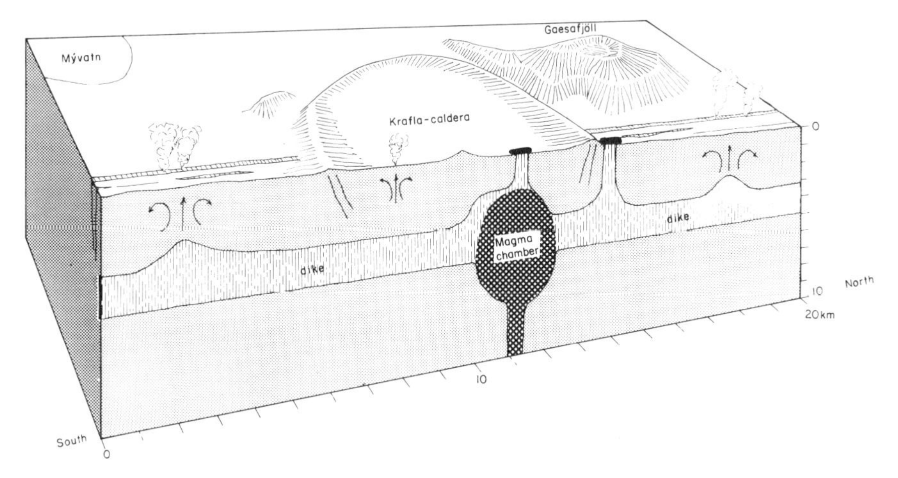

Fig. 15. Schematic model of the Krafla volcano. From Bjornsson et al. (1979).

drillholes outside of geothermal fields. In some wells outside the geothermal areas, it has been observed that temperature does not increase with depth. Figure 17 shows a temperature profile from such a well. A constant temperature of 4°C is observed down to 700 m depth. These circumstances are in an agreement with downward flow of surface water to large depths. In all high temperature geothermal systems in Iceland investigated to date, the hydrothermal fluid is found to be local meteoric water. The recharge areas for the high temperature geothermal systems are therefore the closest vicinity, presumably along the rifting axes, where cold surfact water percolates down to some depth and is then drawn to the upflow zones in the high temperature systems.

The general picture of the cooling process in the axial rift zone of Iceland can be described by the following simplified picture;

The upflow zones of magma are predominantly beneath the central volcanoes, which are located with a mean separation of 12-15 km along the axial rift zones of Iceland. The magmatic heat is, however, dispersed along the rift axes by lateral flow of magma into the fissure swarms. Some of the magmatic heat is released directly to the surface by volcanic eruptions, and some is stored in the crust both below the central volcanoes and in less extent in the fissure swarms along the rift axes outside of the central volcanoes. This heat stored in the crust of the rift zones is transported to the surface by hydrothermal circulation. The major upflow zones of the hydrothermal fluid are located in the central volcanoes as most of the stored heat is concentrated below the central volcanoes, but where large accumulations of magma occur outside of the volcanoes there will be separated upflow zones (examples Námafjall, Gjástykki and Axarfjörður). The heat carried out in the fissure swarms between the cooling spots of the central volcanoes is usually too dispersed to create separate hydrothermal upflow zones. Instead, the heat is transported back to the cooling spots by the recharge water of the hydrothermal systems. A simplified picture of the proposed cooling process along the axial rift zone in Iceland is shown in Figure 18. It should be stressed that the model presented in Figure 18 is only a hypothesis but it has several predictable features which can be tested in the future.

An important point in these considerations is the assumption of heat flow along the spreading axes together with the distinct locations of major cooling spots along the rift axes. There do not seem to be any obstacles which prevent similar processes to take place on submerged spreading axes.

The concept of spacing between cooling spots (central volcanoes) has a different interesting point of view. Vogt (1974)

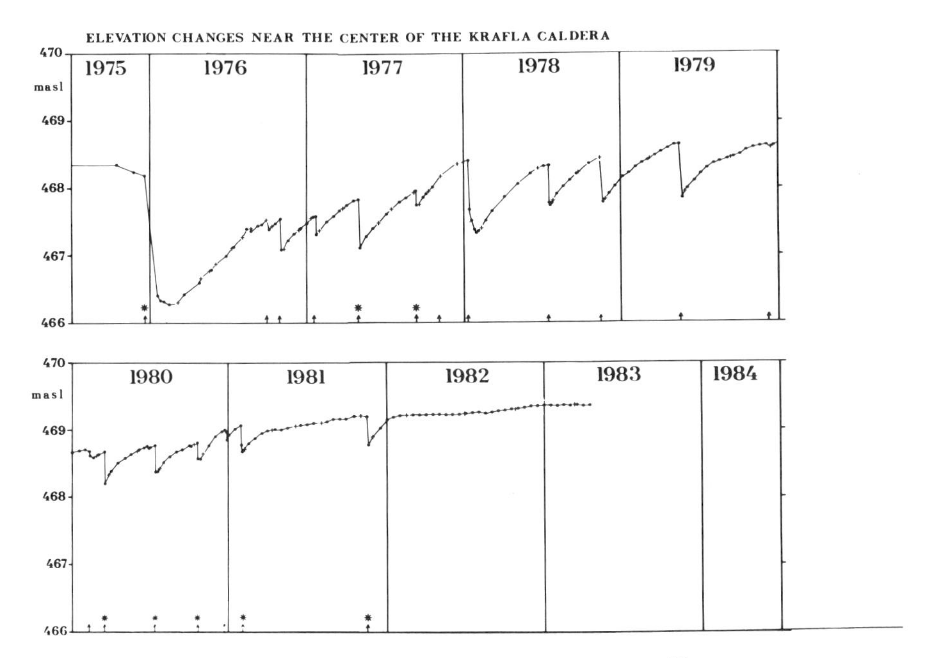

Fig. 16. Changes in land elevation in the Krafla area.

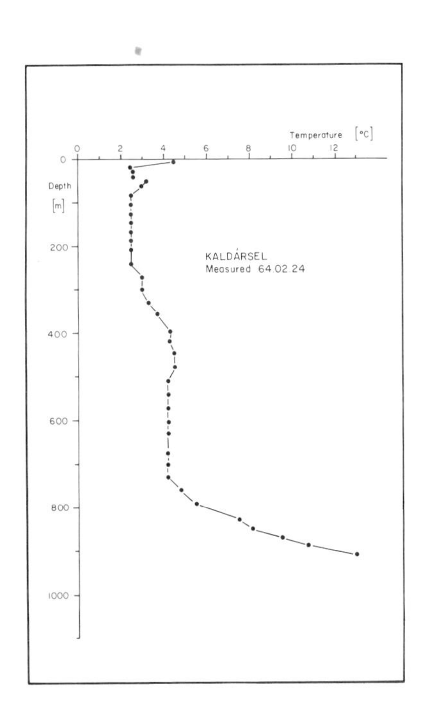

Fig. 17. Temperature profile in the hole in Kaldårsel, Iceland.

has investigated volcano spacing, fractures and the thickness of the lithosphere in various parts of the world. He finds the volcano spacing and the lithosphere thickness to be nearly equal and that volcano or fracture spacing is greater for older than younger oceanic lithosphere. Most mid-ocean ridges are created at an oceanic depth of 2.5 km and subside to depths greater than 6 km (Sclater et al. 1971). For crust younger than 80 Ma the depth increases linearly with the square root of the age (Davis and Lister 1974). Parker and Oldenburg (1973) predicted that the thickness of the oceanic crust should also increase linearly with the square root of the age. These considerations are simply based on isostatic equilibrium over ridge/axes (Talwani et al. 1965).

Spooner and Fyfe (1973) proposed that midoceanic ridge hydrothermal systems are recharged by geographically dispersed, slow moving downflow of cold seawater and discharged by relatively rapid scent of hot fluid through localized fracture systems. In the uppermost kilometer or so of the oceanic crust, significant downflow of cold seawater is indicated by the high $\delta^{18}O$ values obtained from samples of this layer (Muehlenbachs and Clayton 1976). Sleep and Wolery (1978) have examined under what conditions hot water can vent directly into the ocean. They find it unlikely that wide spread porous convection would vent hot water directly into the ocean. Sleep and Wolery modeled spatially restricted flow in planar and tubular crack zones. They found the spacing of high temperature vents to be 3.5 km in young (0-1 My) crust and 35 km for crust older than 1 My but younger than 10 My. In this work, we present a mean value of 12-15 km for the separation of central volcanoes (cooling spots) in the axial rift zone of Iceland. Furthermore, we propose that the recharge of high temperature geothermal systems in the Icelandic rift zones is disperced over the intervals between cooling spots. In this respect, certain similarities exist between observations in Iceland and on submarine rift axis. The crustal thickness within the volcanic zones of Iceland (Bebo and Björnsson 1978, 1980; Thayer et al. 1981, Hermance 1981, Palmason 1971, Bott and Gunnarsson 1980, Evans and Sacks 1979, Gebreude et al. 1980, Sanford and Einarsson 1982) happens to be almost equal to the mean spacing between central volcanoes in the active volcanic zones. The meaning of this coincidence has not been explained so far.

CONCLUSIONS

Hydrothermal systems in Iceland and on other parts of mid ocean ridges have been compared. The two major differences between the physical environment of hydrothermal systems in Iceland and on submerged rifting axes are that they discharge into different fluids and that the absolute pressure is higher in submerged systems than in those located on land. A consequence of this is found to be

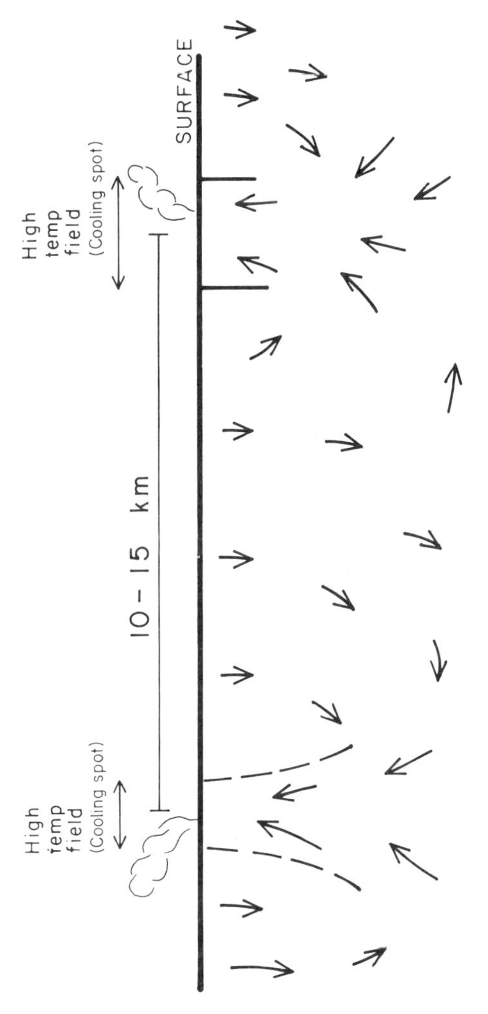

Fig. 18. Schematic cross section along the spreading axis in Iceland.

that topography influences submarine systems in the opposite way to subaerial systems. Furthermore, the high pressure in submarine system prevents phase transitions of the fluid and the chemical output of submarine systems has higher metallic concentrations than observed on the surface in land based hydrothermal systems.

Downward penetration of cold water into hot rock is found to be the only known process which can explain the high heat flux density of some hydrothermal systems in Iceland, and a controlled experiment of magma cooling during the Heimaey eruption of 1973 supports the opinion that this process is actually taking place in nature.

The heat flow through Iceland is compared with the energy released on submarine spreading axes. It is found that the estimated sum of heat released by volcanism and hydrothermal circulation in Iceland is less than the estimate for submarine spreading axes. Extrapolation of knowledge on hydrothermal systems in Iceland to the submarine systems is therefore expected to be realistic.

Cooling of the axial rift zone in Iceland seems to be concentrated to the central volcanoes. The spacing of central volcanos in the rift zones is 12-15 km, which happens to be equal to the estimated thickness of the crust of the axial rift zone. It is proposed that submarine axial rift zones might have similar cooling spots, and their spacing should be related to the spreading rate at each location.

ACKNOWLEDGEMENTS

I am indebted to many colleagues at the National Energy Authority of Iceland for fruitful discussions of the present paper. Prof. S. Björnsson, Dr. G.S. Bodvarsson, Dr. I.B. Fridleifsson, Mr. K. Gunnarsson, Mr. S.Th. Guðlaugsson, Dr. G. Palmason, Dr. K. Saemundsson and Mr. B. Steingrimsson reviewed an earlier version of this paper and pointed out several improvements in the presentation. Special thanks are due to Mr. G. Axelsson who has read the manuscript at different stages and contributed with many constructive comments.

REFERENCES

Anderson, R. N., 1972, Petrologic significance of low heat flow on the flanks of slow spending midocean ridges, Geol. Soc. Am. Bull. 83, 2947-2956, 1972.

Anderson, R.N. and M.A. Hobart, 1976, The relation between heat flow, sediment thickness, and age in the Eastern Pacific, J. Geophys. Res. 81, 2968-2989, 1976.

Anderson, R. N., M. G., Langseth, and J. G. Sclater, 1977. The isms of heat transfer through the floor of the Indian Ocean. J. Geophys. Res. 82, 3391-3409.

Anderson, R. N., 1979, Oceanic heat flow, In: The Sea, Vol. 7, pp 489-523 Wiley-Interscience, New York.

Armannsson, H., G. Gíslason, and T. Hauksson, 1982. Magmatic gases in well fluids aid the mapping of the flow pattern in a geothermal system. Geochim. Cosmochim. Acta 46, 167-177.

Arnold, M., and S. M. F. Sheppard, 1981. East Pacific Rise at Latitude 21°N: isotopic composition and origin of the hydrothermal sulphur. Earth Planet. Sci. Lett. 56, 148-156,

Banwell, C. J., E. R. Cooper, G. E. K. Thompson, and K. J. McCree, 1957. Physics of the New Zealand thermal area. N.Z. Dept. Sci. and Indus. Research Bull. 123, 109 pp.

Banwell, C. J., 1963. Thermal energy from the Earth's crust; Introduction and part I, Natural hydrothermal systems. N.Z.J. Geol. Geophys. 6, 52-69.

Beblo, M., and A. Bjornsson, 1978. Magnetotelluric investigation of the lower crust and upper mantle beneath Iceland. J. Geophys. 45, 1-16.

Beblo, M., and A. Bjornsson, 1980. A model of electrical resistivity beneath NE-Iceland, correlation with temperature. J. Geophys. 47, 184-190.

Bischoff, J. L., and F. W. Dickson, 1975. Seawater- basalt interaction at 200°C and 500 bars; Implications for origin of sea-floor heavy metal deposits and regulation of seawater chemistry. Earth Planet. Sci. Lett. 25, 385-397.

Bischoff, J. L., and W. E. Seyfried, 1978. Hydrothermal chemistry of seawater from 25° to 350°C. American J. Sci. 278, 838-860.

Bischoff, J. L., 1980. Geothermal system at 21°N, East Pacific Rise; Physical Limits on Geothermal fluid and role of adiabatic expansion. Science 207, 1465-1469.

Bjornsson, A., K. Saemundsson, P. Einarsson, E. Tryggvason, and K. Gronvold, 1977. Current rifting episode in North Iceland. Nature, 266, 318-323.

Bjornsson, A., G. Johnsen, S. Sigurdsson, and G. Thorbergsson, 1979. Rifting of the Plate Boundary in North Iceland 1975--1978. J. Geophys. Res. 84, 3029-3038.

Bjornsson, H., 1974. Explanation of jokulhlaups from Grimsvotn, Vatnajokull, Iceland, Jokull, 24, 1-26.

Bjornsson, H., S. Bjornsson, and Th. Sigurgeirsson, 1982. Penetration of water into hot rock boundaries of magma in Grimsvotn, Nature, 295, 580-581.

Bodvarsson, G., 1951. Report on the Hengill thermal area, (in Icelandic with English summary), J. Engin. Assoc. Iceland, 36, 1-48.

Bodvarsson, G., 1961. Physical Characteristics of Natural Heat Resources in Iceland. Jokull, 11, 29-38.

Bodvarsson, G., and R. P. Lowell, Ocean-floor heat flow and the circulation of interstitial waters. 1972, Ocean-floor heat flow and the circulation of interstitial waters. J. Geophys. Res. 77, 4472-4475.

Bodvarsson, G., 1979, Elastomechanical phenomena and the fluid conductivity of deep geothermal reservoirs and source regions. Workshop on Geothermal Reservoir Engineering, Dec. 1979, Stanford University, Stanford, California.

Bodvarsson, G., 1982, Terrestrial energy currents and transfer in Iceland. In: Continental and Oceanic Rifts, ed. G. Palmason, Geodynamics Series vol. 8, pp 271-282, AGU, Washington D.C.

Bonatti, E., and Y. R. Naudu, 1965, The origin of manganese nodules on the ocean floor. American Journal of Science, 263, 17-39.

Bonatti, E., and O. Joensuu, 1966, Deep-sea iron deposit from the South Pacific. Science, 154, 643-645,

Bonatti, E., 1975, Metallogenesis at oceanic spreading centers. Ann. Rev. Earth Planet. Sci., 3, 401-431,

Bonatti, E., 1978, The Origin of Metal Deposits in the Oceanic Lithosphere. Scientific American, 238, 54-61,

Bostrom, K., and M. N. A. Petersen, 1966, Precipitates from hydrothermal exhalations on the East Pacific Rise. Econ. Geology, 61, 1258-1265.

Bott, M. H. P., and K. Gunnarsson, 1980, Crustal structure of the Iceland-Faeroe Ridge, J. Geophys. 47, 221-227.

Bottinga, Y., 1974, Thermal aspects of sea-floor spreading and the nature of the suboceanic lithosphere, Tectonophysics, 21, 15-38,

Castillo B., F., F. J. Bermejo M., B. Domiquez A., C. A. Esquer P., Y. F. J. Navarro O., 1981, Distribucion de temperatures en el campo geotermico de Cerro Prieto. Proceedings Third Synposium on the Cerro Prieto geothermal field, Baja California, Mexico, pp 474-483, Report LBL-11967.

Corliss, J. B., 1971, The origin of Metal-Bearing Submarine Hydrothermal Solutions. J. Geophys. Res. 76, 8128-8138, 1971.

Corliss, J. B., M. Lyle, J. Dymond, and K. Crane, 1978. The chemistry of hydrothermal mounds near the Galapagos rift. Earth Planet. Sci. Lett. 40, 12-24, 1978.

Corliss, J. B., J. Dymond, L. I. Gordon, J. M. Edmond, R. P. Von Herzen, R. D. Ballard, K. Green, D. Williams, A. Bainbridge, K. Crane, T. H. van Audel., 1979, Submarine thermal springs on the Galapagos Rift. Science, 203, 1073-1083.

Crane, K., and W. R. Normark, 1977, Hydrothermal activity and crestal structure of the East Pacific Rise at 21°N, J. Geophys. Res. 82, 5336-5348.

Cronan, D. S., 1976, Implications of metal dispersion from submarine hydrothermal systems for mineral exploration on mid-ocean ridges and in island areas. Nature 262, 567-569.

Cronan, D. S., G. P. Glasby, S. A. Moorby, J. Thomson, K. E. Knedler, and J. C. McDougall, 1982, A submarine hydrothermal manganese deposit from the south-west Pacific Island arc. Nature, 298, 456-458.

Cyamex Scientific Team, 1979, Nature 277, 523-

Davis, E. E., and C. R. B. Lister, 1974, Fundamentals of ridge crest topography. Earth Planet. Sci. Lett. 21, 405-413.

Dunn, J. C., and H. C. Hardee, 1981, Superconvective geothermal zones. J. Volc. Geoth. Res. 11, 189-201, 1981.

Dymond, J., J. B. Corliss, G. R. Heath, C. W. Field, E. J. Dasch, H. H. Veeh, 1973, Origin of Metalliferous Sediments from the Pacific Ocean. Geol. Soc. Am. Bull., 84, 3355-3372,

Edmond, J. M., C. Measures, R. E. McDuff, L. H. Chan, R. Collier, B. Grant, L. I. Gordon, and J. B. Corliss, 1979a. Ridge crest hydrothermal activity and the balances of the major and minor elements in the ocean; the Galapagos data. Earth Planet. Sci. Lett. 46, 1-18, 1979a.

Edmond, J. M., C. Measures, B. Mangum, G. Grant, F. R. Sclater, R. Collier, A. Hudson, L. I. Gordon, and J. B. Corliss, 1979b, On the formation of metal-rich deposits at ridge crests. Earth Planet. Sci. Lett. 46, 19-30.

Einarsson, P., 1978, S-wave shadows in the Krafla caldera in NE-Iceland, evidence for a magma chamber in the crust. Bull. Volcanol. 41, 1-9, 1978.

Einarsson, P., and B. Brandsdóttir, 1980, Seismological evidence for lateral magma intrusion during the July, 1978 deflation of the Krafla volcano in NE-Iceland. J. Geophys. 47, 160-165.

Evans, J. R., and I. S. Sacks, 1979, Deep structure of the Iceland Plateau. J. Geophys. Res. 84, 6859-6866.

Francheteau, J., H. D. Needham, P. Choakroune, T. Juteau, M. Séguret, R. D. Ballard, P. J. Fox, W. Normark, A. Carranza, D. Cordoba, J. Guerrero, C. Rangin, H. Bougault, P. Canbon, and R. Hekinian, 1979, Massive deep-sea sulphide ore deposits discovered on the East Pacific Rise. Nature 277, 523-528.

Friedman, J. D., R. S. Williams Jr., S. Thorarinsson, and G. Pálmason, 1972. Infrared emission from Kverkfjoll subglacial volcanic and geothermal area, Iceland. Jokull, 22, 27-43.

Fyfe, W. S. and P. Lonsdale, 1981. Submarine hydrothermal activity. In. C. Emiliani (ed) The Sea 7, Wiley-Interscience, New York, 589-638.

Gebrande, H., H. Miller, and P. Einarsson, 1980, Seismic structure of Iceland along PRISP-profile I. J. Geophys. 47, 239-249.

Green, K. E., R. P. von Herzen, and D. Williams, 1981, The Galapagos Spreading Center at 86°W: A detailed geothermal field study. J. Geoph. Res. 86, 979-986

Hajash, A., 1975, Hydrothermal Processes along Mid-Ocean Ridges; An Experimental Investigation. Contrib. Mineral. Petrol. 53, 205-226.

Haymon, R. M., and M. Kastner, 1981, Hot spring deposits on the East Pacific Rise at 21°N; preliminary description of mineralogy and genesis. Earth Planet. Sci. Lett. 53, 363-381.

Heath, G. R., and J. Dymond, 1977, Genesis and transformation of metalliferous sediments from the East Pacific Rise, Bauer Deep, and Central Basin, northwest Nazca plate. Geol. Soc. Am. Bulletin, 88, 723-733.

Hekinian, R., M. Fevrier, J. L. Bischoff, P. Picott, and W. C. Shanks, 1980, Sulfite deposits from the East Pacific Rise near 21°N. Science, 207, 1433-1444.

Henley, R. W., and A. McNabb, 1978, Magmatic vapor plumes and ground-water interaction in porphyry copper emplacement. Econ. Geol., 73, 1-20.

Helgason, J., and M. Zentilli, 1982, Stratigraphy and correlation of the region surrounding the IRDP drill hole 1978, Reydarfjordur, Eastern Iceland. J. Geophys. Res., 87, 6405-6417.

Hermance, J. F., 1981, Crustal genesis in Iceland; Geophysical constraints on crustal thickening with age. Geophys. Res. Lett., 8: 203-206.

Hoffert, M., A. Perseil, R. Helkinian, P. Choukroune,, H. D. Needham, J. Francheteau, Y. Le Pichon, 1978, Hydrothermal deposits sampled by diving saucer in Transform Fault "A" near 37°N on the Mid-Atlantic Ridge, Famous area. Oceanol. Acta, 1, 1., 73-86.

Jenkins, W. J., J. M. Edmond, and J. B. Corliss, 1978, Excess ^{3}He and ^{4}He in Galapagos Submarine hydrothermal waters. Nature, 272, 156-158.

Jonsson, V. K., and M. Matthiasson, 1974, Cooling the Heimaey lava with water - Report on the operation (in Icelandic). J. Engin. Assoc. Iceland, 59, 70-83.

Lister, C. R. B., 1972, On the Thermal Balance of a Mid-Ocean Ridge. Geophys. J. R. astr. Soc. 26, 515-535.

Lister, C. R. B., 1974, On the Penetration of water into hot rock. Geophys. J. R. astr. Soc. 39, 465-509.

Lister, C. R. B., 1976, Qualitative theory on the deep end of geothermal systems. U.N. Symp. Development and Use Geothermal Resources, 2nd 1975, Proc. 1, 459-463.

Lister, C. R. B., 1977a, Qualitative models of spreading-center processes, including hydrothermal penetration. Tectonophysics, 37, 203-218.

Lister, C. R. B., 1977b, Estimators for heat flow and deep rock properties based on boundary-layer theory. Tectonophysics, 41, 157-171.

Lister, C. R. B., 1980, Heat flow and hydrothermal circulation. Ann. Rev. Earth. Planet. 8, 95-117.

Lister, C. R. B., 1982, "Active" and "Passive" Hydrothermal Systems in the Oceanic Crust; Predicted Physical Conditions. In: The Dynamic Environment of the Ocean Floor, K. A. Fanning and F. T. Manheim Eds, pp 441-470, Lexington, Mass.

Lonsdale, P. F., J. L. Bischoff, V. M. Burns, M. Kastner, and R. E. Sweenly, 1980, A high-temperature hydrothermal deposit on the seabed at a Gulf of California Spreading Center. Earth Planet. Sci. Lett. 49, 8-20.

Lowell, R. P., 1980, Topographically driven subcritical hydrothermal convection in the oceanic crust. Earth Planet. Sci. Lett., 49, 21-28.

MacDonald, K. C., K. Becker, F. N. Spiess, and R. D. Ballard, 1980. Hydrothermal heat flux of the "black smoker" vents on the East Pacific Rise. Earth Planet. Sci. Lett. 48, 1-7.

Mottl, M. J., and H. D. Holland, 1978, Chemical exchange during hydrothermal alteration of basalt by seawater - I. Experimental results for major and minor components of seawater. Geochim, Cosmochim. Acta, 42, 1103-1115.

Mottl, M. J., H. D. Holland, and R. F. Corr, 1979, Chemical exchange during hydrothermal alteration of basalt by seawater - II. Experimental results for Fe, Mn, and sulfur species. Geochim. Cosmochim. Acta, 43, 869-884, 1979.

Muehlenbachs, K., and R. N. Clayton, 1972, Oxygen isotope geochemistry of submarine greenstones. Can. J. Earth Sci. 9, 471-478, 1972.

Muffler, L. J. P., and D. E. White, 1969, Active metamorphism of Upper Cenozoic sediments in the Salton Sea geothermal field and the Salton Trough, Southeastern California. Geol. Soc. Am. Bull. 80, 157-182.

Oudin, E., P. Picot, and G. Pouit, 1981, Comparison of sulphide deposits from the East Pacific Rise and Cyprus. Nature, 291, 404-407.

Palmason, G., 1981, Crustal structure of Iceland from explosion seismology. Soc. Sci. Islandica, 40, 187 pp, 1971.

Parker, R.L., and D.W. Oldenburg, 1973, Thermal model of ocean ridges. Nature, 242, 137-139.

Reyss, J. L., V. Marchig, and T. L. Ku, 1982, Rapid growth of a deep-sea manganese nodule. Nature, 295, 401-403.

Robinson, P. T., J. M. Hall, N. I. Christensen, I. L. Gibson, I. Fridleifsson, H.-U. Schmincke, and G. Schonharting. 1982, The Iceland Research Drilling Project: Syntheses of results and implications for the nature of Icelandic and oceanic crust. J. Geophys. Res. 87, 6657-6667.

Rona, P. A., 1976, Pattern of hydrothermal mineral deposition; mid-Atlantic Ridge crest at latitude 26°N. Marine Geology 21, 59-66.

Rona, P. A., R. N. Harbison, B. G. Bassinger, R. B. Scott, and A. J. Natwalk, 1976, Tectonic fabric and hydrothermal activity of mid-Atlantic Ridge crest (lat. 26°N). Geol. Soc. Am. Bull. 87, 661-674.

Rona, P. A., 1980, TAG Hydrothermal field; Mid-Atlantic Ridge crest at latitude 26°N. J. Geol. Soc. London, 137, 385-402.

Sanford, A. R., and P. Einarsson, 1982, Magma chambers in rifts. In: Continental and Oceanic Rifts, ed. G. Palmason; pp 147-168, AGU, Washington D.C.
Sclater, J. G., and J. Francheteau, 1970, The implications of terrestrial heat flow observations on current tectonic and geochemical models of the crust and upper mantle of the earth. Geophys. J. R. astr. Soc., 20, 509-542.
Sclater, J. G., R. N. Anderson, and M. L. Bell, 1971, The evaluation of the central eastern Pacific. J. Geophys. Res. 76, 7888-7915.
Sclater, J. G., R. P. Von Herzen, D. L. Williams, R. N. Anderson, and K. Klitgord, 1974, The Galapagos Spreading Centre; Heat-flow low on the North Flank. Geophys. J.R. astr. Soc. 38, 609-626.
Sclater, J. G., C. Jauport, and D. Galson, 1980, The heat flow through oceanic and continental crust and the heat loss of the Earth. Rev. Geoph. Space Phys., 18, 269-311.
Scott, R. B., P.A. Rona, B. A. McGregor, and M. R. Scott, 1974, The TAG hydrothermal field. Nature 251, 301-302.
Scott, R. B., J. Malpas, P. A. Rona, and G. Udintsev, 1976. Duration of hydrothermal activity at an oceanic spreading center, Mid-Atlantic Ridge (lat. 26°N). Geology, 4, 233-236.
Seyfried, W. G., and M. J. Mottl, 1977, Origin of submarine metal--rich hydrothermal solution: Experimental basalt-seawater interaction in a seawater-dominated system at 300°C, 500 bars. Proc. 2nd Int. Sympos. Water-Rock Interaction, IAGC, Strassbourg, France, pp IV 173 - IV 180.
Seyfried, W., and J. L. Bischoff, 1977, Hydrothermal transport of heavy metals by seawater; The role of seawater/basalt ratio. Earth Planet. Sci. Lett. 34, 71-77.
Seyfried, W. E., Jr., and J. L. Bischoff, 1981, Experimental seawater-basalt interaction at 300°C, 500 bars, chemical exchange, secondary mineral formation and implications for the transport of heavy metals. Geochim. Cosmochim. Acta, 45, 135-147.
Seyfried, W. E., and M. J. Mottl, 1982, Hydrothermal alteration of basalt by seawater under seawater-dominated conditions. Geochim. Cosmochim. Acta, 46, 985-1002.
Shanks, W. C., and J. L. Bischoff, 1977, Ore transport and deposition in the Red Sea geothermal system; a geochemical model; Geochim. et Cosmochim. Acta, 41, 1507-1509.
Sigurdsson, H., and R. S. J. Sparks, 1978, Lateral magma flow within rifted Icelandic crust. Nature, 274, 126-130.
Skornyakova, I. S., 1965, Dispersed iron and manganese in Pacific Ocean Sediments. International Geology Review, 7: 2161-2174.
Sleep, N. H., and T. J. Wolery, 1978, Egress of hot water from midocean ridge hydrothermal systems: Some thermal constrains. J. Geophys. Res. 83, 5913-5922.
Sourirajan, S., and G. C. Kennedy, The system H_2O-NaCl at elevated temperatures and pressures. Am. J. Sci. 260, 115-141.

Spiess, F. N., K. C. Mcdonald, T. Atwater, R. Ballard, A. Carranza, D. Cordoba, C. Cox, M. Draz Garcia, J. Francheteau, J. Guerrero, J. Hawkins, R. Haymon, R. Hessler, T. Juteau, M. Kastner, R. Larson, B. Luyendyk, J. D. Macdougall, S. Miller, W. Normark, J. Orcutt, C. Rangin, 1980, East Pacific Rise; Hot Springs and Geophysical Experiments. Science, 207, 1421-1433.

Spooner, E. T. C., and W. S. Fyfe, 1973, Sub-sea floor metamorphism, heat and mass transfer. Contrib. Mineral. Petrol., 42, 278-303.

Stefånsson, V., 1981, The Krafla Geothermal Field, Northeast Iceland. In: Geothermal Systems, Principles and Case Histories, ed. L. Rybach and L. P. L. Muffler, pp. 273-294, John Wiley and Sons.

Strauss, J. M., and G. Schubert, 1977, Thermal convection of water in a porous medium: Effects of temperature - and pressure - dependent thermodynamic and transport properties. J. Geophys. Res. 82, 325-333.

Styrt, M. M., A. J. Brackmann, H. D. Holland, B. C. Clark, V. Pisutha-Arnold, C. S. Eldridge, and H. Ohmoto, 1981, The mineralogy and the isotopic composition of sulfur in hydrothermal sulfide/sulfate deposits on the East Pacific Rise, 21°N latitude. Earth Planet. Sci. Lett. 53, 382-390.

Swanson, D. A., W. A. Duffield, and R. S. Fiske, 1976, Displacement of the south flank of Kilauea Volcano: The result of forceful intrusion of magma into the rift zones. U.S. Geol. Survey Prof. Paper 963, 1-39.

Talwani, M., X. Le Pichon, and M. Ewing, 1965, Crustal structure of the mid-ocean ridges.
2. Computed model from gravity and seismic refraction data. J. Geophys. Res. 70, 341-352, 1965.

Thayer, R. E., A. Bjornsson, L. Alvarex, and J. F. Hermance, 1981, Magma genesis and crustal spreading in the northern neovolcanic zone of Iceland. Telluric-magnetotelluric constrains. J. Geophys. Res. astr. Soc., 65: 423-442.

Vogt, P. R., 1974, Volcano spacing, fractures, and thickness of the lithosphere. Earth Planet. Sci. Lett. 21, 235-252.

Weiss, R. F., P. Lonsdale, J.E. Lupton, A. E. Bainbridge, H. Craig. 1977, Hydrothermal plumes in the Galapagos Rift. Nature, 267, 600-603.

White, D. E., 1957, Thermal waters of volcanic origin. Bull. Geol. Soc. Am. 68, 1637-1657.

White, D. E., 1968, Hydrology, activity, and heat flow of the steamboat springs thermal system, Washoe County, Nevada. U.S.G.S. Prof. Paper 458 - C 109 p.

Williams, D. L., Von Herzen, R. P., J. G. Sclater, and R. N. Anderson, 1974. The Galapagos Spreading Center and hydrothermal circulation. Geophys. J. R. astr., Soc. 38, 587-608.

Williams, D. L., K. Green, T. H. van Andel, R. P. Von Herzen, J. R. Dymond, and K. Crane, 1979, The hydrothermal mounds of the Galapagos Rift; Observations with DSRV Alvin and detailed heat flow studies. J. Geophys. Res. 84, 7467-7484.

GEOCHEMICAL MASS BALANCES AND CYCLES OF THE ELEMENTS

Karl K. Turekian

Department of Geology and Geophysics
Yale University, Box 6666
New Haven, Connecticut 06511

The role of hydrothermal convection associated with volcanism in the chemical transport in the Earth's crust has been known for a long time. The focus until fairly recently, however, had been on processes observed at volcanic sites on land. It was known that the dominant source of water emanating from hot springs and fumaroles was of meteoric origin and that despite the obvious relation to deep sources of magma, juvenile water was virtually impossible to identify at these sites. Precipitation permeated the surface and, as ground water, provided both the reaction solution and the transporting medium for elements released by the hydrothermal alteration of rocks. Many of these chemical species were then deposited around the vents of fumaroles, geysers and hot springs resulting in the characteristic deposits seen in many places. Deposition also took place within the cooler environment of the upper aquifer.

Not until the establishment of the plate tectonic theory was there any reason to believe that hydrothermal processes were active several kilometers below the ocean surface. The identification of oceanic ridges as expressions of the diverging boundaries of the plates and the expectation of thermal processes at these boundaries already implied by the presence of volcanic deposits, led to the proposition that indeed hydrothermal circulation of sea water through the ridges was to be expected just as fresh water circulation occurred on land where volcanic activity was known.

A direct search, using submersibles, for such activity at the ridges began with the FAMOUS project in the Atlantic Ocean. Evidence for hydrothermal activity was discovered in the form of characteristic deposits but active vents were not found. It was the Pacific Ocean, with its more rapid spreading rates, that was to provide the proof of the process. Starting with the Galapagos spreading center and continuing to additional sites in the eastern Pacific, hydrothermal activity was established through chemical, physical and biological observations.

As heat transport by hydrothermal convection at the ocean ridges is a dominant aspect of the heat loss of the Earth, obviously the chemical transports and reactions accompanying the process must be considered in any study of global mass balances or element cycling involving the oceans. The hydrothermal fluid in oceanic systems is not pure fresh water, but obviously sea water, so that chemical reactions in the hydrothermal system can result in chemical losses from the fluid as well as gains. The divergence of the plates occurring at the ocean ridges is the consequence of mantle convection, meaning that both basaltic rock and juvenile volatiles are released to the Earth's surface by the process and the latter may be detected through the use of proper tracers.

In order to determine hydrothermally driven fluxes through the ridge system, the mechanism of supply and removal relative to the heat flux must be known. This is not a simple task as there are three ways in which heat is transferred to the oceans by convection at the ridges: (1) Deep convection brings sea water close to or in contact with the magma chamber after which it returns, modified, to the ocean bottom at about 350°C in spectacular geysers or "smokers" (typically observed at the 21°N EPR site). (2) Entrainment of ambient sea water in the rock system by the rising 350°C water results in less spectacular hot springs and seeps close to the vents (typified by the Galapagos spreading center). (3) Thermally driven convection away from the ridge crest results in water return to the ocean bottom that is not much above ambient sea water temperature but capable of being modified chemically (typified by the mounds area at the Galapagos spreading center and elsewhere). There may be more pervasive sea water convection through major parts of the ocean bottom as the oceanic lithosphere cools as it moves away from the ridge crest, but the evidence for this is still debated.

If the dominant reaction altering the composition of sea water occurs above 350°C then it is possible to determine fluxes of elements by determining the difference between the composition of ambient deep sea water and the water venting at 350°C and multiplying this difference by the total amount of 350°C water released in the worldwide ridge systems. This was done first for the Galapagos spreading center using temperature-element plots which were compatible with conservative mixing of high temperature (∿350°C) water with ambient 2°C water by Corliss et al. (1979) and Edmond et al. (1979a,b) and confirmed for the most part by direct analysis of ∿350°C water at the 21°N EPR site as reported in this volume by Von Damm et al.

The hydrothermal flux was calculated to be about one third of the total oceanic heat flow based on models constructed from heat flow data (Wolery and Sleep, 1976) but the numerical value commonly used by geochemists is based on the T vs. ^3He plot of water from the Galapagos spreading center from Jenkins et al. (1978) which yields a slope of $7.6 \pm 0.5 \times 10^{-8}$ cal/atom ^3He. The currently revised worldwide flux of ^3He across the ocean-atmosphere plane is about 3 atoms/cm^2s (Welhan and Craig, this volume) which, for a total ocean area of 3.62×10^{18} cm^2, yields 2.6×10^{19} cal/y from hydrothermal convection at all ridges of the oceans. This is a value lower than the 4.6×10^{19} cal/y used by Edmond et al. (1979a) in calculating the hydrothermal fluxes of the elements.

The results to date indicate major sources (Si, Ca, K, Li, Rb, Fe, Mn) or sinks (Mg and SO_4^{-2}) at the high temperature reaction sites in the hydrothermal systems (G. Thompson, this volume) that act as important controls on the composition of sea water. The 350°C hydrothermal water is reducing with a high H_2S concentration and this has important consequences as it reaches the oceanic interface. As the hydrothermal water rich in H_2S mixes with the O_2-rich ambient water, sulfur oxidizing bacteria, commonly living in the tissues of organisms, use the chemical energy to provide food for the sustenance of a lush community of worms, clams, crustacea and fish.

Within the rocks near the rock-sea water interface the secondary sea water convection system driven by the rising hot water causes the precipitation of metals as sulfides or reduced oxides (Edmond et al., 1979b). Although trace metals are supplied by the processing of ambient sea water the removal rates for most are smaller than their removal rates in other oceanic sites such as the near-shore environments, deep-sea ferromanganese nodules and deep reducing ocean basins. Metals released from basaltic rocks and magma during the hydrothermal alteration process and transported as chloride complexes stable at high temperatures are also precipitated in this cooler environment. Some precipitation occurs as spectacular chimneys associated with the "smokers", and some within cracks and chambers created in the quenched submarine lavas.

The interaction of basaltic rock and magma with sea water in the ridge hydrothermal systems modifies the isotopic composition of the processed sea water. Water enters the system with $\delta^{18}O \sim 0$ and 350°C water returns with $\delta^{18}O = 1.6‰$ (Welhan and Craig, this volume). In order to maintain the $\delta^{18}O$ of water in the oceans at about the value of SMOW there must be preferential removal of ^{18}O. At low temperatures the deposition of calcium carbonate, silica and authigenic clay minerals all have a higher $\delta^{18}O$ than the water from which they are deposited so the buffering can be maintained by these ongoing processes so long as weathering and diagenetic processes on the continents altering the $\delta^{18}O$ of runoff are compensated as well.

Similarly, sea water strontium has a $^{87}Sr/^{86}Sr$ = 0.709 whereas 350°C water has a value of 0.7031 (Albarede et al., 1981; Von Damm et al. in this volume). Thus the $^{87}Sr/^{86}Sr$ variation with time will reflect the relative importance of strontium supply from sources with a higher-than sea-water $^{87}Sr/^{86}Sr$ and oceanic hydrothermally imprinted strontium. This provides for some of the variation in $^{87}Sr/^{86}Sr$ over time observed in marine carbonates.

Piepgras and Wasserburg (1982) have reviewed their own work and the work of others indicating that the $^{143}Nd/^{144}Nd$ ratio in the Pacific Ocean is strongly influenced by a young basaltic contribution. The issue of whether hydrothermal circulation can provide this imprint has been raised. At the time of this writing there are no published results to affirm or deny this possibility, but F. Albarede in a personal communication in December, 1982, said that his measurements of the rare earth concentration of 350°C water at the 13°N EPR site studied by the French, indicated several orders of magnitude higher concentrations of the rare earth elements than ambient sea water. This would certainly affect the $^{143}Nd/^{144}Nd$ ratio of Pacific sea water and may be the dominant source of young basalt imprint required to explain the data.

Chow and Patterson (1962) have shown that ferromanganese nodules and deep-sea sediments have a range of lead isotopic compositions and Tatsumoto (1978) has shown that deep-sea deposits closest to the ridge axis have lead isotopic compositions most resembling that of mid-ocean ridge basalts (MORB). Obviously there is an input of lead from the hydrothermal processes operating at the ridges. Using ^{210}Pb as a tracer, Chung et al. (1981) argue that 350°C water venting directly into the ocean, as at the 21°N EPR site, can supply the entire flux of Pb to the oceans required by concentration and mean residence time data. If, however, as Krishnaswami and Turekian (1982) argue, most of the Pb released during the hydrothermal alteration is precipitated in the low temperature part of the hydrothermal plumbing system, as at the Galapagos spreading center, only 0.5% of the Pb flux to the oceans will be supplied by hydrothermal processes associated with the ridges. The answer probably lies between these two extremes but only additional future work on lead isotopic systematics in sea water and marine deposits will resolve this question.

In addition to ^{3}He other gases are released to the ocean by the ridge hydrothermal systems. Methane, hydrogen, carbon monoxide and carbon dioxide fluxes have been determined using the ratio of each gas to ^{3}He in the waters. The results are complicated at the low temperature vents, such as the Galapagos spreading center, by the uncertainty of the role of microorganisms at the vents in modifying the composition of the reduced gases (see Welhan and Craig, this volume, and Lilley et al., this volume). The mean residence time of methane and hydrogen in the oceanic reservoir has been calculated to be about 30 years by Welhan and Craig (this volume) which, fortuitously, is also the mean residence time of manganese from hydrothermal sources (Weiss, 1977) and oceanically produced ^{230}Th (Nozaki et al., 1981).

The non-reactive rare gases in addition to ^{3}He must also be released by the hydrothermal processes at the vent. It has been demonstrated that glassy MORB has an excess ^{40}Ar reflecting the mantle source rather than the age of emplacement, whereas holocrystalline basalts, commonly found inside the glass pillows as well as separately, do not have this ^{40}Ar and yield more correct ages (Dymond, 1970; Seidemann, 1978).

Craig and Lupton (1976) show that Pacific glassy MORB samples have $^{3}He/^{20}Ne = 0.002$ and $^{3}He/^{36}Ar = 0.03$. Assuming that no significant fractionation occurs when the rare gases are released during the hydrothermal alteration of basaltic rock or magma, the fluxes of ^{20}Ne and ^{36}Ar from the Earth's interior by scaling them to the ^{3}He flux (3 atoms/cm^{2}s) can be calculated following Craig and Lupton (1981). This exercise yields a ^{20}Ne flux of 1500 atoms/cm^{2}s and an ^{36}Ar flux of 100 atoms/cm^{2}s. For atmospheric reservoirs of 1.8×10^{39} atoms of ^{20}Ne and 1.1×10^{28} atoms of ^{36}Ar the present day fluxes if operating over the age of the earth (4.5×10^9y) would provide 42% of the ^{20}Ne and only 1.5% ^{36}Ar. H. Craig (personal communication) believes that it is not possible to accept the Ne flux as correct because of the wide range in $^{20}Ne/^{36}Ar$ ratios in the basalts analyzed so far. The fluxes of the heavier rare gases, Kr and Xe, can be scaled to ^{36}Ar using the ratios observed by Dymond and Hogan (1973). When this is done these fluxes, like that of ^{36}Ar, are found to be too small to supply the atmosphere if the rates were the same as the present rates.

Hydrothermal processes associated with ocean floor spreading have now been discovered and studied in the Gulf of California. This environment is unlike the open ocean environment because of very rapid sedimentation rates, especially of organic matter as the result of the strong upwelling regime of that body of water. The natural alteration of marine organic debris by 315°C water has resulted in the release of organic compounds which resemble refinery products! The first results of this study on the organics is presented by Simoneit in this volume. Obviously the chemistry of many of the inorganic components associated with this system will be different from that of the open ocean systems thereby contributing to the kaleidoscope of hydrothermal vent chemistry!

ACKNOWLEDGMENTS

Research at Yale on the geochemistry of hydrothermal systems has been supported by NSF grants OCE 79-19254 and OCE 80-17299.

REFERENCES

Albarede, F., Michaud, A., Minster, J.F. and Michaud, G, 1981, $^{87}Sr/^{86}Sr$ ratios in hydrothermal waters and deposits from the East Pacific Rise at 21°N, Earth Planet. Sci. Lett. 55:229-236.

Chow, T.J. and Patterson, C.C.,1962, The occurrence and significance of lead isotopes in pelagic sediments, Geochim. Cosmochim. Acta, 26:263-308.

Chung, Y., Finkel, R. and Kim, K., 1981, ^{210}Pb on the East Pacific Rise at 21°N,Trans. Am. Geophys. Un., Abstract 03-1-C-7, EOS, 62:913.

Corliss, J.B., Dymond, J., Gordon, L.I., Edmond, J.M., von Herzen, R.F., Ballard, R.D., Green, K., Williams, D., Bainbridge, A., Crane, K. and van Andel, T.H., 1979, Submarine thermal springs on the Galapagos Rift, Science, 203:1073-1083.

Craig, H. and Lupton, J.E., 1976, Primordial neon, helium, and hydrogen in oceanic basalts, Earth Planet. Sci. Lett., 31:369-385.

Craig, H. and Lupton, J.E., 1981, Helium-3 and mantle volatiles in the ocean and the oceanic crust, in: "The Sea", (v. 7 The Oceanic Lithosphere) John Wiley & Sons, New York, 391-428.

Dymond, J.R., 1970, Excess argon in submarine basalt pillows, Geol. Soc. Am. Bull., 81:1229-1232.

Dymond, J.R. and Hogan, L., 1973, Noble gas abundance patterns in deep-sea basalts-primordial gases from the mantle, Earth Planet. Sci. Lett., 20:131-139.

Edmond, J.M., Measures, C., McDuff, R.E., Chan, L.H., Collier, R., Grant, B., Gordon, L.I. and Corliss, J.B., 1979a, Ridge crest hydrothermal activity and the balances of the major and minor elements in the ocean: The Galapagos data, Earth Planet. Sci. Lett., 46:1-18.

Edmond, J.M., Measures, C., Mangum, B., Grant, B., Sclater, J.R., Collier, R., Hudson, A., Gordon, L.I. and Corliss, J.B., 1979b, On the formation of metal-rich deposits in the ridge crests, Earth Planet. Sci. Lett., 46:19-30.

Jenkins, W.J., Edmond, J.M. and Corliss, J.B., 1978, Excess ^{3}He and ^{4}He in Galapagos submarine hydrothermal waters, Nature, 272: 156-158.

Krishnaswami, S. and Turekian, K.K., 1982, ^{238}U, ^{226}Ra and ^{210}Pb in some vent waters of the Galapagos spreading center, Geophys. Res. Lett., 9:827-830.

Nozaki, Y., Horibe, Y. and Tsubota, H., 1981, The water column distributions of thorium isotopes in the western North Pacific, Earth Planet. Sci. Lett., 54:203-216.

Piepgras, D.J. and Wasserburg, G.J., 1982, Isotopic composition of neodymium in waters from the Drake Passage, Science, 217:207-214.

Seidemann, D.E., 1978, $^{40}Ar/^{39}Ar$ studies of deep-sea igneous rocks, Geochim. Cosmochim. Acta, 42:1721-1734.

Tatsumoto, M, 1978, Isotopic composition of lead in oceanic basalt and its implication to mantle evolution, Earth Planet. Sci. Lett., 38:63-87.

Weiss, R.F., 1977, Hydrothermal manganese in the deep sea: scavenging residence time and $Mn/^3He$ relationships, Earth Planet. Sci. Lett., 37:257-262.

Wolery, T.J. and Sleep, N.H., 1976, Hydrothermal circulation and geochemical flux at mid-ocean ridges, Jour. Geol., 84:249-275.

PRELIMINARY REPORT ON THE CHEMISTRY OF HYDROTHERMAL SOLUTIONS AT 21° NORTH, EAST PACIFIC RISE

K. L. Von Damm, B. Grant and J.M. Edmond

Department of Earth and Planetary Sciences
Massachusetts Institute of Technology
E34-201, Cambridge, MA 02139 USA

ABSTRACT

A preliminary report is presented, of the information available to date on the chemistry of the 350° C hot springs on the EPR at 21° N. Three distinct fields were sampled by ALVIN in November 1979. The end-member solutions are completely depleted in magnesium and sulphate and are markedly acidic. The alkalies and alkaline earths are similar in their behavior to what was observed at Galapagos; however, there are significant differences. The waters are sulphide dominated with the H_2S:Fe ratio being 3.5. The Fe:Mn ratio is 1.8. The vents are enriched in Zn relative to Cu with Zn:Cu equal to about 7. The strontium concentration in the end member is only a few percent above the normal sea water value. However, the isotopic composition is very similar to basalts. The large resulting flux of non-radiogenic strontium is necessary to the overall isotopic balance of the element in the oceans. The close compositional agreement between the 21°N and Galapagos solutions provides further evidence for the validity of the general calculations developed using the Galapagos data.

INTRODUCTION

The discovery and sampling of hot springs on the crest of the Galapagos Spreading Center in the spring of 1977 by the submersible ALVIN, marked the culmination of over a decade of speculation, theoretical and experimental work, and sea-floor exploration (Corliss et al., 1979). The chemical results changed our perspective on the geochemical cycles in the ocean in a radical way (Edmond et al., 1979a, b, c). However, the springs were cool, less than 20°C above ambient (2.05°C) because of extensive, subsurface mixing between a hydrothermal end-member of estimated temperature 350°C and groundwater with properties indistinguishable from those of the ambient water column. The primary chemical signature for many elements - those forming insoluble sulphides or oxides under cool, mildly acidic, anaerobic conditions - was obliterated in this process. Even for the "soluble" species there were possible effects of low-temperature, retrograde reactions with the wall rocks (for K and Rb) and precipitation as sulphates (for all of the alkaline earths). The precipitation reactions had a serious effect on the pH and alkalinity of the exiting solutions due to proton release by the anions (McDuff and Edmond, 1982). In addition, the estimates of the global hydrothermal fluxes were based, perforce, on the few (4) hot spring fields sampled along the several kilometers of ridge axis explored.

We present here a preliminary discussion of the chemical data from samples collected from the high temperature system discovered on the crest of the East Pacific Rise in 21°N in the spring of 1979 (Rise Project Group, 1980). The samples come from a short reconnaissance which took place in November of 1979 (Edmond, 1980). While the results are incomplete, they go some way towards clarification of the problems encountered in the Galapagos work and consolidate the basis for the schema from these data.

METHODS

On the 1979 dives, the three "black smoker" occurrences found by Speiss et al., (1980) were sampled. Close-up observations of those springs from the submersible established that the fluids exit as clear homogeneous solutions (Edmond, 1980). The thermocouple probes built for this expedition gave temperatures for all the black smokers within a few degrees of 350°C rather than the 380 ± 30°C estimated on the discovery dives from thermistor data (Rise Project Group, 1980). Spurious higher "temperatures" (up to 1180°C!) such as are discussed by Bischoff (1980) are artifacts of case-heating of the frequency converter on the probe by the hydrothermal fluids, as clearly seen on the video monitor. The black smoke is composed predominantly of pyrrhotite (FeS) whose precipitation as fine grained particles was induced within about 10 cm above the vent orifice by the rapid, turbulent entrainment of ambient waters.

The hot springs exist as single or multiple vents associated with constructional features, "chimneys", between a few tens of centimeters to as much as 9 meters tall. The composition and mineralogy of these edifices has been discussed recently by several authors (see Hekinian, this volume). The samples were collected using evacuated, internally gold-plated, stainless steel cyliners developed for down-hole sampling in the Hot Dry Rock Geothermal Program of the

Los Alamos Scientific Laboratories. Valves were actuated electrically or manually using the manipulator arm on the submarine. Because of the weight and design, the samplers could not be deployed using the manipulator but had to be mounted rigidly on the sample "basket" projecting ahead of the boat. The submarine itself therefore had to be maneuvered to position the sampler inlet in the axis of the flow, i.e., to within less than 10 cm. This procedure met with variable success. All samples contained some admixture of ambient water. The temperature at the sampler inlet was monitored using thermocouple probes. On occasion, during maneuvering, they became separated from the inlet yielding spurious measurements.

The effect of sea water entrainment was to induce precipitation of sulphides and sulphates in the samplers as they cooled in the ambient water column before recovery of the boat. Gelatinous silica separated from the samples at a very much slower rate during storage.

Water was expelled from the samplers on the escort ship within twelve hours after collection and aliquoted for shipboard and laboratory analyses. Since the samples were heterogeneous, this procedure introduced some uncertainties in the determinations of the sulphide forming metals. In the laboratory, the samples were filtered and the precipitates redissolved by oxidative reaction with concentrated nitric acid and liquid bromine. The filtrate and filtrand solutions were analysed separately and the results summed to give the original dissolved concentration. It was impossible using the available procedures to separate particulates formed in the sampler from those entrained from the "smoke" forming over the vent orifices. Hence the data for the sulphide forming metals must be taken as preliminary, the estimated uncertainty being in the range of ten to fifteen percent.

Calcium, magnesium, chloride and alkalinity were determined by titration, strontium by isotope dilution, sulphate by polarography, silica by colorimetry and the other cations by flame and flameless atomic absorption spectrophotometry.

DISCUSSION OF DATA

Magnesium, sulphate and silica

The temperature versus Mg plot is shown in Fig. 1 with the least squares fit to the Galapagos data shown for comparison. The effects of displacement and damage to the thermistor probe at the sampler inlet on the temperature values are apparent. Several samples with relatively high measured "inlet temperatures" have essentially ambient Mg concentrations. When plotted versus sulphate, however, the data define a straight line passing through the origin (Fig. 2). As inferred from the Galapagos data, Mg and sulphate have zero concentrations in the 350°C end-member solutions. One point in Fig. 2 falls to high sulphate concentration. This sample has a stoichiometrically equivalent Ca anomaly (see below) indicating dissolution of previously precipitated anhydrite from the chimney wall. The dashed line in Fig. 2 is the Galapagos trend corrected for the effects of H_2S oxidation (McDuff and Edmond, 1982). These authors also discuss the significance of the non-zero intercept of this trend. It results from the precipitation of sulphates of Mg and Ca during sub-surface mixing with groundwater of composition indistinguishable from ambient sea water.

Given its regular behavior and the high relative accuracy of its determination, Mg will be used to normalize for the effects of entrainment of ambient water during sampling, i.e., results will be presented as elements versus Mg plots.

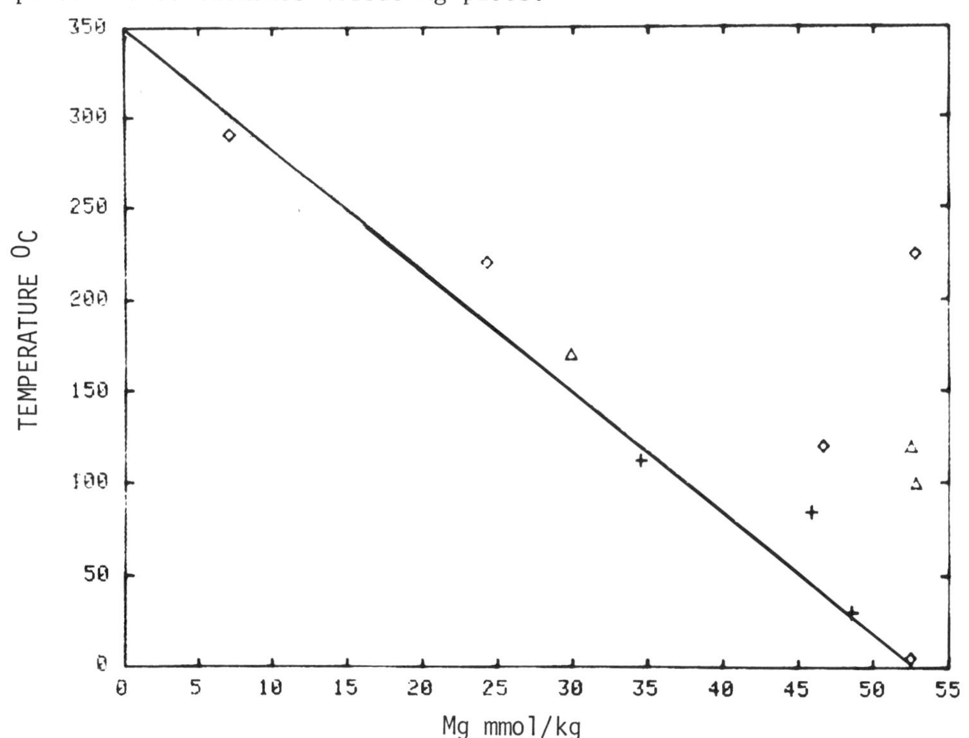

Fig. 1. Temperature vs. Magnesium. Temperature is the measured inlet temperature in degrees celsius. The line is the least squares fit to the Galapagos data. The 3 different fields are identified by symbols: =◇, National Geographic Smoker (NGS), Δ = On-Bottom Seismograph Smoker (OBS), + = South West Smoker (SW).

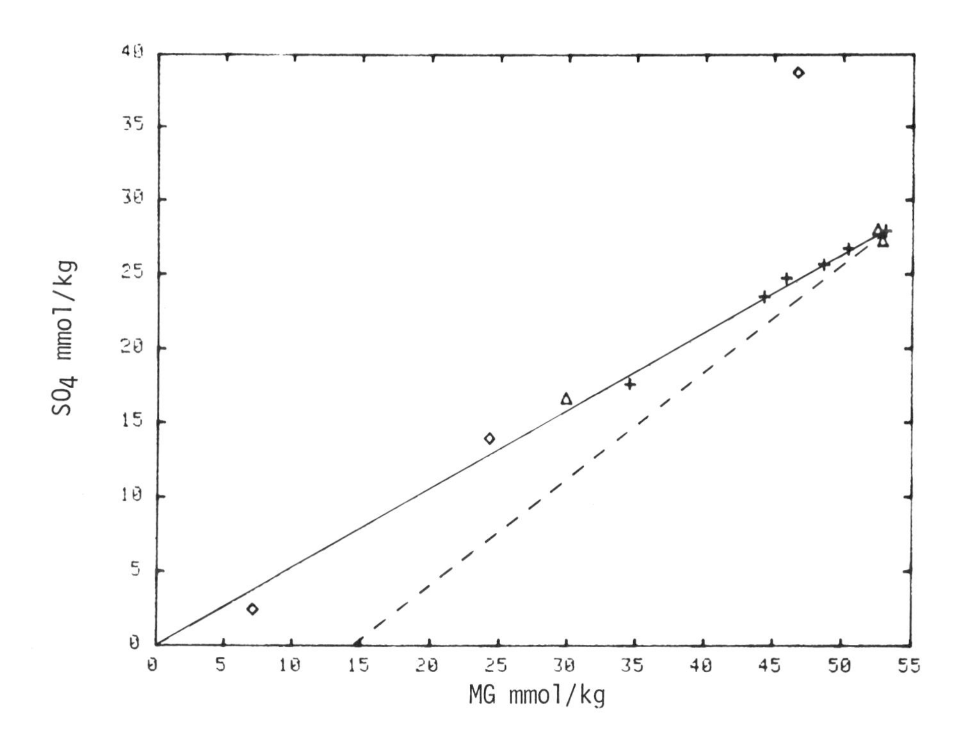

Fi 2. Sulphate vs. Magnesium. The line is the least squares fit to the data. The dashed line is the Galapagos trend corrected for the effects of H_2S oxidation.

The Si-Mg trend (Fig. 3) is linear with a zero-Mg intercept at 21.4 mM Si, identical to the Galapagos estimate. The use of silica "geothermometry" in the latter system is shown to be justified. The residence time of the water in the sub-surface mixing zone at Galapagos must be sufficiently short that the precipitation of amorphous silica is of minor importance. The similarity in the two data sets confirms that the end-member solutions are at equilibrium with quartz.

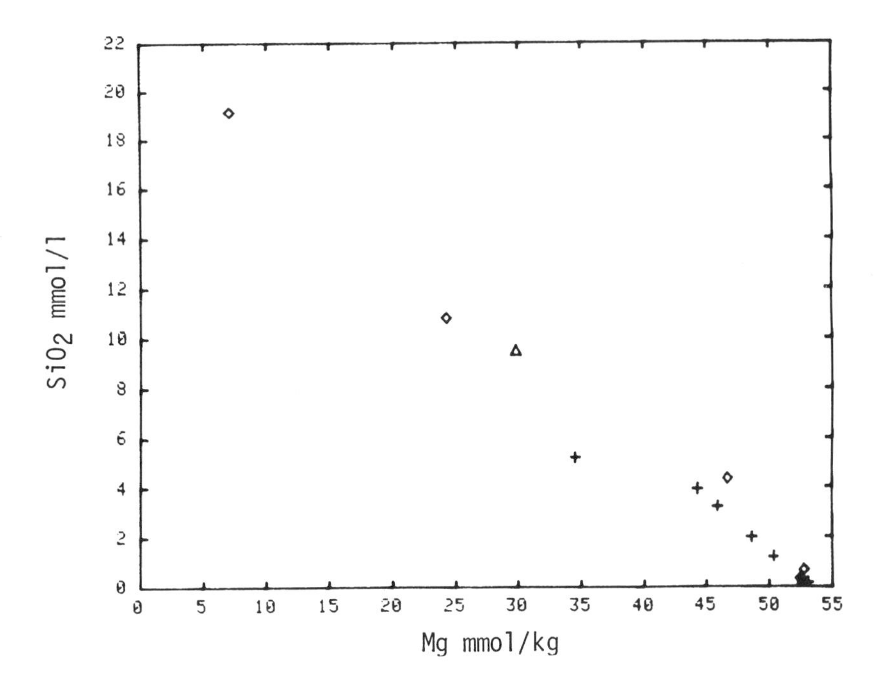

Fig. 3. Silica vs. Magnesium.

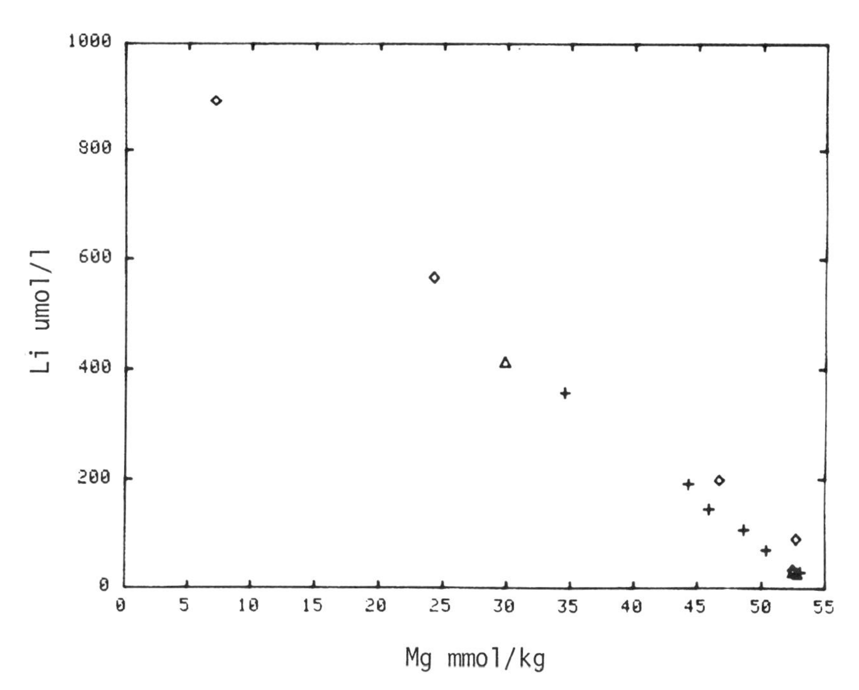

Fig. 4. Lithium vs. Magnesium.

The Alkalies - Li, K, Rb

Lithium (Fig. 4) forms a linear trend with Mg that falls in the lower end of the range observed at Galapagos. The estimated 350°C value of 1.05 mM compares with the previous extrapolations which fall in the range 0.69 to 1.14 mM.

Potassium (Fig. 5) and Rb (Fig. 6) also form linear arrays with Mg. However, the gradients are substantially higher than observed at Galapagos. The extrapolated end-member values 25 mM and 26 uM Rb compare with 18.8 mM and 13.4 to 20.3 uM respectively at Galapagos. Since K and Rb are the two analysed elements with the strongest affinity for basalts at low temperatures, it is possible that retrograde reactions occurred with the wall rocks in the Galapagos sub-surface zone.

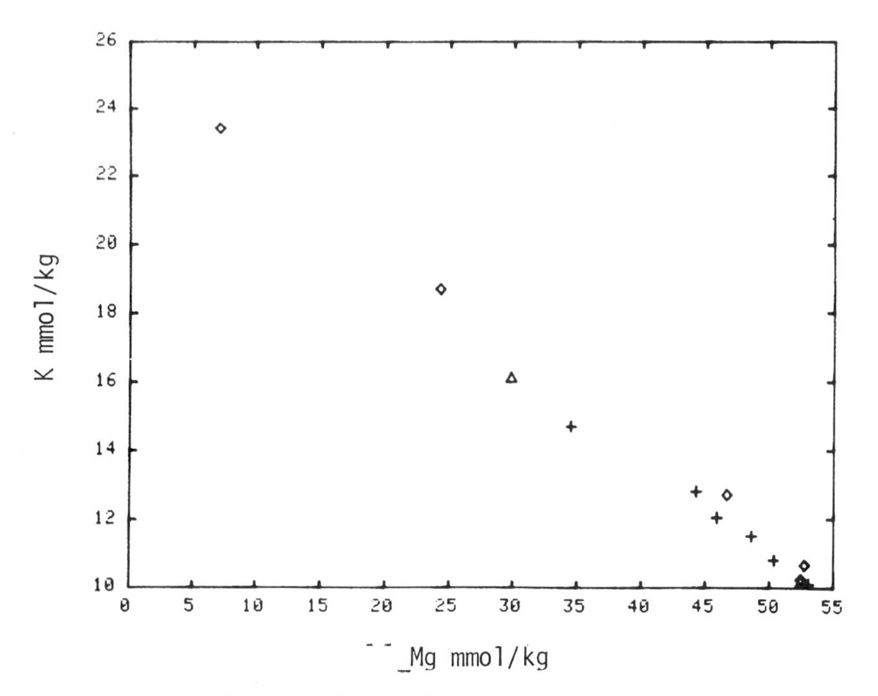

Fig. 5. Potassium vs. Magnesium.

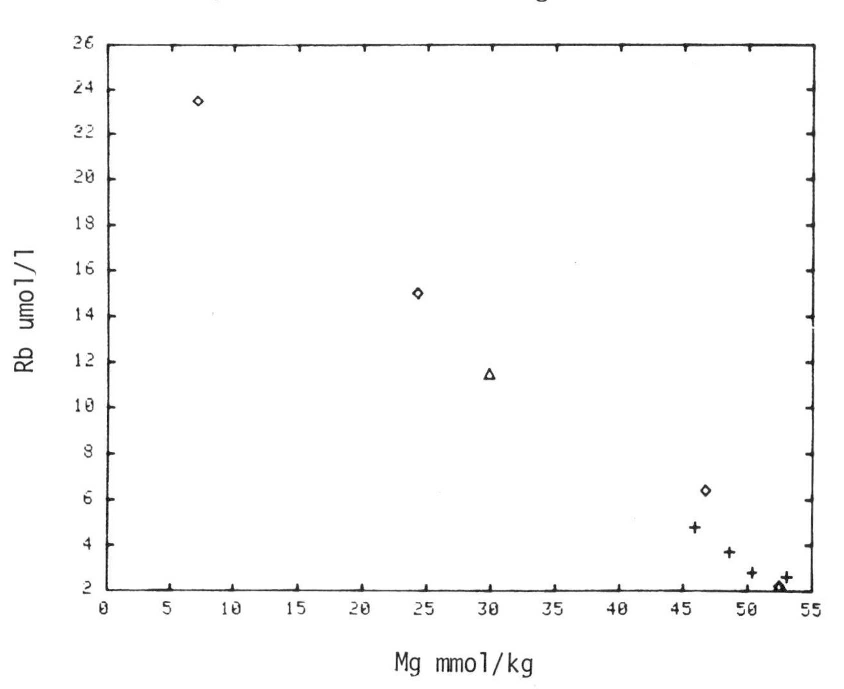

Fig. 6. Rubidium vs. Magnesium.

The Alkaline Earths - Ca, Sr and $^{87}Sr/^{86}Sr$

Calcium (Fig. 7) increases to an estimated value at the end-member of 21.5 mM. This is lower than the lowest Galapagos values (24.6 - 40.2 mM). The high point at 47 mM Mg has been discussed in connection with the sulphate data.

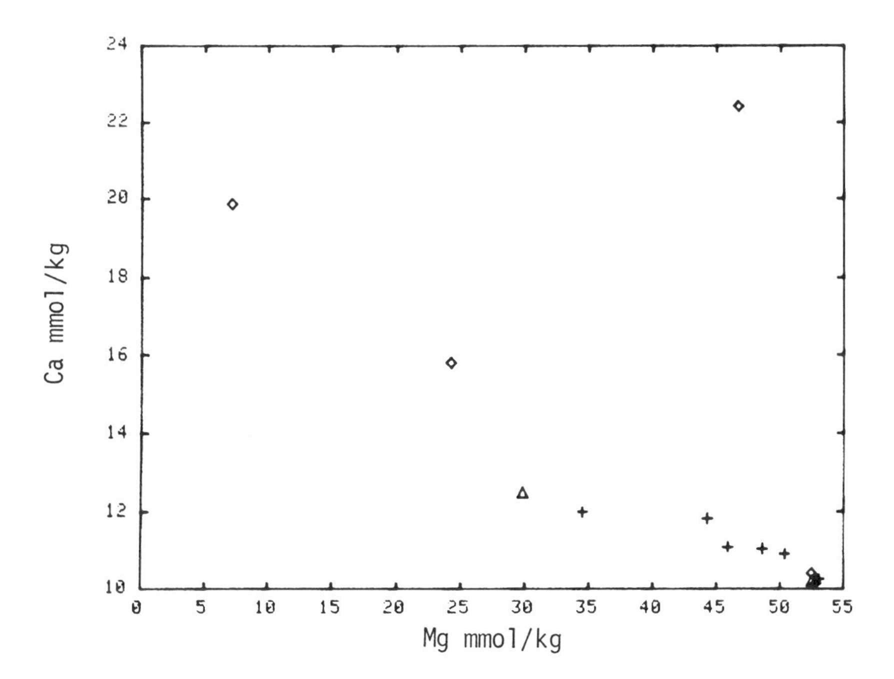

Fig. 7. Calcium vs. Magnesium.

Strontium (Fig. 8) increases by less than 5%. However, the isotopic composition of the end-member, 0.7031, is almost identical to the tholeiitic value. Albarede et al., (1981) reported duplicate measurements made on several of these samples. The flux of Sr from the ridge axis, $13*10^9$ moles per year, compares to the estimated

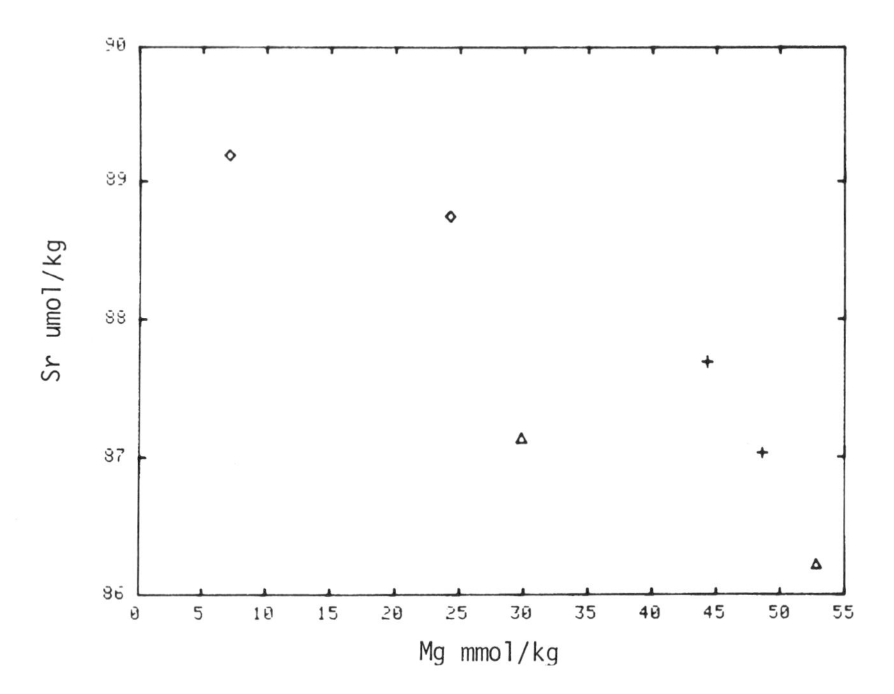

Fig. 8a. Strontium vs. Magnesium.

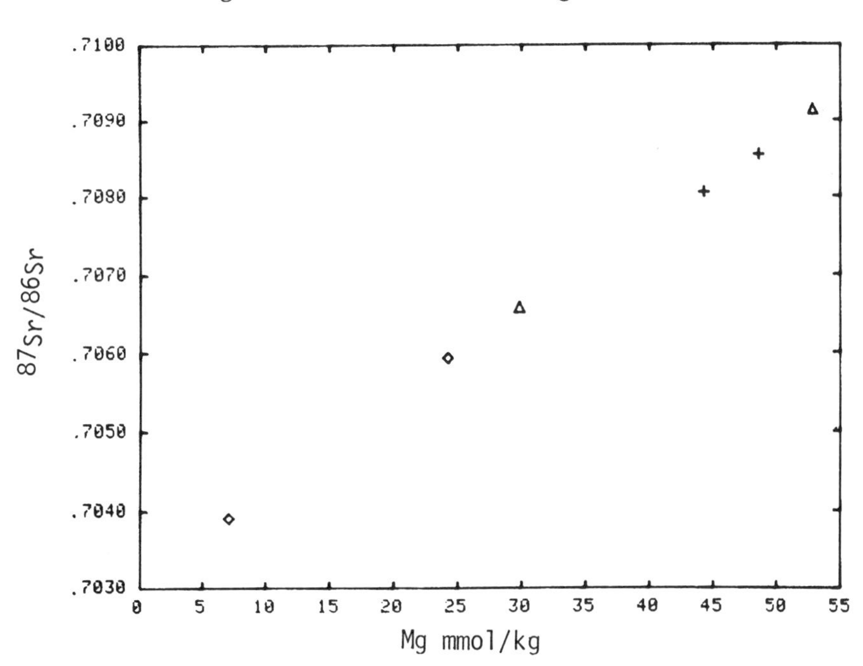

Fig. 8b. $^{87}Sr/^{86}Sr$ vs. Magnesium.

river value (1 uM Sr) of $31*10^9$. However, the net axial flux is only about $0.7*10^9$. One can estimate the relative importance of the flux of non-radiogenic Sr from the ridge on the isotopic composition of sea water Sr using the simple model of Brass (1976);

$$\frac{dR_{sw}}{dt} = \frac{\sum_i \Delta R_i \phi_i}{\Sigma Sr}$$

where R is the 87/86 ratio in a given source _i_, $\Delta R = R_i - R_{sw}$, ϕ_i is the annual flux of Sr and ΣSr is the total amount in the ocean. Considering rivers and hot springs as the only sources;

$$\frac{dR_{sw}}{dt} = \{\phi_{RC} (R_{RC} - R_{sw}) + \phi_R (R_R - R_{sw})\}/T\ (\phi_R + \phi_{RC})$$

where T is the oceanic residence time for Sr. If the present ocean is at steady state then;

$$R_R = R_{sw}\{1 + \frac{\phi_{RC}}{\phi_R}\} - R_{RC} \frac{\phi_{RC}}{\phi_R} = 0.7113$$

This is a reasonable value for contemporary rivers dominated as they are by recycled marine carbonates. The Amazon value, for example, is 0.7107. If the river value were 0.715, then the steady state ocean would be at 0.7118, higher than at any time in the geologic past. In the absence of the ridge crest source, the ocean and fluvial values would be the same at steady state. Given the large scale variations observed in the 87/86 ratio in sea water over geologic time, variations that appear to require that in the Jurassic, 175 my ago, the entire flux of non-carbonate Sr was of basaltic origin (Brass, 1976), it is impossible to sustain the sea water value by continental weathering alone. The ridge crest flux resolves this problem. The limiting factor in establishing the exact value for this input is our presently very inadequate knowledge of the fluvial term. However, the estimate derived from the 21°N data is completely consistent with the existing information and therefore supports the flux calculations in general.

Cl, Na and the Charge Balance

As at Galapagos, two trends are seen in the chloride data (Fig. 9). In one black smoker area, NGS, the concentrations increase to an end-member value of about 565 mM, an increase of about 4.5% over the ambient level. The gradient, 71.5 umol/kcal is about half that observed at Clambake in the Galapagos. The other two fields show a uniform decrease to 480 mM at 350°C. This is a drop of 12.5%, the

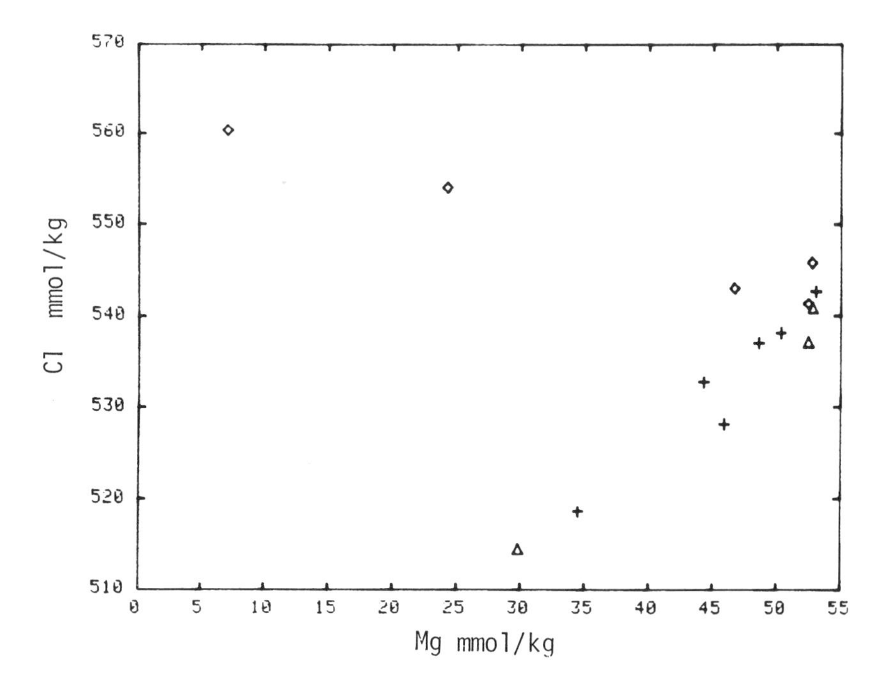

Fig. 9. Chloride vs. Magnesium.

gradient being -171 umol/kcal. This is substantially less than observed at Dandelions (-419) or Garden of Eden (-629).

The Na data (Fig. 10) show parallel effects although, for analytical reasons, the data are much more scattered. Computation of the Na concentrations (Δ Na) from the charge balance of the other major species gives a close coincidence in the trends. Therefore, the processes responsible for the variations in Na and Cl must be closely coupled. Indeed, given the dominant role these two species play in the charge balance, they cannot vary independently.

The increasing trend observed at NGS is most simply related to rock hydration, i.e., removal of water during clay mineral formation. The decreasing trend requires precipitation of an insoluble chloride containing mineral in some abundance. Possible candidates were reviewed in connection with the Galapagos data (Edmond et al., 1979b). There do not appear to have been any important developments in the interim. What can be said is that, since Ca does not form complex chlorides, e.g., of the atacamite type, the cation involved must, on the basis of availability, be iron or magnesium. As pointed out previously (Edmond et al., 1979b), the extent of Cl and Na removal and the stability of the responsible

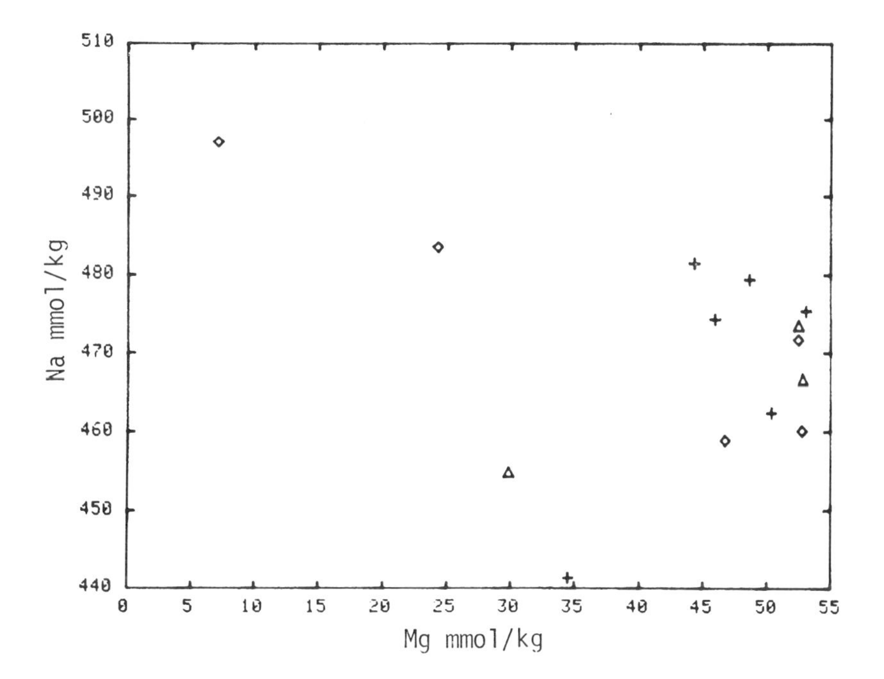

Fig. 10a. Sodium vs. Magnesium. Measured sodium.

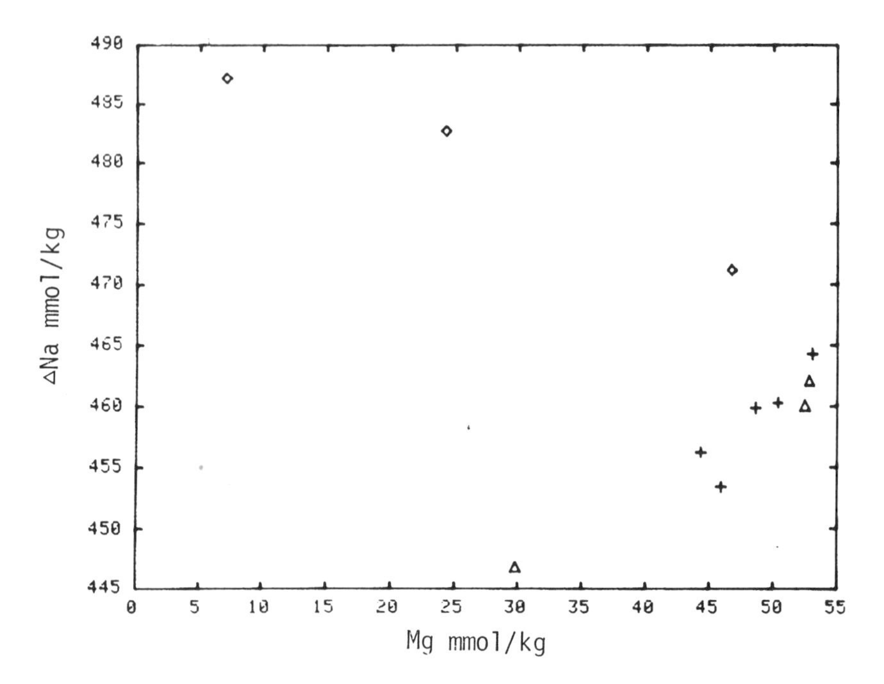

Fig. 10b. Sodium vs. Magnesium. Sodium calculated by difference from the charge balance.

minerals to low temperature processes have important implications both for the geochemical cycles of the two elements and for metallogenic processes involving partial remobilization of subducted oceanic crust.

Reduced Sulphur

The values for reduced sulphur are derived from measurements of both the dissolved H_2S and particulate abundances of sulphide and are subject to the uncertainties discussed above. The data show a general linear trend with Mg to end-member values of 6.5 mM H_2S (Fig. 11). This is equivalent to about 25% of the sea water sulphate concentration. The origin of the H_2S, whether igneous or reductive from sulphate, remains problematic. Recent isotopic measurements on sulphide from solution and from the chimneys give $\delta^{34}S$ values in the range 1 to 5 $^{\circ}/_{\circ\circ}$ indicative of a sea water contribution (Kerridge, 1982). However, the problem will only be resolved by systematic isotopic work on high integrity samples of the fluids and on the distributions of the chemical analogues arsenide and selenide in the fluids and the basalts.

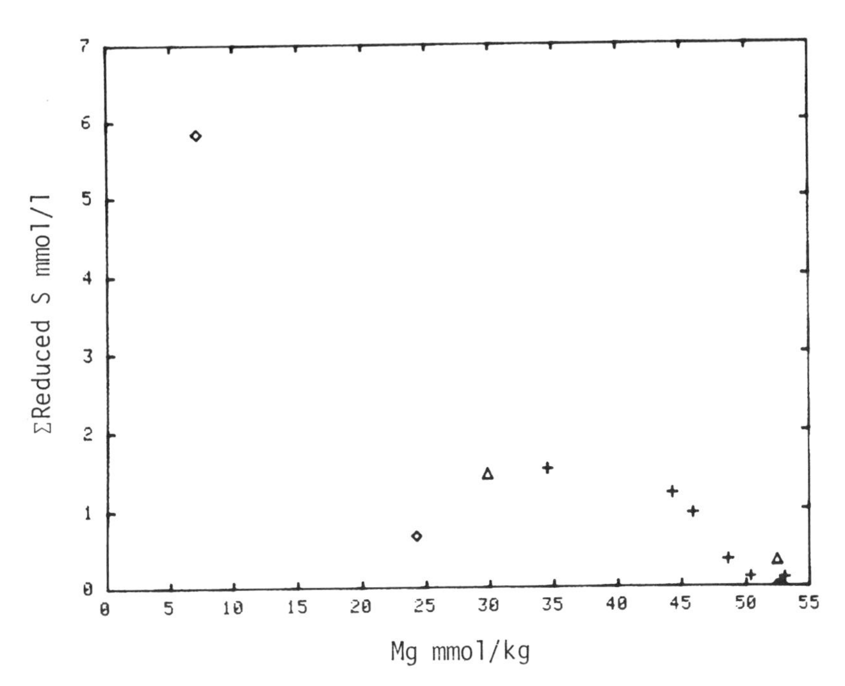

Fig. 11. Total reduced Sulphur vs. Magnesium

pH and Alkalinity

The end-member pH (25°C, 1 atmos) is close to 4 (Fig. 12). Concomitantly, the alkalinity is zero. In this the hydrothermal fluids are distinctly different from what would be expected on the basis of the experimental data. Low pH levels are reported only in the first two or three days of the runs (Bischoff and Dickson, 1975) and are probably the result of the precipitation of magnesium hydroxy sulphates during heating of the sea water rather than water-rock reactions per se (McDuff and Edmond, 1982). In this vein, the alkalinity trends reported from Galapagos (Edmond et al., 1979a) are steeper than would be expected from the present data and again reflect precipitation of hydroxy salts and also sulphides during sub-surface mixing (McDuff and Edmond, 1982).

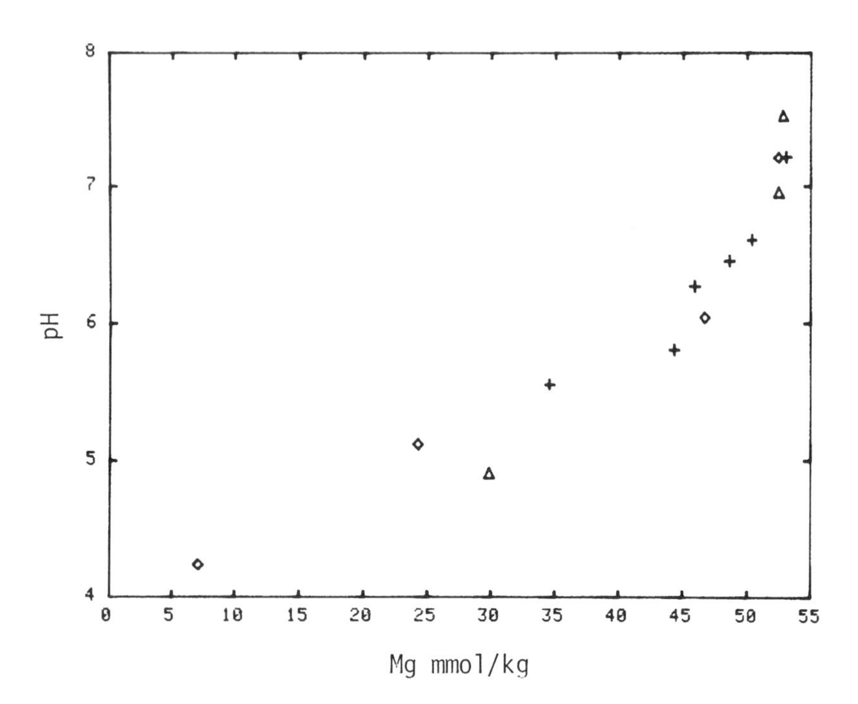

Fig. 12a. pH vs. Magnesium.

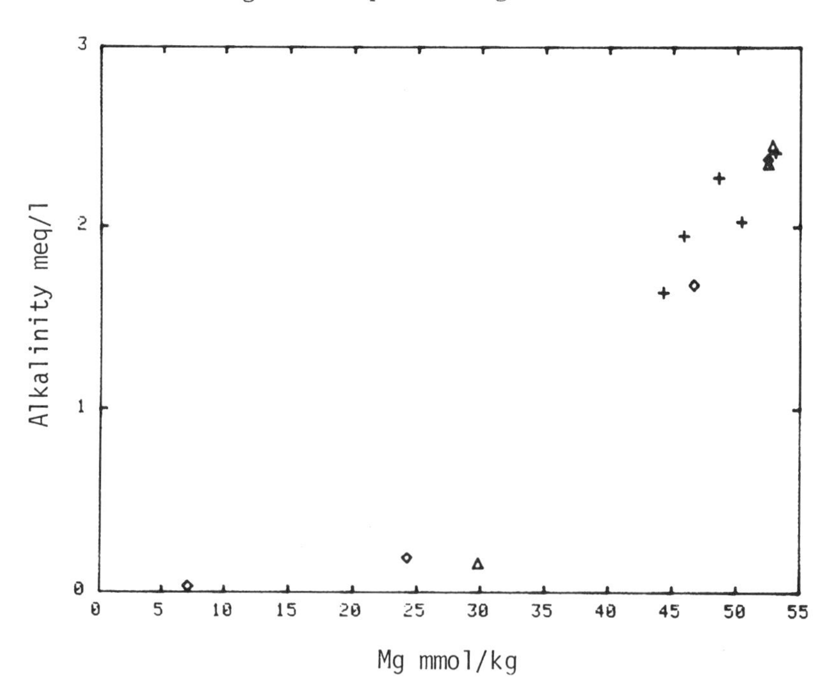

Fig. 12b. Alkalinity (total) vs. Magnesium.

Mn and Fe

The Mn data (Fig. 13) show a general linear trend with Mg to an end-member value of approximately 1 mM. this value is in the upper part of the range extrapolated from the Galapagos data (0.36 -1.14 mM).

The Fe data is most sensitive to the precipitation artifact caused by entrainment during sampling. A rough value of 1.8 mM can be estimated from the scattered data. The low Fe:Mn ratio, as compared to 50, in basalts, suggests that either a large amount of ferrous iron is incorporated in secondary silicates or chlorides or that it is precipitated as insoluble oxides during sulphate reduction. Since ferrous iron and Mn have very similar crystal chemistries, it is unlikely that their hydrothermal fractionation is a primary effect.

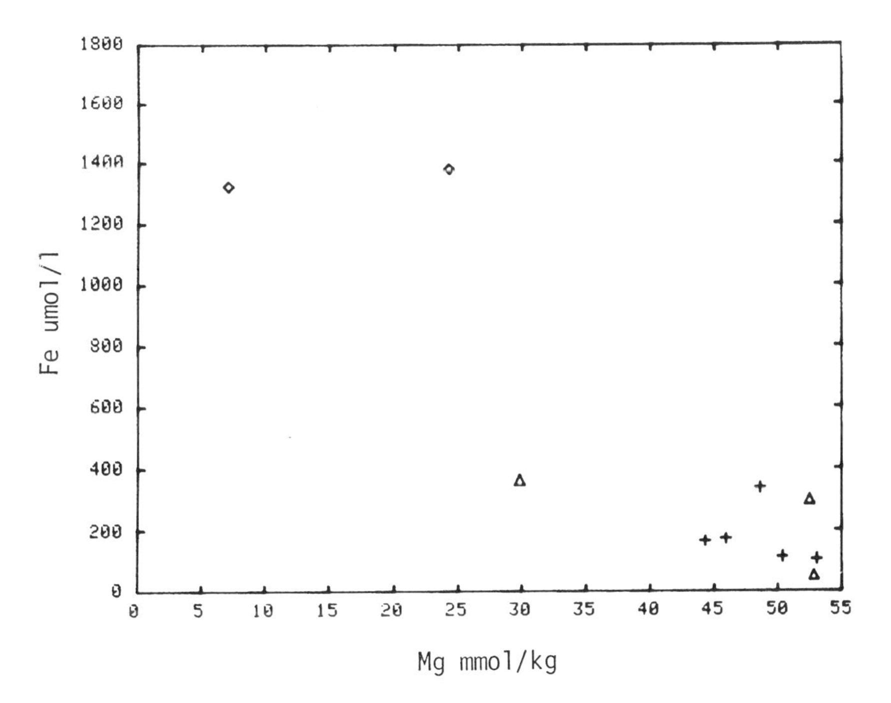

Fig. 13a. Iron vs. Magnesium.

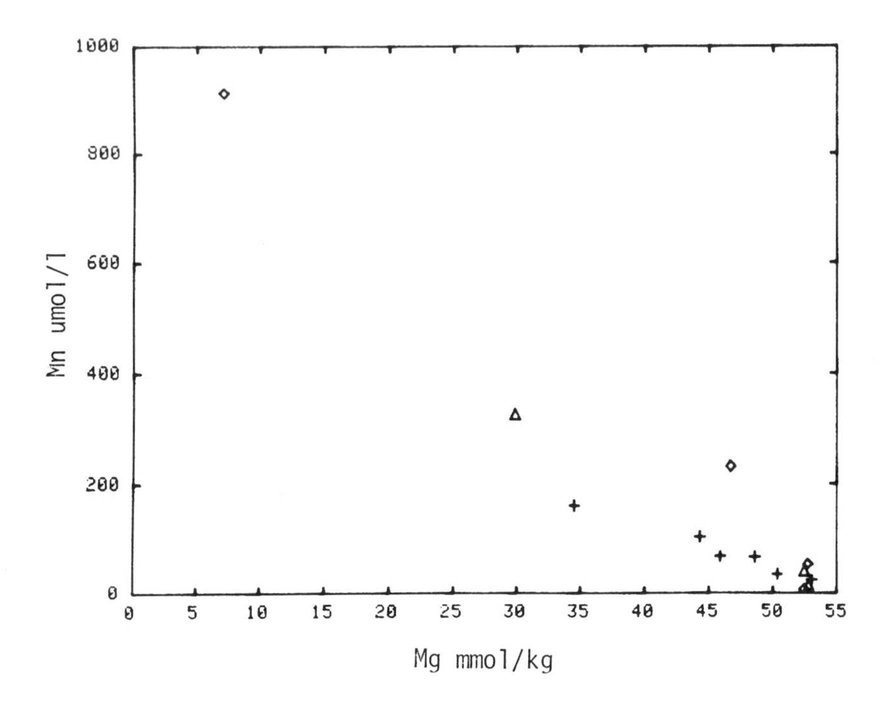

Fig. 13b. Manganese vs. Magnesium.

Zn and Cu

The zinc data (Fig. 14) form a linear trend with Mg to an end-member value of approximately 110 uM. The high point at 46 mM Mg has a corresponding sulphide anomaly, suggesting a ZnS particle was entrained. The data for Cu scatter widely; however, the lowest values place an upper limit on the true vent concentrations. This value is not more than 15uM. Since the ratio in basalts is approximately unity, there must be either preferential mobilization of Zn for reasons that are not obvious, or fractionation of Cu into sulphides precipitating in the conduits.

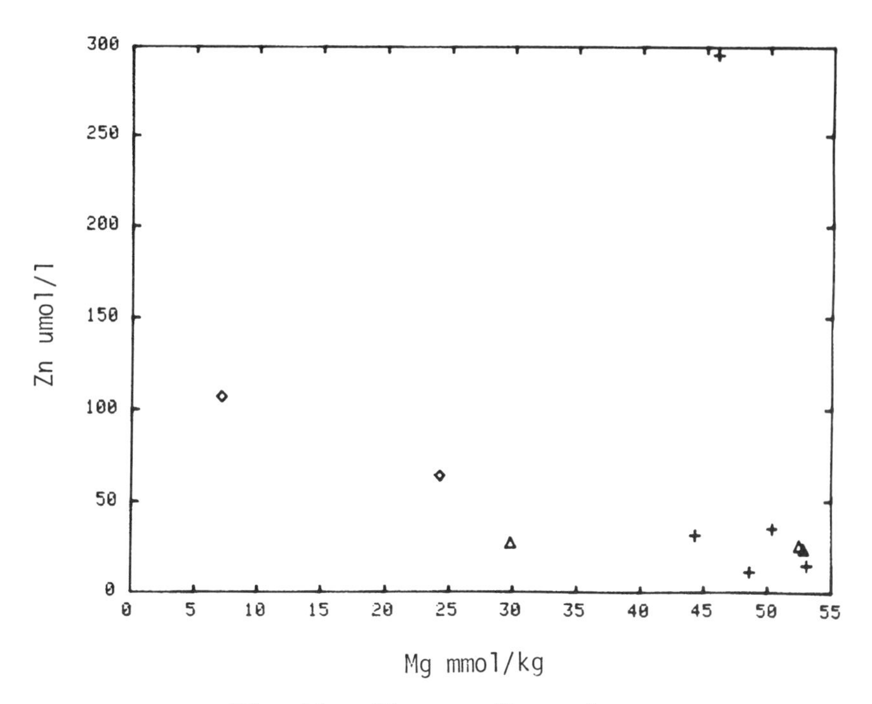

Fig. 14. Zinc vs. Magnesium.

CONCLUSIONS

The first reconnaissance samples of the 350°C "black smokers" on the EPR at 21°N have compositions very similar to those estimated on the basis of the original data from the hot springs on the Galapagos Spreading Center at 86°W. Since the ratio of ^3He to transported heat is identical in the two systems, the validity of the flux computations for the major ions and Mn based on the Galapagos data (Edmond et al., 1979b) is confirmed. The importance of ridge crest hydrothermal processes to the geochemical balance of the oceans is further demonstrated by its role in supplying the flux of non-radiogenic Sr required to maintain the observed $^{87}Sr/^{86}Sr$ ratio observed in sea water.

The undiluted high temperature system at 21°N is a true ore-forming solution. It is acid and it is sulphide dominated and it is particularly enriched in Fe and Zn. However, the interelement ratios, e.g., Fe:Mn and Zn:Cu, are very different from those in the source basalts. High-integrity samples of the undiluted, uncontaminated end-members will be necessary to unravel the chemical dynamics of the high temperature reaction zone.

ACKNOWLEDGEMENTS

We wish to thank the officers, crew, pilots and engineers of ALVIN/LULU for their great efforts in making the sampling program a success. J. Archeuleta of Los Alamos is thanked for his generosity in providing the samplers. We were enchanted by the conversations, helpful suggestions and general high jinks of H. Craig during the expedition. We thank the NSF for generous support.

REFERENCES

Albarede, F., A. Michaud, J. F. Minster & G. Michaud (1981), "$^{87}Sr/^{86}Sr$ ratios in hydrothermal waters & deposits from the East Pacific Rise at 21°N", Earth Planet. Sci. Lett. 55 229-236.

Bischoff, J. L. (1980), "Geothermal system at 21°N, East Pacific Rise: Physical limits on geothermal fluid & role of adiabatic expansion", Science 207 1465-1469.

Bischoff, J. L. & F. W. Dickson (1975), "Seawater-Basalt interaction at 200°C & 500 bars: Implications as to the origin of sea floor heavy metal deposits & regulation of seawater chemistry", Earth Planet. Sci. Lett. 25 385-397.

Brass, G. W. (1976), "The variation of the marine $^{87}Sr/^{86}Sr$ ratio during Phanerozoic time: Interpretation using a flux model", Geochim. Cosmochim. Acta 40 721-730.

Corliss, J. B., J. Dymond, L. I. Gordon, J. M. Edmond, R. P. von Herzen, R. D. Ballard, K. Green, D. Williams, A. Bainbridge, K. Crane & T. J. H. van Andel (1979), "Exploration of submarine thermal springs on the Galapagos Rift", Science 203 1073-1083.

Edmond, J. M. (1980), "Ridge crest hot springs: The story so far", EOS 61 129-131.

Edmond, J. M., J. B. Corliss & L. I. Gordon (1979a), "Ridge crest hydrothermal metamorphism at the Galapagos Spreading Center and reverse weathering" in Ewing Symposium, Vol. 2 (M. Talwani, ed.) 383-390.

Edmond, J. M., C. Measures, R. E. McDuff, L. H. Chan, R. Collier, B. Grant, L. I. Gordon & J. B. Corliss (1979b), "Ridge crest hydrothermal activity and the balances of the major and minor elements in the ocean: The Galapagos data", Earth Planet. Sci. Lett. 46 1-18.

Edmond, J. M., C. Measures, B. Mangum, B. Grant, F. R. Sclater, R. Collier, A. Hudson, L. I. Gordon & J. B. Corliss (1979c), "On the formation of metal-rich deposits at ridge crests", Earth Planet. Sci. Lett. 46 19-30.

Kerridge, J. F. (1982), "Sulphur isotope systematics at the 21°N site, East Pacific Rise", Earth Planet. Sci. Lett. (in press).

McDuff, R. E. & J. M. Edmond (1982), "On the fate of sulphate during hydrothermal circulation at mid-ocean ridges", Earth Planet. Sci. Lett. 57 117-132.

RISE Project Group (1980), "East Pacific Rise: Hot springs & geophysical experiments", Science 207 1421-1433.

METHANE, HYDROGEN AND HELIUM

IN HYDROTHERMAL FLUIDS AT 21°N ON THE EAST PACIFIC RISE

J. A. Welhan and H. Craig

Isotope Laboratory,
Scripps Institution of Oceanography
University of California at San Diego,
La Jolla, CA 92093

ABSTRACT

Methane in 350°C hydrothermal fluids at 21°N on the East Pacific Rise occurs in concentrations greater than 1.1 cc(STP)/kg. Hydrogen concentrations vary from 8 to 38 cc(STP)/kg, showing a considerable range between different vent fields. Helium concentrations exceed 0.021 cc(STP)/kg. The injection rates of methane and hydrogen into the deep ocean indicate replacement times of the order of 30 years, implying that consumption of methane and hydrogen in the water column must be very rapid. Variations of end-member concentrations of methane, hydrogen and possibly helium, as well as $\delta^{13}C(CH_4)$, among vent fields suggests either chemical control of reactive gas abundances and/or variations in gas contents of ridge crest basalts. Measurements of methane and helium in basalt glass from the EPR show $CH_4/^3He$ ratios of 2.5 x 10^6, compared to 3.5 x 10^6 in hydrothermal fluid from the same area. Carbon isotope evidence, CO_2-CH_4 isotope geothermometry, the lack of suitable thermocatalytic sources of organic carbon, and the similarity between $CH_4/^3He$ ratios in these hydrothermal fluids and mid-ocean ridge basalts, point to an abiogenic origin of hydrothermal methane, extracted directly from basalt by circulating seawater.

INTRODUCTION

The discovery of hydrothermal plumes on the Galapagos Ridge (Weiss, Lonsdale, Lupton, Bainbridge and Craig, 1977) confirmed the existence of submarine hydrothermal convection systems along mid-ocean ridge spreading axes, a phenomenon that had been postulated to explain the anomalously low conductive heat flow at ridge crests. Previously, the only geochemical evidence of their existence had been the mid-depth ^3He anomaly in the Pacific Ocean (Clarke, Beg and Craig, 1969), which indicated injection of mantle volatiles into the water column along the ridge crests.

Following detailed sampling of the Galapagos low temperature hydrothermal waters by the ALVIN deep submersible, the impact of hydrothermal seawater circulation on the ocean's major and trace element chemistry was confirmed (Edmond, Measures, McDuff, Chan, Collier and Grant, 1979). Samples of Galapagos hydrothermal fluid were also analyzed for total dissolved inorganic carbon (ΣCO_2), and were found to contain excess carbon with $\delta^{13}C$ = -5.1 to -5.9 ‰ vs PDB (Craig, Welhan, Kim, Poreda and Lupton, 1980). The similarity of the $^{13}C/^{12}C$ composition of this carbon to that of diamonds and carbonatites which is believed to represent juvenile carbon (Craig, 1953), its similarity to CO_2 in basalt vesicles from the East Pacific Rise (Moore, Batchelder and Cunningham, 1977), and its association with ^3He in the hydrothermal fluids (Lupton, Weiss and Craig, 1977), provided the first evidence of mantle carbon being injected directly into the hydrosphere.

After an extensive series of Deep Tow and Angus profiles on the East Pacific Rise at 21°N, dives with ALVIN in April, 1979 led to the discovery of high temperature (350°C) hydrothermal fluids and massive sulfide deposits (RISE group, 1980). Measurements of helium isotope ratios in the 21°N hydrothermal fluids demonstrated the primordial nature of this helium (Lupton, Klinkhammer, Normark, Haymon, MacDonald, Weiss and Craig, 1980). The helium isotope ratio (7.8 x the atmospheric ratio) is very similar to that observed in oceanic basalts (Lupton and Craig, 1975; Craig and Lupton, 1976), and clearly shows that helium is extracted from the ridge crest basalts by circulating hydrothermal fluids.

In addition to helium, these fluids were also found to contain high concentrations of methane, hydrogen, and, like the Galapagos fluids, excess ΣCO_2 (Welhan and Craig, 1979; Craig et al., 1980). Samples of fluids and dissolved gases were obtained during a second series of dives in November, 1979, for gas and isotopic analysis. The $\delta^{13}C$ of excess CO_2 in the hydrothermal fluids was found to be -7.0 ‰ vs PDB (Craig et al., 1980), very similar to that at Galapagos, and provided further evidence for mantle carbon outgassing at ridge crests.

The presence of methane in these hydrothermal fluids has excited interest in abiogenic methane, particularly in light of Gold's (1979) hypothesis that the mantle contains considerable quantities of hydrocarbon gases. In this paper, we summarize the stable isotope and gas concentration data on methane, hydrogen and helium in the EPR hydrothermal fluids, and discuss the stable isotope data in the context of gas origins, particularly the abiogenic origin of methane.

SAMPLING AND ANALYTICAL METHODS

The hydrothermal vents at 21°N on the East Pacific Rise were sampled in November, 1979 with the DSRV ALVIN during five dives. Sampling locations are shown in Figure 1. The vent fields are referred to as the National Geographic Society site (NGS), the Ocean Bottom Seismometer site (OBS), and the Southwest Vents (SWV). A new site, the "Hanging Gardens" (see Frontispiece) was discovered in 1981.

Water samples were collected in evacuated heavy-walled stainless steel cylinders of nominal one-liter volume which were constructed at the Los Alamos Scientific Laboratories. The bottles were fitted with valves at both ends, and were plated inter-

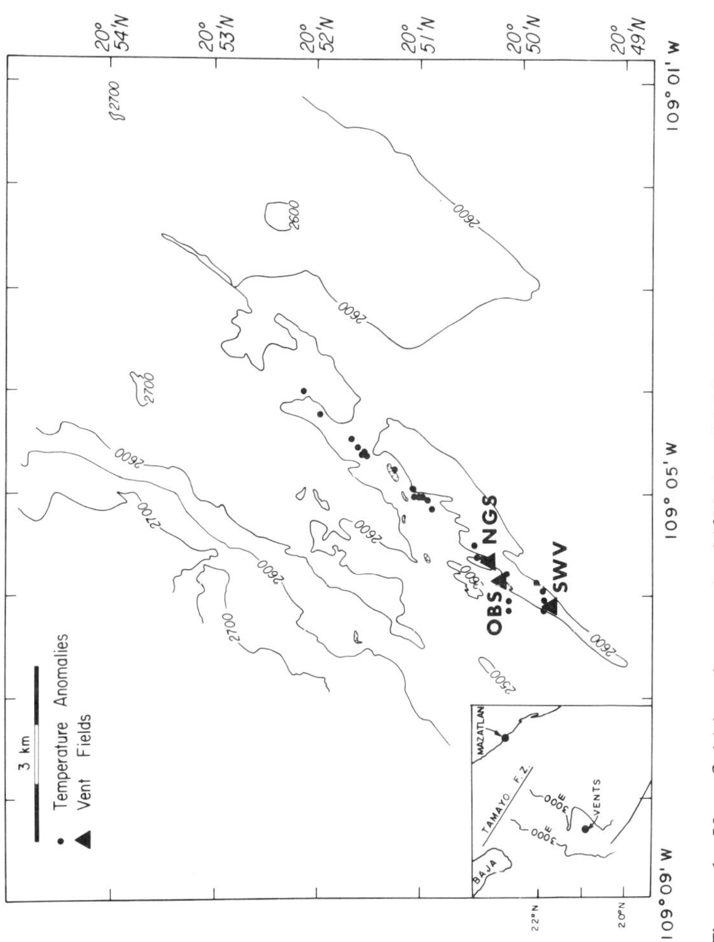

Figure 1. Vent field locations at the 21°N site. NGS = National Geographic Society site; OBS = Ocean Bottom Seismometer site; SWV = Southwest Vents site.

nally with gold to minimize reactions between hydrothermal fluid and the steel. The sampling bottles were evacuated with a two-stage rotary vacuum pump and were strapped to the front equipment cradle of ALVIN. Three bottles were filled on each dive. One bottle's inlet valve was operated by a DC motor designed and built at Los Alamos; the other two bottles were opened with a steel T-bar that mated to the inlet valve handle and was operated by ALVIN's electromechanical arm.

Because the sampling arm was occupied in this manner, the inlet port of each sample bottle was placed into the hydrothermal vent effluent by delicate maneuvering of ALVIN. Vent water temperatures were read by thermocouples strapped to the bottles a few inches from each inlet valve orifice. Temperatures measured by these thermocouples during opening of the bottles were often much higher than the actual temperature of the sampled fluid, due to the turbulence in the vicinity of the vent orifice and entrainment of ambient seawater. For this reason, fluid properties which reflect the degree of mixing between ambient seawater and hydrothermal fluid were used to determine mixing ratios. Magnesium data were obtained from J. Edmond (personal communication); $^{18}O/^{16}O$ ratios in the fluid, expressed as per mil deviations from the SMOW standard, were measured at SIO by the usual CO_2 equilibration technique.

Many samples formed a gas phase in the bottles resulting from fluid contraction during cooling and/or incomplete filling due to premature closure. In order to correct for this effect, both gas and liquid phases were sampled for gas chemistry, to reconstruct the original dissolved gas concentrations. The head-space gases were expanded into a high-vacuum line on shipboard and then Toepler-pumped through a U-trap at the temperature of dry ice-isopropol alcohol. The dry gases were sealed in glass breakseal tubes for later analysis at SIO. Following collection of the head-space gases, the bottle was pressurized slightly with compressed nitrogen and fluid was drawn through Tygon tubing into evacuated 50 ml glass flasks constructed of Corning 1720 aluminosilicate glass, which has low helium permeability. These samples were not poisoned.

Samples were returned to SIO and gases were processed and split into known fractions for analysis of gas compositions and helium isotope ratios by routine procedures (Welhan, Poreda, Lupton and Craig, 1980). Methane, hydrogen, nitrogen and argon were analyzed by gas chromatography, using an ultrasonic detector. Helium and neon concentrations were analyzed by isotope dilution (Craig and Lupton, 1976).

Splits of five of the largest head-space gas samples were processed on a chromatographic combustion system (Welhan, 1981). Methane was separated chromatographically and oxidized on-line over CuO at 800°C in a purified O_2 stream. CO_2 derived from the methane oxidation was analyzed for $^{13}C/^{12}C$ ratio on a 60° double collecting mass spectrometer. Water derived from methane oxidation was reduced to H_2 gas over uranium at 800°C and analyzed for its D/H ratio on a 60° split tube double collecting mass spectrometer. The smallest methane-carbon isotope samples contained less than 0.05 cc(STP) CO_2, resulting in large blank corrections and spectrometer uncertainties. Two of these small volume samples were re-analyzed on a micro-inlet mass spectrometer in the laboratory of Dr. I.R. Kaplan, University of California at Los Angeles, and these data agreed with the SIO analyses to 0.3 and 0.7 ‰. Estimated total uncertainty in the $\delta^{13}C$ values for the smallest samples, corrected for blank contribution, is ± 1 ‰. The two largest samples contained 0.2 to 0.5 cc(STP) CO_2, with an uncertainty in $\delta^{13}C(CH_4)$ of ± 0.3 ‰.

Gas Correction Procedures

Due to the two-phase partitioning of gases within the bottles, the argon concentrations in the liquid phase were used to calculate the original methane and helium concentrations in the fluid. In a few cases, head-space gas ratios were also used to calculate gas partitioning corrections. A detailed description of these procedures is given by Welhan (1981). Briefly, comparison of argon and neon concentrations in the liquid indicated that the fluid samples had degassed in a non-equilibrium manner during the rapid expansion of head-space gases into the vacuum line, so that gas ratios in the fluid remained constant. Similar effects have been reported for rapid degassing of heated water (Hulston and McCabe, 1962). Argon/nitrogen ratios in head-space gases were also used to calculate original gas concentrations on the assumption that the observed head-space gas ratios represented equilibrium gas partitioning between the liquid and gas phase prior to the extraction of head-space gases. This calculation was performed in five cases where a relatively large head-space formed (i.e. $>$ 10-20 ml); samples with smaller head-space volumes yielded inconsistent results, presumably because in the process of extracting the head-space volume the gas phase ratios were affected by liquid phase degassing, the effect being most pronounced in samples with small head-space volumes.

Values for initial argon and nitrogen concentrations were estimated from observed values at the ridge crest depth (Craig, Weiss and Clark, 1967). Methane and helium concentrations corrected for gas partitioning were calculated on the assumption that argon, neon and nitrogen were conservative and air-derived. In three cases where the head-space volume was greater than $\sim$ 50 ml, an additional correction was applied to the liquid phase correction, assuming that prior to the expansion of head-space gases into the vacuum line, the gases had partitioned at equilibrium, proportional to their solubilities. Analytical precision for hydrogen, methane and argon analyses is 3%, and for helium, 1%. Estimated total uncertainty in the corrected concentrations of methane and hydrogen is 6%, and for helium, 4%. Uncertainties in methane concentrations due to possible bacterial effects in unpoisoned samples are discussed in the following section.

RESULTS

The oxygen isotopic composition of the sampled fluid was used as an indicator of the degree of dilution of the hydrothermal end-member with ambient seawater ($\delta^{18}O(H_2O)$ = -0.08 ‰). Figure 2 shows the values of $\delta^{18}O$ vs Mg concentration; extrapolation to zero Mg gives a hydrothermal end-member $\delta^{18}O$ value of +1.6 ‰ vs SMOW (Craig et al., 1980). Mixing ratios of hydrothermal fluid and ambient seawater were then calculated from the $\delta^{18}O$ data.

Table 1 lists the percent hydrothermal end-member, corrected methane and helium concentrations, and stable isotope data on the methane samples. Corrected hydrogen concentrations and hydrogen isotope data on H_2 in the least diluted samples ($>$ 20% hydrothermal fluid) are presented in Table 2. The corrected methane and helium data are plotted vs $\delta^{18}O$ in Figures 3 and 4.

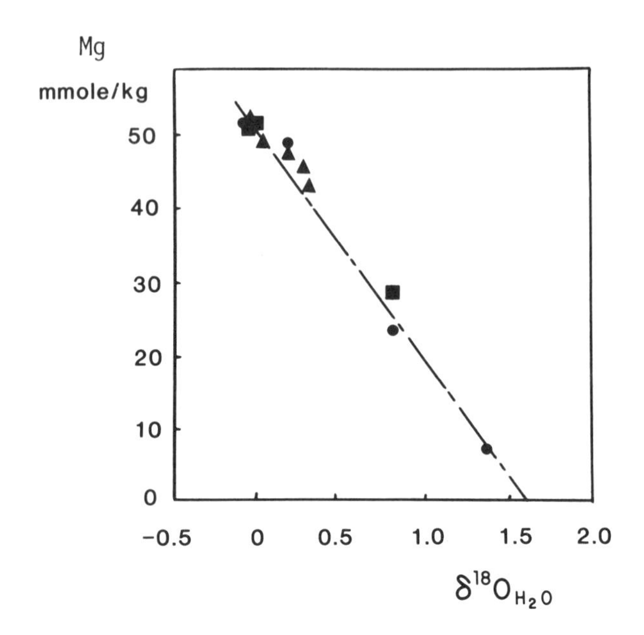

Figure 2. Correlation of Mg vs $\delta^{18}O(H_2O)$ reflects dilution of hydrothermal fluid (Mg = 0; $\delta^{18}O = +1.60$ ‰) with ambient seawater during the sampling process.

Welhan (1981) presented comparisons of H_2, CH_4 and CO_2 concentrations in three liquid samples that were analyzed on shipboard, and duplicate samples that were stored without poison in glass flasks for up to two months. He suggested that bacteria may have been present in the fluid. The possibility of bacterial utilization of methane in the unpoisoned liquid samples upon which some of the corrected methane concentrations are based, implies that some of the corrected data might be inaccurate. However, the general good agreement between gas concentrations calculated from liquid samples and splits of dry head-space gases (which were not subject to bacterial alteration) suggests that the present estimates of end-member concentrations are, in fact, reliable.

Isotopic fractionation of methane and hydrogen due to bacterial activity in samples analyzed for $^{13}C/^{12}C$ and D/H also does not appear to be likely, since (a) these samples represent head-space gases which were extracted and dried on shipboard within 8-48 hours of sample recovery, and (b) decompression and thermal shock accompanying the entry of fluids into the sample bottles would tend to hinder bacterial activity during the time prior to extraction of gases from the head-space of the samples.

Table 1 - Corrected methane and helium concentrations, plus stable isotope data for the five largest methane gas samples.

Sample/Area			% End-member°	He, μcc/g	CH_4, cc/kg	$\delta^{13}C(CH_4)$	$\delta D(CH_4)$
978	Bag	NGS$^{\Delta}$	1.2	–	0.013	–	–
981	1/2	SWV	4.1	0.165	<0.05	–	–
979	1/2	OBS	4.7	0.229	<0.05	–	–
979	11/12	OBS	7.1	0.195	<0.05	–	–
982	5/6	NGS‡	7.1	0.978	0.106	–	–
981	11/12	SWV	8.8	2.33	0.048	–	–
980	5/6	SWV^{+}	17.1	3.10	0.095	–	–
978	7/8	NGS	19.4	2.25	0.147±.007*	–	–
981	3/4	SWV	24.1	6.55	0.301±.012*	-17.6±1.0	–
980	9/10	SWV^{+}	40.0	9.10	0.411±.044*	-17.2±0.8	–
979	3/4	OBS	51.8	15.8	0.548±.048*	-17.4±1.0	–
978	5/6	NGS	54.7	–	–	-16.6±0.3	-126±3
982	7/8	NGS‡	85.9	17.3	1.24±.060*	-15.0±0.3	-102±3

° Calculated from measured $\delta^{18}O$ in H_2O and Mg-$\delta^{18}O$ correlation.
– Not measured.
Δ Low temperature, diffuse vent sampled with plastic bag; all other samples are from "black smoker" vents.
+ Same vent.
‡ Same vent.
* Range represents uncertainty in calculated concentrations, based on liquid and head-space gas data. All other gas concentrations represent corrected concentrations based on liquid data only.

Table 2 - Calculated hydrogen concentrations and hydrogen isotope data for samples with >20% hydrothermal fluid.

Sample/Area			% End-member	End-member H_2, cc(STP)/kg	End-member H_2/CH_4	$\delta D(H_2)$, ‰
981	3/4	SWV	24.1	7.3±5.0	5.7±2.7	-387±20*
980	9/10	SWV	40.0	9.9±5.0	9.2±4.0	-373±3
979	3/4	OBS	51.8	37.5±5.0	37.5±8.0	-401±3
978	5/6	NGS^{+}	54.7	–	–	-396±3
982	7/8	NGS	85.9	30.5±3.5	21.4±3.5	-383±3
					Mean $\delta D(H_2)$ =	-388

* Isotope ratio measurement on very small sample; large spectrometer uncertainty.
+ Corrected gas concentrations not available due to air leakage into bottle.

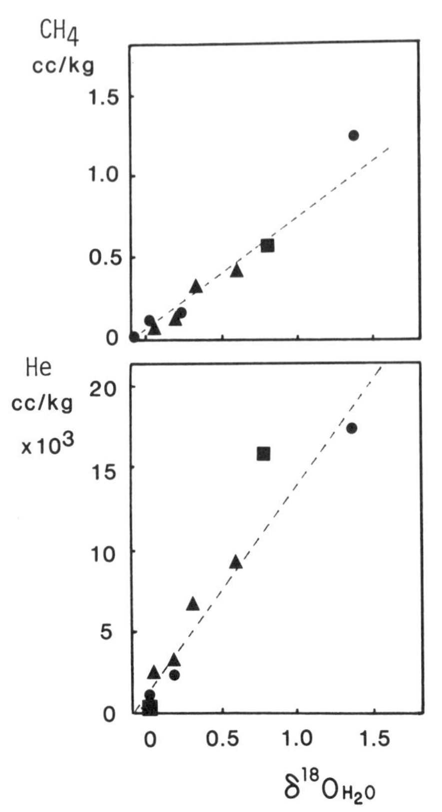

Figure 3. Corrected methane and helium concentrations versus $\delta^{18}O(H_2O)$. Symbols represent different vent fields: ● = NGS; ■ = OBS; ▲ = SWV.

DISCUSSION

End Member Gas Concentrations

Concentrations of methane, hydrogen and helium in the hydrothermal fluid were calculated from their observed correlations with $\delta^{18}O$. Table 3 summarizes the end-member fluid concentrations, together with their isotopic values.

The corrected methane concentrations are approximately linear with $\delta^{18}O$ (Figure 3), although sample 982 7/8 from the NGS site contains up to 30% more methane than expected from the correlation of methane vs $\delta^{18}O$ in other samples. As indicated in Figure 4, methane concentrations relative to helium in all NGS samples appear to be high compared to other areas. Methane/^3He ratios calculated from the two regression lines in Figure 4 range from 3.5 to 6.5 x 10^6 for the NGS and OBS-SWV groups, respectively. In contrast, the end-member helium concentration in the NGS field is 40% lower than in the OBS field (Figure 3). No corresponding difference in $^3He/^4He$ ratios between vent fields was observed.

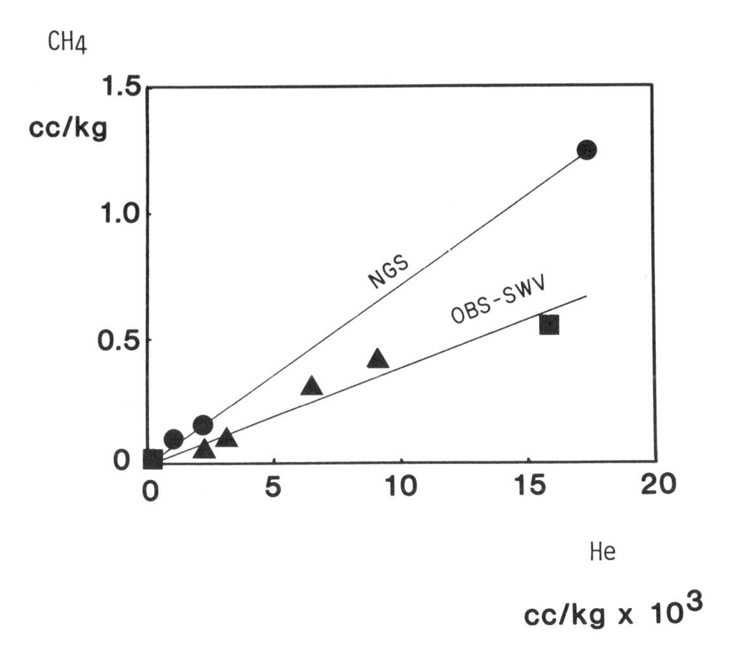

Figure 4. Corrected methane concentration versus corrected helium concentration. Symbols are same as in Figure 3. Calculated $CH_4/^3He$ ratio for NGS site = 6.5 x 10^6; for OBS-SWV sites, = 3.5 x 10^6.

Unlike methane and helium, which are approximately linear vs $\delta^{18}O$, hydrogen concentrations are very non-linear vs $\delta^{18}O$. This effect is clearly demonstrated in the least-diluted samples which contain the highest hydrogen concentrations (Table 2). The possibility of excess hydrogen, produced through reaction of hydrothermal fluid with stainless steel, must be considered, since the gold lining in many of the sampling bottles failed, and corrosion was subsequently noted in some. However, a "black smoker" fluid sample collected in a PVC sample bottle from the NGS site during the April, 1979 dives (Welhan and Craig, 1979), although severely diluted with seawater, had a H_2/CH_4 ratio of 18.2, similar to that observed in sample 982 7/8 from the same vent field (21.4 ± 3.5), suggesting that hydrogen contamination was minor in at least some of the November, 1979 samples.

Furthermore, recent measurements of hydrogen and methane in pure hydrothermal fluid samples collected in all-Titanium bottles during November, 1981, corroborate the 1979 hydrogen data: H_2/CH_4 ratios differ by more than a factor of five among vent fields, primarily due to substantial differences in hydrogen concentrations (variations in methane concentrations of the same magnitude as observed in the 1979 samples were also observed).

The mean and range of five $\delta D(H_2)$ values in Table 2 is -388 ± 15 ‰. There is no apparent correlation of $\delta D(H_2)$ with hydrogen end-member concentrations or H_2/CH_4 ratios. Apparent hydrogen isotope equilibration temperatures calculated for

H_2-H_2O and H_2 - CH_4 exchange (assuming $\delta D(H_2O = 0$ ‰) indicate temperatures of 315 ± 20°C, some 300-450°C lower than temperatures calculated from the CH_4-CO_2 carbon isotope geothermometer (Welhan, 1980; Welhan, 1981). These results suggest that hydrogen isotope reequilibration progresses relatively rapidly as temperature decreases, so that the hydrogen isotope geothermometers essentially reflect exit temperatures. Similar relationships have been observed in continental geothermal fluids (Arnasson and Sigurgeirsson, 1968; Welhan, 1981).

Together, the large variations in hydrogen concentrations and the rapidity of hydrogen isotope equilibration suggest chemical reactivity, in which case the range in hydrogen concentrations and H_2/CH_4 ratios among vent fields may be an indicator of local chemistry (redox state, pH, sulfide equilibria, etc.). Alternatively, differences in hydrogen concentration may reflect fundamental differences in basalt volatile contents.

The variability of hydrogen and methane among vent fields raises the question of whether gas concentrations at 21°N are representative of ridge crest hydrothermal activity in general. Data on methane and helium in hydrothermal plumes from the vicinity of 20°S on the East Pacific Rise demonstrate that $CH_4/^3He$ ratios vary from 5.8 to 17.4 x 10^6 (Craig, 1981); the corresponding ratio in the Red Sea Brines is 0.70 x 10^6. The mean $CH_4/^3He$ mole ratio at 21°N is calculated to be 5.0 ± 1.5 x 10^6. Therefore, although the proportions of these gases vary from one hydrothermal system to another, it appears that 21°N hydrothermal methane and helium abundances are not atypical of ridge crest hydrothermal emanations.

Table 3 - East Pacific Rise 350°C hydrothermal end-member fluid composition.

Component	Concentration, cc(STP)/kg
CH_4	1.15 - 1.45
H_2	8 - 38
He	0.021*

Isotope Ratio	End Member Value
$^3He/^4He$	7.8 x air ratio ▽
$\delta^{13}C(CH_4)$	-15.0 to -17.6 ‰ vs PDB
$\delta D(CH_4)$	-102 to -126 ‰ vs SMOW
$\delta D(H_2)$	-388 ± 15 ‰ vs SMOW
$\delta^{13}C(CO_2)$	-7.0 ± 0.1 ‰ vs PDB‡
$\delta^{18}O(H_2O)$	+1.60 ‰ vs SMOW+

* Based on NGS - SWV data; OBS data indicates 40% higher helium content.

▽ After Lupton et al. (1980); $^3He/^4He$ ratio in air = 1.4 x 10^{-6}.

‡ Represents excess ΣCO_2, over and above seawater ΣCO_2. Isotopic value calculated from observed $\delta^{13}C(CO_2)$ vs $1/\Sigma CO_2$ after Craig et al. (1980).

\+ After Craig et al. (1980).

Assuming that the injection of ^3He at the ridge crests represents the primary source of the mid-depth oceanic ^3He saturation anomaly (Clarke et al., 1969), the calculated global average rate of ^3He outgassing from the upper mantle is 3 atoms $^3He/cm^2 \cdot sec$ (Craig et al. (1975) originally calculated a value of 4 atoms/$cm^2 \cdot$sec, but this value assumed a mean $^3He/^4He$ ratio for injected ridge-crest helium of 11 x the atmospheric ratio, based on measurements of helium in the water column above the ridge-crest. This helium has since been shown to have a ratio of 8 $\pm$ 0.5 x the atmospheric ratio , thereby reducing the calculated mantle ^3He flux proportionately).

If the 21°N $CH_4/^3He$ ratio is representative of mantle outgassing at the ridge crests, then the global average hydrothermal methane injection rate is $2.5 \pm 0.8 \times 10^{-17}$ moles/$cm^2 \cdot$sec. For a deep ocean volume of 10^9 km^3, and a deep water methane concentration of about 2×10^{-6} cc(STP)/kg (Sieler and Schmidt, 1974) this injection rate would supply the entire deep ocean inventory of methane in 30 $\pm$ 10 years. Similar calculations for hydrogen indicate a residence time of 10 to 45 years, based on the observed range of hydrogen concentrations. These calculations agree remarkably well with the previous estimates by Welhan and Craig (1979) based on the first samples of highly-diluted hydrothermal fluid, and indicate very high consumption rates for these compounds in the water column.

Origin of Methane

In discussing the origin of carbon in 21°N methane, three facts must be considered: (a) the ^{13}C-enriched nature of methane carbon; (b) the range in $\delta^{13}C(CH_4)$; and (c) the isotopic differences between 21°N methane and other geothermal methanes, and possible organic carbon sources.

Carbon isotope data on 21°N methane and CO_2 are plotted in Figure 5. Isotopically, 21°N methane is unlike methane from any continental high temperature systems that we have studied. Apparent carbon isotope equilibrium temperatures calculated for CH_4-CO_2 exchange in the 21°N system, using the calculated fractionation factors of Richet, Bottinga and Javoy (1977), range from 620-770°C. These apparent temperatures are the highest yet reported for CH_4-CO_2 isotopic exchange in hydrothermal fluids. However, isotopic exchange temperatures are meaningless unless the methane and carbon dioxide are of related high temperature origins. In continental geothermal systems, where methane is predominantly of thermocatalytic origin (DesMarais, Donchin, Nehring and Truesdell, 1981; Welhan, 1981) and CO_2 is of volcanic, metamorphic or sedimentary origin, carbon isotope geothermometry is of dubious validity (Craig, 1963).

However, unlike continental hydrothermal environments, the 21°N hydrothermal system appears to be relatively free of organic and sedimentary carbon contamination. Possible organic sources for hydrothermal methane are thermocatalysis of organic matter at high temperature and production by methanogenic bacteria in sediment-rich recharge areas. The latter possibility can be ruled out on the basis of the ^{13}C enrichment of 21°N methane relative to biogenic methane (Claypool and Kaplan, 1974; Schoell, 1980). Since sediment thickness is negligible within 1 km of the ridge axis (Ballard, Francheteau, Juteau, Rangan and Normark, 1981), a significant sedimentary carbon input to this system seems unlikely. The concentration of dissolved organic carbon in 21°N bottom water is 1-3 mg/l (Williams, Smith, Druffel and Linick, 1981).

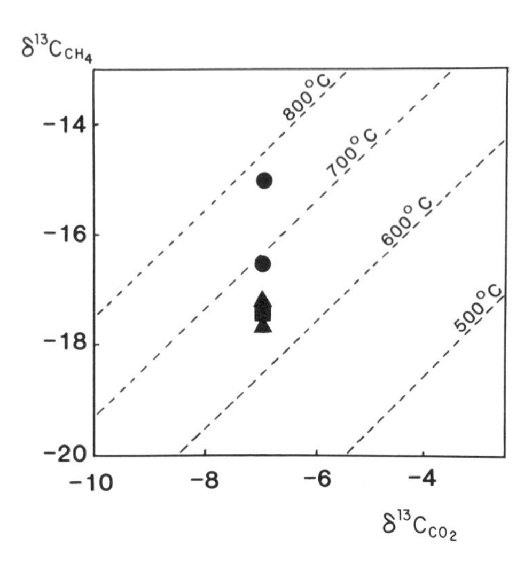

Figure 5. Carbon isotope ratios in CH_4 and CO_2 at 21°N. "CO_2" refers to excess ΣCO_2 in the fluid, derived from basalt. Temperatures refer to carbon isotope equilibrium for CH_4-CO_2 exchange, calculated from Richet et al. (1977).

Conceivably, this organic carbon could be thermocatalytically degraded to produce methane at high temperature within the hydrothermal system, but the kinetic isotope effect accompanying thermocatalytic conversion would produce methane much more ^{12}C-enriched than observed here (Sackett, 1978; Schoell, 1980). Even assuming the presence of a sufficiently large organic carbon source, the observed enrichment of ^{13}C in the methane would require an abnormally ^{13}C-enriched organic substance, which, aside from vestimentiferae tissue, has not been observed in the 21°N area (Williams et al., 1981).

Fluid samples collected during November, 1981, contained very low concentrations of the higher homologues of methane ($C_2H_6/CH_4 < 2 \times 10^{-3}$, with but traces of propane and butane; Welhan and Craig, 1982). This is inconsistent with a thermocatalytic origin, in which appreciable quantities of ethane, propane and butane would be expected (Stahl, 1977). In contrast, the Guaymas Basin hydrothermal vent fluids, which debouch from a thick, organic-rich sedimentary blanket at comparable temperatures (315°C), contain very high concentrations of C_1-C_5 hydrocarbons (100-1000 times higher than at 21°N), with $\delta^{13}C(CH_4)$ values of -40 to -50 ‰ vs PDB (Welhan and Craig, 1982), very typical of thermocatalytically-derived gas. The conspicuous lack of significant amounts of C_2+ hydrocarbons at 21°N and the ^{13}C-enriched character of this methane, argue against a thermocatalytic origin for methane in the 21°N system.

Thus, considering the geological environment as well as the isotopic and chemical characteristics of 21°N methane relative to biogenic and thermogenic methane, the possibility of an organic carbon source for methane can be ruled out.

On the other hand, a basalt sample from the 21°N area contained 176 ppm by weight of carbon (R. Poreda, personal communication). Based on the He contents of basalt and hydrothermal fluid (15 μcc/g and 21 μcc/g, respectively; Craig et al., 1980; Table 3), a water/rock ratio near unity is indicated for this system. The concentration of excess carbon in the fluid corresponds to 40 ppm by weight, relative to 176 ppm of carbon in the basalt. Thus the basalt is capable of supplying all of the excess carbon (CO_2 + CH_4) observed in the hydrothermal fluid.

Methane in Mid-Ocean Ridge Basalt

Hekinian, Chaigneau and Cheminee (1973) and Chaigneau, Hekinian and Cheminee (1980) have carried out detailed studies of gases in mid-ocean ridge (MOR) basalts, in which they found high ratios of CO to CO_2, but only minor amounts of H_2 and CH_4. Unfortunately, their analyses were made by heating crushed samples to 1000°C in vacuum, and it has been well known since the time of Chamberlin (1908) that this procedure merely serves to bring about partial or total chemical equilibrium of the reactive gases contained in the rock. Thus, while heating of basalts to extract gases may tell something about the overall oxidation state of the basalt itself, it tells nothing about the gases contained within the basalt when it was collected.

Pineau, Javoy and Bottinga (1976) analyzed gases in vesicles in the glassy margin of a basalt pillow from the Mid-Atlantic Ridge, by crushing the glass at room temperature in steel tubes. This sample had also been analyzed by Hekinian et al. (1973) by their technique of heating to 1000°C. A comparison of the data from these two papers shows that the CO/CO_2 ratios and H_2/CH_4 ratios in the gases analyzed by Pineau et al. are lower by factors of 100 (!) and 10, respectively. This comparison clearly shows that the gas compositions reported by Hekinian et al. (1973) and Chaigneau et al. (1980) reflect a high-temperature chemical re-equilibration, with much higher CO and H_2 concentrations than occur in the vesicles of the basalts.

Thus the single analysis of gases from a tholeiitic MOR glass by Pineau et al. appears to be the only reasonably reliable literature data on gases in MOR basalt glass. Because of the high gas concentrations in this sample (CH_4 = 2300 μcc/g; H_2 = 44000 μcc/g; H_2/CH_4 = 19), blank problems and gas adsorption effects are probably minor. Zolotarev, Voytov, Sarkisyan and Cherevichnaya (1981) also measured gases in Mid-Atlantic Ridge basalt glass, but at much lower concentrations (CH_4 = 70 μcc/g; H_2 = 600 μcc/g). Although they presented no discussion of blank problems or gas adsorption losses, their H_2/CH_4 ratios (8-9) are similar to those found by Pineau et al., so their data may well be reliable.

The available data on occluded gases in MOR basalts indicate that both hydrogen and methane (as well as CO_2) occur within these rocks, in proportions very similar to those observed in 21°N fluids. Since primordial helium is derived directly from basalt, and since the proportions of methane and hydrogen to primordial helium in 21°N hydrothermal fluid are known, we have measured these gases in several MOR basalts in order to determine whether methane in 21°N fluids is extracted directly from basalt, or whether it is produced within the hydrothermal system, perhaps by high temperature basalt-seawater reactions.

Samples of fresh basalt glass were obtained with DSRV ALVIN in 21°N vent areas, from dredge hauls on the EPR at 8°S, and from the MAR "popping rocks" area (provided by R. Hekinian). Results of these analyses are shown in Table 4, together with data from the 21°N fluids. Considerable caution must be exercised in the interpretation of reactive gas data in rocks. Crushing methods are susceptible to production as well as self-adsorption of both hydrogen and methane within the crushing vessel (Roedder, 1972; unpublished data, this laboratory). The data in Table 4 represent samples that were crushed in a stainless steel ball mill for 5 minutes, and have been corrected for maximum possible blank contribution (the actual blank level depends on the extent of adsorption of the blank on the fresh rock surfaces, and is less than the blank determined with no sample in the ball mill).

Table 4 - Methane, hydrogen and helium concentrations in mid-ocean ridge basalt glass, obtained by crushing in vacuum. Analyses are blank corrected (see text). Hydrogen values are very preliminary because of high blanks.

Location	μcc(STP)/gram					
	CH_4	H_2	He	R/R_a*	$CH_4/^3He$ x 10^{-6}	H_2/CH_4
MAR basalt glass CY74 30-32A	52 (60%)+	n.m.‡	6.4	8.0	0.7	–
MAR "popping rock", DR11-1001A	280 (28%)	2700 (80%)	11.5	8.4	2.1	~9.5
EPR 8°S basalt glass, P6702-44	100 (19%)	n.m.‡	5.2	8.2	1.7	–
EPR 21°N basalt glass, 981-R23	13▽ (19%)	46▽ (85%)	0.46▽	8.2	2.5	~3.5
EPR 21°N, SWV hydrothermal fluid (same location as basalt sample 981-R23)	–	–	–	7.8	3.5	5.7-9.2

* Helium isotope ratio R = $^3He/^4He$; atmospheric ratio R_a = 1.4 x 10^6. The helium concentrations and ratios were measured by R. Poreda and W. Rison.

\+ Magnitude of blank correction indicated in parentheses.

‡ Blank correction = 100%.

▽ Only 20% of a very large rock sample was crushed in ball mill after five minutes. Actual gas concentrations are some five times higher than indicated.

These data represent lower limits on reactive gas concentrations in basalt, since they have not been corrected for self-adsorption. However, the volume of methane adsorbed on fresh silicate surfaces, as determined in separate experiments, indicates that these methane data may be low by a factor of two at most.

Based on these results, the $CH_4/^3He$ ratios found in EPR and MAR basalts are quite uniform and very close to the ratios observed in 21°N fluids, constituting the strongest evidence yet for the abiogenic origin of hydrothermal methane, derived from basalts. Comparing the $CH_4/^3He$ ratios in the 981-R23 glass and hydrothermal fluid from the same location (SWV), at least 70% of hydrothermal methane in this field appears to be basalt-derived.

Methane/^3He ratios as low as 0.7 x 10^6 have been observed in the Red Sea Brines (Welhan and Craig, 1982), and values up to 20 x 10^6 have been observed in hydrothermal plumes over the 21°S hydrothermal field on the EPR (Craig, 1981), indicating that MOR basalts may well contain variable methane concentrations relative to helium. This may explain the variations in CH_4/He found among the 21°N vent fields (Figure 3); similar variations in basalt hydrogen concentrations might also be the cause for the large variations in hydrogen concentrations observed in 21°N fluids, although the data of Table 4 do not permit a quantitative interpretation due to the large blank corrections.

Thus, from the preceding arguments, a significant organic or thermocatalytic source for 21°N methane can be ruled out, and the ^{13}C enrichment of this methane must represent either high temperature isotopic exchange with CO_2 within the basalt, or the original isotopic composition of methane in the upper mantle.

Variability of Carbon Isotope Ratios in Methane

The range in $\delta^{13}C(CH_4)$ indicated in Table 1 may be related to the degree of dilution of the hydrothermal fluid with ambient seawater, or may reflect a spatial trend along the ridge axis. Dilution with dissolved methane in ambient seawater appears unlikely, due to the low methane concentrations in seawater relative to the hydrothermal fluid. Assuming methane in ambient seawater is as light as -70 ‰ vs PDB (for a typical biogenic source; Schoell, 1980), the observed $\delta^{13}C(CH_4)$ vs $\delta^{18}O(H_2O)$ trend indicates that the ambient seawater methane concentration would have to be as high as ~ 0.02 cc(STP)/kg. Analyses of methane in hydrocast samples at the 21°N site, however, show bottom methane concentrations < 10^{-5} cc(STP)/kg (M. Lilley, personal communication).

Alternatively, the variation of $\delta^{13}C(CH_4)$ in these samples may represent a real trend along the ridge axis from north to south: samples 978 5/6 and 982 7/8 are from the NGS vent area, whereas 980 9/10 and 981 3/4 are from the SWV area. Sample 979 3/4 is from the OBS site, intermediate between the NGS and SWV areas. Thus, the most ^{13}C-enriched methane occurs in the NGS area, with lower $^{13}C/^{12}C$ ratios in the OBS and SWV fields. The $^{13}C/^{12}C$ ratio variation in methane and the differences in $CH_4/^3He$ ratios may be related: the NGS site contains the highest methane concentrations as well as the greatest ^{13}C enrichment in methane. A trend in $\delta^{34}S$ of vent sulfides has also been reported (Styrt et al., 1981), with the most primitive sulfur ($\delta^{34}S$ ~ 0 ‰) occurring in the northeastern part of the 21°N area, becoming progressively heavier southwestward along the ridge. One can speculate whether such trends represent local heterogeneity in the basalts, but we need to better establish the magnitude of the $\delta^{13}C(CH_4)$ variation as well as correlations of $\delta^{13}C(CH_4)$ with $\delta^{13}C(CO_2)$ and He in hydrothermal fluids along the ridge axis.

CONCLUSIONS

Methane and hydrogen concentrations in 21°N hydrothermal fluids are greater than 100,000 times normal deep ocean concentrations. This methane is characterized by large enrichments in ^{13}C, with maximum values of -15.0 to -16.6 ‰ vs PDB. Deuterium concentrations in methane and hydrogen are also relatively high, but essentially reflect isotopic equilibrium with hot seawater. The $CH_4/^3He$ mole ratio in the 350°C end-member fluid is $5.0 \pm 1.5 \times 10^6$. Ratios of $H_2/^3He$ range from 35 to 163×10^6. If these concentrations are representative of ridge crest hydrothermal fluids, the calculated average global injection rates of methane and hydrogen into the oceans indicate residence times of the order of 30 years, implying that consumption of these gases in the water column must be very rapid.

Measurements of methane and helium in samples of MOR basalt glass indicate $CH_4/^3He$ ratios of the same order of magnitude as observed in the 21°N hydrothermal fluids ($0.7 - 2.5 \times 10^6$), with similar H_2/CH_4 ratios. A basalt glass sample from the SWV vent field at 21°N had a $CH_4/^3He$ ratio of 2.5×10^6, very similar to the $CH_4/^3He$ ratio observed in SWV fluid (3.5×10^6), indicating that most of the hydrothermal methane is derived directly from the basalt.

The association of ^{13}C-enriched methane with ^3He and CO_2 derived from the ridge crest basalts, the lack of suitable organic or thermocatalytic sources of carbon, both in terms of magnitude and isotopic composition, and the similarity of $CH_4/^3He$ ratios in 21°N basalt and hydrothermal fluids provide conclusive evidence that methane in the 21°N system is of an abiogenic origin. Gas data on MOR basalts indicate that methane is derived directly from the basalts; further work is required to ascertain whether hydrogen may also be so derived. The absence of significant biogenic or thermogenic methane sources lends credibility to the very high isotopic equilibrium temperatures calculated for CH_4-CO_2 exchange, suggesting that carbon isotope exchange between these molecules may be occurring at very high temperatures in the basalt.

ACKNOWLEDGMENTS

We are indebted to the pilots of DSRV ALVIN for their skill in the collection of the hydrothermal vent samples, and to J. Archuleta of Los Alamos Scientific Laboratories for the development of the sampling bottles. We thank K. Kim and M. Lilley for assistance during the shipboard work, and J. Lupton and L. Wetherell for the analysis and tabulation of the He and Ne data. J. Edmond, K. von Damm and C. Measures supplied Mg data on these samples, and assisted in the diving program. This work was supported by National Science Foundation Grant OCE-7921392.

REFERENCES

Arnasson, B., and Sigurgeirsson, T., 1968, Deuterium content of water vapor and hydrogen in volcanic gas at Surtsey, Iceland, Geochim. Cosmochim. Acta, 32:797.

Ballard, R. D., Francheteau, J., Juteau, T., Rangan, C., and Normark, W., 1981, East Pacific Rise at 21°N: the volcanic, tectonic and hydrothermal processes of the central axis, Earth Planet. Sci. Lett., 55:1.

Chaigneau, M., Hekinian, R., and Cheminee, J. L., 1980, Magmatic gases extracted and analyzed from ocean floor volcanics, Bull. Volcanol., 43:241.

Chamberlin, R. T., 1908, The gases in rocks, Carnegie Inst. Wash. Publ., 106:1.

Clarke, W. B., Beg, M. A., and Craig, H., 1969, Excess ^{3}He in the sea: evidence for terrestrial primordial helium, Earth Planet. Sci. Lett., 6:213.

Claypool, G. E., and Kaplan, I. R., 1974, The origin and distribution of methane in marine sediments, in: "Natural Gases in Marine Sediments", I. R. Kaplan, ed., Plenum Press, New York.

Craig, H., 1953, The geochemistry of the stable carbon isotopes, Geochim. Cosmochim. Acta, 3:53.

Craig, H., 1963, The isotopic geochemistry of water and carbon in geothermal areas, in: "Proc. Spoleto Conference on Nuclear Geology", E. Tongiorgi, ed., Spoleto, Italy.

Craig, H., 1981, Hydrothermal plumes and tracer circulation along the East Pacific Rise: 20°N to 20°S, Trans. Am. Geophys. Union (EOS), 62:893.

Craig, H., Clarke, W. B., and Beg, M. A., 1975, Excess ^{3}He in deep water on the East Pacific Rise, Earth Planet. Sci. Lett., 26:125.

Craig, H., and Lupton, J. E., 1976, Primordial neon, helium and hydrogen in oceanic basalts, Earth Planet. Sci. Lett., 31:369.

Craig, H., Welhan, J. A., Kim, K., Poreda, R., and Lupton, J. E., 1980, Geochemical studies of the 21°N EPR hydrothermal fluids, Trans. Am. Geophys. Union (EOS), 61:992.

Craig, H., Weiss, R. F., and Clarke, W. B., 1967, Dissolved gases in the equatorial and South Pacific Ocean, J. Geophys. Res., 72:6165.

DesMarais, D. J., Donchin, J. H., Nehring, N. L., and Truesdell, A. H., 1981, Molecular carbon isotopic evidence for the origin of geothermal hydrocarbons, Nature, 292:826.

Dubrova, N. V., and Nesmelova, Z. N., 1968, Carbon isotope composition of natural methane, Geochem. Int., 5:872.

East Pacific Rise Study Group, 1981, Crustal processes of the mid-ocean ridge, Science, 213:31.

Edmond, J. M., Measures, C., McDuff, R., Chan, L. H., Collier, R., and Grant, B., 1979, Ridge crest hydrothermal activity and balances of the major and minor elements in the oceans: the Galapagos data, Earth Planet. Sci. Lett., 46:1.

Gold, T., 1979, Terrestrial sources of carbon and earthquake outgassing, J. Petrol. Geol., 1:3.

Gunter, B. D., and Musgrave, B. C., 1971, New evidence on the origin of methane in hydrothermal gases, Geochim. Cosmochim. Acta, 35:113.

Hekinian, R., Chaigneau, M., and Cheminee, J. L., 1973, Popping rocks and lava tubes from the Mid-Atlantic rift valley at 36°N, Nature, 245:371.

Hulston, J. R., and McCabe, W. J., 1962, Mass spectrometer measurements in the thermal areas of New Zealand, Part 2. Carbon isotope ratios, Geochim. Cosmochim. Acta, 26:399.

Lupton, J. E., and Craig, H., 1975, Excess ^{3}He in oceanic basalts: evidence for terrestrial primordial helium, Earth Planet. Sci. Lett., 26:133.

Lupton, J. E., Weiss, R. F., and Craig, H., 1977, Mantle helium in hydrothermal plumes in the Galapagos Rift, Nature, 267:603.

Lupton, J. E., Klinkhammer, G. P., Normark, W. R., Haymon, R., MacDonald, K. C., Weiss, R. F., and Craig, H., 1980, Helium-3 and manganese at the 21°N East Pacific Rise hydrothermal site, Earth Planet. Sci. Lett., 50:115.

Moore, J. G., Batchelder, J. N., and Cunningham, C. G., 1977, CO_2-filled vesicles in mid-ocean basalt, J. Volcanol. Geothermal Res., 2:309.

Pineau, F., Javoy, M., and Bottinga, Y., 1976, $^{13}C/^{12}C$ ratios of rocks and inclusions in popping rocks of the Mid-Atlantic Ridge, Earth Planet. Sci. Lett., 29:413.

Richet, P., Bottinga, Y., and Javoy, M., 1977, A review of hydrogen, carbon, nitrogen, oxygen, sulfur, chlorine stable isotope fractionation among gaseous molecules, Ann. Rev. Earth Planet. Sci., 5:65.

RISE Project Group, 1980, East Pacific Rise: hot springs and geophysical experiments, Science, 207:1421.

Roedder, E., 1972, The composition of fluid inclusions, U.S. Geol. Survey Prof. Paper, 440-JJ.

Sackett, W. M., 1978, Carbon and hydrogen isotope effects during thermocatalytic production of hydrocarbons in laboratory simulation experiments, Geochim. Cosmochim. Acta, 42:571.

Schoell, M., 1980, The hydrogen and carbon isotopic composition of methane from natural gases of various origins, Geochim. Cosmochim. Acta, 44:649.

Seiler, W., and Schmidt, U., 1974, Dissolved nonconservative gases in seawater, in: "The Sea", Volume 5, E. D. Goldberg, ed., J. Wiley and Sons, New York.

Stahl, W. J., 1977, Carbon and nitrogen isotopes in hydrocarbon research and exploration, Chem. Geology, 20:121.

Styrt, M. M., Brackmann, A. J., Holland, H. D., Clark, B. C., Pisutha-Arnond, V., Elridge, C. S., and Ohmoto, H., 1981, The mineralogy and the isotope composition of sulfur in hydrothermal sulfide/sulfate deposits on the East Pacific Rise, 21°N latitude, Earth Planet. Sci. Lett., 53:382.

Weiss, R. F., 1970, The solubility of nitrogen, oxygen and argon in water and seawater, Deep Sea Res., 17:721.

Weiss, R. F., 1971, Solubility of helium and neon in water and seawater, J. Chem. Eng. Data, 16:235.

Weiss, R. F., Lonsdale, P., Lupton, J. E., Bainbridge, A. E., and Craig, H., 1977, Hydrothermal plumes in the Galapagos Rift, Nature, 267:600.

Welhan, J. A., 1980, Gas concentrations and isotope ratios at the 21°N EPR hydrothermal site, Trans. Am. Geophys. Union (EOS), 61:996.

Welhan, J. A., 1981, Carbon and hydrogen gases in hydrothermal systems: the search for a mantle source, Ph.D. thesis, University of California at San Diego, 195 pp.

Welhan, J. A., and Craig, H., 1979, Methane and hydrogen in East Pacific Rise hydrothermal fluids, Geophys. Res. Lett., 6:829.

Welhan, J. A., and Craig, H., 1982, Abiogenic methane in mid-ocean ridge hydrothermal fluids, in: "Proc. Deep Source Gas Workshop", W. J. Gwilliam, ed., Morgantown, West Virginia.

Welhan, J. A., Poreda, R., Lupton, J. E., and Craig, H., 1980, Gas chemistry and helium isotopes at Cerro Prieto, Geothermics, 8:241.
Williams, P. M., Smith, K. L., Druffel, E. M., and Linick, T. W., 1981, Dietary carbon sources of mussels and tubeworms from Galapagos hydrothermal vents determined from tissue ^{14}C activity, Nature, 292:448.
Zolotarev, G. I., Voytov, I. S., Sarkisyan, I. S., and Cherevichnaya, L. F., 1981, Doklady Akad. Nauk SSSR, Earth Sci. Sec., 243:207.

REDUCED GASES AND BACTERIA IN HYDROTHERMAL FLUIDS:

THE GALAPAGOS SPREADING CENTER AND 21°N EAST PACIFIC RISE

Marvin D. Lilley, John A. Baross and Louis I. Gordon

School of Oceanography
Oregon State University
Corvallis, OR 97331

ABSTRACT

Hydrothermal fluids at the Galapagos Spreading Center (GSC) and at 21°N on the East Pacific Rise were enriched in methane, hydrogen and carbon monoxide by orders of magnitude over ambient bottom water. Nitrous oxide showed both enrichment and depletion in warm vent waters. Each GSC vent field exhibited unique dissolved gas to silica ratios indicating that complex source - sink mechanisms operated within a small geographic region. The $CH_4/^3He$ ratio at 21°N was 6.1 x 10^6 whereas at the GSC, a range of 12.4 to 42 x 10^6 was seen. At 21°N the $H_2/^3He$ ratio was on the order of 100 times that at the GSC. Microbial data showed as many as 10^9 organisms ml^{-1} in GSC samples and 10^5 ml^{-1} in the hot 21°N waters. These microbial communities are complex and include organisms known to produce and consume the gases discussed here. We conclude that microbial activity in the warm GSC vents is a significant contributing factor in determining the final gas concentrations in the vent waters.

INTRODUCTION

The hydrothermal circulation systems associated with mid-ocean ridges profoundly influence the chemistry of the oceans and the oceanic crust (Corliss et al., 1979; Edmond et al., 1979a, 1979b, 1982). Data have been presented indicating that the chemical properties associated with these ridges may be greatly influenced by microorganisms (Corliss et al., 1979; Jannasch and Wirsen, 1979, 1981; Karl et al., 1980).

The detection of an unusually high $^3He/^4He$ ratio in hydrothermal plumes above the ridge crest at the Galapagos Spreading Center (Lupton et al., 1977) and the subsequent direct sampling of these hot spring emanations with anomalously high ^3He concentrations (Jenkins et al., 1978) indicated that other volatiles may be present in hydrothermal fluids. Subsequently, methane and hydrogen were detected in diluted samples of the hot (350°C) fluids flowing out of the sulfide chimneys at 21°N on the East Pacific Rise (Welhan and Craig, 1979). Samples collected directly from warm water (7-23°C) vents at the Galapagos Spreading Center (GSC) by the research submersible ALVIN showed elevated concentrations of methane, hydrogen, carbon monoxide and nitrous oxide (Lilley and Gordon, 1979; Lilley et al., 1982). Recent work has shown that obligately thermophilic microorganisms associated with the hot vent waters at 21°N can produce H_2, CH_4 and CO (Baross et al., 1982a). Thus microbial production must be added to water-rock interactions, magma equilibria and direct injection from the mantle as possible sources of these gases.

We present here methane, hydrogen, carbon monoxide and nitrous oxide concentration data from the GSC and 21°N along with microbiological data indicating the diversity of the microbial communities associated with these vent waters, including species of organisms capable of producing and consuming all the gases discussed.

SAMPLING AND METHODS

Samples were taken from three vent fields at the Galapagos during dives 898-905 of the DSRV ALVIN in March, 1979. The locations of these vent fields (Rose Garden, Mussel Bed and East of Eden) and

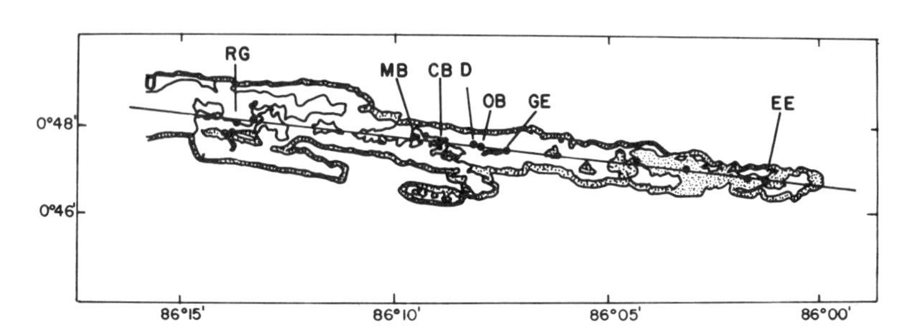

Fig. 1 Locations of the vent fields sampled at the Galapagos Spreading Center. (After Ballard et al., 1982).

CB - Clambake, 1977
D - Dandelions, 1977
OB - Oyster Bed, 1977
GE - Garden of Eden, 1977
RG - Rose Garden, 1979
MB - Mussel Bed, 1979
EE - East of Eden, 1979

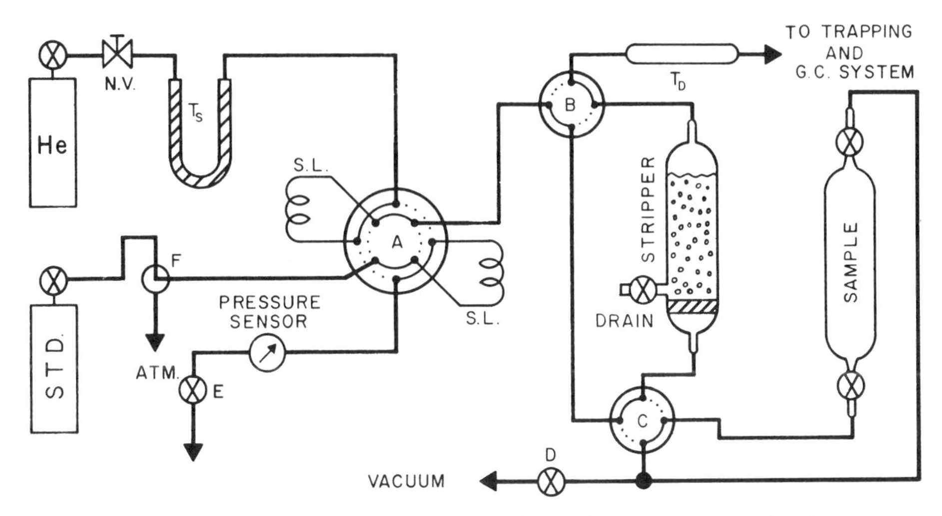

Fig. 2. Stripping system. The labeled components include: (A) Micro volume valve with 0.5 and 2 cc sample loops, Carle; (B & C) Teflon® 4-way valves, Fluorocarbon; (D & E) stainless steel shut-off valves, Whitey; (F) stainless steel 3-way valve, Whitey; (NV) ultra fine metering valve, Nupro; (T_S) liquid nitrogen cooled trap (6.4 mm x 30.5 cm) of Molecular Sieve® 5A to reduce H_2 concentration in He used as stripping gas; (T_D) a series of 3 glass tubes (11 mm x 24.5 cm) containing 1) Drierite to remove water, 2) Molecular Sieve 3A to further reduce the amount of water in the gas stream, 3) Ascarite® to remove CO_2 ; (Pressure Sensor) 0-15 psi semiconductor type, National Semiconductor Corp., calibrated by barometer.

their relationship to the vents sampled in 1977 are shown in Figure 1. The sampling system used for these dives was essentially that described by Corliss et al., (1979). The downstream sensors for temperature, conductivity, dissolved oxygen and pH were not present, however. Since these samplers were constructed of acrylic which in contact with water produces significant quantities of hydrogen (Lilley and Gordon, unpublished data) samples for dissolved hydrogen were taken in newly fabricated PVC samplers. All samples for gas analysis were subsampled into 125 ml tubular Pyrex® flasks with Teflon® stopcocks at either end and poisoned to 0.4 mM with $HgCl_2$.

The 21°N samples were taken during dives 978-982 of the DSRV ALVIN in November, 1979. The locations of the three vent fields sampled during these dives are given by Welhan and Craig (this volume). The water samples were collected in evacuated, gold-plated stainless steel cylinders fitted with valves at both ends. The samplers were constructed at the Los Alamos Scientific Laboratories. Subsamples of the hydrothermal fluids were taken in 20 ml tubular Pyrex® flasks with Teflon® stopcocks at either end and poisoned to 3 mM with $HgCl_2$. For both the 21°N and GSC samples, the analysis procedure consisted of stripping out the dissolved gases with helium, cryogenic trapping, then desorption and detection by gas chromatography. A schematic of the stripping system is shown in Figure 2. The stripping vessel was made of Pyrex® with a fritted glass disk at the bottom to disperse the stripping gas (ultra high purity He - U.S. Bureau of Mines).

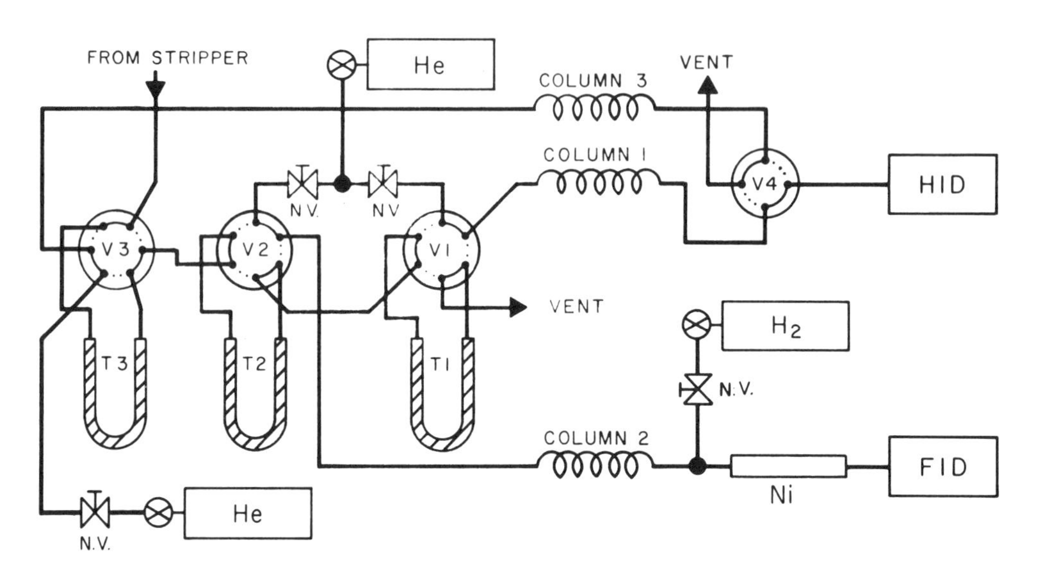

Fig. 3. Configuration of the gas chromatograph. The labeled components are:
(V1-V4) Mini Volume GC valves, Carle; (T1) liquid nitrogen cooled trap of Molecular Sieve 5A, 3 mm x 30 cm stainless steel; (T2) dry-ice/acetone cooled trap half filled with both Molecular Sieve 5A and activated charcoal to ensure retention of both CH_4 and CO, 3 mm x 18 cm stainless steel; (T3) ice water cooled trap filled with Molecular Sieve 13X, 3 mm x 18 cm stainless steel; (Column 1) 3 mm x 6 m stainless steel column filled with Molecular Sieve 5A, 25°C; (Column 2) 3 mm x 2 m stainless steel column filled with Molecular Sieve 5A, 100°C; (Column 3) 3 mm x 46 cm stainless steel column filled with Molecular Sieve 5A, 250°C; (Ni) nickel oxide on firebrick, 3 mm x 10 cm stainless steel, 360°C; (H_2) hydrogen gas for reducing CO to CH_4 over Ni catalyst for detection by FID; (He) helium carrier gas, U.S. Bureau of Mines ultra high purity; (HID) Beckman helium ionization detector; (FID) Fisher flame ionization detector.

The connecting tubing through which the sample flowed and that portion connecting the stripper exit to valve B and T_D was high pressure nylon (0.32 cm o.d.). All other connecting tubing was stainless steel (either 0.16 or 0.32 cm o.d.) and all connecting fittings were stainless steel or nylon (Swagelok®). With Valve C in the position shown in Figure 2, a sample could be connected to the system while maintaining He flow through the stripper. Once the sample was in place, the connecting tubes were evacuated by opening Valve D. During this time period the stripping vessel, still under He pressure, was drained. The new sample could then be transferred to the stripper by closing Valve D, rotating Valve C to put the stripping gas in line with the sample flask and then opening the stopcocks on the sample flask. This pushed the sample into the stripper, bubbled He through the sample and carried the stripped gases to the subsequent drying and trapping stages. The samples were stripped for 10 min. at a flow rate of 125 ml min^{-1}.

The configuration of the gas chromatographic system used on the 21°N Cruise is shown in Figure 3. This multitrap approach separated N_2O, CH_4 and CO, and H_2 onto traps 3, 2, and 1, respectively. The gases were then manually valved into the appropriate column and detector. Hydrogen analysis was accomplished by removing the liquid nitrogen dewar and switching V1 to connect T1 in series with Column 1 and the HID for 90 sec. This was sufficient time for H_2 to elute from the trap but prevented any O_2 or N_2 from reaching the HID. The detection of CH_4 and CO was done simultaneously with that of H_2 by switching V2 to connect T2 in series with Column 2 and heating T2 to 250°C with an aluminum block heater. Carbon monoxide was reduced to CH_4 over the Ni catalyst and detected by the FID. After the H_2 peak was recorded V4 and V3 were switched to connect T3 with Column 3 and the HID. The trap was then heated to 250°C with the block heater to elute and detect N_2O. The gas chromatograph was a Beckman GC-55; the peaks were recorded on a Leeds and Northrup Speedomax® recorder and integrated with Hewlett Packard 3373B digital integrators.

The 1979 Galapagos gas data were collected using a two trap version of the system described above for H_2, CH_4 and CO while N_2O was determined on a separate system by the method of Cohen (1977).

Many of the 21°N samples formed a gas phase within the metal sample containers which made it necessary to apply corrections to the fluid phase data to compensate for partitioning into the gas phase. The manner in which this was done is described by Welhan and Craig (this volume). Since we did not ourselves analyze for species which could be considered conservative (N_2, Ar, Ne), we made these corrections using Ar and Ne data provided by J. Welhan (pers. comm.). We consider this to be acceptable since our samplers and his were connected in series during the sampling process and duplicate samplers of Welhan's connected similarly in series showed N_2/Ar ratios which varied by no more than 3% (J. Welhan, pers. comm.).

This does not rule out, of course, the possibility that random bubble formation in the sampler train could have affected our results on occasion. Bubble formation was noted during the sampling process but care was taken to dislodge the bubbles and flush the samplers prior to enclosing the sample. However, this process may account for some of the variability between our 21°N hydrogen and methane values and those of Welhan and Craig (this volume). Due to the high solubility of nitrous oxide, its concentration should be only minimally affected by bubble formation.

Methods and reagents used to enumerate bacteria from water samples, as well as procedures for measuring the ability of bacteria to denitrify, nitrify, fix N_2, and produce and consume various trace gases have been described (Baross, et al., 1982a; 1982b). Methane oxidizing bacteria from the water column at 21°N were enumerated using the spread plate technique for dilutions of 0.1 ml and 0.01 ml whereas 1, 10, and 50 ml samples were concentrated on 0.2 μm Nuclepore® membrane filters using sterile techniques. The filters were then placed on the surface of agar plates consisting of a minimal salts-trace element mix with KNO_3 as a nitrogen source (see Baross, et al., 1982a for details of this medium). The plates were placed in vented anaerobic jars containing 50% air and 50% methane and incubated at 5°C. Representative isolates from each water sample were tested for their ability to oxidize CH_4 using ^{14}C-CH_4 and the procedures already described (Baross et al., 1982a).

Samples prepared for electron microscopy were fixed in sterile artificial seawater containing 2.5% electron microscope grade gluteraldehyde (Tousimis) within minutes after the Alvin surfaced. Samples prepared for scanning electron microscopy were dried by the critical point method, mounted on aluminum stubs and coated under a vacuum with a layer of gold 10-20 nm in thickness. The samples were viewed with an AMR model 1000 scanning electron microscope. Samples prepared for transmission electron microscopy were fixed with gluteraldehyde, and after fixation resuspended in 0.2 M cacodylate buffer and postfixed with 1% uranyl acetate. These samples were then sectioned, stained with 1% OsO_4 and examined with a Philips Electron Microscope Model 300.

RESULTS

A summary of the 1979 GSC data is presented in Table 1. Due to uncertainties in the measured sample temperatures, the temperatures listed were calculated with the Si-T relationship ($T = -0.15 + 0.0160$ Si) determined from the 1977 dive data (Edmond et al., 1979a). The ^3He data were also calculated; ^3He analyses were not done on the 1979 samples. For this calculation we assumed that ^3He in these vents did not differ appreciably from the values obtained by Jenkins et al., (1978) for the vents sampled in 1977. We used the ^3He-T relationship

Table 1. Properties of the 1979 Galapagos vent samples.

Dive/ Sampler	Temp (°C)	Silicic Acid (μM)	O_2 (μM)	$NO_3^- + NO_2^-$ (μM)	H_2S (μM)	N_2O (nM)	CH_4 (μM)	CO (nM)	^3He (10^{-14}M)	H_2 (nM)
					A. Mussel Bed					
898/B7	4.1	275	84	31.0	0	33.5			3.8	
898/B4	3.9	259	114	32.4	10	31.6	0.429		3.5	
898/A3	6.3	402	48	22.0	18	67.9	0.953		7.7	
898/B2	6.2	400	60	23.2	15	44.7	1.145		7.5	
898/B3	7.2	457	36	19.3	0				9.3	
898/B6	2.9	201	106	36.3	3	25.5	0.183		1.7	
902/B4	9.6	605	0	10.2	30				13.5	
902/B3	10.0	626	0	9.9	35	88.0	2.236	37.1	14.2	
902/A4	7.7	490	43	18.5	20				10.1	
902/B7	9.2	575	0	13.3	30	85.1	2.040	39.8	12.8	
902/B2	7.7	486	13	18.4	20	71.9	1.637	28.1	10.1	
902/B5	7.7	486	10	17.8	20	77.4	1.566	48.2	10.1	
902/A3	6.9	443	55	20.8	15	73.0		31.2	8.7	
904/B5	9.4	588	0	9.8	27	109.5	2.040	35.7	13.1	
904/B7	8.9	561	3	12.1	24	99.5	1.959	46.5	12.2	
904/B4	8.0	507	15	17.7	21	85.3	1.709	23.2	10.7	
904/B3	8.1	514	35	17.1					10.8	
904/B2	9.5	595	0	11.2	32	89.7	2.044	30.8	13.3	
904/A4	3.6	243	140	32.5					3.0	
904/H3	9.1	574	11	13.1	27	80.1	1.941		12.6	71.8
904/H1	2.6	185	138	37.5		18.7	0.094		1.2	9.6
					B. East of Eden					
901/A9	2.2	161	117		>1	19.0	0.031	11.2	0.5	
901/A2	2.2	161	119		0.5	19.9	0.049	48.2	0.5	
901/A7	2.2	160	115		0.9	20.7	0.112	72.4	0.5	
901/A12	2.3	167	109			19.9	0.103	16.5	0.7	
901/A6	5.2	336	0		142	14.5	1.990	51.8	5.8	
901/A5	5.6	360	0		182	13.0	3.230	76.8	6.5	

(Continued)

Table 1. (Continued)

Dive/ Sampler	Temp (°C)	Silicic Acid (μM)	O_2 (μM)	$NO_3^-+NO_2^-$ (μM)	H_2S (μM)	N_2O (nM)	CH_4 (μM)	CO (nM)	^3He (10^{-14}M)	H_2 (nM)
					C. Rose Garden					
899/A5	9.9	619	0	20.7	130	22.8			14.0	
899/A6	9.5	596	0	21.5	130				12.3	
899/A7	9.4	587	0	21.9	140	23.7	1.364		13.1	
899/A1	8.1	513	4	24.8	75				10.8	
899/A9	8.7	550	0	22.4	130	36.7	1.288		11.9	
899/A13	7.6	479	0	25.7	70				10.0	
899/A12	6.1	395	19	20.1	50	20.6	0.787		7.3	
899/A2	4.6	303	58	32.8	55	28.1		59.8	4.7	
900/B2	13.8	854	0	7.8	150	30.0	2.849	60.7	20.8	
900/B6	13.7	846	0	7.5	230				20.6	
900/A3	10.0	625	0	14.9	160	24.5	1.838	57.6	14.2	
900/B7	5.4	348	34	24.5	90	29.7	0.742	48.2	6.1	
900/B4	9.7	608	0	13.1	140	35.2	1.736	51.4	13.6	
900/B3	4.9	322	54	27.9	90	26.7	0.662	52.7	5.2	
900/B5	4.2	279	66	28.3	80	21.3	0.425	42.9	4.0	
903/A2	8.9	558	0	19.5	120	37.4	1.463		12.2	
903/A7	13.5	837	0	11.9	120	25.0	2.568		20.3	
903/A12	16.2	994	0	5.3					25.0	
903/A6	15.0	927	0	7.8	195	24.3	3.024		22.9	
903/A1	10.2	636	0	16.2	120	31.5	1.776		14.5	
903/A9	11.6	724	0	13.2	140	38.8	2.192		7.0	
903/H2	6.6	420	26	26.7	110	26.7			8.2	10.9
905/A13	16.4	1007	0	2.7	250	53.1	3.449		25.4	
905/A2	12.4	769	0	12.7					18.4	
905/A6	3.5	238	33	26.4	30	28.9	0.662		2.8	
905/A12	5.8	372	28	27.4	85	21.6	0.850		6.8	
905/A1	10.0	627	0	19.9	160	23.4	1.919		14.2	
905/A9	9.5	596	0	21.3	135	23.2	1.776		13.3	
905/H4	5.3	343	317	31.6	0				5.9	
Ambient	2.2	160	120	39.1	0	20.8	0.0003			0.4

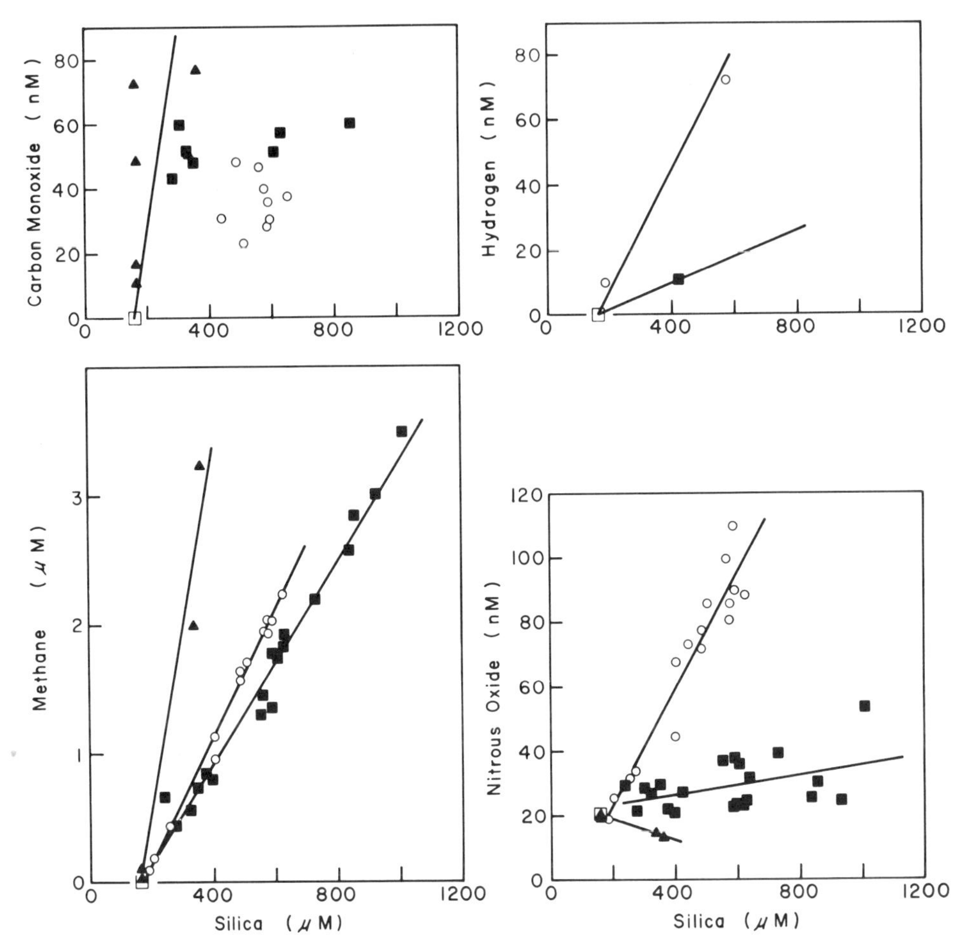

Fig. 4 Dissolved gas concentrations vs Si for the 1979 GSC vent fields. The fields are represented by symbol: ■ = Rose Garden, O= Mussel Bed, ▲ = East of Eden. This figure reproduced from Lilley et al. (1982) by courtesy of Nature.

from their Figure 2 (as East of Eden was near Garden of Eden, we used the slope which included Garden of Eden) and the temperature calculated from Si to produce the estimated ^3He values for our samples. The Si, O_2, NO_3^- + NO_2^- and H_2S data were provided by J. Edmond (pers. comm.).

Since silica provides a good measure of the degree of dilution of the hydrothermal end-member with ambient seawater (Corliss et al., 1979), the dissolved gas concentrations from Table 1 are plotted against silica in Figure 4.

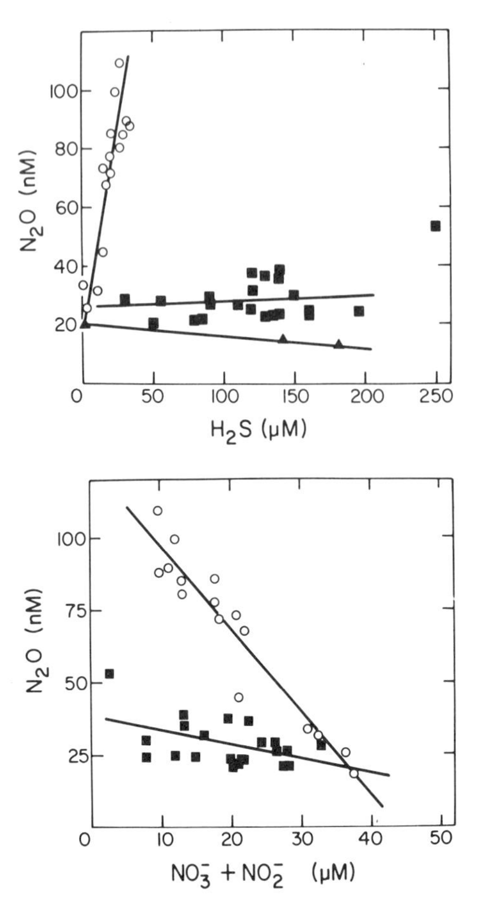

Fig. 5 Nitrous oxide at the GSC vs (a) hydrogen sulfide; (b) nitrate and nitrite. Symbols as in Figure 4.

The GSC methane and nitrous oxide data (Fig. 4) showed clearly that each of the vent fields sampled followed a unique dilution line. The few hydrogen samples taken with noncontaminating PVC samplers also indicated considerably different dilution lines for the two vent fields sampled. The carbon monoxide data showed considerable scatter but also indicated differences among the vent fields. Note that the Rose Garden data did not follow a dilution line between a high temperature end-member and ambient seawater. This may indicate that the carbon monoxide concentration in this vent field is not governed by a hydrothermal end-member but is produced in more shallow regions of the convection system. The concentration of nitrous oxide in the marine environment is closely related to the redox conditions and the concentrations of other nitrogen species (Goering, 1978). Figure 5 depicts the relationships between N_2O and H_2S and N_2O and nitrate + nitrite.

Chemical and microbiological data obtained during the 1977 diving expedition to the GSC are included here in Tables 2 and 3 for comparison purposes and to aid in later discussion. These samples were analyzed for molecular hydrogen; however, the acrylic samplers used in 1977 resulted in high and uncertain amounts of hydrogen being added to the samples. An indication of elevated hydrogen levels associated with the Garden of Eden vent area was detected, however, by transponder navigation of a bottom-contact hydrocast with a non-contaminating "kamikaze" near-bottom sampler similar to that described by Schink et al., (1966). The results of this cast are shown in Figure 6.

The microbial counts in these water samples were approximately 10^9 ml^{-1} (Table 3). The difference between these counts and the much lower (10^6 ml^{-1}) levels of organisms seen at the GSC by Karl et al., (1980) may be due to differences in sampling techniques. The 1977 samples were taken by extending the inlet tube as far into the vent opening as possible, thus increasing the likelihood of dislodging and collecting microbial mats with the water samples. Rock and animal surfaces near the vents also showed heavy microbial colonization as indicated for the surfaces of a mussel shell in Figure 7 and a limpet in Figure 8.

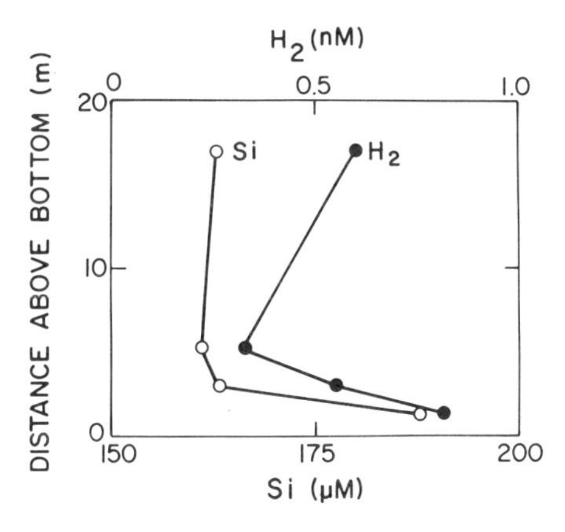

Fig. 6 Near bottom gradient in H_2 and Si seen by hydrocast at the Garden of Eden site.

Table 2. Properties of the 1977 Galapagos vent samples.

Dive Number	Sample	T (^{0}C)	Si (μM)	O_2 (μM)	H_2S (μM)	NO_3^- (μM)	NO_2^- (μM)	NH_3 (μM)
				A. Clambake				
715	A5	7.3	461	16	7	10.1	0.79	0.9
	A6	6.9	426	26	6	13.6	0.64	0.9
	A7	7.9	516	14	8	10.4	0.69	0.1
	B1	5.4	334	68	4	24.1	0.24	0.8
	B2	4.6	317	62	4	24.4	0.32	0.7
	B3	4.0	255	81	2	30.3	0.15	0.9
	B4	7.5	470	12	7	8.1	0.86	1.0
717	B1	8.5	564	9	11	11.3	0.86	>1.7
	B2	8.5	597	16	9	12.9	0.78	>1.9
	B3	6.9	464	25	7	16.4	0.78	1.7
	B4 *	2.2	173	104	0	37.2	0.06	0.4
	B6 *	2.2	157	104	0	39.2	0.03	0.0
719	A1	8.0	505	18	2	12.1	0.82	2.1
	B4 *	2.0	163	108	0	38.5	0.02	0.0
722	B1	8.6	580	4	8	6.4	0.81	1.9
	A4	8.6	581	3	9	6.5	0.92	1.9
727	B1	9.3	590	0	10	6.2	0.80	1.0
	B2	9.3	582	3	10	7.6	0.78	1.0
	B5	-	424	39	6	19.9	0.47	0.9
				B. Oyster Bed				
723	A1	6.8	401	0	59	4.1	0.35	0.3
	B7	6.8	409	0	82	1.9	1.12	0.0
726	A5	8.3	491	0	125	4.0	-	0.0
	B4	8.3	484	0	100	5.9	2.51	0.0
	B8	6.6	386	6	55	14.8	0.37	0.0
	B7	9.1	549	0	145	-	-	0.0
	A3	6.1	353	16	50	17.8	0.52	0.0
	A6	9.1	548	0	145	0.4	-	0.0

Dive Number	Sample	T (^0C)	Si (μM)	O_2 (μM)	H_2S (μM)	NO_3^- (μM)	NO_2^- (μM)	NH_3 (μM)
				C. Garden of Eden				
728	A2	2.9	208	26	3	35.3	0.00	0.5
	A8	2.9	208	28	0	34.4	0.00	0.2
	B8	9.7	631	0	67	9.8	0.00	0.0
	A6	9.7	631	0	60	9.7	0.00	0.0
	A7	10.8	632	0	64	9.2	0.00	0.0
	B7	10.8	639	0	64	9.1	0.00	0.0
	A3	12.0	677	0.4	71	7.7	0.00	0.0
	B4	12.0	714	0	75	5.8	0.00	0.0
730	B4	-	698	0	68	8.2	1.23	0.0
	B1	-	480	30	38	19.6	0.26	0.0
	B2	6.0	471	33	37	23.3	0.24	0.0
735	B6	10.4	615	0	55	9.2	1.50	0.0
	A4	10.4	636	0	55	36.5	1.40	0.0
	B3	12.0	703	0	56	2.6	1.05	0.0
	A7	12.0	709	0	58	12.4	1.11	0.0
				D. Dandelions				
716	A2	6.0	327	10	0	11.3	1.28	0.9
	A3	6.6	372	48	0	21.9	0.31	0.5
	A4	6.4	364	52	0	22.5	0.30	0.5
	A8 *	2.1	161	110	0	39.5	0.03	0.0
	B5	6.6	370	52	0	21.9	0.31	0.5
	B8	6.6	372	51	0	21.9	0.32	0.5
723	A3	6.2	325	13	0	10.8	1.39	0.5
	A5	4.0	224	83	0	33.7	0.07	0.0
	A6	4.8	265	64	0	28.7	0.11	0.2
	A8	4.0	229	81	0	32.5	0.08	0.0

* denotes an "ambient" sample

Table 3. Summary of bacterial counts found in, and cultured from, warm vent waters of the Galapagos Spreading Center (1977 Cruise).

	Number of Bacteria ml^{-1}					
		Aerobic		Anaerobic		
Sample Source	Total Direct counts	Heterotrophs[1]	S/Mn Oxidizers[2]	Grown in $H_2 + CO_2$[3]	SO_4 Reducers[4]	N_2 fixers[5]
Clambake	approx. 1×10^9	not detected	not detected	1.1×10^5	3.6×10^4	not detected
Garden of Eden	approx. 1×10^9	2.0×10^7	1.2×10^7	not tested	not tested	not detected
Garden of Eden	4.0×10^8	8.0×10^6	6.8×10^6	2.2×10^6	1.0×10^2	$<10^3$
Sediment mounds[6]	8.0×10^7	1.1×10^7	7.8×10^6	5.8×10^4	5.0×10^2	approx. 10^4

[1] Many of the S/Mn oxidizing bacteria can grow heterotrophically.

[2] Medium as described in Baross et al. (1982a).

[3] Medium was MS (Baross et al., 1982a) with NO_3^- incubated in an atmosphere of $H_2 + CO_2$.

[4] Sulfate reduction indicated by FeS formation.

[5] N_2 fixing bacteria from heterotrophic and chemolithotrophic S/Mn medium; confirmed by acetylene reduction.

[6] This sample was taken by shoving the inlet of the pumping system described by Corliss et al. (1979) beneath the surface of a sediment mound.

Fig. 7. SEM of the surface of a GSC mussel shell showing large bacterial forms and a dense population of small rod shaped bacteria. Bar is 3 µm.

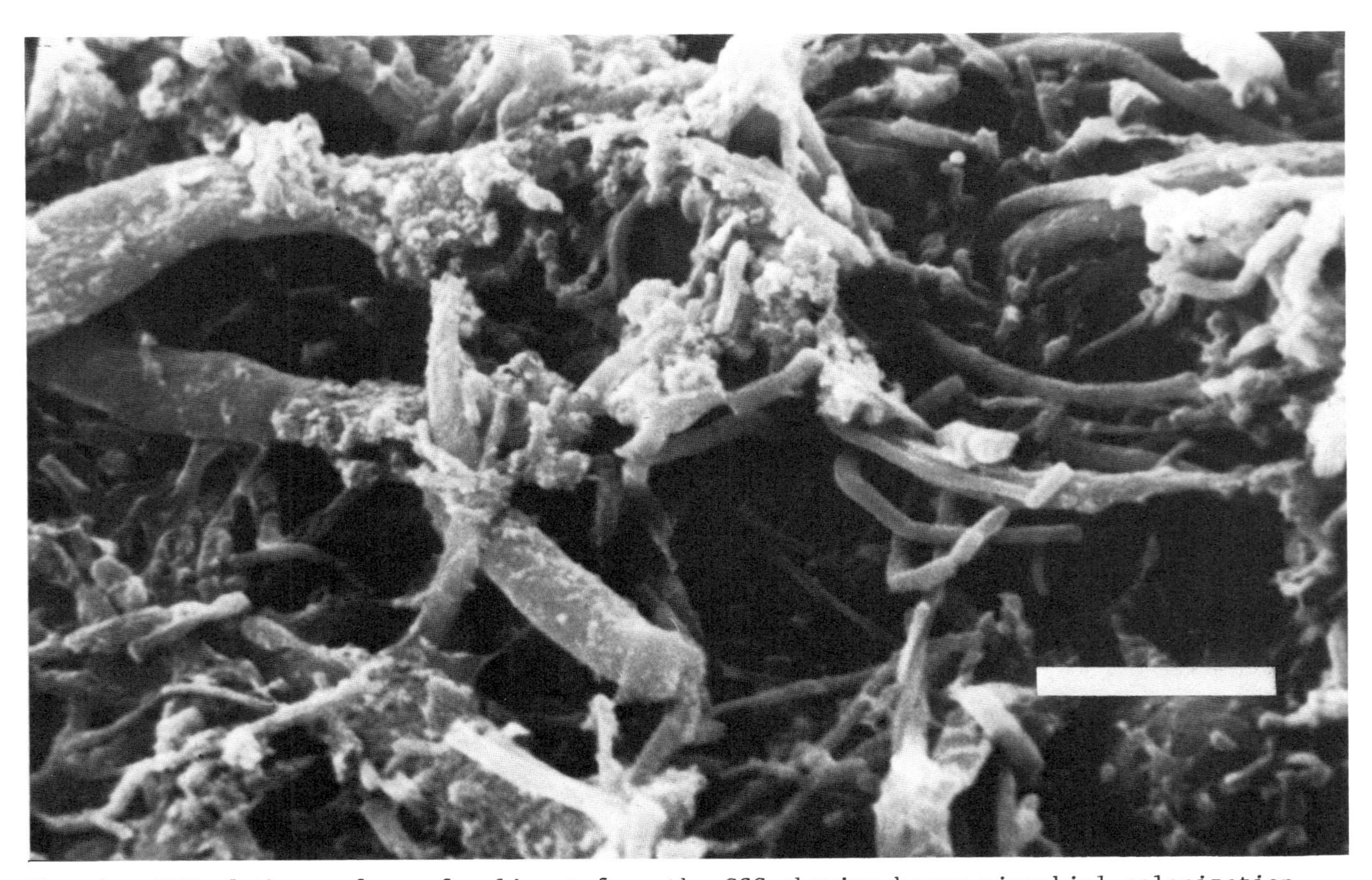

Fig. 8. SEM of the surface of a limpet from the GSC showing heavy microbial colonization. Bar is 5 μm.

Table 4. Chemical properties of the 21°N samples.

Dive/Sample		Vent Field[1]	Temp.[2] (°C)	Mg^2 (mM/kg)	CH_4 (μM)	H_2 (μM)	CO (nM)	N_2O (nM)
978	Bag[3]	NGS	2	52.44	0.5	0.02	3	18
978	7/8	NGS	40	46.67	8.8	-	56	nd
979	3/4	OBS	152	29.83	30.2	639	134	nd
979	1/2	OBS	2	52.47	0.1	13.5	8	27
979	11/12	OBS	2	52.84	0.2	(5.3)[4]	26	22
980	9/10	SW	121	34.52	22.0	54.3	8	13
980	5/6	SW	27	48.59	6.7	23.1	16	41
981	3/4	SW	55	44.31	12.7	(10.3)[4]	102	9
981	1/2	SW	2	53.03	0.2	3.1	-	171
981	11/12	SW	15	50.35	3.5	50.0	32	44
982	7/8	NGS	304	7.02	53.4	562	95	nd
AMBIENT			1.7	52.31	0.3nM	0.36nM	0.3	20

[1] NGS = National Geographic Society site.
OBS = Ocean Bottom Seismometer site.
SW = Southwest site.

[2] These temperatures were calculated from the relationship:
$Tmg = Ta + \Delta\ Mg/0.15\ \frac{m\ mol\ Mg}{°C}$; where Tmg = sample temperature, Ta = ambient temperature (1.7°C), Δ Mg = Mg ambient - Mg measured, Mg ambient = 52.31 mM. This relationship and the Mg ambient value were provided by Russell McDuff and John Edmond. The Mg data were taken from McDuff and Edmond (1982).

[3] Niskin sterile bag sampler filled at a "Galapagos type" warm water vent.

[4] These are minimum values due to GC electrometer saturation.

The 21°N dissolved gas data are presented in Table 4. Precipitation of silica within the samplers caused uncertainty in the 21°N silica concentrations. Thus in Figure 9 the dissolved gas concentrations are plotted against $\delta^{18}O$ as a measure of dilution of the hydrothermal end-member with ambient seawater. Methane showed less variation between vent sites than was the case at the Galapagos vents. Nitrous oxide was anomalously high at the SW site and hydrogen appeared to follow different trends at the three sites. Carbon monoxide showed a general tendency toward increasing concentration with increasing fraction of hydrothermal fluid.

Based upon the 21°N data, Welhan and Craig (this volume) calculate that the hydrothermal methane flux is sufficient to replace the

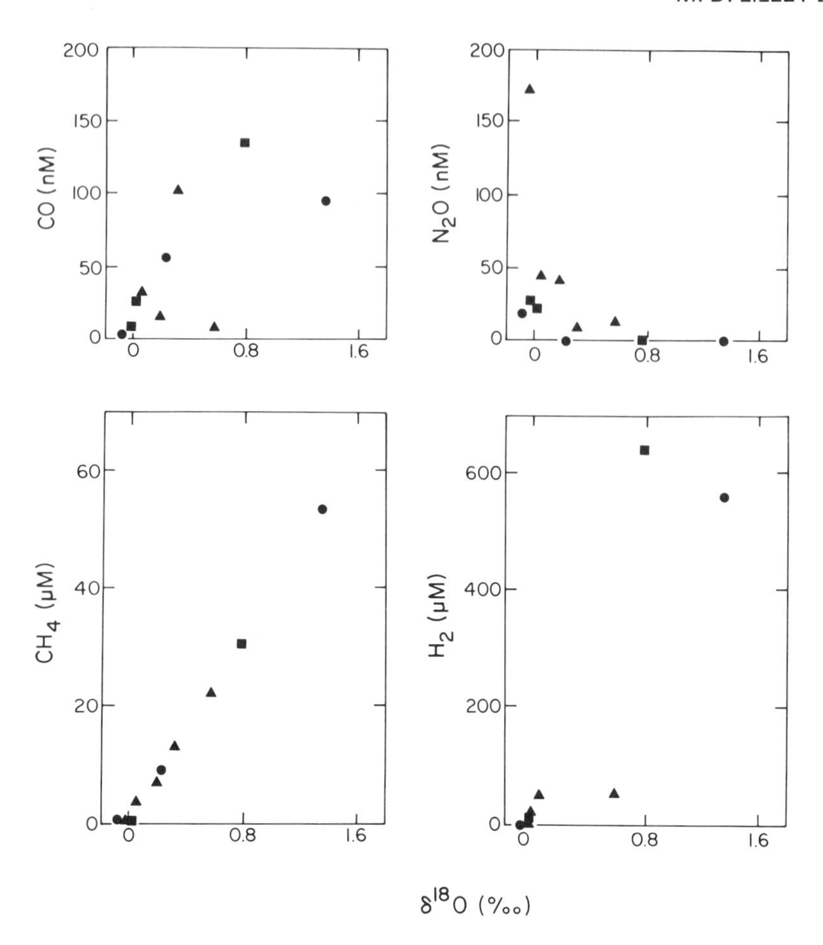

Fig. 9 Dissolved gas concentrations vs $\delta^{18}O$ of water from the 21°N vents. The fields are represented by symbol: ● = NGS, ■ = OBS, ▲ = SW. The $\delta^{18}O$ data were provided by John Welhan and Harmon Craig (pers. comm.).

deep-sea methane in 25 years. As they state, this strongly implies rapid bacterial methane oxidation below the thermocline. Craig (1981) also reported that methane was unobservable 300 km west of the East Pacific Rise at 20°S whereas directly above the ridge crest plumes containing approximately 9 nM were detected. In Figure 10 we show that significant numbers of methane oxidizing bacteria were

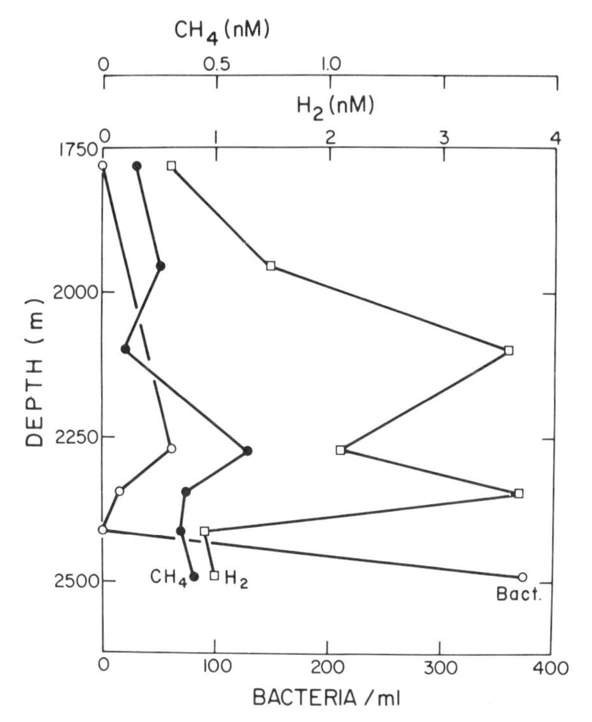

Fig. 10 Water column concentrations of H_2, CH_4 and CH_4-oxidizing bacteria directly over the ridge crest at 21°N.

found in the hydrothermal plumes directly above the ridge crest at 21°N. It is likely that hydrogen oxidizing organisms are also associated with such plumes and that both hydrogen and methane are rapidly oxidized as the plume moves downstream.

These 21°N vent samples contained up to 10^5 microorganisms ml^{-1} and thermophilic organisms capable of growth at 85°C were detected in all samples except one taken from a low temperature vent (Table 5). The nutritional versatility of organisms from 21°N samples is shown in Table 6. Bacteria were found which could nitrify with either NH_4^+ or NO_2^-, oxidize methane, couple denitrification with sulfur oxidation, and exhibit mixotrophic growth. The morphological diversity of organisms found at 21°N is indicated by Figures 11-15. Considerable variation was seen in the degree and location of microbial colonization of chimney rock material. Figure 16 depicts an EDAX scan of a region near the throat of an active "black smoker" showing the

Table 5. Chemical* and microbiological properties of hot water venting from sulfide chimneys at 21°N along the East Pacific Rise.

Sample	Magnesium Temp (°C)	pH	Si(μM)	Concentration of Biologically Important Constituents (μM)			Total numbers of Bacteria (ml^{-1})			Thermophiles Present		
								Viable Ct at 25°C on inorganic medium		inorganic medium		organic medium
				H_2S	Mn	Fe	Direct Ct	Aerobic	Anaerobic	ph 5	pH 7	pH 7
978 (Bag)	2	(sample not analyzed chemically)					8.6×10^5	5.3×10^3	<1	-	-	-
979 11/12	2	7.5	200	1.2	17	56	1.8×10^5	<1	25	+	+	-
979 1/2	2	7.0	390	3.4	46	14	3.6×10^5	5	40	+	+	-
978 7/8	40	6.0	1300	600	120	20	7.4×10^5	<1	310	+	+	-
980 7/8	44	6.3	2839	475	68	-	3.5×10^5	<1	<1	+	+	-
979 3/4	152	4.9	9600	1300	286	206	2.5×10^5	<1	<1	+	+	-
978 5/6	189	5.1	11700	7.5	1055	153	1.4×10^4	<1	80	+	+	-
982 7/8	304	4.2	18320	4300	800	-	4.7×10^5	<1	<1	+	+	-

* Chemical data provided by John Edmond (personal communication).

Table 6. Nutritional versatility of mesophiles and thermophiles isolated from hot water and sulfide chimney animals taken at 21°N (East Pacific Rise).

Organism	Source	Incubation Temp(°C)	Nitrification NH_4^+	Nitrification NO_2^-	Methane[1] oxidation	Denitrification[2] with sulfur	Mixotrophic[3]
10	Polychaete worm gut	40	+	+	+	+	+
978 5/6 M	220°C water	50	+	-	+	+	+
15	Polychaete worm gut	40	+	+	+	+	+
14	120°C water	40	+	-	+	+	+
1	2°C water	50	+	+	+	+	+

1 Methane oxidation using ^{14}C-methane.
2 Denitrification using acetylene blockage procedures.
3 Mixotrophic growth tested using formate, acetate, methanol, glucose and amino aicds.

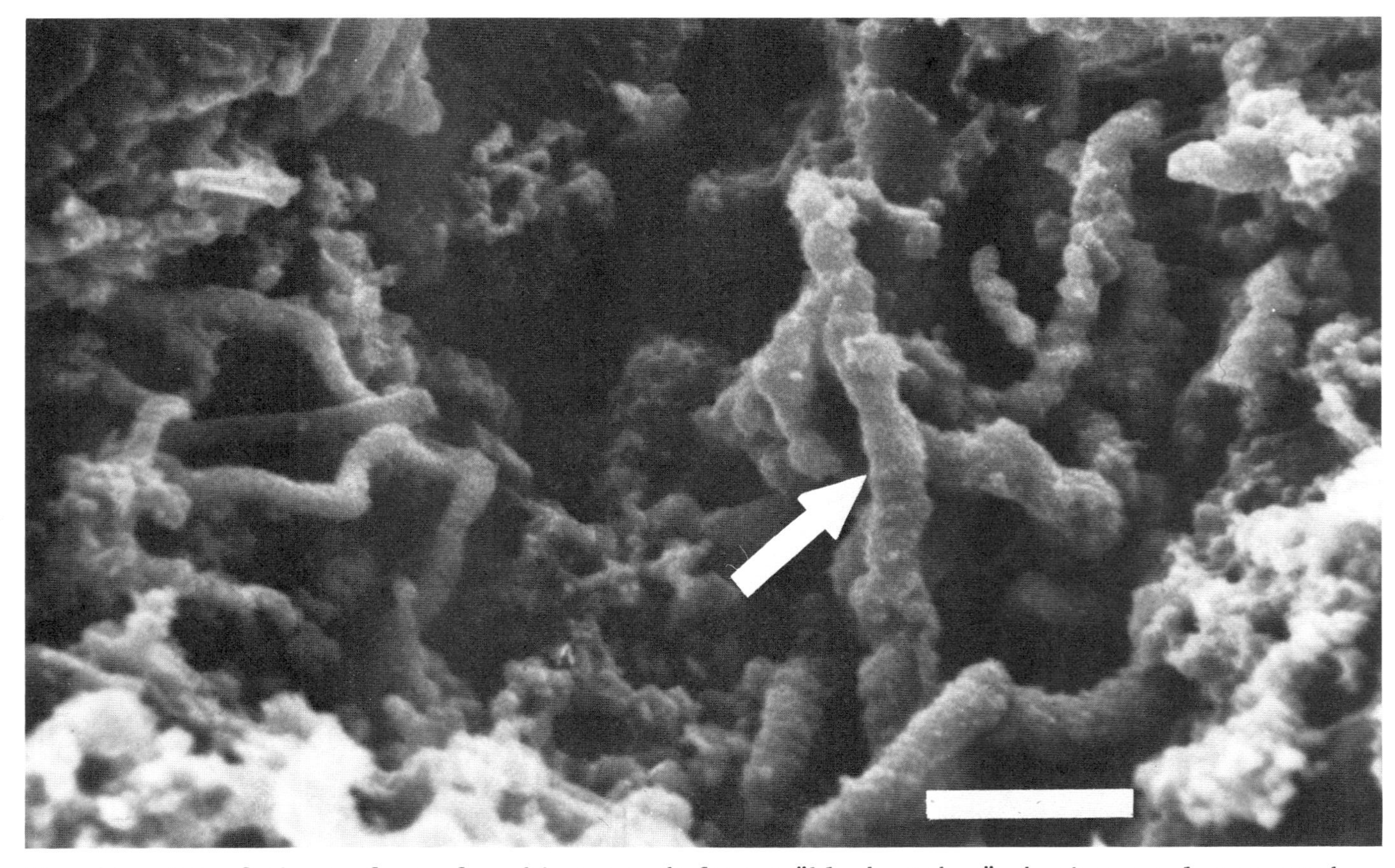

Fig. 11. SEM of the surface of a chimney rock from a "black smoker" showing metal encrusted bacteria. Bar is 5 μm.

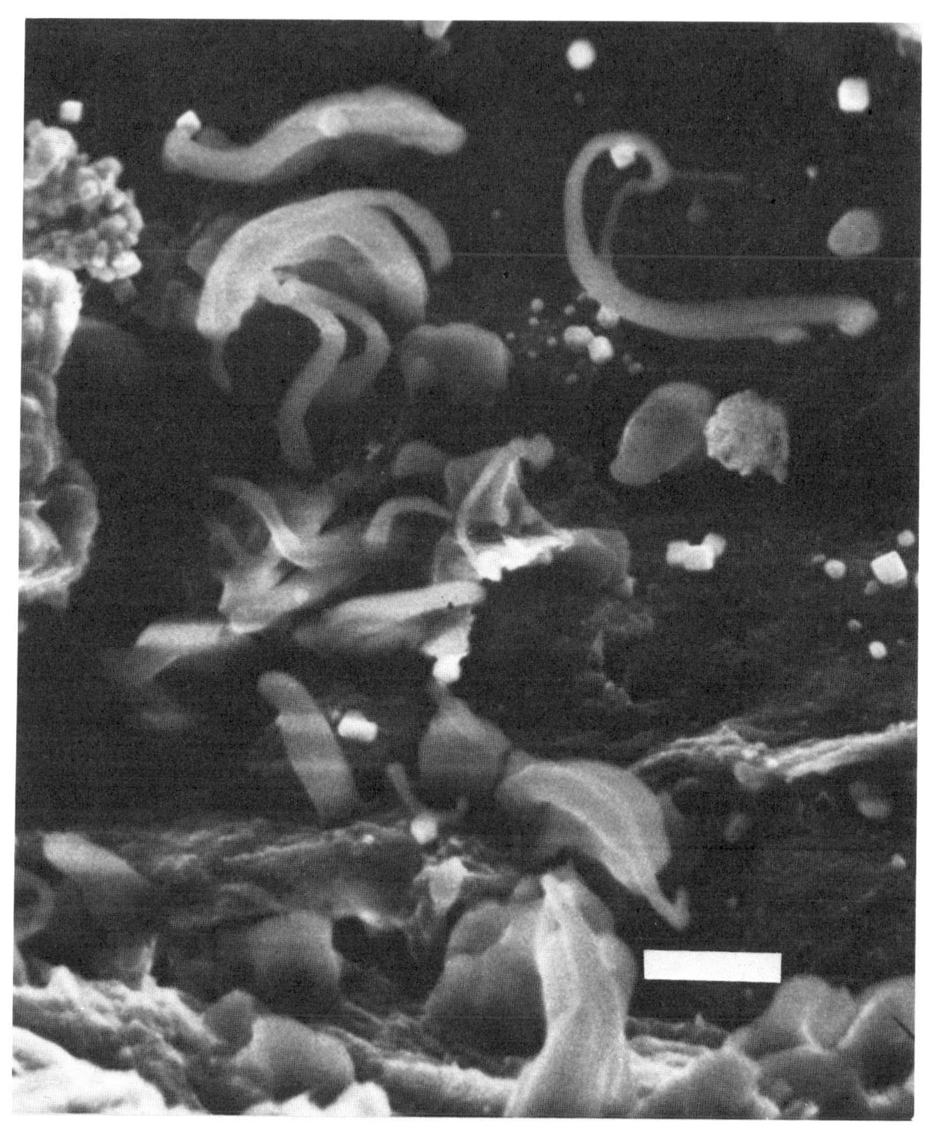

Fig. 12. SEM of the surface of a chimney rock from a "black smoker" showing a microcolony of a previously undescribed fan-shaped bacteria. Bar is 5 μm.

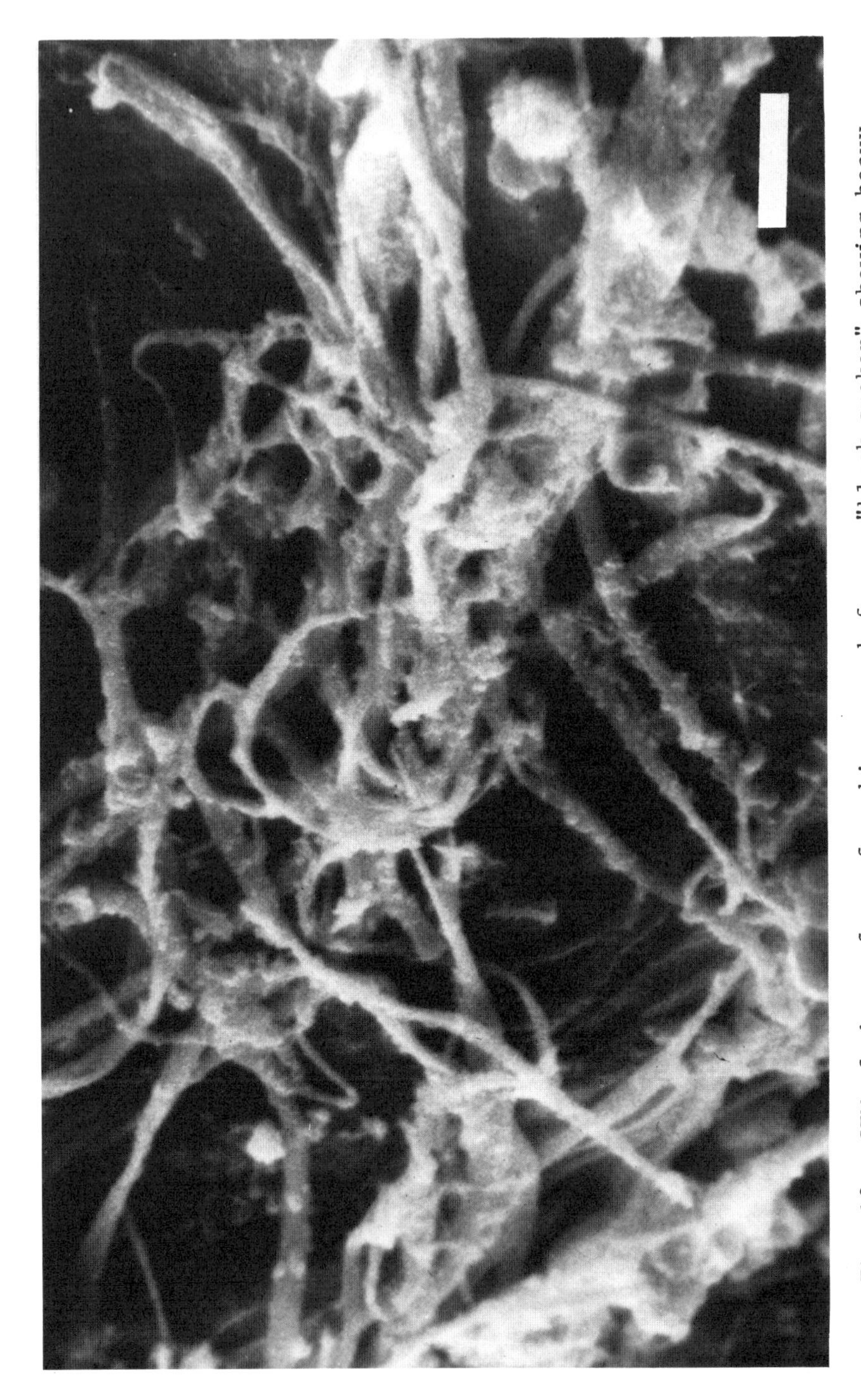

Fig. 13. SEM of the surface of a chimney rock from a "black smoker" showing heavy colonization by filamentous bacteria. Bar is 5 μm.

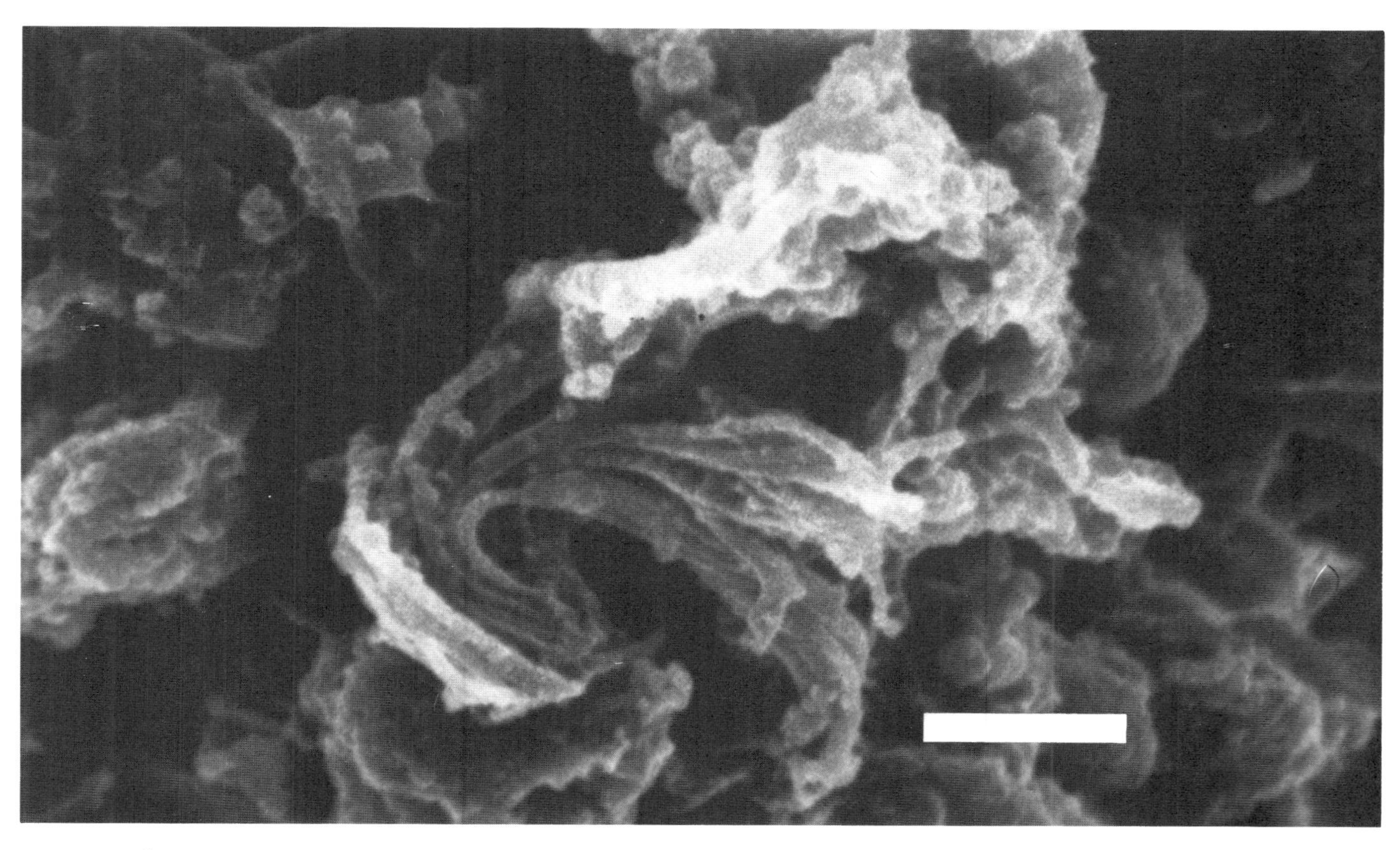

Fig. 14. SEM of the surface of a limpet shell taken from a "black smoker" chimney showing clusters of thin rod shaped bacteria. Bar is 3 μm.

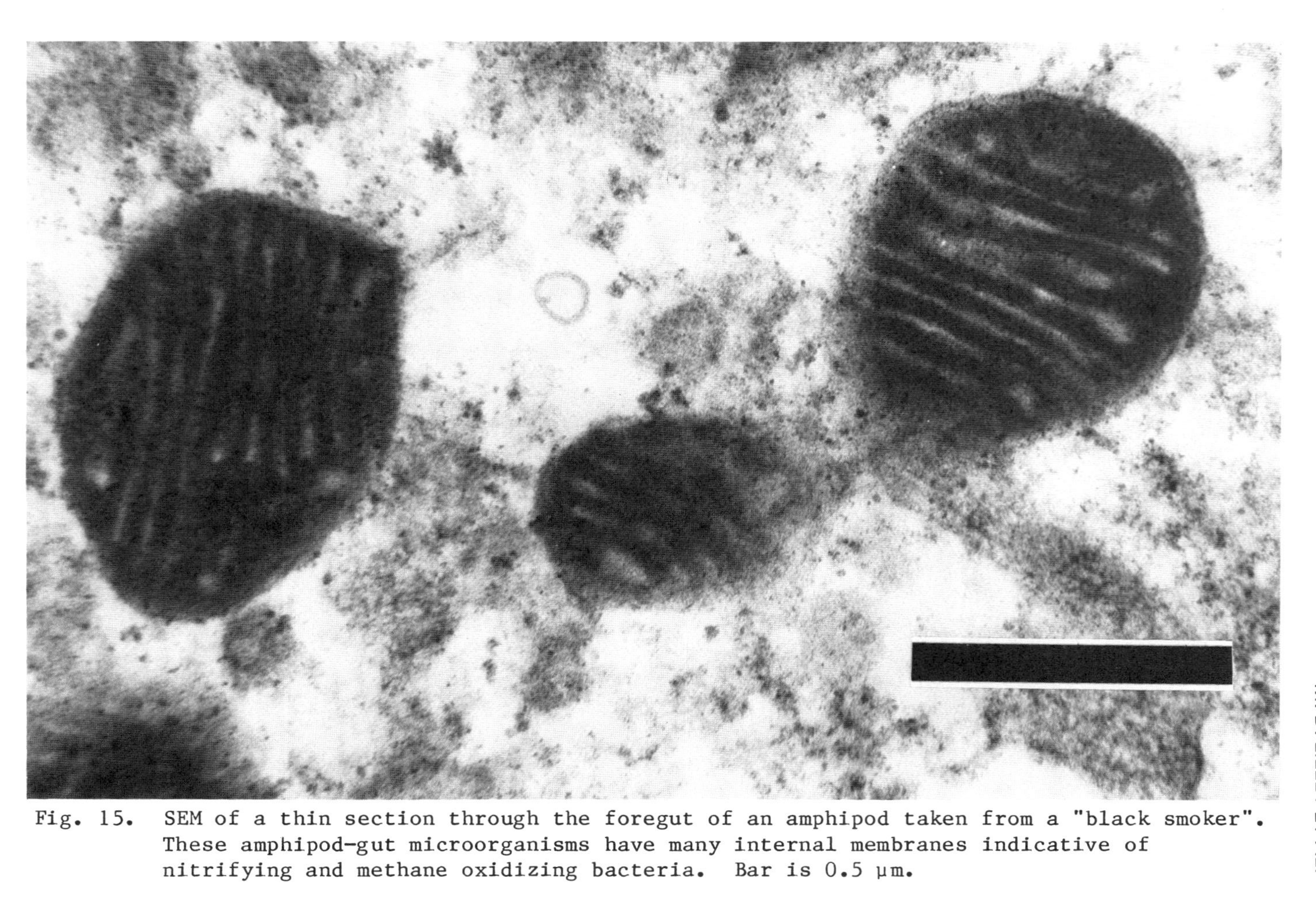

Fig. 15. SEM of a thin section through the foregut of an amphipod taken from a "black smoker". These amphipod-gut microorganisms have many internal membranes indicative of nitrifying and methane oxidizing bacteria. Bar is 0.5 μm.

Fig. 16. SEM of a section of chimney rock. The material at the lower left is pyritic and the large crystals are a Ca compound. The dot scan shows the location of Ca atoms. Bar is 250 μm.

Table 7. Physiological groups of microorganisms identified at 21°N and at the Galapagos Spreading Center.

Physiological group	GSC	21°N	Comments
1. Sulfur oxidizers	+	+	Dominant in water, isolates from 21°N water and chimney animals can also oxidize C_1 to C_4 compounds and NH_4^+.
2. Metal oxidizers	+	+	Cooxidize with sulfur, dominant in water.
3. Sulfate reducers	+	+	Only viable organism isolated from "Clambake" water at the GSC. Low numbers in other water samples and in animal guts.
4. Methanogens	+	+	Methanogens in 21°N water samples, presumptive association with animal guts.
5. N_2-fixers	+	-	Associated with animals and particularly with the vestimentiferan tube worm; heterotrophic
6. Denitrifiers	+	+	Associated with sulfur oxidation. Most abundant at 21°N.
7. Nitrifiers	+	+	Most abundant at 21°N. Many strains tested can cooxidize methane and sulfur compounds.
8. Heterotrophs	+	+	High numbers associated with animals. Mostly fermentative or mixotrophic.
9. Methane oxidizers	Not tested	+	In hydrocast samples and hot vent waters at 21°N. Presumptive (presence of multiple membranes) in guts of animals associated with chimneys.

presence of Ca. Other major elements detected were Si, S, Fe, Zn and Cu. A summary of the diverse physiological groups of organisms found at the GSC and 21°N is shown in Table 7.

DISCUSSION

Methane, hydrogen, carbon monoxide and nitrous oxide can be produced both abiotically and microbially. Possible abiotic sources of methane, carbon monoxide and hydrogen include injection from the mantle and extraction from basalt by hot fluids. Additional abiotic sources include the reduction of carbon dioxide to form methane and the oxidation of magnetite to hematite,

$$2Fe_3O_4 + H_2O = 3Fe_2O_3 + H_2$$

which has been postulated as the major source of volcanic hydrogen (Arrhenius, 1981). Nitrous oxide could be produced abiotically by the reduction of nitrate in the descending seawater or by the oxidation of ammonium ascending with the hydrothermal fluids. Ammonium is known to substitute for alkali metal ions in silicate minerals and to be stable at high temperatures and pressures (Hallam and Eugster, 1976), and nitrous oxide has been detected in a few volcanic gas samples (Cicerone et al., 1978). Carbon monoxide is a common constituent of volcanic gases (Gerlach, 1980a,b) and could originate from C-O-H-S equilibria within the magma (Gerlach and Nordlie, 1975).

An alternative source of these gases in the hydrothermal fluids could be biological production. Methane, carbon monoxide, hydrogen and nitrous oxide are produced and consumed by microorganisms (Cole, 1976). Hydrogen can be produced by many microbial species, methane by only one group, the methanogens. Little is known about the production of carbon monoxide by microorganisms but there is evidence that carbon monoxide may be an intermediate in methane oxidation (Baross et al., 1982c). Considerable microbiological information related to the hydrothermal systems at the GSC and 21°N has been amassed in recent years. Some of this information will be summarized and its relevance to vent chemistry in general and dissolved gases in particular will be discussed.

Microbiological analysis of water samples from the 1977 GSC expedition showed that large numbers of sulfur-oxidizing bacteria were associated with the vents and led to two hypotheses: that significant numbers of bacteria colonize fractures and fissures to considerable depth in the basalt and influence the chemistry of the system, and that the chemolithotrophic sulfur-oxidizing bacteria provide the primary food source for the animal communities associated with the vents (Corliss et al., 1979). Additional microbiological information obtained from these same water samples (Table 3) gave an indication that the vent waters harbor a diverse group of microorganisms including aerobic heterotrophs, sulfur-manganese oxidizers, anaerobic sulfate-reducers and nitrogen fixing organisms. In addition, sulfur-oxidizing bacteria that could only grow anaerobically with nitrate as the electron acceptor were found. The lack of aerobic sulfur-oxidizers at Clambake may be the result of low levels of H_2S emitted by this vent system (Table 2). This early work indicated the presence in the vents of a large and physiologically diverse population of bacteria and suggested that the dominant bacterial groups in each vent may be distinct and reflect the differences in the chemical properties of each vent.

Subsequent microbiological work at the GSC demonstrated a diverse population and many bacterial clumps indicating that the organisms were being sloughed off the vent walls. *In situ* CO_2 uptake

Table 8. Rate of gas production at 100 ± 2°C by bacterial communities from 21°N samples.§

Sample	Magnesium Temp (°C)	Inorganic medium (basal salts and trace elements) supplemented with:			Gas Production nmol/ml/hr		nmol/ml/9 hr	presence of
		Energy Source	Nitrogen Source	Carbon Source	H_2	CH_4	CO	N_2O (nmol/ml)
978 5/6* (water)	195	$S_2O_3^{2-}$ Mn^{2+} Fe^{2+}	$(NH_4)_2SO_4$	$NaHCO_3$	102	42	10	0.15
		Mn^{2+} Fe^{2+}	$(NH_4)_2SO_4$	formate	31	27	11	0.12
		none	none	none	0.3	0.3	0.6	nd
982 7/8† (water)	306	$S_2O_3^{2-}$ Mn^{2+} Fe^{2+}	$(NH_4)_2SO_4$	$NaHCO_3$	54	22	14	0.025-0.1
		Mn^{2+}	$(NH_4)_2SO_4$	formate	110	46	22	0.025-0.1
		none	none	none	0.2	nd	nd	nd
3140 (sulfide chimney rock)‡		$S_2O_3^{-2}$ Mn^{+2} Fe^{+2}	KNO_3	$NaHCO_3$	24 23	24 22	7.5 10	0.025-0.1 nd
		Mn^{+2} Fe^{+2}	$(NH_4)_2SO_4$	formate	26	25	10	
		none	none	none	0.3	nd	nd	nd

* Inoculum was 1.3×10^6 per 30 ml.

† Inoculum was 3×10^5 per 30 ml.

‡ Inoculum was 6.1×10^5 per 30 ml.

§ No gas production using formalin-killed cells as a control for all three samples.

experiments showed that these communities fixed CO_2 at rates exceeding those at the H_2S - O_2 interfaces of the Black Sea and Cariaco Trench (Karl et al., 1980). The $^{13}C/^{12}C$ and ^{14}C data from vent animals have strengthened the idea that chemoautotrophic bacteria form the base of the vent community food chain (Rau and Hedges, 1979; Rau, 1981a; Williams et al., 1981).

At 21°N in 1979 large numbers of bacteria were found in the hot water samples. Diverse physiological groups were present as had been found at the GSC (Tables 5-7). Note in Table 5 that although direct counts indicated 10^5 organisms ml^{-1} in the vent samples, only the sample from a warm-water "Galapagos type" vent showed appreciable growth when incubated at 25°C. This, coupled with growth of organisms at 85°C in all samples except that from the warm water vent, was a strong indication that substantial numbers of bacteria were associated with the hot waters. Table 8, from Baross et al., (1982a) shows that bacteria isolated from the two hottest water samples and from a sulfide chimney could produce methane, hydrogen and carbon monoxide and oxidize methane at 100°C when supplied with only inorganic carbon, nitrogen and energy sources.

This experiment indicates the potential for microorganisms associated with the high temperature hydrothermal vents to produce hydrogen, methane and carbon monoxide. Whether these organisms produce appreciable levels of these gases *in situ* is yet to be proven but the high methane levels at the East of Eden site indicate that this type of vent system would be a good site to test for methanogenic activity.

Methane at 21°N has a $\delta^{13}C$ value of -15.0‰ vs PDB (Welhan and Craig, this volume). This is much heavier than methane that is typically considered to be of biogenic origin (Schoell, 1980). However, following the approach of LaZerte (1981) a plot of the data available from determinations of the $^{13}C/^{12}C$ fractionation factors of pure cultures of methanogenic bacteria (Fig. 17) indicates a temperature dependence that is considerably higher than that predicted by thermodynamic considerations. The high temperature data points (65°C) represent the reduction of CO_2 to CH_4 with H_2 by a thermophilic methanogen, *Methanobacterium thermoautotrophicum* (Games et al., 1978; Fuchs et al., 1979). Methanogens associated with the vents may have a similar physiology. These data indicate that at 65°C the CH_4 produced would be 25‰ to 34‰ lighter in ^{13}C than the CO_2 substrate. It may not be strictly correct to combine data for physiologically different organisms supplied with different substrates (CO_2, CH_3OH), but the sense of the plot would not be significantly altered if the data of Rosenfeld and Silverman (1959) were excluded. If this temperature trend were applicable to vent methanogens, the available data would imply that at approximately 86°C, the reduction of CO_2 to CH_4 would not result in fractionation of the carbon isotopes. Data for higher temperatures does not exist nor is it known what effect hydrostatic

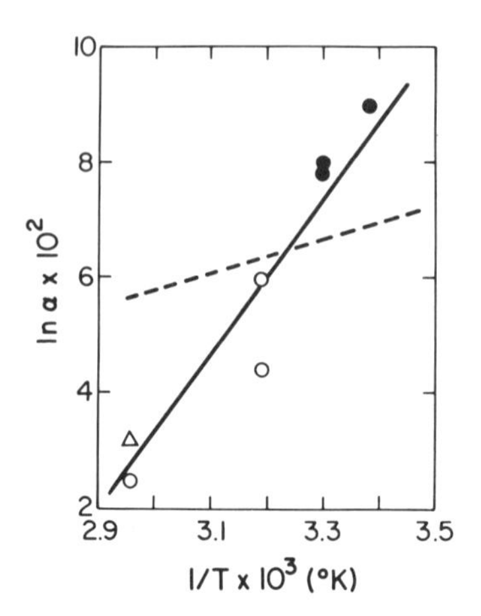

Fig. 17 Plot of ln α vs $^1/T$ determined from pure cultures of methanogens. These data represent the fractionation between ^{13}C and ^{12}C during the reduction of CO_2 to CH_4 (Games et al., 1978 and Fuchs et al., 1979; O and Δ in the figure, respectively) and the reduction of methanol to CH_4 (Rosenfeld and Silverman, 1959; ● in the figure). See text. The intercept, slope and correlation coefficient for the regression are -0.407, 146.2 and 0.96, respectively. The dashed line represents the relationship between the equilibrium thermodynamic fractionation factor and $^1/T$ derived from the data of Richet et al., (1977).

pressure has on the fractionation factor exhibited by these organisms. Until such information is available any inference about the source of methane based on stable isotope data must be made with caution.

An important and as yet unanswered question concerning the microbiology and chemistry of hydrothermal vents is, what principal nitrogen sources are being utilized by the complex communities of bacteria found in the vents? The $^{15}N/^{14}N$ data (Rau, 1981b) indicated that the nitrogen found in vent animals was depleted in ^{15}N relative to deep-sea sources and its assimilation into animal tissue must have been preceded by bacterial assimilation and fractionation if the source is "deep-sea" rather than mantle. Organisms capable of nitrification, denitrification and nitrogen fixation have been found in the vents (Tables 6 and 7) and it is possible that all these processes occur at some point in the convective circulation system wherein ambient seawater descends through the heated rocks, is altered, and mixes with additional descending water as it rises to the sea floor. Ammonia was detected in some 1977 GSC vent samples

(Table 2) and its direct assimilation by organisms may be an important process in some vents.

Nitrous oxide was found to vary considerably at the GSC (Fig. 4) and to be present at higher than ambient levels at the SW site at 21°N (Fig. 9). Nitrous oxide can be produced microbially through both nitrification and denitrification. As ambient seawater penetrates into the warm rocks, oxygen and nitrate will be made available to the resident chemolithotrophic population to be used as electron acceptors in biological oxidations. During denitrification, nitrite accumulates until the nitrate levels are low and then the nitrite is further reduced to N_2O and N_2 (Goering, 1978). If nitrite does not become limiting then N_2O may accumulate. If nitrite is depleted then N_2O will be used as the terminal electron acceptor and reduced to N_2. This sequential process has been demonstrated to occur in low-oxic and anoxic marine environments (Cohen and Gordon, 1978; Cohen, 1978) and is the likely process controlling the nitrous oxide concentrations in the GSC vents. As shown earlier in Table 3, denitrifying organisms have been found in the GSC vents including obligately anaerobic types having physiological characteristics similar to *Thiobacillus denitrificans* which can reduce nitrate to nitrous oxide using sulfide as the electron donor (Aminuddin and Nicholas, 1973).

Rose Garden showed only slight nitrous oxide production. This is consistent with Rose Garden being a leaky system with appreciable seawater penetration and therefore considerable nitrate in the upper zone of the system where denitrifiers would likely be found. Nitrate extrapolates to zero at about 17.5°C. Note in Figure 5 that although Rose Garden has high levels of H_2S (H_2S/Si = 0.232), relatively high levels of NO_3^- are also present and little accumulation of N_2O is noted. Mussel Bed, although it has much less sulfide (H_2S/Si = 0.072), appears to be less leaky; nitrate extrapolates to zero at about 11.8°C and appreciable N_2O accumulation is evident. East of Eden has the highest H_2S/Si value (0.853) and shows net consumption of N_2O which indicates that the available NO_3^- and NO_2^- is quickly utilized in this vent system.

The SW site at 21°N showed high levels of nitrous oxide in the most diluted sample (981 1/2) and higher levels than the NGS and OBS sites for less diluted samples. This indicates that a source for nitrous oxide must exist near the chimney outlet, possibly within the outer margin of the chimney wall itself or from a nearby warm water vent.

Using the GSC hydrothermal end-member Si concentration of 21.9 mM (Edmond et al., 1979a) and the 21°N end-member water $\delta^{18}O$ value of +1.6 ‰ vs SMOW (Craig et al. 1980), extrapolated end-member concentrations of methane and hydrogen were derived for each vent system. These concentrations along with the ratios of these constituents to ^3He for each vent area are summarized in Table 9. The extrapolations

Table 9. Inferred properties of the hydrothermal end-member fluids.

Vent field	CH_4 (μM)	H_2 (μM)	$CH_4/^3He$* (mole ratio)	$H_2/^3He$* (mole ratio)
Galapagos†				
East of Eden	302		$42.0 \pm 10.9 \times 10^6$	
Mussel Bed	107	3.7	$14.8 \pm 1.8 \times 10^6$	$0.7 \pm .2 \times 10^6$
Rose Garden	85	0.9	$12.4 \pm 1.1 \times 10^6$	0.1×10^6
East Pacific Rise				
21°N, 109°W	65	227-1275	6.3×10^6 ($5.0 \pm 1.5 \times 10^6$)‡	$22\text{-}124 \times 10^6$ ($35 - 163 \times 10^6$)‡
20°S, 114°W§			$5.8 - 17.4 \times 10^6$	

* ^3He values for the Galapagos are those derived in Table 1. The ^3He value used for 21°N is from Welhan and Craig (this volume).

† If, as suggested in this paper, intermediate biological sources and sinks of H_2 and CH_4 are shown to be active in the low temperature vents then these extrapolations to "end-member" values result in a set of compositions other than the 350°C hydrothermal fluid.

‡ Welhan and Craig (this volume).

§ This data taken from Craig (1981) for comparison.

for the low temperature GSC vents are made in order to examine the potential magnitude of microbial effects upon the more slowly ascending fluids. These data constitute the only basis for comparison of dissolved gases in warm water vent systems such as the GSC with the high temperature 21°N system.

The end-member methane concentration derived from our data for all three 21°N vent areas (65 μM) is close to the value of Welhan and Craig (this volume) for the NGS site (1.45 cc/kg). This results in a $CH_4/^3He$ value roughly half that of the lowest value seen at the GSC and demonstrates that these low temperature vents may be significantly influenced by microbial activity.

The most striking difference between the GSC and 21°N data is the large difference seen in H_2 concentrations. If the hydrothermal end-member solutions were in equilibrium with the quartz-fayalite-magnetite (QFM) buffer, they should contain approximately 400 mM H_2 at 350°C and 500 bars (Wolery and Sleep, 1976). This would imply a $H_2/^3He$ ratio on the order of 10^{10} which is orders of magnitude higher than those observed (Table 9). Either the QFM buffer system does not control the H_2 concentration in these solutions or significant H_2 consumption occurs as the fluids ascend to the sea floor.

Molecular hydrogen is an excellent energy source for several species of chemolithotrophic bacteria (Cole, 1976). In the vent systems it is surely utilized by methanogens, sulfate reducers and denitrifiers within the anaerobic zones and by hydrogen oxidizing bacteria (e.g. Pseudomonas, Alcaligenes) in the aerobic zones of warm water vents. Thus it is not surprising that molecular hydrogen is found in much lower concentrations relative to ^3He in the warm water vents at the GSC than in the high temperature vents at 21°N.

Systematic variations in chemical species among the hydrothermal vents have been ascribed to differences in vent field age, substrate composition and water/rock ratio (Edmond et al., 1979a). We have presented microbiological data here which indicate the presence of bacteria in these systems which may also influence the chemistry of the exiting vent water. It has been shown that some of these organisms can grow at 100°C and produce the reduced gases discussed here (Baross et al., 1982a). Organisms have been isolated which can oxidize methane, ammonia, reduced sulfur, Mn^{2+} and Fe^{2+}. Sulfate reducing and denitrifying bacteria have been found as well. Many of these reactions can be coupled by a single organism, for instance sulfide oxidation with denitrification, and the cooxidation of reduced sulfur and iron, i.e. both components of pyrite.

Although reduced sulfur is probably very important as an energy source to the microbial community in the vents, the finding of high levels of CH_4, H_2 and reduced metals along with organisms capable of using these additional energy sources indicates that the complex biological communities found near the vents have a more diverse food source than sulfur oxidizing bacteria alone.

It is expected that microbial influences on the exiting vent waters will be most significant in the warm water vents where the hot hydrothermal fluid mixes with appreciable quantities of descending sea water below the sea floor thus forming energy-rich mixtures of oxidized and reduced compounds. However, the finding of bacterial gas production at high temperature (Baross et al., 1982a) and the recent discovery by Karl et al., (1983) that high levels of ATP are emanating from "black smokers" indicates that microbial activity may be prevalent in the high temperature vents as well. It is important to our understanding of hydrothermal vents, and the chemical history of sea-water, that the degree of microbial influence upon both the liquid and solid phases within hydrothermal systems be quantitatively evaluated.

ACKNOWLEDGMENTS

We thank the officers and crews of the R/Vs KNORR, GILLISS, LULU and ALVIN for their assistance in sample collection. Carlos Lopez was also instrumental in this regard. Marie de Angelis performed the N_2O analyses in 1979 and provided editorial assistance. We thank John Edmond, John Welhan, and Harmon Craig for generously supplying unpublished data. Discussions with, and data supplied by, John Welhan were essential for the corrections applied to the 21°N gas data. We also thank John Welhan and Harmon Craig for providing us with an advance copy of their contribution to this volume. This work was supported by the Office of Naval Research, Marine Chemistry and Biology Program under Contract N00014-79-C-0004.

REFERENCES

Aminuddin, M. and Nicholas, D. J. D., 1973, Sulphide oxidation linked to the reduction of nitrate and nitrite in Thiobacillus denitrificans, Biochim. Biophys. Acta, 325:81-93.

Arrhenius, G. (1981) Interaction of ocean-atmosphere with planetary interior. Adv. Space Res., 1:37-48.

Ballard, R. D., van Andel, T. H. and Holcomb, R. T., 1982, The Galapagos Rift at 86°W 5. Variations in volcanism, structure, and hydrothermal activity along a 30 - kilometer segment of the rift valley. J. Geophys. Res., 87:1149-1161.

Baross, J. A., Lilley, M. D. and Gordon, L. I., 1982a, Is the CH_4, H_2 and CO venting from submarine hydrothermal systems produced by thermophilic bacteria? Nature, 298:366-368.

Baross, J. A., Dahm, C. N., Ward, A. K., Lilley, M. D. and Sedell, J. R., 1982b, Initial microbial response in lakes to the Mt. St. Helens eruption. Nature, 296:49-52.

Baross, J. A., Lilley, M. D., Dahm, C. N., and Gordon, L. I., 1982c, Evidence for microbial linkages between CH_4 and CO in aquatic environments. EOS, Trans. Am. Geophys. Union, 63:155.

Cicerone, R. J., Shetter, J. D., Stedman, D. H., Kelly, T. J. and Liu, S. C., 1978, Atmospheric N_2O: measurements to determine its sources, sinks, and variations. J. Geophys. Res., 83:3042-3050.

Cohen, Y., 1977, Shipboard measurement of dissolved nitrous oxide in seawater by electron capture gas chromatography, Anal. Chem., 49:1238-1240.

Cohen, Y., 1978, Consumption of dissolved nitrous oxide in an anoxic basin, Saanich Inlet, British Columbia. Nature, 272: 235-237.

Cohen, Y. and Gordon, L. I., 1978, Nitrous oxide in the oxygen minimum of the eastern tropical North Pacific: evidence for its consumption during denitrification and possible mechanisms for its production. Deep-Sea Res. 25:509-524.

Cole, J. A., 1976, Microbial gas metabolism. Adv. Microbial Phys. 14:1-92.

Corliss, J. B., Dymond, J., Gordon, L. I., Edmond, J. M., von Herzen, R. P., Ballard, R. D., Green, K., Williams, D., Bainbridge, A., Crane, K. and van Andel, T. H., 1979, Submarine thermal springs on the Galapagos Rift. Science, 203:1073-1083.

Craig, H., Welhan, J. A., Kim, K., Poreda, R. and Lupton, J. E., 1980, Geochemical studies of the 21°N EPR hydrothermal fluids. EOS, Trans. Am. Geophys. Union, 61:992.

Craig, H., 1981, Hydrothermal plumes and tracer circulation along the East Pacific Rise: 20°N to 20°S. EOS, Trans. Am. Geophys. Union, 62:893.

Edmond, J. M., Measures, C., McDuff, R. E., Chan, L. H., Collier, R., Grant, B., Gordon, L. I. and Corliss, J. B., 1979a, Ridge crest hydrothermal activity and the balances of the major and minor elements in the ocean: the Galapagos data. Earth Planet. Sci. Lett., 46:1-18.

Edmond, J. M., Measures,C., Mangum, B., Grant, B., Sclater, F. R., Collier, R., Hudson, A., Gordon, L. I. and Corliss, J. B., 1979b, On the formation of metal-rich deposits at ridge crests. Earth Planet. Sci. Lett., 46:19-30.

Edmond, J. M., Von Damm, K. L., McDuff, R. E. and Measures, C. I., 1982, Chemistry of hot springs on the East Pacific Rise and their effluent dispersal. Nature, 297:187-191.

Fuchs, G., Thauer, R., Ziegler, H. and Stichler, W., 1979, Carbon isotope fractionation by Methanobacterium thermoautotrophium. Archiv. Microbiol. 120:135-139.

Games, L. M., Hayes, J. M., and Gunsalus, R. P., 1978, Methane-producing bacteria: natural fractionations of the stable carbon isotopes. Geochim. Cosmochim. Acta, 42:1295-1297.

Gerlach, T. M., 1980a, Evaluation of volcanic gas analyses from Kilauea volcano. J. Volcanol. Geotherm. Res., 1:295-317.

Gerlach, T. M., 1980b, Evaluation of volcanic gas analyses from Surtsey volcano, Iceland, 1964-1967. J. Volcanol. Geotherm, Res., 8:191-198.

Gerlach, T. M. and Nordlie, B. E., 1975, The C-O-H-S gaseous system, part II: temperature, atomic composition, and molecular equilibria in volcanic gases. Am. J. Sci., 275:377-394.

Goering, J. J., 1978, Denitrification in marine systems. in: "Microbiology-1978", D. Schlessinger, ed., American Society for Microbiology, Washington D.C. p. 357-361.

Hallam, N. and Eugster, H. P., 1976, Ammonium silicate stability relations. Contrib. Miner. Petrol., 57:227-244.

Jannasch, H. W. and Wirsen, C. O., 1979, Chemosynthetic primary production at East Pacific sea floor spreading centers. BioSci. 29:592-598.

Jannasch, H. W. and Wirsen, C. O., 1981, Morphological survey of microbial mats near deep-sea thermal vents. Appl. Environ. Microbiol., 41:528-538.

Jenkins, W. J., Edmond, J. M. and Corliss, J. B., 1978, Excess ^{3}He and ^{4}He in Galapagos submarine hydrothermal waters. Nature, 272:156-158.

Karl, D. M., Wirsen, C. O. and Jannasch, H. W., 1980, Deep-sea primary production at the Galapagos hydrothermal vents. Science, 207:1345-1347.

Karl, D. M., Burns, D. J., and Orrett, K., 1983, Biomass and in situ growth characteristics of deep sea hydrothermal vent microbial communities, Abs. annual meeting Am. Soc. Microbiol., p. 235, Abs. N 69.

LaZerte, B. D., 1981, The relationship between total dissolved carbon dioxide and its stable carbon isotope ratio in aquatic sediments. Geochim. Cosmochim. Acta, 45:647-656.

Lilley, M. D. and Gordon, L. I., 1979, Methane, nitrous oxide, carbon monoxide, and hydrogen in the hydrothermal vents of the Galapagos Spreading Center. EOS, Trans. Am. Geophys. Union, 60:863.

Lilley, M. D., de Angelis, M. A. and Gordon, L. I., 1982, Methane, hydrogen, carbon monoxide and nitrous oxide in submarine hydrothermal vent waters. Nature, 300:48-50.

Lupton, J. E., Weiss, R. F. and Craig, H., 1977, Mantle helium in hydrothermal plumes in the Galapagos Rift. Nature, 267:603-604.

McDuff, R. E. and Edmond, J. M., 1982, On the fate of sulfate during hydrothermal circulation at mid-ocean ridges. Earth Planet. Sci. Lett., 57:117-132.

Rau, G. H., 1981a, Hydrothermal vent clam and tube worm $^{13}C/^{12}C$: further evidence of nonphotosynthetic food sources. Science, 213:338-339.

Rau, G. H., 1981b, Low $^{15}N/^{14}N$ in hydrothermal vent animals: ecological implications. Nature, 289:484-485.

Rau, G. H. and Hedges, J. I., 1979, Carbon-13 depletion in a hydrothermal vent mussel: suggestion of a chemosynthetic food source. Science, 203:648-649.

Richet, P., Bottinga, Y. and Javoy, M., 1977, A review of hydrogen, carbon, nitrogen, oxygen, sulfur and chlorine stable isotope fractionation among gaseous molecules. Ann. Rev. Earth Planet. Sci., 5:65-110.

Rosenfeld, W. D. and Silverman, R. S., 1959, Carbon isotope fractionation in bacterial production of methane. Science, 130:1658-1659.

Schink, D. R., Fanning, K. A. and Piety, J., 1966, A sea-bottom sampler that collects both bottom water and sediment simultaneously. J. Mar. Res., 24:365-373.

Schoell, M., 1980, The hydrogen and carbon isotopic composition of methane from natural gases of various origins. Geochim. Cosmochim. Acta, 44:649-661.

Welhan, J. A. and Craig, H., 1979, Methane and hydrogen in East Pacific Rise hydrothermal fluids. Geophys. Res. Lett., 6:829-831.

Williams, P. M., Smith, K. L., Druffel, E. M. and Linick, T. W., 1981, Dietary carbon sources of mussels and tubeworms from Galapagos hydrothermal vents determined from tissue ^{14}C activity. Nature, 292:448-449.

Wolery, T. J. and Sleep, N. H., 1976, Hydrothermal circulation and geochemical flux at mid-ocean ridges. J. Geol., 84:249-275.

EFFECTS OF HYDROTHERMAL ACTIVITY ON SEDIMENTARY ORGANIC MATTER: GUAYMAS BASIN, GULF OF CALIFORNIA - PETROLEUM GENESIS AND PROTOKEROGEN DEGRADATION

Bernd R. T. Simoneit

School of Oceanography
Oregon State University
Corvallis, Oregon 97331

ABSTRACT

Guaymas Basin is an actively-rifting oceanic basin, consisting of two tectonically active rift valleys, the Northern and Southern Troughs, separated by a transform fault area. The sediment blanket accumulates rapidly, keeping the basin floor covered (geologic age - Quaternary), and the process of ocean plate accretion occurs by dike and sill intrusions into the unconsolidated diatomaceous muds, resulting in localized contact metamorphism of both organic and mineral matter. Also, deep seated magmatic heat flow causes extensive thermal alteration, especially at DSDP Site 477 in the southern rift. This high conductive heat flow results in hydrothermal circulation with concomitant migration of organic pyrolysate.

The interstitial gas, gasoline, bitumen and protokerogen geochemistry of these sediments was examined, as well as how these carbonaceous fractions were affected by the hydrothermal activity. A thermogenic component is admixed with the biogenic interstitial gas (the CH_4) at depth in the DSDP sites and CO_2 is derived from thermal decomposition of carbonates at those depths.

The bitumen at depth of the DSDP sites contains thermogenic components such as primary olefins, thermodynamically stabilized molecular markers [17α(H)-hopanes], and elemental sulfur, with traces of polynuclear aromatic hydrocarbons (PAH). The major components are n-alkanes with no carbon number predominance and unresolvable complex branched and cyclic hydrocarbons (hump), which are typical of mature petroleum and were generated by the heat from the intrusives.

Dredge samples recovered from a mound on the southern rift floor have a characteristic petroliferous odor and the bitumen of some fragments was analyzed. The thermogenic origin of this petroleum was confirmed by the presence of gasoline range hydrocarbons; a broad distribution of n-alkanes (C_{13}-C_{33}, no carbon number predominance), hump, pristane and phytane; of olefins and PAH; and of large amounts of aromatic/naphthenic and asphaltic material. Similar petroleum was subsequently recovered from other active mounds in the southern rift by the D.S.R.V. *Alvin*. This petroleum is derived from the immature, primarily marine organic matter at depth by thermal alteration and rapid quenching by hydrothermal removal, followed by condensation at the seabed as part of the sulfide formations. Exterior, exposed oil samples are degraded and leached, whereas interior samples are essentially unaltered.

GEOLOGIC SETTING

The Guaymas Basin is an actively-spreading oceanic basin, which is part of the system of spreading axes and transform faults that extend from the East Pacific Rise to the San Andreas fault (Curray *et al*., 1979, 1982). The processes of ocean plate accretion result in high conductive heat flow (locally exceeding 1.2 Wm^{-2}) and dike and sill intrusions into the unconsolidated sediments (Williams *et al*., 1979; Einsele *et al*., 1980; Curray *et al*., 1982). The sediments accumulate at a rate of more than 1 m/1000 yrs and have covered the rift floors to a depth of up to 400 m (Curray *et al*., 1979; 1982).

Organic matter of these Recent hemipelagic sediments is comprised of various operational fractions as defined by the analytical procedures. The fractions that have been analyzed here are interstitial gas, lipids, fulvic and humic substances and detrital carbon (protokerogen). All of these carbonaceous fractions are very sensitive to thermal stress, as for example from an intrusion or a deep-seated heat source and are thus easily pyrolized, which has been found to be the case in Guaymas Basin, Gulf of California (Fig. 1), especially in the southern rift.

The organic geochemistry of the Guaymas Basin sediments will be discussed here, as well as how the organic matter is affected by the hydrothermal activity and how the data correlates with the petroleum observed at the seabed and in shallow sediments.

EXPERIMENTAL

The samples described in this summary are derived from gravity coring (30G, Simoneit *et al*., 1979), Deep Sea Drilling Project, Leg 64 coring (Curray *et al*., 1979, 1982), dredging operations (7D, Simoneit and Lonsdale, 1982), piston coring (LaPaz 9P, 13P and 15P) and sampling with the submersible Alvin (Fig. 1). The salient analytical results for all samples are collected in Table 1.

Analyses for gasoline range (C_4-C_9) hydrocarbons were carried out on canned or bagged samples by the methods described (Simoneit et al., 1979; Whelan and Hunt, 1982). Interstitial gases in vacutainers were analyzed for composition and stable isotope contents (Simoneit, 1982a; Galimov and Simoneit, 1982a,b). The extractable bitumen and protokerogen analyses were carried out by the well defined organic geochemical practice (Simoneit and Philp, 1982; Jenden et al., 1982) and the petroleum was analyzed by the same methods after extractive separation from the minerals (Simoneit and Lonsdale, 1982).

The various organic fractions were analyzed by capillary gas chromatography (GC) and gas chromatography/mass spectometry (GC/MS). The GC/MS analyses were conducted on a Finnigan Model 4021 quadrupole mass spectrometer interfaced directly with a Finnigan Model 9610 GC and equipped with a 30 m x 0.25 mm i.d. fused silica capillary column (wall-coated with SE-54). The operating conditions for the GC and GC/MS were as reported earlier (Simoneit et al., 1979). The MS data were acquired and processed with a Finnigan-Incos Model 2300 data system. Compound assignments were made from individual mass spectra and GC retention times, with comparison to authentic standards where possible.

The protokerogen from sediments was separated and then analyzed by Curie point pyrolysis, electron spin resonance spectrometry (ESR), and for stable isotope and elemental compositions (Simoneit and Philp, 1982; Jenden et al., 1982).

RESULTS AND DISCUSSION

Gravity coring by Scripps Institution of Oceanography (SIO) in 1972 recovered a 3.5 m section (site 30G, 27°23.0'N, 111°26.9'W, 1970 m water depth, Fig. 16) with a strong petroliferous odor. In view of the high heat flow of the area, the organic matter of this core was analyzed to assess any potential petroleum genesis within the basin (Simoneit et al., 1979). The bulk of the organic matter (humates and kerogens) was of an unaltered marine origin. The major hydrocarbons, fatty acids, ketones and fatty alcohols, with the molecular markers were also of an autochthonous marine origin and minor amounts of allochthonous terrestrial lipids were detectable. Methane was principally of a biogenic origin throughout, however, the lower sections of the core contained significant concentrations of thermogenic C_2 to C_8 hydrocarbons. This was the first indication that thermogenic products were diffusing to the seabed from depth in Guaymas Basin.

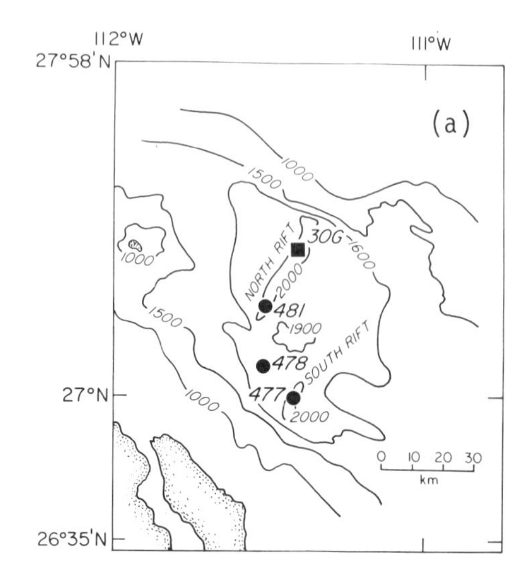
112°W
111°W
27°58'N
(a)
1000
1500
1000
30G
1600
NORTH RIFT
2000
481
1500
1900
478
SOUTH RIFT
27°N
477
2000
1000
0 10 20 30
km
26°35'N

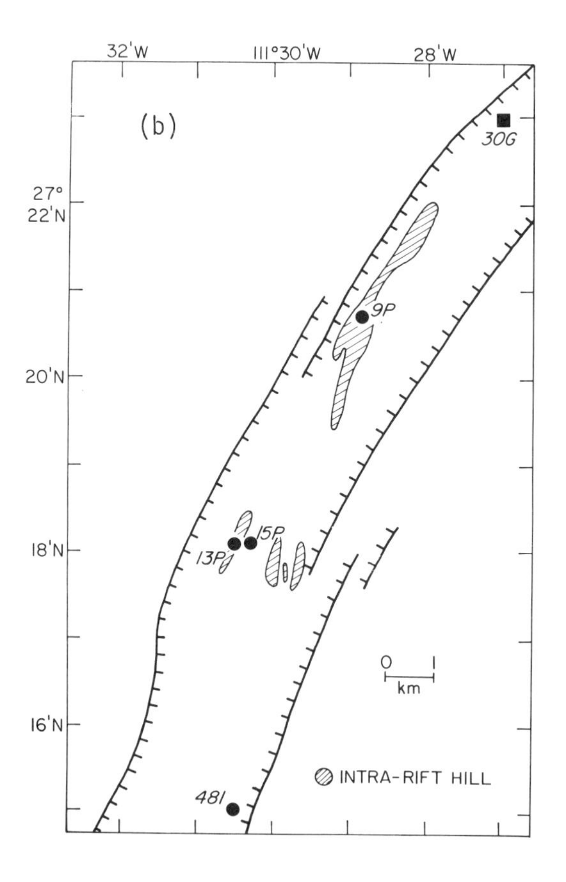
32'W
111°30'W
28'W
(b)
30G
27°
22'N
9P
20'N
15P
18'N
13P
0 1
km
16'N
INTRA-RIFT HILL
481

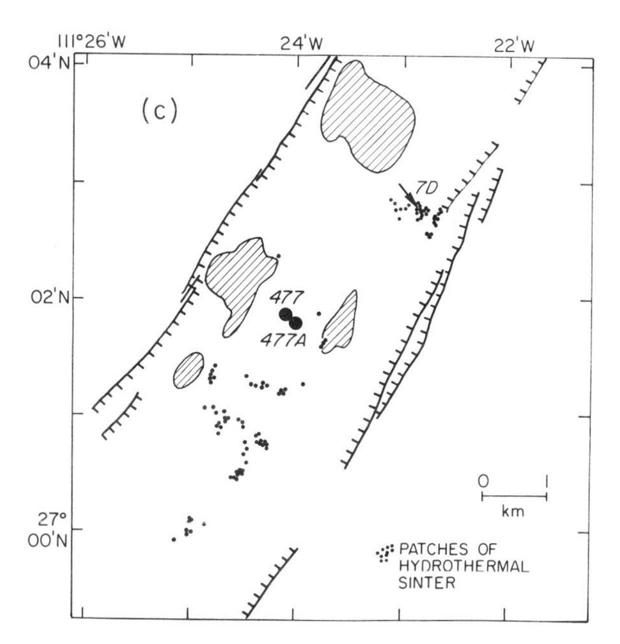

Fig. 1. Location maps of the Guaymas Basin area in the Central Gulf of California, showing the sampling locations. (a) Overall basin with DSDP sites, (b) Northern Trough with locations of DSDP Site 481, piston coring sites and gravity core 30G (Simoneit et al., 1979), and (c) portion of Southern Trough showing locations of DSDP Site 477, dredge 7D and the mound patches visited by DSRV Alvin.

Table 1. Summary of analytical results for sedimentary bitumen and hydrothermal petroleum from Guaymas Basin, Gulf of California.

Samples (cf. Fig. 1)	Total Organic Carbon (%)	Total Extract Yield (μg/g dry weight of sample)[1]	Total Hydrocarbon Yield (μg/g dry sample)[2]	CPI[3]	Reference
30G (∿3 m)	1.7-2.5	n.d.	1-24 (60-1000)	1.4-6.0	Simoneit _et al._ (1979)
64-477-5-1	2.6	2160 (83,000)	78 (3000)	1.9	Simoneit and Philp (1982)
64-477-17-3	0.8	1230 (153,750)	430 (53,700)	1.03	Simoneit and Philp (1982)
64-477A-9-1	0.4	2380 (600,000)	14 (3500)	1.02	Simoneit and Philp (1982)
7D-2B	n.a.	1075	27	1.03	Simoneit and Lonsdale (1982)
7D-4A,B	n.a.	70,400	4850	1.2	Simoneit and Lonsdale (1982)
9P (12.9-14.4 m)	n.d.	2110	41	1.05	----
13P (10.5-12.4 m)	n.d.	3040	17	1.7	----
15P (11.2-11.9 m)	n.d.	5000	16	1.2	----
172-2 (DSRV _Alvin_)	n.a.	76,300	n.d.	1.0	----
1172-4 (DSRV _Alvin_)	n.a.	552,000	n.d.	n.d.	----
1177-3 (DSRV _Alvin_)	n.a.	55,000	n.d.	0.88	----

1 Values include entrapped bitumen liberated after mineral removal with HF; values in parentheses are μg/g C_{org}.

2 Values in parentheses are hydrocarbon yield in μg/g C_{org}.

3 Carbon preference index (Simoneit, 1978).

n.d. - not determined

n.a. - not applicable

Deep Sea Drilling Project Samples, Leg 64

A thermogenic gas component could be identified at greater subbottom depths for DSDP Sites 477, 478 and 481 (Simoneit, 1982a), analogous to other areas (e.g., Kvenvolden and Field, 1981). Stable carbon isotope data were determined for methane (CH_4) from all sites (Fig. 2) (Galimov and Simoneit, 1982a,b) and at shallow depths, the data indicated a typically biogenic pattern (Sites 481 and 478, and also 30G, cf. Simoneit et al., 1979). However, with increasing depth, the $\delta^{13}C$ values became heavier indicating the removal (by diffusion and/or distillation) of the lighter $^{12}CH_4$ due to the thermal stress from intrusive and conductive heat sources. The CH_4 at Site 477 was heaviest, reflecting the highest temperature effects, and the data for Site 481 between the sills indicated various thermal stresses.

The total hydrocarbon fractions of the lipids were analyzed to evaluate the thermal effects and two examples are shown in Fig. 3, one of unaltered biogenic lipids and one of thermogenic petroliferous bitumen. The unaltered sample exhibits n-alkanes ranging from C_{14}-C_{35}, with a strong odd carbon number predominance, especially $>C_{23}$ (terrestrial plant wax, e.g., Simoneit, 1978), and subordinate amounts of C_{20} and C_{25} natural cyclic olefins and triterpenes. The thermally altered sample exhibits n-alkanes with essentially no carbon number predominance and a range from C_{15}-C_{31}. Primary olefins and elemental sulfur are also dominant components.

The lipids in the sediments from shallow depths of Sites 477, 478 and 481 are primarily of an autochthonous marine origin, with a minor influx of terrestrial plant wax (Simoneit and Philp, 1982). The paleoenvironmental conditions of sedimentation were partially euxinic, probably as a result of the high deposition rates. Lipids are thermally altered close to and below the sills (Simoneit and Philp, 1982; Ishiwatari and Fukushima, 1979). This is indicated by the loss of the carbon number predominance of the n-alkanes, the appearance of a broad hump of unresolvable complex material (cf. Table 1) and the thermodynamic equilibration of certain stereoisomers of the hopane molecular markers. Also, large amounts of primary olefins are found close to sills in the altered samples and significant elemental sulfur is present. This thermal alteration of organic matter is most severe at Site 477 and intermediate at Sites 478 and 481.

Kerogen, the isoluble, detrital organic matter is an excellent in situ indicator of the effects of thermal stress. For example, the organic nitrogen content at Site 477, expressed as the carbon to nitrogen ratio in Fig. 4 can be used as an illustration. A C/N value of 11-14 is typical for immature, marine organic matter (Simoneit, 1982a). At depths exceeding 150 m, the C/N ratio

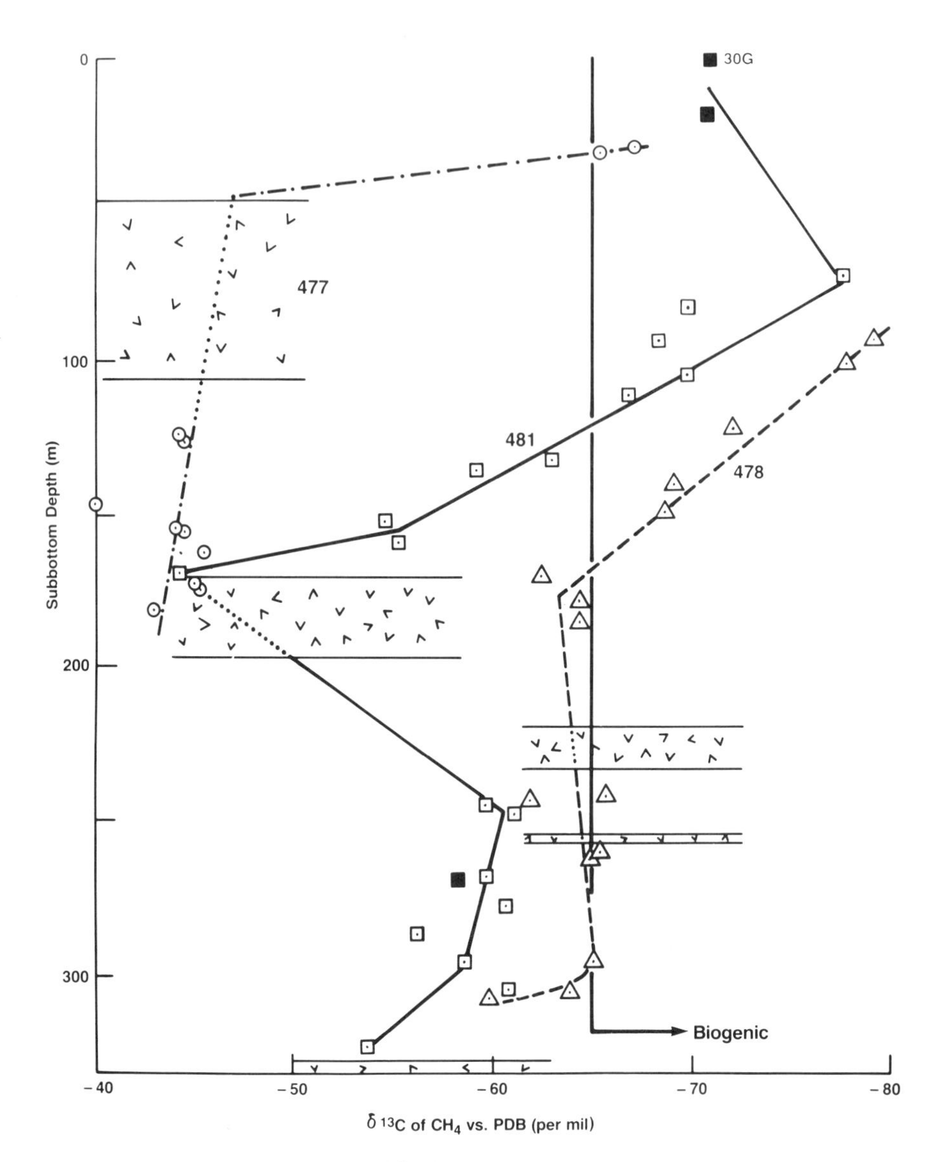

Fig. 2. Trends of carbon-12 depletion versus depth in interstitial methane at DSDP Sites 477 (—○—), 478 (—△—) and 481 (—□—), ■ = canned or frozen samples) ($\delta^{13}C$ data reported versus PDB standard, the intrusions are indicated at the appropriate depth for each site).

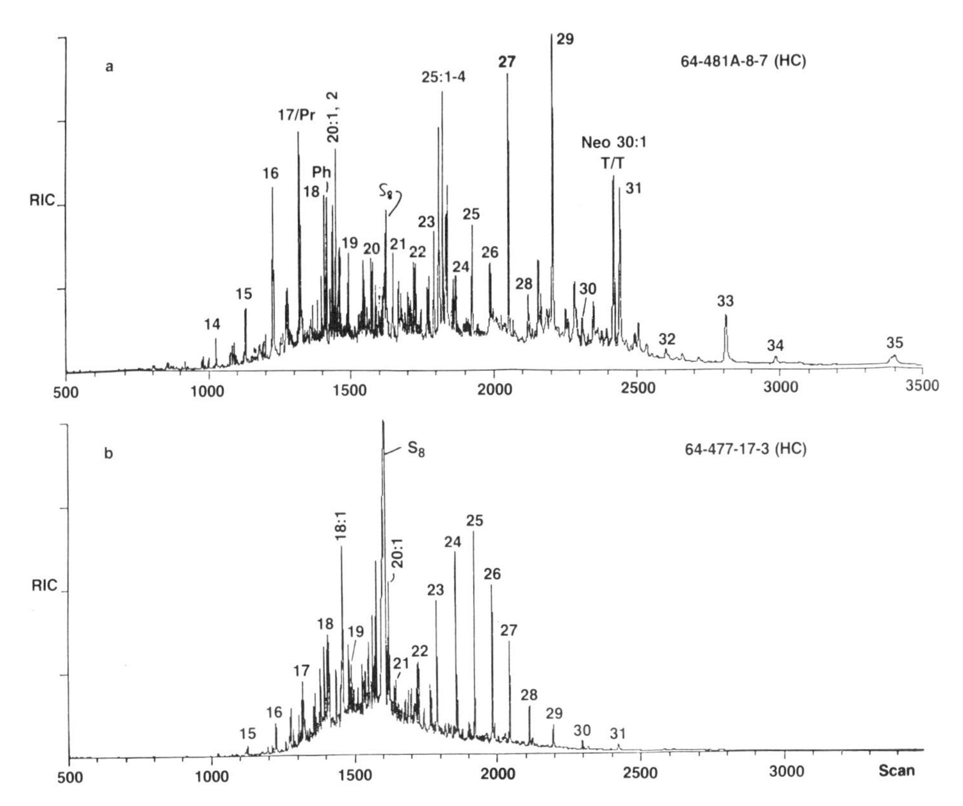

Fig. 3. Total ion current traces for total hydrocarbon fractions from (a) an unaltered, immature sample and (b) a thermally altered sample (a: 64-481A-8-7; b: 64-477-17-3; numbers indicate n-alkanes).

increases to infinity, indicating all organic nitrogen has been removed by pyrolysis (under laboratory simulation this requires temperatures in excess of 500°C). An increase in the C/N is also observed above and below the upper sill, confirming that it is an intrusion and not a flow, i.e., the organic matter was thermally stressed on both sides of the sill.

Pyrolysis-GC and pyrolysis-GC/MS has been utilized to characterize kerogens (van de Meent et al., 1980) and the DSDP samples were analyzed by these same techniques to assess both their composition and degree of thermal catagenesis (Simoneit and Philp, 1982). Examples of pyrograms for unaltered and thermally altered kerogen from Site 477 are given in Fig. 5. The shallow sample exhibits a pyrogram that is virtually identical to those of surface samples from all the other sites and is representative of typical unaltered marine organic matter (Simoneit and Philp, 1982). The other (deepest) sample reflects a pattern of essentially complete

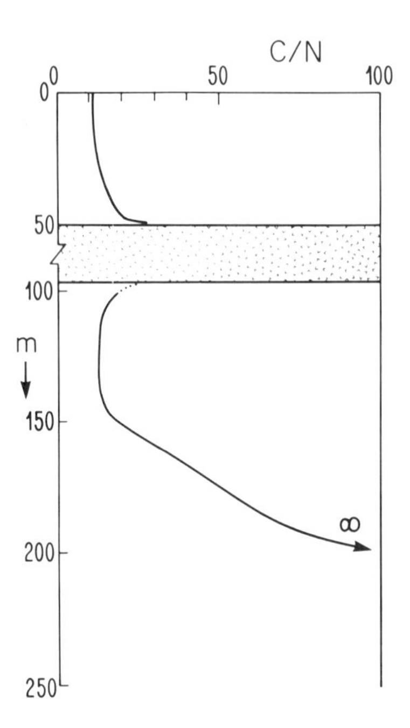

Fig. 4. Atomic ratios of organic carbon to nitrogen versus depth at DSDP Site 477 (includes data for 477A) (Simoneit, 1982a).

expulsion of pyrolysate. The *in situ* appearance of the kerogen in the zones of high thermal stress (at depth of Sites 477 and 478) resembles activated amorphous carbon (Simoneit, 1982a) and is representative of the spent kerogen.

Seabed Shallow and Surface Samples

Piston coring by the R.V. *Melville* of SIO (LaPaz Leg 2, July-August, 1980) recovered some sediments with a petroliferous odor in the Northern Trough of Guaymas Basin. Three composited samples (about six intervals of 2 cm each per sample) were analyzed from core 9P (12.9-14.4 m), 13P (10.5-12.4 m) and 15P (11.2-11.9 m) (cf. Fig. 1b). Site 9P is located on a large intra-rift hill and the core lithology consists of stiff, low-porosity mud, with possible pieces of hydrothermal crust. Sites 13P and 15P are located on a narrow ridge and coring recovered gas charged, stiff

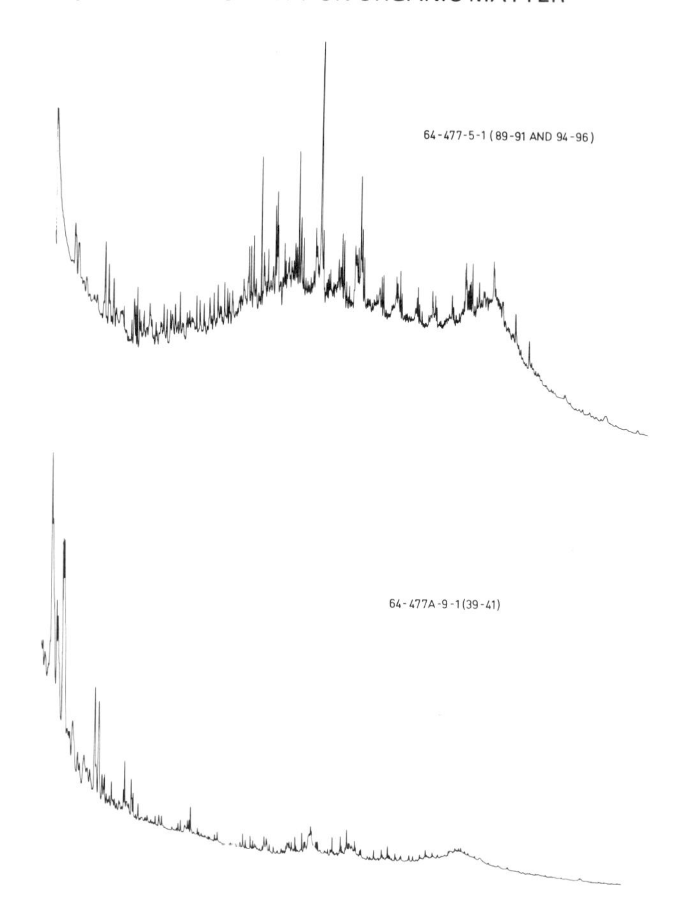

Fig. 5. Examples of Curie point pyrolysis - GC traces for kerogen concentrates from Site 477 (subbottom depths are given in parentheses) (Simoneit and Philp, 1982):
a) 64-477-5-1 (unaltered, 30 m)
b) 64-477A-9-1 (strong hydrothermal alteration, 240 m).

mud with a strong petroliferous odor. The n-alkane distributions of these samples are given in Fig. 6 and they can be compared with some examples from DSDP Site 481 (Simoneit and Philp, 1982). The n-alkanes of sample 9P range from C_{13}-C_{33}, with a maximum at C_{18} and essentially no carbon number predominance to C_{26} (CPI_{13-33} = 1.05), typical of many petroleums. A complex, unresolved mixture of branched and cyclic hydrocarbons (i.e., hump) ranges from C_{12}-C_{31}, also typical of petroleum. These hydrocarbons, coupled with the abundant pristane and phytane, are characteristic of a thermogenic component, representing the volatile boiling range of the petroliferous components (Hunt, 1979; Tissot and Welte, 1978) and are migrating upward. A minor higher plant wax component ($>C_{27}$)

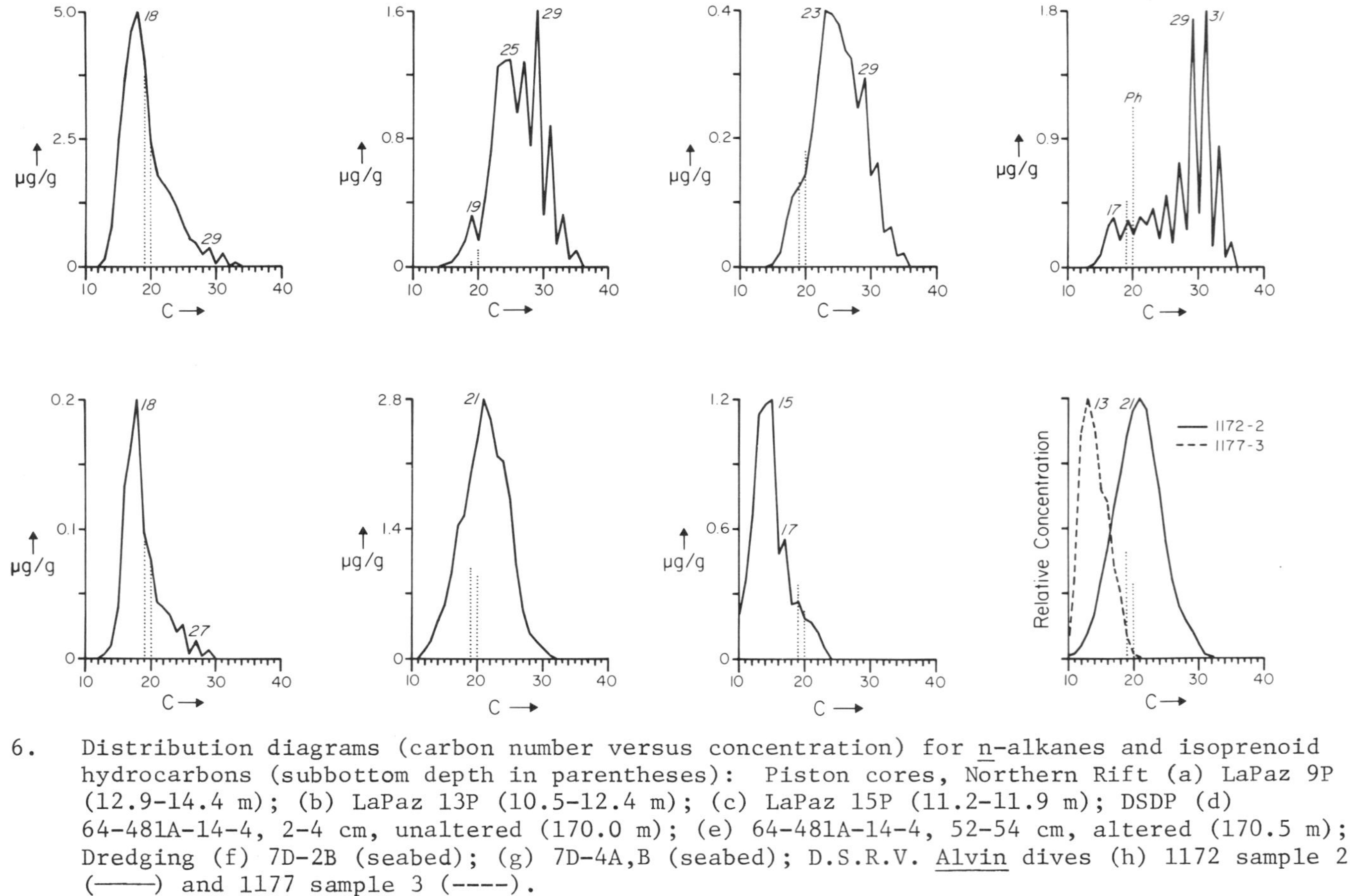

Fig. 6. Distribution diagrams (carbon number versus concentration) for n-alkanes and isoprenoid hydrocarbons (subbottom depth in parentheses): Piston cores, Northern Rift (a) LaPaz 9P (12.9-14.4 m); (b) LaPaz 13P (10.5-12.4 m); (c) LaPaz 15P (11.2-11.9 m); DSDP (d) 64-481A-14-4, 2-4 cm, unaltered (170.0 m); (e) 64-481A-14-4, 52-54 cm, altered (170.5 m); Dredging (f) 7D-2B (seabed); (g) 7D-4A,B (seabed); D.S.R.V. Alvin dives (h) 1172 sample 2 (——) and 1177 sample 3 (----).

is also present in this sample, reflecting some terrigenous influx.

Samples 13P and 15P have similar n-alkane distributions ranging from C_{14} to C_{35} (CPI_{14-35} = 1.7 and 1.2, respectively), with a lesser petroliferous component maximizing at slightly higher boiling range (C_{23}-C_{25}). Normal alkanes from higher plant wax are major components, especially for sample 13P. The hump ranges from C_{12}-C_{30} for both samples and the overall thermogenic fraction is highest in 9P > 15P > 13P (cf. Table 1). The thermogenic nature of the petroliferous components in these samples is supported further by the molecular indicators present in minor amounts. This material appears to have migrated from depth by advection in pore fluids or by diffusion, since no major conduits or oil droplets were visible in the cores (cf. analogies - Sayles and Jenkins, 1982; Maris and Bender, 1982).

Numerous hydrothermal mounds rising 20-30 m above the basin floor (water depth about 2000 m) have been mapped in the Southern Trough and some have active hydrothermal plumes (Lonsdale, 1980). A dredge haul (7D, Fig. 1c) across a 50 m-wide patch of sinter deposits (forming a mound) recovered claystones, massive sulfides, barite, talc and other hydrothermal minerals, together with tubeworm specimens. Many of the fragments in the dredge were stained with a petroleum-like oil and had a strong odor similar to diesel fuel.

Gasoline range hydrocarbons (C_5-C_{10}) were analyzed in two dredge samples (Simoneit and Lonsdale, 1982). The relative distributions of all hydrocarbons in this range were very similar for both samples, and their large amounts with the associated structural diversity confirmed their origin by thermal generation from the sedimentary organic matter (Whelan and Hunt, 1982).

The total bitumen (lipid) extract of sample 7D-2B is light amber in color and black for sample 7D-4A,B (cf. Table 1); both extracts of the samples exhibit blue fluorescence in solution typical for heavy oil condensates. These two bitumens are clearly different (Simoneit and Lonsdale, 1982). The aliphatic fraction (F1) is lowest for both samples and most of the bitumen is comprised of asphaltic (F3) and then aromatic/naphthenic (F2) material. The aromatic to aliphatic ratio for sample 7D-2B is 9.3 and it is 4.7 for sample 7D-4A,B, typical for many crude oils (Hunt, 1979).

The GC traces of the aliphatic hydrocarbons (Fig. 7) are dramatically different. Sample 7D-2B exhibits a pattern typical of petroleum, where the dominant n-alkanes range from C_{12}-C_{33} with no carbon number predominance (CPI_{12-33} = 1.03) and a maximum at n-C_{21} (cf. Fig. 6). Pristane and phytane are about equal

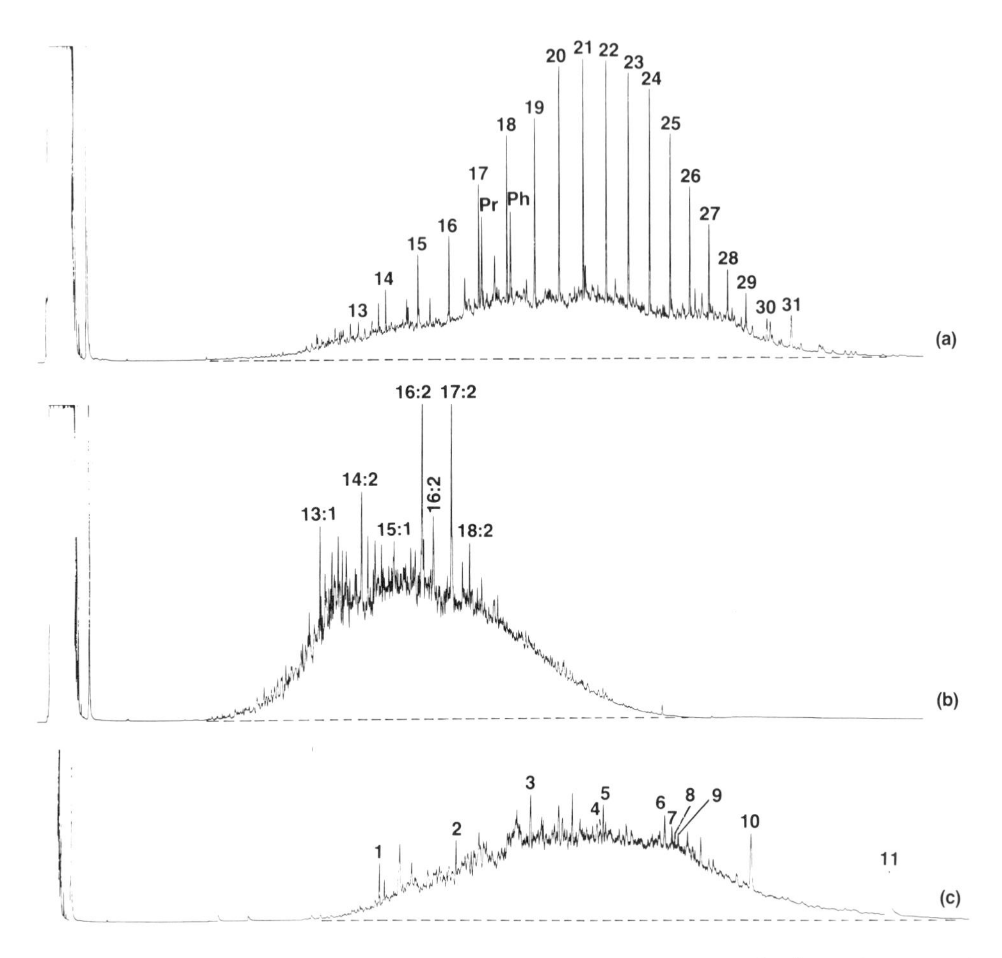

Fig. 7. Capillary gas chromatograms of aliphatic (F1) and aromatic (F2) hydrocarbon fractions for dredge samples (a and c) 7D-2B and (b) 7D-4A,B (Simoneit and Lonsdale, 1982): (a) sample 7D-2B, F1 (the carbon chain length of the *n*-alkanes is indicated by the arabic numerals, Pr = pristane, Ph = phytane); (b) sample 7D-4A,B, F1 (olefins are indicated by chain length: double bond equivalent); (c) sample 7D-2B, F2 (1 = tetramethylbenzene, 2 = phenanthrene, 3 = pyrene, 4 = benz(a)anthracene, 5 = chrysene, 6 = benzofluoranthenes, 7 = benzo(e)pyrene, 8 = benzo(a)-pyrene, 9 = perylene, 10 = benzoperylene, 11 = coronene).

(Pr/Ph = 1.06). The mixture of branched and cyclic hydrocarbons (i.e., hump) ranges from C_{11} to C_{31}, also typical of petroleum. On the other hand, sample 7D-4A,B exhibits a much narrower GC profile, skewed to lower carbon numbers. The dominant resolved peaks are various mono- and diolefins ranging from C_{12} to C_{19} and the hump ranges from C_9 to about C_{21}. This is very much unlike typical petroleum, as olefins are not found in mature crudes (Hunt, 1979; Tissot and Welte, 1978).

The n-alkanes of sample 7D-4A,B were determined by GC/MS analysis to be present as minor components, ranging from C_{10} to C_{23}, with a minor odd carbon number predominance (CPI_{10-23} = 1.20) and a maximum at C_{15} (Fig. 6). Pristane is more abundant than phytane (Pr/Ph = 1.6). The GC/MS data for both samples confirmed the presence of mono- and diolefins ranging from C_{12} to C_{19}. These compounds are not terminal olefins as were identified near the sills of DSDP Site 481A (Simoneit and Philp, 1982). They are slightly more stable "in-chain" olefins with methyl branching, similar to those in Bradford crude oil (Hoering, 1977).

The major diagnostic molecular markers in the hydrocarbon fraction (7D-2B,F1) consist of triterpenoids, extended tricyclic diterpanes and steranes with their rearranged analogs (diasteranes) (Simoneit and Lonsdale, 1982) and are discussed as an example. The extended tricyclic diterpanes range from C_{20} to C_{29} in a similar distribution pattern as observed for other mature petroleum samples (Simoneit and Kaplan, 1980). They are, however, not present in the unaltered surface sediments. The triterpenoids (Fig. 8) are essentially "mature"; they are for the most part in their thermodynamically more stable form, completely different from those in unaltered lipids of surface sediments (Simoneit and Philp, 1982; Simoneit et al., 1979). The triterpenoids are comprised primarily of the 17α(H),21β(H)-hopane series ranging from C_{27} (no C_{28}) to C_{35} and the homologs from C_{31}-C_{35} are present as 22-S and R diastereomeric pairs in a ratio of about unity (Fig. 8). This series constitutes the stable mature form of these compounds (Dastillung and Albrecht, 1976; Ensminger et al., 1974) and appears to have been generated by the hydrothermal activity. The compounds are not found in the unaltered sediments, where the biogenic markers with the 17β(H),21β(H) stereochemistry and various triterpenes predominate. Minor amounts of 17β(H),21α(H)-moretanes (C_{29}, C_{30} and C_{31}), 17β(H),21β(H)-hopanes (C_{27} and C_{30}), iso-hop-13(18)-ene and 17β(H)-moret-22(29)-ene are also present in sample 7D-2B. These are precursor relics and intermediates from the thermal conversion process.

The steroidal markers consist of primarily 5α(H),14α(H),17α(H)-steranes (20 R), with lesser amounts of 5β(H),14α(H),17α(H)-steranes (20 R) and diasteranes [mainly the 13β(H),17α(H)-20 R or S series]. The steranes ranged from C_{26} to C_{30}, with cholestane as the major homolog. The distribution pattern indicates a marine autochthonous origin and fits best with similar data for DSDP Site 477 (Simoneit and Philp, 1982). The large 5α(H)-sterane concentration is a result of the elevated thermal stress which probably converted

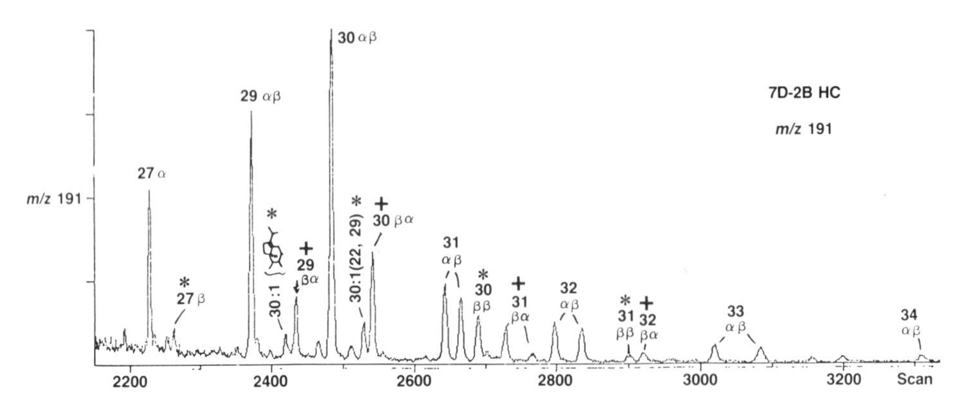

Fig. 8. Mass chromatogram (m/z 191) showing the triterpenoid distribution for dredge sample 7D-2B (Simoneit and Lonsdale, 1982). The predominant 17α(H)-hopanes are indicated by the arabic numerals followed by α, moretanes by βα, 17β(H)-hopanes by β and olefins by 30:1 (* indicates not present in mature crude oils, + not common in mature crude oils).

other steroidal compounds to these hydrocarbons. The extended diterpenoidal, triterpenoidal and steroidal indicator compounds are not detectable by GC/MS in the hydrocarbon fraction of sample 7D-4A,B.

An example of a GC trace of the aromatic/naphthenic fraction (F2) of sample 7D-2B is shown in Fig. 7c. The GC/MS data indicate that the major resolved peaks are polynuclear aromatic hydrocarbons (PAH), another group of compounds uncommon in petroleums but ubiquitous in higher temperature pyrolysis residues (Geissman *et al.*, 1967; Hunt, 1979; Blumer, 1975; LaFlamme and Hites, 1978; Ishiwatari and Fukushima, 1979). The dominant analogs for both samples are the pericondensed aromatic series as for example pyrene, benzopyrenes, perylene, benzoperylene and coronene. A further indication for a pyrolytic origin is the presence of five-membered alicyclic rings (e.g., acenaphthene, fluorene, fluoranthene, etc.), which are found in all pyrolysates from organic matter, since once formed they do not easily revert to pericondensed aromatic hydrocarbons (Blumer, 1975, 1976; Scott, 1982). It should also be noted that this fraction contains significant amounts of toxic PAH. The benzopyrenes are major components and the highly toxic benzo(a)pyrene is present in the two samples at levels of 25 and 16 ng/g of total bitumen, respectively. Perylene is present in these fractions and it is the predominant PAH in the unaltered sedimentary lipids of samples deposited in the Gulf from oxygen-minimum environments (Simoneit and Philp, 1982; Simoneit, 1982b). Thus, the chemical composition of this fraction indicates a source from pyrolysis with rapid quenching by hydrothermal removal and subsequent condensation at the seabed.

This process of pyrogenesis, removal and transport to the seabed appears to be operative at most of the mounds explored by subsequent diving with the deep submersible R.V. _Alvin_ during January, 1982. Most of the areas with lower temperature regimes that were sampled at the mounds consisted of hydrothermal minerals cemented by solid petroleum. This material liquified upon warming on deck, thus liberating the characteristic volatile and odorous components. Preliminary data indicate that hydrothermal petroleum has become incorporated into seven of the nine areas sampled by the D.S.R.V. _Alvin_ and one vent was discharging a water-oil emulsion. The boiling ranges and compositions of the various oil samples are quite variable (e.g., Fig. 6h). The oil on the exterior surfaces of the mounds is weathered, leached and/or degraded, whereas interior samples are relatively unaltered.

OVERVIEW AND IMPLICATIONS

The rifts of the Guaymas Basin are areas of active petroleum genesis, primarily due to deep seated heat sources acting on the recent, immature organic matter. Migration of petroleum occurs with the pore fluids to the vents and by advection/diffusion through the sedimentary cover to the seabed. This overall process should be considered in petroleum exploration for areas with deep heat sources and a massive sedimentary cover (e.g., Northern Gulf of California with sediments in excess of 1-2 km thickness). In such areas the presence of potential traps (impermeable lithologies) could aid the accumulation of hydrothermal petroleum.

The sediments at depth (DSDP Sites 477, 478 and 481) contain large amounts of thermogenic (admixed with some original biogenic) interstitial gas, thermally-derived bitumen (again admixed with original biogenic lipid residues), and residual, spent protokerogen (hydrogen-poor carbonaceous detritus). This occurs in close proximity to major sill intrusions, in temperature regimes of 150-200°C (the oil window, Tissot and Welte, 1978), and especially at depth in Site 477 on approach to the deeper high heat source, where the probable temperature regime exceeds 300°C. These temperatures are also in accord with the inorganic geochemistry of the hydrothermal mineralogy (Gieskes _et al._, 1982), where the deep seated heat source in the Site 477 area has been interpreted as being active for an extensive time period to develop a greenschist facies. The temperatures in excess of 300°C would result in early expulsion of the petroleum products and subsequent high temperature synthesis of the PAH.

The pyrolysates have migrated by advection, diffusion, distillation and hydrothermal circulation away from these heat sources and upward to the seabed. There the petroleum condenses according to the ambient temperatures in the conduits and vugs of the hydrothermal mineral mounds. PAH and sulfur condense in the hot vents; waxes crystallize in intermediate temperature regions (∿20-80°C); and the volatile petroleum partially collects in cold areas (0°C) and emanates into the sea water.

The thermogenic origin of the petroleum sampled at the seabed (e.g., the dredge samples, piston core samples and dive samples) has been confirmed. This is based upon the following chemical parameters: 1) presence of gasoline range hydrocarbons (also the odorous compounds), 2) broad distribution of hydrocarbons (hump and $\underline{n}$-C_{13}-C_{33}, with no carbon number predominance) and the presence of pristane and phytane, 3) the relative amounts of aromatic/naphthenic (F2) and asphaltic (F3) material versus the aliphatic hydrocarbons (F1), and 4) the presence of olefins and PAH.

The presence of petroliferous material at the seabed indicates that the plume and interstitial waters must contain large amounts of methane and higher molecular weight organic matter. This represents a carbon source to microbes at the vent site, as evidenced in part by the degradation of the exposed oil samples. The microbiota could thus in turn be consumed by the higher order fauna. This is a major difference compared to hydrothermal systems without a sediment blanket, where the carbon source of the ecosystem is primarily carbon dioxide, although some methane appears also to be metabolized. The presence of toxic PAH in the petroleum is of interest in terms of its effect on the biota. The macrofauna does not appear to be detrimentally influenced by the presence of the PAH nor the oil.

ACKNOWLEDGEMENTS

I thank the Deep Sea Drilling Project and the National Science Foundation for my participation on DSDP Leg 64, the NSF for my participation on the D.S.R.V. Alvin, Pluto 6 cruise; Dr. P. Lonsdale for the dredge and piston core samples, and Dr. E. M. Galimov, Dr. R. P. Philp, Mr. P. Jenden, Mr. C. Jackson, and Mr. E. Ruth for data and assistance. Funding from the National Science Foundation, Division of Ocean Sciences (Grant OCE81-18897) and from the OSU Oceanography Industrial Cooperative Program is gratefully acknowledged.

REFERENCES

Blumer, M., 1975, Curtisite, idrialite and pendletonite, polycyclic aromatic hydrocarbon minerals: Their composition and origin, Chem. Geol., 16:245-256.

Blumer, M., 1976, Polycyclic aromatic compounds in nature, Scient. Amer., 234(3):34-45.

Curray, J. R., Moore, D. G., Aguayo, J. E., Aubry, M. P., Einsele, G., Fornari, D. J., Gieskes, J., Guerrero, J. C., Kastner, M., Kelts, K., Lyle, M., Matoba, Y., Molina-Cruz, A., Niemitz, J., Rueda, J., Saunders, A. D., Schrader, J., Simoneit, B. R. T. and Vacquier, V., 1979, Leg 64 seeks evidence on development of basins, Geotimes, 24(7):18-20.

Curray, J. R., Moore, D. G., Aguayo, J. E., Aubry, M. P., Einsele, G., Fornari, D. J., Gieskes, J., Guerrero, J. C., Kastner, M., Kelts, K., Lyle, M., Matoba, Y., Molina-Cruz, A., Niemitz, J., Rueda, J., Saunders, A. D., Schrader, H., Simoneit, B. R. T. and Vacquier, V., 1982, Initial Reports of the Deep Sea Drilling Project, Vol. 64, Parts I and II, U.S. Govt. Printing Office, Washington, D.C., 1314 pp.

Dastillung, M. and Albrecht, P., 1976, Molecular test for oil pollution in surface sediments, Mar. Poll. Bull., 7:13-15.

Einsele, G., Gieskes, J., Curray, J., Moore, D., Aguayo, E., Aubry, M. P., Fornari, D. J., Guerrero, J. C., Kastner, M., Kelts, K., Lyle, M., Matoba, Y., Molina-Cruz, A., Niemitz, J., Rueda, J., Saunders, A., Schrader, H., Simoneit, B. R. T. and Vacquier, V., 1980, Intrusion of basaltic sills into highly porous sediments and resulting hydrothermal activity, Nature, 283:441-445.

Ensminger, A., Van Dorsselaer, A., Spyckerelle, C., Albrecht, P. and Ourisson, G., 1974, Pentacyclic triterpenes of the hopane type as ubiquitous geochemical markers: origin and significance, in: "Advances in Organic Geochemistry 1973," B. Tissot and F. Bienner, eds., Editions Technip, Paris, pp. 245-260.

Galimov, E. M. and Simoneit, B. R. T., 1982a, Geochemistry of interstitial gases in sedimentary deposits of the Gulf of California, Leg 64, in: "Initial Reports of the Deep Sea Drilling Project," Vol. 64, J. R. Curray, D. G. Moore et al., U.S. Govt. Printing Office, Washington, D.C., pp. 781-788.

Galimov, E. M. and Simoneit B. R. T., 1982b, Variations in the carbon isotope compositions of CH_4 and CO_2 in the sedimentary sections of Guaymas Basin (Gulf of California), Geokhimya, Acad. Nauk SSSR, 7:1027-1034.

Geissman, T. A., Sim, K. Y. and Murdoch, J., 1967, Organic minerals. Picene and chrysene as constituents of the mineral curtisite (idrialite), Experientia, 23:793-794.

Gieskes, J. M., Kastner, M., Einsele, G., Kelts, K. and Niemitz, J., 1982, Hydrothermal activity in the Guaymas Basin, Gulf of California: A synthesis, in: "Initial Reports of the Deep Sea Drilling Project," Vol. 64, J. R. Curray, D. G. Moore, et al., U.S. Govt. Printing Office, Washington, D.C., pp. 1159-1167.

Hoering, T. C., 1977, Olefinic hydrocarbons from Bradford, Pennsylvania, crude oil, Chem. Geol., 20:1-8.

Hunt, J. M., 1979, "Petroleum Geochemistry and Geology," W. H. Freeman and Company, San Francisco, 617 pp.

Ishiwatari, R. and Fukushima, K., 1979, Generation of unsaturated and aromatic hydrocarbons by thermal alteration of young kerogen, Geochim. Cosmochim. Acta, 43:1343-1349.

Jenden, P. D., Simoneit, B. R. T. and Philp, R. P., 1982, Hydrothermal effects on protokerogen of unconsolidated sediments from Guaymas Basin, Gulf of California, elemental compositions, stable carbon isotope ratios and electron spin resonance spectra, in: "Initial Reports of the Deep Sea Drilling Project," Vol. 64, J. R. Curray, D. G. Moore, et al., U.S. Govt. Printing Office, Washington, D.C., pp. 905-912.

Kvenvolden, K. A. and Field, M. E., 1981, Thermogenic hydrocarbons in unconsolidated sediment of Eel River Basin, offshore Northern California, Amer. Assoc. Petrol. Geol. Bull., 65:1642-1646.

LaFlamme, R. E. and Hites, R. A., 1978, The global distribution of polycyclic aromatic hydrocarbons in recent sediments, Geochim. Cosmochim. Acta, 42:289-304.

Lonsdale, P., 1980, Hydrothermal plumes and baritic sulfide mounds at a Gulf of California spreading center (Abstract), EOS Trans. Am. Geophys. Union, 61:995.

Maris, C. R. P. and Bender, M. L., 1982, Upwelling of hydrothermal solutions through ridge flank sediments shown by pore water profiles, Science, 216:623-626.

Sayles, F. L. and Jenkins, W. J., 1982, Advection of pore fluids through sediments in the equatorial east Pacific, Science, 217:245-248.

Scott, L. T., 1982, Thermal rearrangements of aromatic compounds, Acc. Chem. Res., 15:52-58.

Simoneit, B. R. T., 1978, The organic chemistry of marine sediments, in: "Chemical Oceanography," Vol. 7, J. P. Riley and R. Chester, eds., Academic Press, London, pp. 233-311.

Simoneit, B. R. T., 1982a, Shipboard organic geochemistry and safety monitoring, Leg 64, Gulf of California, in: "Initial Reports of the Deep Sea Drilling Project," Vol. 64, J. R. Curray, D. G. Moore, et al., U.S. Govt. Printing Office, Washington, D.C., pp. 723-728.

Simoneit, B. R. T., 1982b, Organic geochemistry of laminated sediments from the Gulf of California, in: "Coastal Upwelling - Its Sediment Record," E. Suess and J. Thiede, eds., NATO - Adv. Res. Inst., in press.

Simoneit, B. R. T. and Kaplan, I. R., 1980, Triterpenoids as molecular indicators of paleoseepage in Recent sediments of the Southern California Bight, Mar. Environ. Res., 3:113-128.

Simoneit, B. R. T. and Lonsdale, P. F., 1982, Hydrothermal petroleum in mineralized mounds at the seabed of Guaymas Basin, Nature, 295:198-202.

Simoneit, B. R. T. and Philp, R. P., 1982, Organic geochemistry of lipids and kerogen and the effects of basalt intrusions on unconsolidated oceanic sediments: Sites 477, 478 and 481, Guaymas Basin, Gulf of California, in: "Initial Reports of the Deep Sea Drilling Project," Vol. 64, J. R. Curray, D. G. Moore, et al., U.S. Govt. Printing Office, Washington, D.C., pp. 883-904.

Simoneit, B. R. T., Mazurek, M. A., Brenner, S., Crisp, P. T. and Kaplan, I. R., 1979, Organic geochemistry of recent sediments from Guaymas Basin, Gulf of California, Deep-Sea Res., 26A:879-891.

Tissot, B. P. and Welte, D. H., 1978, "Petroleum Formation and Occurrence, A New Approach to Oil and Gas Exploration," Springer-Verlag, Berlin, 538 pp.

van de Meent, D., Brown, S. C., Philp, R. P. and Simoneit, B. R. T., 1980, Pyrolysis-high resolution gas chromatography and pyrolysis gas chromatography-mass spectrometry of kerogens and kerogen precursors, Geochim. Cosmochim. Acta, 44:999-1013.

Whelan, J. K. and Hunt, J. M., 1982, C_1-C_8 in Leg 64 sediments, Gulf of California, in: "Initial Reports of the Deep Sea Drilling Project," Vol. 64, J. R. Curray, D. G. Moore, et al., U.S. Govt. Printing Office, Washington, D.C., pp. 763-779.

Williams, D. L., Becker, K., Lawver, L. A. and von Herzen, R. P., 1979, Heat flow at the spreading centers of the Guaymas Basin, Gulf of California, J. Geophys. Res., 84:6757-6769.

GENESIS OF FERROMANGANESE DEPOSITS-

DIAGNOSTIC CRITERIA FOR RECENT AND OLD DEPOSITS

Kurt Boström

Department of Geology
University of Stockholm
10691 Stockholm, Sweden

ABSTRACT

Marine iron manganese deposits show a greater chemical variability than was realized only 10 years ago. As a consequence correct conclusions regarding the genesis of a given deposit frequently require knowledge about several geochemical parameters and preferably about accumulation rates and spatial relations vis-a-vis e.g. spreading centers.

For old deposits the analogous interpretation is considerably more difficult, since evidence may be missing for a nearby spreading center in the past and since elevated accumulation rates may not be detectable. Hence petrographic and geochemical characteristics may be the only ones available for a genetic interpretation. I will here show that a combination of geochemical relations such as Fe/Ti - Al/(Al+Fe+Mn), Fe/Mn, Fe-Mn-(Co+Ni+Cu), U/Th and REE (rare earth element) data may yield conclusive genetic information.

INTRODUCTION

The metal enriched sedimentary deposits on the ocean floor fall largely into two major categories:

a) manganese nodules, which occur in areas of slow accumulation and

b) hydrothermal deposits which occur at or near spreading centers.

a. Manganese nodules

The manganese nodules were discovered in the deep sea already during the Challenger expedition 1873-1876 (Murray and Renard, 1891), but not until after World War II was there a rapid growth in the interest in the geology, geochemistry, origin and possible exploation of the nodules. A number of excellent monographs and reports have appeared during the last decade (Horn, 1972; Horn et al., 1973; Glasby, 1977; Bischoff and Piper, 1979; Lalou 1979; Sorem and Fewkes, 1979; Cronan, 1980; Varentsov and Grasselly 1980; and Roy 1981) which have considerably deepened our knowledge about bulk chemistry, internal structure, mineralogy etc. Economic aspects are touched upon in several of these publications but additional discussions of this topic are given in Govett and Govett (1976), Ross (1978) and Earney (1980). Glasby and his coworkers have also published a number of useful literature compilations (Meylon et al., 1981); an analytical data-set has furthermore been published by Monget, et al., (1976). However, the origin of these deposits is not well known in spite of over 100 years of studies. Early opinions invoked among other mechanisms also volcanic activity (von Gümbel, 1878), but during the last few decades the nodules have increasingly been regarded as of hydrogenic or biogenic origin. Some of these controversies and problems are being treated by C. Lalou and Fleet in the papers in this section. The paper by Lalou is addressed to the enigmatic problem of nodule ages and growth rates. Most studies (Ku, 1977, Ku and Glasby, 1972) suggest very small growth rates; generally various methods yield consistent results. However, these growth rates are about 1000 times smaller than the accumulation rates of the sediments on which the nodules rest. Many solutions have been proposed for this problem, but none so far seems to be widely accepted. The paper by Fleet treats the mixing problem - i.e. to what extent nodules are essentially hydrogenous and hydrothermal and how these inputs can be identified and quantified, using REE-data. Another article of interest is the report on the Manop-project by Bender (1983). As is obvious from the reports discussed here there is at present a tendency to interpret the nodules as products of combined biogenic, diagenic, volcanic etc. processes.

b. Hydrothermal deposits

During the Challenger expedition also deposits rich in colloidal iron and manganese oxides in the carbonate

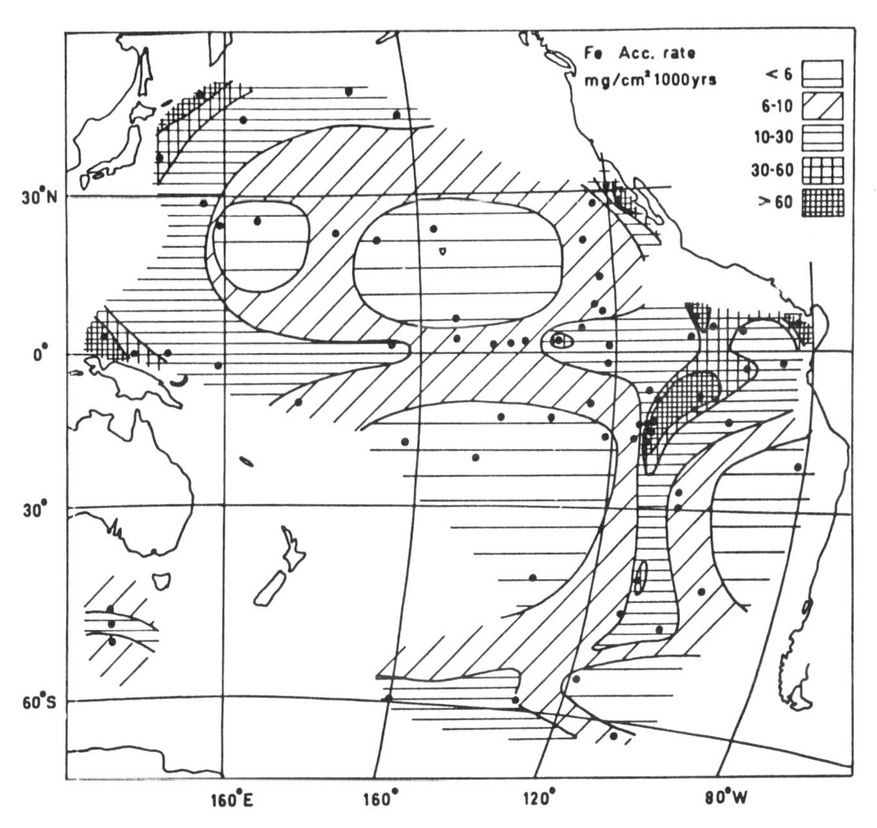

Figure 1. Accumulation rates of Fe in Pacific Ocean. Areas with low accumulation rates at 10-30°N and 10°-50°S largely coincide with the manganese nodule rich regions; high accumulation rates near the continents are caused by detrial iron. The high accumulation rates on the East Pacific Rise coincide with very low ones for Al, less than 2 mg/cm² 1000 years. (After Boström et al., 1973).

free fraction were obtained at stations 292 and 293 on the East Pacific Rise, but the significance of these finds was not realized then. Not until the early sixties would scientists turn their interest to such deposits and recognize them as sediments of volcanic origin (Skornyakova, 1964; Arrhenius and Bonatti, 1965; Boström and Peterson, 1966;

1969). Subsequently many more papers have been published on these sediments; reviews have been presented by Bonatti (1975), Boström (1980) and Rona (1983).

The origin of these sediments had been a controversial question before 1970, mainly because the early studies were insufficiently quantified as to rate of formation, but the volcanic hypothesis was finally accepted after accumulation rate maps conclusively demonstrated that Fe and Mn deposited fastest near the spreading center (Boström 1970; Bender et al., 1971) (See Fig 1). However, the most sensational support for the volcanic theory eventually came from several in situ studies whereby increased temperatures were registered at sites where hydrotheraml deposits form (Rona et al; 1975) or by actual observation of hot springs and associated sulfide deposition (Corliss et al., 1979; Francheteau et al., 1979).

These findings also suggested a connection with the earlier finds of sulfide rich hot brine deposits in depressions in the Red Sea; these had earlier been ascribed an evaporite-leaching origin by several authors, but Bischoff (1969) had advocated a hydrothermal origin, which at present appears as the most likely explanation.

Such hydrothermal rock-water reactions may proceed at highly variable conditions. In deep sea hydrothermal systems we can as a rule assume a deficiency in oxygen whereas shallow hydrothermal systems may involve much oxygen and hence form very acid solutions, which are found on volcanic islands (Boström and Widenfalk, 1983). The volcanic sediments formed near volcanic islands therefore differ much from those in the deep sea.

It is now clear that the hydrothermal deposits on the ocean floor show a variety of appearances, which, however, can be brought into a coherent classification that well explains the sequence of mineralforming events, etc., as is shown by the paper by Bonatti in this section. Depending on the circumstances various criteria may furthermore be required to recognize different hydrothermal deposits (Rona, 1978).

c. Old metalliferous oceanic deposits

Metalliferous deposits have been encountered as basal layers at several deepsea drilling sites (DSDP) as was to be expected (Boström and Petersen, 1969) e.g. during DSDP leg 5 in NE Pacific (von der Borch and Rex 1970). More difficult to interpret are old deposits that have lost their association with the plate or when the latter cannot be identified as a substratum that has originated at a spreading center. Also other basaltic rocks, e.g. seamounts, may be covered with iron-manganese pavements. After such deposits have been involved in orogenic events and associated metamorphism at a continental margin collision center the primary nature of the Fe-Mn deposit may indeed be hard to unveil. A search for conclusive petrographical and geochemical criteria is therefore of interest.

DATA SOURCES

The geochemical relations discussed below refer to numerous different deposits. Therefore a condensed data account is given here in order to avoid several repetitous quotations.

Epicontinental nodules: Manheim, 1965 (Baltic Sea); Calvert and Price, 1970 (Loch Fyne); Ku and Glasby, 1972 (Jervis inlet); Boström et al., 1982 (N. Baltic Sea, 8 mean values of 97 analyses); Manheim, 1961; 1982 ($MnCO_3$--nodules).

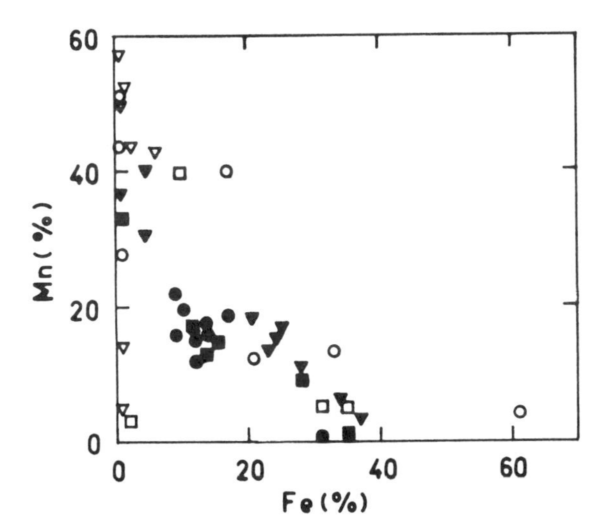

Figure 2. Distribution of Fe and Mn in various fossil nodules (open symbols) and Recent ones (filled symbols). Circles represent pelagic environments, squares shallow marine banks, horsts or sea mounts and triangles represent littoral and epicontinental sedimentation regimes.

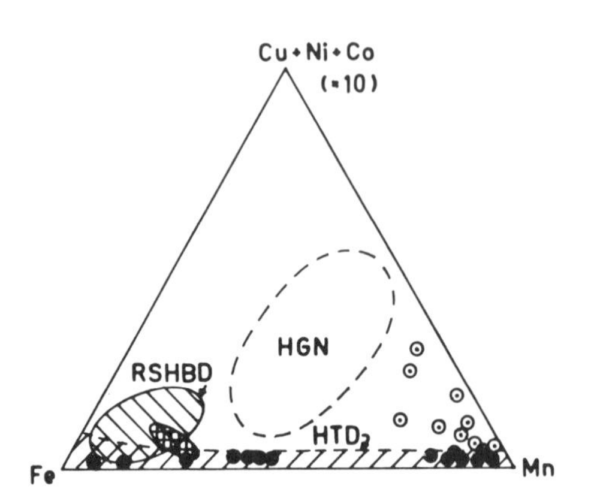

Figure 3. Distribution of Mn, Fe and (Cu+Ni+Co) in various Fe-Mn rich deposits. RSHBD= Red Sea hot brine deposits, HGN= hydrogenous nodules, and HTD (shaded field)= hydrothermal deposits (crusts).
Small shaded envelope at about 70% Fe, 25% Mn represents unconcolidated East Pacific Rise deposits. Circles represent diagenetic nodules; filled circles = epicontinental nodules (Baltic Sea, Loch Fyne); open circles= nodules from E. Pacific Ocean. Note overlap between diagenetic and hydrothermal depposits.

Shallow water marine nodules and crusts: Mero, 1965 (Off Japan; Blake Plateau); Bonatti et al., 1972 b (Thera; Stromboli).

Deep Sea nodules: Mero, 1965 (World- wide); Cronan 1980 (Pacific nodules re Fe, Mn, Al and Ti content).

Fossil nodules: Jenkyns, 1977.

Unconsolidated volcanic sediments: Boström and Peterson, 1969; Boström and Rydell, 1979 (East Pacific Rise); Bischoff, 1969 (Red Sea hot brine deposits); Scott et al., 1972 (S. Indian Ocean); Boström and Widenfalk,

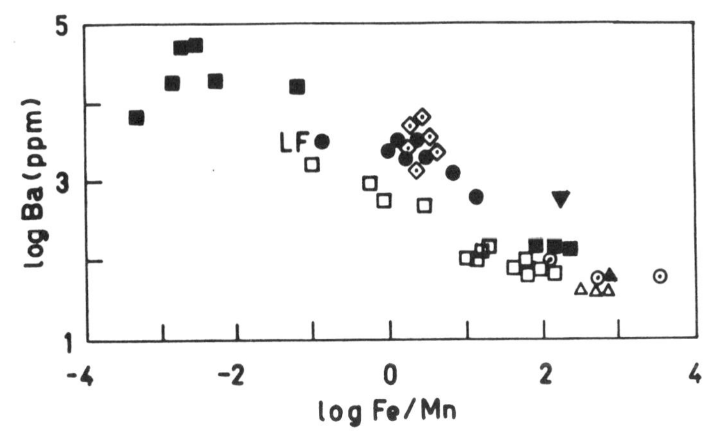

Figure 4. Distribution of Ba, Fe and Mn in various metal rich deposits. Filled squares = hydrothermal deposits from En Kafala; Open squares = hydrothermal deposits from Stromboli, Thera and East Pacific Rise (solid crusts). Open diamonds = hydrothermal (unconsolidated) deposits from the East Pacific Rise. Open circles = hydrothermal muds from Santorini. Filled circles = nodules from epicontinental environments; ordinary pelagic nodules (not shown) cluster near log Fe/Mn = 0 and log Ba = 3. Open triangles = rocks from Santorini. Filled triangles (small) = mean terrigenous matter; d:o (large) = mean lateritic matter).

1983 (Santorini, iron rich muds).

Consolidated volcanic deep sea sediments (crusts): Bonatti and Joensuu, 1966; Veeh and Boström, 1971, Rydell and Bonatti, 1973 (all about East Pacific rise deposits); Scott et al., 1974 (Mid Atlantic Ridge crust); Moore and Vogt, 1976 (Galapagos Rise crust).

Ordinary Pacific pelagic sediments: Boström 1980; Boström et al., 1976.

Laterites, bauxites and other weathering products: Valeton 1972; Loughnan 1969.

Fossil exhalative deposits: Bonatti et al., 1972 a (En Kafala); Boström et al., 1979 (Långban).

Mean values Boström 1976 (Basaltic matter, biological matter, terrigenous matter). Boström and Widenfalk, 1983

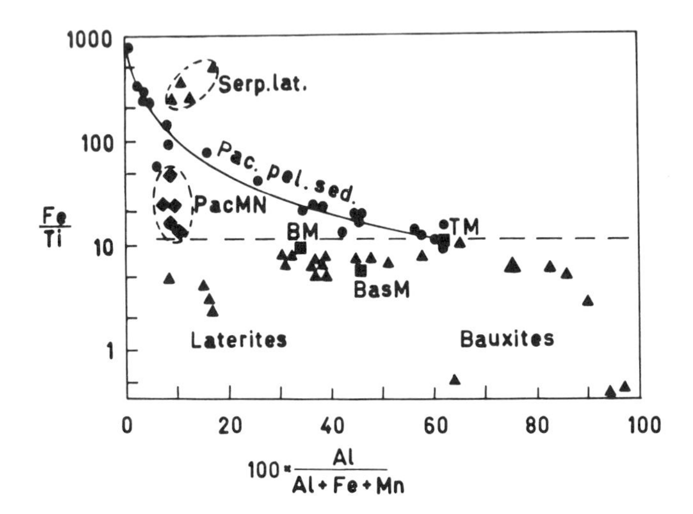

Figure 5. Relations between Fe, Ti, Al and Mn in various Fe - Mn - Al rich deposits. Bas M = = basaltic matter, TM = terrigenous matter and BM = biological matter (all shown as large squares). Small triangles = laterites and bauxites; large triangle = mean bauxite. Serp.lat. = laterites formed on serpentinite.
Circles = Pacific pelagic unconsolidated sediments. Solid curve represents the theoretical mixing line between East Pacific Rise deposits (top left corner) and TM (Boström 1970, Boström et al 1976); the real and theoretical distributions agree well. Diamonds = Pacific manganese nodules.

(volcanics rocks from Santorini)

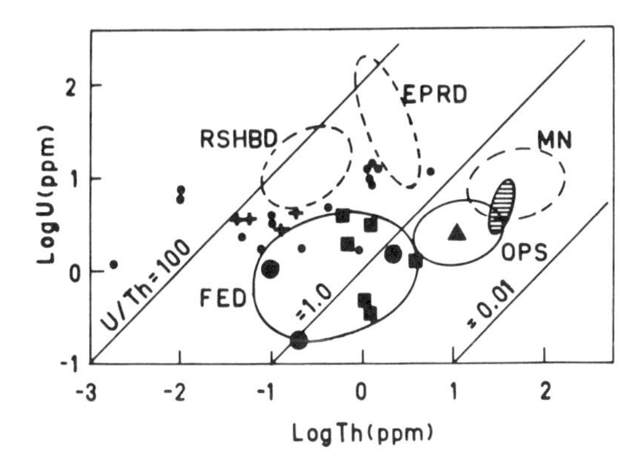

Figure 6. Relations between U and Th in MN = manganese nodules; OPS = ordinary pelagic sediments; filled triangle = terrigenous matter; small shaded field = bauxites; RSHBD = Red Sea hot brine deposits; EPRD = East Pacific Rise deposits; small circles = hydrothermal crusts near spreading centers.
FED = Fossil exhalative deposits, involving the En Kafala deposit (large squares) and the Långban deposit (large circles). Several more swedish Precambrian iron and manganese ores show U/Th ratios larger than 1.0 (Boström and Loberg, unpubl. results).
Crosses represent ophiolitic Fe - Mn deposits, probably of an exhalative - sedimentary origin.

PETROGRAPHIC CHARACTERISTICA

The various deposits (concretionary nodules and unconsolidated Fe-Mn sediments) will tend to recrystallize during metamorphism, a process that will be controlled by the pressure and temperature conditions and the major element geochemistry of the deposit. Deposits of different origins but with similar major element compositions will hence develop the same petrographical features, i.e. petrographic information may be of restricted value.

GEOCHEMICAL CHARACTERISTICS

Fossil nodules have been described in the literature, particularly by Jenkyns 1977, who also analyzed several of the nodules for Fe, Mn and some other elements (see Fig 2). Most of the nodules he studied were furthermore little altered; the nature of the associated strata could hence be ascribed a pelagic, littoral etc. nature. It is clear, however, from the data in Fig 2 that little could be inferred from the Fe-Mn relations had the alterations been more severe.

The relations from Fe - Mn - (Cu+Ni+Co) and for Fe/Mn-Ba are shown in Figs 3 and 4. Several groups of deposits with very different origins coalesce in the plots. If the general geological setting is well known these plots can be of use in genetic discussions, but are of little value only for the discussion of strongly metamorphosed sediments that have been separated from their associated rocks.

The relations for Fe/Ti - Al/(Al+Fe+Mn) and U/Th are shown in Figs 5 and 6. It is apparent that all hydrothermal sediments with Fe/Ti over 20 and Al/(Al+Fe+Mn) values less than 0.35 should be well recognizable. Thus almost all laterites and bauxites have Fe/Ti ratios distinctly lower than 11.0, the only rare exception being laterites formed on serpentinites. Furthermore, manganese nodules fall distinctly below the curve describing the Pacific pelagic sediment series; however, only few nodules have been analyzed for both Al and Ti, additional studies of the distribution of these elements in nodules should therefore be performed.

The relation U/Th is shown in Fig.6. The plot clearly demonstrates that hydrogenous manganese nodules and ordinary pelagic sediments have U/Th ratios near 0.3; the same is also true for bauxites. In contrast to this pattern most hydrothermal deposits show U/Th ratios well exceeding 1.0 and in several cases even exceed 100.

Two fossil (or supposedly fossil) exhalative deposits (FED) are shown in Fig. 6, namely the En Kafala deposit (Bonatti et al., 1972 b) and the Långban deposit (Boström et al., 1979). Even their lowest U/Th ratios are somewhat higher than is normal for mean "normal" oceanic sediments and their highest ratios reach well into the "hydrothermal" region. This is perhaps most interesting for the Långban deposit, where greenshist metamorphism probably has caused some loss of the fairly mobile U, i.e. the original U/Th ratios were probably even higher than today. The conclusion that the Långban deposit is of an exhalative-sedimentary origin has support also in many other observations (Brotzen, 1955).

U/Th relations furthermore strongly suggest that certain ophiolites are indeed oceanic products, as reviewed in Bonatti (1955).

Rare earth element studies have shown that hydrothermal deposits show relative REE abundances that are virtually identical to those in seawater (Bender et al., 1971; Fleet 1983), whereas hydrogenous nodules show relative enrichments in the lighter REE (La, Nd and Sm) and show a large positive Ce-anomaly. However, some non-oceanic nodules (Black Sea) lack the Ce-anomaly and have much lower total REE-abundance than ordinary oceanic nodules (Fomina and Volkov, 1969; Piper, 1974). Indeed, shallow water nodules show hardly any preferential uptake of REE relative to associated sediments. This variability in REE-abundances due to the genetic conditions (Fleet 1983) make REE data less simple to use in a straightforward manner when old deposits are being studied.

CONCLUSIONS

It appears likely that the relations between Fe/Ti - Al/(Al+Fe+Mn) and U/Th will be useful genetic indicators also for metamorphosed iron-manganese deposits. As long as surrounding strata are not rich in reducing matter there is a tendency for the iron-manganese oxides to show a considerable redox buffering effect as is demonstrated by the Långban deposit (see above) and only with considerable increase in temperature and reducing conditions will the highly oxidized manganese compounds eventually disappear. These redox capacity-phenomena may explain the tenacious preservation of manganese in fossil nodules as observed by Jenkyns, 1977, and may also help preserve the original composition of the deposit.

These effects of the redox capacities are often underestimated in geological studies (Breck, 1974; Boström, 1981), but do perhaps make iron-manganese deposits better geochemical paleoenvironment indicators than commonly realized.

ACKNOWLEDGEMENTS

Funds for support of this research were provided by the Swedish Board for Technical Development (STU) and the Swedish Natural Science Research Council (NFR).

REFERENCES

Arrhenius, G.O.S., and Bonatti, E., 1965. Neptunism and volcanism in the ocean. In Progress in Oceanography, vol 3, M. Sears ed., Pergamon Press, New York, 7-21.

Bender, M.L., 1983. The manganese nodule program EOS, 64(5) 42-43.

Bender, M., Broecker, W., Gornitz, V., Middel, U., Kay, R., Sun, S.S., and Biscay, P., 1971. Geochemistry of three cores from the East Pacific Rise. Earth and Planetary Sci. Letters, 12; 425-433.

Bischoff, J.L., 1969. Red Sea geothermal brine deposits: their mineralogy, chemistry and genesis. In:Hot Brines and Recent Heavy Metal Deposits in the Red Sea, E.T. Degens and D.A. Ross eds., Springer-Verlag, New York, 368-401.

Bischoff, J.L., and Piper, D.Z., 1979., eds, Marine Geology and Oceanography of the Pacific Manganese Nodule Province, 842 p. Plenum Press, New York.

Bonatti, E., 1975. Metallogenesis at oceanic spreading centers. In: Annual Review of Earth and Planetary Sciences, F.A. Donath, F.G. Stehli and G.W. Wetherill eds. Annual Reviews, Inc., Palo Alto, 401-431.

Bonatti, E., 1983; Hydrothermal Metal Deposits from the Oceanic Rifts: a Classification. In: Hydrothermal Processes at seafloor spreading centers, P.A. Rona, K. Boström, L. Laubier and K. Smith, eds. NATO Conference series, Plenum Press.

Bonatti, E., and Joensuu, O., 1966. Deep Sea iron deposits from the South Pacific. Science, 154, 643-645.

Bonatti, E., Fischer, D.E., Joensuu, O., Rydell, H.S., and Beyth, M.,1972 a. Iron-manganese-barium deposits from the Northern Afar Rift (Ethiopia). Econ. Geol. 67, 717-730.

Bonatti, E., Honnorez, J., Joensuu, O., and Rydell, H.S., 1972 b. Submarine iron deposits from the Mediterranean Sea. In: Symposium on the sedimentation in the Mediterranean Sea.VIII Internat. Sedim. Congr., D.J. Stanley, ed., Heidelberg.

Bonatti, E., Kraemer, T., and Rydell, H., 1972 c.Classification and genesis of submarine iron-manganese deposits. In: Ferromanganese deposits on the Ocean Floor, 149-161. D. Horn, ed., National Science Foundation,

IDOE, Washington, D.C.

Boström, K., 1970. Submarine volcanism as a source for iron. Earth and Planet Sci.: Letters, 9, 348-354.

Boström, K., 1976. Particulate and dissolved matter as sources for pelagic sediments. Stockholm Contr. Geology 30:15-79.

Boström, K., 1980. The origin of ferromanganoan active ridge sediments. In: Seafloor spreading Centers: Hydrothermal systems, P.A. Rona and R.P. Lowell eds., Dowden. Hutchinson & Ross, Inc., Stroudsburg, 288-332.

Boström, K., 1981. On the formation of magnetite-hematite-apatite "magmas" from metalliferous sediments at subduction zones. Stockh. Contrib. Geology, 37, 21-42.

Boström, K., and Peterson, M.N.A., 1966. Precipitates from hydrothermal exhalations on the East Pacific Rise. Econ. Geology, 61, 1258-1265.

Boström, K., and Peterson, M.N.A., 1969. Origin of aluminum-poor ferro-manganoan sediments in areas of high heat flow on the East Pacific Rise. Marine Geology, 7, 427-447.

Boström, K., Kraemer, T. and Gartner, S., 1973. Provenance and accumulation rates of opaline silica, Al, Ti, Fe, Mn, Cu, Ni, and Co in pelagic sediments. Chem. Geol., 11, 123-148.

Boström, K., Joensuu, O., Valdés, S., Charm, W. and Glaccum, R., 1976. Geochemistry and origin of East Pacific sediments sampled during DSDP leg 34. In: Initial Reports of the Deep Sea Drilling Project, 34,559-574. Yeats et al., eds., U.S. Government. Print. Office, Washington.

Boström, K., and Rydell, H., 1979. Geochemical behavior of U and Th during exhalative sedimentary processe. In: La Genese des nodules de manganese. 151-166. Colloq. Internat. CNRS No. 289, C. Lalou, ed., CNRS, Paris.

Boström, K., Rydell, H. and Joensuu, O., 1979. Långban-An exhalative sedimentary deposit? Econ. Geol., 74, 1002-1011.

Boström, K., Wiborg, L. and Ingri, J., 1982. Geochemistry and origin of ferromanganese concretions in the Gulf of Bothnia. Marine Geology, 50, 1-24.

Boström, K. and Widenfalk, L., 1983. The origin of iron-rich muds at the Kameni Islands, Santorini. Chem. Geol.(Accepted for publication)

Breck, W.G., 1974. Redox levels in the sea. In: The Sea, 5, E.D. Goldberg ed., Wiley Interscience, New York. 153-179.

Brotzen, O., 1955. Some microstructures in jasper from the Långban mine. GFF, 77, 275-283.

Calvert, S.E., and Price, N.B., 1970. Compositions of manganese nodules and manganese carbonates from Loch Fyne, Scotland. Contrib. Mineral. Petrol., 29, 215-233.

Corliss, J.B., Dymond, J., Gordon, L.I., Edmond, J.M., von Herzen, R.P., Ballard, R.D., Green, K., Williams, D., Bainbridge, A., Crane, K. and van Andel, T.H., 1979. Submarine thermal springs on the Galapagos Rift. Science, 203, 1073-1083.

Cronan, D.S., 1980. Underwater Minerals. Academic Press, London. 362 p.

Degens, E.T. and Ross, D.A., Eds., Hot Brines and Recent Heavy Metal Deposits in the Red Sea. Springer Verlag.

Earney, F.C.F., 1980. Petroleum and Hard Minerals from the Sea. Edward Arnold, London, 291 p.

Fleet, A.J., 1983. Hydrothermal and hydrogenous ferro-manganese deposits: Do they from a continuum? The rare earth element evidence. In: Hydrothermal processes at seafloor spreading centers, P.A. Rona, K. Boström, L. Laubier and K. Smith, eds. NATO Conference series, Plenum Press.

Fomina, L.S., and Volkov, I.I., 1969. Rare earths in iron-manganese concretions of the Black Sea. Dokl. Akad. Nauk SSSR, 185, 188-191.

Francheteau, J., Needham, H.D., Chouckroune, P., Juteau, T., Seguret, M., Ballard, R.D., Fox, P.J., Normark, W., Carranza, A., Cordoba, D., Guerrero, J., Rangin, C., Bougalt, H., Cambon, P. and Hekinian, R., 1979. Massive deep sea sulfide ore deposits discovered on the East Pacific Rise. Nature, 277, 523-530.

Glasby, G.P., 1977, ed., Marine Manganese Deposits. Elsevier Scientif. Publ. Co, Amsterdam. 523 p.

Govett, G.J.S. and Govett, M.H., 1976. World Mineral Supplies - Assessment and perspective. Developments in Economic Geology, 3. Elsevier Scientif. Publ. Co, Amsterdam, 472 p.

Horn, D.R., 1972 ed., Ferromanganese Deposits on the Ocean Floor. The Office for the International Decade of Ocean Exploration, National Science Foundation, Washington D.C. 293 p.

Horn, D.R., Delach, M.N. and Horn, B.M. 1973. Metal Content of Ferromanganese Deposits of the Oceans. Technical report no 3, NSF GX 33616. National Science Foundation, Washington D.C. 51 p.

Jenkyns, H.C., 1977. Fossil nodules. In: Marine manganese deposits (Ed. G.P. Glasby) 87-108, Elsevier Scientif. Publ. Co, Amsterdam.

Ku, T.L., 1977. Rates of accretion. In: Marine Manganese Deposits (Ed: G.P. Glasby) 249-267. Elsevier Scientific. Publ. Co, Amsterdam.

Ku, T.L., and Glasby, G.P., 1972. Radiometric evidence for rapid growth rate of shallow-water continental margin manganese nodules. Geochim. Cosmochim. Acta, 36, 699-704.

Lalou, C., 1979, Ed., La Genese des Nodules de Manganese. Colloq. Internat. CNRS no 289, CNRS, Paris.

Lalou, C., Genesis of ferromanganese deposits. Hydrothermal origin. In: Hydrothermal processes at seafloor spreading centers. P.A. Rona, K. Boström, L. Laubier and K. Smith, eds., NATO Conference series, Plenum Press.

Loughnan, F.C., 1969. Chemical Weathering of the Silicate Minerals. Amer. Elsevier Publ. Co., Inc, New York, 154 p.

Manheim, F.T., 1961. A geochemical profile in the Baltic Sea. Geochim. Cosmochim. Acta, 25, 52-70.

Manheim, F.T., 1965. Manganese-iron accumulations in the shallow marine environment. In: Symposium on marine geochemistry Occas. publ. 3, D.R. Schink and J.T. Corliss, eds., 217-276, Mar. Lab., Univ. of Rhode Island.

Manheim, F.T., 1982. Geochemistry of manganese Carbonates in the Baltic Sea. Stockholm Contributions in Geology, 37, 145-159.

Mero, J.L., 1965. The mineral resources of the Sea. Elsevier Scientif. Publ. Co., Amsterdam.

Meylan, M.A., Glasby, G.P. and Fortin, L.I., 1981. Bibliography and Index to Literature on Manganese Nodules (1861-1979). Dept of planning and economic development, State of Hawaii, Honolulu, Hawaii, 530 p.

Monget, J.M., Murray, J.W. and Mascle, J., 1976. A worldwide compilation of published, multicomponent analyses of ferromanganese concretions. Techn. Rep. No 12, NSF-IDOE, Washington, D.C.

Moore, W.S., and Vogt, P.R., 1976. Hydrothermal manganese crusts from two sites near the Galapagos spreading axis. Earth and Planet. Sci. Letters, 29, 349-356.

Murray, J. and Renard A.F., 1891. Report on Deep Sea Deposits. Challenger Exped. Reports, 3, Her Majesty's Stationary office, London.

Piper, D.Z., 1974. Rare earth elements in ferromanganese nodules and other marine phases. Geochim. Cosmochim. Acta, 38, 1007-1022.

Rona, P.A., 1978. Criteria for recognition of hydrothermal mineral deposits in oceanic crust. Econ. Geol., 73, 135-160.

Rona, P.A., 1983. Hydrothermal mineral deposits at seafloor spreading centers. Earth Science Reviews (in press).

Rona, P.A., McGregor, B.A. Betzer, P.R. Bolger, G.W. and Krause, D.C., 1975. Anomalous water temperatures over the Mid-Atlantic Ridge Crest at 26°N latitude. Deep-Sea Research, 22, 611-618.

Rona, P.A., and Lowell, R.P., eds., 1980 Seafloor Spreading centers, Hydrothermal Systems. Benchmark Papers in Geology, 56. Dowden, Hutchinson and Ross. Inc. Stroudsburg, Pennsylvania. 425 p.

Ross, D.A., 1978. Opportunities and uses of the ocean. Springer Verlag, New York 320 p.

Roy, S., 1981. Manganese deposits. Academic Press, London. 458 p.

Rydell, H.S., and Bonatti, E., 1973. Uranium in submarine metalliferous deposits. Geochim. Cosmochim. Acta, 37, 2557-2565.

Rydell, H.S., Kraemer, T., Boström. K., and Joensuu, O., 1974. Postdepositional injections of uranium-rich solutions into East Pacific Rise sediments. Mar. Geol., 17, 151-164.

Scott, M.R., Osmund, J.K., and Cochran, J.K., 1972. Sedimentation rates and sediment chemistry in the South Indian Basin. In: Antarctic Oceanology II, D.E. Hayes, ed., Antarctic Res. Serie 19., Amer. Geophys. Union.

Scott, M.R., Scott, R.B., Rona, P.A., Butler, L.W. and Nalwalk, A.J., 1974. Rapidly accumulating manganese deposit from the median valley of the Mid-Atlantic Ridge. Geophys. Res. Lett., 1, 355-358.

Skornyakova, I.S., 1964. Dispersed iron and manganese in Pacific Ocean sediments. Int.Geology Rev. 7: 2161-2174.

Sorem, R.K. and Fewkes, R.H., 1979. Manganese Nodules. IFI/Plenum, New York. 723 p.

Valeton, I., 1972. Bauxites. Elsevier Publ. Co., Amsterdam. 226 p.

Varentsov, I.M., and Grasselly, G., 1980. Geology and Geochemistry of Manganese, vol 3: Manganese on the bottom of recent basins. E. Schweizerbart'sche Verlagsbuchhandlung Stuttgart. 357 p.

Veeh, H.H., and Boström, K., 1971. Anomalous $^{234}U/^{238}U$ on the East Pacific Rise. Earth Planet. Sci. Letters, 10, 372-374.

von der Borch, C.C., and Rex, R.W., 1970. Amorphous iron oxide precipitates in sediments cored during leg 5. In: Initial Reports of the Deep Sea Drilling Project, 5, D.A. Mc Manus et al., eds., U.S. Government Print. Office Washington.

von Gümbel, G., 1878. Ueber die im Stillen Ozean auf dem Meeresgrunde vorkommenden Manganknollen. Sitz. Berichte d.K. Bayerischen Akademie d. Wissenschaften Münschen. Matem - Physik Klasse, 189-209.

HYDROTHERMAL METAL DEPOSITS FROM THE OCEANIC RIFTS: A CLASSIFICATION

Enrico Bonatti

Lamont Doherty Geological Observatory
of Columbia University
Palisades, N.Y. 10964

INTRODUCTION

Much ground has been covered since the early suggestions by Skornyakova (1964), Arrhenius and Bonatti (1965), Bonatti and Joensuu (1966), and Bostrom and Peterson (1966) that metal concentrations are formed along mid-ocean ridges due to hydrothermal activity. The most spectacular of these developments have been, of course, the direct observation and sampling by submersible of the Galapagos Rift warm springs and related metal concentrations (Corliss et al., 1979), and of the 21°N East Pacific Rise hot vents and associated metal sulfide deposits (Francheteau et al., 1979).

It appears now that subseafloor hydrothermal circulation, particularly active along accretionary plate boundaries, can give rise to different types of metal deposits, depending on depth below the sea floor reached by the circulating solutions, types of rocks encountered during this circulation, temperature reached by the circulating waters, and a number of other factors. In this paper I attempt to review briefly and classify the different types of oceanic hydrothermal deposits on the basis of the aforementioned factors.

HYDROTHERMAL MODEL OF METALLOGENESIS IN THE OCEAN BASINS

The classification discussed in this paper is based on a model of sub-seafloor hydrothermal circulation which, since its earliest formulation by Elder (1965), has been supported and refined by a wealth of geophysical and geochemical data, and is now widely accepted. Briefly, seawater penetrates within the crust in the vicinity of a hot, highly fissured zone of crustal accretion, that is, along the axial, crestal zone of active mid-ocean ridges. The depth of penetration of the water depends on permeability and thermal gradient in the crust, and ranges probably from several km's to hundreds of meters. Both permeability (caused by faulting, fissuring and fracturing of the crust) and thermal regimes are probably quite different in fast (whole rate of seafloor spreadng: > 6 cm/y) and slow (< 4 cm/y) spreading ridges. The former (East Pacific Rise) are characterized by relatively shallow (1 to a few km) and large, permanent or semipermanent magma chambers and by sharp thermal gradients (Kusznir, 1980). The latter (mid-Atlantic Ridge), though highly fractured, has probably intermittent and smaller magma chambers and weaker thermal gradients (Kusznir, 1980). Accordingly, the sub-seafloor hydrothermal circulation will tend to be different in the two cases, with a higher probability of waters reaching rapidly high temperatures (> 350°C) in the case of fast-spreading ridges.

In order for the hydrothermal fluids to reach temperatures close to 400°C, as seen in the EPR 21°N system, the fluids must be heated close to the roof of a magma chamber (Bishoff, 1980; Strens and Cann, 1982). The fluids can then rise rapidly and adiabatically (as probably is the case at the EPR 21°N "black smoker" sites) or non-adiabatically (as is probably the case at the Galapagos Rift deposits).

Reactions of the circulating waters with the host rocks result in significant elemental exchanges which depend primarily on temperature, chemistry and mineralogy of the rock, chemistry of the water and water/rock ratio. The host rocks are mainly of basaltic composition, particularly in fast-spreading crust as in the East Pacific, while in a slow-spreading ocean (Atlantic) rocks of different composition (i.e., ultramafics) may also be involved to some extent.

The chemical exchanges between crust and seawater during the hydrothermal convective circulation can give rise to concentrations of a number of elements (among them, several metals, sulfur and silica) in solution. Some of these elements can then be deposited, either within the crust during different stages of the sub-sea floor circulation; or just before and during the injection of the solutions through the sea floor; or on the sea floor after discharge of the hydrothermal fluids.

A GEOLOGICAL AND CHEMICAL-MINERALOGICAL CLASSIFICATION OF MARINE HYDROTHERMAL DEPOSITS

Among the various metal deposits of hydrothermal origin observed and/or sampled in the ocean basin, a distinction can be made (table 1) between (a) deposits formed within the igneous crust before discharge of the solutions through the sea floor (<u>hydrothermal pre-dis-</u>

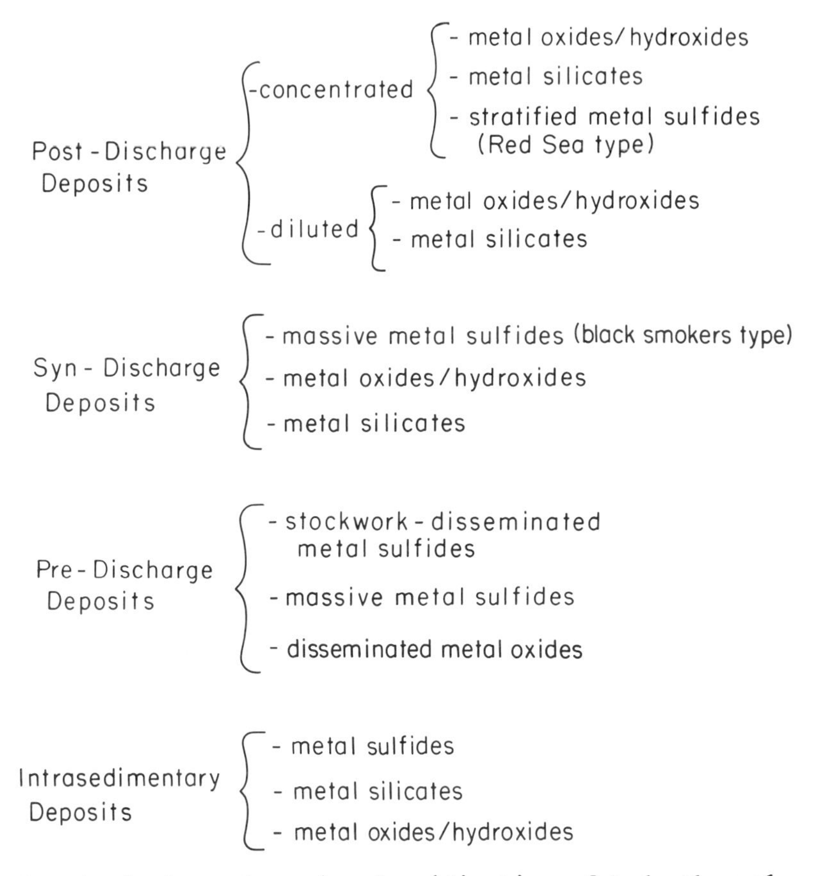

Table 1. A simple, schematic classification of hydrothermal metal deposits from oceanic rifts.

charge deposits); (b) deposits formed at the site of hydrothermal discharge through the sea floor (hydrothermal syn-discharge deposits); and (c) deposits formed by precipitation on the sea floor of hydrothermally-derived elements after some residence time in bottom seawater (hydrothermal post discharge deposits). Fig. 1 summarizes this classification.

An additional class of deposits is that found when the hydrothermal solutions are discharged from the igneous crust into sediments with metal concentrations forming within the sediment column (hydrothermal intra-sedimentary deposits).

I will consider each of these groups separately.

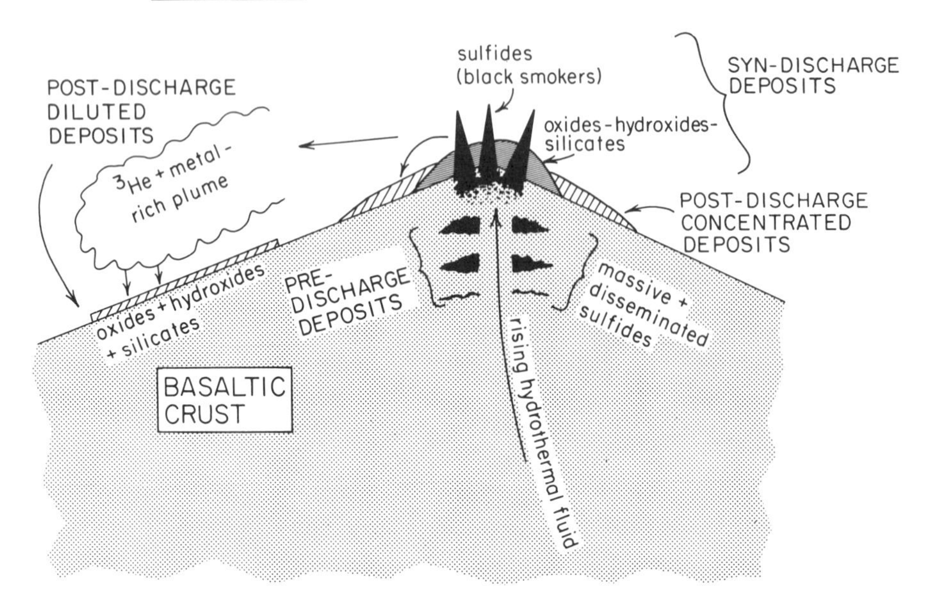

Fig. 1. Schematic representation of different types of metal deposits resulting from sub-sea floor hydrothermal circulation in the axial zone of accretion. Intra-sedimentary deposits are not represented.

HYDROTHERMAL PRE-DISCHARGE METAL DEPOSITS

The main direct evidence to date that metals can be deposited by the hydrothermal solutions within the crust, that is, before the solutions are discharged through the sea floor into sea water, is the observation of stockwork and disseminated type metal sulfides found in metabasalts sampled at or below the bottom. Moreover, metal oxides and hydroxides associated with silicates and carbonates of hydrothermal origin have been frequently observed in veins cutting through metabasalts, particularly those drilled by the DSDP program. However, it is likely that also sizeable massive sulfide deposits can be formed within the crust, particularly during the upward return flow of the hot fluid. Pre-discharge sulfide deposits can be formed by the following processes.

(1) In the case of fast spreading ridges (East Pacific Rise) where a shallow (~2 km) magma chamber may be present, the hot (>400°C) solutions probably rise adiabatically and rapidly in their return flow towards the sea floor, after having extracted H_2S, silica, Fe and base metals from the rock close to the roof of the magma chamber. Adiabatic expansion of the fluid during ascent will

cause strong supersaturation of sulfide minerals, which will tend to be deposited within any available space, particularly in the upper several hundred meters of the flow (Bischoff, 1980).

(2) It is possible that the hydrothermal fluids after having reached at depth a temperature close to 400°C, will ascend non-adiabatically but with significant heat loss to the wall rocks. This may happen when the upward path towards exit is tortuous and the return flow is relatively slow. Drop of temperature of the metal-rich hydrothermal fluids during ascent may result in deposition of sulfides within the crust.

(3) Boiling of the hydrothermal fluid at decompression during the ascending return flow does not occur at the 21°N East Pacific Rise "black smoker" site (Welhan and Craig, 1979) because the hydro thermal system operates there at pressures >260 bars (depth below sea level at discharge is roughly 2,800 meters). However, where hydrothermal systems discharge on the sea floor at pressures well below the critical pressure of sea water (that is, where the sea floor is shallower than roughly 2,000 meters below sea level), boiling might occur (Delaney and Cosens, 1982). Boiling could take place if and when in P/T space the adiabatic curve for the ascending hot hydrothermal fluids would intersect the liquid/vapor curve for sea water (Delaney and Cosens, 1982). Separation of a vapor and a liquid phase triggered by boiling will tend to partition salts (and base metals) into the liquid and may thus cause deposition of some of the metals before discharge of the solutions through the sea floor (Delaney and Cosens, 1982).

(4) If during the shallowest stages of the return flow the hot fluids mix with descending cold sea water, a drop in temperature of the solution is expected. Precipitation of metals (possibly as oxides in addition to sulfides) as well as silica can be expected to take place.

The type of deposits to be expected in the pre-discharge stages of a subseafloor hydrothermal system are the following (table 1):

<u>Stockwork-Disseminated Metal Sulfides.</u> This type of deposit has been recognized in metabasalts from the mid Atlantic Ridge (Bonatti et al., 1976), and in several basaltic sections drilled by DSDP, where chalcopyrite, pyrite and other phases fill veins and patches criss-crossing the rock and disseminated throughout it. This type of mineralization appears to be similar to the stockwork-disseminated deposits spatially associated with massive sulfide deposits in ophiolite complexes.

<u>Massive metal sulfides.</u> Though not sampled yet, it is likely that massive sulfide deposits are formed in empty spaces within

the upper part of the crust by precipitation from the ascending hot solutions. The mineralogy of these deposits is probably similar to that of the syn-discharge, black smoker-type deposits, except for absence or paucity of phases produced by interaction of the hydrothermal fluids with oxygenated sea water (i.e., barite, etc.).

Metal oxides. If the hydrothermal solutions undergo some mixing with colder sea water during the ascending shallow stages of their circulation, it is possible for Fe-Mn oxides and hydroxides to precipitate, such as those observed lining veins within basalts.

SYN-DISCHARGE HYDROTHERMAL METAL DEPOSITS

A preferred locus of metal deposition is where the hot or warm hydrothermal fluids are discharged through the sea floor at the end of their convective circulation within the crust. In the case of the 350°C East Pacific Rise hot springs, the rapid decrease of temperature and the pH change which are bound to occur at discharge tend to favor rapid sulfide precipitation from solutions which are already supersaturated by adiabatic expansion during ascent. Moreover, both in these high exit temperature springs and in the low (~17°C) exit temperature Galapagos-type springs, oxidation reactions taking place at discharge determine precipitation of Fe and Mn oxides and hydroxides. The following is a simple classification of the syn-discharge hydrothermal metal deposits (table 2).

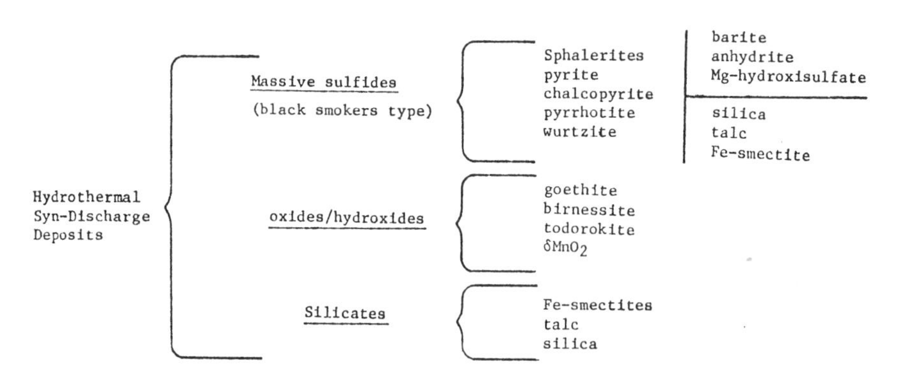

Table 2. Simple classification of hydrothermal syn-discharge metal deposits, with the major mineral components of each group.

HYDROTHERMAL METAL DEPOSITS FROM THE OCEANIC RIFTS

Massive metal sulfide deposits ("black smoker" ty Sulfide deposits on the East Pacific Rise at 21°C of t smoker" type first discovered by Francheteau et al. (sist of a complex phase assemblage (Haymon and Kastn The dominant sulfide minerals are sphalerite, pyrit pyrite. Sulfates (anhydrite, barite and Mg-hydrox silica, and Fe-Mn oxide-hydroxides, together with abundant phases, are also present either directl of discharge or in the immediate vicinity (table

Metal oxide-hydroxides. Oxidation during fraction of the Fe and Mn dissolved in the hyd ids causes deposition of Fe-Mn oxide-hydroxides i y of the vents.

Metal silicates. Precipitation of Fe- ay take place directly from the hydrothermal solu discharge, by reactions of silica with Fe, both amor r constituents of the solution.

POST-DISCHARGE HYDROTHERMAL METAL DEPOS

Those elements which were not depos n the pre-discharge subseafloor circulation stage and du g the injection of the solutions through the sea floor, will be carried into bottom sea water. Of these elements those which are highly soluble in sea water (for instance, Ca) tend to be kept in solution during mixing and dilution of the hydrothermal solutions with sea water, and to augment the ocean budget for that given element. The elements which are instead relatively insoluble in sea water (for instance, Fe and Mn will tend to precipitate on the sea floor after some residence time in sea water).

The residence time for a given element depends on its solubility and its concentration in the solution, on the temperature of the solution and on the rate of admixture and dilution of the hydrothermal liquid with the water. In the case of Mn, Weiss (1977), using a one-dimensional scavenging model of Craig (1974), has estimated a residence time of roughly 50 years for Mn injected into sea water by the Galapagos hydrothermal springs. This implies that hydrothermal Mn plumes may be able to travel horizontally for distances up to 1,000 km from the source before all the excess hydrothermal Mn is deposited (Weiss, 1977).

Depending on the residence time of the elements of interest, it is convenient to separate "concentrated" metal deposits, formed close to the hydrothermal vents when the residence time is relatively short and consisting prevalently (say >50%) of hydrothermally-derived components; and the "diluted" deposits, with longer residence times and dilution of the hydrothermal component by other components.

Concentrated Deposits

Metal oxide/hydroxide type. After discharge of the hydrothermal solutions through the sea floor and their dilution by sea water, partial or total oxidation of some of the dissolved hydrothermal metals, such as Fe and Mn, will occur in most cases. Gradual oxidation of these elements, due primarily to oxygen dissolved in sea water, results generally in relatively insoluble compounds, i.e., metal oxides and hydroxides, and/or silicates. Fractionation of Fe from Mn may occur during their gradual precipitation, due to their different solubilities, and resulting in deposits ranging from Fe-rich/Mn-free to Mn-rich/Fe-free. Table 3 gives data on the mineralogy of this class of deposits, and reference to some examples.

Metal silicates. Fe-silicates, generally a Fe-rich smectite, have been found as a major component of several "post-discharge" hydrothermal deposits (Table 3). These compounds may originate either by direct precipitation of smectites from the hydrothermal liquids, or by post-depositional, diagenetic reactions involving hydrothermally-derived silica and iron hydroxides.

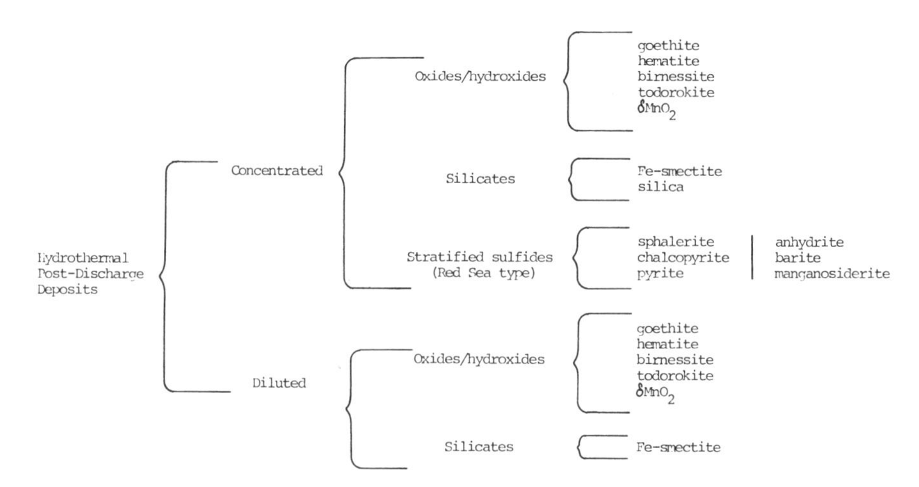

Table 3. Simple classification of hydrothermal post-discharge metal deposits, with the major mineral components of each group.

Stratified metal sulfides (Red Sea type). In the case of sub-sea floor hydrothermal circulation occurring below the axial troughs of the Red Sea, the hydrothermal solutions acquire high salinity (up to >30%) and high density, probably because they circulate within thick evaporite deposits. The ensuing dense brines tend upon discharge to stagnate within topographic lows, without mixing with oxygenated sea water. Under these conditions sulfides can precipitate (table 3) forming deposits (which include sphalerite, pyrite, chalcopyrite and with concentrations of Zn averaging 12.2%, and of CuO 4.5%) which are interstratified with Fe and Mn-rich oxide/hydroxide and Fe-silicate layers and with terrigenous-biogenous sediment units (Bishoff, 1969).

The Red Sea-type stratified sulfide deposits are clearly different as far as structural setting, facies association and genesis from the black-smoker-type sulfide deposits found at the axis of fast spreading ridges.

Diluted Deposits

Not all the chemical species mobilized by the sub-seafloor hydrothermal circulation are deposited before or during discharge into sea water of the thermal solutions. Some elements, for instance Mn, can be kept in solution for considerable time and therefore travel considerable distance before being deposited on the bottom. Dilution with sea water during transport, and co-deposition with other terrigenous, hydrogenous and biogenour phases, will result in deposits where the hydrothermal components constitute only a fraction (say <50% by weight) of the total.

Metalliferous sediments from the Nazca plate and the Bauer Deep (Bostrom et al., 1975; Dymond et al., 1973) provide examples of diluted hydrothermal deposits, where the hydrothermal components derive primarily from the crest of the East Pacific Rise. Their geochemical signature (Al/Al+Fe+Mn <0.4, relatively high concentrations of Mn, Ba, Si, Hg, U, Fe, Cu, etc.) discriminate them from South Pacific pelagic sediment with negligible hydrothermal constituents. The mineralogy of the diluted deposits can include Fe-hydroxides, Fe-smectites, various Mn-hydroxides and oxides (todorokite, birnessite δMnO_2) and barite. These phases are diluted to various extents by biogenous carbonates and silica, and by terrigenous and authigenous silicates.

INTRA-SEDIMENTARY HYDROTHERMAL METAL DEPOSITS

Generally axial zones of spreading are sediment-free; therefore, the hydrothermal fluids after their sub-seafloor circulation exit directly from the basaltic crust into sea water. However, in areas where the axis of accretion is located close to land and terrigenous sedimentation rates are very high, the axis of spreading can be buried by sediment and the hydrothermal fluids may be discharged into the sediment. Possible examples of this situation may occur in areas of the Gulf of Aden at the Red Sea and of the Gulf of California. Where this situation obtains, metal deposits can be formed either within the sediment column or at the sediment-water interface. They can include metal sulfides, oxide-hydroxides and silicates. The composition of these intra-sedimentary hydrothermal deposits may be influenced by reactions between sediment and the hydrothermal fluids.

One example of this type of deposit is from the Gulf of California, where above axial segments of spreading buried by sediment, mounds of barite-metal sulfide have been found associated with zone of high heat flow and with high temperature (~300°C) fluids (Simoneit and Lonsdale, 1982).

CONCLUSIONS

Already from this brief survey, it is apparent that hydrothermal circulation at oceanic axial zones of accretion is responsible for the formation of a variety of types of metal deposits. Moreover, to ridge-axis deposits one has to add deposits formed by hydrothermal activity in marginal basins, island arcs, intraplate volcanic seamounts, etc.

It is increasingly clear that hydrothermal metal deposits from the modern sea floor can provide models to understand the mode of formation of a variety of metal deposits found in the geological record, ranging from massive sulfides and manganiferous sediments of ophiolite complexes to Kuroko-type and Besshi-type ores.

ACKNOWLEDGMENTS

Research supported by ONR Contract Contribution No. 0000 from Lamont-Doherty Geological Observatory, Columbia University, Palisades, N.Y. 10964.

REFERENCES

Arrhenius G. and Bonatti E. 1965. Neptunism and volcanism in the ocean. In Progress in Oceanography, Vol. 3, M. Sears, Ed. Pergamon Press, New York, pp. 7-21.

Bishoff J.L. 1969. Red Sea geothermal brine deposits: their mineralogy,chemistry and genesis. In Hot Brines and Recent Heavy Metal Deposits in the Red Sea. Springer-Verlag, Berlin/New York, pp. 368-401.

Bishoff J.L. 1980. Geothermal system at 21°N, East Pacific Rise: Physicallimits on geothermal fluid and role of adiabatic expansion. Science, 207, 1465-1469.

Bonatti E. and Joensuu O. 1966. Deep sea iron deposits from the South Pacific. Science, 154, 643-645.

Bonatti E., Guerstein-Honnorez B.M. and Honnorez J. 1976. Copper-iron sulfide mineralizations from the equatorial Mid Atlantic Ridge. Econ. Geol., 71, 1515-1525.

Bostrom K. and Peterson M.N.A. 1966. Precipitates from hydrothermal exhalations on the East Pacific Rise. Econ. Geol. 61, 1258-1265. Bostrom K. 1975. The origin and fate of ferromanganoan active ridge sediments. Stockholm Contrib. Geol., 27, 149-243.

Corliss J.B., Dymond J., Gordon L.I., Edmond J.M., VonHezen R.P., Ballard R.D., Green K., Williams D., Bainbridge A., Crane K., and Van Andel T.H. 1979. Submarine thermal springs on the Galapagos Rift. Science, 203, 1073-1083.

Craig H. 1974. A scavenging model for trace elements in the deep sea. Earth Planet Sci. Lett., 23, 149-159.

Delaney J.R. and Cosens B.A. 1982. Boiling and metal deposition in sub-marine hydrothermal systems. Marine Techn. Soc. Jour., 16, 62-65.

Dymond J., Corliss J.B., Heath G.R., Dasch C.W. and Veeh H.H. 1973. Origin of metalliferous sediments from the Pacific Ocean. Geol. Soc. Amer. Bull., 84, 3355-3372.

Dymond J. and Eklund W. 1978. A microprobe study of metalliferous sediment components. Earth Planet Sci. Lett., 40, 243-251. Elder J.W. 1965. Physical processes in geothermal areas. Am. Geophys. Union Monograph 8, pp. 211-239.

Francheteau, J., Needham H.D., Chouckroune, P., Juteau T., Seguret, M., Ballard R.D., Fox P.J., Mormark W., Carranza A., Cordoba D., Guerrero J., Rangin C., Bougalt H. Cambon P., and Hekinian R. 1979. Massive deep sea sulfide-ore deposits discovered on the East Pacific Rise. Nature, 277, 523-530.

Kusznir N.S. 1980. Thermal evolution of the oceanic crust: its dependence on spreading rate and effect on crustal structure. Geophy. Jour. Roy. Astron. Soc., 61, 167-181.

Simoneit B.R.T. and Lonsdale P.F. 9182. Hydrothermal petroleum in mineralized mounds at the seabed of Guaymas Basin. Nature, 295, 198-202.

Skornyakova I.S. 1964. Dispersed iron and manganese in Pacific ocean sediments. Int. Geol. Rev., 7, 2161-2174.

Streus M.R. and Cann J.R. 1982. A model of hydrothermal circulation in fault zones at mid ocean ridge crests. Geophys. Jour. Roy. Astr. Soc., 71, 225-240.

Wehlen J.A. and Craig H. 1979. Methane and hydrogen in East Pacific Rise hydrothermal fluids. Geophys. Res. Lett., 6, 829-831.

Weiss R.F. 1977. Hydrothermal manganese in the deep sea; scavenging residence time and Mn/^{3}He relationship. Earth Planet. Sci. Lett., 37, 257-262.

GENESIS OF FERROMANGANESE DEPOSITS:

HYDROTHERMAL ORIGIN

Claude Lalou
Centre des Faibles Radioactivités
Laboratoire mixte CNRS-CEA
91190 Gif sur Yvette, France

ABSTRACT

The recent discovery of active hydrothermal vents initiated a revolution in knowledge of the chemistry of the oceans and in the concept of a steady state ocean. The literature on known hydrothermal deposits, including metalliferous sediments, sulfides, barite, talc, iron silicates and manganese oxides is reviewed. Special consideration is given to studies of growth rates of manganese oxides in light of the new knowledge of hydrothermal processes.

The problem of the genesis of manganese nodules is also considered. It is shown that, owing to the flux of manganese presently delivered to the oceans by active hydrothermal vents, it is possible that the manganese forming the nodules has an hydrothermal origin. An hydrothermal source of manganese of the manganese nodules would answer such unresolved problems as why the nodules stay at the surface of sediments which accumulate at rates 1000 times faster than nodules are considered to form.

INTRODUCTION

Although the first reference to a possible submarine hydrothermal activity and to an eventual link between this activity and manganese nodules is attributed to Murray and Renard (1891), it has been necessary to wait for about 75 years to find new arguments in support of this link. The new arguments are primarily based on the close association of Mn oxide precipitation and volcanic debris (e.g. Arrhenius and Bonatti, 1965; Bonatti and Nayudu, 1965; Bonatti, 1967). Unfortunately, these ideas did not really develop for an additional 10 years. Another discovery curtailed discussion of the origin of manganese nodules: the measurement of their growth rates. After more or less precise measurements of ^{226}Ra, very slow growth rates were established by the decrease of the ^{230}Th excess with depth (Bender et al., 1966; Ku and Broecker, 1967; Barnes and Dymond, 1967) and confirmed later on by ^{10}Be measurements (Krishnaswami et al., 1972). Consequently, it was unnecessary to find a source other than normal seawater for manganese and other metals, even though discrepencies in some nodules led Lalou and Brichet (1972) and Lalou et al. (1973 a and b) to question the validity of the interpretation of the radiochemical data in nodules which are open chemical systems.

Bonatti et al. (1972) proposed a chemical classification based on the relative proportions of Mn, Fe and (Co+Ni+Cu) in which the hydrothermal deposits were limited to iron-rich deposits. This classification has been modified (Bonatti, 1975) to take into account the newly discovered manganese rich hydrothermal deposits (Fig.1). A break was then established between manganese nodules and hydrothermal Fe-Mn precipitates, defined either by their chemical composition or by their tectonic setting. Some models were established, which can be schematically presented as shown in Fig. 2. These models imply a different origin for the elements forming either hydrogeneous deposits (manganese nodules) or hydrothermal

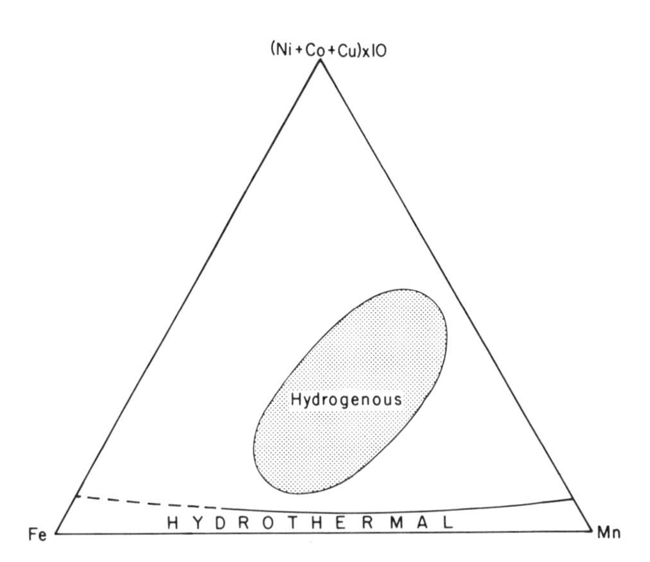

Fig. 1. Ratio Fe/Mn/(Ni + Co + Cu x 10) in oceanic metal rich deposits showing how hydrothermal and hydrogeneous deposits plot in two different fields (Bonatti, 1981).

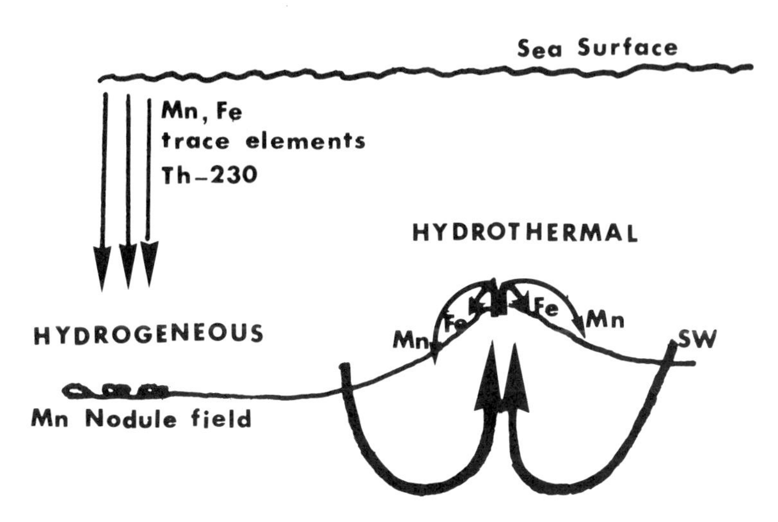

Fig. 2. Model for the formation of hydrothermal deposits, in contrast to hydrogeneous deposits.

deposits. Following a review of known hydrothermal deposits, including locations, principal characteristics and, when measured, growth rates, an attempt is made to establish a link between these recognized hydrothermal deposits and the manganese nodules.

HYDROTHERMAL DEPOSITS: A REVIEW.

In 1966, only 15 years ago, metalliferous sediments associated with mid-oceanic ridges were described near the East Pacific Rise, by Boström and Peterson (1966). The same year, samples dredged at station AMPH D2, on the flank of a seamount close to the crest of the East Pacific Rise (10°38'S, 109°36'W) were attributed to deposition by hydrothermal solutions of volcanic origin (Bonatti and Joensuu, 1966).

The first review paper on hydrothermal deposits is attributed to Rona (1978). In this paper, 17 locations where either metalliferous sediments or encrustations or concretions have been found were cited and shown on a map. The present inventory of hydrothermal deposits, made four years later, recognizes some 70 locations (Table 1, Fig. 3) situated in the Pacific, Atlantic and Indian oceans, as well as in the Mediterranean and Red Seas and in the Afar Rift.

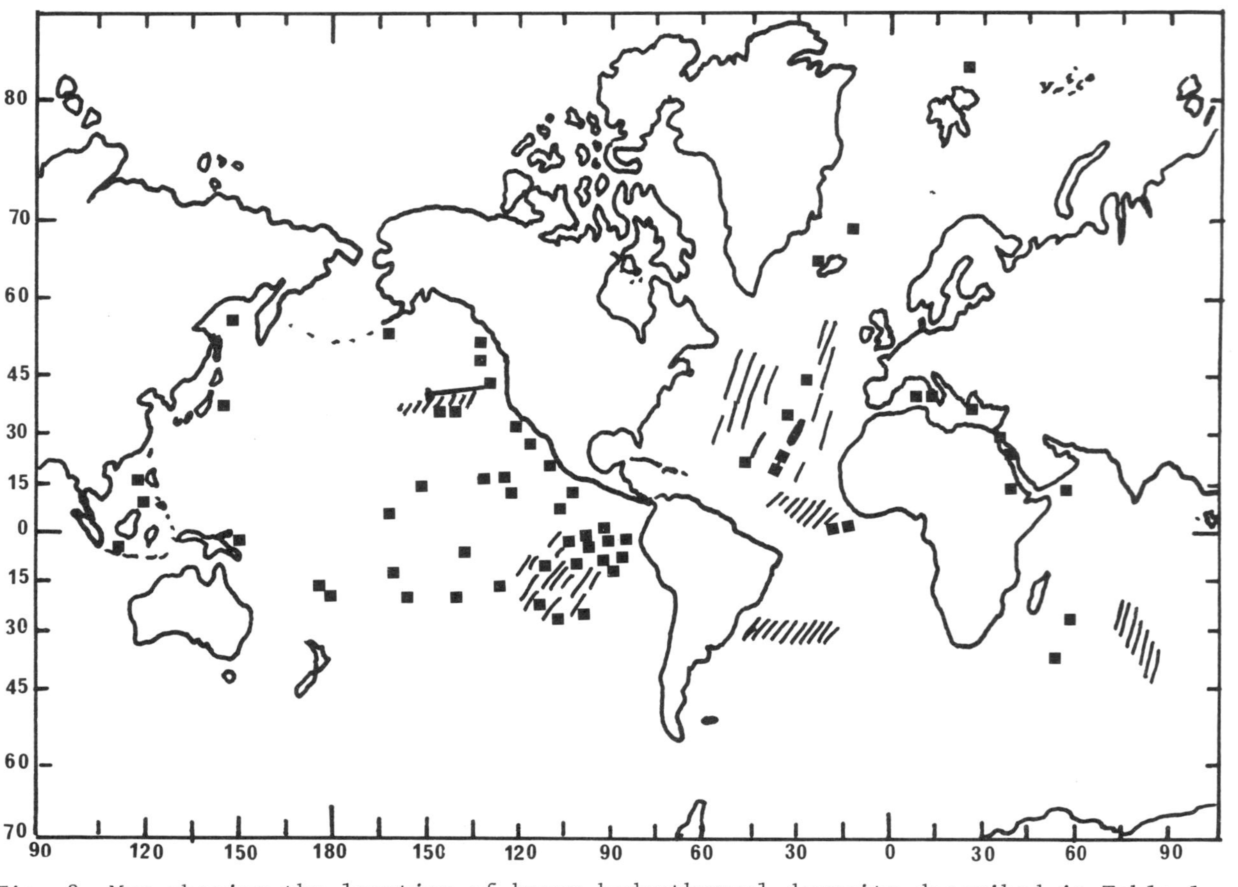

Fig. 3 Map showing the location of known hydrothermal deposits described in Table 1.

The diverse nature of these deposits and differences in their chemical composition, mineralogy and modes of occurence contrasts with the uniform appearance of manganese nodules. Four broad classes of deposits can be distinguished, as follows:

(1) Metalliferous Sediments: These sediments are essentially characterized by enrichment in iron and manganese, relative to aluminium (Aluminium poor, ferromanganoan sediments of Boström and Peterson, 1969 and Boström et al., 1969). Figure 4 shows the

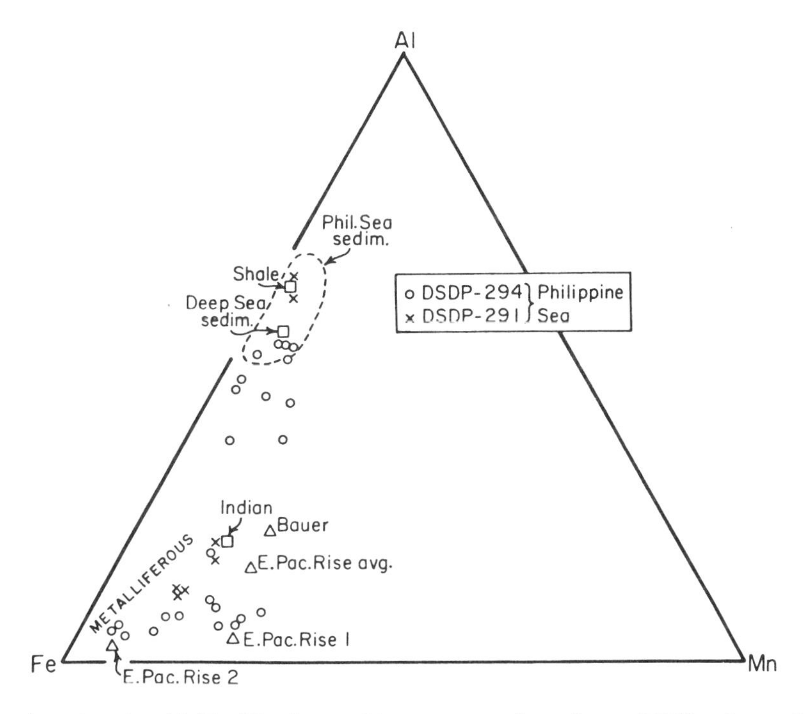

Fig. 4. Ratio Al/Fe/Mn in sediment samples from DSDP sites 291 and 294-295, and in various oceanic metal-rich hydrothermal deposits. The ratio for average shale and average deep-sea sediments is also shown. The field enclosed by a dashed line marks modern sediments from about 80 sites in the Philippine Sea (from Bonatti et al., 1979)

position of such metalliferous sediments relative to shales and deep sea sediments in a Al-Fe-Mn diagram (Bonatti et al., 1979). These metalliferous sediments have been attributed to hydrothermalism as they were initially found in the vicinity of active ridges. They were later recognized at the base of numerous DSDP cores, at the contact with basalt, which also implies their formation near active ridges. In the Pacific ocean, it is probably during Deep Sea Drilling Project Leg 7 that the first basal metalliferous sediments were described by Dreven (1971).

These metalliferous sediments have been extensively studied from a chemical point of view and are generally enriched in various metallic elements (e.g. Sayles and Bishoff, 1973; Boström et al., 1973 for trace element enrichments, Fischer and Boström, 1969; Boström and Rydell, 1979 for uranium and thorium content and Dash et al., 1971 for isotopic analysis).

The metalliferous sediments are considered to have formed at discharge zones of hydrothermal springs at spreading centres as the result of mixing of hydrothermal solutions with cool, oxygenated seawater. In anoxic deeps, as in the Red Sea, the same phenomenon produces metal-rich brines overlying sulfide deposits. Two alternate origins have been proposed for the metalliferous sediments of the Bauer Deep within the Nazca plate: either the metalliferous sediments originate from the East Pacific Rise and accumulate in the deep after transit in water masses, or they are autochthonous, which implies the existence of a hydrothermal source in the deep (Dymond and Veeh, 1975; Mc Murty and Burnett, 1975).

In the Atlantic Ocean, metalliferous sediments are not as extensively known as in the Pacific Ocean. Cronan (1972) has analyzed sediments from the crestal region of the Mid Atlantic Ridge at 45°N and compared them to sediments cored on both sides of the median valley. He found that the sediments from the valley are highly enriched in iron, arsenic and mercury. Analyzing sediments from the TAG Hydrothermal Field, Scott et al.,(1978) found that the non-detrital accumulation rates for Mn, Fe and trace metals exceed the background values for the Atlantic Ocean, but values of Mn and Fe are lower than on the East Pacific Rise; accumulation rates of Ni, Co, Cu and Cr are higher at the TAG Hydrothermal Field than on the East Pacific Rise. These authors attribute these differences in metal content to differences in hydrothermal systems between fast and slow spreading ridges. Such Fe and Mn enriched sediments have also been cored on the Iceland plateau (Varentsov et al., 1980). Analyzing 9 DSDP sites in the North Atlantic, Horowitz and Cronan (1976) have shown that basal sediments generally contain higher concentrations of Fe, Mn, Mg, Pb and Ni than the material overlying them and that these enrichments have been roughly constant since Cretaceous times.

(2) Sulfide deposits: Sulfide deposits were not considered as a major component of the hydrothermal contribution to the oceans and sediments until the discovery of massive sulfide deposits on the East Pacific Rise at 21°N (CYAMEX, 1978; Hekinian et al., 1980; RISE, 1980) and at numerous other sites along the East Pacific Rise between about 21°N and 20°S. Sulfides were more frequently encountered in disseminated and vein forms in the basaltic basement or exposed in large fracture zones as the Romanche, Vema and Atlantis (Bonatti et al., 1976a and b). Sulfides were also found in special environments such as the Red Sea brines (Stephens and Wittkopp, 1969), and in such restricted areas as Santorini caldera (Butuzova, 1966) and the island of Vulcano (Honnorez, 1969). Massive sulfide deposits older than middle Miocene have been cored at DSDP site 471, at the contact with basalt (Devine and Leinen, 1981). Recently, Malahoff et al., (1982) described sulfide deposits in the eastern part of the Galapagos Spreading Center.

This discovery is an important contribution not only to the understanding of metallogenesis processes, but has an important impact on the chemistry of the ocean. The active vents are a source for mantle volatiles such as helium with a higher $^3He/^4He$ than atmospheric helium, which is a good indicator of the presence of hydrothermal plumes, a source for hydrogen and methane and, of special interest to this paper, a source for large quantities of manganese which are introduced in the ocean (see the review paper by Craig and Lupton, 1981)

(3) Miscellaneous deposits: Hydrothermal deposits other than metalliferous sediments, sulfides and Mn-Fe oxides have been reported in the literature. Barite has been found in the Afar Rift, associated with opal, gypsum, nontronite, goethite and Mn-oxide (Bonatti et al., 1972). Barite has also been found in the Laue basin, where it is also associated with opal and with volcanic detritus (Bertine and Keene, 1975), and in the San Clemente fault, where it is associated, like the 21°N sulfides, with dense colonies of large benthic animals (Lonsdale, 1979).

Talc deposits have been reported as high temperature hydrothermal deposits by Lonsdale et al., (1980). They were observed from the manned submersible DSV-4 Seacliff during dives onto the spreading axis of the Guayamas Basin where the deposit is associated with terraces and ledges and clusters of large white clam shells. The talc deposit is partly covered with ferromanganese oxide and is rich in iron sulfide. Study of the sulfur and oxygen isotopes indicates precipitation at about 280°C. In the Atlantic Ocean, another important talc deposit has been described by Bonte (1981) in the Romanche fracture zone, between 4700 and 4650 m depth. It consists of plate-like samples of Fe-Mn oxide overlaying a reddish deposit which is a mixture of talc and hematite, associated with dolomite and serpentine. The isotopic composition of oxygen indicates a temperature of formation of about 250°C, which is similar to the temperature determined for the talc of Guayamas basin.

(4) Ferromanganese oxide deposits and development of the criteria to differentiate hydrothermal deposits from other oxide deposits: It is unusual to discuss ferromanganese deposits in a paper dealing with hydrothermalism. Theoretically, in classical hydrothermal deposits, iron and manganese are well separated during precipitation, forming an iron-rich phase and a manganese-rich phase. Nevertheless, the two phases are often found in the same area and frequently the iron-rich phase is represented by the iron-rich smectite, nontronite.

The principal areas where iron or manganese oxides of hydrothermal origin have been found are shown on Fig. 5. The stars represent samples which have been described only from their mineralogical or chemical composition; the asterisks represent samples which have been submitted to more extensive studies, including radiochemical analysis, which indicate an age or a growth rate for these samples. With their chemical composition and geological setting near active vents, the fast growth rates of these deposits is one of the main characteristics used presently to distinguish between hydrogeneous or hydrothermal origin.

(a) Fe-Mn hydrothermal deposits characterized by their chemical composition or their geologic setting. In the Pacific ocean the three stars at location 1 (Fig.5) are in the vicinity of the Mendocino fracture zone where Nayudu (1966) dredged rocks with encrustations of ferromanganese oxide minerals and manganese nodules with nuclei of palagonite tuff. In nearby sediments, manganese micronodules are associated with palagonite grains. From the close association of the manganese oxides with the palagonite tuffs and breccias, Nayudu (1966) suggested a volcanic origin for the former. At the present time this association would probably not constitute a proof; the Mn oxide of nodules and encrustations would be considered as hydrogeneous in origin; no chemical analysis was given.

Star 2 indicates the location of a seamount on the Pacific Rise where Bonatti and Joensuu (1966) dredged samples of poorly crystallized goethite and fragments of black porous manganese oxide. In this case, it is the segregation between iron and manganese that led the authors to infer a hydrothermal origin.

Star 3 indicates the location of Delwood seamount, the site of a dredge in which Piper at al., (1975) recovered a sample

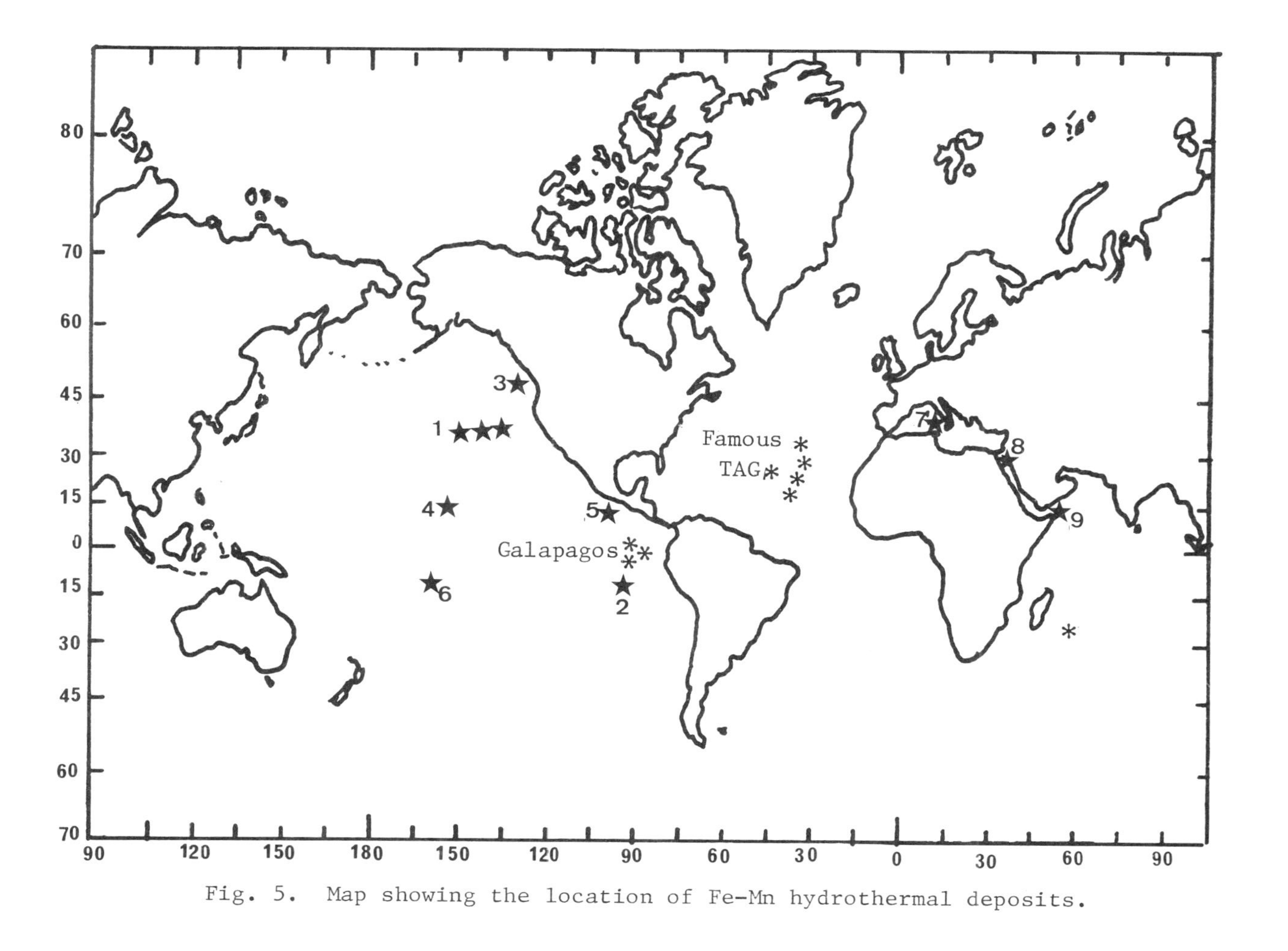

Fig. 5. Map showing the location of Fe-Mn hydrothermal deposits.

consisting of amorphous hydrated iron oxide. Their criterion for an hydrothermal origin was the very low thorium content, the very low Th/U ratio, the high $^{234}U/^{238}U$ ratio and an anomalous REE pattern.

Star 4 indicates the location of a Fe/Mn oxide crust, found in a box grab sampler, SW of Hawaii, associated with an anomalously high temperature in the overlying water column. Despite this coincidence, Grundlach et al., (1976) inferred a hydrogeneous origin, due to the Mn/Fe/Ni/Cu/Co relations which are in the same ranges as in abyssal manganese nodules. In this case, the chemical composition prevailed over the environment in classifying the origin of the deposit.

The East Seamount (star 5) is a young seamount ($\ll 0.6 \times 10^6$ y) where a field of nodules (2 to 10 cm diameter of individual nodules) has been photographed in the caldera. The nodules are composed of pure, rather well crystallized birnessite that was rapidly precipitated (Lonsdale, 1980). The nodules have subsequently been coated with a hydrogeneous ferromanganese crust. The difference of origin for the two parts of the nodules is established from their chemical and mineralogical compositions: the hydrothermal central part is Mn-rich and poor in iron and other trace metals and is composed of birnessite; the surrounding hydrogeneous layer is richer in Ni and composed of δ MnO_2. Hyaloclastite samples with either iron-rich or Mn-rich coatings have been dredged in association with these nodules.

On the Tonga Kermadec ridge, Cronan et al., (1982) have recently dredged a hydrothermal Mn-oxide deposit (star 6) characterized by the purity of its manganese and its textural similarity with the TAG deposit.

In the Mediterranean Sea, hydrothermal Fe-Mn deposits are essentially found around active volcanoes including Eolian Island and Santorini Island (Thera), or around submarine seamounts. Generally, these deposits are represented more by Mn or Fe micronodules in the sediment (see for example Kidd and Armannson, 1979) than by thick encrustation. Nevertheless, two well-defined types of specimens have been dredged.

The first type of specimen, dredged on the eastern slope of the Tyrrhenian sea (star 7) is reported by Castellarin and Sartori (1978) as the first case of a submarine hydrothermal sediment not directly connected to an active ridge or volcano. This occurence consists of oxides and hydroxides of iron and mangano-siderite nodules. The criterion for hydrothermal origin is the diversity and the segregation of materials present. The second type of specimen, dredged on the volcanic seamounts Enarete and Eolo in the submerged portion of the Eolian arc and on the top of Mt Lametino, in the eastern portion of the Eolian arc, are reported by Rossi et al., (1980) as hydrothermal since Fe and Mn are extremely fractionated. The specimens from Mt Enarete and Eolo exhibit a high Fe/Mn ratio; the specimens from Mt Lametino exhibit a very low Fe/Mn ratio.

In the Red Sea, the only well defined Mn deposit (star 8) that has been described is the one reported in Sudan on the northern coast of the Red Sea where coral reef material is replaced by manganese oxide (El Rabaa and Basher, 1975). The authors consider that geothermal activity comparable to the one leading to the formation of hot brines occurred a few tens of thousands of years before present. In the Gulf of Aden, a manganese oxide deposit associated with nontronite has been described by Cann et al., (1975). This deposit (star 9) was found in a region of small transform faults. From its Mn-rich Fe-poor chemical composition a hydrothermal origin is inferred. A maximum age of 300,000 years is inferred from the age of the underlying lava flow; a probable age of 100,000 to 200,000 years is favored, resulting in a rate of deposition greater than 50-100 cm/My. The hydrothermal origin at this location is attested by all known criteria including chemical composition, segregation between Fe and Mn, geologic setting, and rapid growth. Mn oxide coating the basalts somewhat farther from the spreading centre is presumed of hydrogeneous origin. Cann et al., (1977) suggest that the smectite (nontronite) presumably crystallized first out of the ascending solution and was followed and covered by the Mn oxide; as will be considered, this is the succession found in the Galapagos Mounds area.

(b) Fe-Mn hydrothermal deposits characterized by their fast growth rates.The principal difference between hydrothermal precipitation and hydrogeneous precipitation is the rate of deposition due to the availability of the elements. In the first case, a rapid rate may be postulated while in the second case, due to the low concentrations of elements in seawater, very slow growth rates are evident. This has been verified on some well-established hydrothermal sites such as TAG, FAMOUS and GALAPAGOS mounds.

In TAG area, the first attempt to date undoubtedly hydrothermal deposits is attributed to Scott et al., (1974). Using the hypothesis that no thorium 230 or protactinium 231 enters the crystal lattice at the moment of precipitation, they calculate ages from the ratio $^{230}Th/^{224}U$, and growth rates from the increase of this age determined in successive layers of the sample. The growth rates so obtained are 130 mm/10^3 years with ^{230}Th and 250 mm/10^3 years with ^{231}Pa. Manganese crusts dredged in the nearby Atlantis fracture zone, either in the median valley or at the eastern end of the fracture zone, exhibit, as measured by the decrease of excess

^{230}Th, classical slow growth rates. The sample dredged in the median valley, the nearest from the TAG Hydrothermal Field, shows a growth rate of 4.9 mm/10^6 years and the farthest a growth rate of 0.88 mm/10^6 years.

Some features of these data should be discussed:

1) in the hydrothermal crust, ^{232}Th is relatively high, which would not be the case if the precipitation occurred without thorium,

2) the increase in age with depth is due more to a decrease in uranium content than to an increase of ^{230}Th activity,

3) the observation that hydrogeneous crusts exhibit higher growth rates in the vicinity of the hydrothermal field than farther away.

These points will be considered later.

In the FAMOUS area, an oriented specimen recovered by the diving saucer CYANA was studied (Lalou et al., 1977) In this sample, anomalies other than the one measured in the TAG sample occur. The upper layer contains an excess of thorium 230, due to its exposure after formation to ^{230}Th fallout from seawater. The lower layers are depleted in thorium 230. When the ^{230}Th/^{234}U ratio is interpreted as an age, the ages first decrease with depth, reaching a minimum value of 47,000 years, then increase again. This anomalous variation is clearly due to the variations of U content, while ^{230}Th is quite constant. The structure of this sample shows that this deposit was formed by consolidation of preexisting sediment, without any evidences of growth structures, and has probably formed in a very short time, about 50 years ago.

Between the Kane and Atlantis fracture zones, Bonte et al., (1980) studied three cores in which Mn deposits in the form of encrustations and nodules are interbedded in the sediments. All the Mn deposits are closely associated with palygorskite. While the normal calcareous sediment under and above the palygorskite contains ^{230}Th excess, no ^{230}Th excess is found in the palygorskite which may be as thick as 70 cm, supporting the inference of a very rapid deposition. A hydrothermal origin was proposed by the authors for the palygorskite and, by consequence of its close association, to the Mn oxide as well. The episodic character of the deposits and the lack of ^{230}Th excess in the palygorskite show that the hydrothermal activity occured during relatively short periods and, from the rate of accumulation of the normal sediment, that the last episode probably took place about 100,000 years ago. Slightly north of the coring area, the MADCAP submarine volcano is covered with Mn nodules with nuclei of palygorskite or phillipsite (Hoffert et al., 1975). In spite of their normal slow growth rate (Lalou and Brichet unpublished results), this close association of the manganese nodules with palygorskite tentatively relates the nodules to hydrothermalism.

The first samples of hydrothermal crusts were dredged in the GALAPAGOS field in 1972, before the discovery of the Galapagos spreading center or the Galapagos Mounds area (Moore and Vogt, 1976). From their radiometric and chemical measurements and observation of the samples, the authors deduced a rate of deposition several orders of magnitude faster than the more common hydrogeneous nodules. In the case of the sample farthest from the spreading axis, they identified a 3 mm thick second component of hydrogeneous origin on the outer surface of the hydrothermal material. This conclusion is based on the excess of ^{230}Th found in this layer, but it must be noted that in this sample, such an excess is also present deeper in the sample, in the hydrothermal part. The authors also characterize the two parts by an enrichment in trace elements in the hydrogeneous portion. However, the deepest layer analyzed (20-22 mm) that is presumed to be hydrothermal is richer in Cu, Zn and Ni than the 3 upper mm. The only distinct chemical characteristic is the very low Fe/Mn ratio of the hydrothermal portion (0.0011) compared to the hydrogeneous portion (0.65) of the sample.

In the Galapagos mounds area, a comprehensive sequence has been obtained by use of a hydraulic piston corer on Deep Sea Drilling Project Leg 70. Figure 6 is a schematic section of a mound (Honnorez et al., 1981). A complete segregation between iron and manganese is present, iron being localized in a Fe-rich smectite (nontronite) capped in some holes by a blanket of pure todorokite. A chronology has been established (Lalou et al., in press) indicating that nontronite represents an episodic transformation of a calcareous ooze that occured during deposition of the ooze between 300,000 years and 90,000 years BP. Thereafter, manganese precipitation has occured (DSDP hole 509B), from 80,000 to 20,000 years BP. However, these results cannot be translated into growth rates as the formation is not continuous. The structure of the Mn deposit is that of a layered cake: dense, hard plates about 3mm thick, depleted in ^{230}Th, are embedded in a powdery material in which planctonic organisms are present and which exhibits an excess of ^{230}Th. Pulses of Mn-rich solution produce the very rapid formation of the Mn plates at the sediment-water interface and when the Mn flow decreases, the solution slowly impregnates the settling sediment as it reaches the oceanic floor. This is the only way to explain the such well-defined structure and the differences of ^{230}Th activity, low in the plates and in excess in the powdery material in between (Lalou et al., 1983).

The final example is located in the Indian Ocean in the Crozet and Madagascar basins. Eight box cores have been recovered each containing beds of nodules interstratified with the normal calcareous sedimentation. From a biostratigraphic study, Denis-Clochiatti et al.,(1979) have shown that the beds may be correlated at the scale of the basins; from a radiochemical study of one of the cores, containing 5 beds of nodules, Reyss et al.,

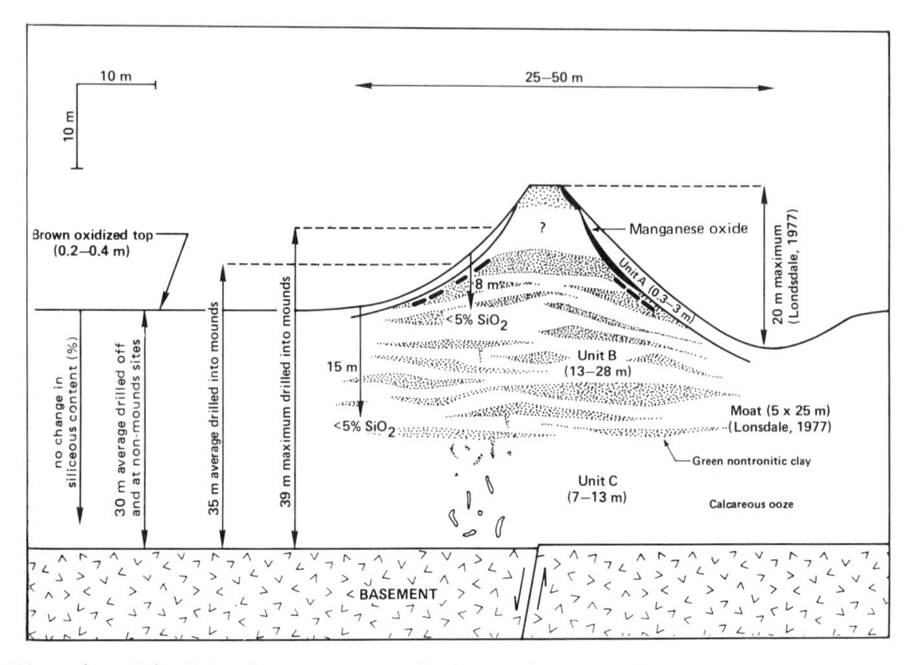

Fig. 6. Idealized geometry of the Galapagos hydrothermal mounds (from Honnorez et al., 1981)

(1981) have shown that the carbonate-free sedimentation rate is $1mm/10^3$ years, without any interruption at the levels of the nodules beds, while nodules at the surface and at 23 cm depth show apparent growth rates of 1 $mm/10^6$ years.

The different inconsistancies found in all the cases presented leads directly to the crucial problem of the significance of ^{230}Th measurements used for growth rate calculations of manganese nodules.

POSSIBLE RELATIONS BETWEEN Mn NODULE FORMATION AND HYDROTHERMALISM

The many inconsistencies that exist between slow growth rates and the characteristics of manganese nodules have been summarized by Lalou et al., (1976, 1981). The principal inconsistencies are as follows: 1) Radioactive disequilibrium in the cores of nodules presenting apparent slow growth rates and in which, as a consequence, all uranium-series nuclides must be in equilibrium. Furthermore, in such nodules, a ^{14}C activity has been determined in the nucleus, while, between the nucleus and the Mn precipitate, a postglacial (<10,000 y.BP) fauna was present. 2) Distribution of elements in discrete layers and abrupt changes in the structure, incompatible with a slow accumulation process in a homogeneous environment. 3) Expansion of the nucleus material in the oxide. 4) Large difference in ^{230}Th activity between the upper surface of the nodule (more active) and its lower surface, showing that ^{230}Th does not arrive with the oxy-hydroxides. 5) The very low integrated ^{230}Th activity within the nodules. This phenomenon is sometimes found in sedimentary cores, but the inverse is also found, in contrast, in nodules it is always much lower than the value calculated for the theoretical flux. 6) The fact that the ^{230}Th activity is normal for recent sediments underlying the nodules. 7) ^{232}Th gradients, similar to radioactive decreases, but that cannot be radioactive decreases; such gradients have also been found by Burnett and Piper (1977) in a sample from the Hess deep. 8) The fact that when different dating methods are applied, some of which do not use radioactive gradients, such as K/Ar, hydration rinds etc... the different dating methods do not give the same ages (Burnett and Piper, 1977; Lalou et al., 1979). 9) Finally, the fact that nodules lie at the surface of sediment accumulating 1000 times faster than the nodules are supposed to form.

Different hypotheses have been proposed to explain this unconformity between nodules and their sedimentary substrate. These hypotheses contend that either nodules would be constantly pushed up by benthic fauna, or microseisms would maintain large objects at the top of sediments, or nodules would float in the boundary layer. All the hypotheses call for a deus ex machina that cannot be proven, and none of them can account for the fact that nodules occur in beds, sometimes crowded one against the other, or that they also occur in beds in the sedimentary sequence and not randomly distributed. Apart from the radiometric measurements and taking into account the possibility of an hydrothermal origin, many of these characteristics can be explained.

As discussed, anomalies in the distribution of radionuclides are often observed in hydrothermal as well as in hydrogeneous deposits in the form of U and ^{232}Th gradients that cannot be attributed to a decrease in activity with time. Other gradients such as those obtained for ^{230}Th may also be attributed to a cause other than radioactive decay. Such gradients indicate either a chemical gradient during rapid precipitation, or, particularly for ^{230}Th, a gradient due to the porosity of the nodule material with the major portion of the ^{230}Th arriving after the formation of the nodule.

The Hydrothermal Hypothesis.

In the case of an hydrothermal origin, specially for manganese, it may be postulated that the manganese which is delivered at spreading centers, as at the present time at the East Pacific Rise between 21°N and 20°S, may be transported far from its source (Weiss, 1977; Klinkhammer, 1980 and Edmond et al., 1982). In the light of this new discovery, it is tempting to complete the model of figure 2 with a long arrow going from spreading centers towards abyssal plains and representing the manganese flux (Figure 7). This manganese flux is probably not omnidirectional, but the hydrothermal plume is preferentially transported by the general circulation, and the major accumulation is to be found under the lee of the vents.

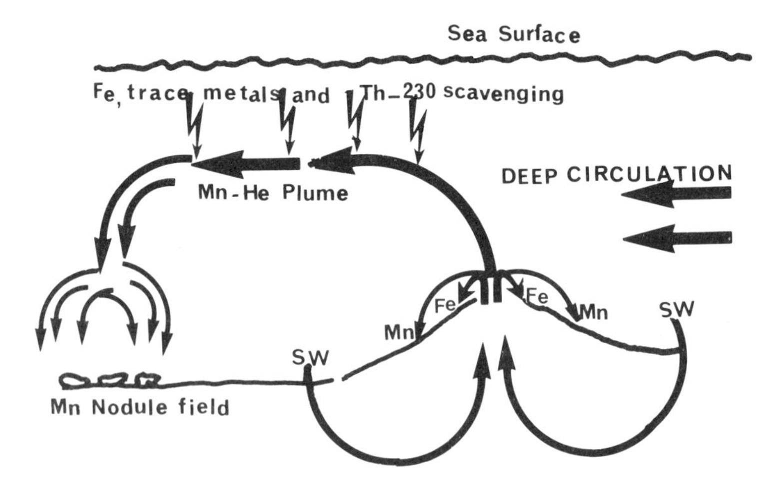

Fig. 7. Model, modified from Figure 2, showing the possible relationship between hydrothermal active vents and manganese nodule formation.

This manganese, in the form of minute hydroxide particles, is one of the best scavengers for trace elements in the ocean. Therefore, during its transport, it is enriched in Co, Cu, Ni and probably also in other trace elements. Moreover, a portion of these trace elements may also have their origin directly in the hydrothermal fluid. Until the present time, the total list of metals which may be introduced into the ocean in this way is not well known. The first evaluations obtained are those by Edmond et al., (1982). They indicate that for Li and Mn, this hydrothermal source is probably as great or greater than the source of rivers and that for Ca, Ba and Si it contributes substantially. During its transport, Mn certainly also scavenges ^{230}Th and other radionuclides. At the termination of formation of the nodule, or a hiatus in its formation, the nodule is exposed to the oceanic rain of ^{230}Th. This exposure may explain the very high activities in the outermost millimeter of the nodule.

The hydrothermal hypothesis has the advantage to explain most of the characteristics observed for nodules, as follows: 1) The layered structure and abrupt changes may be more easily explained by a pulsed phenomenon than in a steady-state ocean. 2) The abrupt changes in chemical composition are probably due to differences in intensity of the hydrothermal discharge and, consequently, to different duration of scavenging. 3) The difference in activity of upper and lower surfaces of nodules is explained by the fact that only the upper surface, which has always been the upper one, has been exposed after its formation to the rain of ^{230}Th. 4) The fields of nodules would have been formed during periods of increasing submarine hydrothermalism.

This may also perhaps explain some chemical gradients found in the nodule fields. For example, in the Sonne basin, nodules richer in manganese are found in the northern part of the basin. The authors (Halbach et al., 1980) attribute this gradient to a different origin for Mn, however, they propose pore water for the northern part and seawater for the southernmost part of the Sonne basin. The growth rates of nodules in the northernmost part of the basin have been determined as very high by Reyss et al., (1982). An alternate explanation may be advanced: This repartition, at a larger scale probably because of a more intense hydrothermal injection, may be compared to the TAG Hydrothermal Field. There, when moving away from the true hydrothermal field, manganese encrustations exhibit decreasing apparent growth rates. The apparent growth rates are more rapid near the spreading axis than farther away, as indicated in the paragraph dealing with the TAG Hydrothermal Field.

Is Hydrothermal Formation of Nodules Quantatively Possible ?

An attempt is made to determine whether the hypothesis proposed is realistic in terms of manganese supply. In the Pacific ocean, the amount of Mn in nodules has been estimated at 400×10^9 tons by Mero (1965). From the data of the active vents at 21°N, a Mn flux for these vents is calculated using a diameter orifice of 15 cm, a rate of flow for the fluid of a few meters per second (assumed equal to 5 m/sec) (Edmond, 1981) and a concentration of total dissolvable manganese of (310 x 50) microgram/kg as evaluated from samples 50 fold diluted with ambient seawater by Lupton et al., (1980). The life of a chimney is between 100 and 30 years (Lalou and Brichet 1981, 1982), so 50 years is assumed taking into account a possible fluctuation in the activity. The Mn escape so calculated is 9×10^9 g of manganese for a chimney, or 9×10^3 tons. Therefore, about 4×10^6 active chimneys are needed to form the nodule field in 50 years. This is certainly unrealistic, but:

1) the Mn values used in the calculations are probably underestimated, since new values of 30 to 50 mg/kg are obtained at 21°N by Edmond et al.,(1982) and at 13°N, the first results available deduced from ^{3}He data show a higher activity.

2) from the layered structure of the nodules it may be assumed that they are not formed during a single event,

3) the time span to form nodules in our hypothesis is approximately equivalent to the time needed for the sediment to bury the nodules. Considering a sedimentation rate of 1 mm/10^3 years, this allows between 50,000 to 100,000 years to form the nodules, so the theoretical number of chimneys may be lowered by a factor 10^3 to 2×10^3.

4) All the calculations made until now are made within an uniformitarian theory, as opposed to a catastrophic theory, and it is evident that in such phenomenon as hydrothermalism, it cannot be considered that the level of ongoing activity necessarily represents the level of past activity.

The notion of mean spreading rate is certainly true at the scale of millions years, but certainly not true at shorter time scale. Direct evidence of this have been measured in the Afar rift where the mean spreading rate is 1.5 cm/year and where, in 1978, in less than a week, the fault margins shifted apart, in some places, more than one meter (Allard et al., 1979) accompanied by a fissure eruption of large quantities of lavas.

As it does not seem that the present is a specially active period, the estimated flux values may be low. Perhaps the next formation of a manganese field, after a quiet period when present nodules are buried, will begin when Southern California will become an island.

ACKNOWLEDGEMENTS

Thanks are due to many people without whom this work would not have been possible: Evelyne Brichet, who has collaborated with me for the last 10 years, Héloïse Leclaire, Célestine Jehanno, Philippe Bonte and Jean Louis Reyss. During these 10 years fruitful discussions have been conducted with Dr J. Labeyrie, J.C. Duplessy and L. Labeyrie Jr. I thank R. Chesselet for a critical review of this paper. Finally, all my thanks go to Peter Rona for his work as organizer of the Conference, for his kind invitation to participate and for his work of rewriting the manuscript to give it a correct english form.

The financial support of this study is due to CNRS and CEA, and to an ATP Chemical Oceanography grant which is gratefully acknowledged.

Table 1. Location and nature of known hydrothermal deposits

Geographic area	Position	Nature	Reference
	PACIFIC OCEAN		
1) Metalliferous sediments			
East Pacific rise	Between 12° and 14° South		Boström & Peterson 1966,1969; Veeh & Boström, 1971.
Okhotsk Sea	c. 60°N 143°W		Strakhov & Nesterova 1968
DSDP Leg VII site 66	2°24N, 166°7W	Basal metalliferous sedim.	Drever, 1971
DSDP Leg XVI			
Sites 159	12°19N,122°17W	Ferrugineous sediments at contact basalt-sediment.	Cronan,1972 Cronan & Garret,1973
160	11°42N,130°52W		
161	10°50N,139°57W		
162	14°52N,140°02W		
DSDP Leg XIX			
Site 183	52°34N,161°12W		Natland,1973
192	53°00N,164°42W		
DSDP Leg XXXIV			
Site 319	13°01S,101°31W	Fe and Mn rich Al poor sedim.	Boström & al.1976, Dymond & Corliss, 1976,1977.
320	9°00S, 83°31W		
321	12°01S, 81°54W		
E.P.R. Bauer deep	South of 10°S to 25°S,103°W		Heath & Dymond,1977
NE Pacific between Clarion & Clipperton.	15°12N,126°58W	Metal.sed may be relict	Bishoff & Rosenbauer, 1977.
E.P.R.	3°25S	Metal.sed & abundant biol. activity.	Lonsdale, 1977
DOMES site	c. 15°N,126°W		Bishoff & Rosenbauer, 1977;Bishoff 1979
Philippine Sea			
DSDP Leg XXXI			
site 291	12°48N,127°49E	Eocene metal rich sedim.	Bonatti & al 1979
site 295	12°34N,131°23E		
EPR	9°S, 108°W	Fe-rich,Mn & metal rich	Cronan & Varnavas 1981

Sulfides and active vents			
DSDP Leg XXXI			
Site 297	30°52N, 134°09E	Sulfides	Hosking,1975
EPR	21°N	Sulfides & active vents	CYAMEX,1978 RISE, 1980
DSDP Leg LXIII			
Site 471, off California	23°28N, 112°29W	Metallic sulfides.	Devine & Leineir,1981
China sea, near Luzon island		Giant shells	Bouchet pers.comm.
Mariana Trench	18°13N, 144°44E	^{3}He & methane anomalies	Craig (oral comm.1981)
Galapagos Rift	86°W	Fossil sulfid. & Mn deposits	Malahoff & al.,1982
Mn and Fe encrustations			
Banu Wuhu	Indonesia	Fe and Mn in submarine exhalations	Zelenoff, 1965
Mendocino F.Z.	32-41°N, 117-149°W & 40°N,125-128°W	Nodules, palagonite, tuff & breccias with Fe-Mn encrust.	Nayudu,1965
AMPH D2	10°38S, 109°36W	Poorly cristallized goethite and Mn oxide	Bonatti & Joensuu,1966
DSDP Leg V			
Site 37	40°58N, 140°43W	Amorphous Fe deposits	von der Borsh & Rex 1970, Dymond & al.1973
Site 38	38°42N, 140°21W		
Site 39	32°48N, 139°31W		
DSDP Leg VIII			
Site 74	6°14S, 136°05W	Fe- and Mn-rich deposits	von der Borsh & al. 1971
Site 75	12°31S, 134°16W		
DSDP Leg XVI			
Site 156	1°40N, 82°24W	Crusts of Mn 50cm beneath sediment surface.	Cronan,1972
Site 159	12°19N, 122°17W		
Site 160	11°42N, 130°52W		
Site 161	10°50N, 139°57W		
Site 162	14°52N, 14°02W		
South Marquesa fracture zone	c. 18°S, 155°W	Fe-rich smectite	Aoki & al. 1974
Galapagos rift	0°48N, 86°W	Active vents, biol.activity	Detrick & al 1974
		Mn on pillows	Crane & Ballard,1980
Delwood seamount	c. 51°N, 131°W	Fe-rich depos.	Piper & al. 1975

(continued)

Table 1 (cont.2)

Galapagos	c. 2°60N, 95°W	Mn crusts	Moore & Vogt 1976
SW Hawaii	14°N, 153°W	Mn-Fe crusts (non-hydro-thermal for authors)	Grundlach & al, 1976
Hess depression	2°17N, 101°01W	16mm of Fe-Mn crust	Burnett & Piper,1977
Galapagos mounds	0°36N, 86°07W	Mn oxide & nontronite	Corliss & al.1978
DSDP Leg LIV	-	-	Dymond & al 1980
Leg LXX	-	-	Honnorez & al.1981
Siqueiroz FZ	9°08N, 105°12W	Mn encrust. on fresh basalt	Lalou & al. 1979
Seamount NW Clipperton FZ	8°48N, 103°53W	Nodule field (birnessite)	Lonsdale & al. 1980
Seamount EPR	19°S	Fe-rich enc.	Toth, 1980
Juan da Fuca Ridge	48°N, 129°W	Fe-rich enc.	Toth, 1980
Explorer Ridge	50°05N, 130°W	Mn oxide & nontronite	Grill & al 1981
Tonga Kermadec Ridge		Mn crust	Cronan & al 1982
Miscellaneous			
New Guinea, Matupi harbour	2°40S, 148°E	Metal rich brines	Fergusson & Lambert,1972
Australia, Laue Basin	16°55S, 176°49W	Barite & Opale	Bertine 1974 Bertine & Keene,1975
San Clemente Basin	32°16N, 117°43W	Barite	Lonsdale 1979
Guayamas Basin	Gulf of Calif.	Talc encrus-ted with Mn and Fe	Lonsdale & al,1981
ATLANTIC OCEAN			
Metalliferous sediments			
West flank of MAR	Between 28°S and 31°S	Metalliferous sediments	Boström & al 1972
Mid Atlantic Ridge	near 45°N	Iron rich sed.rich in As, Hg.	Cronan,1972
Atlantic Basin	Between 20°N and 60°N	Metalliferous sediments	Horowitz & Cronan,1976
TAG area	26°N to 30°N on MAR	Metal enriched sediments	Scott & al 1978, Cronan & al.1979, Rona,1980

Iceland Plateau	69°01N, 10°43W	Metalliferous sediments	Varentsov & al 1980
Sulfide deposits			
Romanche FZ	0°S	Sulfides, pyrite	Bonatti & al 1976a
Vema and Atlantis FZ	MAR between 0° and 20°N	Metallic sulfides	Bonatti & al 1976b
Kolbeinsen Ridge		Hydrothermal vents	Stefansson (this volume)
Fe-Mn encrustations			
TAG area	26°N to 30°N	Mn encrust.	Scott & al 1974, Rona & al 1976, Rona 1980
FAMOUS area	MAR at 37°N	Mn oxide & nontronite	ARCYANA 1975
MAR	23°N	Mn encrust.	Thompson & al 1975
Kane & Atlantis FZ		Mn oxide & paligorskite	Bonte & Lalou 1980
Romanche FZ	0°S	Mn encrust.	Bonte 1981
	MEDITERRANEAN SEA		
Metalliferous sediments			
Santorini Island	c.36°N 27°W	Iron rich sed. sulfides, phosphates	Butuzova 1966
Santorini Island	-	Iron rich sed. & opal	Bonatti & al 1972
Stromboli Island	c.39°N 15°E	Iron rich sed. & Mn deposit	Bonatti & al 1972
South Tyrrhenian	39°30N, 15°E	Micronodules of iron-rich smectite or Mn	Kidd & Armanson 1979
Sulfides			
Vulcano	c.39°N, 15°E		Honnorez 1969
Fe-Mn encrustations			
Tyrrhenian sea	c.40°N, 14°E	Iron and mangano siderite	Castellarin & Sartori 1978,
Eolian Island arc		Mn deposit	Rossi & al 1980
	RED SEA		
		Brines in axial thought sulfides	Degens & al 1969

(continued)

Table 1 (cont.4)

	c. 22°N, 36°E	Mn oxide replacing corals	El Rabaa & A'Basher 1975
	INDIAN OCEAN		
Indian Ridge system	bet. 16°S and 40°S	Metalliferous sediments	Boström & al 1969
SW Indian ridge			
DSDP XXV site 245	31°22S, 52°18E	Metalliferous sediments	Warner & Gieske 1974
site 248	c. 30°S, 36°E	Metalliferous	Marchig &
Site 249	24°S, 60°E	sediments	Vallée 1974
Gulf of Aden	12°31N, 47°39E 12°35E, 47°39E	Mn oxide & nontronite	Cann & al 1977
Madagascar basin	24°S, 60°E	Mn nodules	Reyss & Lalou 1981

REFERENCES

Allard,P., Tazieff, H. and Dajlevic, D. (1979) Observation of sea floor spreading in Afar during the november 1978 fissure eruption. Nature 279, 30-33.

Aoki,S., Koyama, N. and Sudo,T. (1974) An iron-rich montmorillonite in a sediment core from the northeastern Pacific. Deep Sea Res. 21, 865-875.

ARCYANA. (1975) Transform fault and rift valley from bathyscaph and diving saucer. Science 190, 108-116. (1965)

Arrhenius, G.and Bonatti, E. (1965) Neptunism and vulcanism in the ocean: In Progress in oceanography 3, Pergamon press, 7-22.

Barnes, S.S. and Dymond, J.K. (1967) Rates of accumulation of ferromanganese nodules. Nature, 213, 1218-1219

Bender, M.L., Ku, T.L. and Broecker, W.S. (1966) Manganese nodules, their evolution. Science 151, 325-328.

Bertine, K.K. (1974) Origin of Laue basin rise sediments. Geochim. Cosmochim. Acta 38, 629-640.

Bertine, K.K. and Keene, J.B. (1975) Submarine barite opal rocks of hydrothermal origin. Science, 188, 150-152.

Bischoff, J.L., Piper, D.Z. and Quintero, P. (1979) Nature and origin of metalliferous sediments in Dome site C, Pacific manganese nodule province. In: Colloque Intern. CNRS, 289; C. Lalou Ed: Paris, 119-138.

Bischoff, J.L. and Rosenbauer, R.J. (1977) Recent metalliferous sediments in the North Pacific manganese nodule area. Earth. Planet. Sci. Lett. 33, 379-388.

Bonatti, E. (1967) Mechanisms of deep sea volcanism in the South Pacific In: Researches in Geochemistry P.H. Abelson Ed. John Willey and Sons. 2, 453-491.

Bonatti, E. (1975) Metallogenesis at oceanic spreading centers. Annual Rev. Earth. Planet.Sci. 3, 401-431.

Bonatti, E. (1981) Metal deposits in the oceanic lithosphere. Chapt.17 In: The Sea, vo 17, C. Emiliani Ed. John Wiley and Sons, 639-686.

Bonatti, E., Fisher, D.E., Joensuu, O., Rydell, H.S. and Beyth, M. (1972) Iron manganese barium deposits from the northern Afar rift (Ethiopia). Econ. Geol. 67, 717-730.

Bonatti, E., Guerstein-Honnorez, B.M. and Honnorez, J. (1976b) Coper-iron sulfide mineralizations from the equatorial mid-Atlantic ridge Econ. Geol. 71, 1515-1525.

Bonatti, E., Guerstein-Honnorez, B.M., Honnorez, J. and Stern, C. (1976a) Hydrothermal pyrite concretions from the Romanche Trench (Equat. Atl). Metallogenesis in oceanic fracture zones. Earth Planet. Sc. Lett. 32, 1-10.

Bonatti, E., Honnorez, J., Joensuu, O. and Rydell H.S. (1972) Submarine iron deposits in the Mediterranean Sea. In: The Mediterranean Sea Stanley D.J. Ed. Dowden, Hutchinson and Ros Strondsberg Pennsylvania, 701-710.

Bonatti, E. and Joensuu, O. (1966) Deep sea iron deposit from the South Pacific. Science, 154, 643-645.

Bonatti, E., Kolla, V., Moore, W.S. and Stern, C. (1979) Metallogenesis in marginal basins: Fe-rich basal deposits from the Philippine Sea. Marine Geol 32, 21-37.

Bonatti, E., Kraemer, T. and Rydell, H.S. (1972) Classification and genesis of submarine iron-manganese deposits. In: Ferromanganese deposits on the ocean floor. IDOE, Columbia Univ. N.Y. 149-166.

Bonatti, E. and Nayudu, Y.R. (1965) The origin of manganese nodules on the ocean floor. Amer. Journ. Sc. 263, 17-39.

Bonte, Ph. (1981) Relations entre l'environnement et les caractéristiques des concrétions polymétalliques marines dans la Fosse de La Romanche. Dr.Sc.Thesis Paris. Note CEA 2237, 245pp.

Bonte, Ph., Lalou, C. and Latouche, C. (1980) Ferromanganese deposits in cores from the Kane and Atlantis fracture area: possible relationships with hydrothermalism. J. Geol. Soc. London. 137, 373-377.

Borch, Van der C.C., Nesteroff, W.D. and Galehouse, J. (1971) Iron rich sediments cored during Leg VIII of Deep Sea Drilling Project. Initial Rept. DSDP US Govt Printing Off. VIII, 829-837.

Borch, Van der C.C. and Rex, R.W. (1970) Amorphous iron-oxide precipitation in sediments cored during Leg V, DSDP. Initial. Rept. DSDP US Govt Print. Off., V, 541-544.

Boström, K., Joensuu, O., Valdes, S., Charm, W. and Glaccum, R. (1974) Geochemistry and origin of East Pacific sediments sampled during DSDP leg XXXIV. Init. Rept. DSDP US Govt Print. Off. XXXIV, 559-574

Boström, K. Joensuu, O., Valdes, S. and Riera, M. (1972) Geochemical history of South Atlantic ocean sediments since late Cretaceous. Marine. Geol. 12, 85-121.

Boström, K., Kraemer, T. and Gartner, S. (1973) Provenance and accumulation rates of opaline silica, Al, Ti, Fe, Mn, Cu, Ni and Co in pacific pelagic sediments. Chemical Geol. 11, 123-148.

Boström, K. and Peterson, M.N.A. (1966) Precipitates from hydrothermal exhalations on the East Pacific Rise. Econ. Geol. 61, 1258-1265.

Boström, K. and Peterson, M.N.A. (1969) Origin of aluminium poor ferromanganoan sediments in areas of high heat flow on the East Pacific Rise. Marine Geol. 7, 427-447.

Boström, K., Peterson, M.N.A., Joensuu, O. and Fisher, D.E. (1969) Aluminium-poor ferromanganoan sediments on active ocean ridges. J. Geophys. Res. 74, 3261-3270.

Boström, K. and Rydell, H. (1979) Geochemical behavior of U and Th during exhalative sedimentary processes. Coll. Intern. CNRS, 289, C.Lalou Ed. CNRS Paris, 151-166.

Burnett, C.and Piper, D.Z. (1977) Rapidly-formed ferromanganese deposit from the eastern Pacific Hess Deep. Nature 265, 596-600.

Butuzova, G.Y. (1966) Iron-ore sediments of the fumarole field of Santorin volcano, their composition and origin. Dokl. Akad. Sci. USSR. Earth Science sect. 168, 215-217.

Cann, J.R., Winter, C.K. and Pritchard, R.G. (1977) A hydrothermal deposit from the floor of the Gulf of Aden. Mineral. Mag. 41, 193-199

Castellarin, A. and Sartori, R. (1978) Quaternary iron manganese deposits and associated pelagic sediments (radiolarian clays and chert, gypsiferous mud) from Tyrrhenian sea. Sedimentology, 25, 801-821

Corliss, J.B., Lyle, M., Dymond, D.J. and Crane, M. (1978) The chemistry of hydrothermal mounds near the Galapagos rift. Earth Planet. Sci. Lett. 40, 12-24.

Craig, H., and Lupton, J.E. (1981) Helium 3 and mantle volatiles in the ocean and the oceanic crust. In:The oceanic lithosphere C. Emiliani Ed. The Sea, vol.VII, John Wiley and Sons, 391-428.

Crane, K. and Ballard, R.D. (1980) The Galapagos rift at 86°W: Structure and morphology of hydrothermal fields and their relationships to the volcanic and tectonic processes of the rift valley. Jour Geophys Res. 85, B-3, 1443-1454.

Cronan, D.S. (1972) Basal ferrugineous sediments cored during Leg XVI DSDP.Init. Rept. DSDP. US Govt. Print. Off., 601-604.

Cronan, D.S. (1972) The mid Atlantic ridge near 45°N, XVII. Al, As, Hg and Mn in ferrugineous sediments from the median valley. Canadian J. Earth. Sc. 9, 319-323.

Cronan, D.S. and Garret, D.E. (1973) Distribution of elements in metalliferous Pacific sediments collected during DSDP. Nature Phys.Sc. 242, 88-89.

Cronan, D.S., Glasby, G.P., Moorby, S.A., Thomson, G., Knedlar, K. and McDougall, J. (1982) A submarine hydrothermal manganese deposit from the SW Pacific Island arc. Nature 298, 456-458.

Cronan, D.S., Rona, P.A. and Shearme, S. (1979) Metal enrichments in sediments from the TAG hydrothermal field. Marine Mining 2, 78-89.

Cronan, D.S. and Varnavas, S.P. (1981) Hydrothermal and dissolution related geochemical variations in sediments from an East Pacific Rise fracture zone at 9°S. Coll. C4, Geologie des Oceans. Congrès Intern. Geol. Paris. Special Publ. Ocenologica Acta, 4. 47-58.

CYAMEX, équipe scientifique (1978) Découverte par submersibles de sulfures polymétalliques massifs sur la dorsale du Pacifique oriental par 21°N (Projet RITA). C.R. Acad. Sc. Paris 287, 1365-1368.

Dash, E.J., Dymond, J.R. and Heath, G.R. (1971) Isotopic analysis of metalliferous sediments from the East Pacific Rise. Earth Planet.Sc. Lett. 13, 175-180.

Degens, E.T. and Ross, D.A. (1969) Hot brines and recent heavy metal deposits in the Red Sea. Springer Verlag N.Y. Inc. 600pp.

Denis-Clocchiati, M., Leclaire, L. and Giannesini, P.J. (1979) Vitesse de croissance des nodules de manganèse dans l'océan Indien. In: Coll. Intern. CNRS, 289 C. Lalou, Ed. CNRS Paris, 241-250.
Detrick, R.S., Williams, D.L., Mudie, J.D. and Sclater, J.C. (1974) The Galapagos spreading center: bottom water temperature and the significance of geothermal heating. Geophys. J. Roy.Astr. Soc. 38, 627-637.
Devine, J.D. and Leinner, M. (1981) Chemistry of the massive sulfide deposits cored at site 471.Init. Rept. DSDP US. Gvt. Print. Off. LXIII, 679-681.
Drever, J.I. (1971) Chemical and mineralogical studies, site 66. Init. Rept. DSDP US Gvt. Print.Off. VII (2) 965-976.
Dymond, J., Corliss, J.B., Cobler, R., Muratli, C.M., Chou, C. and Conard, R. (1980) Composition and origin of sediments recovered by deep drilling of the sediment mounds, Galapagos spreading center. Init. Rept. DSDP US Gvt. Print. Off. LIV, 377-386.
Dymond, J., Corliss, J.B. and Heath, G.R. (1977) History of metalliferous sediments, DSDP site 319 in the Southern Pacific. Geochim. Cosmochim. Acta. 41, 741-753.
Dymond, J., Corliss, J., Heath, G.R., Field, C.W., Dasch, E.J. and Veeh, H.H. (1973) Origin of metalliferous sediments from the Pacific ocean. Geol.Soc. Amer. Bull. 84, 3355-3372.
Dymond, J., Corliss, J.B. and Stillinger, R. (1976) Chemical composition and metal accumulation rates of metalliferous sediments from sites 319, 320B and 321.Init. Rept DSDP US Gvt. Print. Off. XXXIV, 575-588
Dymond, J. and Veeh, H.H. (1975) Metal accumulation rates in the Southeast Pacific and the origin of metalliferous sediments. Earth Planet. Sc. Lett. 28, 13-22.
Edmond, J.M. (1981) Hydrothermal activity at mid-ocean ridge axes. Nature, 290, 87-88.
Edmond, J.M., Von Damm, K.L., McDuff, R.E. and Measures, C.I. (1982) Chemistry of hot springs on the East Pacific rise and their effluent dispersal. Nature, 297, 187-191.
Ferguson, J. and Lambert, T.T.B. (1972) Volcanic exhalations and metal enrichments at Matupi Harbor, New Britain. TPNG. Econ. Geol. 67 25-27.
Fisher, D.E. and Boström, K. (1969) Uranium rich sediments on the East Pacific rise. Nature 224, 64-65.
Grill, E.V., Chase, R.L., Mac Donald, R.D. and Murray, J.W. (1981) A hydrothermal deposit from Explorer ridge in the northeast Pacific Earth Planet. Sci. Lett. 52, 142-150.
Gundlach, H. and Beiersdorff, H. (1976) Heated bottom water and associated Mn/Fe oxide crusts from the Clarion fracture zone, SE of Hawaii. J.O.A. Edimburg (last abstracts)
Halbach, P., Marchig, V., and Scherhag, C. (1980) Regional variations in Mn, Ni, Cu and Co of ferromanganese nodules from a basin in the Southeast Pacific. Marine Geol. 38, M1-M9.

Heath, G.R. and Dymond, J. (1977) Genesis and transformation of metalliferous sediments from the East Pacific rise, Bauer Deep and central basin, Northwest Nacza plate. Geol. Soc. Amer. Bull. 88, 723-733.

Hekinian, R., Fevrier, M., Bischoff, J.L., Picot, P. and Shanks, W.C. (1980) Sulfide deposits from the East Pacific rise near 21°N. Science, 207, 1443-1444.

Hoffert, M., Lalou, C., Brichet, E. and Bonte, Ph. (1975) Presence en Atlantique nord de nodules de manganèse à noyaux d'attapulgite et de phillipsite authigène. C.R. Acad. Sc. Paris 281, 231-233.

Honnorez, J. (1969) La formation actuelle d'un gisement sous marin de sulfures fumerolliens à Vulcano (mer thyrrhénienne) Partie I: les minéraux sulfurés des tufs immergés à faible profondeur. Mineral. Deposit. 4, 114-131.

Honnorez, J., Von Herzen, R.P., Barret, T.J., Becker, K., Bender, M.L., Borella, P.E., Hubberten, H.W., Jones, S.C., Karato, S., Laverne, C., Levi, S., Migdison, A.A., Moorby, S.A. and Schrader, E.L. (1981) Hydrothermal mounds and young ocean crust of the Galapagos: preliminary deep sea drilling results, leg LXX. Geol. Soc. Amer.Bull., 92, 457-472.

Horowitz, A. and Cronan, D.S. (1976) The geochemistry of basal sediments from the north atlantic ocean. Marine Geol. 20, 205-228.

Hosking, K.G.F. (1975) The sulfides of certain deep sea sediment samples collected on DSDP Leg XXXI. Init. Rept. DSDP US Gvt Print. Off. XXXI 515-517.

Kidd, R.B. and Armannson, H. (1979) Manganese and iron micronodules from a volcanic seamount in the Tyrrhenian sea. J. Geol. Soc. London, 136, 71-76.

Klinkhammer, G.P. (1980) Observations of the distribution of manganese over the East Pacific rise. Chemical Geol. 29, 211-226.

Krishnaswami, S., Somayajulu, B.L.K. and Moore, W.S. (1972) Dating of manganese nodules using beryllium 10. In: Horn Ed. Ferromanganese deposits on the oceans Papers from a conference held at Lamont Doherty Geological Obs. IDOE-NSF, 117-122.

Ku, T.L. and Broecker, W.S. (1967) Uranium, thorium and protactinium in a manganese nodule. Earth Planet. Sc. Lett. 2, 317-320.

Lalou, C. and Brichet, E. (1972) Signification des mesures radiochimiques dans l'évaluation de la vitesse de croissance des nodules de manganèse. C.R. Acad.Sc. Paris, 275, 815-818.

Lalou, C. and Brichet, E. (1981) Possibilité de datation des dépôts de sulfures metalliques hydrothermaux sous-marins par les descendants à vie courte de l'uranium et du thorium. C.R. Acad.Sc. Paris 293, 821-826.

Lalou, C. and Brichet, E. (1982) Ages and implications of East Pacific Rise sulphide deposits at 21°N. Nature, 300, 169-171.

Lalou, C., Brichet, E. and Bonte, Ph. (1976) Some new data on the genesis of manganese nodules. 25th Intern.Geol.Congr. 3, 779-780.

Lalou, C., Brichet,E. and Bonte, Ph. (1981) Some new data on the genesis of manganese nodules. In: Geology and geochemistry of Manganese Hungarian Acad of Sc. Publ. 31-90.

Lalou, C., Brichet, E., Jehanno, C. and Leclaire, H. (1983). Hydrothermal manganese deposits from Galapagos mounds, DSDP leg LXX, Hole 509B and Alvin dives 729 and 721 Earth Planet. Sc.Lett. 63, 63-75.

Lalou, C., Brichet, E., Ku, T.L.,and Jehanno, C. (1977) Radiochemical, scanning electron microscope (SEM) and X ray dispersive energy (EDAX) studies ot a Famous hydrothermal deposit. Marine Geol. 24, 245-258.

Lalou, C., Brichet, E., Leclaire, H. and Duplessy, J.C. (1983) Uranium series disequilibrium and isotope stratigraphy in hydrothermal mounds samples from DSDP sites 506-509, leg LXX and site 424, leg LIV: an attempt at chronology. Init. Rept. DSDP U.S. Govt. Print. Off. LXX 303-314.

Lalou, C., Brichet, E., Poupeau, G., Romary, P and Jehanno, C. (1979) Growth rates and possible ages of a North Pacific manganese nodule. In:Marine geology and Oceanography of the Pacific manganese nodules province. J.L. Bischoff and D. Piper Ed. Plenum publ. Corp. 815-833.

Lalou, C., Brichet, E. and Ranque, D. (1973) Certains nodules de manganèse trouvés en surface des sédiments sont-ils des formations contemporaines de la sédimentation ? C.R. Acad. Sc. Paris 276, 1661-1664.

Lalou, C., Delibrias, G., Brichet, E. and Labeyrie, J. (1973) Existence de carbone 14 au centre de deux nodules de manganèse, âges C-14 et Th-230 de ces nodules. C.R. Acad. Sc. Paris 276, 3013-3015.

Lonsdale, P. (1977) Clustering of suspension feeding macrobenthos near abyssal hydrothermal vents at oceanic spreading centers. Deep Sea Res. 84, 857-863.

Lonsdale, P. (1979) A deep sea hydrothermal site on a strike slip fault Nature 281, 531-534.

Lonsdale, P.F., Bischoff, J.L., Burns, V.M., Kastner, M. and Sweeney, R.E. (1980) A high temperature hydrothermal deposit on the seabed at the Gulf of California spreading center. Earth Planet. Sci. Let. 49, 8-20.

Lonsdale, P., Burns, V.M. and Fisk, M. (1980) Nodules of hydrothermal birnessite in the caldera of a young sea-mount. Jour. Geol. 88, 611-618.

Lupton, J.E., Klinkhammer, G.P., Normark, W.R., Haymon, R., MacDonald, K.C., Weiss, R.F. and Craig, H. (1980) Helium 3 and manganese at the 21°N East Pacific rise hydrothermal site. Earth Planet. Sc. Lett. 50, 115-127.

Malahoff, A., Cronan, D.S., Skirrow, R. and Embly, W. (1982) Submarine hydrothermal mineralization from the Galapagos rift

at 86°W. Marine Mining (in the press).
Marchig, V.S. and Vallier, T.L. (1974) Geochemical studies of sediments and interstitial water, site 248 and 249, Leg XXV, DSDP. Init. Rept. US Govt. Print. Off. XXV, 405-415.
Mc Murty, G.M. and Burnett, W.C. (1975) Hydrothermal metallogenesis in the Bauer Deep of the South eastern Pacific. Nature, 254, 42-44
Mero, J.L. (1965) The mineral ressources of the sea. Elsevier oceanography series, 312 pp.
Moore, W.S. and Vogt, P.R. (1976) Hydrothermal manganese crust from two sites near the Galapagos spreading axis. Earth Planet. Sc. Lett. 29, 349-356.
Murray, J. and Renard, A.F.(1891) Deep sea deposits. Rept. Sci. Results. Explor. voyage Challenger 525 pp.
Natland, J.H. (1973) Basal ferromanganoan sediments at DSDP site 183, Aleutian abyssal plain and site 192, Meiji guyot, Northwest Pacific, Leg XIX. Init. Rept. DSDP, US Govt. Print. Off. XIX, 629-636.
Nayudu, Y.R. (1965) Petrology of submarine volcanics and sediments in the vicinity of the Mendocino fracture zone.Progress Oceanogr. 3, 207-220.
Piper, D.Z. (1973) Origin of metalliferous sediments from the East Pacific Rise. Earth Planet. Sc. Lett. 19, 75-82.
Piper,D.Z., Veeh, H.H., Bertrand, W.G. and Chase R.L. (1975) An iron rich deposit from the Northeast Pacific. Earth Planet. Sc. Lett. 26, 114-120.
Rabaa, El M. and Basher, R.H.M. (1975) Replacement of coral reefs by manganese deposit. IX th Congrés Int. Sedim. Nice, 9, 38-43.
Reyss, J.L. and Lalou, C. (1981) Nodules and associated sediments in the Madagascar basin. Chemic. Geol. 34, 31-41.
Reyss, J.L., Marchig, V. and Ku, T.L. (1982) Rapid growth of deep-sea manganese nodule. Nature, 295, 401-403.
RISE project group. (1980) East Pacific Rise: Hot springs and geophysical experiments. Science 207, 1421-1442.
Rona, P.A. (1978) Criteria for recognition of hydrothermal mineral deposits in oceanic crust. Econ. Geol. 73-2- 135-160.
Rona, P.A. (1980) TAG hydrothermal field: Mid Atlantic ridge crest at latitude 26°N. Geol. Soc. London, 137, 385-402.
Rona, P.A., Harbison, R.N., Bassinger, B.G., Scott, R.B. and Nalwak, A.J. (1976) Tectonic fabric and hydrothermal activity of the Mid-Atlantic ridge crest (Lat. 26°N.) Geol Soc. Amer. Bull. 87, 661-674.
Rossi, P.L., Bocchi, G. and Lucchini, F. (1980) A manganese deposit from the South Tyrrhenian region. Oceanologica Acta 3-1, 107-113.
Sayles, F.L. and Bischoff J.L. (1973) Ferromanganoan sediments in the equatorial East Pacific.Earth Planet. Sci. Lett. 19, 330-336.
Scott, M.R., Scott, R.B., Morse, J.W., Betzer, P.R., Butler, L.W.,

and Rona, P.A. (1978) Metal enriched sediments from the TAG hydrothermal field. Nature, 276, 811-813.
Scott, M.R., Scott, R.B., Rona, P.A., Butler, L.W. and Nalwak, A.J. (1974) Rapidly accumulating manganese deposit from the median valley of the Mid-Atlantic ridge. Geophys. Res. Lett. 1, 355-358.
Stephens, J.D. and Wittkopp, R.W. (1969) Microscopic and electron beam microprobe study of sulfide minerals in Red Sea mud samples. In: Hot brines and recent hydrothermal deposits in the Red Sea. E.T. Degens and D.A. Ross Ed. Springer Verlag, N.Y. Inc. 441-447.
Strakhov, N.M. and Nesterova, J.L. (1968) Effect of volcanism on the geochemistry of marine deposits in the sea of Okhotsk. Geochem.Intern., 5, 644-666.
Thompson, G., Woo, C.C. and Sung, G.W. (1975) Metalliferous deposits on the Mid-Atlantic ridge. Abst. Geol. Soc. Amer. 7, 1297-1298.
Toth, J.R. (1980) Deposition of submarine crusts rich in manganese and iron. Geol. Soc. Amer. Bull. 91, 44-54.
Varentsov, I.M., Ryabushkin, D.A., Kazimirov, D.A., Koreneva, Y.E., Gendler, T.S., Udinstev, G.B., Gradusov, B.P., Chizhirova, N.N., Saidova, K.M., Belyayeva, N.V. and Dzhinoridze, R.N. (1980)Metalliferous sediments of the Iceland Plateau, North-Atlantic, geochemical characteristics of formation. Trans. from Geokhimya 10, 1528- 1541 (Geochem. Intern., 140-153).
Veeh, H.H. and Boström, K. (1971) Anomalous $^{234}U/^{238}U$ on the East Pacific rise. Earth Planet. Sc. Lett. 10, 372-374.
Warner, J.B. and Gieskes, J.M. (1974) Iron-rich basal sediments from the Indian Ocean: site 245 DSDP. Init.Rept. DSDP US Govt. Print. Off.XXV, 395-403.
Weiss, R.F. (1977) Hydrothermal manganese in the deep-sea, scavenging residence time and $Mn/^{3}He$ relationships. Earth Planet. Sc. Lett. 37, 257-262.
Zelenof, K.K. (1965) Iron and manganese in exhalations of the submarine Banu Wuhu volcano (Indonesia). Doklady Akad. Nauk. SSR. 155, 1317-1320 (English translation, 94-96).

HYDROTHERMAL AND HYDROGENOUS FERRO-MANGANESE DEPOSITS: DO THEY FORM A CONTINUUM? THE RARE EARTH ELEMENT EVIDENCE

A.J. Fleet

Exploration and Production Division, BP Research Centre
Chertsey Road, Sunbury-on-Thames
Middlesex, U.K., TW16 7LN

ABSTRACT

Ferro-manganese deposits in the deep-sea are not all likely to have just formed by either hydrothermal processes or hydrogenous ones. Some will contain material contributed by both these processes so that such deposits probably form a continuum. In this paper the rare earth element (REE) evidence for this supposition is appraised and the REE contents of marine ferro-manganese deposits are discussed. The hydrothermal and hydrogenous "endmembers" of the hypothetical continuum generally have distinct REE contents: the former being relatively depleted in cerium and the latter enriched. Ferro-manganese deposits with REE contents derived by both hydrogenous and hydrothermal processes apparently exist, but only those with predominantly "hydrothermal REE" can be easily recognised. This is because hydrogenous deposits generally contain higher REE abundances than hydrothermal ones so that even a small "hydrogenous REE" content swamps the "hydrothermal REE". In the geologic record it should be possible to distinguish between ferro-manganese deposits formed by these different processes and ones formed in shallow water largely as a result of diagenesis. Hydrogenous deposits with apparently atypical REE contents (e.g. those of the Red Sea and Santorini) may also be identified.

INTRODUCTION

Marine ferro-manganese deposits may be formed by either hydrogenous or hydrothermal processes (e.g. Bonatti, 1981; Cronan, 1976). Hydrogenous deposits form by direct precipitation from seawater and by interaction at the seafloor of seawater and surface sediments. Hydrothermal deposits are also precipitated from seawater, but from

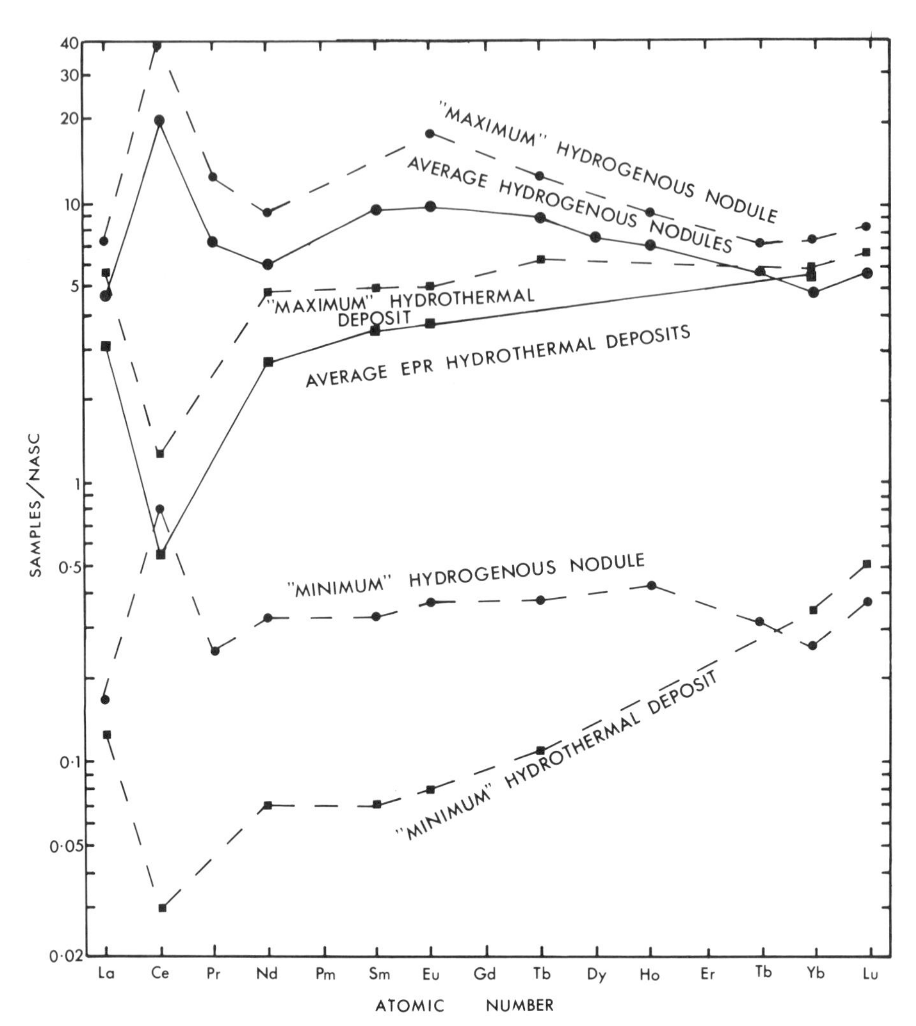

Fig. 1. REE abundances in ferro-manganese deposits compared to a composite of North American shales (NASC) (Haskin et al., 1968).Data for the hydrogenous nodules are from Ehrlich (1968):the "maximum" plot is for his Indian Ocean sample and the "minimum" plot is for his Pacific Area B nodule. The average values for East Pacific Rise (EPR) hydrothermal deposits are from Elderfield and Greaves (1981):they are on a carbonate-free basis and corrected for REE in foraminiferal calcite.The "maximum" hydrogenous plot is for sample A-21 of Piper and Graef (1974) on a carbonate-free basis.The "minimum" hydrogenous plot is for sample 14A FAMOUS of Bonnot-Courtois (1981).

seawater which has circulated through the igneous lithosphere and reacted with it at elevated temperatures.Marine ferro-manganese deposits are not necessarily formed solely by either hydrothermal or hydrogenous processes.Both processes may contribute to the formation of such deposits,so that a continuum between the two types of "end-member" ferro-manganese deposit might be expected (Cronan,1980,p.249-250).This review assesses the rare earth element (REE) evidence for such a continuum and discusses the REE contents of marine ferro-manganese deposits.

The REE generally behave similarly to one another in the geologic environment.However,processes which fractionate one or more of these elements from the others do occur and are significant.The REE contents which result from such fractionation provide an important line of evidence for characterising those deposits which have the same or similar geneses (Fleet,in press).

THE REE CONTENTS OF "ENDMEMBER" HYDROTHERMAL AND HYDROGENOUS FERRO-MANGANESE DEPOSITS

The REE contents of the "endmember"marine ferro-manganese deposits are distinct.Hydrogenous ferro-manganese deposits have high REE contents compared with sediments in general (Table 1) and are enriched in Ce relative to other REE (Fig. 1).In Figure 1,as is common practice,the REE contents of the deposits are normalised relative to those of a North American shale composite (NASC) (Haskin et al.,1968). This normalisation is carried out to allow easy comparison of different analyses.The NASC is used to represent sediments in general because all continental lithogenous material has similar REE abundances:any fractionation of the REE which occurs during weathering seems to be negated during transport so that aluminosilicate materials entering the oceans have similar REE contents (e.g. Fleet, in press). Figure 1,therefore, shows that hydrogenous nodules are generally enriched about five-tenfold in the REE relative to lithogenous material in general.In contrast to hydrogenous ferro-manganese deposits,hydrothermal ones, which are also generally enriched in REE relative to sediments in general (Table 1),are depleted in Ce and are somewhat enriched in the heavy REE (HREE) relative to the light (LREE) and other REE (Fig. 1). The REE contents of ferro-manganese deposits, especially as reflected by their normalised plots, can, therefore,be used to give clues to whether a deposit formed from either a hydrothermal source or a hydrogenous source or both,by showing that the deposit has received REE from one or both these types of source.The extent to which this is possible is discussed below.

REE FRACTIONATION IN THE MARINE ENVIRONMENT

The REE mainly enter the marine environment incorporated in

Table 1. REE abundances in ferro-manganese deposits and other sedimentary materials (ppm)

	La	Ce	Pr	Nd	Sm	Eu	Gd	Tb	Dy	Ho	Er	Tm	Yb	Lu
HYDROGENOUS DEPOSITS (Ehrlich, 1968)														
Average	150	1460	57	200	55	12		7.5	44	7.3		2.8	15	2.7
"Maximum" (Indian Ocean Sample)	237	2800		306	77.5	22	22		10.5			3.6	23	3.9
"Minimum" (Pacific Area B Sample)	5.4	60		11	1.88	0.46		0.32				0.16	0.82	0.18
HYDROTHERMAL DEPOSITS														
Average EPR crest (Piper and Graef, 1974, corrected by Elderfield *et al.*, 1981a).	100	40		91	20	4.6							16	
"Maximum" (Sample A-21 of Piper and Graef, 1974 on a carbonate-free basis).	182.1	93.7		158.9	28.3	6.16		5.42					18.3	3.21
"Minimum" (Sample 14A FAMOUS of Bonnot-Courtois, 1981)	4.0	1.87		2.4	0.40	0.105		0.09					1.10	0.25
LITHOGENOUS MATERIAL														
North American shales composite (Haskin *et al.*, 1968)	32	73	7.9	33	5.7	1.24	5.2	0.85	5.8	1.04	3.4	0.50	3.1	0.48
PLANKTONIC FORAMINIFERA														
(Elderfield *et al.*, 1981a)	1.28	0.355		1.0	0.199	0.0601	0.29		0.33		0.27		0.28	
SEAWATER ($\times 10^{-7}$)														
Average of 11 samples of North Atlantic Deep Water (Høgdahl *et al.*, 1968	34	12	6.4	28	4.5	1.3	7.0	1.4	9.1	2.2	8.7	1.7	8.2	1.5

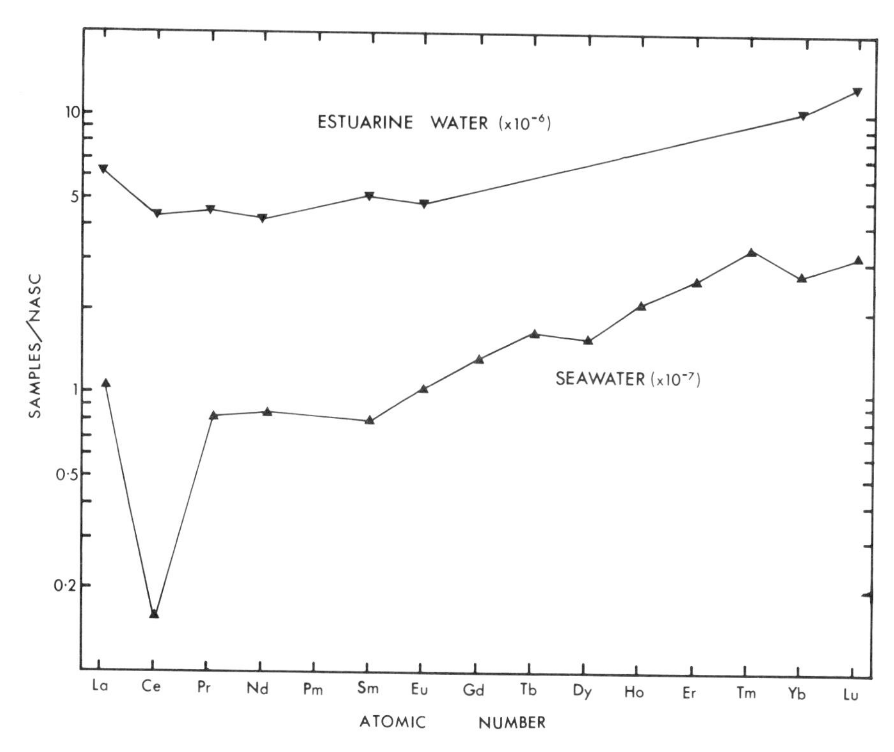

Fig. 2. Average REE abundances dissolved in Gironde estuary water of 0.42‰ salinity (Martin et al.,1976) and in eleven samples of North Atlantic Deep Water (Høgdahl et al.,1968) compared to a composite of North American shales (Haskin et al.,1968).

particulate materials,only a few percent of the supply are dissolved (Martin et al.,1976).The dissolved REE entering the oceans and the REE in seawater (Fig. 2) are slightly enriched in both the LREE and the HREE relative to the intermediate REE.Unlike riverwater,though, seawater is markedly depleted in Ce (Fig. 2).This fractionation of Ce relative to the other REE is due to Ce being removed rapidly, relative to the other REE,from the oceans.Goldberg (1961) was the first to suggest an explanation,which has not been superceded,for this Ce fractionation.He proposed that Ce^{3+} in the oceans is oxidised to Ce^{4+} and is precipitated from solution as CeO_2,while the other REE remain in the 3+ state and are lost from solution without discernible fractionation of other individual REE.This Ce removal from seawater probably occurs in the open ocean rather than in estuaries or shelf waters (e.g.Fleet,in press).The other REE in seawater are also very efficiently removed from solution (Turner and Whitfield, 1979).

Hydrogenous ferro-manganese nodules act as a major repository for the REE and,in particular,for Ce.Elderfield et al. (1981a) suggest that Ce might be incorporated directly into nodules,unlike the other, trivalent REE which Elderfield et al. consider probably undergo some cycling through surface sediments.Hydrothermal ferro-manganese deposits,on the other hand,accumulate the REE without significant fractionation of the REE one from another.They take up REE from seawater which has previously lost Ce during its residence time in the oceans but which has undergone no marked change in contents during hydrothermal circulation. Normalised REE plots of the hydrothermal deposits,therefore,parallel those of seawater (cf.Figs. 1 and 2).Indeed the REE contents of ocean-ridge metalliferous were first used as contributory evidence to show that these sediments form as precipitates from hydrothermal solutions of seawater which has circulated through the oceanic crust;in particular,the negative Ce anomalies exhibited by the normalised plots of these sediments were taken as indicating a seawater source at least for the REE of the sediments (e.g.Bender et al.,1971;Dymond et al.,1973;Piper and Graef, 1974).Evidence from ophiolites has shown that REE scavenging by ferro-manganese deposits can go on within the crust where conditions are suitable for oxyhydroxides to form (e.g.Robertson and Fleet,1976).

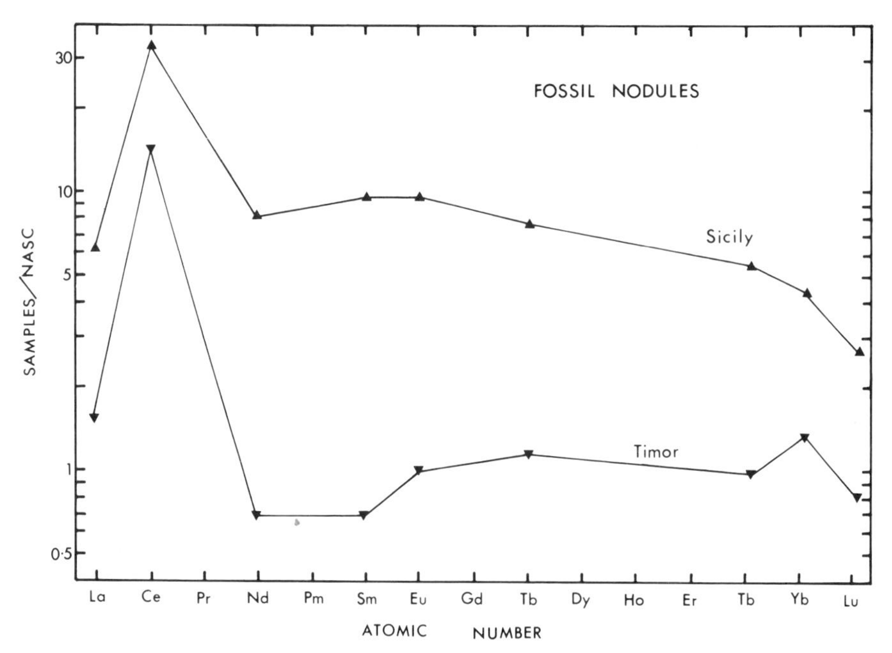

Fig. 3. REE abundances in fossil ferro-manganese nodules from Sicily and Timor (Fleet,unpublished data) compared to a composite of North American shales (Haskin et al.,1968).

REE VARIABILITY IN FERRO-MANGANESE DEPOSITS

The range of REE contents which hydrogenous ferro-manganese deposits have is illustrated in Figure 1.These data (Table 1) are all from Ehrlich (1968),but other analyses of deep-sea nodules,which are apparently hydrogenous,fall within this range (e.g.Volkov and Fomina, 1967;Piper,1974;Rankin and Glasby,1979;Courtois and Clauer,1980; Elderfield et al.,1981a),as do ones from micronodules (Addy,1979) and for fossil nodules (Fig. 3).Piper (1974) suggested that there is some systematic variation of REE content with water depth but stressed that there were too few data available to him to ascertain any regional variations. More recently Enderfield et al. (1981b) have demonstrated that the variation in the REE contents within individual nodules is compatible with the variation between different nodules. Those hydrogenous deposits which are not enriched in the REE relative to sediments (i.e.have values falling below 1.0 on NASC-normalised plots) tend to the exceptions but even they have distinct positive Ce anomalies.

Although discussion here is limited to deep-sea deposits it is

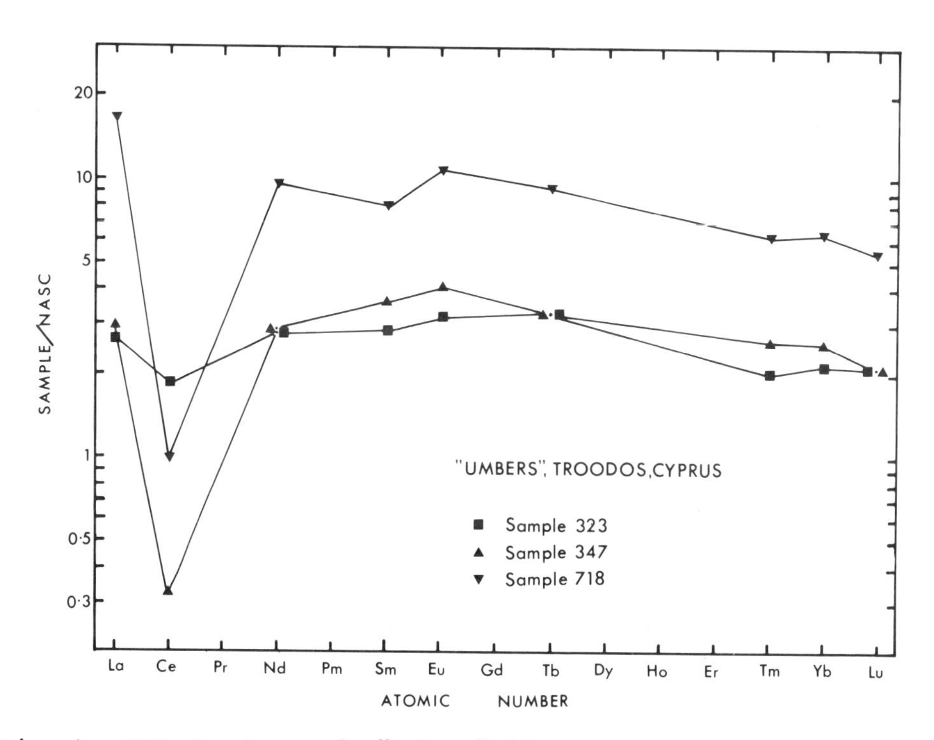

Fig. 4. REE abundances in "umbers" (ferro-manganese hydrothermal deposits) from Troodos,Cyprus (Robertson and Fleet,1976) compared to a composite of North American shales (Haskin et al.,1968).Sample 323 is rich in lithogenous material.

important to note that shallow-water,continental-margin nodules are enriched neither in the REE generally nor in Ce specifically (Calvert and Price,1977).This must be because the genesis of these nodules is very different from that of their deep-sea equivalents.The shallow-water nodules grow rapidly,mainly as a result of diagenetic mobilisation of Fe and Mn in the underlying sediments.They,therefore,have little opportunity to incorporate REE,and,in particular,no preferential build-up of Ce can occur in them.In contrast,in the deep sea, hydrogenous nodules grow slowly and,as discussed above,probably accumulate the trivalent REE after some cycling through surface sediments and Ce directly from seawater (Elderfield et al.,1981a). The interpretation of the REE contents of shallow-water nodules from the seafloor should present few problems but other geochemical and sedimentary criteria should be taken into account when considering the origin of fossil nodules.Only deep-sea deposits are considered below.

Ferro-manganese deposits which are believed to be hydrothermal are generally enriched in the REE,except for Ce,relative to sediments in general (Fig. 1).Those ferro-manganese deposits from ophiolites which are believed to be hydrothermal have similar REE contents (Fig. 4) (e.g.Bonatti et al.,1976;Robertson and Fleet,1976).Hydrothermal nontronites contain lower REE abundances but are also depleted in Ce (Fig. 5) (Barrett et al.,in press).Hydrothermal sulphides,like sulphides in general,are unable to accommodate significant

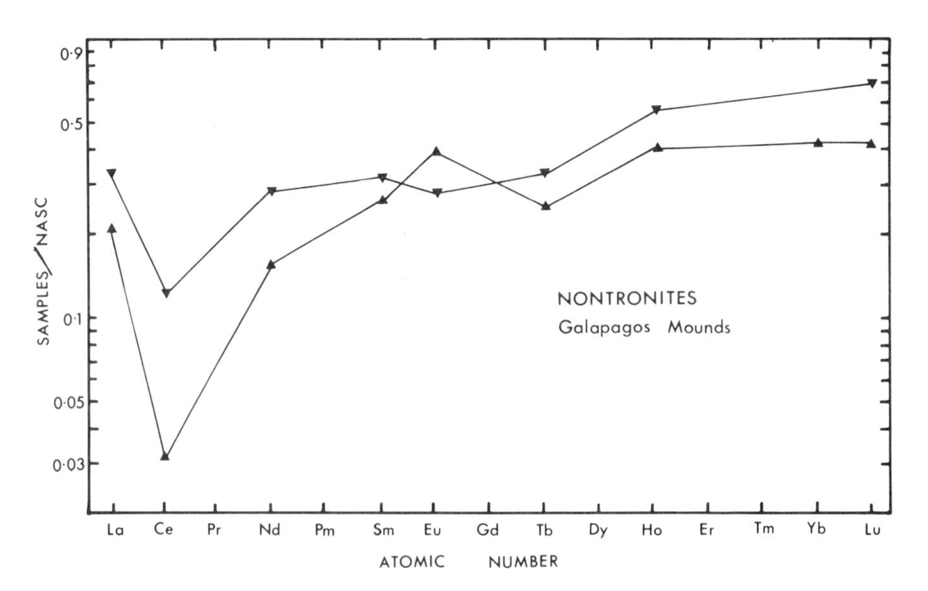

Fig. 5 REE abundances in nontronites from the Galapagos mounds of the eastern Equatorial Pacific (Barrett et al.,in press) compared to a composite of North American shales (Haskin et al.,1968).

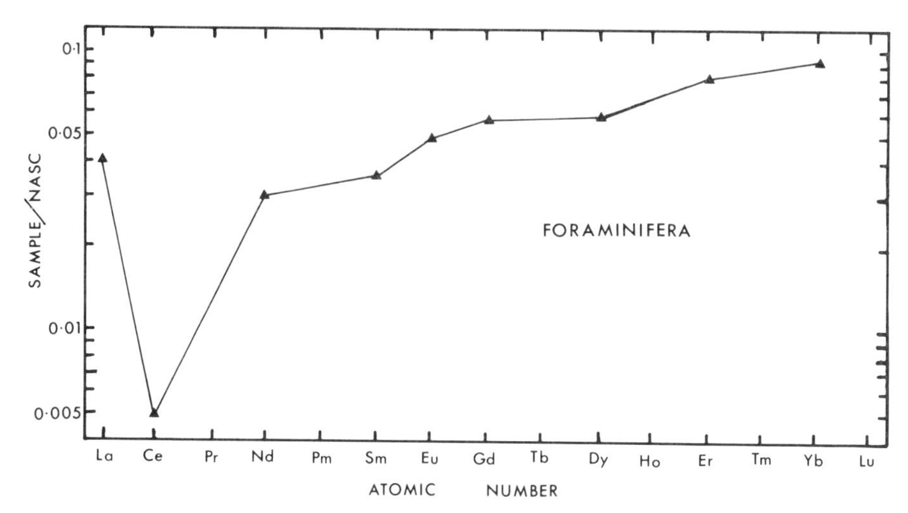

Fig. 6. REE abundances in planktonic foraminifera (Elderfield et al., 1981a) compared to a composite of North American shales (Haskin et al.,1968).

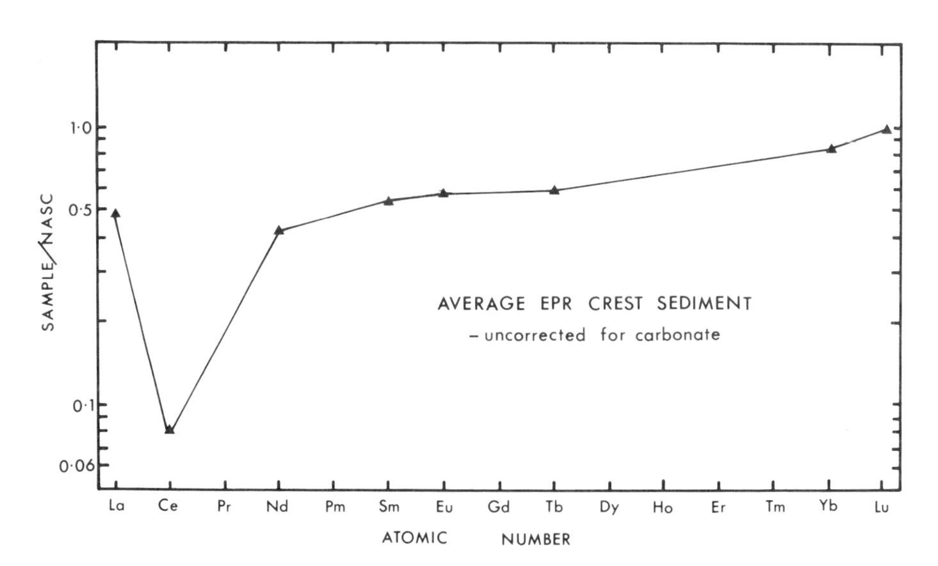

Fig. 7. Average REE abundances in metalliferous sediments from the crest of the East Pacific Rise (Piper and Graef, 1974), uncorrected for carbonate content, compared to a composite of North American shales (Haskin et al., 1968).

amounts of REE (e.g.Robertson and Fleet,1976).

Because both hydrothermal and hydrogenous deposits are generally enriched in the REE relative to sediments in general,the prescence of biogenic or lithogenous material dilutes the REE contents of most deposits.This is particularly true of hydrothermal deposits which usually occur on spreading ridges and other volcanic edificies above the calcite compensation depth and often contain 70-90% or more $CaCO_3$.The REE contents of biogenic materials are poorly known.Calcareous foraminifera apparently have very low REE contents which on a normalised plot parallel those of seawater (cf.Figs. 2 and 6) (Spirn,1965;Elderfield et al.,1981a).Diatoms and radiolaria also seem to contain lower REE abundances than sediments in general (e.g. Elderfield et al.,1981a).The presence of biogenic material would, therefore be expected to lower the bulk REE contents of ferro-manganese deposits.In the case of calcareous material,if the REE contents of foraminiferal are typical,this will just lower the normalised plot of hydrothermal deposits as the plots of both foraminifera and the deposits parallel that of seawater (cf.Figs. 1,6 and 7). Calcareous material in hydrogenous nodules would also tend to lower the REE contents of nodules and depress their Ce enrichment.This is likely to be significant in very few hydrogenous deposits,though, as they usually contain very little calcareous material and their REE contents are usually very high compared with biogenic calcite.

The presence of lithogenous material,as typified by NASC,will tend to dilute and flatten the normalised plots of most hydrothermal and hydrogenous deposits (Fig. 9);though,rarely,it will enrich the REE contents of those deposits with REE contents less than those of NASC.An actual example of the effects of lithogenous dilution is seen in Figure 4.Because of their generally lower REE contents the normalised plots of hydrothermal deposits may be expected to be more affected than those of hydrogenous ones (Fig. 8).The presence of lithogenous and biogenic material in ferro-manganese deposits should,of course,be detectable from geochemical and mineralogical data.

Particular care should be taken when interpreting the REE contents of ferro-manganese crusts on basaltic material from spreading ridges.Because such basaltic material is depleted in the LREE relative to the HREE (e.g.Haskin et al.,1968),appreciable quantities of it will dilute the REE abundances of the crusts.Also it must be borne in mind that such ferro-manganese deposits might in part be halmrolysates.It is unlikely,though,that the REE will be mobilised from the basalt into precipitated ferro-manganese deposits as selective REE uptake seems to be the process occurring during basalt weathering (e.g.Fleet,in press).

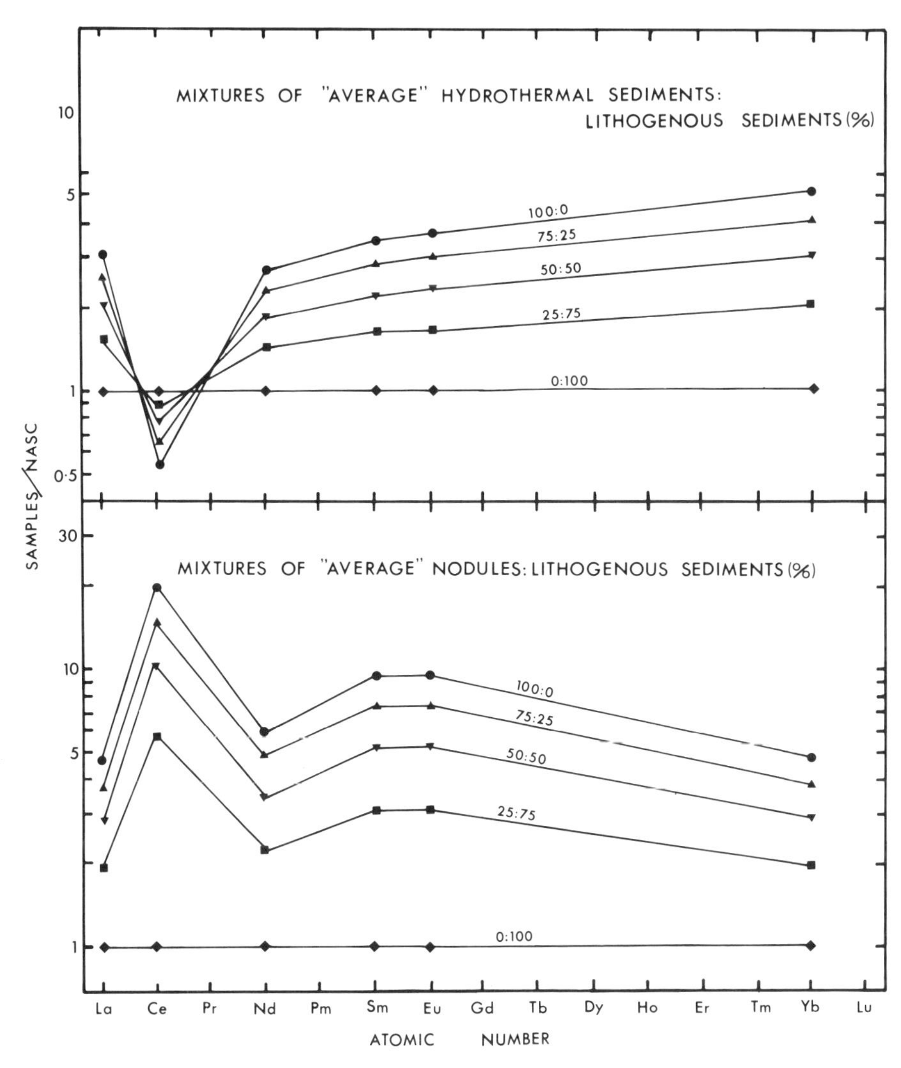

Fig. 8. Calculated REE abundances in hypothetical mixtures of average East Pacific Rise hydrothermal metalliferous sediments (Piper and Graef, 1974, as corrected by Elderfield et al., 1981a) and average ferro-manganese hydrogenous nodules (Ehrlich, 1968) mixed with average lithogenous sediment (NASC), compared to North American shales (Haskin et al., 1968).

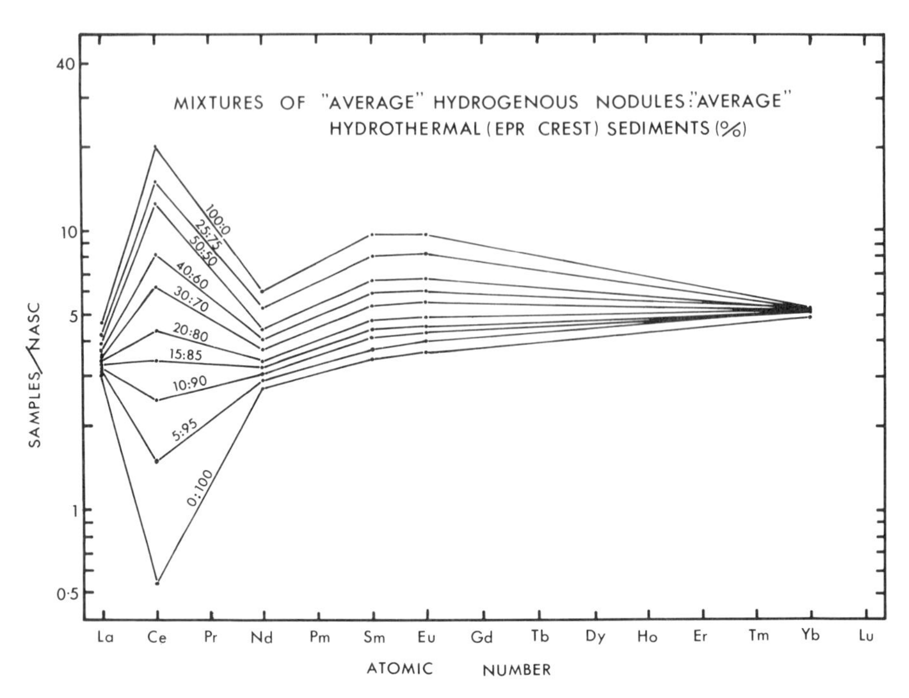

Fig. 9. Calculated REE abundances in hypothetical mixtures of average East Pacific Rise hydrothermal metalliferous sediments (Piper and Graef,1974,as corrected by Elderfield et al.,1981a) mixed with average ferro-manganese hydrogenous nodules (Ehrlich, 1968), compared to North America shales (Haskins et al., 1968).

MIXED HYDROGENOUS AND HYDROTHERMAL DEPOSITS

Some ferro-manganese deposits contain REE which were apparently derived from both hydrogenous and hydrothermal sources.The REE contents which would result from mixing different proportions of "average" East Pacific Rise crest sediment and "average" manganese nodule (Table 1; Fig. 1) are given in Figure 9. The relative proportion of "hydrothermal" REE has to be high for the deposit to be depleted in Ce. Figure 9 suggests 90% or more of the REE must be "hydrothermal" for this to occur.In actual examples,though,the likely REE contents of local hydrothermal and hydrogenous deposits must be considered when estimating the relative proportions.Elderfield and Greaves (1981) have described ferro-manganese nodules from the Bauer Deep with marked Ce deficiencies (Fig.11).They have estimated,using average REE contents for north Equatorial Pacific nodules and for East Pacific Rise crest sediment expressed on a carbonate-free basis,that the REE in the nodules are about 80% "hydrothermal" and 20% "hydrogenous".

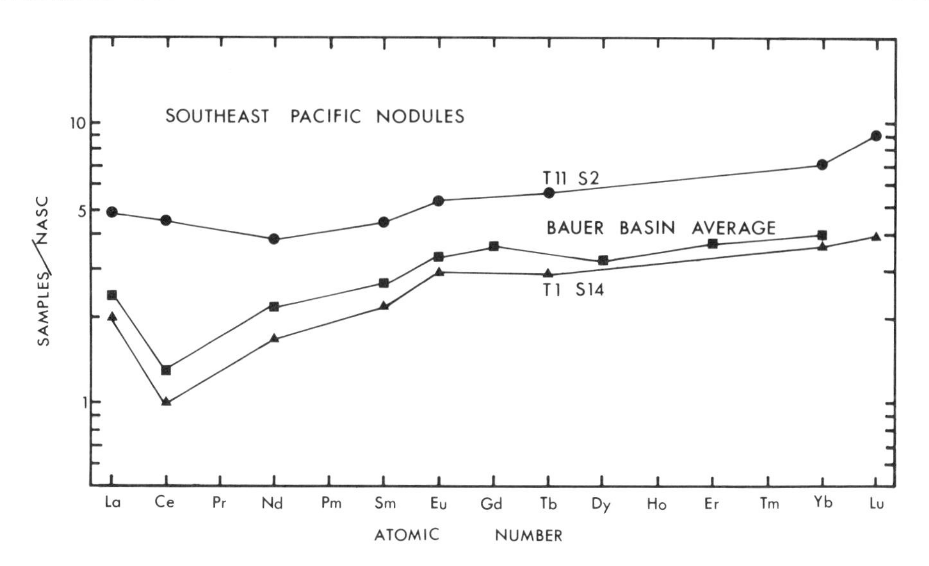

Fig. 10. REE abundances in ferro-manganese nodules from the southeast Pacific compared to North American shales (Haskin et al., 1968).The average for Bauer Basin nodules is from Elderfield and Greaves (1981);the others are from Courtois and Clauer (1980).

Some other nodules from the southeast Pacific and elsewhere in the Pacific have small Ce deficiences or no Ce anomalies on normalised plots (e.g.Fig.10) (Piper,1974;Courtois and Clauer,1980).Most of these are from relatively near active ridges suggesting that deep-water advection,of the type identified by Craig (this volume) and his co-workers using ^{3}He,may have contributed significant quantities of

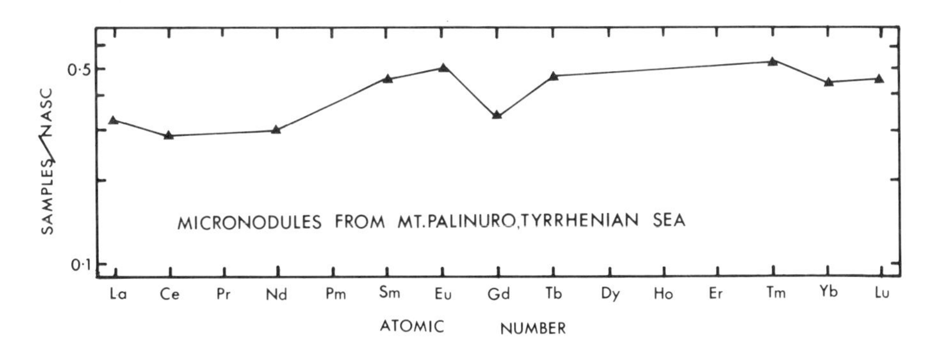

Fig. 11. REE abundances in ferro-manganese micronodules from Mount Palinuro,Tyrrhenian Sea (Fleet,unpublished data) compared to North American shales (Haskin et al.,1968).

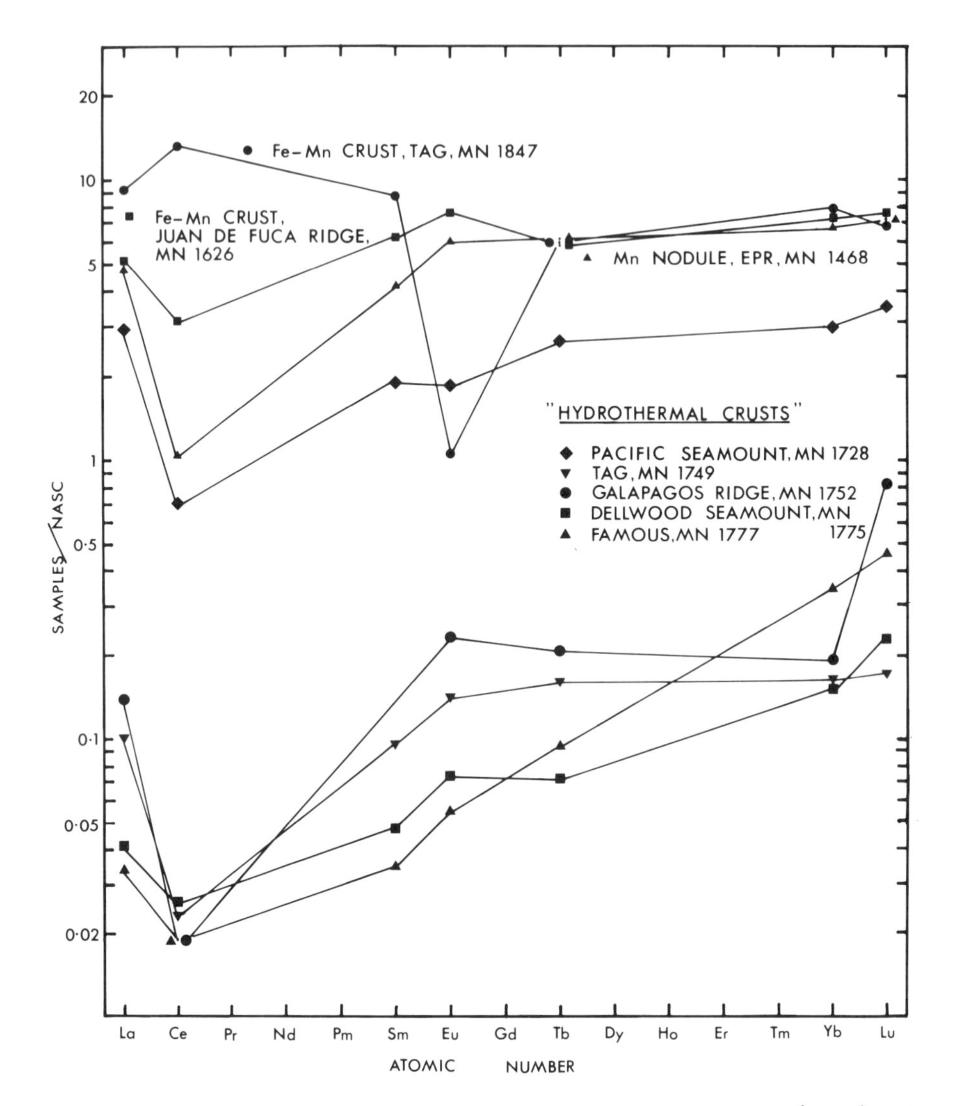

Fig. 12. REE abundances in various ferro-manganese deposits (Toth, 1980) compared to North American shales (Haskin et al. 1968).

hydrothermal REE to these nodules.

Micronodules from Mount Palinuro in the Tyrrhenian Sea, described by Kidd and Ármannson (1979) and believed by these authors to be hydrothermal, are probably a further example of this kind of nodule with REE mainly from a hydrothermal source but also a minor contribution of "hydrogenous" REE (Fig.11). There is the possibility that this might be due to the REE contents of Tyrrhenian seawater differing from what is taken to be "typical "seawater (cf.Morten et al.,1980). No REE data on Tyrrhenian seawater are available and,indeed,our knowledge of REE abundances in seawater is based on very few data (e.g.Fleet,in press).Results of Elderfield and Greaves (1982),though, show that the REE contents of Mediterranean outflow water in the Atlantic are very much like those of what is generally considered "typical" seawater in the literature.Local and past variations in the REE abundances in seawater must always be borne in mind when interpreting REE data from ferro-manganese deposits,but far more data on seawater are needed before any new conclusions can be drawn.

Various ferro-manganese crusts analysed and discussed by Toth (1980),which were thought to have a hydrothermal origin,also seem to represent some kind of intermediate deposit on any hypothetical ferro-manganese hydrogenous-hydrothermal continuum.Some of these deposits have low REE abundances relative to other ferro-manganese deposits (Fig. 12).On normalised plots they all have small negative Ce anomalies or,in the case of one TAG sample,a very small positive anomaly.This suggests,as Toth concluded,that small proportions of "hydrogenous" REE are present in these samples.Analyses of ferro-manganese deposits from the mounds near the Galapagos Rift,where hydrothermal solutions have passed through sediments,have yielded similar results (Fig. 13) (Corliss et al.,1978;Barrett et al.,in press).Ferro-manganese crusts from seamounts in the south Tyrrhenian Sea,reported by Morten et al.(1980),may be analogous,although the considerations advanced by these authors and discussed above in relation to the Mount Palinuro micronodules,must be considered when evaluating these samples.A fossil equivalent of these kind of deposit has been reported by Barrett (1981) (Fig. 14). Lithogenous

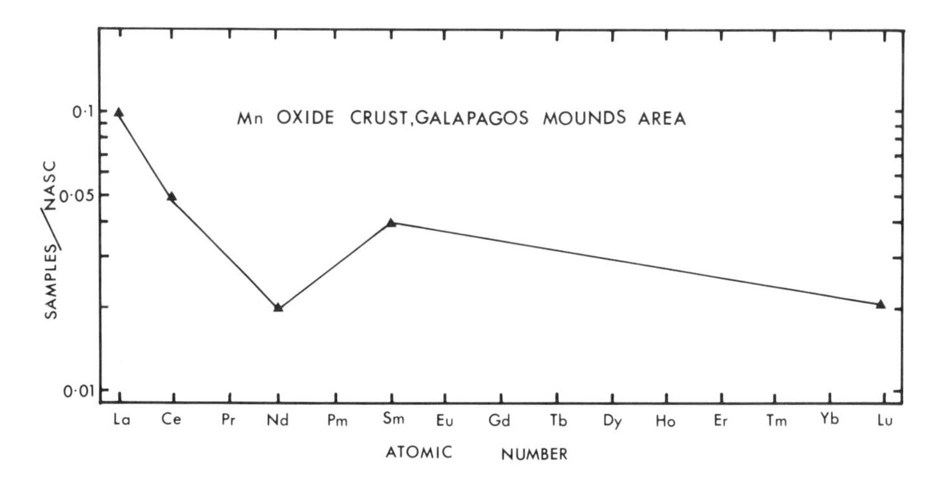

Fig. 13. REE abundances in a "manganese oxide crust" from the Galapagos mounds area (Barrett et al.,in press) compared to North American shales (Haskin et al.,1968).

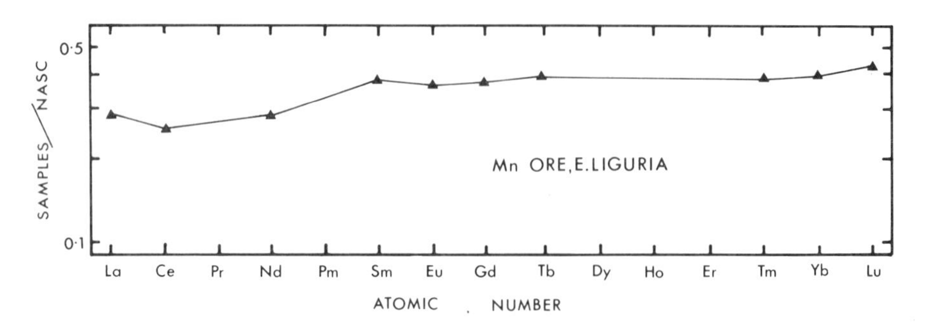

Fig. 14. REE abundances in a manganese ore from East Liguria (Barrett, 1981) compared to North American shales (Haskin et al.,1968).

dilution might be responsible for the REE contents of this sample but other geochemical evidence cited by Barrett suggests that the sample is predominantly hydrothermal.

OTHER FERRO-MANGANESE DEPOSITS

For the sake of completeness it is important to mention other ferro-manganese deposits which have been analysed for the REE. These are from off the island of Santorini in the Aegean and from the Atlantis II and Chain Deeps of the Red Sea.These are clearly hydrothermal deposits (e.g.Degens and Ross,1969;Smith and Cronan, 1976) but have very distinct REE contents (Fig. 15).Courtois and Treuil (1977) noted that the normalised plots for their Red Sea samples are similar to those of ultrabasic igneous rocks and suggested that the sediments were deposited from hydrothermal solutions which had reacted with peridotite etc.at depth.They also suggested that redox conditions and complexation at depth might be responsible. The normalised patterns of the Santorini samples are not like those of the igneous rocks of Santorini reported by Puchelt (1978).It seems more feasible that the REE contents of the Santorini ferro-manganese deposits result from the seawater in the hydrothermal system of Santorini undergoing severe and/or lengthy reducing conditions.Such conditions could have released soluble trivalent Ce into the seawater and led to the uptake of Eu^{2+} from the seawater as a result of rock alteration. The participation of acid and intermediate igneous rocks,rather than just basic ones as along spreading ridges,in the Santorini hydrothermal system may also be of importance.Further work is needed to elucidate this problem.

CONCLUSIONS

1. The ferro-manganese deposits formed in the oceans derive

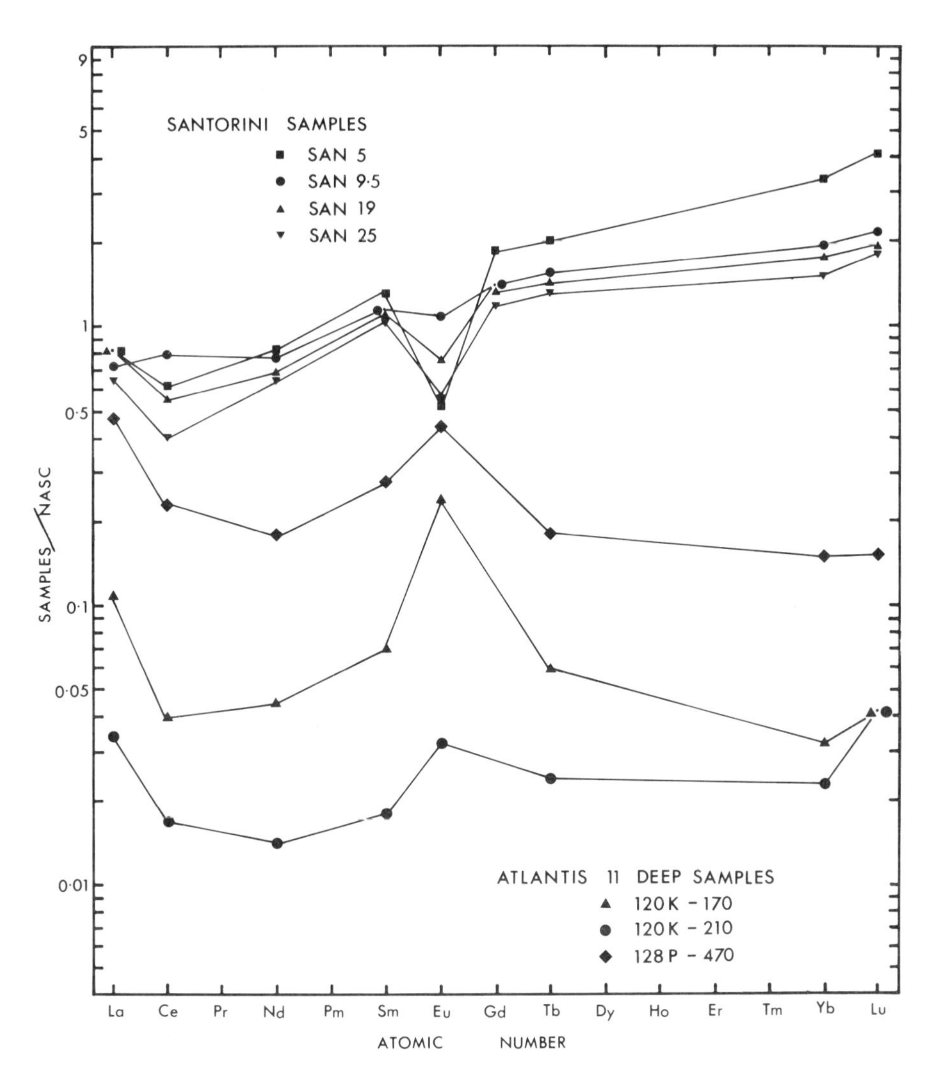

Fig. 15. REE abundances in ferro-manganese hydrothermal deposits from Santorini (Fleet,unpublished data) and Atlantis II Deep,Red Sea (Courtois and Treuil,1977) compared to North American shales (Haskin et al.,1968).Santorini samples 5 and 25 are from the inner exhalative zones of Palaea and Nea Kameni,and samples 9.5 and 19 are from the outer exhalative zones (Smith and Cronan,1976).

their REE contents from seawater.In the case of hydrogenous deposits the REE are probably either precipitated directly from seawater or cycled through surface sediments.Some fractionation of the REE occurs during these processes,in particular Ce is preferentially incorporated in the deposits. In contrast the REE contents of hydrothermal deposits seem to be incorporated in deposits from seawater without significant fractionation of the REE from each other.

2. Oceanic ferro-manganese deposits are generally enriched in the REE relative to sediments in general but hydrothermal deposits are depleted in Ce and hydrogenous ones are enriched in Ce.

3. Ferro-manganese deposits with REE contents derived by both hydrogenous and hydrothermal processes apparently exist, but only those with predominantly "hydrothermal REE" can be easily recognised. This is because hydrogenous deposits generally contain higher REE abundances than hydrothermal ones so that even a small "hydrogenous REE" content swamps the "hydrothermal REE".

4. The REE contents of hydrothermal deposits can be markedly distinct from the "norm" (e.g. the deposits of Santorini). This suggests that the hydrothermal seawater giving rise to these deposits undergoes redox and other reactions at depth not normally experienced by the seawater in hydrothermal systems at oceanic spreading axes.

ACKNOWLEDGEMENTS

When this paper was presented verbally a number of people, in particular Drs. A.H.F. Robertson and S.D. Scott, usefully commented on it. Helpful reviews of the draft manuscript came from Drs. P. Henderson and C.P. Summerhayes. My own work in this field has been carried out over a number of years, while at Chelsea and Goldsmiths' Colleges, University of London, and the Open University, utilising the facilities of the University of London Reactor Centre. The help of the Director and staff of the Centre is gratefully acknowledged, as is the help and guidance of many colleagues, in particular Drs. P. Henderson, S.J. Parry and P.J. Potts. The samples from Mount Palinuro, Santorini, Sicily and Timor, analyses of which have not been previously reported, were kindly provided by Drs. R.B. Kidd, P.A. Smith, H.C. Jenkyns and A.J. Barber. Finally I thank Mrs. Doreen Norman for speedy typing of the paper. The paper is published with the permission of the management of the British Petroleum Co. Ltd..

REFERENCES

Addy, S. K., 1979, Rare earth element patterns in manganese nodules and micronodules from northwest Atlantic, Geochim. Cosmochim. Acta, 43:1105-1115.

Barrett, T. J., 1981, Chemistry and mineralogy of Jurassic bedded chert overlying ophiolites in the North Apennines, Italy, Chem. Geol., 34:289-317.

Barrett, T. J., Fleet, A. J., and Friedrichsen, H., in press, Major element,rare earth element,and O- and H- isotopic composition of metalliferous and pelagic sediments from the Galapagos mounds area,Leg 70, Initial Rep. Deep Sea drill. Proj., 70.

Bender, M. L.,Broecker, W.,Garnitz, V.,Middel, V.,Kay, R.,Sun S. S., and Biscaye, P., 1971, Geochemistry of three cores from the East Pacific Rise, Earth planet Sci. Lett., 12:425-433.

Bonatti, E., 1981, Metal deposits in the oceanic lithosphere, in "The Sea, Volume 7," C.Emiliani, ed., Wiley Interscience.

Bonatti, E.,Zerbi, M.,Kay, R., and Rydell, H., 1976, Metalliferous deposits from the Apennine ophiolites:Mesozoic equivalents of modern deposits from oceanic spreading centres, Bull. geol. Soc. Amer., 87:83-94.

Bonnot-Courtois, C.,1981, Distribution des terres rares dans le depots hydrothermaux de la zone FAMOUS et des Galapagos - comparison avec les sédiments metalliferes, Mar. Geol., 39:1-14.

Calvert, S.E., and Price, N.B., 1977, Shallow water,continental margin and lacustrine nodules:distribution and geochemistry, in "Marine manganese deposits," G.P.Glasby, ed., Elsevier,pp.45-86.

Corliss, J.B.,Lyle, M.,Dymond, J., and Crane, K., 1978, The chemistry of hydrothermal mounds near the Galapagos Rift, Earth planet Sci. Lett., 40:12-24.

Courtois, C., and Clauer, N.,1980, Rare earth elements and strontium isotopes of polymetallic nodules from southeastern Pacific Ocean, Sedimentology, 27:687-695.

Courtoiu, C., and Treuil, M., 1977, Distribution des terres rares et de quelques éléments en trace dans les sédiments récents des fosses de la Mer Rouge, Chem. Geol., 20:57-72.

Cronan, D. S., 1976, Manganese nodules and other ferro-manganese oxide deposits, in "Chemical oceanography,2nd Edition,Volume 5," J.P.Riley and R.Chester, eds., Academic Press, pp. 217-263.

Cronan, D. S., 1980, "Underwater minerals," Academic Press, 400 pp.

Degens, E. T., and Ross, D. A., 1969, "Hot brines and recent heavy metal deposits in the Red Sea," Springer-Verlag, 600 pp.

Dymond, J.,Corliss, J. B.,Heath, G. R.,Field, C. W.,Dasch, E. J., and Veeh, H. H., 1973, Origin of metalliferous sediments from the Pacific Ocean, Bull. geol. Soc. Amer., 84:3355-3372.

Ehrlich, A. M., 1968, Rare earth element abundances in manganese nodules, Unpubl. Ph.D. Thesis, Massachusetts Institute of Technology, 225 pp.

Elderfield, H., and Greaves, M. J., 1981, Negative cerium anomalies in the rare earth element patterns of oceanic ferromanganese nodules, Earth planet Sci. Lett., 55:163-170.

Elderfield, H., and Greaves, M. J., 1982, The rare earth elements in seawater, Nature, 296:214-219.

Elderfield, H.,Hawkesworth, C. J.,Greaves, M. J., and Calvert, S. E., 1981a, Rare earth element geochemistry of oceanic ferromanganese nodules and associated sediments, Geochim. Cosmochim. Acta, 45:1231-1234.

Elderfield, H.,Hawkesworth, C. J.,Greaves, M. J., and Calvert, S. E., 1981b, Rare earth element zonation in Pacific ferromanganese

nodules, Geochim. Cosmochim. Acta, 45:1231-1234.
Fleet, A. J., in press, Aqueous and sedimentary geochemistry of the rare earths, in "Rare earth element geochemistry," P.Henderson, ed., Elsevier.
Glasby, G. P., 1973, Mechanisms of enrichment of the rarer elements in marine manganese nodules, Mar. Chem., 1:105-125.
Goldberg, E. D., 1961, Chemistry in the oceans, in "Oceanography," Am. Assoc. Adv. Sci. Publ., 67:583-597.
Haskin, L. A.,Haskin, M. A.,Frey, F. A., and Wildeman, T. R., 1968, Relative and absolute terrestrial abundances of the rare earths, in "Origin and distribution of the elements," L. H. Ahrens, ed., Pergamon, New York, pp. 889-912.
Høgdahl, O. T.,Melsom, S., and Bowen, V. T., 1968, Neutron activation of lanthanide elements in seawater, Adv. Chem. Ser., 73:308-325.
Kidd, R. B., and Armannson, H., 1979, Manganese and iron micronodules from a volcanic seamount in the Tyrrhenian Sea, Jl. geol. Soc. Lond., 136:71-76.
Martin, J. M.,Høgdahl, O. T., and Phillpott, S.C., 1976, Rare earth element supply to the ocean, J. geophy. Res., 81:3119-3124.
Morten, L.,Landini, F.,Bocchi, G.,Mottana, A., and Brunfelt, A. O., 1980, Fe-Mn crusts from the Tyrrhenian Sea, Chem. Geol., 28: 261-278.
Piper, D. Z., 1974, Rare earth elements in ferromanganese nodules and other marine phases, Geochim. Cosmochim. Acta, 38:1007-1022.
Piper, D. Z., and Graef, P. A., 1974, Gold and rare-earth elements in sediments from the East Pacific Rise, Mar. Geol., 17:287-297.
Puchelt, H., 1978, Geochemical implications for the Santorini Island Group (Aegean Sea, Greece), in "Alps,Apennines and Hellenides," Closs, H.,Roeder, D., and Schmidt, K., eds., Int. Union Geodyn. Scientific Report No. 38,Stuttgart, Schweizerbart, pp. 489-493.
Rankin, P. C., and Glasby, G. P., 1979, Regional distribution of rare and minor elements in manganese nodules and associated sediments in the southwest Pacific and other localities, in "Marine geology and oceanography of the Pacific manganese nodule province," Bishoff, J. L., and Piper, D. Z., eds., Marine Science Volume 9, Plenum, pp. 681-697.
Robertson, A. H. F., and Fleet, A. J., 1976, The origin of rare earths in metalliferous sediments of the Troodos Massif,Cyprus, Earth planet Sci. Lett., 28:285-394.
Robertson, A. H. F., and Fleet, A. J., 1977, REE evidence for the genesis of metalliferous sediments of Troodos,Cyprus, Spec. Publ. geol. Soc. London, 7:78-79.
Smith, P. A., and Cronan, D. S., 1976, The dispersion of metals assoc- with an active submarine exhalative deposit, Oceanology International, 75:111-114.
Spirn, R. V., 1965, Rare earth distributions in the marine environment, Unpubl. Ph.D. Thesis, Massachusetts Institute of Technology, 165 pp.
Toth, J. R., 1980, Deposition of submarine crusts rich in manganese and iron, Bull. geol Soc. Amer., 91:44-54.

Turner, D. R., and Whitfield, M., 1979, Control of seawater composition, Nature, 281:468-469.
Volkov, I. I., and Fomina, L. S., 1967, Rare-earth elements in sediments and manganese concretions of the ocean, Lithol. Mineral. Resour., 5:579-595.

SUBMARINE VOLCANIC EXHALATIONS THAT FORM MINERAL DEPOSITS: AN OLD IDEA NOW PROVEN CORRECT

Brian J. Skinner

Department of Geology and Geophysics
Yale University
New Haven, Connecticut 06511

ABSTRACT

Submarine volcanism, the circulation of seawater through rocks of the seafloor and the formation of mineral deposits on and in rocks on the seafloor are all processes that were realized during the 19th Century. The possibility that the three processes might be connected and that formation of many kinds of mineral deposit might result, is a realization of the 20th Century, with the principal ideas coming from Japan and Norway. Observations of modern ore-forming systems in operation in the Mediterranean, the Red Sea and the Pacific have proven the process but leave open the question of equivalence in tectonic settings between modern deposits and older ores in the geological record.

INTRODUCTION

When the distinguished geologist and historian Thomas Crook reviewed the history of ideas concerning the genesis of mineral deposits, he was led to conclude that "the active speculation that took place during the nineteenth century on the process involved in ore genesis left very little scope for originality to twentieth century workers . . .". We can now see that Crook (1933) was much too pessimistic. The second half of the twentieth century has been and still is a time of great intellectual ferment and originality where ideas concerning the origin of mineral deposits are concerned. Nevertheless, while new ideas and new understandings have appeared in abundance, it is certainly true that the roots of some of today's most exciting ideas and discoveries can be traced to obeservations and speculations by workers of the last - and sometimes even earlier - centuries. The realization that submarine hot springs can form mineral deposits on or in rocks on the sea floor is one such - a modern discovery or tremendous importance that has roots of considerable antiquity.

Where the origins of mineral deposits are concerned, the great debates of the 19th century and first four decades of the 20th tended to focus on two key questions. First, whether mineral deposits are epigenetic, meaning that the mineralizing materials were introduced sometime later than the time of formation of their enclosing host rocks, or whether some or all deposits are syngenetic, and formed contemporaneously with the host rocks. Proponents for the more extreme aspects of the two alternatives sometimes allowed their imaginations to decide their stances, rather than relying on careful observation and as a result the debate had become somewhat dogmatic and sterile by the end of the nineteenth century. For those who remained unbiased, however, field evidence was convincing that both epigenetic and syngenetic deposits occur.

All students of mineral deposits are eventually led to conclude that the chemical constituents of the ore minerals found in most syngenetic and epigenetic deposits must have been transported in solution. The second key question, therefore, and the one that sparked the most heated and even fanciful dialogue, concerned the origin or origins of the transporting fluids. A small and sometimes persuasive group argued that the fluid was a magma. Even as late as the 1920's, there were a few adherents to the idea that most mineral deposits, and even quartz veins, were all part of a magmatic continuum (Spurr, 1923). The great majority of observers were quickly led to conclude, correctly, that aqueous fluids must be the medium by which most ore constituents are moved. During the first half of the nineteenth century students of mineral deposits were troubled by a problem of aqueous chemistry - ore minerals are practically insoluble in water, so how could they move in solution? Debate on the chemical problem became muted during the later years of the nineteenth century when the French geologist Daubrée demonstrated that ore minerals of tin, tungsten, and other elements could be readily moved in solution if the water contained suitable "mineralizing agents" such as fluorine. But what was the origin of the "mineralizers" and the hot aqueous solutions? There are two obvious possibilities. The first possibility attracted many influencial and persuasive adherents - called the magmatic hydrothermalists - who came to believe that all mineralizing fluids must be juvenile exudations released by crystallizing bodies of igneous rock. Their most compelling evidence came from observations of ore minerals deposited around fumaroles and from the close temporal and spatial relationships between many ore deposits and igneous rocks.

The second possibility, meteoric waters, for which there were also many proponents, had first been suggested by Agricola. Surface water circulates deep into the crust, becomes heated, reacts with rocks, slowly gathers mineralizers and then picks up a dissolved mineral load that is eventually deposited in some local setting to form an ore deposit. The process came to be called lateral secretion and the clearest observational evidence for its operation derived from springs, aquifers, wells, water in mines and similar features prove that sub-surface circulation systems are ubiquitous. The most difficult criticism for the secretionists to answer was the scarcity mineral deposits. If sub-surface circulations are so common, why are mineral deposits not more abundant?

Ideas of the 19th Century

It is now apparent that almost all of the ideas proposed during the 19th century are correct in some measure but that none is entirely correct. This is certainly true for the origin of the family of deposits we now call volcanic-hosted massive sulfide deposits. Indeed, recognition that several kinds of deposits have a family relationship and that massive sulfide deposits even exist as a class is one of the important advances of the second half of the 20th century. So too is the realization that volcanic-hosted massive sulfide deposits can be subdivided into several subgroups and that all such deposits are formed by precipitation from submarine, volcanic exhalations.*

Even though prior to the 1970's no one seems to have recognized the diversity of volcanic-hosted massive sulfide deposits and to have deciphered all the steps by which they form, individual fragments of the story were deciphered long ago. The first person who seems to have suggested a relationship between any class of mineral deposit and submarine volcanic activity is often reported to have been the English geologist Henry de la Bêche (see, for example, Stanton, 1959.) I have not discovered the specific reference, but de la Bêche is reputed to have suggested that certain of the British sedimentary iron formations might have derived their iron from submarine volcanic sources. De la Bêche's idea was aired during the 1830's and does not seem to have been especially influential in its day, but it was revived with great effect some 80 years later. In one of the classic publications of geology, Van Hise and Leith (1911) suggested that the siliceous banded-iron formations of the Lake Superior region had received their iron and silica through submarine volcanism. From today's vantage it seems possible that neither the British iron beds nor the Lake Superior formations are volcanic in origin, but regardless of the correctness of a submarine-volcanic origin for the specific deposits studied either by de la Bêche or Van Hise and

*The term exhalation is usually reserved for a gaseous emanation, and was no doubt so meant when originally used by authors discussing submarine volcanic processes. Use of the term by students of mineral deposits has now been broadened to include liquid as well as gaseous emanations.

Leith, it is clear that the chemistry suggested by them does indeed occur and that a specific class of banded iron formations common and widespread in submarine volcanic sequences of many geological ages - called the Algoma type by the Canadian geologists - owes its origin to submarine volcanism. It is also clear that many of the manganese-rich layers found in submarine volcanic sequences of all ages owe their origin to the same process (Krauskopf, 1957).

The idea that submarine hot springs can deposit silica and iron oxide minerals is easier to accept than is the idea that sulfide minerals might be deposited by the same mechanism because one can see siliceous sinter and iron oxides around most fumaroles. The conclusion that iron sulfide minerals might also be formed by such a process was not drawn for a long time after there was strong observational evidence that pointed the way to a connection between certain sulfide ores and submarine volcanism. Had he not been enticed to spend much of his life considering zoological matters, that perspicacious observer Charles Darwin might well have been the first to articulate the correct relation between mineral deposits and submarine volcanism. He almost did, anyway, because he recognized that some of the sulfide ores he observed in Chile were emplaced in submarine volcanics and as he puzzled over their origin he was led to wonder why similar rocks on volcanic islands did not also contain ores.

Soon after Darwin published the geological observations he made during the voyage of the H.M.S. Beagle, one of the seminal papers on the geology of mineral deposits appeared - a paper by Elie de Beaumont (1847) discussing metallic mineral deposits and volcanic emanations. De Beaumont was, in his day, the most perceptive and persuasive member of the magmatic hydrothermalist school. He argued strongly that all mineralizing solutions had a juvenile, igneous origin, but he also realized that because certain sulfide and sulfate deposits - such as the kuperschiefer of Germany - are bedded and enclosed by sedimentary strata, the deposits are presumably syngenetic in origin. Syngenetic ores, he realized, must also have been deposited from solution because ore minerals are not abundant constituents of clastic sediments. True to his magmatic hydrothermal beliefs de Beaumont argued that the depositing solutions must be released by the cooling magma body, must rise from depth, mix with seawater, and then deposit their dissolved mineral load in the sediment pile accumulating on the seafloor. He seems to have envisioned a widespread change in the composition of the seawater rather than a localized change. Elie de Beaumont drew his evidence from European examples and from deposits within sedimentary rocks. He apparently did not perceive the process arising from submarine volcanism with deposition in volcanic rocks - the igneous sources he visualized were deep and remote from the seafloor, they fed their derivative solutions pervasively upward through fissures. Following de Beaumont, no major pieces of the story fell into place until the 20th century.

The First Half of the 20th Century

A great deal of scientific attention given mineral deposits during the later years of the 19th and early years of the 20th Century drew evidence from deposits in the Cordillera of the Americas. So many of the newly discovered deposits owed their origins to subaerial volcanism or to associated igneous intrusions, that the tide of opinion ran strongly in favor of a magmatic hydrothermal origin for essentially all mineral deposits. This is not to say that essential observations were not being made elsewhere, for they were. But the most vocal and persuasive writers, such as Waldemar Lindgren, J.F. Kemp, L.C. Graton, and F. Posepny dominated the international literature. They may have heard of, but they did not heed, bodies of work that produced the next step forward in our understanding of submarine hot springs and associated ore deposits.

One group of workers were conducting their studies in Japan but because most of their papers were in Japanese the fact that their significance was not widely appreciated is not difficult to understand. The studies concerned the copper ores at the Kosaka Mine in the Hokuroko District, Akita Prefecture, Japan. The Kosaka mine is one of the largest and oldest of those working the kuroko, which are massive, black, lead-zinc ores enclosed within Tertiary volcanic breccias, tuffs and lavas, of dacitic and rhyolitic composition. Kuroko deposits are now accepted as one of the typical examples of a massive sulfide deposit formed by submarine volcanic exhalations. The beginnings of the recognition of the origin of kuroko deposits go back to the first years of the 20th century, and the teachings of Wataru Watanabe who had studied in Germany with A.W. Stelzner and A. von Grodeck during the closing years of the 19th century. While in Germany, Watanabe was exposed to the idea that some of the markedly different bedded sulfide ore bodies in Spain and Portugal might be sediments. Upon his return, Watanabe observed that many of the ores in Japan are also bedded and in his teaching he started referring to the kuroko and other deposits as ore beds. It was a logical step to consider the possibility that if an ore bed enclosed by volcanic rocks is syngenetic, might not ores have a volcanic origin? The first to consider the deposits to be due to submarine volcanism were apparently N. Fukuchi and K. Tsujimoto (1904) whose paper I cannot read because it is in Japanese, but who are credited by Ohashi (1920) as proposing ideas in accord with his own. To Ohashi must go the credit for the first explicit statement of the process. Ohashi published earlier works in Japanese, but in 1920 he published a short paper in English in which he discussed the origin of the kuroko deposits. The deposits, he said, are syngenetic and they formed while the enclosing muds, pyroclastics and volcanic rocks were accumulating. They are a direct result of rhyolitic, submarine volcanism and "they are none other than sinter formed on the sea-floor by hot springs." Ohashi was not explicit as to the origin of the spring waters but for the rest nothing could be more clearly stated. According to T.

Watanabe (1970), the ideas of Ohashi and others were recognized as being important by Japanese geologists and played a part in mineral exploration, but the international community of geologists seems not to have recognized the importance of kuroko for deciphering the origin of massive sulfide deposits until the 1960's.

While Ohashi was conducting his study of kuroko deposits, an equally important study related to submarine hot spring deposits was being made in Norway by C.W. Carstens (1919). Again language might have been something of a barrier to dissemination of ideas because Carstens wrote in Norwegian. In a discussion of the geology of the Norwegian Caledonides in the Trondhjem area Carstens described numerous thin but very extensive and apparently bedded pyritic horizons. The deposits were concordant and invariably associated with rocks which even in their metamorphosed state he recognized as having originally been volcanic or pyroclastic. The pyritic horizons were formed, he suggested, through a submarine, volcanic-exhalative process in which the iron was leached from the volcanic rocks. This was a variation on the ideas of Van Hise, Leith and de la Bêche; the deposits described by Carstens, were iron formations but this time they were clearly iron sulfide deposits. Unlike the earlier workers Carstens seems to have had lateral secretion in mind. Unfortunately Carsten's work did not make the impact it deserved to do among students of mineral deposits. It was not until his work was referenced and expanded by a later Norwegian volcanologist, C. Oftedahl, in 1958 that the significance of his observations became apparent.

Through the 1920's, 1930's and 1940's the idea that certain ores might form through the process of submarine volcanic exhalation grew slowly. Most of the attention, such as it was, was given to the possibility that iron formations might form by such a means. Niggli (1925) included deposits formed in this fashion in his classification of ores with magmatic affinities, and when Hans Schneiderhöhn published his text Erzlagerstatten in 1944, he included a chapter on the ore-type, using as examples, the Devonian-aged iron formations of the Lahn and Dill basins in Germany.

During the 1930's two field discoveries of considerable significance were made. Behrend (1934) reported the existence of iron oxide-rich sediments formed through submarine fumarolic activity which arose as a result of historic eruptions in Santorini caldera, Komeni Islands, Greece. More importantly, Bernauer (1935) reported the iron sulfides, pyrite and marcasite, to be forming as a result of submarine fumarolic activity associated with a modern eruption of La Fossa, a volcano on the island of Vulcano in the Eolian Archipelago. The deposits described in both cases are small and though they served to establish the existence of the process in submarine environments, the observations at Santorini and Vulcano apparently did not convince people that large ore deposits could form by the same means.

The potential importance of submarine volcanic exhalations was slowly realized so that by the time hostilities ceased at the end of World War II the idea was firmly established, though the real significance of the process was still not appreciated and in some quarters a great deal of resistance still existed. Many important questions were barely touched; for example, the compositions of the volcanic exhalations, how they were generated, what kind of volcanism caused them and even the diversity of ore deposits so formed remained to be explored. Papers by European geologists such as F. Hegeman (1948) and H. Carstens (1955), by the Australian geologist, R.L. Stanton (1959) and by many others added further descriptions and developed some of the ideas, but the paper that seems to have had the greatest impact was that by C. Oftedahl (1958).

The Second Half of the 20th Century

In my opinion, Oftedahl's 1958 paper marked a turning point for several reasons, one being the prosaic fact that it is in English and more widely accessible than some of its predecessors, another that he was an able volcanologist, but the most important reason I believe was that he showed that several different types of mineral deposits could be re-evaluated as exhalative-sedimentary ores associated with volcanic rocks. Oftedahl's paper seems to be the first to suggest that the process might be a much more widespread and important one than had formerly been realized. Oftedahl suggested that both oxide and sulfide iron-formations could derive their iron from submarine igneous sources and though many of his examples were drawn from Scandinavia he included other, somewhat better known, European deposits such as the great Rio Tinto ores of Spain. He was apparently unaware of the work of Ohashi and he also chose to include a couple of less than convincing examples, such as Mississippi Valley-type lead-zinc ores. The only product of modern submarine volcanism cited by Oftedahl was the shallow fumarolic activity along the north shore of Vulcano and he seems to have considered this to be of minor significance. The possibility of modern, widespread, deep sea volcanism apparently did not occur to him. Despite certain weaknesses in the paper and the fact that several Scandinavian colleagues immediately challenged some of the evidence he cited (Marmo, 1958; Kautsky, 1958; Landergren, 1958; and Kullerud et al., 1959) most of Oftedahl's ideas have stood the test of time.

Most of the deposits cited by Oftedahl are associated, at least in part, with pyroclastic rocks that are felsic rather than mafic in composition. Magmas giving rise to felsic rocks generally contain dissolved water as a result of their formative melting processes so Oftedahl returned to the classic magmatic-hydrothermal stance and concluded that the most likely source of mineralizing fluids was in the intrusive from which the pyroclastic rocks were derived. Now that there is much contrary evidence from stable isotope studies this is one of the points on which most geologists would disagree with Oftedahl. The second point with which we must disagree is one for which Oftedahl can be excused because the great discovery days for oceanic volcanic processes had not started in 1958 - Oftedahl concluded that "submarine ore formations are quite rare as a present day phenomenon."

Despite Oftedahl's spirited answers (1959) to his critics, his ideas were not accepted by all. For example, one of the most articulate and widely read geologists of his day, C.F. Davidson (1964), strongly refuted the whole idea of submarine exhalations. Davidson based his opposition on one of the central creeds of geology - that the present is the key to the past - and argued that evidence was not in favor of ores forming through such a seemingly bizarre medium as submarine volcanic exhalations. Davidson's closing words are pure bombast; "it can thus reasonably be deduced that the various ore deposits attributed to an actualistic environment such as abnormal atmospheres, extraordinary volcanic exhalations, rivers of strange chemical composition, and other departures from uniformitarianism can all be accounted for, within the limits of contemporary experience, as being deposited at length from saline groundwaters with or without juvenile additives."

Davidson's was the last protest of any real significance because the very year his words appeared - 1964 - saw the first publications by Charnock (1964) and Miller (1964) announcing the presence of ponds of hot, dense brines along the axis of the Red Sea. These papers were soon followed by a flood of papers describing not only brines, but also the sulfide minerals of zinc, lead, copper and iron deposited by the brines in the underlying sediments. When the comprehensive volume discussing all aspects of the Red Sea brines and their associated metal deposits (Degens and Ross, 1969) appeared, all opposition to the notion of ore deposits formed by submarine volcanic exhalations disappeared. The sulfide deposits in the Red Sea are as large or larger than many of the ore deposits that were being mined in the 1960's so it was no longer possible to brush the evidence aside as being insignificant. The importance of the Red Sea brine deposits in focussing and changing ideas can hardly be overestimated. When it became apparent that the brines are a product of a circulations system within rocks of the sea floor, that the source of heat is apparently magmatic while the dissolved metallic constituents come from the fractured and porous rocks through which the circulating meteoric fluids pass, geologists immediately realized they had been misinterpreting obvious evidence for years.

It is difficult to say just who should be given remaining credit for finally completing the circle by recognizing that lateral secretion of seawater through hot oceanic crustal rocks is far more important than is primary magmatic water in generating submarine ore-forming fluids, that magma serves as a heat source, and that ore constituents can come from both the volcanic rocks and the magma. It is not difficult, however, to identify the places where the significance of deposits being formed by submarine exhalations took root most strongly - Canada and Japan. From 1960 onward, Canadian and Japanese authors developed the concept as they described and discovered new deposits. Most of the Canadian work dealt with Archean deposits, the Japanese work with the younger kuroko deposits. Following the lead of the Canadians and Japanese, geologists elsewhere have been reevaluating deposits so that by 1982, more than a thousand massive sulfide deposits have been recognized as having an origin through submarine, volcanogenic processes. When modern ore deposits were found to be forming in the Pacific, the discovery came more as a confirmation than a revelation.

Massive sulfide deposits have a great variety of shapes and sizes and occur in many tectonic settings, but they are all pyritic, with or without associated sulfide minerals of Cu, Pb and Zn. Given an adequate heat source and an adequate volume of permeable volcanic or pyroclastic rocks through which seawater can circulate, it is now apparent that several kinds of mineral deposits can form. Not all of the deposits are stratiform or even very massive. Some, for example, are stockworks enclosed in highly altered volcanic rocks and were apparently once part of the conduit up which a rising plume of heated and pregnant ore-solutions rose toward the sea floor.

The best definition of volcanogenic massive sulfide deposits is the succinct statement by Franklin, Lydon and Sangster (1981), authors of the most comprehensive review paper to date concerning massive sulfide deposits. Massive sulfide deposits, they state, "are stratabound and in part stratiform accumulations of sulfide minerals which are normally composed of at least 60 percent sulfide minerals in their stratiform portions. The stratiform portion may comprise up to 100 percent of the total sulfide present, but many deposits have a substantial component of discordant vein-type sulfide mineralization, the stringer zone, mainly in the footwall strata. Massive sulfide deposits can occur in virtually any supracrustal rock type, although volcanic rocks and pelitic to semipelitic strata are the predominant host rocks."

Many active submarine exhalative systems have now been discovered and they are being studied intensively. Several of the recently discovered submarine hot springs as reported in this volume by Hekinian et al. (1983) are, at this very moment, depositing sulfide ore-minerals on the sea floor. But all of the modern ore-forming systems discovered to date have been in the same tectonic setting - along spreading ridges. The Red Sea deposits are forming in the sediments that still coat a young and very slow spreading ridge, those in the Pacific are deposited directly on mafic lavas extruded along a fast and much older spreading ridge. Both the Red Sea and the Pacific deposits are, however, the result of the same kind of volcanism - basaltic. Considering the great lateral extent of modern spreading ridges it is probable that hundreds, and possibly even thousands, of active sites of sulfide mineral deposition on submarine basaltic substrates remain to be discovered, and that within the ocean basins, thousands of fossil sites must exist. Some of the deposits, like those along the East Pacific Rise, will be found on the sea floor itself, because fast-spreading rates and the consequent shallow magma chambers put the deposition on the sea floor. Other deposits - associated with slow spreading ridges where magma chambers are much deeper - will probably be found within the rocks of the sea floor.

The large number of modern ore-forming systems immediately suggests that identical deposits should be common in the geological record. Unfortunately this conclusion seems to be incorrect. The reasons for the apparent discrepancy are few. The fact that massive sulfide deposits do occur in old rocks proves that the process has worked in the past and that oxidation and resolution of deposits on the sea floor does not remove them all. The most likely reason for the discrepancy is that spreading ridges and their associated ore deposits do not often "jump" a subduction trench to be thrust on to continental rocks, so we rarely if ever see examples of an identical tectonic setting on continents. There does not, in fact, seem to be a single mineral deposit in the geological record that one can point to and confidently assert that it is identical in both geologic setting and petrological characteristics to modern ridge deposits. The question, obviously, is why not?

There are, certainly, a large number of massive sulfide deposits in ophiolites of many ages and these deposits have clearly been formed by the same processes forming deposits along modern spreading ridges. But are ophiolites fragments of true oceanic crust and were the deposits in them formed above ancient spreading ridges? For a number of reasons, many workers now accept the suggestions by Armstrong and Dick (1974) that ophiolites are actually formed in back-arc basins and are not fragments of uncontaminated oceanic crust at all. Two such centers of ophiolite hosted deposits - the Troodos deposits in Cypress and those in the Semail Nappe of Oman, are discussed in this volume by Robertson and Boyle (1983). Both the Troodos and Oman deposits are Cretaceous in age and both were formed in the Mesozoic Tethys Sea. Both seem to fit the true oceanic crust model as well as any known deposit. In both cases, however, there are felsic as well as mafic volcanic rocks in the section so that even the Cyprus and Oman ophiolites do not exactly fit the tectonic setting of the East Pacific Rise. The dilemma can possible be explained by the recent suggestions of Uyeda and Nishiwaki (1979, 1980) concerning differences in subduction zones. According to Uyeda and Nishiwaki there are two different kinds of subduction margins, one, a Chilean type, is characterized by large earthquakes and compressional tectonics because the plates are coupled along the subduction zone. A second, the Marianas type, is characterized by extensional tectonics and small earthquakes due to plate decoupling along the subduction zone. Back-arc basin formation and the eruption of submarine basalts and associated felsic rocks is associated with Marianas-type subduction and Uyeda and Nishiwaki suggest that essentially all massive sulfide deposits so far discovered in the geological record have been formed in such back-arc basins. They specifically cite kuroko-type and Cyprus-type deposits as examples.

We must conclude, therefore, that the process of deposit formation through submarine volcanic exhalations is a well-established and widespread process that has operated as far back in geological time as we have rocks to provide the necessary evidence - for at least 3.5 billion years. The exact tectonic settings within which the deposits have formed and the associated volcanism occurred remains, in most cases, uncertain. The answers, if they are to be found in modern settings, will probably lie along the compression or subduction margin of plates. Thus, while students of mineral deposits follow the present exploration of mid-ocean ridges with great interest they await exploration of the submarine regions of the subduction zones of plates with even greater interest. Only part of the story has so far been revealed, and if the geological record has any validity as a guide to the present, the richest and mineralogically most diversified deposits on the modern sea floor remain to be found.

REFERENCES

Armstrong, R. L., and Dick, H., 1974, A model for the development of thin overthrust sheets in crystalline rocks, Geology, 2:35-40.

Behrend, F., 1936, Eisen and Schwefel fördernde Gasquellen auf den Kameni Inseln, in "Santorin; Der Werdegang eines Inselvulcans und sein Ausbruch 1925-1928," Band II, Kap. XII, H. Reck, ed., Dietrich Reiner/Andres and Steiner, Berlin.

Bernauer, F., 1935, Rezente Erzbildung auf der Insul Volcano, Teil I, Neues Jahrb. Miner., 69:60-91.

Bernauer, F., 1939, Rezente Erzbildung auf der Insul Volcano, Teil II, Neues Jahrb. Miner., 75:54-71.

Carstens, C.W., 1919, Oversight over Trondhjemsfeltests bergbygning, Det. Kgl. norske Videnskabers. Selsk. Skr.

Carstens, H., 1955, Jermalmeni i det vestlige Trondhjemsfelt og forholdet til Kisforekomstene, Norske Geol. Tidsk., 35:211-220.

Charnock, H., 1964, Anomalous bottom water in the Red Sea, Nature, 230:590.

Crook, T., 1933, "History of the theory of ore deposits," Thomas Murby and Co., London.

Davidson, C. F., 1964, Uniformitarianism and ore genesis, Mining Magazine (London), 110:176-185 and 244-253.

de Beaumont, E., 1849, Note sur les emanations volcanique et metallifères, Bull. de la Soc. Géol. de France (Seance du 5 Juillet, 1847), 4(2):1249.

Degens, E. T., and Ross, D. A. (eds.), 1969, "Hot Brines and Recent Heavy Metal Deposits in the Red Sea. A Geophysical and Geochemical Account," Springer-Verlag Inc., New York.

Franklin, J. M., Lydon, J. W., and Sangster, D. F., 1981, Volcanic-massive sulfide deposits, in "Economic Geology, 75th Anniversary Volume," B. J. Skinner, ed., Economic Geology Publishing Co., New Haven, CT.

Fukuchi, N. and Tsujimoto, K., 1904, Ore beds in Misaka Series, Jour. Geol. Soc., Tokyo, 11:393 (in Japanese).

Hegeman, F., 1948, Über sedimentäre lagerstatten mit submarines Vulcanischen Stoffzufuhr, Fortschr. Mineral., 27:54-55.

Hekinian, R., Renard, V., and Cheminee, J. L., 1983, Hydrothermal deposits and the East Pacific Rise near 13°N: Geological setting and distribution of active sulfide chimneys, this volume.

Kautsky, G., 1958, The theory of exhalative-sedimentary ores proposed by Chr. Oftedahl, Geöl. Foren. Stockholm Förh., 80:283-287.

Kinkel, A. D., Jr., 1966, Massive pyritic deposits related to volcanism and possible methods of emplacement, Econ. Geol., 61:673-694.

Krauskopf, K. B., 1957, Separation of manganese from iron in sedimentary processes, Geochim. et Cosmochim. Acta, 12:61-84.

Kullerud, G., Vokes, F. M., and Barnes, H. L., 1959, on exhalative-sedimentary ores, Geöl. Foren. Stockholm Föhr., 81:145-148.

Landergren, S., 1958, Comments to "A theory of exhalative-sedimentary ores", Geöl. Foren. Stockholm Förh., 80:288-290.

Marmo, V., 1958, On the theory of exhalative-sedimentary ores, Geöl. Foren Stockholm Förh., 80:277-282.

Miller, A. R., 1964, Highest salinity in the world ocean?, Nature, 230:590.

Niggli, P., 1929, "Ore deposits of magmatic origin: Their genesis and natural classification," Thomas Murphy and Co., London.

Oftedahl, C., 1958, A theory of exhalative-sedimentary ores, Geöl. Foren. Stockholm Förh., 80:1-19.

Oftedahl, C., 1959, On exhalative-sedimentary ores. Replies and discussions, Geöl. Foren. Stockholm Förh., 81:139-144.

Ohashi, R., 1920, On the origin of the Kuroko of the Kosaka Copper Mine, Northern Japan, Journal of the Akita Mining College, 2:11-18.

Robertson, A. H. F., and Boyle, J., 1983, Tectonic setting and origin of metalliferous sediments in the Mesozoic Tethys, this volume.

Spurr, J.E., 1923, "The Ore Magmas. A Series of Essays on the Ore Depositions," McGraw Hill Book Co., New York.

Stanton, R. L., 1959, Mineralogical features and possible mode of emplacement in the Brunswick mining and smelting ore bodies, Glouster County, New Brunswick, Canad. Mining Metall. Bull., 52:631-643.

Stanton, R. L., 1972, "Ore Petrology," McGraw Hill Book Co., New York.

Uyeda, S. and Nishiwaki, C., 1979, Back-arc opening and mode of subduction, Jour. Geophys. Res., 84:1044-1961.

Uyeda, S. and Nishiwaki, C., 1980, Stress field, metallogenesis and mode of subduction, in "Geol. Soc. Canada, Spec. Paper 20," Strangeways, D. W., ed.

Van Hise, R., and Leith, C. K., 1911, "The Geology of the Lake Superior Region," U.S. Geological Survey Monograph Vol. 52.

HYDROTHERMAL DEPOSITS ON THE EAST PACIFIC RISE NEAR 13° N: GEOLOGICAL SETTING AND DISTRIBUTION OF ACTIVE SULFIDE CHIMNEYS

Roger Hekinian[+], Vincent Renard[+] and Jean L. Cheminée[++]

[+] Centre Océanologique de Bretagne
B.P. 337, 29273 BREST CEDEX FRANCE
[++] C.N.R.S. Ecole Normale Supérieure, Laboratoire de
Géologie, 46, rue d'Ulm, 75230 PARIS CEDEX 05 FRANCE
Contribution n° 799 du Centre Océanologique de Bretagne

ABSTRACT

The study of a fast spreading segment (12 cm y^{-1}, full rate) of the East Pacific Rise (E.P.R.) near 12°50'N explored in detail by a surface ship in 1981 and by a manned submersible in 1982 revealed the existence of intense hydrothermal activity on the Rise crest. In the area of study (600 Km^2), the Rise crest (regional depth 2600 m) is about 1500 m wide and is occupied in its center by a linear graben (> 300 m wide and 50 m deep) bordered by vertical scarps. The Rise flank topography (near the 2800 m contour line) is limited by linear fault scarps and by four volcanoes located at about 1 Km, 6 Km and 18 Km from the accreting plate boundary. The general orientation of the Rise and graben is N 165°.

The central graben made up essentially of lobated basaltic flows and remnants of collapsed structures (lava lake type) is the site of most recent extrusive volcanism. Lobated flows made up of olivine basalts were found to occur near sulfide deposits in the graben axis. Intense hydrothermal activity is found along a 20 Km segment of the Rise axis. More than eighty hydrothermal deposits, twenty four of which were active, were observed at an average interval of 100 to 200 meters. The deposits consist of copper, zinc and iron in sulfide and sulfate phases similar to those found near 21° N on the E.P.R. The active chimneys responsible for the building of large size edifices show compositional and textural zonation across their walls. The inner sections of the orifices consist mainly of chalcopyrite, while the outer zones are formed with a mixture of anhydrite, hydrated silica and iron-oxide and hydroxide complexes.

Repeated observations of an active chimney have demonstrated its rapid growth (40 cm in 5 days) : it is likely that all sulfide deposits from this segment of the Rise axis were formed rapidly, perhaps within a few decades.

INTRODUCTION

After the discovery of hydrothermal fields in the Eastern Pacific Ocean in 1977 (Lonsdale et al., 1977 and Cyamex team et al., 1979), the accreting plate boundary of the East Pacific Ocean has become a preferential target for further investigations by several sea-going expeditions. Sulfide edifices were found in the Guaymas basin near 27° N in the Gulf of California along a 6 km segment of sedimented bottom (Lonsdale et al., 1980 ; Simoneit and Lonsdale, 1982). Several active sites were found along a 6 Km segment of the E.P.R. near 21°N (Spiess et al., 1980 ; Ballard et al., 1981). Fossil sulfide deposits and low temperature vents containing only animal life were reported from a 24 Km long segment of the Galapagos Spreading Center near 86° W (Ballard et al., 1982).

A large scale regional exploration of the East Pacific Rise crest was made in 1980 (Searise cruise, Francheteau, 1981). One of the areas examined near 13° N on the E.P.R. showed indications of a high concentration of manganese and helium (Merlivat et al., 1981 ; Francheteau, 1981). A more detailed study of the East Pacific Rise area near 13° N and 11°30' N took place in 1981 with the R/V Jean Charcot (Clipperton cruise) (Hekinian et al., 1981, 1983). During this cruise the main emphasis was to map detailed bathymetry using a multi-channel narrow-beam echo sounder (SEABEAM), to take color photographs with a deep towed camera (RAIE), and to carry out water sampling, rock dredging and sediment sampling. It was also during this cruise (Clipperton) that the study of intra-plate volcanic structures (seamounts) was initiated in the area.

In January, February and March of 1982, a submersible investigation (Cyatherm cruise) took place with the diving saucer CYANA and its support ship "Le Suroit". Hydrothermal processes were studied on two segments of the Rise crest near 13° N, 11°30' N and on off-axis seamounts (Fig. 1). The Cyatherm cruise was divided

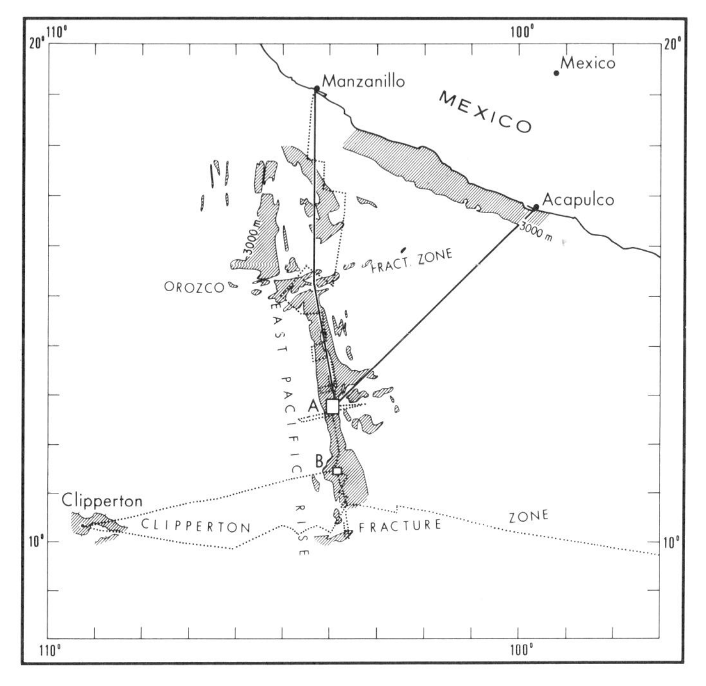

Fig. 1. Generalized map of Eastern Pacific Ocean showing the ship tracks and the detailed zones (A) and (B) of study during the 1981 cruise of N.O. Jean Charcot and the 1982 cruise of the N.O. Le Suroit. The shaded areas indicate depths less than 3000 meters (courtesy of J. Mammerickx).

into three geological legs (Geocyatherm) and one biological leg (Biocyatherm). During the Geocyatherm cruise, in addition to 32 dives of CYANA, 4 deep towed camera profiles, 14 hydrocasts and 5 sediment traps were acquired.

The present paper is mainly intended to review previous studies based on data obtained during the Clipperton and the Geocyatherm cruises (Hekinian et al., 1981, 1983 in press). In addition, some new preliminary observations concerning the distribution and composition of the hydrothermal deposits along the Rise crest near 13° N are incorporated. More detailed work on the regional structure, the microtectonics and the petrology of the basement rocks associated with the hydrothermal material from the Rise crest and from the off-axis seamounts is now underway.

GEOLOGICAL SETTING

The main area selected for detailed work (zone "A") lies about half-way between the two major fracture zones Orozco and Clipperton, between which the Rise crest shows and extremely linear and continuous structure for at least 400 km (Fig. 1) (Hekinian et al., 1981). Relief (< 300 m in height) along the Rise crest is gentle and the depth at the axis varies only between 2800 and 2500 m. The detailed study of zone "A" located between 12°38'N and 12°54' N covers about 600 km^2 centered around the Rise crest and includes off-axis seamounts.

A general pattern of fault structures is prominent on the western side of the Rise where the tectonic grain is parallel to the strike of the accreting plate boundary. From bathymetric observations it is observed that the scarps located on the western side of the Rise axis have both east and west facing slopes forming graben and horst like-structures (Figs. 2, 3). A parallel ridge located 2 km west of the axial zone shows abundant sediment cover with isolated pillow-lavas as seen from a deep towed camera station. Structural asymmetry of the Rise crest is shown by the absence of a typical "linear fabric" of the accreting plate boundary towards the east (Fig. 2). This "fabric" is obliterated by the presence of seamounts. The fact that the smallest volcanic cone is found near the Rise crest and the larger one the farthest away (18 km) might suggest a rapid growth away from the accreting plate boundary. However, the number of seamounts mapped so far is insufficient to make any inferences on their distribution and evolution. It is thought that these seamounts started to form near/or on the Rise axis and have evolved as independent structures away from the central graben.

The Rise crest is oriented N 165° and its axial zone defined by the 2700 m contour line has a width of about 1500 m and a general regional relief of 150 m above the surrounding floor (Fig. 2). Submersible observations (dives CY 82-17 and CY 82-20) made on westerly traverses normal to the Rise axis for a distance of 10 km showed a tectonically narrow (4 km wide) active zone (Hekinian et al., in press). This interpretation is based on the presence of freshly slumped talus piles at the foot of the scarps exposing pillow flow cross sections bared of any sediment cover. Similar observations were previously made near 21° N on the E.P.R. where the tectonically active zone reaches up to 25 km in width (Cyamex team, 1981).

At the center of the Rise crest there is a central depression (graben) about 20-40 m deep and more than 300 m wide (Hekinian et al., 1981, 1983 ; Francheteau, 1981) (Fig. 2). The central graben is bounded on each side by vertical scarps about 10 to 30 m in height. At a few places along the Rise axis, because of the dis-

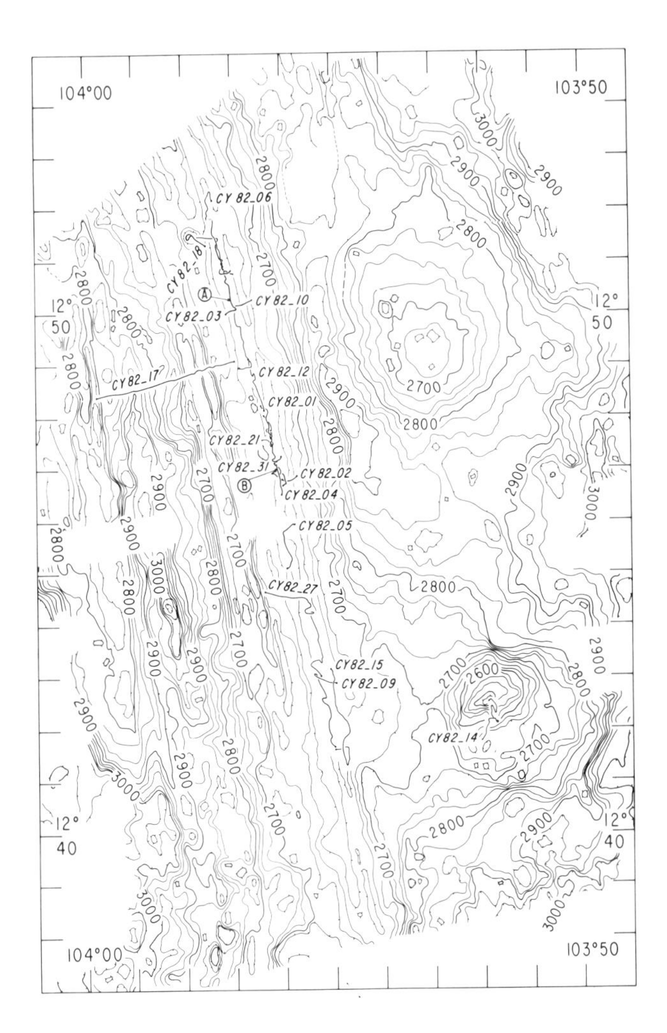

Fig. 2. Detailed bathymetric map (contour lines every 25 meters) of zone "A" made by the N.O. Jean Charcot using bottom moored network and multi beam echo-sounder (SEABEAM) (Hekinian et al., 1981, 1983). The tracks of the diving saucer CYANA on the sea floor are shown (CY = CYANA, 82 = year 1982 followed by the dive number e.g. CY 82-01. In the central graben (A) includes dives : CY 82-13, 16, 19, 22, 23, 24, 25, 30 and (B) CY 82-06, 26, 28 and 32.

continuity of the marginal scarps, the central graben is not as well defined (Fig. 2). Small and isolated topographic highs (> 10 m in height) occur at the margins of the graben, on top of the vertical scarps (Fig. 2). These highs consist primarily of tubiform and bulbous pillow flows and contain small isolated or interconnecting sediment pockets suggesting an older crustal age than that of the graben. Using similar terminology to that of Ballard et al., (1979) and Cyamex (1981), the lava flows of the Rise axis in zone "A" were grouped into three main categories : 1) lobated sheet flows, 2) flat sheet flows and 3) pillow flows. The most abundant are the lobated flows found essentially in the central graben associated with a minor occurrence of flat lying and wrinkled flows (Fig. 3). The lobated flows, about 0.5 to 1 m in relief, have flattened tops which are often collapsed displaying empty cavities at several levels. They form depressions with isolated columns (pillars) and linear ridges typical of the "lava lake" type of structure already described from other Eastern Pacific regions (Francheteau et al., 1979 ; Ballard et al., 1979).

Extensional fissures with a N 165° trend and varying in size between less than 1 m up to about 15 m in width cut through the sheet flows. The linearity of the central graben is disrupted between 12°47' N and 12°42' N where the southern segment of the Rise is offset eastward with respect to its northern prolongation (Fig. 2). This is seen when following the 2625 m contour line of the detailed bathymetric map of zone "A" (Hekinian et al., 1983). The southern offset segment (at 12°47' N - 12°42' N) also differs from the northern rise segment by the preponderance of fissures which have also affected the eastern marginal high (dive CY 82-15 ; CY 82-27) (Hekinian et al., in press) (Fig. 2).

The central graben is the site of the most recent extrusion where fresh basaltic rocks were recovered. Fragments of lobated flows (CY 82-21-04) collected on a hydrothermal site (near 12°48' N at about 5 m from an active vent) showed a thin (≃ 1 mm thick) coating of iron-manganese hydroxide (table 1, Fig. 4). Another rock sample (CY 82-21-06) was taken at about 60 m north of the same site and did not show any surface coating but had a fresh pristine glassy crust. These two rocks have a similar composition they consist of low TiO_2 (< 1.2 %) content and a forsterite rich olivine (Fo_{87-89}). Similar types of primitive olivine basalts, that is to say the least differentiated rocks, were also recovered in association with active sulfide deposits along the accreting plate boundary (at age zero) near 21° N on the E.P.R. (Hekinian, in preparation). It is interesting to note that Fe-Mn deposition started early during the stage of crustal formation and caution has to be exercised when using as reference the thickness of manganese crust for determining the relative age of volcanic flows.

Table 1. Microprobe analyses of early formed phases set in a matrix of basaltic glass recovered from a lobated type of flow located at about 5 meters away from an active hydrothermal site visited during dive CY 82-21 (Fig. 2., 4). The area surrounding the site was coated by a thin (< 3 mm thick) film of iron-manganese hydroxide.

	Glass CY 82-21-04 AV.3	Olivine AV.3	Chromium Spinel	Fe-Mn coating CY 82-21-04 AV.2
SiO_2	48.98	40.22	-	12.16
Al_2O_3	16.70	0.07	33.60	0.78
FeO_T	7.98	10.71	15.38	24.84
MnO	0.19	0.15	-	29.48
MgO	9.03	48.81	17.77	3.32
CaO	12.18	0.28	-	3.99
Na_2O	2.67	-	-	1.53
K_2O	0.07	-	-	0.36
TiO_2	1.18	-	-	-
Cr	N.D	N.D	33.29	N.D
Total	98.98	100.17	100.04	76.46

The off-axis seamounts in zone "A" occur at variable distances from the Rise axis. The closest to the accreting plate boundary (centered at about 1 km west of the central graben axis) consists of a small circular edifice about 80 m in height and 800 m in base diameter (Fig. 2). This small seamount has a central caldera (< 50 m in depth made up of pillow and sheet flows (CY 82-18). Two other larger seamounts are located further away, about 6 km to the east of the Rise axis (Fig. 2). The first one to the northeast has gentle slopes and stands about 250 m above the surrounding floor. The second one found to the southeast shows steeper slopes and rises from a regional depth of 2700 m up to 2440 m. The summit and the southern flank of this volcanic edifice was explored by both deep towed camera and by submersible. Basaltic pillow-lava fragments and sulfide material were recovered (Figs. 2, 3 ; Hekinian et al., 1983). Isolated outcrops of massive sulfides made up essentially of pyritic material exposed on steep scarps are partially covered by a reddish brown and orange powdery product made up essentially of Fe-hydroxide and clay-like material coated with a manganese crust. This ocker-colored material abounds and forms a continuous blanket for at least 500 m on a north-south crossing of the seamount (Fig. 2) ; it is believed that such material represents the altered top layer of massive sulfides buried underneath. Low-temperature altered products similar in appearance and composition to that encountered on this southeastern seamount were previously reported from transform fault "A" (FAMOUS area, Mid-Atlantic Ridge) and from the Galapagos Spreading Center (Hoffert et al., 1979 ; Fevrier, 1981).

A fourth edifice discovered named the "Clipperton seamount" occurs at about 18 km the west of the Rise axis near 12°36' N (Hekinian et al., 1983). It rises about 1000 m (1900-2000 m) high above the surrounding floor and its summit is about 4 km in diameter at the 2200 m contour line. The summit consists of five constructional highs or subcircular peaks, 40-150 m in height and 20-200 m in width. One dive (CY 82-29) took place on the summit of the seamount and revealed the existence of thin sediment patches intercalated with fresh glassy sheet flows covered by Fe-Mn material. One of the small subcircular volcanic peaks (< 50 m in height) explored in detail by submersible has slopes inclined about 50-60°. The slopes consist entirely of elongated pillow flows oriented downslope. The top of this constructional hill is made up of bulbous pillows associated with a mixture of reddish brown Fe-Mn and silica hydroxide materials. Coral-like organisms

(15-20 cm in diameter) coated with iron and manganese were also found scattered between the pillow flows. A dredge haul (CL DR 02) made accross the larger constructional hill, brought up a variety of basalts and fractionated rocks together with small chips of sulfides and unusually thick Fe-Mn slabs (1-2 cm thick).

DISTRIBUTION AND COMPOSITION OF THE HYDROTHERMAL DEPOSITS ON THE RISE CREST

For the purpose of this study and in order to avoid confusion an appropriate terminology is used to designate the hydrothermal fields discovered along the Rise crest near 13° N (E.P.R.). The term "hydrothermal deposit" is used to indicate any concentration of hydrothermal material forming mounds, columns, tubiform chimneys or any other irregularly shaped sulfide edifices or sulfide concentrations recognized either by bottom towed instruments or manned submersibles whose relationships with the surroundings have not been mapped. The term "hydrothermal site" is given to a sulfide deposit whose boundaries with the surrounding geology are well outlined from submersible observations. A chimney (either hydrothermally active or inactive) has a more or less cylindrical and/or conical shape conduct and is set either on large sulfide edifices or is found standing up directly on the basaltic basement as an isolated entity.

Most of the hydrothermal deposits were found to occur within a narrow band (< 100 m wide) near the center of the graben along the mean N 165° direction of the East Pacific Rise axis, between 12°41' N and 12°52' N (Fig. 3, 4). Other deposits were found at the foot and on the top of scarps at the margins of the axial graben (Fig. 4). On the basis of both deep towed camera (RAIE) and submersible observations (CYANA), more than eighty hydrothermal deposits were recognized within a 500 m wide band extending about 20 km along the central graben (Figs. 3, 4), except for the southern (south of latitude 12°47') offset segment of the Rise where hydrothermal activity appears to have switched eastward from the central graben to the marginal high (Hekinian et al., in press) (Figs. 2, 4). The dimension of a hydrothermal site was estimated during several passes with the submersible across individual sites. In general, the individual sites are made up of conical columnar and irregularly shaped edifices (up to 26 m in height and an average diameter of about 5 m) with overgrown small active and inactive chimneys (< 2 m in height). Most active sites consist of chimneys having both white and dark colored exiting fluids. On several occasions during sampling, breaking up a chimney which had white colored fluid gave rise to the outpouring of dark fluid.

The flow rate of hydrothermal exiting fluids varies considerably. The white colored fluids are rapidly dispersed above their orifices while the plumes of black fluids were seen to rise up to about 40-50 m above their vents. The velocity at which the dark fluids are spewing out from their orifices was estimated by timing the motion on the television screen of suspended particles rising in the column of hot fluid and was estimated to vary between 0.5 and 2 m/sec. From this velocity estimate and from measurements made on sections of active chimneys (3 cm in average diameter) sampled by CYANA, the flow rate was found to be on the order of 3.5 to 14 liters/sec. (Hekinian et al., in press).

Most of the intensely active vents occur in the northerm segment of the central graben, that is above 12°47' N (Fig. 4). Only a few sites showed diffuse low temperature (< 100° C) hydrothermal activities. These are located near 12°49' N in the central graben (dives CY 82-10 and CY 82-12) (Fig. 4) and are characterized by the abundance of animal life such as vestimentiferan pogonophorans (tube worm), brachyuran crabs, galatheid crabs, and pinkish white eel-like fish, a fauna similar to that discovered near 21° N on the E.P.R. However, in addition, light brown colored mussels (5-10 cm in length) were also seen associated with the tube worms and limpets. Further geological observations and sampling were carried out on the twenty-four active sites visited by the submersible. Successful temperature measurements were made on eight active sites and, hydrothermal fluid sampling operations were performed on one active site near 12°51' N (Hydro 1) and on two other sites near 12°47' N (Hydro 2 and 3 ; Fig. 4).

Most of the detailed work was concentrated in the northern area near 12°50' N and in the middle area near 12°47' N of the Rise crest explored (Figs. 2, 4). Two high temperature hydrothermal sites located about 7 km from each other in the central graben which can be considered as typical examples of the newly discovered hydrothermal fields (Hekinian et al., in press) will be described. The most northerly site ("A" on Fig. 2) was revisited during several dives which enabled the divers to observe and measure for the first time the growth rate of a sulfide chimney. A chain made up of a stainless steel alloy was attached along the side of a

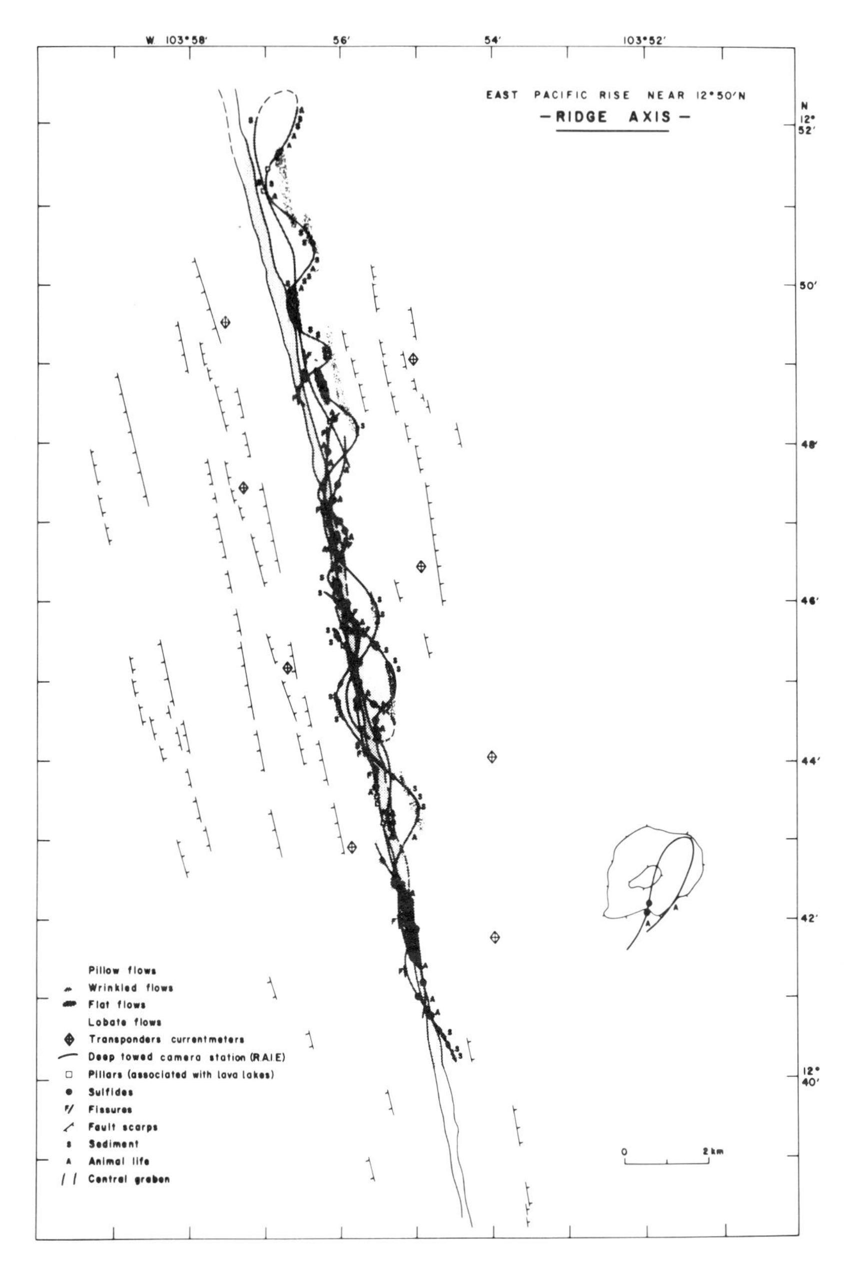

Fig. 3. Schematic representation of the main morphological features interpreted from bathymetric and deep towed navigated photographic coverage (RAIE) made during the Clipperton cruise (1981) (Hekinian et al., 1983). The fault scarps away from the rise axis (constructional zone) were drawn from bathymetric data. Remnants (pillars) of lava lake and hydrothermal deposits are mainly found in the central graben.

Fig. 4. Distribution of the hydrothermal deposits along the northern portion of the Rise Crest explored near 13° N. The 2650 and 2700 m contour line were chosen to delineated the axial zone with a mean N 165° orientation. The hachured lines represent the highs (2600 m) located at the margins of the central graben which is shown by a 2640 m contour lines. The black triangles indicate the hydrothermally active sites ; the black dots are inactive hydrothermal sites seen by CYANA and deep towed camera station (RAIE) during the GEOCYATHERM cruise (1982), while the black squares represent hydrothermal deposits (the majority inactive) found during the Clipperton cruise of 1981 (RAIE stations) (Hekinian et al., 1981). The particular sites and sample stations mentioned in the text are shown. The rate growth of a sulfide edifice was measured at the "chain site". The pogo-10 and pogo-12 sites discovered during dives CY 82-10 and CY 82-12 consist of abundant animal life (vestimentiferan, pogonophorans, etc...) in association with low temperature sulfide vents. 21-04 showns an active hydrothermal site where a fresh basaltic rock (sample CY 82-21-04) was obtained from the vicinity of an active vent (see text and table 1).

Figure 5 will be found in the color insert following page 18.

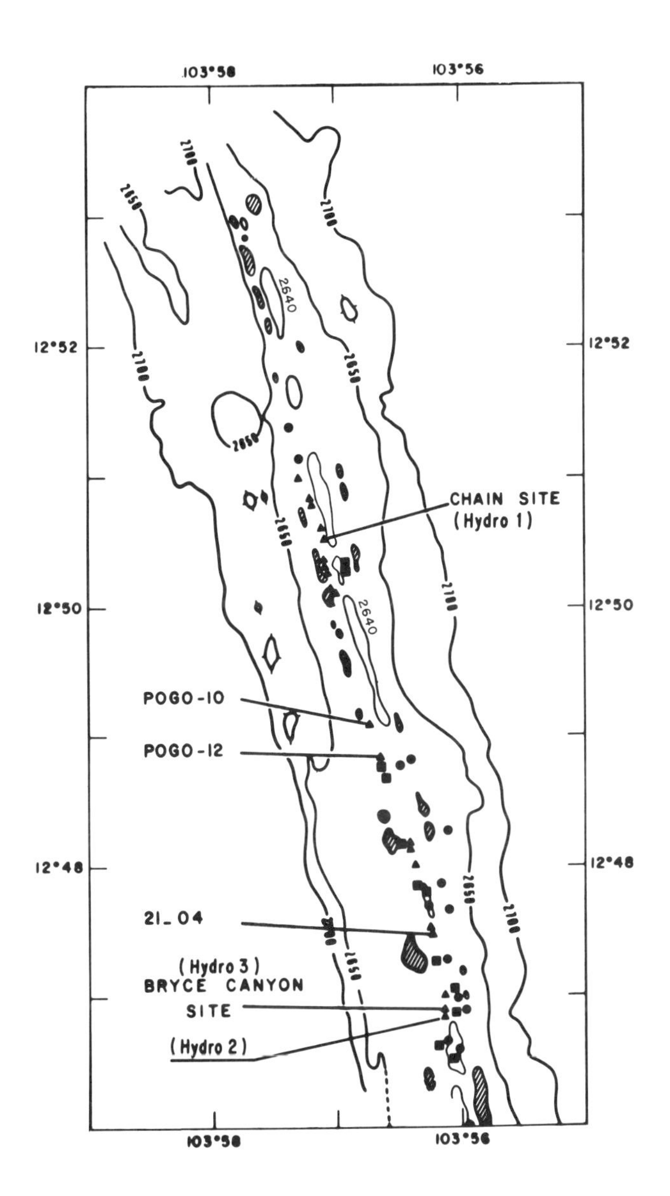
103°58
103°56
12°52
12°50
12°48
2640
2650
2700
CHAIN SITE
(Hydro 1)
POGO-10
POGO-12
21_04
(Hydro 3)
BRYCE CANYON
SITE
(Hydro 2)

chimney and permitted a measurement of the growth rate (Hekinian et al., in press). This site, located at a depth of 2635-2637 m was then called the "chain site". The other hydrothermal site ("B" on Fig. 2) made up of gullies and of numerous isolated and variegated sulfide peaks was found at a depth of 2629-2630 m, near 12°47' N (Fig. 2, 4). Because this particular landscape reminded the observer of Bryce canyon in Utah, (U.S.A.), site B was named (Fig. 4) the "Bryce canyon site".

The "chain site" was revisited during several dives (CY 82-16, CY 82-19, CY 82-24, CY 82-25, CY 82-30, CY 82-37) in order to make detailed geological observations and to sample the exiting fluids (Hekinian et al., in press). The site is located on top of lobated flows bounded to the east by a lava lake structure and to the west by a scarp which represents the eastern wall of a fissure parallel to the graben N 165° trend cutting through the lava lake. This site consists of a tall edifice, about7-8 m high topped with several active chimneys less than 50 cm tall. At the foot of this tall edifice, active chimneys with black exiting fluids are observed (Fig. 5). These chimneys occur at less than 1 m from the base of the edifice and the setting is comparable to that found on other sites, notably the "Bryce canyon" one.

The visual observations which permitted a measurement of the growth rate of the chimney started during dive CY 82-25, on the 19th of February 1982 (13 H 25) when a chain with markers every 10 cm was hung along the remaining trunk of a small active chimney after its top was broken down cy CYANA's mechanical arm (Figs. 5A, B). The site was revisited on the 24th of February at 11 H 00 and the divers witnessed the new growth of the chimney. A cylindral tube about 40 cm high and 10 cm in diameter had grown above the marker chain (Fig. 5B). Also it was noticed that all the neighboring chimneys which had previously been broken down (dive CY 82-25) were rebuilt (Fig. 5A). The last visit payed to the "chain site" on the 4th of March (dive CY 82-37) showed that the chimney had grown still taller and wider (Fig. 5B). However, by this time, the chain was fully corroded due to the acid nature of the exiting fluid, the chimney which had originally one single orifice, showed three vents with exiting black fluids (Hekinian et al., in press).

From previous work (Macdonald et al., 1980) on the E.P.R. near 21°N, the relatively fast growth rate of active vents had also been inferred but no quantitative measurements were made. It is likely that all the sulfides seen in the central graben and the immediate vicinities (marginal highs) were formed rather rapidly, probably within a few decades. Indeed, it is believed that off-axis activities, except for the localized volcanic structure, might be inexistant or may be of very low temperature types. Several submersible crossings and deep towed camera surveys made on the normal flank region with linear fabric of the Rise crest did not reveal any indication of the presence of hydrothermal deposits.

Fig. 6. Schematic representation of the "chain site" as it looked during dive CY 82-30. The sample localization is shown (Hekinian et al., 1983). (3) indicates sample CY 82-25-02 taken at the foot of the edifice with the chain. (1) and (2) are samples CY 82-30-01 and CY 82-30-02. Sample CY 82-30-03 (4) is a "thumb" like fragment of an inactive edifice (table 2). The graduated chain shows (C) horizontal bars (10 cm in length) every 10 cm intervals. (A) indicates the portion of the overgrown chimney over the marker. A pulverulent cone (B) with concentric rims is also seen on the left of the diagram.

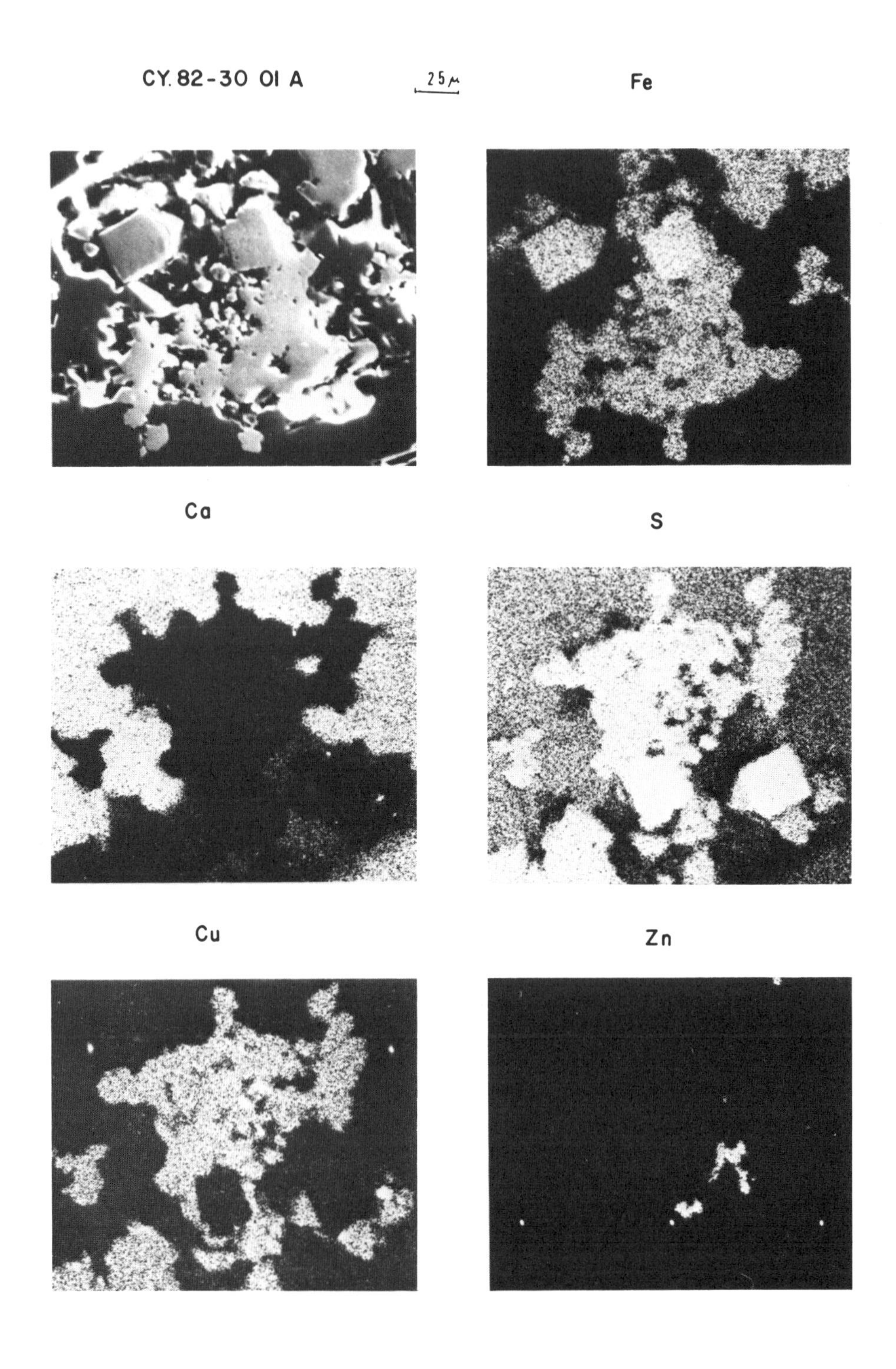
CY. 82-30 01 A
25μ
Fe
Ca
S
Cu
Zn

Fig. 7. Scanning electron micrographs of chalcopyrite-pyrite and anhydrite association in the intermediate zone across the wall of an active chimney sampled (CY 82-30-01 Ax 420) on the "chain site" near 12°50' N in the central graben (area"A", Fig. 2, 3, Table 2). Small inclusions of sphalerite occur in Chalcopyrite. The exposure time of the sample to the secondary electron beam was 320 sec. for all the elements except for Ca for which it was 160 seconds.

Figure 8 will be found in the color insert following page 18.

From the four samples taken at the "chain site", two are from active vents broken by CYANA's mechanical arm (CY 82-30-1 and 30-2), and one from the foot of the chimney with the chain probably represents a fragment of the newly built edifice (CY 82-25-02) (Fig. 6). The fourth sample is from an inactive chimney with low relief (CY 82-30-03) (Figs. 5, 6). Except for sample CY 82-25-02 which is structureless, all samples show concentric outward zonation of chalcopyrite, chalcopyrite-pyrite association, pyrite-anhydrite and iron-silica hydroxide association (Table. 2). While the innermost part of the chimneys are coated with massive chalcopyrite, the exterior is more friable and heterogeneous in composition. Anhydrite is usually the most common constituent forming the outer layer of the chimneys (Fig. 7). Traces of sphalerite (Zn, Fe, sulfide) were found as inclusions in chalcopyrite near the outer margin of the chimney (CY 82-30-1) (Fig. 7).

The "Bryce canyon site" was visited during dive CY 82-31. It is set on the roof of a lava lake made up of lobated flows and bordered by a linear fissure oriented N 165°. Both active and inactive edifices, some attaining heights of 13-15 m, stand among many smaller and skeletal sulfide edifices. Slumped blocks of sulfides and slabs of lobated flows covered with a dark colored powdery material were seen in the immediate surroundings. One active chimney less than 2 m high was found at the foot of a tall sulfide edifice. This active chimney is slightly bent and had three orifices. Only one of the orifices discharged black fluid. The other two orifices were sampled and found partially clogged (Fig. 8A). The orifices of the chimney are about 3 cm inside diameter (sample CY 82-31-09) and the temperature of the exiting fluid measured at the mouth of the orifice was 329° C.

About 5 m slightly downslope, on the southwestern side of the previous chimney, another active edifice with at least three chimneys (2-7 m high) was seen (Fig. 8B). One of the chimneys was sampled (CY 82-31-10) and a temperature measurement made (327° C ; Fig. 8B). These samples consist mainly of copper-rich sulfide material and are comparable to those found in the "chain site". Both chimneys (CY 82-31-09 ; CY 82-31-10) consist of a zoned structure with the inner walls made up of coarse grained chalcopyrite (5-1.5 mm thick) surrounded externally by a grey colored layer of anhydrite-pyrite-chalcopyrite association (Table 2). The most external part of the chimney consists of anhydrite-pyrite and a mixture of iron-silica material. Traces of chlorine associated with copper were detected on the outer margin of the chimney (sample CY 82-31-09) (Figs. 8, 9). Copper chlorate phases (atacamite)were also identified in sulfide deposits from 21° N on the E.P.R. (Février, 1981) Chlorine (CL up to 1.2%) also occurs in some clay-like minerals made up essentially of Si-Al-Mg-Fe assemblages found as interstitial products associated with pyrite (CY 82-31-09) (Fig. 9). The origin of these hydrated silicates could either be attributed to sea-water sulfide interaction or early low temperature deposition from hydrothermal fluid.

Table. 2. General description of coexisting phases across the walls of hydrothermal chimneys sampled from the central graben of the East Pacific Rise Crest near 13° N. The inside diameter of the orifices is less than 4 cm. Samples CY 82-25-02, 30-01, 30-02, and 31-(5+6) are fragmented chimneys (Figs. 7, 9).

Sample CY 82	Inner Zone (3-8 mm)	Intermediate Zone (3-15 mm)	Outer Margin (< 2 mm)	Temp. t°C	Remark
21-02	cp	cp, bn, anh, py	anh, py, Si, Fe-ox	N.D.	actif
25-02	?	?	anh, py cp	232-257	actif
30-01	cp	cp, py, anh,	anh, py, Fe-ox	319	actif
30-02	cp	cp, py, anh,	anh, Fe-ox	319	actif
30-03	cp	?	Si, Fe-ox	-	dead
31-(5+6)	cp	py, sp, anh	anh, sp Fe-ox	N.D.	actif
31-09	cp	cp, py, anh,	anh, py, Si, Fe-ox	330	actif
31-10	cp	cp, py,	anh, Si, Fe-ox	320	actif

Abbreviations are : cp : chalcopyrite ; py : pyrite ; sp : sphalerite ; bn : bornite ; anh : anhydrite ; Si : amorphous hydraded silica complexes ; Fe-ox : iron oxide-hydroxide complexes.

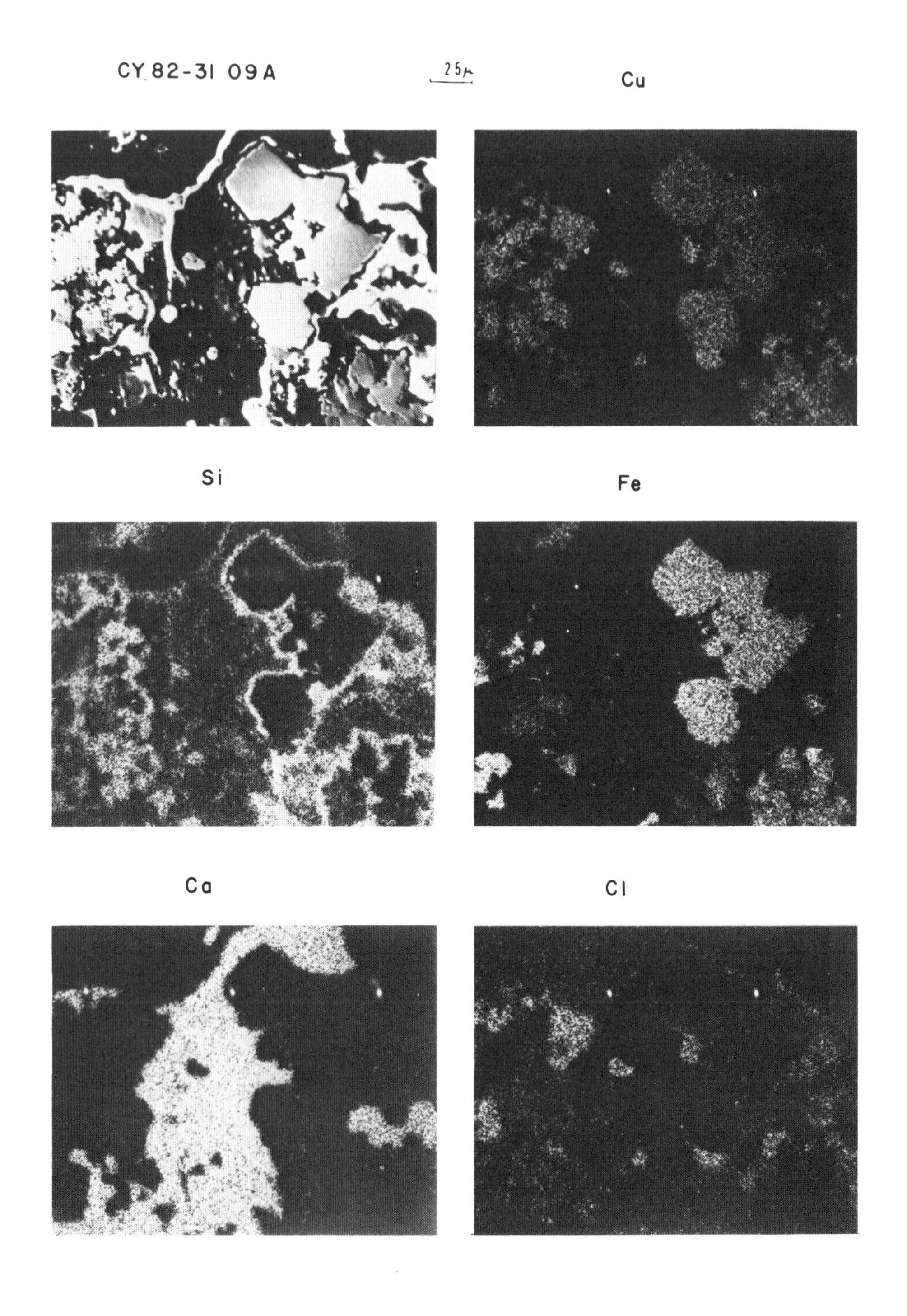
CY 82-31 09A
25μ
Cu
Si
Fe
Ca
Cl

Fig. 9. Scanning electron photomicrographs of chalcopyrite-anhydrite-pyrite association in the intermediate zone across ← the wall of an active chimney sampled (CY 82-31-09 A) at the "Bryce canyon" hydrothermal site located in the central graben near 12°47'N (area"B", Fig. 2, 3, 4, Table 2). Notice the distribution of a Si-Mg hydroxide compound around the chalcopyrite and in the groundmass. Traces of copper-chlorate bearing phases occur. The exposure time to the secondary electron beam was 160 seconds for all the elements.

SUMMARY AND CONCLUSIONS

A detailed study of a fast spreading ridge segment (12 cm y^{-1}) was carried out near 13° N, located between the Orozco transform fault and another small fracture zone near 12° N on the East Pacific Rise. The axial zone is less than 1500 m wide and the crest is occupied by a narrow central graben (300 m average width).

This Rise segment represents the most hydrothermally active portion of an accreting plate boundary so far discovered. More than eighty active and inactive hydrothermal deposits were found along a 20 km long segment of the Rise crest.

The average extent of an individual hydrothermal site was estimated to be about 50 m in diameter. The average occurence of hydrothermal sites was estimated to be about one every 100-200 m along a narrow band (< 100 m wide) within the Rise crest.

The sulfide edifices found at the Rise crest are believed to have been formed rather rapidly, probably within a few decades. The rate of growth of one chimney was measured and it was found that in five days a cylindrical edifice of about 10 cm in diameter grew about 40 cm in height. From this observation, it was estimated (Hekinian et al., in press) that a cylindrical shaped chimney having an internal and external diameter of 3 cm and 10 cm respectively (with an average density of 2.9 gr/cm^3) will increase its mass by about 1.6 kg per day. Assuming that the individual chimneys coalesce and form large conical shaped sulfide edifices averaging 6 m in height and 3 m in diameter and with a weight of 41 tons, these edifices could be built in about seventy years. However, these estimations made on the volume of sulfide material formed at the Rise axis have to be taken with caution since the observed sulfide edifices are irregularly shaped (columns, mounds, filling cavities, etc...) and their activity is probably discontinuous in time. The active chimneys found on the sulfide edifices are relatively small (<2-3m in height and less than 20 cm external diameter). Most of the active chimneys sampled showed a zonal compositional variation. The most inner part consists mainly of chalcopyrite surrounded by a mixture of lower temperature phases such as anhydrite and clay-like material.

Other sulfide deposits were found on hydrothermally inactive off-axis seamounts located at 18 and 6 km from the Rise crest. From preliminary observations made essentially on the southeastern seamount (6 km from the Rise axis), it seems that the hydrothermal deposits are more abundant and form more continuous deposits than those encountered on the Rise crest. It is likely that off-axis seamounts are formed rather rapidly from individual eruptive centers (as opposed to the axial graben fissural eruption) enabling the formation of large sulfide deposits. However, it is still premature to make speculations on off-axis hydrothermal and volcanic processes until more detailed work is carried out in order to understand the distribution and formation of intraplate volcanic structures.

ACKNOWLEDGMENTS

The data presented here were obtained during the Clipperton 1981 and the GEOCYATHERM 1982 cruises organised and sponsored by CNEXO-COB. This study represents a collective effort and incorporates previous published and unpublished work done in collaboration with colleagues participating in the Clipperton and the CYATHERM cruises.

The captains of the N.O Jean Charcot, G. Paquet and of the N.O Le Suroit, J. Keranflec'h, as well as the officers and the crew members participating in the Clipperton (1981) and the Cyatherm (1982) cruises are gratefully acknowledge. We are also thankful to J. Roux, the CYANA diving group and the navigation team for their contribution during the GEOCYATHERM cruise. The preliminary description of the sulfide was performed by M. Fevrier The electron microprobe determinations were done by COB, by M. Bohn using the "CAMEBAS de l'Ouest". The thin sections and polished sections were made by G. Floch (COB) and the photographic work was performed and supervised by G. Vincent. Drafting of the illustrations was done by D. Carre and J.P. Maze. C. Ollivier typed the manuscript.

CITED REFERENCES

Ballard, R.D., Holcomb , R.T., and Van Andel, T.H., 1979, The Galapagos Rift at 86°W : sheet flows, collapsed pits and lava lakes of the Rift Valley, J. Geophys. Res., 84, 5407-5422.

Ballard, R.D., Francheteau, J., Juteau, T., Rangin, C., and Normark W., 1981, East Pacific Rise at 21° N : the volcanic tectonic and hydrothermal processes of the Central axis, Earth. Planet. Sci. Lett., 55, 1-10.

Cyamex team and Bougault, H., Cambon, P., and Hekinian R., 1979, Massive deep-sea sulfide ore deposits discovered on the East Pacific Rise., Nature., 277 (5601), 523-528.

Cyamex Team., 1981, First manned submersible dives on the East Pacific Rise at 21° N (Projet RITA) : general results., Marine Geophys. Res., 4, 345-379.

Fevrier, M., 1981, Hydrothermalisme et minéralisation sur la dorsale Est Pacifique à 21° N. Thèse, Université de Bretagne Occidentale et C.O.B., Brest, France, 270 p.

Francheteau, J., Juteau T., and Rangin C., 1979, Basaltic pillars in collapsed lava pools on the deep ocean floor., Nature., 281 (5728), 209-211.

Francheteau, J., 1981, Characteristics of the axial region of the East Pacific Rise (abstract), in AGU Chapman Conference. The creation of the Oceanic Lithosphere, April 6-10, Warrenton, Virginia.

Francheteau, J., and Ballard, R.D., (in preparation), The Pacific Rise near 21° N, 13° N and 20° S. Inference for along strike variability of axial processes of the Mid-Oceanic Ridge.

Hekinian, R., Fevrier, M., Needham, H.D., Avedik, F. and Cambon, P., 1981, Sulfide deposits : East Pacific Rise near 13° N (abstract) EOS, 62 (45), 913.

Hekinian, R., Fevrier, M., Avedik, F., Cambon, P., Charlou, J.L., Raillard, J., Needham, H.D., Boulegue, J., Moinet, A., Merlivat, L., Manganini, S., and Lange, J., 1983, East Pacific Rise near 11-13° N : geology of new hydrothermal fields, Science, (in press).

Hekinian, R., Francheteau, J., Renard, V., Ballard, R.D., Choukroune, P., Cheminée, J.L., Albarede, F., Minster, J.F., Marty, J.C., Boulegue, J., and Charlou, J.L., Intense hydrothermal activity at the axis of the East Pacific Rise near 13° N : submersible witnesses the growth of sulfide chimney, Marine Geophys. Res. (in press).
Hoffert, M., Perseil, A., Hekinian, R., Choukroune, P., Needham, H.D., Francheteau, J., and Le Pichon X., 1978, Hydrothermal deposits sampled by diving saucer in transform fault "A" near 37° N on the Mid-Atlantic Ridge, Famous area, Oceanol. Acta 1, 1, 73-86.
Lonsdale, P.F., 1977, Deep-town observations at the mounds abyssal hydrothermal field Galapagos Rift., Earth. Planet. Sci. Lett., 36, 92-110.
Lonsdale, P.F., Bischoff, J.L., Burns, V.M., Kastner, M., and Sweeney, R.E., 1980, A high temperature hydrothermal deposit on the seabed at the Gulf of California spreading center. Earth. Planet. Sci. Lett., 49, 8-20.
Macdonald, K.C., Becker, K., Spiess, F.N., and Ballard, R.D., 1980, Hydrothermal heat flux of the "black smoker" vents on the East Pacific Rise. Earth. Planet. Sci. Lett., 48, 1-7.
Merlivat, L., Boulegue, J., and Dimon, B., 1981, Helium isotopes and manganese distribution in the water column at 13° N on the East Pacific Rise (abstract) AGU Fall Meeting, EOS, 62, (45), 913.
Simoneit, B.R.T., and Lonsdale, P.F., 1982, Hydrothermal petroleum in mineralized mounds at the seabed of Guaymas Basin. Nature, 295, 198-202.
Spiess, F.N. et al., Rise Project Group, 1980, East Pacific Rise : Hot springs and geophysical experiments., Science, 207, 1421-1433.

TECTONIC SETTING AND ORIGIN OF METALLIFEROUS SEDIMENTS IN THE MESOZOIC TETHYS OCEAN

A.H.F. Robertson and J.F. Boyle

Department of Geology
University of Edinburgh
Edinburgh EH9 3JW
Scotland

ABSTRACT

Plate tectonic reconstruction of the Mesozoic Tethyan realm shows that metalliferous-oxide and sulphide-sediments are associated with Late Triassic continental rift volcanism, Triassic-Jurassic seamounts and subsiding carbonate platforms, and Jurassic to Lower Cretaceous passive margins adjacent to ocean basins. Cupriferous sulphide- and oxide-sediment precipitation accompanied seafloor spreading in Upper Jurassic-Lower Cretaceous and Upper Cretaceous time. The field relations, mineralogy and chemistry (particularly of Rare Earth Elements) is summarised for each sediment type.

The Middle Triassic initial stages of continental rifting (Periadriatic Line) were marked by deposition of shallow marine carbonates and related Pb-Zn-rich brine exhalations. Hydrothermal leaching was driven by elevated geothermal gradients in areas of thinned continental crust. Widening rifts during Upper Triassic time were accompanied by extensive mostly alkalic mafic volcanism and deep water Mn- and Fe-Mn oxide-sediments (Oman, Antalya). Ferruginous pisoliths accumulated slowly on seamounts and disintegrating carbonate platforms adjacent to rifts (Sicily). After volcanism ended, hydrogenous accumulation produced Mn- and trace metal-rich crusts above the rift lavas (Mamonia, Antalya), and also metal-rich horizons on the adjacent deep water clastic rift margins (Antalya). During continental break-up and the passive margin phase, Fe-Mn crusts and nodules accumulated slowly on subsiding seamounts (Sicily). The pink, often nodular, ammonitico rosso or knollenkalke is the product of condensed pelagic carbonate deposition. The passive margins comprise predominantly distal terrigenous and radiolarian sediments, deposited below the calcite compensation depth. Manganiferous intercalations in the Upper Jurassic-Lower Cretaceous signal renewed volcanism along the margins (Antalya, Oman, Mamonia, Baër-Bassit).

Upper Jurassic ocean floor formed at a rifted spreading axis possibly in an elongate strike-slip controlled basin (Ligurian Apennines), with stratiform cupriferous sulphides at several levels, and trace element-poor manganese ores along the lava-pelagic sediment interface. Later radiolarian deposition was pelagic with a distal terrigenous component (Elba). During the Upper Cretaceous, seafloor spreading, in small ocean basins, produced major cupriferous sulphides, and both Fe-rich (ochres) and Fe-Mn-rich (umbers) oxide-sediments (Troodos Massif, Semail Nappe, Baër-Bassit).

Comparisons of modern oceanic sediments highlight the key controls of Tethyan metallogenesis. High-temperature discharge from less-rifted fast-spreading axes produced major stratiform cupriferous sulphide orebodies and Fe-Mn oxide-sediments (umbers). Rifting and slower spreading allowed greater seawater penetration and favoured formation of small stratiform cupriferous sulphides and Fe-poor,Mn-oxide sediments (Ligurian Apennines). The metals of the Tethyan rifts and passive margin precipitated from more dilute lower-temperature thermal springs, with varying degrees of trace element scavenging from seawater. The end-product was the condensed Fe-Mn nodules and crusts which slowly accumulated on sediment-starved seamounts and subsiding platforms.

INTRODUCTION

This paper synthesises information on metalliferous, mostly oxide-and cupriferous-sulphide, sediments from the Mesozoic Tethyan ocean, in the light of modern oceanic processes. Stress is placed on ancient deposits for which the tectonic setting, field relations and chemistry are relatively well known. Data from the literature are supplemented by our own fieldwork and by new analyses of major and trace elements of oxide-sediments from Oman, and of Rare Earth Elements (REE) for a wide range of Tethyan deposits. The sediments are discussed in relation to the tectonic history of the Tethys, which involves continental rifting, passive margins and several modes of ocean crust genesis. Active margins are excluded. The review concludes with comparisons with modern oceanic counterparts.

OUTLINE TETHYAN PALAEOGEOGRAPHY

Continental reconstructions stemming from palaeomagnetic data for Permo-Triassic time reveal a major ocean open to the east, narrowing westwards towards the present West Mediterranean (Briden et al. 1970; Smith 1971; Dewey et al. 1973). This is the palaeo-Tethys, which is mostly recorded by metamorphic rocks, for example, in the Caucasus, and Black Sea area, and possibly the Pontides of Northern Turkey

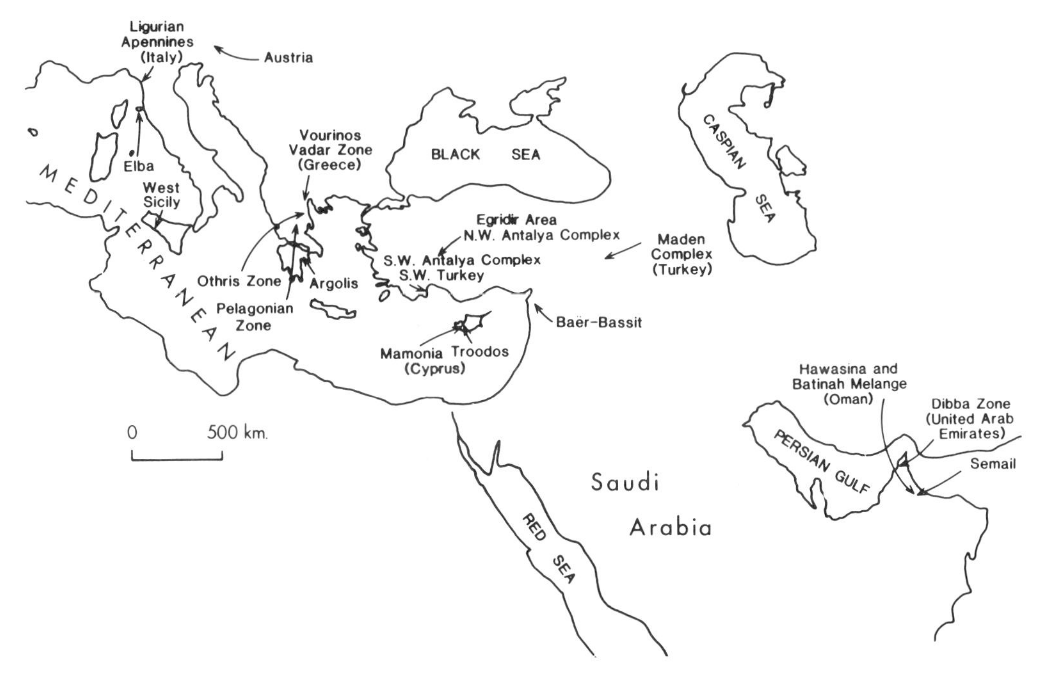

Fig. 1. Outline map to show the locations of metalliferous sediments mentioned in the text

(Laubscher and Bernoulli, 1977; Sengör, 1979; Jenkyns, 1980). None of the metal deposits discussed here come from this older ocean (Fig. 1).

During Triassic time the margins of Gondwanaland underwent continental rifting, seen for example in the Mediterranean, Atlantic and East African areas (e.g. Cox, 1970). In the Mediterranean, these rifts can be considered in a general way to have developed adjacent to the palaeo-Tethys and thus in several cases could be back-arc basins produced by intra-continental rifting (Nisbet, in Smith et al., 1975; Pe-Piper, 1982). Irregular fragmentation produced a complex array of horst-grabens and micro-continental slivers of various scales. During Upper Triassic time substantial alkalic mafic volcanism occurred in many areas; e.g. Dolomites, Italy (Bernoulli and Jenkyns, 1974; Laubscher and Bernoulli, 1977), Pindos, Greece (Aubouin et al., 1970), Mamonia, Cyprus (Lapierre and Rocci, 1976). Some of the rifts remained inactive (aulacogens), while others grew to form the ocean basins of the Neo-Tethys. In the East Mediterranean, rifting well into the southern margin of Gondwanaland initiated a southern branch of the Neo-Tethys, often termed the Mesogea (Biju-Duval et al., 1977, Fig. 2). In the West Mediterranean, Middle and Late Triassic rifting

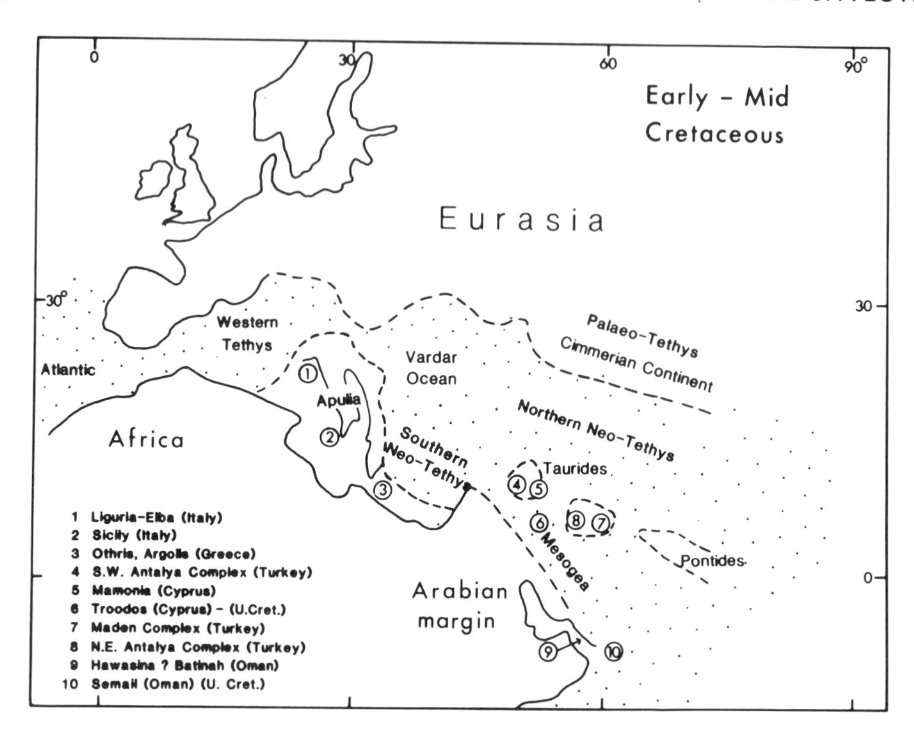

Fig. 2. Palaeogeographic sketch of the Tethys of Early-Mid Cretaceous time. The Palaeozoic Tethys has virtually closed, with a single Western Neo-Tethys and several branches of the Eastern Neo-Tethys. The longitudes of various continental blocks are arbitrary. Inferred locations discussed in the text are shown. Sources of data: Smith and Briden, 1977; Lauer, 1981.

was followed by renewed extension and final continental break-up reportedly during Upper Jurassic time (Oxfordian, de Graciansky et al., 1979). The Apennine ophiolite deposits, discussed here, formed in this ocean. Further east, in the Greek area the timing of initial ocean-floor spreading has been taken as Upper Triassic (Smith et al., 1975). In central Greece, the Othris ophiolite was emplaced westwards prior to Lower Cretaceous time (Cemomanian to Santonian age of sedimentary cover), but whether it originated in a local ocean basin (Hynes et al., 1971; Barton, 1975), or was thrust from the north-east from a distant Vardar ocean (Aubouin et al., 1970; Bernoulli and Laubscher, 1972) remains controversial.

Eastwards, the northern branch of the Neo-Tethys apparently extended into the Pontides of Northern Turkey. To the south lay a Neo-Tethys strand in central western Turkey (Ankara-Izmir Zone, Sengor and Yilmaz, 1981), then south again was the Mesogea, represented by the Antalya Complex (S.W. Turkey, Brunn et al., 1971; Dumont et al., 1972; Woodcock and Robertson, 1982), the Mamonia Complex (S.W. Cyprus, Lapierre, 1975; Robertson and Woodcock, 1979) and Baër-Bassit (Syria, Delaune-Mayere et al., 1976, 1977).

If ocean-floor spreading in the Eastern Mediterranean Neo-Tethyan areas began soon after Upper Triassic time, almost all trace of Triassic oceanic crust has been subducted. Alternatively, ocean crust formation could have been delayed until the Upper Jurassic, as in the Western Mediterranean or even later. During the Upper Cretaceous (Delaloye et al., 1977; Rocci et al., 1980) the Troodos, Antalya, Baër-Bassit, Hatay, and Guleman ophiolites were created in the Neo-Tethys (e.g. Gass and Masson-Smith, 1963; Parrot, 1977; Robertson and Woodcock, 1980). Trace element chemistry of the Troodos Upper Pillow Lavas suggests genesis above a subduction zone, possibly in a small back-arc basin (Pearce, 1975; Pearce et al., 1980), but supporting field evidence is generally lacking; certainly the faunal and palaeomagnetic results oppose the concept of a wide southern Neo-Tethys (e.g. Sengör and Yilmaz, 1981). The complexity of the Tethyan belt is further illustrated by recent palaeomagnetic data which imply that during the Mesozoic, Turkey was split into three blocks sited off the Arabian continental margin near the equator (Lauer, 1981, Fig. 2).

Well south of the East Mediterranean in the Oman area (Figs. 1, 2), continental rifting in the Permian and Upper Triassic (Glennie et al., 1973, 1974) was followed by outpouring of Upper Triassic alkalic to tholeiitic basalts of the Haybi Complex (Searle et al., 1980) which was a precursor to ocean-floor spreading adjacent to the Hawasina continental margin. It has been proposed that the Semail ophiolite formed in the Upper Cretaceous above a north-east dipping subduction zone in which Tethyan oceanic crust was being consumed (Pearce et al., 1981). The palaeogeography of the area between Oman and the East Mediterranean can not at present be specified in detail (Zagros crush zone, e.g. Ricou, 1971).

Summarising, the metalliferous sediments treated here relate to various tectonic settings in the Neo-Tethys:

(i) Middle to Late Triassic continental rifting (Periadriatic, Antalya, Mamonia, Othris, Argolis, Oman, Fig. 1.), pre- and syn-rift condensed pelagic deposition on horsts and continental platforms (Sicily, Austria), (ii) Jurassic-Lower Cretaceous ocean floor spreading (Apennines), (iii) Jurassic-Cretaceous passive margins (Antalya, Mamonia, Baër-Bassit, Oman), (iv) Late Cretaceous spreading (Troodos, Baër-Bassit, Semail).

UPPER TRIASSIC-EARLY JURASSIC CONTINENTAL RIFTING

Continental rifting (Fig. 3) during Middle to Upper Triassic time is documented by block-faulting, accelerated clastic input, and onset of deep water pelagic deposition within subsiding rift zones.

In the vicinity of the Periadriatic Line, in Northern Yugoslavia and Southern Austria, the earlier stages of continental rifting in Middle Triassic time were accompanied by the deposition of stratiform carbonate-hosted Pb-Zn sulphide deposits of Mississippi Valley-type (Brigo et al., 1977). The ores, which are still being mined, are located in limestones and dolomites of lagoonal facies, and are then overlain by deeper water clastic and pelagic dolomites. The genesis of the Pb-Zn ores is thought to relate to elevated geothermal gradients during the earlier stages of continental rifting and crustal thinning. With downward penetrating convection, Pb and Zn (also Mn, Cu)were leached then precipitated in an overlying saline sea(Russell et al., 1981).

During the later stages of continental extension, the basin deepened and became pelagic. For example, in the Antalya Complex (S.W. Turkey) red nodular iron-rich pelagic limestones with ammonites (ammonitico rosso facies, see below) appear from Middle-Triassic time (Ladinian, Marcoux, 1970, 1974). Volcanism was initiated with localised acidic tuffs, followed by extrusion of over 1500 m of mafic alkalic lavas mostly in deep water (Juteau,1975). Subaerial erosion of both the parent platform and the isolated horsts in the rift produced large volumes of terrigenous clastics. Flourishing carbonate build-ups liberated large volumes of carbonate clastics and peri-platform ooze (Robertson and Woodcock, 1981a, b, Fig. 3). Similar rift-relationships are seen in Oman (Glennie et al., 1973, 1974).

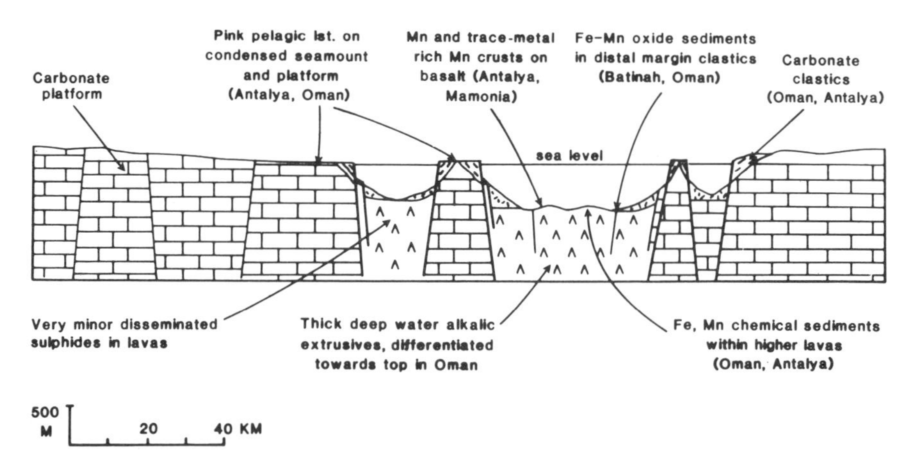

Fig. 3. Setting of metalliferous and pelagic sedimentation associated with Triassic rift volcanism (e.g. Antalya, Mamonia, Pindos, Oman, Dolomites).

Rift Margins

The margins of the rift zones, during the later stages, as seen in the Antalya (Turkey), Mamonia (Cyprus) and the Batinah (Oman) areas (Fig. 1) are mostly dominated by rapidly deposited terrigenous and carbonate clastics with low metal content. In Antalya, red and brown, metal-enriched, argillaceous mudstones occur at the top of the Upper-Triassic succession of *Halobia*-bearing fine grained pelagic limestones. These mudstones mark the top of the calcareous succession below thick non-calcaraeous radiolarian sediments of Jurassic and Cretaceous age (see below). They contain goethite and are markedly enriched in Fe, also in Co, Cu, Ni, Y and Zr (Table 1, analysis 1). Alone of the rift margin and passive margin sediments, they show relatively high REE values and a marked positive Ce anomaly indicating relatively slow hydrogenous accumulation from seawater (Fig. 10a; cf. Haskin et al., 1966). All the REE plots which follow are in chondrite-normalised form. These metal-enriched mudstones correspond to a hiatus in clastic sedimentation following volcanism when the area subsided below the calcite compensation depth (Table 2).

Distal margin successions in the upper Batinah Complex (Oman) include dark grey manganiferous mudstones intercalated with bentonitic clays, radiolarites, fine grained carbonate and terrigenous turbidites. The Batinah Complex structurally overlies the Semail Nappe in contrast to the Hawasina and associated volcanic rocks (Haybi Complex, Searle et al., 1980), which represent the continental margin stacked beneath the Semail Nappe (Glennie et al., 1973; Graham, 1980). According to Woodcock and Robertson (1982c), the upper Batinah Complex was either derived from a north-easterly margin to the Oman ocean basin, or possibly represents part of the rifted Arabian margin. As shown in Table 1 (analysis 6), these metalliferous sediments are highly ferruginous and manganiferous, virtually non-calcareous with high values of Ba and Sr, elements which correlate with Mn. Levels of other trace elements (e.g. Cu, Ni, Cr, V) are relatively low. These manganiferous, and to a lesser extent ferruginous, chemical sediments reflect hydrothermal discharge during rift volcanism. The precipitates drifted to the adjacent rift margins, scavenged trace elements from seawater (Ba, Sr), then accumulated during breaks in distal turbidite input. These metal-rich intercalations do not extend into the more proximal clastic successions (Woodcock and Robertson, 1982a,c).

Upper Triassic Rift Volcanism

Metal-oxide sediments related to the Late Triassic rift volcanism have been described in most detail from Antalya (Turkey), Mamonia (Cyprus) and the Batinah (Oman) (Fig. 1, 2, 3). In the Antalya Complex, metalliferous oxide sediments within the mafic lavas are generally restricted to minor interstitial ferruginous sediments, but near the top of the lava pile local intercalations of ferruginous-, manganiferous- or ferromanganiferous-oxide sediments are seen, plus minor volumes of red pelagic limestone (Robertson, 1981; Fig. 4).

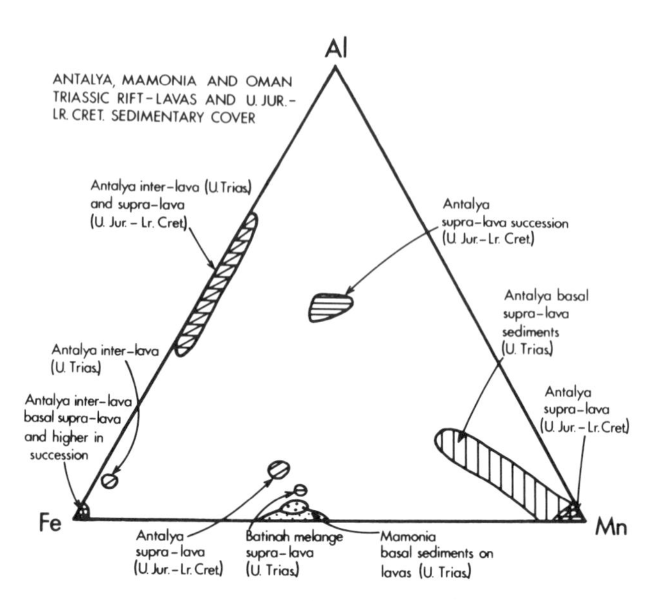

Fig. 4 Al, Fe, Mn plots of metalliferous sediments associated with the Upper Triassic rift lavas and Upper Jurassic-Lower Cretaceous sedimentary cover (Antalya and Mamonia Complexes).

Adjacent to carbonate build-ups the highest lavas are overlain by manganese crusts and pink, locally nodular, pelagic limestones (Table 1, analysis 3). As shown in analysis 4 (Table 1), overlying mudstones are strongly enriched in manganese, relative to hemipelagic muds (also in Ba, and Sr), but are relatively depleted in iron. REE plots (Fig. 5a) for the interlava mudstones show a small negative Ce anomaly, possibly indicative of rapid accumulation from seawater (cf. Haskin et al., 1966), but the overlying ferruginous pelagic carbonate, (Table 2, analysis 3) exhibits low REE values with slight heavy REE enrichment.

Very similar Upper Triassic mafic alkalic lavas, mostly pillow lavas, again crop out in the Mamonia Complex, Cyprus (Lapierre, 1975). Particularly notable are grey pink, to black pelagic limestones up to several tens of metres thick, locally overlying the mafic lavas (Dhiarizos Group, Swarbrick and Robertson, 1980). As shown in analyses 5 (Table 1), these overlying carbonates (Fig. 4) are strongly enriched in Mn, Fe, also in Cu, Ni, Zn and Zr. A sample just above the mafic lavas shows a pronounced negative Ce anomaly, while another from 3 m higher in the succession shows a well-marked positive Ce anomaly (Table 2, analyses 4, 5; Fig. 5b.).

More extensive interlava and supra-lava metalliferous oxide-sediments are seen in Upper Triassic pillow lavas of the lower Batinah Complex, Oman. During its Upper Cretaceous tectonic emplacement, the Semail ophiolitic nappe split into blocks, allowing the

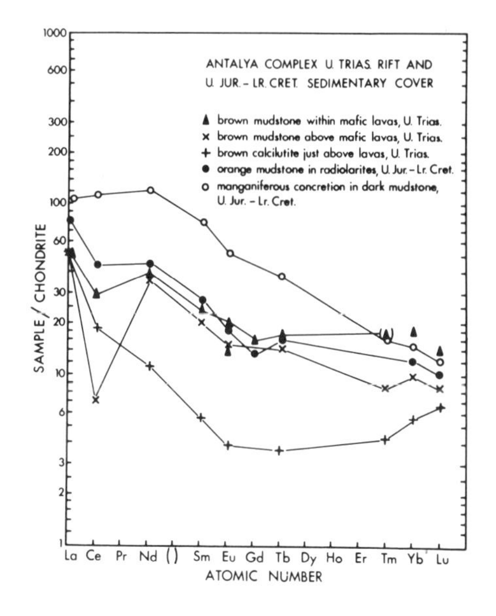

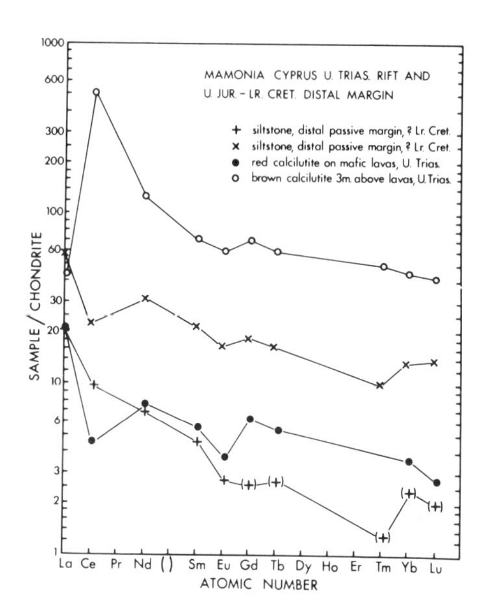

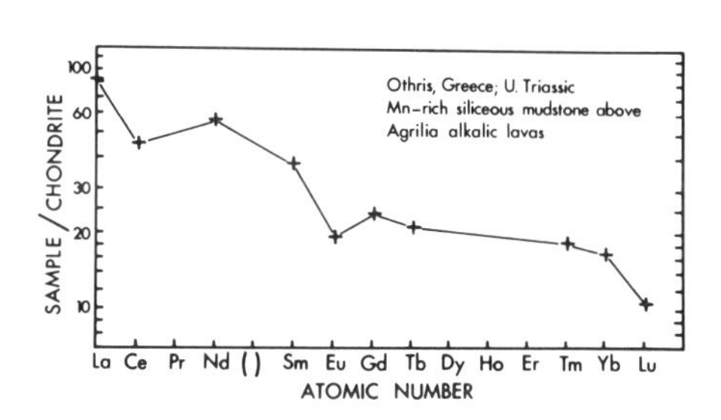

Fig. 5 Rare Earth Elements expressed as chondrite-normalised plots for metalliferous oxide-sediments: a. Upper Triassic rift lavas and the Upper Jurassic-Lower Cretaceous sedimentary cover, Antalya Complex (Turkey); b. Late Triassic rift and late Jurassic to early Cretaceous distal passive margin (Mamonia Complex, S.W. Cyprus); c. Late Triassic, Othris (Greece).

underlying sedimentary and igneous rocks of the Hawasina and Haybi distal margin units to move upwards in corridors and reach high levels above the Semail Nappe (Woodcock and Robertson, 1983). The lower Batinah Melange thus records the most distal preserved part of the Triassic rift lavas and associated sediments. Numerous intercalations of red and brown siliceous mudstones appear towards the top of the lavas, which include trachytes and other differentiated extrusives. The overlying sediments, also dated Upper Triassic by Radiolaria (E.A. Pessagno, personal communication, 1980) comprise alternations of redeposited Halobia-bearing limestones, radiolarites and radiolarian mudstones, often reddish brown to black with metal-impregnation. Some siliceous mudstones are strongly Fe-enriched relative to Mn (analysis 7, Table 1), but Fe/Mn ratios vary greatly (Robertson and Fleet, unpublished data; Fig. 4).

The metalliferous oxide-sediments within and above the mafic lavas of the Upper Triassic rift zones (Fig. 3) were locally hydrothermally precipitated during and immediately after volcanism ended. The relatively small volume of interlava chemical sediment in the Antalya and Mamonia lavas probably reflects rapid extrusion, rather than absence of hydrothermal activity. By contrast, the higher lavas in these areas, and in Oman, were erupted more spasmodically, allowing greater metal accumulation. In the Mamonia, the negative Ce anomaly of the sediments above the lavas points to initial rapid ferromanganese precipitation. This was followed by slower accumulation to produce a positive Ce anomaly (c.f. Goldberg et al., 1963). The setting was probably on a topographic high which was isolated from clastic input (Fig. 3).

Table 1. Selected representative analyses of Tethyan metal-oxide sediments associated with rifts and passive margin.

1. Mudstone, rift margin succession; 2. Pelagic carbonate within pillow lavas; 3. Pelagic limestone overlying lavas; 4. Calcareous mudstone above alkalic lavas; 1-4. Upper Triassic S.W. segment of Antalya Complex (Turkey); 5. Mudstone overlying lavas, Upper Triassic, Mamonia (Cyprus); 6. Mudstone, rift margin succession, Upper Triassic, Oman (Batinah sheets); 7. Mudstone succession above mafic lavas, Upper Triassic, Oman (lower Batinah melange); 8. Chamosite-goethite-haematite pisoliths, Lower Jurassic, Sicily; 9. Goethite-haematite pisoliths, Lower Jurassic (W. Sicily); 10. Ferromanganese nodule, Middle Jurassic (W. Sicily); 11. Siltstone, proximal part of passive margin, Upper Jurassic-Cretaceous, Antalya Complex (Turkey);12,13.Siltstone, Lower Cretaceous, Antalya Complex (Turkey); 14. Mudstone of the distal passive margin, Upper Jurassic-Lower Cretaceous, Mamonia (Cyrpus): 15. Radiolarite above peralkaline lavas, ?Lower Cretaceous, Baër-Bassit (Syria); 16. Mudstone of the passive margin succession, Upper Jurassic to Lower Cretaceous, lower Batinah melange (Oman); 17. Radiolarite, Upper Jurassic to Lower Cretaceous, Lower Batinah melange (Oman).

Sources of data: a. Robertson, 1981; b. Swarbrick, 1979; c. A.H.F. Robertson, unpublished; d. Jenkyns, 1970a; f. Delaune-Mayere (1978); Dritterbass (1979). Estimated average from semi-quantitative microprobe scan.
N.B. Major element oxide anlayses in wt.% to one decimal place.

Table 1.

Analysis No.	1	2	3	4	5	6	7	8	9	10	11	12	13	14	15	16	17
Sample No.	124	213	265	185	198	79-52	79-61	-	-	-	222	428	433	151	117/08	79-45	79-46
SiO_2	55.7	21.0	11.4	71.5	27.1	8.7	66.2	~13.0	~1.5	3.64	86.7	63.4	85.8	12.4	93.9	70.8	92.1
Al_2O_3	7.8	4.8	1.3	0.6	3.9	1.7	2.9	~3.5	~1.5	1.89	2.0	1.4	2.6	0.9	1.9	17.5	3.5
Fe_2O_3	29.2	3.4	0.7	6.7	28.8	31.4	29.9	~30	95	12.79	10.5	1.0	10.3	39.4	1.0	5.5	1.8
MgO	1.4	2.0	1.5	0.3	0.9	1.8	1.7	-	-	2.58	0.2	0.9	0.5	3.3	0.5	1.4	0.9
CaO	0.7	66.3	76.0	14.3	16.2	4.4	0.3	~50	~50	26.18	0.2	4.6	0.4	6.8	<0.1	1.0	0.7
Na_2O	0.2	b.d.	b.d.	-	<0.1	0.3	<0.1	-	-	0.27	0.1	0.4	0.1	<0.1	<0.1	0.4	0.2
K_2O	2.0	1.4	0.2	0.1	<0.1	0.2	0.1	-	-	0.06	0.4	0.3	0.4	1.3	0.1	1.3	0.5
TiO_2	0.6	0.5	<0.1	<0.1	0.2	0.1	0.1	~0.8	~1.5	0.42	0.1	0.7	0.1	0.2	0.1	0.8	0.1
MnO	0.2	<.1	6.5	5.9	20.0	53.4	0.3	~1.5	~0.5	13.75	0.02	27.0	0.06	33.4	0.01	0.06	0.06
P_2O_5	0.2	0.2	0.2	0.1	0.13	0.1	<0.1	~0.1	.15	0.46	<0.1	<0.1	<0.1	<0.1	0.07	<0.1	<0.1
Total	100.0	99.8	97.8	99.5	-	102.1	101.39	-	-	59.04	100.3	99.0	100.4	-	-	99.1	100.8
Ba	64	25	1150	156	50	5851	36	b.d.	b.d.	270	41	93	97	1119	<20	150	124
Co	185	29	87	19	58	-	-	1000	~2000	1900	4	1934	29	114	15	-	-
Cr	29	21	15	19	23	13	19	~400	~400	20	21	13	20	7	5	115	21
Cu	100	21	77	79	436	28	22	b.d.	~300	20	20	900	25	11	11	50	30
Ni	523	53	81	63	408	90	31	500	~750	1730	77	119	52	127	<10	49	16
Pb	62	<5	2	51	148	-	14	-	-	670	43	13	28	43	-	23	10
Sr	25	147	230	141	146	2709	9	-	-	450	32	889	23	69	11	147	54
V	280	47	30	135	143	67	76	~500	~500	-	45	313	130	69	33	154	21
Zn	-	-	-	-	270	218	67	1000	1000	360	-	-	-	-	52	65	37
Zr	297	174	75	48	322	37.3	18	-	-	-	57	43	47	318	56	208	33

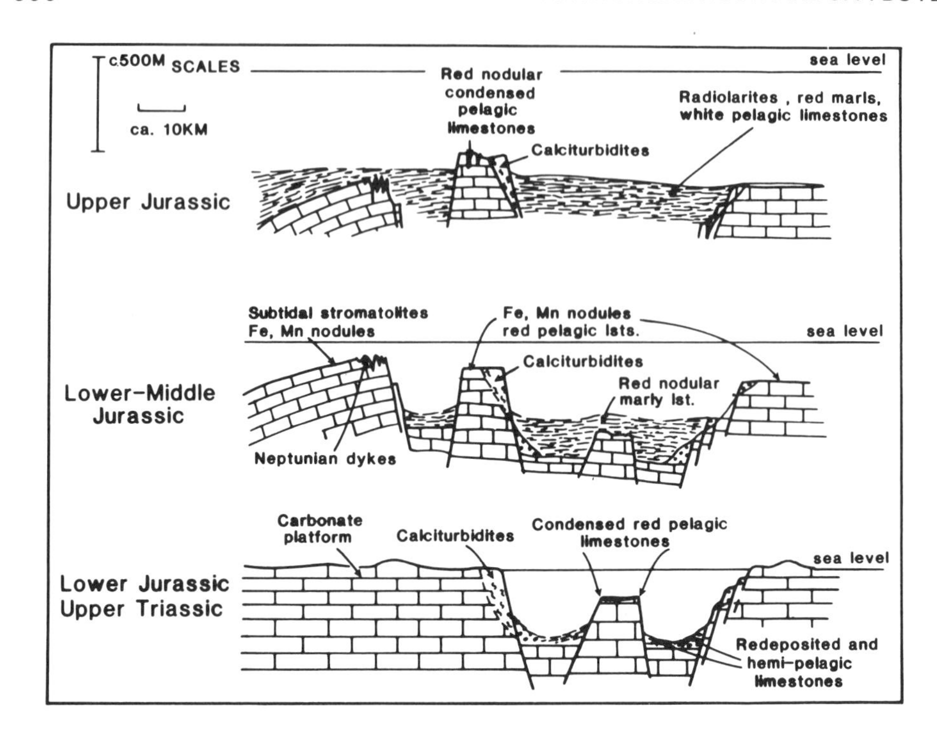

Fig. 6 Sites of metallogenesis associated with disintegration and subsidence of a Tethyan carbonate platform in Western Sicily; modified after Bernoulli and Jenkyns (1974).

Greek Manganese Ores

Highly manganiferous oxide-sediments, also associated with mostly Triassic volcanism, are reported by Panagos and Varnavas (1980), from the Othris and Argolis areas of Greece (Figs. 1, 2). Few field relations are given. According to Spathi (1964), the mineralogy is mostly goethite, haematite, pyrolusite and quartz. Typical analyses (Table 3, analyses 16, 17; Fig. 1) show high Mn values relative to Fe, Al and Mg, which are strongly depleted relative to pelagic clays (e.g. Turekian and Wedepohl, 1961). Fe and Mn typically vary inversely. Trace metals are markedly enriched relative to pelagic clays, particularly Co.

Panagos and Varnavas (1980) relate the ferromanganese oxides to rapid precipitation from thermal springs, with scavenging of trace elements from sewater. Co enrichment is attributed to the ready substitution of Co for Mn^{4+}, relative to Fe^{3+}, which has a much larger ionic radius (Burns and Burns, 1977). The Greek chemical sediments show similar Fe/Mn ratios to the Antalya Upper Jurassic-Lower Cretaceous passive margin manganiferous oxide-sediments and the Apennine manganese ores, but trace elements are relatively enriched, more in the range of ferromanganiferous Cyprus umbers (see below). One sample of manganiferous mudstone collected by A.G. Smith from above the Upper Triassic Agrilia lavas in Othris (Smith et al., 1975) shows a small negative Ce anomaly in chondrite-normalised plots (Table 2, analysis 6, Fig. 5c).

Lower Jurassic Ferruginous Pisoliths

Throughout much of the Tethys, major carbonate platforms grew then disintegrated during Triassic and Jurassic time. For Sicily the stages are illustrated in Fig. 6 as are the sites of pelagic, metalliferous and clastic deposition (Bernoulli and Jenkyns, 1974). The rifting appears to have been in two separate phases, Upper Triassic to Liassic and Middle Jurassic (e.g. de Graciansky et al., 1979). During the Liassic, successions on the carbonate horsts include up to 0.4 m thick horizons of brown iron-rich pisoliths which are concentrically laminated sub-spherical bodies up to several centimetres in diameter. Jenkyns (1970b) states that the pisoliths are composed of calcite, goethite, haematite and rarely chamosite. The associated fauna includes Tethyan ammonites, belemnites, gastropods, rare brachiopods, fish teeth and crinoid ossicles. The nuclei of the pisoliths are often clasts of differentiated lavas (Fig. 11a). Electron microprobe scans over the concentric laminations show high values of Fe and many trace metals. Small volumes of ferromanganiferous nodules are also locally present. Average elemental values are given in Table 1 (analyses 8,9) and Fig. 7.

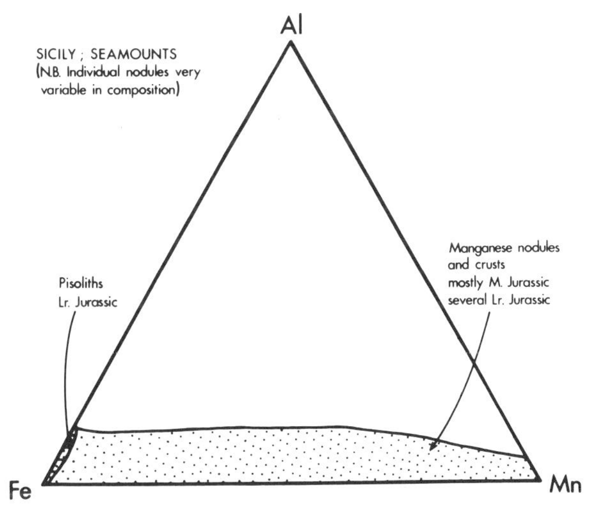

Fig. 7. Triangular plot of major elements in Early and Middle Jurassic pisoliths (Toarcian) and manganese nodules and crusts (Middle Jurassic), Western Sicily, data from Jenkyns (1970a, b).

In the light of the regional setting and modern sediments, Jenkyns (1970b) concluded that much of the Fe and, to a lesser extent, the Mn and trace metals, were derived from submarine exhalations related to coeval volcanism. The pisoliths apparently formed within the photic zone (few tens of metres) on flat-topped seamounts undergoing minimal net pelagic carbonate deposition. The highly ferruginous pisoliths contrast with the other Tethyan metalliferous sediments related to rift volcanism, which are mostly more manganiferous. Depositional conditions were clearly fully oxidising from the open marine fauna; algal and fungal matter in the pisoliths (Fig. 11b) was possibly sufficiently abundant to locally depress Eh and thus exclude manganese from the sediment.

JURASSIC-CRETACEOUS PASSIVE MARGINS

The timing of final continental break-up and ocean-floor spreading is reported to be early Upper Jurassic (Oxfordian) in the Western Tethys (de Graciansky et al., 1979), but evidence is more equivocable in the Eastern Tethys where the preserved ophiolites are mostly Upper Cretaceous. At least some seafloor spreading is believed to have taken place during the Jurassic-Lower Cretaceous interval (e.g. Hynes et al., 1972; Biju-Duval et al., 1977; Searle et al., 1980). Many of the ocean-floor spreading axes followed previous Triassic rifts (e.g. Oman, Mamonia, Antalya). These older rifts thus became the deeper more proximal parts of passive margins during the ocean genesis which ensued. In other cases the new spreading axes cut the older Triassic rifts.

Clastic Margins

In the Eastern Tethys, the former Upper Triassic rift successions pass up into deep water non-calcareous radiolarian pelagic sediments, deposited below the calcite compensation depth. These pelagic sediments are intercalated with calcareous and terrigenous clastic sediments which become finer grained, thinner bedded and more siliceous further from the parent continental margin (Glennie et al., 1973, 1974; Robertson and Woodcock, 1981a). While most of the successions are not metal-enriched, locally, more proximal parts of the Antalya former passive margin, dated Upper Jurassic to Lower Cretaceous, are dominated by goethite-rich siltstones and mudstones which are strongly enriched in Fe, but show normal shale values of most other elements (Table 1, analysis 11; Fig. 8). This ferruginous accumulation records relatively condensed deep water accumulation on a proximal part of a starval margin bordered by a carbonate platform. REE patterns of mudstones and claystones from the Jurassic and Lower Cretaceous intervals of the Mamonia and Baër-Bassit (Syria) margins all show shale patterns characteristic of continentally-derived sediments (Table 2, analyses 11-18; Fig. 5a, b; Fig. 10a, b).

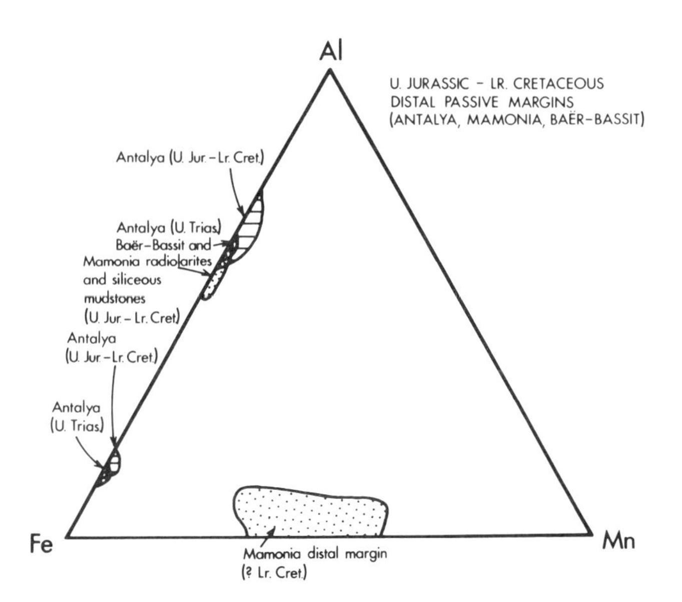

Fig. 8 Triangular plots to show major element compositions of Upper Triassic to Lower Cretaceous metal-rich sediments deposited during passive margin subsidence phase.

Sequences stratigraphically above the Upper Triassic rift lavas are generally highly deformed (e.g. Hawasina, Oman; Glennie et al., 1973), but at least in the Antalya Complex unbroken Upper Triassic to Middle Cretaceous successions are known (Robertson and Woodcock, 1981b). These sequences are dominated by deep water non-calcareous radiolarian mudstones and radiolarites with minor clastics, generally without metal enrichments. Clay minerals, elemental ratios, trace elements and REE values again point to terrigenous source materials (e.g. Table 1, analysis 14; Fig. 9).

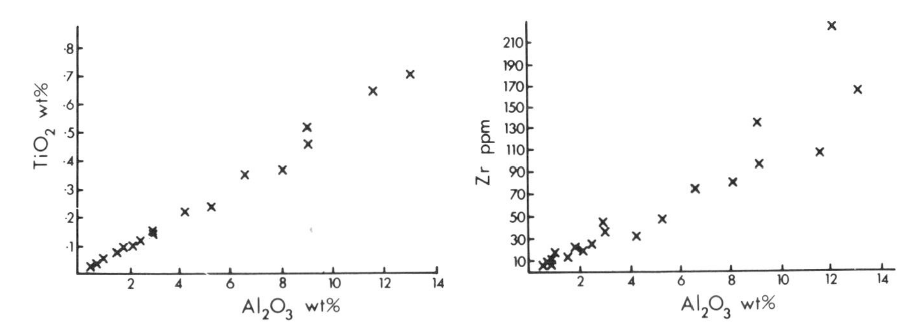

Fig. 9 Correlation of major and trace elements in metal-rich and associated sediments from the Jurassic passive margin phase of the N.W. segments of the Antalya Complex (Turkey), data from Waldron, 1981. a. TiO_2 versus Al_2O_3; b. Zr versus Al_2O_3.

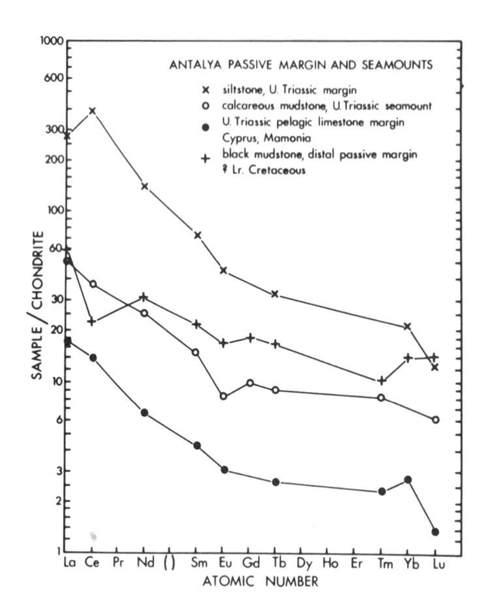

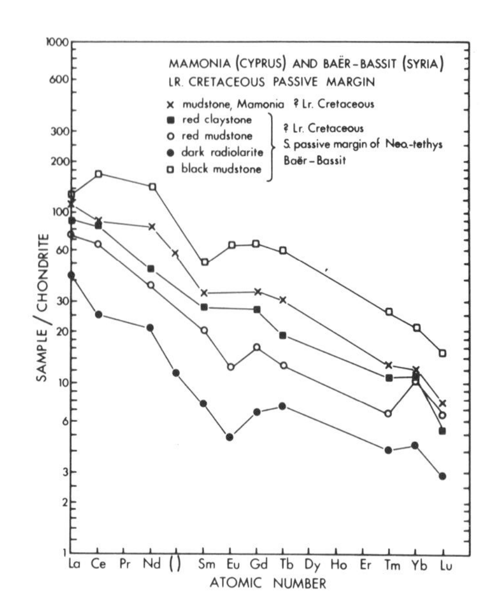

Fig. 10 Rare Earth Elements distribution in Upper Triassic, Jurassic and Lower Cretaceous, passive margin sediments. a. Antalya Complex (S.W. Turkey); b. Mamonia (S.W. Cyprus) and Baër-Bassit (Syria). Previously unpublished data, A.J. Fleet and A.H.F. Robertson

Manganese Deposits

In some areas intercalations of manganese ore and oxide-sediments appear in the dominantly clastic margin successions, in the Late Jurassic-Early Cretaceous.

In the Antalya Complex, manganiferous ores, which were once mined, appear as en echelon lenses up to several metres thick within the higher levels of the radiolarian successions overlying the Triassic alkalic lavas. The mineralogy is mostly goethite, pyrolusite and quartz. These ores, which are seen only above the lavas and not in the adjacent former rift-margin successions, are particularly strongly enriched in Mn, Co, Ni and Cu relative to the intercalated radiolarian sediments (Table 1, analyses 12, 14). The deposits include small manganiferous concretions up to several centimeters in diameter which are exceptionally enriched in trace metals, particularly Co, Sr, Cu and Mo (Robertson , 1981). Several samples of the manganese oxide-sediments show small negative Ce anomalies, but others exhibit shale patterns (Table 2, analysis 19; Fig. 5a). In the Mamonia Complex, no long intact sequences above the Upper Triassic lavas are known, but distal margin sequences above the former Triassic rift margins (see above, Robertson and Woodcock, 1979; Swarbrick and Robertson, 1981) consist of thin bedded mixed terrigenous and bioclastic turbidites interbedded with claystones. In the ?Lower Cretaceous many of the turbidites are black due to impregnation with manganese oxide. These deposits and similar metal-rich intercalations in the distal part of the Baër-Bassit margin (Syria, Delaune-Mayere, 1978; Delaune-Mayere et al., 1977) show shale patterns in normalised plots of REE (Fig. 10b).

Similar manganiferous mudstones and cherts are reported from the Upper Jurassic-Lower Cretaceous interval in Oman (Glennie et al., 1974, p. 199).

The Upper Jurassic to Lower Cretaceous chemical sediments in the margin successions of Oman, Antalya, Mamonia and Baër-Bassit are strongly relatively enriched in Mn but not in trace elements, except in some diagenetic phases. Renewed alkalic volcanism is known from this time, from Baër-Bassit (Parrot, 1974), possibly also from Antalya (Robertson and Woodcock, 1981b) and the Northern Oman Mountains (A.H.F. Robertson and A.J. Kemp, unpublished work).

Two hypotheses can be invoked to explain these hydrothermal manganiferous oxides within otherwise passive margin successions. First, if after Late Triassic rifting ocean-floor spreading did take place throughout much of the Jurassic, then the hydrothermal activity could relate to onset of subduction and marginal basin formation (Pearce et al., 1981). Alternatively, if little or no spreading took place until the Upper Jurassic, as in the Western Tethys, as seems more likely, the hydrothermal activity could record the final separation of the southern branch of the Neo-Tethys in the Eastern Mediterranean (Robertson and Woodcock, 1980).

Condensed Pelagic Limestones

Pink stratigraphically condensed pelagic limestones, the ammonitico roso or knollenkalke facies, are a characteristic feature of Tethyan carbonate platforms and seamounts which underwent slow pelagic accumulation from Upper Triassic to Upper Jurassic time. In some areas (e.g. Antalya, Mamonia, Oman exotics), the pink limestones of Upper Triassic age are closely related to coeval volcanism. Elsewhere, pink condensed pelagic limestones on carbonate platforms and seamounts are widely known from Sicily, Austria, Greece, Albania and Turkey (Aubouin, 1964; Bernoulli and Jenkyns,1974). Facies range from shallower (Jenkyns, 1974) to deeper water (Garrison and Fischer, 1969), and may be nodular to varying extents (Wendt, 1970; Ogg, 1981).

More chemical data, particularly of the non-calcareous component, is needed to evaluate any possible role of volcanism but condensed hydrogenous accumulation alone may be sufficient to form these pink pelagic limestones.

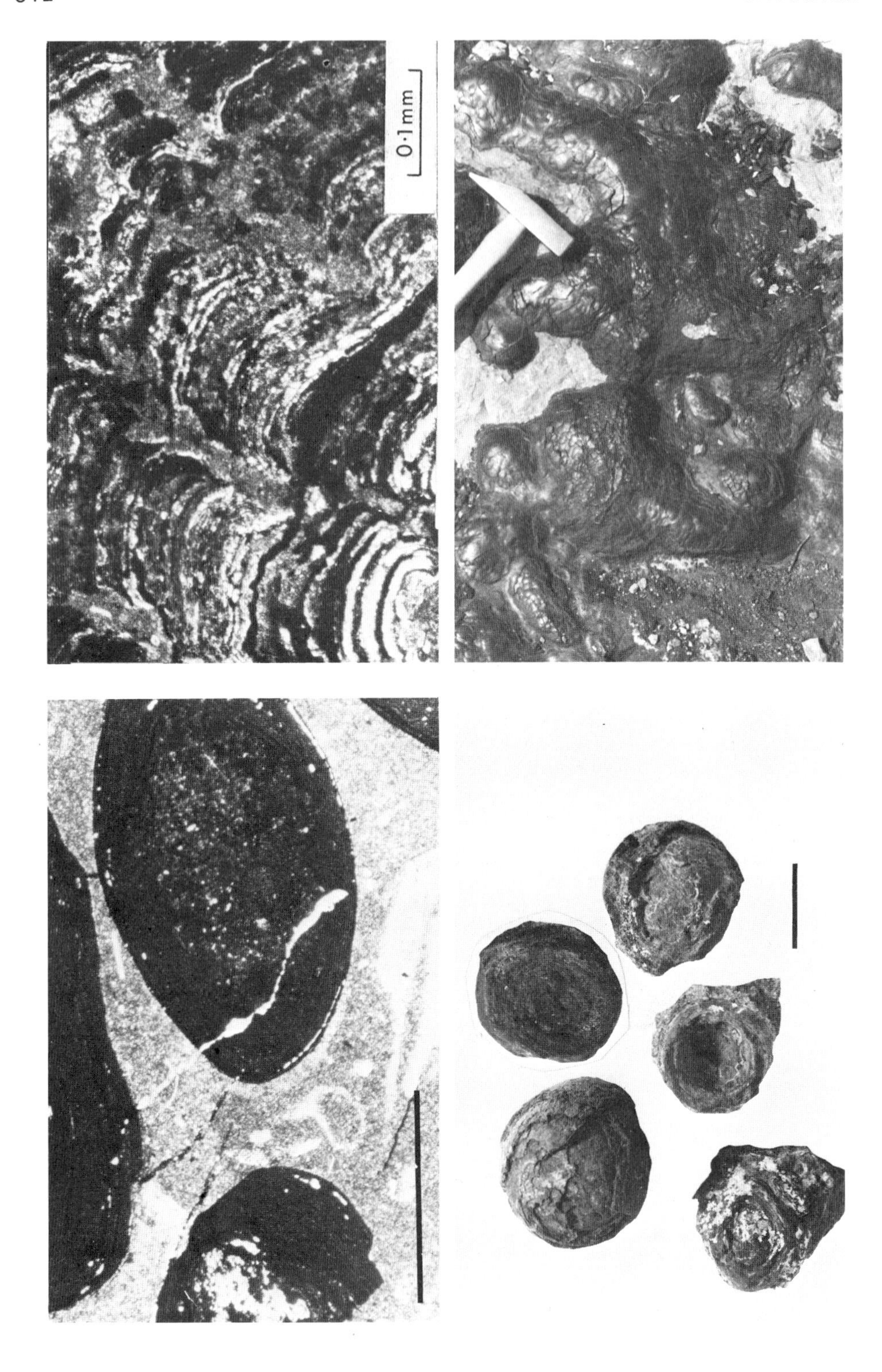
0·1mm

Fig. 11 Field photographs and micrographs of metalliferous oxide-sediments from Italy and Austria. a. Concentrically laminated Lower Jurassic iron pisolith. The core is trachyte with flow structure. Rocca Busambra, western Sicily. Thin section. Scale bar = 1 mm; b. Columnal stromatolithic microfabric in ferromanganese nodules of Early to Middle Jurassic age. Thin section, Unken Syncline, northern calcareous Alps, Austria; c. Fossil ferromanganese nodules from various localities of Middle Jurassic age in western Sicily. Scale bar = 3cm; d. Thick ferromanganese crust with limestone interstices. Bedding surfaces. Monte Kumeta, western Sicily. Hammer handle is 28 cm long. a, from Jenkyns 1970a, b, from Jenkyns 1977; c,d, from Jenkyns 1970b.

Manganese Nodules and Pavements

Manganese crusts, nodules and pavements have been described in detail from stratigraphically condensed pelagic successions on seamounts in Western Sicily (Jenkyns, 1970a, 1977; Figs. 6, 11c, d), and are also present in other slowly-deposited Neo-Tethyan margin successions (Bernoulli and Jenkyns,1974). According to Jenkyns (1970a), the nodules range from scattered to a continuous pavement. Intraclasts or shells often form the nuclei. The ferromanganese overlies differentiated extrusives in places. Mineralogically, the nodules contain goethite, haematite, todorokite and calcite. Electron-microprobe scans reveal considerable compositional variation both within individual nodules and between those from different locations. Average compositions are given in Table 1 (analysis 10).

Many of the nodules exhibit high values of Fe, Ni and V but tend to be less enriched in Cu and Zn. New REE results for manganese nodules and crusts from Sicily and Austria show a spectacular positive Ce anomaly in normalised plots (Table 2, analyses 7, 8, 9, 10; Fig. 12).

Jenkyns (1970a) concludes that the Sicilian nodules grew slowly in shallow water, largely within the photic zone. He infers an important metal contribution from coeval Mid-Jurassic volcanism, which probably records the final phase of continental break-up in the Western Tethys followed by ocean floor spreading in the early Upper Jurassic.

JURASSIC-LOWER CRETACEOUS SPREADING AXES

Mesozoic ophiolites associated with cupriferous sulphides tend to be Jurassic and Lower Cretaceous in the Western Mediterranean as far as Greece, and Upper Cretaceous in the remainder of the Eastern Mediterranean and in the Middle East (Figs. 2, 3).

The two age groups can be further distinguished in that the earlier has a generally mid-ocean ridge chemical character and contain mostly relatively small sulphide bodies. Here we concentrate on the well preserved and extensively studied Apennine ophiolites. Other 'Alpine' ophiolites, which are variously metamorphosed and deformed, are too numerous to be individually described here, but include associated metal deposits in the Western Alps, West Liguria and Calabria (Zuffardi, 1977), the Eastern Alps (Derkmann and Klemm, 1977). Albania (Aleksandër, 1980), Rumania (Gioflica et al., 1980);Turkey (Griffiths et al., 1972), and several parts of Greece, including Pindos (Sideris et al., 1980) and Macedonia (Kelepertsis, and Andrulakis 1980). All these deposits can be distinguished from the

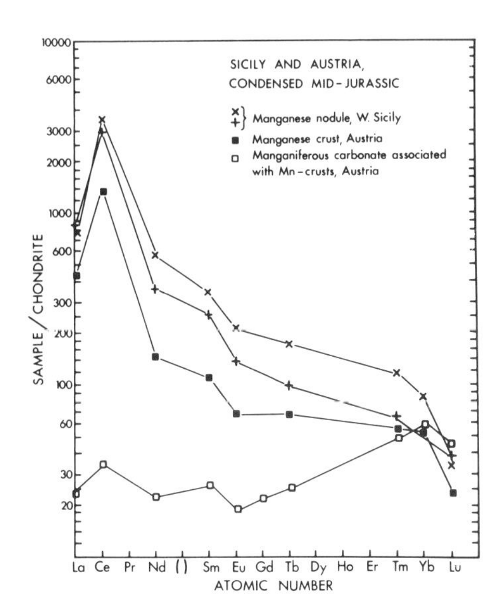

Fig. 12. Rare Earth Elements and Mid-Jurassic manganese nodules from Sicily and Austria. Previously unpublished data of A.J. Fleet and A.H.F. Robertson. Samples collected by H.C. Jenkyns and J.D. Hudson.

cupriferous sulphides of the Upper Cretaceous Semail Nappe, Oman, the Troodos Massive, Cyprus and other Upper Cretaceous ophiolites in the Eastern Mediterranean and the Middle East which contain very much larger cupriferous sulphides with extrusives which are reported to show subduction-related (high Mg low TiO_2) chemistry (Pearce, 1980)

Apennine Sulphides

An extensive literature has been recently summarised by Ferrario and Garuti (1980); field relations are shown in Fig. 13c. Massive and disseminated cupriferous sulphides occur at four main horizons in the ophiolite stratigraphy. The lowest stratiform sulphides appear in 'basal' breccias (Barrett and Spooner, 1977; Cortesogno et al., 1980), dominated by ultramafic rock clasts. The ores occur between the intrusive rocks and the overlying mafic extrusive and lava breccias. Anomalously high Ni in these lower sulphides is from the ultramafic rocks. A ferruginous crust above the breccias marks a significant hiatus according to Ferrario and Garuti (1980), but could alternatively be interpreted as a relatively rapid hydrothermal precipitate. Ophicalcite, the distinctive carbonate-veined and brecciated serpentinite, was also formed by a hydrothermal process

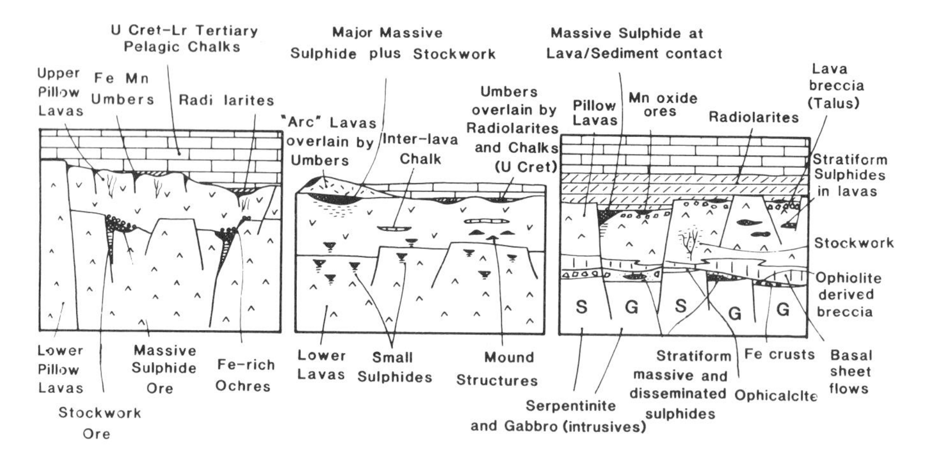

Fig. 13 Schematic cross-section to show the relationships of metalliferous deposits in Tethyan ophiolites. a. Troodos (Cyprus); b. Apennines (Italy); c. Semail (Oman).

Table 2. Rare Earth Elements (REE) for Tethyan metalliferous and pelagic sediments from rift, seamount and passive margin settings.

Analysis No.	1	2	3	4	5	6	7	8	9	10	11	12	13	14	15	16	17	18	19
Sample No.	124	213	265	310	313	71-10	1	39549	A1	A2	75-11	79-14	71-143	75-8	75-28	75-31	75-36	75-44	428
La	82.4	7.7	15.1	5.4	48.9	21.7	190.6	199.1	105.7	6.0	5.9	26.6	6.5	10.4	21.0	19.2	25.7	22.6	12.0
Ce	332.4	17.5	15.5	2.9	364.2	30.2	2012	238.7	955	23.3	19.0	55.5	8.4	16.0	46.7	44.2	117.2	57.6	34.0
Nd	83.7	6.6	6.7	3.6	62.3	26.3	160.9	27.0	69.7	10.7	18.3	38.5	4.0	9.7	15.4	16.0	68.5	22.2	6.9
Sm	15.2	1.3	1.2	0.88	11.6	5.9	38.4	54.0	17.6	4.1	4.7	9.1	0.97	1.8	2.6	3.2	7.9	4.3	1.7
Eu	3.21	0.30	0.29	0.23	3.63	1.12	8.0	11.8	3.7	1.02	1.16	1.94	0.70	0.45	0.62	0.73	3.73	1.09	0.45
Tb	1.58	0.1	0.18	0.10	2.38	0.76	3.1	6.6	2.5	0.95	0.76	1.13	0.13	0.26	0.38	0.44	2.26	0.66	0.25
Tm	0.69	[0.06]	0.14	-	1.33	0.47	1.6	2.8	1.4	1.28	0.34	0.4	0.04	0.1	0.32	0.2	0.67	0.3	0.17
Yb	3.6	0.45	0.98	0.65	9.72	2.68	7.9	13.8	8.4	9.0	2.51	2.31	0.38	0.77	1.68	1.66	3.40	1.95	0.83
Lu	0.38	[0.05]	0.20	0.11	1.78	0.42	-	1.3	0.96	1.69	0.48	0.33	0.06	0.11	0.25	0.25	0.63	0.02	0.15
Ce/Nd	4.0	2.6	2.3	0.8	5.8	1.1	12.5	8.8	13.7	2.2	1.05	1.4	2.1	1.6	3.0	2.7	1.7	2.5	4.9

Details of sediments: 1. Mudstone rift margin succession; 2. Pelagic carbonate within pillow lavas; 3. Mudstone above lavas; 1-3. Upper Triassic, Antalya Complex (S.W. Turkey); 4. Pink pelagic calcilutite on lavas; 5. Manganese crust above lavas; 4-5, Upper Triassic Mamonia Complex (S.W. Cyprus); 6. Mn-rich mudstone above agrilia lavas, Othris (Greece); 7. Manganese nodule (W. Sicily); 8. Manganese crust (W. Sicily); 9-10. Manganese crusts in condensed pelagic crust (W. Sicily); 9-10. Manganese crusts in condensed pelagic succession, Austria; 11. Black siltstone; 12. Red shale; 13. Radiolarite; 11-13. ? Lower Cretaceous, passive margin succession, Mamonia Complex (S.W. Cyprus); 14. Dark chert; 15. Orange mudstone;
16. Orange mudstone; 17. Black mudstone; 18. Red Clay; 14-18. ? Lower Cretaceous passive margin succession, Baër Bassit (Syria); 19. Black Mn-rich siltstone within radiolarian cherts, Upper Jurassic Lower Cretaceous, Antalya (Turkey).
Source of data: a. Robertson, 1981; A.J. Fleet and A.H.F. Robertson unpublished.
Full analysis given in Table 1.

involving alteration of ultramafic rocks exposed on the seafloor (Abbete et al., 1972). Above the breccias, sulphide-bearing stockworks are located in the overlying mafic pillow lavas and lava breccias. Similar sulphides, some with intact stockworks, lie along the lava-sediment surface. Mineralogy, although variable, is dominated by pyrite, pyrrotite, chalcopyrite, with minor sphalerite, marcasite, magnetite, haematite, arsenopyrite, mackinawite, bornite, gold and cubanite.

Apennine Manganese Ores

Along the ophiolite-pelagic sediment contact, or slightly higher in the succession, are manganese ores which were once economic (Cortesogno et al., 1979). As summarised by Bonatti et al. (1976c), two types of deposit are found (e.g. at Gambetesa and Molinello). The first type is massive ore in lenses up to 30m thick and 10-200m long, often quartz-veined, fractured or brecciated, grading laterally into the second type. This is banded ore deposits forming mineralised layers interstratified with radiolarian cherts up to several metres thick and persistent laterally from 10-200m. Some massive manganese ores are also reported from the axial zones of major fold hinges.

Mineralogically, according to Bonatti et al. (1976c), the manganese ores consist of variable amounts of quartz and braunite, but small cross-cutting veins are much more heterogenous, including manganite, pyrolusite, calcite, rhodonite, rhodochrosite, barite, tinzenite, parsettensite and piedmonite. These ores are slightly more metamorphosed than many counterparts in the Eastern Tethys (see below).

Chemically, most of the Apennine manganiferous ores show high values of Mn and Si, but low values of almost all other elements, relative to shale or deep sea sediments. As shown in analysis 13 (Table 3, Fig. 14), Ba is very enriched, also to some extent Cu, but other trace elements exhibit low values. Bonatti et al. (1976c) give U and Th data. While U in seawater is relatively abundant, Th is very depleted because of low solubility. The Apennine manganese ores are relatively depleted in Th. Consistent with this the REE trends (Table 4, analysis 8: Fig. 15) show a marked negative Ce anomaly, indicative of precipitation from seawater with little fractionation.

Sedimentary Cover

Above the ophiolite, non-calcareous radiolarian sediments up to 300 m thick (Decandia and Elter, 1972; Folk and McBride, 1978) are composed of two alternating lithologies: siliceous mudstones and radiolarites (Barrett, 1981). The radiolarites are interpreted as siliceous turbidites redeposited from oceanic highs (Nisbet and Price, 1974), while the siliceous mudstones are interturbidite deposits which accumulated slowly in topographical lows with near-surface dissolution of radiolaria.

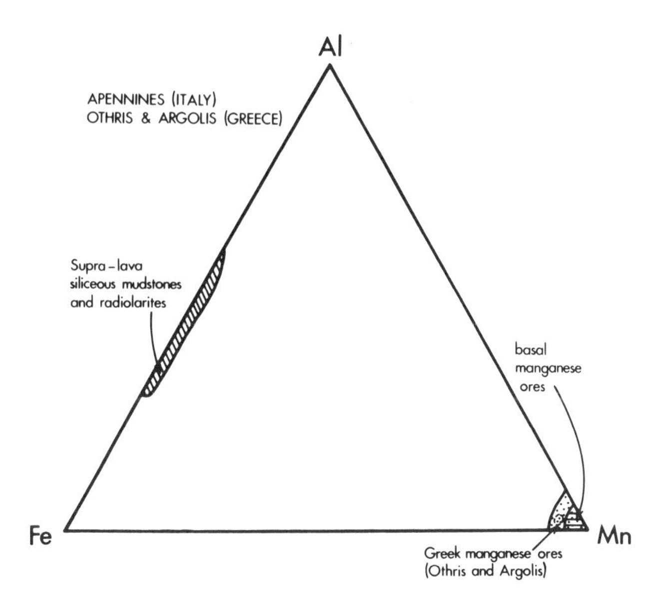

Fig. 14. Triangular plot to show major elements of metalliferous sediments associated with the Apennine ophiolite and Greek manganese ores.

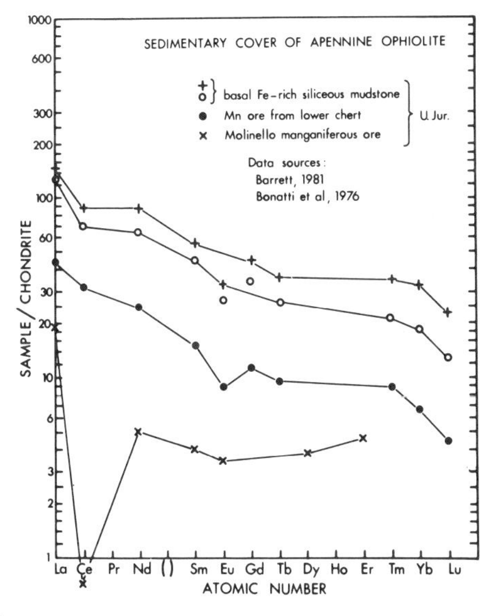

Fig. 15. Rare Earth Elements in metalliferous sediments associated with the Apennine ophiolite.

Barrett (1981) also shows that, while silica in the radiolarian succession remains relatively constant, the distribution of Fe is clearly related to the ophiolitic basement (Table 3, analyses 14). Increased Al and Ti values in some basal sediments are attributed to a volcaniclastic component again derived from the ophiolite. Higher Mg values lower in the succession reflect the presence of chlorite partly derived by alteration of basalt. Higher in the sequence, correlations of major and trace metals with Al imply a fine grained terrigenous component. Trace metals show low and relatively constant values except locally close to the ophiolite (Table 3, analysis 15; Table 4, analyses 9, 10). Pb isotope data (Barrett, 1980) confirm that the pelagic sediments contain varying proportions of basalt-derived Pb and more radiogenic Pb which could have been derived from fine grained detritus, radiolarian silica, or from seawater.

Metallogenesis on a Rifted Spreading Axis

An origin as a transform fault zone has been invoked to explain the highly faulted nature of the Ligurian Apennine ophiolites (Cortesogno et al., 1980). On the other hand Barrett and Spooner (1977) and Barrett (1981) stress the absence of ophiolite-derived clastics above the basal pelagic sediments as favouring formation on a strongly rifted spreading ocean ridge, rather than a fracture zone (Fig. 13). In either case spreading probably took place in an elongate strike- slip -controlled basin, analogous to the Gulf of California (Kelts, 1981). The intense faulting, 'basal breccias' and exposure of intrusives, including serpentinite, reflect a rifted setting (see also Garrison, 1974). Stratiform massive and disseminated sulphide

---

Table 3. Tethyan ophiolite-related metal-oxide sediments and Greek manganese ores.

1. Interstitial sediment in lavas, Margi; 2. Ochre; 3. Basal pale brown umber; 4. Higher dark umbers; 1-4. Skouriotissa; 5. Mudstone, Arakapas transform fault; 6. Radiolarite, Kalavassos; 1-6 Upper Cretaceous, Troodos Massif, Cyprus; 7. Interlava umber, Wadi Jizi; 8. 'Mound' gossan, Yanbu; 9. Black umber around 'Mound', Yanbu; 10. Supra-lava umber, Suhaylah; 11. Supra-'arc' umber, Mulayyinah; 7-11. Upper Cretaceous, Semail Nappe, Oman; 12. Umber, ? Upper Cretaceous, Baër-Bassit (Syria); 13. Mn mineralisation, Molinello; 14. Lower siliceous mudstone, Elba; 15. Radiolarite, Elba; 13-15. Upper Jurassic, Apennines; 16-17. Average of 21 samples from Argolis, Palaeo-Epidauros (Greece).
Sources of data: a. Robertson and Hudson, 1973; b. Robertson, 1976; c. Robertson, 1978; d. J.F. Boyle, unpublished, e. A.H.F. Robertson, unpublished; f. Bonatti et al., 1976; g. Barrett, 1981; h. Parrot, 1977; i. Panagos and Varnavas 1980.

Major oxide analyes to one decimal place only.

Table 3

Analysis No.	1	2	3	4	5	6	7	8	9	10	11	12	13	14	15	16	17
Sample No.	26A	351	1	2	30	10	74	244	226	15	44	7312 -801	2-17	1c	1b	-	-
SiO_2	39.6	8.8	13.6	15.1	34.7	93.1	57.6	83.5	40.0	50.9	69.6	33.0	31.0	65.5	92.5	-	-
Al_2O_3	14.5	6.2	3.7	1.8	7.5	2.4	5.1	0.3	1.6	7.1	9.1	10.3	<.1	18.6	2.8	2.6	1.3
Fe_2O_3 (tot)	27.3	53.3	62.2	43.9	46.8	2.0	13.6	9.5	15.8	26.3	9.8	25.3	<.1	16.4	2.4	1-2	2.8
MgO	13.0	1.2	0.8	0.8	3.0	0.3	0.8	0.1	0.4	2.4	1.4	3.8	<.1	2.8	0.7	6.9	0.92
CaO	1.8	6.2	5.5	10.0	2.3	0.5	16.2	6.8	1.2	1.7	2.9	1.5	2.9	0.7	<.1	0.95	2.5
Na_2O	1.7	-	-	-	0.1	-	0.1	0.1	0.1	0.3	0.7	0.5	<.1	<.1	<.1	-	-
K_2O	0.9	0.9	6.8	0.7	0.9	0.4	0.1	<0.1	0.2	1.7	2.3	1.2	0.1	2.5	0.5	-	-
TiO_2	0.2	0.2	2.1	1.5	0.3	<0.1	0.1	<0.1	1.2	0.3	0.4	0.3	<.1	0.4	0.8	0.04	0.04
MnO	0.7	0.9	2.0	18.5	0.6	0.9	5.5	1.2	26.2	7.3	2.6	4.3	59.5	0.2	0.1	34.2	63.0
P_2O_5	0.4	-	5.9	5.0	1.0	0.1	0.3	<0.1	0.1	0.2	0.3	0.9	-			-	-
Total	101.1	-	-	-	97.1	-	99.7	100.6	95.9	98.4	99.1		-	-	-		
Ba	32	81	756	1190	147	40	80	13	80	216	275	638	3500	140	58	-	-
Co	-	162	129	133	148	16	-	-	-	-	-	133	80	42	42	110	493
Cr	4	28	8	9	135	24	17	6	63	38	45	112	<10	41	10	26	29
Cu	1183	2098	803	1400	10	67	336	15	1361	527	216	-	310	55	44	188	422
Ni	319	368	336	368	204	24	233	17	160	303	94	268	39	130	42	138	192
Pb	15	360	179	190	60	51	99	11	211	236	95	334	-	33	3	46	29
Sr	57	204	-	-	18	-	217	33	325	493	425	488	-	17	20	-	-
V	985	529	886	689	597	38	240	80	59	468	193	1481	-	-	-	-	-
Zn	314	844	467	381	586	134	233	5	142	299	127	185	50	130	69	93	144
Zr	55	766	1102	798	200	31	48	<5	43	94	96	232	-	78	21	-	-

Table 4. Rare Earth Elements (REE) of Tethyan ophiolite-related oxide sediments.

Analysis No.	1	2	3	4	5	6	7	8	9	10
Sample No.	2052	303	347	718	30	341	74	Z-17	V-3	M-4 87
La	11	55.0	95.5	535	50.2	78.6		1.72	29.8	25.5
Ce	4.7	17.2	24.3	72.7	17.7	138.2		5.0	48.0	55.9
Nd	11.5	66.1	95.2	319	66.3	73.0		2.74	30.7	22.4
Sm	2.5	14.4	20.7	45.7	14.7	14.3		0.68	6.6	4.5
Eu	0.7	3.5	5.1	13.6	3.6	3.5		0.173	1.51	0.9
Tb	0.5	2.0	2.8	7.9	2.4	2.4		-	0.94	0.5
Tm	0.4	1.1	1.3	3.1	1.2	0.9		-	0.53	0.3
Yb	2.6	5.9	7.9	19.6	7.1	5.9		0.23	3.10	1.9
Lu	0.4	1.0	1.0	2.6	1.2	0.9		0.029	0.49	0.3
Ce/Nd	0.3	0.2	0.2	0.2	0.2	1.3		1.8	1.5	2.5

1. Interstitial sediment in lavas; 2. Volcaniclastic ochre, 1,2 Mathiati; 3. Basal umber, Skouriotissa; 4. Typical umber, Drapia; 5. Mudstone, Arakapas fault zone; 6. Bentonitic clays above lavas; 1-6. Upper Cretaceous, Troodos Massif; 7. Typical supra-lava umber, Upper Cretaceous, Semail Nappe, Oman; 8. Mn-mineralisation, Molinello; 9. Basal siliceous mudstone, Elba; 10. Upper siliceous mudstone, Elba; 8-10. Upper Jurassic, Apennine ophiolite (Italy).
Sources of data: a. Robertson and Fleet, 1976; b. A.J. Fleet and A.H.F. Robertson, unpublished; c. Bonatti et al., 1976; d. Barrett, 1981.

ores were precipitated from hydrothermal solutions at several stratigraphical levels in the extrusives and ophiolite-derived sediments. The manganese ores could have formed from hydrothermal solutions in two ways. The manganese could have separated from iron during massive sulphide deposition, followed by ponding as precipitates in hollows in the lava surface; alternatively, the ores could reflect deposition from cooler more dilute solutions released well after the end of massive sulphide formation in the area. The manganese ores which overlie radiolarites near the base of the pelagic succession could be of the second type. The overlying pelagic siliceous muds, which were deposited below the calcite compensation depth, show an upwards decrease in volcanic exhalative components and an increase in distal terrigenous matter. Much later, some of the manganese was apparently remobilised and concentrated in major fold hinges.

UPPER CRETACEOUS TROODOS SPREADING AXIS

The stratigraphic settings of the cupriferous sulphides and oxide-sediments of the Troodos Massif, Cyprus, are summarised in Fig. 13a.

Stratiform Cupriferous Sulphides

Most of the major massive sulphide orebodies and the associated ferruginous ochres, are located along the contact between the Lower and the Upper Pillow Lavas (Constantinou and Govett, 1972). By contrast, the ferromanganiferous umbers occur both as dispersion halos around the sulphides, and as small deposits ponded in hollows in the surface of the Upper Pillow Lavas (Robertson, 1975) (Fig. 13a). Minor porphyry copper-gold deposits are also known (Kambos, Laona), one associated with a quartz-diorite plug (Hollister, 1982).

2cm
0·5 m

Fig. 16. (i) Field photographs of Cyprus metalliferous deposits. a. Typical field relationships of umbers in small hollows in the lava surface. Pillowed and thin sheet flows with abundant inter-pillow Fe-oxide sediment are overlain by black umber (middle left) Bentonitic clays and radiolarian mudstone behind are tree-covered. In the distance the chalk cover forms a low scarp (Margi, Nicosia District); b. Umber has been mined out of a major fault-controlled hollow in the former ocean floor topography. Along the right wall lavas became detached and slid down originally almost vertical fault scarps; c. Crudely graded submarine scree-breccia with abundant orange Fe-oxide interstitial sediment. Wall of umber pit in b; d. Disintegrated pillow lava segment in spalled hyaloclastite. Note pale altered chilled margin of original larger pillow; b-d Lymbia umber pit, Nicosia District.

3 cm
0.1 m
0.2 m

Fig. 16. (ii) Field photographs of Cyprus umber and ochre deposits; a. Fe-Mn rich dark umber in typical small hollow floored by pillow lava breccias near Kalavassos, Limassol District; b. Detail of umber showing two thin siliceous clay-rich bands altered volcaniclastic silt (pale), Troulli umber pit, Laranca District; c. Detail of finely laminated umber; b-c Troulli umber pit, Larnaca District; d. Bedded ochre composed mostly of Fe-oxide with detrital pyrite grains, Skouriotissa opencast, Nicosia District.

The massive cupriferous ores, dominated by pyrite and chalcopyrite, are mostly conglomeratic and formed in seafloor depressions underlain by downward-tapering stockworks of disseminated ore and otherwise mineralised lava (Searle, 1972; Constantinou and Govett 1973; Constantinou, 1976). The compositions and temperatures of the sulphide-precipitating solutions have been determined by H, O, Sr, Pb and S isotope data, supplemented by fluid-inclusion work (Spooner et al., 1974; Heaton and Sheppard, 1977; Spooner, 1977; Spooner and Bray, 1977; Chapman and Spooner, 1977; Spooner and Gale, 1982).

Detailed studies of the south-Troodos Kalavassos orebody have revealed a series of elongate massive orebodies and stockwork zones controlled by major normal-faults (Adamides, 1980). Many small stockworks in the Lower Pillow Lavas lie along relatively continuous fault lineaments (e.g. Margi area, Boyle and Robertson, in press) and can be related to the extensional tectonics of a spreading axis (Adamides, 1975).

Ferruginous Ochres

The ochres are brightly coloured ferruginous, manganese-poor, metalliferous oxide-sediments restricted to immediately around and above the massive sulphide ore-bodies (Constantinou and Govett, 1972). Ochres are distinct from numerous gossans, which formed by subaerial oxidation of sulphide ore. Four distinct ochre types exist (Robertson, 1976).

First, there is massive and pseudo-conglomeratic ochre, as formerly exposed at Mathiati (East Troodos) and Mousoulos (South Troodos). This ochre formed a thin carapace up to several metres thick above the massive sulphides and was interpreted by Constantinou and Govett (1972) as a product of submarine weathering prior to extrusion of the overlying unmineralised Upper Pillow Lavas. The pseudo-conglomeratic texture suggests some degree of sedimentary transport in addition to collapse upon oxidation.

The second form of ochre comprises brown to grey or orange ochreous metalliferous siltstones, mudstones, clays and volcaniclastics, with nodules of chalcedonic-chert formed by replacement (Skouriotissa). This ochre, which contains detrital sulphide, shows sedimentary structures indicative of transport over a surface of mineralised lavas and sulphides (Fig. 16 (ii) d). Mineralogically, the sediment contains crystalline goethite, minor quartz, feldspar (probably authigenic), mixed-layer clays and some opal-CT. Apatite and palygorskite are present in some samples. In addition to Fe, these ochres are very strongly enriched in Cu and Zn and many other trace elements (Table 3, analysis 2), relative to normal pelagic clays.

The third type of ochre comprises very finely laminated brown or greyish oxide-sediments, often with veins of selenitic gypsum. These mudstones lie above massive sulphides (Mathiati, east Troodos), or are interbedded with mineralised lavas adjacent to the orebody (Skouriotissa). Minerals are mainly crystalline goethite, gypsum, minor quartz and smectite. In these oxide-sediments, manganese and trace element values are higher, closer to those of ferromanganiferous umbers (Mn to 1.2%, Skouriotissa).

The fourth ochre type consists of orange or red ferruginous veins and interstitial oxide-sediments within mineralised pillow lavas (e.g. Skouriotissa). Mineralogically, goethite and haematite predominate; Mn and trace metals exhibit low values.

Each of the ochre types, which is composed of ferruginous oxide-sediments, is enriched in light relative to heavy REE and have normalised trends with distinct negative Ce anomalies (Robertson and Fleet, 1976). Absolute values fall between the higher values typical of umbers and the lower values in vein and interstitial sediments within unmineralised lavas (Table 4, analysis 2; Fig. 22).

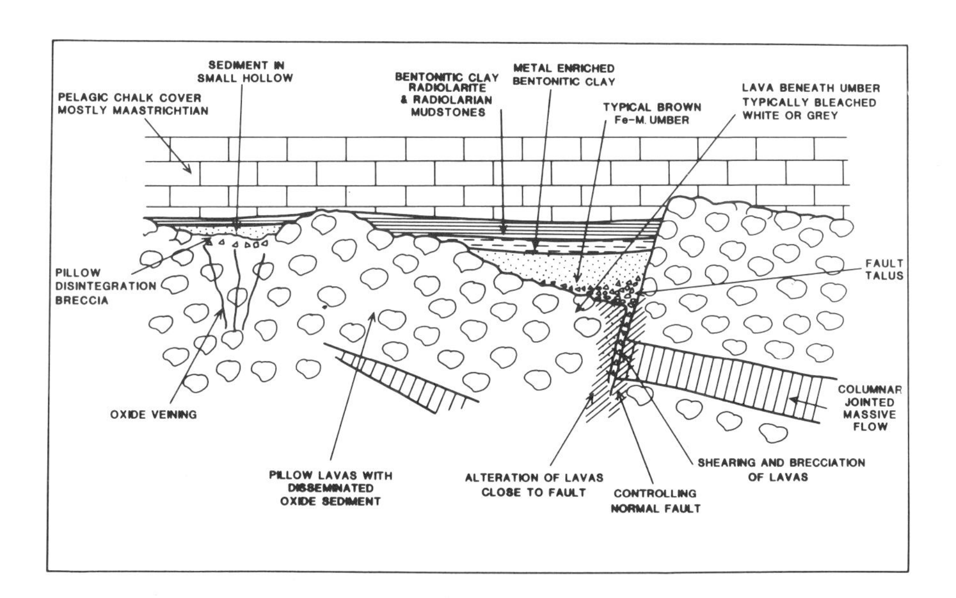

Fig. 17. Reconstruction of the field relationships of a typical small umber hollow related to seafloor faulting, Troodos Massif.

Disseminated Oxide-Sediments

Orange often calcareous ferruginous oxide-sediments, interstitial to pillow lava, are irregularly distributed throughout both the Lower and the Upper Pillow Lavas. Greater abundances are found in areas of sulphide mineralisation. For example, the abundance of metalliferous oxide-sediment in the mineralised Margi area (east Troodos) contrast with virtual absence in the unmineralised Akaki River section (north Troodos). Ferruginous oxide-sediment between pillows and in veins is also widespread in the highest levels of the Upper Pillow Lavas, particularly below umber deposits (see below). Mineralogically, these oxide-sediments contain calcite, goethite, minor mixed-layer clays, principally smectite and analcite; Mn and trace elements generally exhibit low and variable values (Table 3, analyses 1), but extreme Cu and Ba enrichments are observed. REE plots again show a distinct negative cerium anomaly (Table 4, analysis 1; Fig. 22). Locally, the metalliferous sediments are silicified and haematite has replaced goethite.

Ferromanganiferous Umbers

The typical Fe, Mn umbers (Elderfield et al., 1972; Robertson, 1975; Guillemot and Nesteroff, 1980) are finely laminated oxide-sediments distributed in small hollows in the surface of the unmineralised Upper Pillow Lavas (Figs. 16, 17). Detailed mapping (Boyle and Robertson, in press) reveals field relations as summarised in Fig. 17. The umbers typically lie in small fault-bounded hollows in the lava surface. The lavas immediately beneath the umbers, often comprised of fault-talus, are generally highly decomposed chemically altered and impregnated with iron oxide-sediment. The

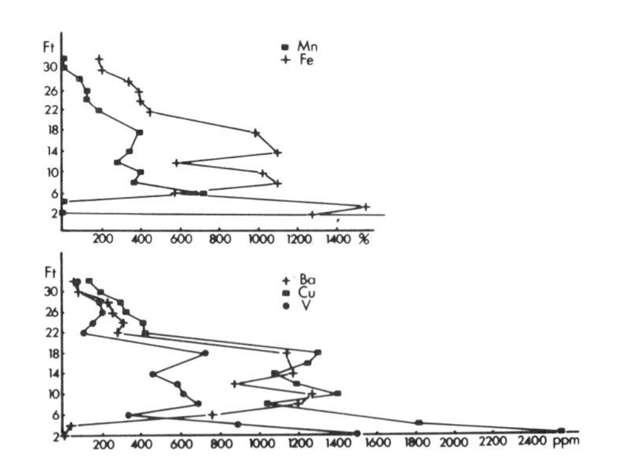

Fig. 18. Vertical variation in element content of ferruginous and ferromanganiferous umbers from Skouriotissa, W. Troodos (Cyprus), after Robertson (1976).

basal umbers, up to several centimeters thick, are typically bright orange, then overlain by finely laminated darker umber. Higher in the succession, which is usually several metres thick, but locally up to 30 m, the umbers sometimes contain siliceous mudstone intercalations, small black manganiferous concretions and nodules of chalcedonic quartz chert of replacement origin.

The umbers generally pass conformably upwards into several metres of pink to grey thin bedded 'ribbon' radiolarian cherts and mudstones of Campanian age (Robertson and Hudson, 1974; Table 3, analysis 6). Elsewhere (e.g. Skouriotissa), the umbers become more argillaceous upwards, then pass into grey bentonitic clays, dated Campanian -Maastrichtian (Fig. 17). Exceptionally, at Skouriotissa, where the umbers overlie sulphides and mineralised lavas directly, the basal ferruginous umbers up to several metres thick contain chert and volcaniclastic mudstone, then pass conformably up into very manganiferous umber (Robertson, 1976).

Mineralogically, the umbers are generally poorly crystalline, dominated by goethite and quartz with subordinate mixed-layer clays and occasional opal-CT (Desprairies and Lapierre, 1972). Current studies (J. Boyle) also reveal minor feldspar (possibly authigenic) and veins of palygorskite and MnO_2.

The typical chemistry of umbers (Table 3, analysis 4) is ferromanganiferous with strong trace metal enrichment relative to pelagic clays (c.f. Turekian and Wedepohl, 1961), particularly in Ba, Co, Cu, Ni, Pb, V and Zr (Robertson and Hudson, 1973). The basal Mn-poor orange ochres are intermediate in composition between the overlying umbers and true ochres associated with the sulphides (Table 3, analysis 3). At Skouriotissa, the metal content wanes upwards as shown in Fig. 18 (Robertson, 1976). Close correlations exist, for example between Fe and Cu, Mn and Ba and Al_2O_3 versus Fe_2O_3 (Fig. 19a, b, c). Comparisons of the major and trace element content with other Troodos metalliferous oxide-sediments are plotted in Fig. 20. The most notable feature of the umber REE chemistry is the high absolute values (Table 4, analysis 4), coupled with a very marked negative cerium anomaly. The magnitude of this Ce anomaly varies inversely with the detrital content (Robertson and Fleet, 1976; Fig. 22 a, b).

Varnavas (1980) investigated the partitioning of metals in the Fe-poor basal umbers (Skouriotissa) and the overlying Fe-Mn umbers (Skouriotissa, Archangelos), by selective leaching with acetic acid, and hydroxylamine-HCl. The chief conclusion was that in typical Fe-Mn umbers Ba, Co and Pb associate with the manganese mineral phase, while Zn, Cu, Al and Mg relate to iron oxy-hydroxides and clay minerals. In the localised basal Mn-poor umbers above the sulphides (Skouriotissa), Ni, Co, Pb, Zn, Cu, Al, Mn all associate with the ferric oxy-hydroxide and clay-mineral phases. Ba in the insoluble residue implies the existence of barite, or a barian feldspar. Associations

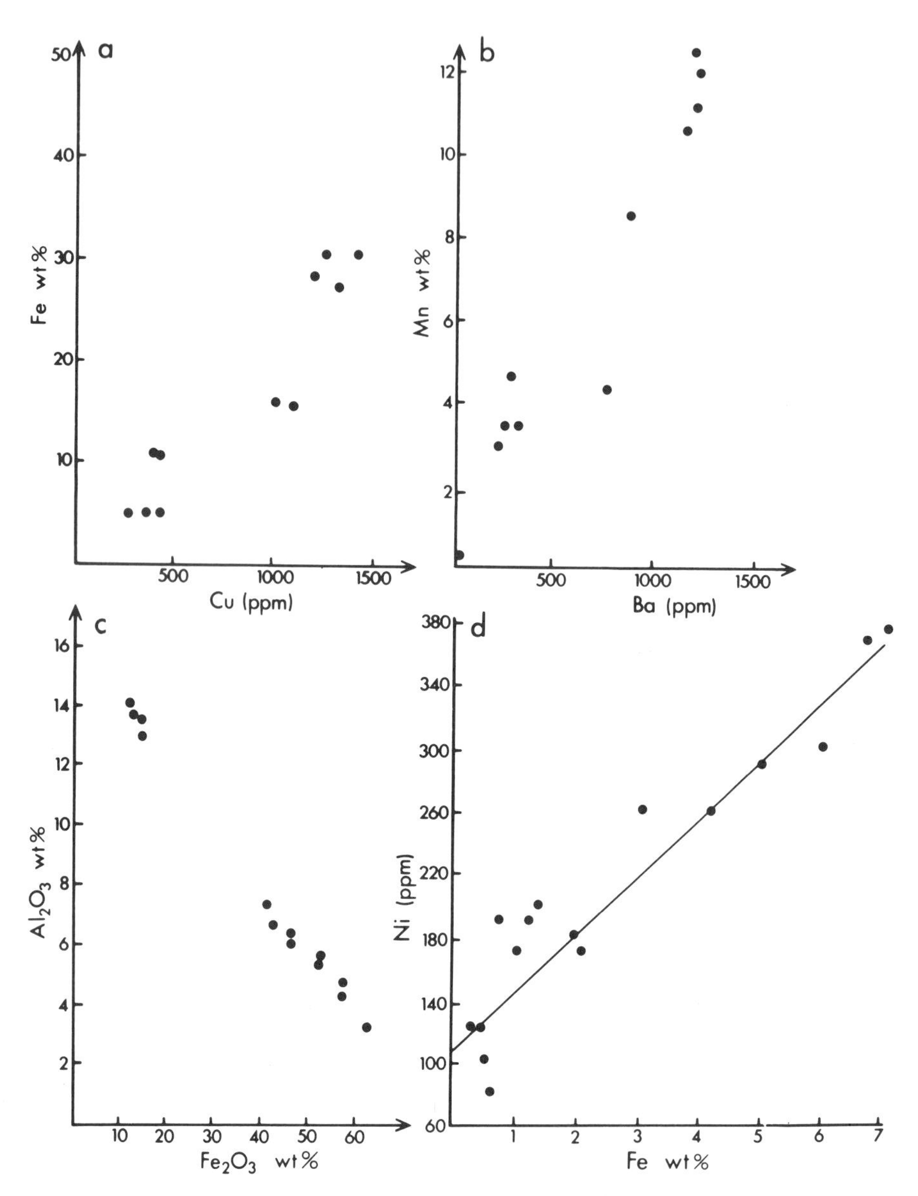

Fig. 19. Variation diagrams of major and trace elements in Greek manganiferous ores and Cyprus umbers; a. Fe verus Cu; b. Mn versus Ba; c. Al_2O_3 versus Fe_2O_3; a-c Cyprus umbers; d. Ni versus Fe, average of 19 analyses from Ermioni, Greece; data sources: a-b Robertson and Hudson, 1973; c. J. Boyle unpublished; d. Panagos and Varnavas, 1980.

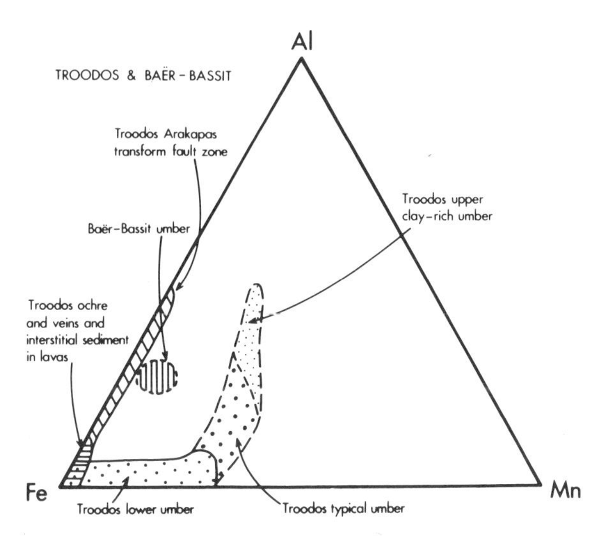

Fig. 20. Triangular plots of Troodos (Cyprus) and Bäer-Bassit (Syria) metalliferous oxide-sediments.

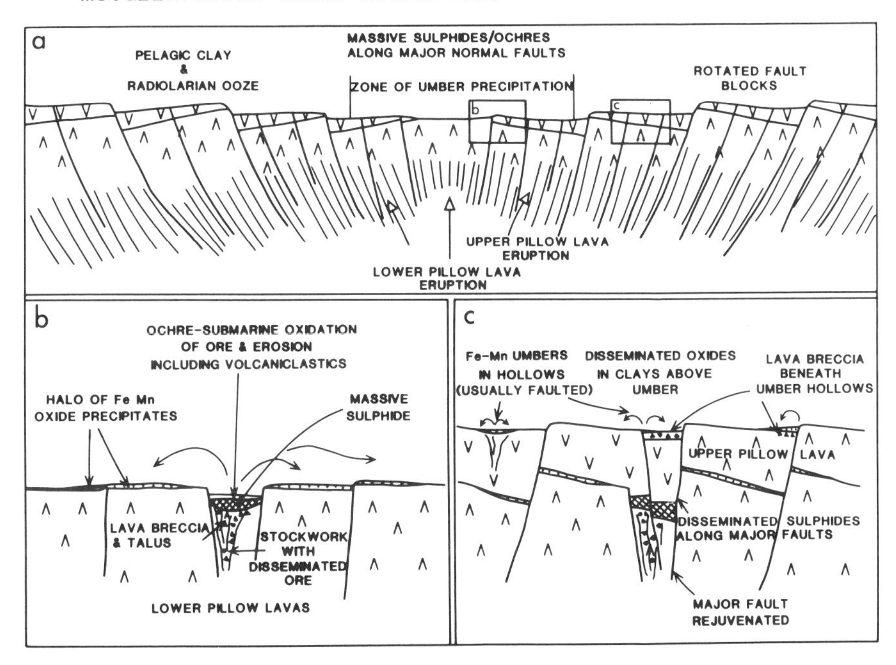

Fig. 21. Reconstruction of the setting and origin of the various Troodos (Cyprus) metalliferous sediments. a. Inferred sites of genesis of the umbers, ochres and cupriferous sulphides; b. formation of cupriferous sulphides with oxide dispersion halos at the contact between the Upper and Lower Pillow Lavas; c. Setting of the ferromanganiferous umbers formed in the highest levels above the Upper Pillow Lavas.

correspond generally to correlations of major and trace elements reported by Robertson and Hudson (1973; Fig. 19 a, b).

Arakapas Transform Fault

Distinctive metalliferous oxide-sediments are intercalated with the volcaniclastic fill of the Arakapas Fault Zone of South Troodos, interpreted as a fossil transform fault (Simonian and Gass, 1978). Red, purple to brownish-grey finely laminated oxide-sediments are intercalated with redeposited volcaniclastic sands and silts within the transform fault depression. Relative to Arakapas basalt, the local detrital material, the oxide-sediments are strongly enriched in Fe, Si, K, Mn and P while Al, Ba, Mg, Ca, Cr, Sr and Ni are all depleted. However, relative to the overlying typical ferromanganiferous umbers, all these oxide-sediments from the Arakapas fault belt are strongly depleted in Mn and trace metals (Table 3, analysis 5). Although varied, several samples show negative cerium anomalies in normalised plots (Robertson, 1978; Table 4, analysis 5).

Origins of Troodos Oxide-Sediments

The combined field, isotope and fluid-inclusion data show that the sulphide ores were derived by deep leaching of mafic ophiolitic rocks, with an important component derived from seawater. The solutions were essentially Cretaceous seawater with 3.5 wt % total dissolved solids, heated to ca. 300-370°C (Spooner and Bray, 1977). Sites of metalliferous deposition relative to the Troodos spreading axis are summarised in Fig. 21.

The sulphides formed from solutions released along major faults located close to sites of lava extrusion near the spreading axis. Ochres formed by seafloor oxidation of sulphides, by erosion of the ores and adjacent lavas, and as ferruginous manganese-depleted precipitates from hydrothermal solutions released during and soon after the sulphide formation. Any manganese precipitated at this time was remobilised during later volcanism and hydrothermal activity. Soluble manganese was oxidised more slowly than iron, so that manganese tended to be dispersed and accumulated slowly as a halo around the sulphide ore bodies. The Upper Pillow Lavas were subsequently erupted on the flanks of the spreading axis, still in the zone of intense hydrothermal activity. The umbers precipitated within and above the highest levels of these flank lavas. Solutions were discharged along numerous normal faults, often reactiviting structures in the Lower Pillow Lavas. Iron precipitated first to form the basal ferruginous umbers, then as manganese oxidised the ferromanganiferous umbers accumulated rapidly with little fractionation of REE's from seawater. The Pb and Sr isotope data also show that much of the trace element content was incorporated from seawater (Gale et al., 1981). Locally, along the Arakapas transform fault the ferruginous oxide-sediments were rapidly precipitated from hydrothermal solutions during short breaks in clastic input. Manganese was oxidised more

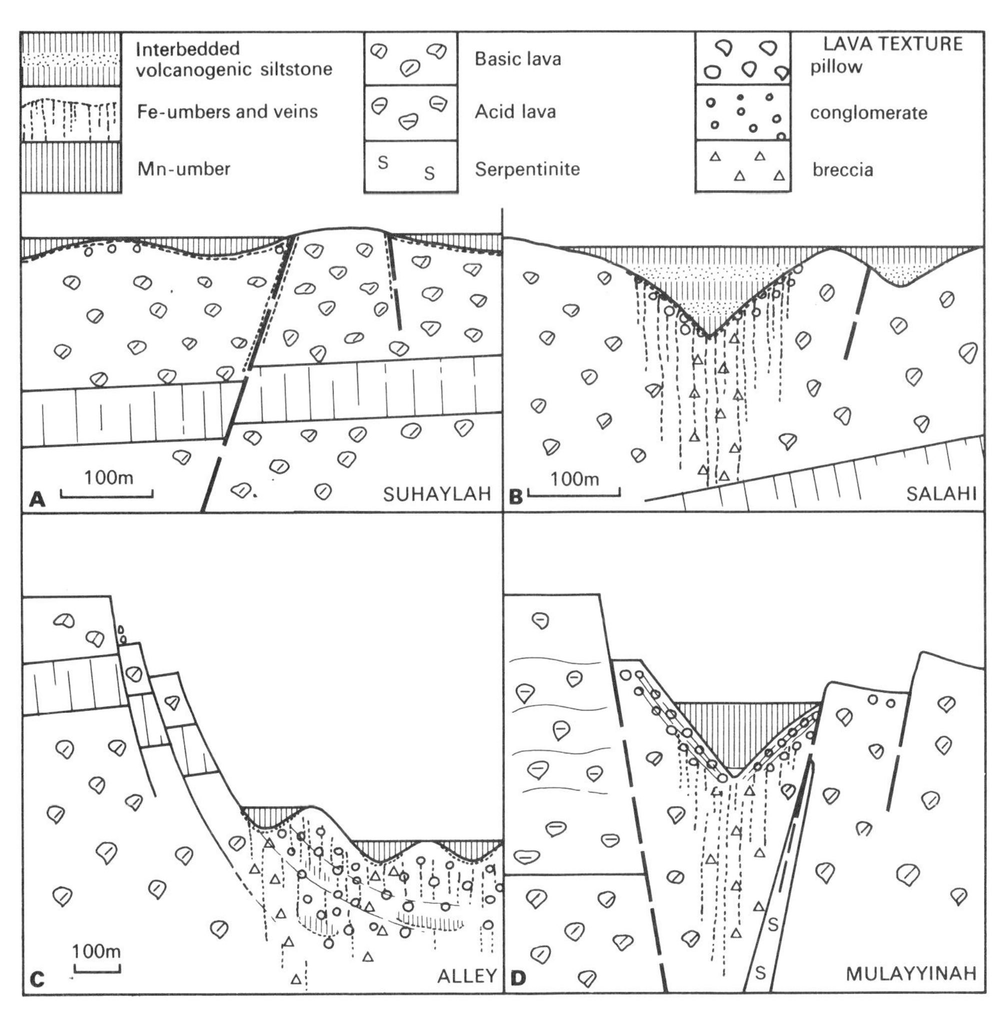

Fig. 24. Setting of ferromanganiferous umbers in the Semail Nappe Oman. a. as a relatively continuous thin blanket above the lower lavas; b. as fault-controlled hollows above the upper lavas; c. as thick ponded deposits in the major N-S trending Alley fault zone, interpreted as a major seafloor fault zone parallel to the spreading axis; d. as ponded deposits above the latest 'arc lavas' which formed restricted volcanic edifices. After Fleet and Robertson (1980)

slowly than iron and so was retained in solution, and only precipitated in the overlying umbers after volcanism ended in the area.

UPPER CRETACEOUS SEMAIL NAPPE, OMAN

Metalliferous sediments, including cupriferous-sulphides, ferromanganese and manganiferous oxide-sediments occur at various levels within and above the extrusives of the Semail Nappe, Oman (Fleet and Robertson, 1980). The Semail Nappe formed at a Late Cretaceous spreading axis, and was then thrust over the Hawasina continental margin onto the Arabian platform (Reinhardt, 1969; Glennie et al., 1973, 1974; Smewing, 1981). The general field relations of metalliferous sediments are illustrated in Figs. 13b, 24.

Sulphides and Oxides in the Axis Lavas

The lower lavas, which are mostly tholeiitic (Alabaster et al., 1980), include small scattered cupriferous sulphides, both massive and disseminated, associated with ferruginous ochres and minor ferromanganiferous umbers. The small massive sulphides are underlain by stockworks of veined, silicified, sulphide- and oxide-impregnated pillow lavas. Many of the small sulphides, now oxidised to gossan material, are conglomeratic, with intercalations of volcaniclastic silt, impure grey calcilutite and bedded ochre. In the north-south trending "Alley" zone (Smewing et al., 1977) numerous small sulphide bodies lie along mineralised fault-planes. In many areas, small sulphides are overlain by substantial columnar-jointed sheet-flows, rather than merely lavas.

Ferruginous Mound Structures

The lower levels of the upper lavas locally contain distinctive

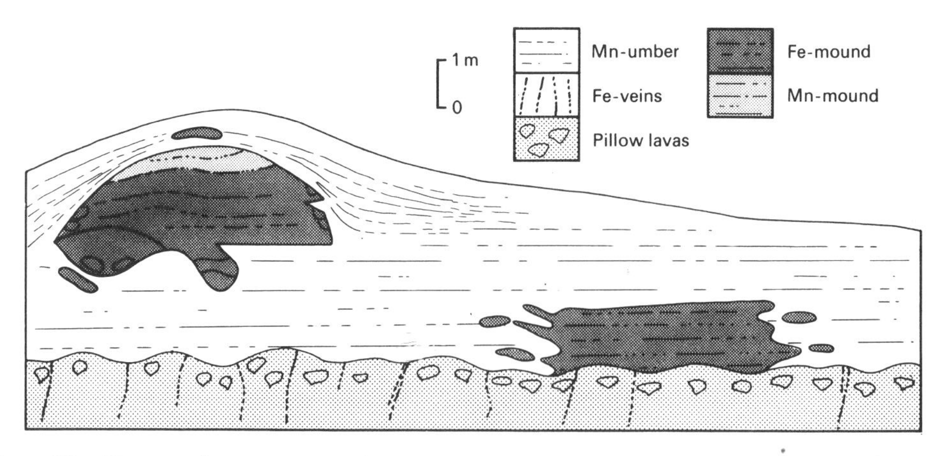

Fig. 23. Ferruginous mound structures high in the upper lavas, Semail Nappe, Oman, after Fleet and Robertson, 1980.

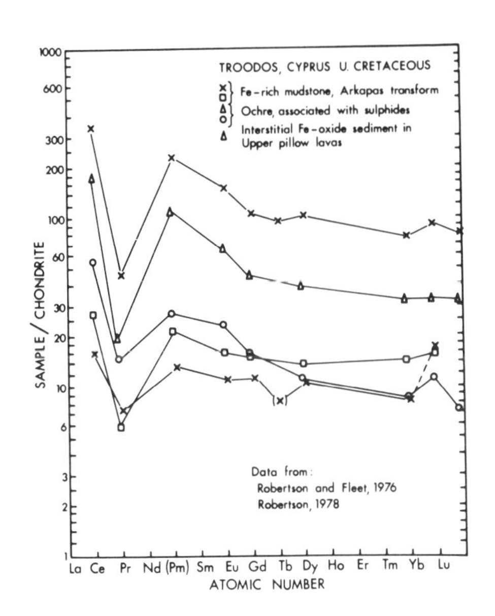

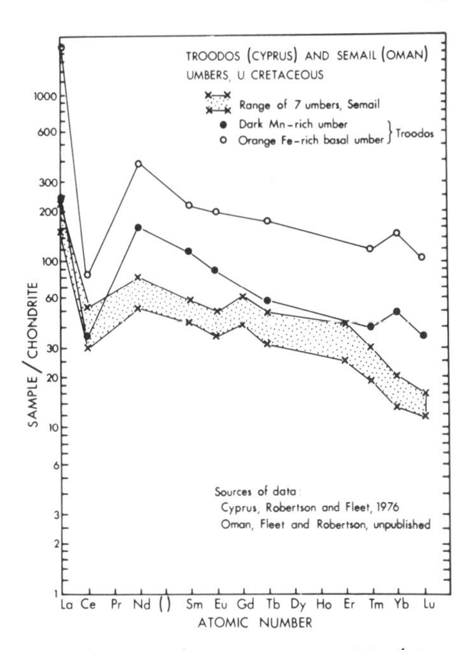

Fig.22 Rare Earth Elements in Troodos (Cyprus) and Semail (Oman) metalliferous sediments: a. Troodos; b. Troodos and Semail. Data from Robertson and Fleet (1976) and unpublished.

ferruginous mound-shaped structures surrounded by highly manganiferous umbers. As shown in Fig. 23 (Fleet and Robertson, 1980), most of the mounds are up to 3 m in diameter and 2 m high and rest directly on mineralised lavas, but in some cases the mounds are enveloped in umber. Steeper-sided mounds exhibit flanking talus admixed with umber. Otherwise mound material and umber interfinger. Internally, the mound substance is crudely stratified, sugary-textured, composed mostly of Fe-oxyhydroxide (haematite and goethite). Chemically, the mounds are strongly enriched in Fe, but highly depleted in Mn and trace metals, relative to pelagic clays (c.f. Turekian and Wedepohl, 1961; Table 3, analysis 8). The surrounding black umber is manganiferous and ferruginous with higher trace element values (Table 3, analysis 9). Normalised REE plots of these umbers show negative cerium anomalies.

Interlava Metalliferous and Pelagic Sediments

Numerous *en echelon* lenses of ferromanganiferous umbers, not spatially associated with sulphides, occur at various levels in the lavas. The lavas beneath these deposits, often include talus-breccia, which is pervasively decomposed and impregnated with iron oxide. Locally, umbers are strongly burrowed. Pelagic carbonates are present as lenses up to 2 m thick between pillowed lavas.

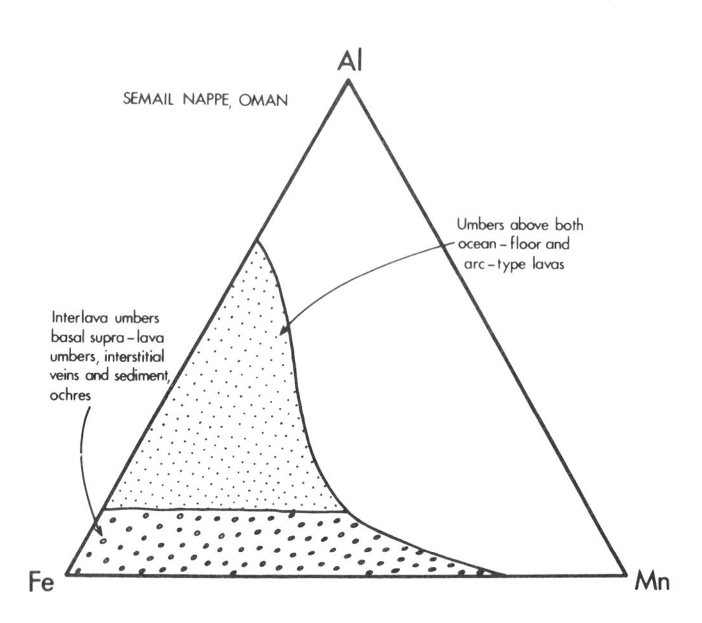

Fig. 25. Triangular plot of metalliferous sediments in the Semail Nappe, Oman. Based on unpublished analyses by A.H.F. Robertson and A.J. Fleet.

Ferromanganiferous Umbers above the Lavas

Small ponded umber deposits, similar to those in Cyprus, occur above the upper lavas. Occasional major fault-controlled hollows contain volcaniclastic sediments up to 18 m thick, in addition to umber (Fig. 24). Fe, Mn umbers above the lavas are thickest (up to 10 m) and extensive in the "Alley", a major north-south trending fault zone formed by seafloor faulting parallel to the spreading axis (Smewing et al., 1977). Locally, the umbers lie directly on the lower lavas (Suhaylah), and there siliceous umbers with nodules of quartzitic chert, formed by replacement, pass upwards through thin radiolarian mudstones, into several metres of pelagic chalk (Fleet and Robertson, 1980).

Chemically, the supra-lava umbers are comparable to the typical Troodos umbers, but are more siliceous and show generally lower trace element values (Table 3, analysis 10; Fig. 25). Normalised REE plots again show negative Ce anomalies (Fig. 22b.).

Metallogenesis Related to Differentiated Edifices

In the Semail Nappe the 'axis' lavas are overlain by discrete volcanic edifices up to several kilometers in diameter, dominated by intermediate to acidic extrusives (Alabaster et al., 1980). Any long time-break between 'axis' and 'edifice' volcanism is ruled out by the absence of significant pelagic sediments between the lava series. The differentiated edifices are underlain by several large potentially economic sulphide orebodies (Aarja, Bayda, Le Sail). Field relations of the sulphides, which are dominated by pyrite-chalcopyrite ore, are similar to those in Cyprus (Alabaster et al., 1980). Scattered umbers in the lavas around the Le Sail orebody may be dispersion halos (J.A. Pearce, personal communication, 1982), like those in Cyprus.

Metalliferous-oxide and pelagic sediments are seen in hollows in the surface of the differentiated edifice lavas (Mulayyinah). Chocolate-brown mudstones with chert and volcaniclastics pass upwards into non-calcareous radiolarian sediments largely obscured by wadi gravel. Chemical analyses (c.f. Table 3, analysis 11) show an upward decrease in Fe, Mn and trace elements, tending to the composition of normal pelagic clays (Robertson and Fleet, unpublished data).

Origin of the Semail Metalliferous Deposits

Although mineralogically similar, the Semail cupriferous sulphides differ from those of Troodos in several respects. Small sulphide bodies occur widely in the lower levels of the Semail lavas without any concentration between the lower and upper lavas as in Cyprus. Secondly, the largest sulphides are spatially linked with the differentiated volcanic edifices, which show distinctive calc-alkaline chemistry, considered by Pearce et al. (1981) to indicate genesis above a subduction zone. Intense hydrothermal discharge immediately preceded construction of these edifices apparently on older 'axis' seafloor.

The small cupriferous sulphides low in the lava pile formed from small scattered hydrothermal vents, located in broad depressions on the seafloor, which were subsequently flooded by sheet-flows. As in Cyprus, the ferromanganiferous umbers are seen both as dispersion halos around sulphide bodies and also as precipitates from hydrothermal solutions released away from the loci of sulphide deposition. The mound structures formed cemented ferruginous bodies on the seafloor surrounded by soft manganiferous-oxide sediments; bacterial cementation is a strong possibility (Fleet and Robertson, 1980). The interlava umbers are generally more siliceous than their Cyprus counterparts possibly due to slower net accumulation. Intercalated pelagic carbonates, not seen in the Troodos, indicate spasmodic eruption. The lower concentrations of Fe, Mn and most trace metals in the Oman supra-lava umbers relative to Cyprus is consistent with generally less intense hydrothermal discharge. The mudstones above the late differentiated, edifice lavas approach the composition of average pelagic clays. Progressive upward decrease in trace element content in the Oman supra-lava umbers records waning hydrothermal discharge and relative increase in background pelagic material away from a spreading axis.

COMPARISON WITH MODERN SEDIMENTS

A comparison with modern oceanic chemical sediments helps in assessment of the origins of Tethyan chemical sediments (Fig. 26). Conversely, the well exposed field relations of ophiolite-related deposits clarify many features, particularly the sub-seafloor stock-works and hydrothermal feeder systems. Also, the hydrothermal sediments of continental margin successions provides clues to the timing and location of oceanic crust which has since been destroyed by subduction. Useful comparisons can also be made with metal-rich sediments in active margin settings and orogenic belts not considered here.

Rift Metallogenesis

The Tethyan metalliferous oxide-sediments which formed during continental rifting have few modern counterparts. Both the Red Sea (Coleman, 1974) and the Gulf of California (Curray and Moore, 1982) have already progressed to oceanic crust genesis. To some extent comparable are the Fe- and Mn-rich deposits of the Afar rift, an incipient oceanic spreading axis. There, encrustations on subaerial basalt are composed of goethite, nontronite, pyrolusite, birnessite, todorokite, strontiobarite, rhodochrosite, opal and gypsum (Bonatti et al., 1972). A closer comparison might be made with the Red Sea in the Miocene (Whitmarsh, 1974), but there alkalic rift volcanism coincided with the Messinian desiccation of the Mediterranean, producing vast volumes of evaporite (Hsü et al., 1973) not seen in the

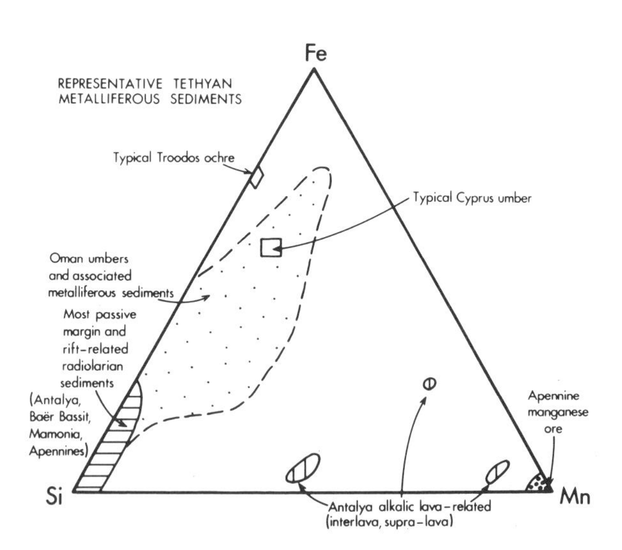

Fig. 26. Fe, Si, Mn triangular plot of representative Tethyan metal-liferous oxide-sediments.

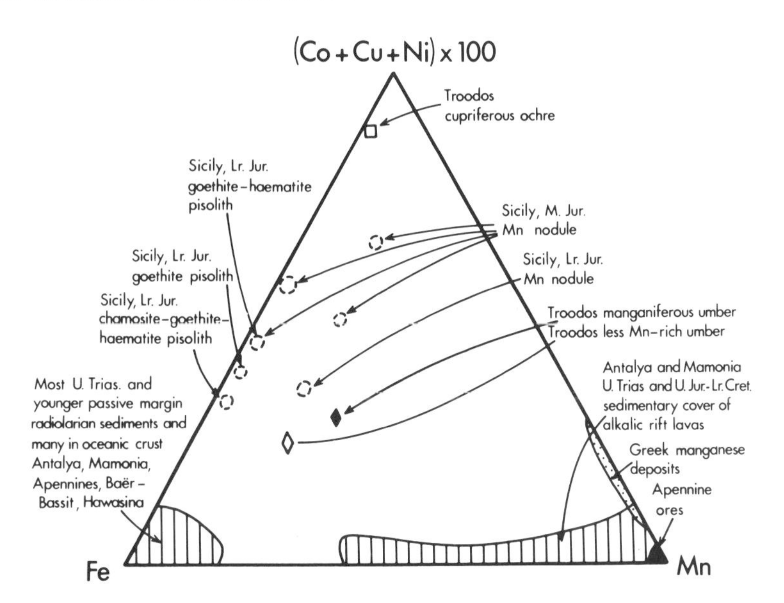

Fig. 27. Triangular plot to illustrate the trace element content of selected Tethyan metalliferous oxide-sediments.

Tethyan Triassic rifts. For this reason counterparts of the Red Sea brine pools (Craig, 1969; Bignell et al., 1976) are unknown in the Mediterranean Tethys. Manganese dispersion halos are present around the Red Sea brine pools and are also a feature of the Oman, Troodos and Apennine metalliferous deposits.

Passive Margin Metallogenesis

Most of the Tethyan passive margin slope deposits are dominated by silicate and carbonate clastic matter with little metal enrichment. Ratios of Fe_2O_3 to Al_2O_3 can be used to assess the terrigenous and pelagic clay versus hydrothermal components (Fig. 28a). The Antalya, Mamonia, Baër-Bassit and Oman (Batinah Melange) radiolarian sediments all exhibit a dominantly terrigenous detrital fraction (Fe/Al 0.16-0.6; Fig. 28a), (Price, 1977; Martin et al., 1981), contrasting with the sedimentary cover of the Apennine (Barrett, 1981), or Vourinos (Steinberg and Mpozodis-Marin, 1978), ophiolite-related sediments, which are closer to normal pelagic clays in composition (Fe/Al ca.1, Fig. 28a, b). Fe/Al ratios of these more oceanic radiolarian sediments (Fig. 28b) are comparable with the diatomites of the Gulf of California, a small ocean basin characterised by high biogenic siliceous productivity (Fe/Al ca. 0.6 Bramlette, in Cressman, 1962; Calvert, 1966). Much higher Mn/Al (0.35 cf. 60) ratios in some Gulf of California diatomites reflect important hydrothermalism as in the Guaymas Basin (Curray and Moore, 1982). Fe_2O_3-Al_2O_3 ratios of other oceanic sediments are shown in Fig. 28c

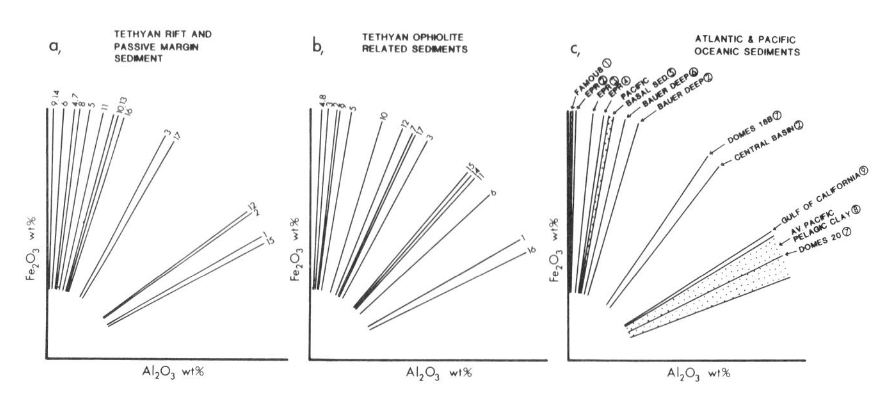

Fig. 28 Comparative Al_2O_3-Fe_2O_3 ratios for Tethyan and oceanic sediments. Pelagic sediments plots with ratios close to 1:1 Fe_2O_3-Al_2O_3. Fe-fractionated metalliferous sediments plot well above this line, and strongly Mn-fractionated metalliferous oxides sediments below the line. a. Upper Triassic to Lower Cretaceous Tethyan rift, passive margin and platform deposits, see Table 1 for details of sediments numbered; b. Tethyan ophiolite-related metal-oxide sediments, see Table 3 for details of sediments; c. Atlantic and Pacific oceanic sediments, both modern and cored by DSDP. Sources of data: 1. Hoffert et al. (1978); 2. Heath and Dymond (1977); 3. Dymond and Veeh (1975) 4. Boström and Peterson (1969); 5. Cronan (1976); 6. Sayles and Bischoff (1973); 7. Bischoff and Rosenbauer (1977); 8. Goldberg and Arrhenius (1958); 9. Bramlette, in Cressman (1962).

Within the limits of semi-quantitative data available (Jenkyns, 1970a), the Sicilian Middle Jurassic ferromanganese deposits are more comparable to some Atlantic manganese nodules and crusts in the TAG area of the Mid-Atlantic Ridge (e.g. Ni ca. 2000 ppm), than to Pacific or Indian Ocean oceanic nodules (Ni 5-6000 ppm; Fig. 28). Chemical data are thus consistent with an important local hydrothermal component, rather than, for example, the ferromanganese associated with the sediment-starved Pleistocene crusts and nodules of the Blake Plateau, western Atlantic (Mero, 1965). The widespread Tethyan pink pelagic limestone facies (ammonitico rosso, knollenkalke) can be compared with the Upper Jurassic pink pelagic claystones and limestones in the Atlantic Upper Jurassic (Tithonian) which owe their colour and iron content to slow pelagic accumulation in strongly oxidising bottom waters without an identifiable volcanic exhalative component (Murdmaa et al., 1978; Jansa et al., 1979).

Ocean Floor Metallogenesis

The manganiferous ores in the highly faulted Ligurian Apennine ophiolite can most closely be compared with the TAG hydrothermal field of the Mid-Atlantic Ridge at 26°N (Scott et al., 1974; Rona, 1980), and also to the Galapagos hydrothermal springs (Corliss et al., 1979; Figs. 26, 29).

In the TAG hydrothermal field manganese crusts up to 4.2 cm thick precipitate on volcanic breccias along the east wall of the rift valley. Diffuse low-temperature hydrothermal solutions, 0.11^{o}C above ambient seawater temperature are exhaled from an area of closely spaced faults. Chemically, levels of Co are similar to the Apennine manganese ores, but both Ni and particularly Cu are higher, probably reflecting nearby massive sulphide deposition. Although inferred at depth in the extrusives, sulphides have not been confirmed in the Mid-Atlantic Ridge TAG field (Rona, 1980). Manganese crusts 50 km from the TAG area have apparently had more time to scavenge metals from seawater and show much higher levels of Fe and trace elements (Scott et al., 1974). These are compositionally close to the Middle Jurassic Sicilian nodules (Jenkyns, 1970a). The Greek manganese ores (Panagos and Varnavas, 1980) show similar Fe_2O_3/Al_2O_3 relations (Fig. 29b) to the TAG deposits, but higher values of trace elements trending towards the off-axis Atlantic ferromanganese values.

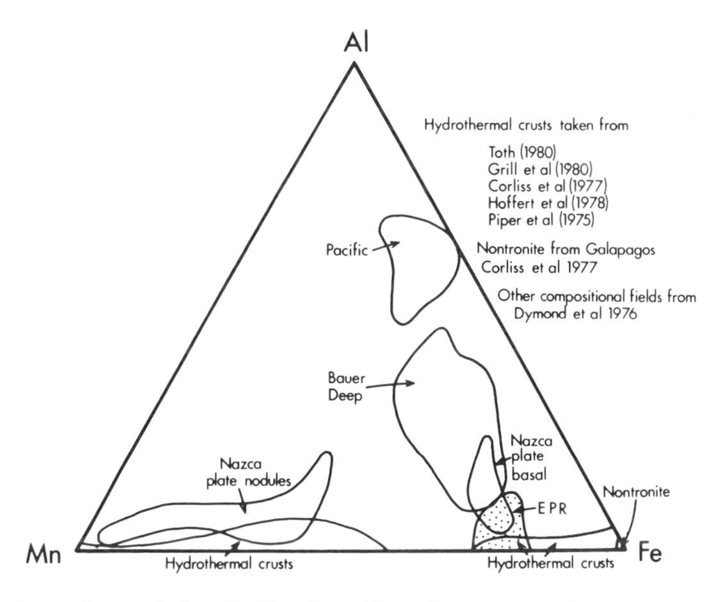

Fig. 29. Triangular plot of Al, Fe, Mn of comparative oceanic metalliferous-oxide sediments.

The less faulted character of the Upper Cretaceous Troodos and Semail ophiolites is similar to the East Pacific Rise Galapagos in Juan de Fuca spreading axes where spectacular discoveries of sulphide- and oxide-sediments have been made (e.g. CYAMEX, 1979; Hekinian, this volume). This, thus, opposes the earlier opinion that 'Cyprus-type' orebodies could only form in peculiar small ocean basins, for example marginal basins formed above subduction zones. Of the recent discoveries, most similar to the Troodos ores are the cupriferous sulphides and oxides of 00°45N, 86°07'W on the Galapagos spreading axis (Malahoff et al., 1983). Sulphide bodies up to 1 km long x 200m wide x 20 m high are located there along major faults near the wall of the median valley. The orebody, which is composed mostly of pyrite and chalcopyrite, forms numerous giant coalescing stacks up to 35 m high and 5-30 m in diameter; sulphides in places cement lava-breccias. Iron oxide-depositing vents are also reported, as well as large volumes of dispersed Fe, Mn-oxide precipitates. The iron-oxide 'vents' are apparently similar to the dispersed Fe-oxide ochre in the Troodos ophiolite.

The most spectacular discovery on the East Pacific Rise at 10° to 30°N was the sulphide chimneys precipitated from both "black-smokers" and "white-smokers" (Hekinian, this volume). S.D. Scott (oral communicaton at meeting) recalls similar structures from the Troodos Mavravouni orebody. Ochreous conglomerates associated with sulphides low in the Semail lavas could be collapsed oxidised chimneys. The numerous small sulphides and oxide deposits in the Semail lavas are similar to Cu-Ni-Zn sulphides and ochres which form mounds in depressions on the seafloor on the East Pacific Rise at 21°N (Francheteau et al., 1979). Also, the ochreous mound-structures within the Semail lavas could be considered superficially similar to the Galapagos hydrothermal mound structures but these were precipitated from dilute Fe, Mn and Si-rich solutions which percolated through pelagic sediments 20-30 m thick, 20 km off the Galapagos Rift (Corliss et al., 1977)m, while the Semail mounds formed much closer to the spreading axis during brief breaks in lava extrusion.

Chemical sediments most similar to the Troodos ferromanganiferous umbers come from high heat-flow areas close to the crest of the East Pacific Rise, particularly at 12-14°S (Boström and Peterson, 1966; Boström, 1969). Average levels of Fe, Mn, Al, Cu, Si, Cu, Cr, V and other trace metals (as well as Fe/Al ratios; Fig. 29c), are strikingly similar when considered on a carbonate-free basis. Active ridge sediments from crestal areas of lower heat-flow show lower Fe, Mn and trace element values similar to many of the supra-lava umbers of the Semail Nappe, Oman and from the Baër-Bassit ophiolite (Syria, Table 3). 'Boström-type' dispersed oxide sediments are reported from many other oceanic settings, including intra-plate seamounts (reviewed by Rona, 1978; see also Hekinian, this volume) and have recently been encountered in S W Pacific back-arc basins (Cronan et al., 1982).

Metallogenesis has also been reported from fracture zones, but this is generally restricted to minor sulphide veins and concretions (e.g. Vema and Romanche Fracture Zones, Bonatti et al., 1976a, b). Although ferruginous oxide precipitates are present along the south Troodos Arakapas transform fault, major sulphides are absent, probably because easy access of seawater cooled the oceanic crust well below the ca. 375°C needed for sulphide discharge.

The Pacific Ocean sulphides and ochres occur on fast-spreading ridge segments (> 6 cm/yr half-rate), yet spreading rates based on lava chemistry imply very slow spreading for the Troodos Massif (Nisbet and Pearce, 1974). On the other hand, the large sulphide bodies, high percentage of sheet-flows and the absence of interbedded pelagic carbonates all tend to suggest fast spreading for Troodos. Spreading rates for the Semail with many smaller sulphides and pelagic interbeds was presumably slower, while the rifted Apennine ophiolite probably formed at even slower rates similar to the Atlantic (> 3cm/yr half-rate).

CONCLUSIONS: CONTROLS OF TETHYAN METALLOGENESIS

The form taken by the various metal deposits was strongly influenced by the tectonic setting in the evolution of the Tethyan oceanic basins (Figs. 30a, b, c). The exhalative sedimentary Pb-Zn deposits associated with the Middle Triassic earlier stages of continental rifting were apparently powered by elevated geothermal gradients in areas of thinned continental crust. Most of the other Tethyan oxide- and sulphide-sediments are directly related to volcanism. The size of the heat engine available to drive hydrothermalism is largely determined by whether the whole oceanic crust or merely rift volcanics are involved. The extent of faulting controls the access of seawater to the hot crust and the degree of dilution by non-hydrothermal sediments is governed by the overall oceanographical setting. Edmond (this volume) models precipitation at ocean floor hot springs in terms of the extent to which a high-temperature end-member solution (c. 350°) interacts with cold seawater in the subsurface environment. Where the undiluted end-member reaches the sea floor, sulphides precipitate near the vent. Conversely, if subsurface dilution is great then sulphide is deposited within the volcanic basement, and only cool manganese-rich fluids emerge. Chalcophile elements, occurring abundantly in the high-temperature end-member fluid, are extracted by sulphide precipitation in the case of high subsurface dilution, and are consequently depleted in the low-temperature, manganese-rich, springs. Much of the trace element content is scavenged from seawater as manganese oxidises more slowly than iron.

The Troodos (Fig. 30c) and Apennine sulphides (Fig. 30b) record high-temperature emanations (ca. 350°C). Dispersed Fe, Mn "Boström-

Fig. 30. Block diagram summarising tectonic settings of Tethyan metalliferous oxide-sediments: a. associated with Upper Triassic rift volcanism (Antalya, Mamonia, Hawasina and Batinah, Oman); b. associated with slow-spreading rifted oceanic spreading axis (Ligurian Apennines); c. associated with a less strongly rifted faster spreading axis (Troodos, Baër-Bassit, Semail).

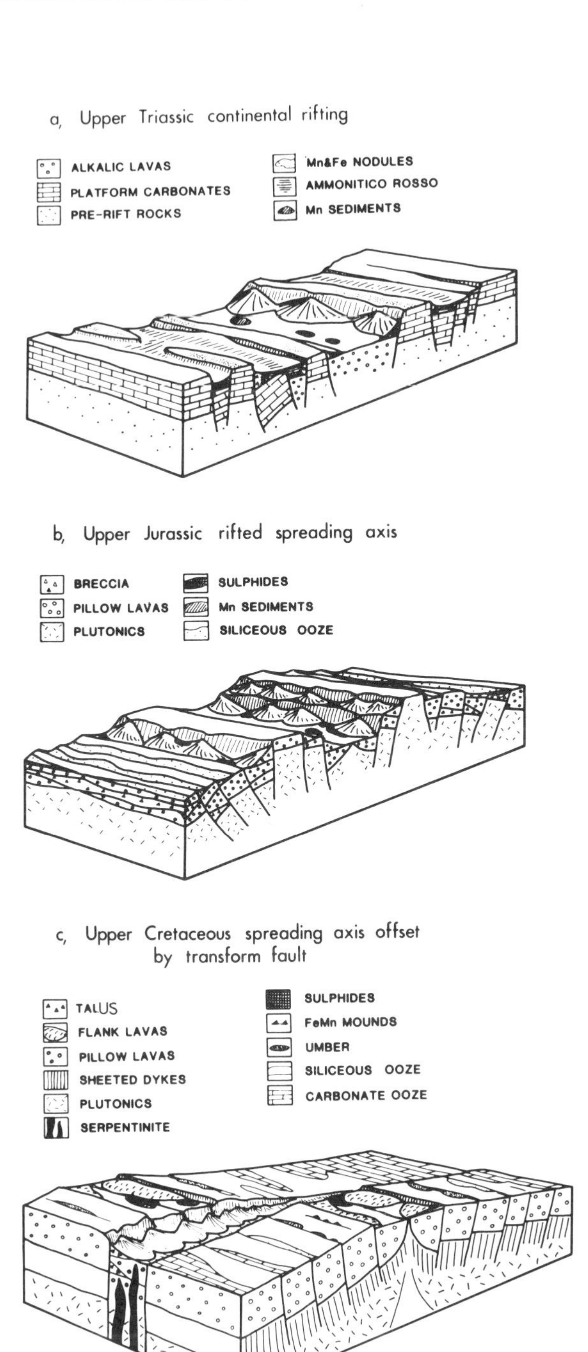
a, Upper Triassic continental rifting
ALKALIC LAVAS
PLATFORM CARBONATES
PRE-RIFT ROCKS
Mn&Fe NODULES
AMMONITICO ROSSO
Mn SEDIMENTS
b, Upper Jurassic rifted spreading axis
BRECCIA
PILLOW LAVAS
PLUTONICS
SULPHIDES
Mn SEDIMENTS
SILICEOUS OOZE
c, Upper Cretaceous spreading axis offset by transform fault
TALUS
FLANK LAVAS
PILLOW LAVAS
SHEETED DYKES
PLUTONICS
SERPENTINITE
SULPHIDES
FeMn MOUNDS
UMBER
SILICEOUS OOZE
CARBONATE OOZE

type" sediments of the Troodos, Baër-Bassit and Oman relate to intense hydrothermalism and ponding of ferromanganiferous sediments on the spreading ridge flanks, similar to parts of the East Pacific Rise. Sulphide formation is inhibited where seawater rapidly cools the crust, as along the South-Troodos Arakapas transform fault. The major Semail sulphides record localised high-temperature discharge immediately prior to construction of a differentiated edifice above the 'axis' extrusives. Increased mixing with pelagic clay produced the less metal-rich chocolate-brown mudstones above the latest lavas in Oman (Fig. 30c).

A greater degree of subsurface mixing prevents undiluted end-member solutions reaching the seafloor, and thus increases the relative proportion of cool manganese-rich emanations. Sea-water easily percolated downwards through the rifted Apennine ocean crust, generating localised sulphide solutions, and large volumes of cooler spring waters from which the manganese ores were precipitated. The Upper Jurassic to Lower Cretaceous manganese sediments of the Antalya margin record waters of similar composition, but with a significant hydrogenous component resulting from slow accumulation away from discharge sites as suggested by a reduced negative cerium anomaly (Fleet, this volume). With further seawater dilution, and thus a greater hydrogenous contribution, the more trace-element-rich Greek manganese ores were generated. Some of the Antalya, Oman and Mamonia deposits formed along the margins of Triassic rifts and above the alkalic rift lavas (Fig. 30a). With further dilution the hydrothermal contribution is masked. The Sicilian ferrogmanganese nodules and crusts, with a marked positive Ce anomaly, accumulated slowly on sediment-starved seamounts and carbonate platforms and represent the distal end-member of the hydrothermal-hydrogenous series.

Finally, the Tethyan red condensed pelagic sediments and some ferruginous oxide-sediments of clastic margins (Antalya) reflect condensed hydrogenous accumulation without any significant hydrothermal contribution (Fig. 30a). In the case of Oman, Antalya, and Mamonia, red radiolarian sediments accumulated below the calcite compensation depth along a passive continental margin; the cover of the Apennine ophiolite formed further out in a Mesozoic ocean basin.

ACKNOWLEDGEMENTS

This work was largely funded by a N.E.R.C. grant to A.H.F. Robertson and N.H. Woodcock, 'The Ophiolite-related geology of the East Mediterranean'. J.F. Boyle acknowledges a N.E.R.C. Research Studentship. Fieldwork in Oman was under the auspices of the Open University Oman ophiolite project. A.J. Fleet helped produce the previously unpublished REE data. For analytical assistance we thank G. Fitton, Mrs D. James and G.R. Angel. Mrs D. Baty drafted the diagrams, Mrs M Wright and Mrs F. Verth typed the draft manuscript and Mrs H. Hooker prepared the final copy.

REFERENCES

Abbate, E., Bortolotti, V. and Passerini, I. 1972: Studies on mafic and ultramafic rocks. 2-Palaeogeographic and tectonic considerations on the ultramafic belts in the Mediterranean area. Boll. Soc. geol. Ital., 91, 239-282.

Adamides, N.G. 1975: Geological history of the Limni concession in the light of the plate tectonic hypotheses. Trans. Inst. Min. Met., 84,B 17-33.

Adamides, N.G. 1980: The form and environment of the Kalavassos ore deposits, in: Ophiolites, Proc. Internat. Ophiolite Symp. A Panayiotou, ed., Nicosia, Cyprus, 1980, 117-129.

Alabaster, T., Pearce, J.A., Mallick, K.I.J. and Elboushi, I.M. 1980: The volcanic stratigraphy and location of massive sulphide deposits in the Oman ophiolite, in: Ophiolites, Proc. Internat. Ophiolite Symp., A. Panayiotou, ed., Nicosia, Cyprus, 1979, 751-758.

Aleksandër, C., 1980: The influence of rocks bearing hydrothermal veinous deposits of copper ore of the ophiolite belt of the Albanides on the mineralogical properties of the deposits. UNESCO Int. Symp. Metallogeny of mafic complexes: Vol. 2. The Eastern Mediterranean-Western Asia, and its comparison with similar metallogenic environments in the world, Athens, I.G.C.P. Project 169, Athens, 39-55.

Aubouin, J., 1964: Réflexions sur le faciès "Ammonitico Rosso". Ann. Soc. géol. Nord 6, 475-501.

Aubouin, J., Bonneau, M., Celet, P., Charvet, J., Clement, B., Degardin, J.M., Dercourt, J., Ferriere, J., Fleury, J.J., Guernet, C., Maillot, H, Mania, J., Mansy, J.L., Terry, J., Thiebault, P., Tsoflias, P., and Verriex, JJ. 1970: Contribution à la géologie des Hellenides: le Gavrovo, le Pinde et la zone ophiolitique subpelagonienne. Ann. Soc. geol. Nord 90, 277-306.

Barrett, T.J., 1980: The Pb isotopic composition of Jurassic cherts overlying ophiolites in the North Apennines. Earth Planet. Sci. Lett., 49, 193-204.

Barrett, T.J., 1981: Chemistry and mineralogy of Jurassic bedded chert overlying ophiolites in the North Apennines, Italy. Chemical Geology, 34, 289-317.

Barrett, T.J., and Spooner, E.T.C., 1977: Ophiolitic breccias associated with allochthonous oceanic crustal rocks in the Eastern Ligurian Apennines, Italy - a comparison with observations from rifted oceanic ridges. Earth Planet. Sci. Lett., 35, 79-91.

Barton, C.M., 1975: Mount Olympus, Greece: new light on an old window. J. geol. Soc. Lond., 131,389-369.

Bernoulli, D., and Laubscher, H., 1972: The palinspastic problem of the Hellenides: Eclogae. Geol. Helvetiae, 65, 107-118.

Bernoulli, D., and Jenkins, H.C., 1974: Alpine, Mediterranean and Central Atlantic facies in relation to the early evolution of the Tethys, in: "Modern and ancient geosynclinal sedimentation", R.H. Dott, Jr. and R.H. Shaver, eds., Spec. Publ. Soc. Econ. Paleontol. Mineral. Tulsa, 19, 129-160.

Bignell, R.D., Cronan, D.S. and Tooms, J.S.,1976: Metal dispersion in the Red Sea as an aid to marine geochemical exploration. Trans. Inst. Min. Metall. Sect. B: Appl. Earth Sci., 85, B274-B278.

Biju-Duval, B., Dercourt, J. and Le Pichon, X., 1977: From the Tethys Ocean to the Mediterranean Seas: A plate tectonic model of the evolution of the western Alpine system, in: Editions Technip., B. Biju-Duval, and L. Montadert, eds., Paris, 143-164.

Bischoff, J.L. and Rosenbauer, R.J., 1977: Recent metalliferous sediments in the North Pacific manganese nodule area. Earth Planet. Sci. Lett., 33, 379-388.

Bonatti, E., Fisher, D.E., Joensuu, O., Rydell, H.S. and Beyth, M., 1972: Iron-manganese-barium deposit from the Northern Afar Rift (Ethiopia): Econ. Geol., 67, 717-730.

Bonatti, E., Guerstein-Honnorez, B.-M., and Honnorez, J., 1976a: Copper-iron sulfide mineralizations from the equatorial Mid-Atlantic Ridge. Econ. Geol., 71, 1515-1525.

Bonatti, E., Guerstein-Honnorez, B.-M., Honnorez, J., and Stern, C., 1976b: Hydrothermal pyrite concretions from the Romanche Trench (equatorial Atlantic): Metallogenesis in oceanic fracture zones. Earth Planet. Sci. Lett., 32, 1-10.

Bonatti, E., Zerbi, M., Kay, R., and Rydell, H., 1976c: Metalliferous deposits from the Apennine ophiolites: Mesozoic equivalent of modern deposits from oceanic spreading centers. Geol. Soc. America Bull., 87, 83-94.

Boström, K., and Peterson, M.N.A., 1966: Precipitates from hydrothermal exhalations on the East Pacific Rise. Econ. Geol., 61, 1258-1265.

Boström, K. and Peterson, M.N.A., 1969: The origin of aluminium-poor ferromanganoan sediments in areas of high heat flow on the East Pacific Rise. Marine Geol., 7, 427-447.

Boyle, J.F. and Robertson, A.H.F., 1983: Structural controls of extrusion and hydrothermal metallogenesis in the Troodos spreading axis, Cyprus. Spec. Pub. J. geol. Soc. Lond. (in press).

Briden, J., Smith, A.G. and Sallomy, J.T., 1970: The geomagnetic field in Permo-Triassic time. Geophys. J.R. astr. Soc., 23, 101-117.

Brigo, L., Kostelka, L., Omenetto, P., Schneider, H.J., Schroll, E., Schultz, O. and Štrukcl, I., 1977: Comparative reflections on four Alpine Pb-Zn deposits, in: "Time- and Strata-Bound Ore Deposits", D.D. Klem. and H.-J. Schneider, eds., Springer-Verlag.

Brunn, J.H., Dumont, J.F., De Graciansky, P.C., Gutnic, M., Juteau, T., Marcoux, J., Monod, O. and Poisson, D., 1971: Outline of the geology of the western Taurides, in: "Geology and History of Turkey", A.S. Campbell, ed., Petroleum Exploration Society of Libya, Tripoli, 225-255.

Burns, R.G. and Burns, V.M., 1977: Mineralogy of marine manganese deposits, 185-248, in: "Marine Manganese Deposits", G.P. Glasby, Elsevier, 532pp.

Calvert, S.E., 1966: Accumulation of diatomaceous silica in the sediments of the Gulf of California. Geol. Soc. Am. Bull., 77, 569-596.

Chapman, H.J. and Spooner, E.T.C., 1977: ^{87}Sr enrichment of ophiolitic sulphide deposits in Cyprus confirms ores formed by circulating seawater. Earth Planet. Sci. Lett., 35, 71-78.

Coleman, R.G.. 1974: Geological background of the Red Sea, in: "The Geology of Continental Margins", Burk, C.A. and Drake, C.L., eds., Springer-Verlag, 743-753.

Constantinou, G., 1976: Genesis of the conglomerate structure, porosity and collomorphic textures of the massive sulphide ores of Cyprus, in: "Metallogeny and plate tectonics", D.F. Strong, ed., Geol. Assoc. Canada, Spec. Publ., 14, 187-211.

Constantinou, G. and Govett, G.J.S., 1972/73: Genesis of sulphide deposits, ochre and umber of Cyprus. Trans. Instn. Min. Metall. (Sect. B: Appl. earth sci.), 81, B34-36.

Constantinou, G. and Govett, G.J.S., 1973: Geology, geochemistry and genesis of Cyprus sulfide deposits. Econ. Geol. 68, 843-858.

Corliss, J.B., Lyle, M., Dymond, J. and Crane, K., 1977: The chemistry of hydrothermal mounds near the Galapagos Rift. Earth Planet. Sci. Lett. 40, 12-24.

Corliss, J.B., Dymond, J., Gordon, L.I., Edmond, R.T., von Herzen, R.D., Ballard, R.D., Green, K. Williams, D., Bainbridge, A.E., Crane, K. and van Andel, Tj.H., 1979: Submarine thermal spring on the Galapagos Rift. Science,203, 1073-1083.

Cortesogno, L., Lucchetti, G. and Penco, A.M., 1979: Le mineralizzazioni a manganese nei diaspri delle ofioliti Liguri: mineralogia e genesi. Rend. Soc. Ital. Mineral. Petrol., 35,151-197.

Cortesogno, L., Galbiati,B.,Principi,G.,Venturelli, G., 1980: Eastern Liguria ophiolitic breccias: new data and discussion on paleogeographic models. Ofioliti, 3, 99-160. (In Italian with an extended summary in English).

Cox, K.G., 1970: Tectonics and volcanism of the Karoo period and their bearing on the postulated fragmentation of Gondwanaland, in: "African magmatism and tectonics", T.N. Clifford, and I.G. Gass, eds., Oliver and Boyd, Edinburgh, 211-236.

Craig, H., 1969: Geochemistry and origin of the Red Sea brines, in: "Hot brines and recent heavy metal deposits in the Red Sea", E.T. Degens, and D.A. Ross, eds., Springer-Verlag, New York, 208-242.

Cressman, E.R., 1962: Data of Geochemistry, Ch.T: Non-detrital siliceous sediments (6th ed.). U.S. Geol. Surv. Prof. Pap. 440-T, 23pp.

Cronan, D.S., 1976: Basal metalliferous sediments from the Eastern Pacific. Geol. Soc. Am. Bull., 87, 928-934.

Cronan, D.S., Glasby, G.P., Moorby, S.A., Thomson, J., Knedler, K.E. and McDougall, J.C., 1982: A submarine hydrothermal manganese deposit from the south-west Pacific Island arc. Nature, 298, 456-458.

Curray, J.R., Moore, D.G., 1982: Init. Repts. DSDP, 64, Washington (U.S. Govt. Printing Office). Parts 1 & 2, 1313pp.

Cymex Scientific Team, 1979: Massive deep sulphide ore deposits discovered by submersible of the East Pacific Rise: Project Rita, Nature, 277, 523-528.

Decandia, F.A. and Elter, P., 1972: La 'zona' ofiolitifera del Bracco del nel settore compreso fra Levanto e Val Graveglia Apennino ligure, Memorie Soc. geol. Ital., 11, 503-530.

Delaloye, M., Vuagnat, M. and Wagner, J.J., 1977: K-Ar ages from the Kizil Dagh ophiolitic complex (Hatay, Turkey) and their interpretation, in: "Structural history of the Mediterranean Basins", B. Biju-Duval, and L. Montadert, eds., Editions Technip. Paris, 73-78.

Delaune-Mayere, M.. 1978: Cherts Mesozoique du bassin Tethysien oriental: mineralogie et geochemie des sediments siliceux du secteur de Taminah (NW Syrien). Cah. OSTROM Ser. Geol., 10, 191-202.

Delaune-Mayere, M. and Parrot. J.-F., 1976: Evolution du Mesozoique de la marge continentale meridionale du bassin Tethysian oriental d'apres l'etude des series sedimentaires de la region ophiolitique de NW Syrien. Cah. OSTROM, 8, 173-183.

Delaune-Mayere, M., Marcoux, J., Parrot, J.-F. and Poisson, A., 1977: Modele d'evolution Mesozoique de la paleomarge Tethysienne au niveau des nappes radiolaritiques et ophiolitiques de Taurus Lycien., d'Antalya et de Baër-Bassit, in: "Structural history of the Mediterranean basins", B. Biju-Duval, and L. Montadert, eds., Editions Technip. Paris. 79-94.

Derkmann, K. and Klemm, D.D., 1977: Strata-bound Kies-ore deposits of the "Tauernfenster" (Eastern Alps, Austria/Italy), in: Time-and Strata-bound ore deposits", D.D. Klemm, and H.-J. Schneider, eds., Springer-Verlag.

Desprairies, A. and Lapierre, H., 1972: Les argiles liees au volcanisme du massif du Troodos (Chypre), et leur remaniement dans sa couverture. Rev. Geogr. Phys. Geol. Cyn. 15, 499-510.

Dewey, J.F., Pitman, W.C.,Ryan, W.B.F. and Bonnin, J., 1973: Plate tectonics and the evolution of the Alpine system. Geol. Soc. Amer. Bull., 84, 3137-3180.

Dritterbass, von W., 1979: Sedimentologie und Geochemie von eisenmanganführenden Knoller und Krusten im Jura der Trento-Zone (östliche Südalpen, Norditalien). Eclogae geol. Helv. 72, 313-345.

Dumont, J.F., Gutnic, M., Marcoux, J., Monod, O. and Poisson,A., 1972: Les Trias des Taurides occidentales (Turquie). Définition du bassin pamphylien: Un nouveau domaine à ophiolithes à la marge externe de la chaine taurique. Z. dtsch. Ges., 123, 385-409.

Dymond, J., Corliss, J.B. and Stillinger, R., 1976: Chemical composition and metal accumulation rates of metalliferous sediments from sites 319,320, 321, in: DSDP Init. Rep., Project 34, 575-588.

Dymond, J. and Veeh, H.H., 1975: Metal accumulation rates in the south-east Pacific and the origin of metalliferous sediments, Earth Planet. Sci. Lett., 28, 13-22.

Edmond, J.M., Measures, C., Magnum, B., Grant, B., Sclater, F.R., Collier, R., Hudson, A., Gordon, L.I. and Corliss, J.B., 1978: On the formation of metal-rich deposits at ridge crests. Earth Planet. Sci. Lett., 46, 19-31.

Elderfield, H., Gass, I.G., Hammond, A. and Bear, L.M., 1972: The origin of ferromanganese sediments associated with the Troodos Massif of Cyprus. Sedimentology, 19, 1-19.

Ferrario, A. and Garuti, G., 1980: Copper deposits in the basal breccias and volcano-sedimentary sequences of the Eastern Ligurian ophiolites (Italy). Mineral. Deposita, 15, 291-303.

Fleet, A.J. and Robertson, A.H.F., 1980: Ocean-ridge metalliferous and pelagic sediments of the Semail Napope, Oman. J. geol. Soc. Lond., 137, 403-422.

Folk, R.L. and McBride, E.F., 1978: Radiolarites and their relation to subjacent "oceanic" crust in Liguria, Italy. J. Sediment. Petrol., 48, 1069-1102.

Francheteau, J., and others, 1979: Massive deep-sea sulphide ore deposits discovered on the East Pacific Rise. Nature, 277, 523-528.

Gale,N.H., Spooner, E.T.C. and Potts, P.J., 1981: The lead and strontium isotope geochemistry of metalliferous sediments associated with Upper Cretaceous ophiolitic rocks in Cyprus, Syria, and the Sultanate of Oman. Can. J. Earth Science, 18, 1290-1302.

Garrison, R.E., 1974: Radiolarian cherts, pelagic limestones and igneous rocks in eugeo-synclinal assemblages, in: "Pelagic Sediments: On Land and Under the Sea", K.J. Hsü. and H.C. Jenkyns, eds., Spec. Publ. Int. Assoc. Sediment., 1, 367-399.

Garrison, R.E. and Fischer, A.G., 1969: Deep water limestones and radiolarites of the Alpine Jurassic. Spec. Publ. Soc. Econ. Palaeontol. Mineral. Tulsa., 14, 20-56.

Gass,I.G. and Masson-Smith, D., 1963: The geology and gravity anomalies of the Troodos Massif, Cyprus. Roy. Soc. London, Phil. Trans., A255, 417-467.

Gioflica, G., Vlad, S., Nicolae, I., Vlad, C. and Bratosin, I., 1980: Copper metallogenesis related to Mesozoic ophiolites from Romania. UNESCO Int. Simp. Metallogeny of Mafic Complexes: Vol. 2. The Eastern Mediterranean - Western Asia, and its comparison with similar metallogenic environments in the world, Athens, I.G.C.P. Project 169, 156-171.

Glennie, K.W., Boeuf, M.G.A., Hughes Clarke, M.W., Moody-Stuart, M., Pilaar, W.F.H. and Reinhardt, 1973: Late Cretaceous nappes in the Oman Mountains and their geologic evolution. Bull. Am. Ass. Petrol. Geol., 57, 5-27.

Glennie, K.F., Boeuf, M.G.A., Clark, M.W.H., Moody-Stuart, M., Pilaar, W.F.H. and Reinhardt, B.M., 1974: Geology of Oman Mountains. Konikl.Nederlands Geol. Mijnbouwkundig Genoot., Verh., 31, 1-423.

Goldberg, E.D. and Arrhenius, G.O.S., 1958: Chemistry of Pacific pelagic sediments. Geochim. Cosmochim. Acta., 13, 153-212.

Goldberg, E.D., Koide, M., Schmitt, R.A. and Smith, R.H., 1963: Rare earth distributions in the marine environments. J. Geophys. Res., 68, 4209-4217.

Graciansky, P-C. de, Bourbon, M., de Charpal, O., Chennet, P-Y. and Lemoine, M., 1979: Genèse et évolution comparée de deux marges continentales passives: marge ibérique de l'Océan Atlantique et marge européenne de la Téthys dans les Alpes occidentales. Bull. Soc. géol. France, 21, 663-674.

Graham, G., 1980: Evolution of a passive margin and nappe emplacement in the Oman Mountains, in: "Ophiolites", "Proc. Internat. Ophiolite Symp.", Panayiotou, A., ed., 1979, 414-424.

Griffiths, W.R., Albers, J.P. and Oner, O., 1972: Massive sulphide deposits of the Ergani Maden area, S.E. Turkey. Econ. Geol., 67, 701-713.

Grill, E.V., Chase, R.L., Macdonald, R.D. and Murray, J.W., 1981: A hydrothermal deposit from the Explorer ridge in the Northeast Pacific ocean. Earth Planet. Sci. Lett., 52, 142-150.

Guillemot, D., and Nesteroff, E.D., 1980: Les dépôts métallifères crétacés de Chypre: comparison avec leurs homologues actuels de Pacifique, in: "Ophiolites", A.Panayiotou, ed., Proc. Internat. Ophiolite Symp., Cyprus, 1979, 139-146.

Haskin, L.A., Wildeman, T.R., Frey, F.A., Collins, K.A., Keedy, C.R. and Haskin, M.A., 1966: Rare earths in sediments. J. Geophys. Res., 71, 6091-6105.

Heath, G.R. and Dymond, J., 1977: Genesis and transformation of metalliferous sediments from the East Pacific Rise, Bauer Deep and Central Basin, northwest Nazca plate. Geol. Soc. Am. Bull., 88, 723-733.

Heaton, T.H.E. and Sheppard, S.M.F., 1977: Hydrogen and oxygen isotope evidence for seawater-hydrothermal alteration and ore deposition, Troodos complex, Cyprus, in: "Volcanic Processes in Ore Genesis", Spec. Publ. Geol. Soc. London, 7, 42-58.

Hoffert, M., Perseil, A., Hekinian, R. Choukroune, P., Needham, H.D., Francheteau, J. and Le Pichon, X., 1978: Hydrothermal deposits sampled by diving saucer in Transform Fault "A" near 37°N on the Mid-Atlantic Ridge, FAMOUS area. Oceanol. Acta, 1, 73-86.

Hollister, V.F., 1982: Porphyry copper-gold deposits of Cyprus. Geol. Soc. Am. Abstracts with programs, New Orleans, 1982, p. 516.

Hsü, K.J., Cita, M.B. and Ryan, W.B.F., 1973: The origin of the Mediterranean evaporites, in: "Initial Reports of the Deep Sea Drilling Project", W.B.H. Ryan, and K.J. Hsü, eds., U.S. Government Printing Office, Washington, D.C., 13, Part 2, 1447 pp.

Hynes, A.J., Nisbet, E.G., Smith, A.G., Welland, M.J.P. and Rex, D.C., 1972: Spreading and emplacement ages of some ophiolites in the Othris region, Eastern Central Greece (Proc. 4th Aegean Symposium Hannover). Z. dtsch. geol. Ges., 123, 455-468.

Jansa, L.F., Enos, P., Tucholke,B.E., Gradstein,F.M. and Sheridan, R.E. 1979: Mesozoic-Cenozoic sedimentary formations of the North American Basin; Western North Atlantic, in: Deep drilling results in the Atlantic ocean: contintental margin and paleoenvironment, M. Talwani, W. Hay and W.B.F. Ryan, eds., Maurice Ewing Series, Amer. Geophys. Union, 3, 1-58.

Jenkyns, H.C. 1970a: Fossil manganese nodules from the west Sicilian Jurassic. Eclogae Geol. Helv., 63, 741-774.

Jenkyns, H.C. 1970b: Submarine volcanism and the Toarcian iron pisolites of western Sicily. Eclogae Geol. Helv., 63, 549-572.

Jenkyns, H.C. 1974: Origin of red nodular limestones (Ammonitico Rosso, Knollenkalke) in the Mediterranean Jurassic: a diagenetic origin, in: Pelagic sediments on land and under the sea, K.J. Hsü and H.C. Jenkyns, eds., Spec. Publ. Internat. Assoc. Sedimentol, 1, 249-273.

Jenkyns, H.C. 1977: Fossil nodules, in: Marine manganese deposits, G.B. Glasby, ed., Elsevier, 532pp.

Jenkyns, H.C. 1980: Tethys: past and present. Proc. Geol. Ass. 91, 107-118.

Jenkyns, H.C. and Hardy R.G., 1975: Basal iron-titanium-rich sediments from hole 315A (Line Islands, Central Pacific), in: S.O. Schlanger and E.D. Jackson et al., Initial Reports of the Deep Sea Drilling Project, 33 (U.S. Government Printing Office, Washington, D.C., 1975) 833-836.

Juteau, T. 1975: Les ophiolites des nappes d'Antalya (Taurides occidentales, Turquie). Mém. Sciences Terre, 32, 692pp.

Kelepertsis, A.E. and Andrulakis, I., 1980: The mineralogy of the stratiform cupriferous Pyrite in Lagadi-Kardylia (Madedonia -Greece) and its genesis problem. UNESCO Int. Symp. Metallogeny of Mafic Complexes: Vol. 2, The Eastern Mediterranean - Western Asia, and its comparison with similar metallogenic environments in the world, Athens, I.G.C.P. Project, 169, 197-218.

Kelts, K. 1981: A comparison of some aspects of sedimentation and translational tectonics from the Gulf of California and the Mesozoic Tethys, Northern Penninic margin, Eclogae geol. Helv., 74, 317-338.

Lapierre, H. 1975: Les formations sédimentaire et éruptives des nappes de Mamonia et leurs relations avec le Massif de Troodos (Chypre occidentale). Mém. Soc. géol. France, 123, 132pp.

Lapierre, H. and Rocci, G. 1976: Le volcanisme du sud-ouest de Chypre et le problème de l'ouverture des region Tethysiennes au Trias. Tectonophysics, 30, 299-313.

Laubscher, H. and Bernoulli, D. 1977: Mediterranean and Tethys. in: The ocean basins and margins, 4A, eds., A.E.M. Nairn, W.H. Kanes and F.G. Stehli, Plenum Press, New York, 1-28.

Lauer, J-P. 1981: L'évolution géodynamique de la Turquie et de Chypre deduite de l'étude paléomagnetique, Thèse du doctorat és-sciences, Université Louis Pasteur de Strasbourg, 291pp.

Malahoff, A., Cronan, D.S., Skirrow, R. and Embley, R.W. 1983: The geological setting and chemistry of hydrothermal sulfides and associated deposits from the Galapagos Rift at 86°W. Marine Mining, 4, 122-137.

Marcoux, J. 1970: Age carnien des termes éffusives du cortège ophiolitique des Nappes d'Antalya (Taurus Lycien oriental, Turquie). C. r. Séances Acad. Sci. Paris, 271, 285-287.

Marcoux, J. 1974: Alpine type Triassic of the upper Antalya nappe (western Taurides, Turkey), in: Die Stratigraphie der alpin-mediterranean Trias, H. Zapfe, ed., Vienna, 145-146.

Martin, A., Marcoux, J., Poisson, A. and Steinberg, M. 1981: Gradients géochimique dans une série radiolaritique mésozoique des Taurides occidentales (Turquie). Implications palaeo-oceanographique. C.r. Acad. Sci. Paris, 292, Ser. 11, 525-530.

Mero, J.L. 1965: The mineral resources of the sea. Amsterdam-London-New York, Elsevier, 312p.

Murdmaa, I.O.,Gordeev, V.V., Bazilevskaya, E.S. and Emelyanov, E.M. 1978: Inorganic geochemistry of the Leg 44 sediments, in: Initial Reports of the Deep Sea Drilling Project, W.E. Benson, R.E. Sheridan et al., eds., Washington (U.S. Government Printing Office), 44, 463-476.

Nisbet, E.G. and Pearce, J.A. 1974: TiO_2 as a possible guide to past oceanic spreading rates. Nature, 246, 468-470.

Nisbet,E.G. and Price, I. 1974: Siliceous turbidites: bedded cherts as redeposited ocean ridge-derived sediments, in: Pelagic sediments on land and under the sea, K.J. Hsü, and H.C. Jenkyns, eds., Int. Assoc. Sedimentol. 1, 351-366.

Ogg, J. 1981: Middle and Upper Jurassic sedimentation history of the Trento Plateau (Northern Italy), in: Rosso Ammonitico Symposium Proceedings, A. Farinacci and S. Elmo eds., Edizioni Tecnoscienza, Rome, 479-503.

Panagos, A.G. and Varnavas, S.P., 1980: Preliminary observations on manganese deposits in the areas of Othris and Argolis (Eastern Greece). UNESCO Int. Symp. Metallogeny of Mafic Complexes: Vol. 2. The Eastern Mediterranean - Western Asia and its comparison with similar metallogenic environments in the world, Athens, I.G.C.P. Project, 169, 257-280.

Parrot, J.-F. 1974: Le secteur de Taminah (Tourkamannli): étude d'une séquence volcano-sédimentaire de la région ophiolitique du Baër-Bassit (North-ouest de la Syrie) Cah. ORSTOM. 6, 127-146.

Parrot, J.-F. 1977: Assemblage ophiolitique du Baër-Bassit et termes effusifs du volcano-sédimentaire, Travaux et Documents de L'ORSTOM, 72, 333 pp.

Pe-Piper, G., 1982: Geochemistry, tectonic setting and metamorphism of Mid-Triassic volcanic rocks of Greece, Tectonophysics, 85, 253-272.

Pearce, J.A. 1975: Basalt geochemistry used to investigate past tectonic environments on Cyprus. Tectonophysics, 25, 41-67.

Pearce, J.A. 1980: Geochemical evidence for the genesis and eruptive setting of lavs from the Tethyan Ophiolites, in, "Ophiolites,"Proc. Internat. Ophiolite Symp., A. Panayiotou, ed., Nicosia, Cyprus, pp 261-272.

Pearce, J.A., Alabaster, T., Shelton. A.W. and Searle, M.P. 1981: The Oman ophiolite as a Cretaceous arc-basin complex: evidence and implications. Phil. Trans. R. Soc. Lond. Ser. A., 300, 299-317.

Piper, D.Z., Veeh, H.H., Bertrand, W.G., and Chase, R.L. 1974: An iron-rich depositfrom the northeast Pacific, Earth Planet. Sci. Lett., 26, 114-120.

Price, I. 1977: Facies distinction and interpretation of primary cherts in a Mesozoic continental margin succession, Othris (Greece). Sediment. Petrol., 18, 321-35.

Reinhardt, BM. 1969: On the genesis and emplacement of ophiolites in the Oman Mountains geosyncline. Schweiz. Mineral. Petrogr. Mitt. 49, 1-30.

Ricou, L.E. 1971: Le croissant ophiolitique peri-Arabe, une ceinture de nappes mises en place au Cretacé supérieur. Revue Géog. Phys. Géol. dyn., 13, 327-349.

Robertson, A.H.F. 1975: Cyprus umbers: basalt-sediment relationships on a Mesozoic ocean ridge. J. geol. Soc. Lond., 131, 511-531.

Robertson, A.H.F. 1976: Origins of ochres and umbers: evidence from Skouriotissa, Troodos Massif, Cyprus. Trans. Instn. Ming. Metall., 85, B245-251.

Robertson, A.H.F. 1978: Metallogenesis along a fossil oceanic fracture zone, Arakapas fault belt, Troodos Massif, Cyprus. Earth Planet. Sci. Lett. 41, 317-329.

Robertson, A.H.F. 1981: Metallogenesis in a Mesozoic passive continental margin, Antalya Complex, S.W. Turkey. Earth Planet. Sci. Lett. 54, 323-345.

Robertson, A.H.F. and Hudson, J.D. 1973: Cyprus umbers: chemical precipitates on a Tethyan ocean ridge. Earth Planet. Sci. lett. 18, 93-101.

Robertson, A.H.F. and Hudson J.D. 1974: Pelagic sediments in the Cretaceous and Tertiary history of the Troodos Masif, Cyprus. Spec. Publ. Int. Assoc. Sedimentol. 1, 403-4367.

Robertson, A.H.F. and Fleet A.J. 1976: The origins of rare earths in Metalliferous sediments of the Troodos Masssif, Cyprus. Earth Planet. Sci. Lett. 28, 385-394.

Robertson, A.H.F. and Woodcock, N.H. 1979: The Mamonia Complex southwest Cyprus: the evolution and emplacement of a Mesozoic continental margin. Geol. Soc. Am. Bull. 90, 651-665.

Robertson, A.H.F. and Woodcock, N.H. 1980: Tectonic setting of the Troodos massif in the East Mediterranean, in: Ophiolites. Proc. Int. Ophiolite Symp., A. Paniotou, ed., Cyprus 1979, 36-49.

Robertson, A.H.F. and Woodcock, N.H. 1981a: Alakır Cay Group, Antalya Complex, S.W. Turkey: a deformed Mesozoic carbonate margin. Sediment Geol., 30, 95-131.

Robertson, A.H.F. and Woodcock, N.H. 1981b: Gödene Zone, Antalya Complex, S.W. Turkey: Volcanism and sedimentation on Mesozoic marginal ocean crust. Geol. Rdsch., 70, 1177-1214.

Robertson, A.H.F. and Woodcock, N.H. 1983: Genesis of the Batinah melange above the Semail Ophiolite, Oman. J. Struct. Geol. 5, 1-17.

Rocci, G., Baroz, F., Bebien, J., Desmet, A.,Lapierre, H., Ohnenstetter, D., Ohnenstetter,M., and Parrot, J.-F. 1980: The Mediterranean ophiolites and their related Mesozoic volcano-sedimentary sequences, in: Ophiolites. Proc. Int. Ophiolite Symp., A. Paniotou, ed., Cyprus 1979, 272-287.

Rona, P.A. 1978: Criteria for recognition of hydrothermal mineral deposits in oceanic crust. Econ Geol. 73, 135-160.

Rona, P.A. 1980: TAG hydrothermal field: mid-Atlantic Ridge crest at latitude 26°N. J. geol. Soc. London, 137, 385-402.

Russell, M.J., Solomon, M., and Walshe, J.L. 1981: The genesis of sediment-hosted, exhalative zinc+lead deposits, Mineral Deposita, 16, 113-127.

Sayles, F.L. and Bischoff, J.L. 1973: Ferromanganoan sediments in the equatorial East Pacific. Earth Planet. Sci. Lett. 19, 330-336.

Scott, M.R., Scott, R.B., Rona, P.A., Butler, L.W. and Nalwalk, A.J. 1974: Rapidly accumulating manganese deposit from the median valley of the Mid-Atlantic Ridge. Geophys. Res. letters, 1, 355-358.

Searle, D.L. 1972: Mode of occurrence of the cupriferous pyrite deposits of Cyprus. Trans. Instn.Min. metall. (Sect. B: Appl. earth sci.), 81, B189-97.

Searle, M.P., Lippard, S.J., Smewing, F.D. and Rex, D.C. 1980: Volcanic rocks beneath the Semail Nappe in the northern Oman mountains and their significance in the Mesozoic evolution of the Tethys. J. geol. Soc. London, 137, 589-605.

Sengör, A.M.C. 1979: Mid-Mesozoic closure of Permo-Triassic Tethys and its implications. Nature, 279, 590-593.

Sengör, A.M.C. and Yilmaz, Y. 1981: Tethyan evlution of Turkey: a plate tectonic approach. Tectonophysics, 75, 181-241.

Sideris, C., Scounakis, S. and Economou, M. 1980: The ophiolite complex of Edessa area and the associated mineralisation, UNESCO Int. Symp. Metallogeny of mafic complexes: Vol. 3. The Eastern Mediterranean - Western Asia, and its comparison with similar metallogenic environments in the world, Athens, I.G.C.P. Project 169, 142-153.

Simonian, K.O, and Gass, I.G. 1978: Arakapas fault belt Cyprus: A fossil transform belt. Geol. Soc. Am. Bull. 89, 1220-1230.

Smewing, J.D. 1981: Regional setting and petrological characteristics of the Oman ophiolite in North Oman. Ofioliti, 335-376.

Smewing, J.D., Simonian, K.O., Elboushi, I.M. and Gass, I.G. 1977: Mineralized fault zone parallel to the Oman ophiolite spreading axis. Geology, 5, 534-538.

Smith, A.G. 1971: Alpine deformation and the oceanic areas of the Tethys, Mediterranean and Atlantic. Geol. Soc. Am. Bull. 82, 2039-2070.

Smith, A.G. and Briden, J.C. 1977: Mesozoic Cenozoic palaeocontinental maps. Cambridge University Press, 63pp.

Smith, A.G., Hynes, A.J., Menzies, M.A., Nisbet, E.G., Price, I., Welland, M.J.P. and Ferrière, J. 1975: The stratigraphy of the Othris Mountains, eastern central reece: a deformed Mesozoic continental margin sequence. Ecologae. Geol. Helv. 68, 463-481.

Spathi, C. 1964: The mineralogical composition of the Greek manganese deposits. Ph.D. thesis, University of Saslonika, 121pp (in Greek with English summary).

Spooner, E.T.C. 1977: Hydrodynamic model for the origin of the ophiolitic pyrite ore deposits of Cyprus. In Volcanic Processes in Ore Genesis, Spec. Publ. Geol. Soc. London, 7, 58-72.

Spooner, E.T.C., Beckinsale, R.D., Fyfe, W.S. and Smewing, J.D. 1974: O^{18} enriched ophiolitic metabasic rocks from E. Liguria (Italy), Pindos (Greece), and Troodos (Cyprus). Contr. Mineralogy Petrology, 47, 41-62.

Spooner, E.T.C. and Bray, C.J. 1977: Hydrothermal fluids of seawater salinity in ophiolitic sulphide ore deposits in Cyprus. Nature, 266, 808-812.

Spooner, E.T.C. and Gale, N.H. 1982: Pb isotopic composition of ophiolitic volcanogenic sulphide deposits, Troodos Complex, Cyprus. Nature, 296, 239-242.

Steinberg, M. and Mpodozis Marin, C. 1978: Classification geochimique des radiolarites et des sediments siliceux oceaniques, signification paleo-oceanographique. Oceanol Acta, 1,

Swarbrick, R.E. and Robertson, A.H.F. 1980: Revised stratigraphy of Mesozoic rocks of southern Cyprus. Geol. Mag. 117, 547-563.

Swarbrick, R.E. 1979: The sedimentology and structure of south west Cyprus and its relationship to the Troodos Complex. Unpublished University of Cambridge, Ph.D. thesis, 277pp.

Toth, J.R. 1980: Deposition of submarine crusts rich in manganese and iron, Geol. Soc. Am. Bull., 91, 44-54.

Turekian, K.K. and Wedepohl, K.H. 1961: Distribution of the elements in some major units of the earth's crust. Bull. geol. Soc. Am. 72, 175-191.

Varnavas, S.P. 1980: Partition geochemical investigations of ferromanganese deposits from the Troodos Massif, Cyprus. UNESCO Int. Symp. metallogeny of mafic complexes: Vol. 2. The Eastern Mediterranean - Western Asia and its comparison with similar metallogenic environments in the world, Athens, I.G.C.P. Project 169, 391-410.

Waldron, J.W.F. 1981: Mesozoic sedimentary and tectonic evolution of the northeast Antalya Complex, Egridir, S.W. Turkey. Unpublished University of Edinburgh Ph.D thesis, 249 pp.

Wendt. J. 1980: Stratigraphische Kondensation in triadischen und jurassischen Cephalopodenkalken der Tethys. Neues Jahrb. Geologie und Palaontologie, Monats. 1970, 433-448.

Whitmarsh, R.B. 1974: Summary of the general features of the Arabian Sea and Red Sea Cenozoic history based on Leg 23 Cores. In R.B. Whitmarsh, O.E. Weser, D.A. Ross et al. Initial Reports of the Deep Sea Driling Project 23 (U.S. Government Printing Office), 1115-1125.

Woodcock, N.H. and Robertson, A.H.F. 1982a: Stratigraphy of the Mesozoic rocks above the Semail Nappe, Oman. Geol. Mag. 119, 67-116.

Woodcock, N.H. and Robertson, A.H.F. 1982b: Wrench and thrust tectonics along the Mesozoic-Cenozoic continental margin: Antalya Complex, S.W. Turkey. J. geol. Soc. Lond. 139, 147-163.

Woodcock, N.H. and Robertson, A.H.F. 1982c : Structure and sedimentology of the Batinah Sheets, above the Semail Nappe, Oman. Can.J. Earth Sci., 19, 1635-1656.

Zuffardi, P., 1977: Ore/mineral deposits related to the Ophiolites in Italy, in,"Times- and strata-bound ore deposits", D.D. Klemm, and H.-J. Schneider, eds., Springer-Verlag.

INTRODUCTION TO THE BIOLOGY OF HYDROTHERMAL VENTS

J. Frederick Grassle

Woods Hole Oceanographic Institution
Woods Hole, Massachusetts 02543

A search for hydrothermal vents forming metal deposits led to the unexpected discovery of dense beds of clams, mussels, and vestimentiferan worms on the Galapagos Rift in 1977. The clams were first seen in photographs taken by Deep Tow (Lonsdale, 1977) and the whole vent community was later observed from ALVIN (Corliss and Ballard, 1977; Corliss et al., 1979). Subsequent expeditions in 1979 to the Galapagos Rift and in 1982 to the Guaymas Basin, 11-13°N and 21°N on the East Pacific Rise included ecologists, microbiologists, systematists, physiologists, and biochemists.

The visually satisfying presence of so many strange creatures clustered against a sparsely-inhabited background will be remembered by many as the essence of the discovery. The biological importance concerns adaptations to a previously unknown facet of the biosphere. Most of the large, visible animals clustered around the vents thrive on chemosynthetic products ultimately derived from energy contained in the heated interior of the planet. The list in Jannasch's review of reduced inorganic compounds (H_2S, S°, $S_2O_3^{2-}$, NH_4^+, Fe^{2+}, NO_2^-, and possibly Mn^{2+}) likely to be used as sources of energy by chemosynthetic microorganisms underscores the complexity of microbial processes. Although the vent community is not particularly diverse, species unique to the vent ecosystem include at least a dozen new families or subfamilies (four have been described: Desbruyères and Laubier, 1981; Jones, 1980; McLean, 1981; Williams, 1980) and many more new genera (e.g., Burreson, 1979; Humes and Dojiri, 1980; Maciolek, 1981; Newman, 1979). As discussed in Jannasch's review, microorganisms are important in providing food for most of the vent animals (Cavanaugh et al., 1981, Corliss et al., 1979; Felbeck, 1981; Hessler, this volume; Jannasch and Wirsen, 1979; Karl, 1980; Ruby et al., 1981), depositing metals (Jannasch and Wirsen, 1981; Ruby et al., 1981; Ehrlich, 1982), and changing the chemical composition of vent water (Baross et al., 1982). Environments similar to those at present-day vent ecosystems may have been important in the origin of life (Corliss et al., 1981; Jannasch, this volume).

Hessler and Smithey's contribution provides the first thorough analysis of the distributions of the megafaunal vent animals and their ecological relationships at the Galapagos Rift vents. The animals closest to vents appear to depend on the supply of warm water and food rather than elevated temperatures alone, and the differences between the communities have more to do with the configuration and chemistry of the venting water than the barriers presented by the between-vent distances of 100 km or less along the Galapagos Rift.

Desbruyères and Laubier discuss the biology of the sites at 11-13°N on the East Pacific Rise (Desbruyères and Laubier, this volume; Desbruyères et al., 1982). The four areas visited by biologists in ALVIN and CYANA (Galapagos Rift, 11-13°N, 21°N, and Guaymas Basin on the East Pacific Rise) are separated by roughly 1000 km intervals. These distances may, in part, explain zoogeographic differences in the faunas. Mussels have not been found north of the 11°-13° site and the pink bythitid fish has only been observed at the Galapagos vents. A different species of the clam, *Calyptogena*, occurs at the northernmost site in the Guaymas Basin (Grassle, 1982) and no *Calyptogena* have been found at 11°-13°N. The pompeii worm, *Alvinella pompejana*, was found only at 21°N and 11-13°N, whereas the other alvinellid, *Paralvinella grasslei*, was collected at all sites but 21°N (Desbruyères and Laubier, this volume). *Riftia* occurs at all four sites and is found with the other family of Vestimentifera, Lamellibrachiidae, only at the Galapagos, 11-13°N and 21°N sites.

Hydrothermal fluid venting through pelagic sediments in the Guaymas Basin provides an opportunity to compare the animal communities in these sediments with extensive studies of soft sediments elsewhere in the deep sea. Most deep-sea sediments, including those in the Galapagos Mounds region (Grassle, in preparation), support a highly diverse community. Communities in Guaymas Basin sediments include a moderately diverse community peripheral to vents, a community with high biomass dominated by single species of bivalves in sediments smelling of sulfide (Grassle, 1982), and low biomass communities consisting of a small number of species in the petroliferous sediment described by Simoneit and Lonsdale (1982). The sediments with sulfide are covered with mats of the filamentous microorganism, *Beggiatoa* (Fig. 1). (Figure 1 will be found in the color insert following page 18).

Three additional vent areas are known only from photographs.

Lamellibrachiid Vestimentifera and Calyptogena are present on the Juan de Fuca Ridge off Oregon (Normark et al., 1982); an unknown animal thought to be an enteropneust is found with Calyptogena at 20°S (Francheteau and Ballard, in press). Riftia were not seen at either site. No vent animals were seen in photographs of hydrothermal vents off Hawaii (Malahoff et al., 1982), nor were they observed from ALVIN at 10°-15° C vents on seamounts near 21°N on the East Pacific Rise (Lonsdale et al., 1982).

The age of active vents is measured in years or decades. Estimates of heat loss at 350° C vents indicate a life span of 1-10 years (MacDonald et al., 1980). The age of hydrothermal sulfides at one area on the East Pacific Rise is estimated to be 20-60 years (Lalou and Brichet, 1981, 1982). The dissolution of clam shells in less than 15 years (Killingley et al., 1981; Lutz, 1982) attests to the recent extinction of a number of vent fields seen in large-scale surveys (Ballard et al., 1982; Desbruyères and Laubier, this volume). Desbruyères and Laubier discuss the species life history features needed for dispersal and colonization of newly-formed vents. These include: 1) rapid growth to maturity (Rhoads et al., 1981, 1982; Turekian et al., 1979; Turekian and Cochran, 1981), 2) high respiration rates of vent animals compared with other deep-sea species (Childress and Mickel, 1982; Mickel and Childress, 1982) (the plentiful food supply and fluctuating conditions during the life of vents may also be important in the evolution of these high metabolic rates — K. L. Smith in Hiatt, 1980; Grassle, 1982), 3) continuous reproduction (Desbruyères and Laubier, this volume), and 4) dispersal ability. The larvae of most vent animals have not been collected but inferences from shell morphology of bivalves and gastropods indicate both planktotrophic and lecithotrophic modes of development (Lutz et al., 1980; Boss and Turner, 1980; Desbruyères and Laubier, this volume). Thus far it is not possible to predict length of larval life with certainty from these results.

Feeding relationships are not fully worked out but radioisotope studies support the view that bacteria are the primary food source for clams, mussels, and vestimentiferans (Rau, 1981a, b; Rau and Hedges, 1979; Williams et al., 1981). As Jannasch (this volume) points out, chemosynthetic production occurs "(1) within the subsurface vent system (2) in microbial mats in the immediate surrounding of vents and (3) in various symbiotic associations with invertebrates." The importance of the last mode of production to the large vent animals has led workers to reexamine shallow water species and discover bacterial symbionts in shallow-water bivalves, oligochaetes and pogonophorans (Felbeck et al., 1981; Giere, 1981, Southward et al., 1981). Metabolic pathways and other biochemical characteristics of vent organisms are only beginning to be understood (Felbeck et al., 1981). The unusual features of blood function have been studied in Bythograea, Calyptogena and Riftia (Arp and Childress, 1981a, b, 1983; Terwilliger et al., 1980; Whittenberg et al., 1981).

The time that biologists have spent at vents is measured in days. The biology papers that follow in this volume are as important for identifying gaps in our understanding as for summarizing current findings. The results thus far have influenced our thinking about trophic relationships in benthic communities, the environmental determinants of rates of metabolism, growth, reproduction and mortality, and the means for larval dispersal to sites less than 100 m across separated by kilometers and, in some cases, hundreds of kilometers.

The studies of vent fauna have been made possible by developments not only in the ability to discover vents but the means to relocate sites in order to conduct experiments. The ability to revisit and collect, after a 10-month interval, a group of mussels marked in situ at 2.5 km depth, is a technological feat we have come to expect as routine. Return to the vents already known will add greatly to our knowledge of the dynamics of these systems.

REFERENCES

Arp, A. J., and Childress, J. J., 1981a, Blood function in the hydrothermal vent vestimentiferan tubeworm, Science, 213: 342-344.

Arp, A. J., and Childress, J. J., 1981b, Functional characteristics of the blood of the deep-sea hydrothermal vent brachyuran crab, Science, 214:559-561.

Arp, A. J., and Childress, J. J., 1983, Sulfide binding by the blood of the hydrothermal vent tubeworm *Riftia pachyptila*, Science, 219:295-297.

Ballard, R. D., 1977, Notes on a major oceanographic find, Oceanus, 20(3):35-44.

Ballard, R. D., and Grassle, J. F., 1979, Return to Oases of the Deep, Nat. Geogr., 156(5):689-705.

Ballard, R. D., Francheteau, J., Juteau, T., Rangan, C., and Normark, W., 1981, East Pacific Rise at 21°N: the volcanic, tectonic, and hydrothermal processes of the central axis, Earth Planet. Sci. Let., 55:1-10.

Ballard, R. D., Holcomb, R. T., and van Andel T. H., 1979, The Galapagos Rift at 86°W: 3. Sheet flows, collapse pits, and lava lakes of the rift valley, Jour. Geophys. Res., 84(B10): 5407-5422.

Ballard, R. D., van Andel, T. H., and Holcomb, R. T., 1982, The Galapagos Rift at 56°W 5. Variations in volcanism, structure, and hydrothermal activity along a 30 km segment of the Rift Valley, Jour. Geophys. Res., 87(B2):1149-1161.

Baross, J. A., Lilley, M. D., and Gordon, L. I., 1982, Is the CH_4, H_2 and CO venting from submarine hydrothermal systems produced by thermophilic bacteria?, Nature, 298:366-368.

Boss, K. J., and Turner, R. D., 1980, The giant white clam from the Galapagos Rift, Calyptogena magnifica species novum, Malacologia, 20(1):161-194.

Burreson, E. M., 1981, A new deep-sea leech, Bathybdella sawyeri gen. et sp. n. from thermal vent areas on the Galapagos Rift, Proc. Biol. Soc. Wash., 94: 483-491.

Cavanaugh, C. M., Gardiner, S. L., Jones, M. L., Jannasch, H. W., and Waterbury, J. B., 1981, Procaryotic cells in the hydrothermal vent tubeworm, Riftia pachyptila Jones: possible chemoautotrophic symbionts, Science, 213:340-342.

Childress, J. J., and Mickel, T. J., 1982, Oxygen and sulfide consumption rates of the vent clam, Calyptogena pacifica, Mar. Biol. Let., 3:73-79.

Corliss, J. B., and Ballard, R. D., 1977, Oases of life in the cold abyss, Nat. Geogr., 152(4):441-453.

Cohen, D. M., and Haedrich, R. L., The fish fauna of the Galapagos thermal vent region, Deep-Sea Res., (in press).

Corliss, J. B., Baross, J. A., and Hoffman, S. E., 1981, An hypothesis concerning the relationship between submarine hot springs and the origin of life on earth, Oceanol. Acta SP:59-69.

Corliss, J. B., Dymond, J., Gordon, L. I., Edmond, J. M., von Herzen, R. P., Ballard, R. D., Green, K., Williams, D., Bainbridge, A., Crane, K., and van Andel, T. H., 1979, Submarine thermal springs on the Galapagos Rift, Science, 203: 1073-1083.

Crane, K., and Ballard, R. D., 1980, The Galapagos Rift at 86°W: 4. Structure and morphology of hydrothermal fields and their relationship to the volcanic and tectonic processes of the Rift Valley, Jour. Geophys. Res., 85(B3):1443-1454.

Desbruyères, D., and Laubier L., 1980, Alvinella pompejana gen. sp. nov., Ampharetidae aberrant des sources hydrothermales de la ride Est-Pacifique, Oceanol. Acta, 3:267-274.

Desbruyères, D., and Laubier, L., 1982, Paralvinella grasslei, new genus, new species of Alvinellinae (Polychaeta: Amphareti- dae) from the Galapagos Rift geothermal vents, Proc. Biol. Soc. Wash., 95:484-494.

Desbruyères, D., Crassous, P., Grassle, J., Khripounoff, A., Reyss, D., Rio, M., and Van Praet, M., 1982. Donnees ecologi- ques sur un nouveau site d'hydrothermalisme actif de la ride du Pacifique Oriental, C.R. Acad. Sci. Paris, Ser. III 295: 489-494.

Ehrlich, H., 1982, Manganese oxidizing bacteria from a hydrothermal active area on the Galapagos Rift. Ecol. Bull., 35, (in press).

Enright, J. T., Newman, W. A., Hessler, R. R., and McGowan, J. A., 1981, Deep-ocean hydrothermal vent communities, Nature, 289: 219-221.

Fatton, E., and Roux, M., 1981a, Modalites de croissance et micro- structure de la coquille de Calyptogena (Vesicomyidae, Lamelli- branches), en relation avec les sources hydrothermales oceani- ques, C.R. Acad. Sci. Paris, 292:55-60.

Fatton, E., and Roux, M., 1981b, Etapes de l'organisation micro- structurale chez Calyptogena magnificia Boss et Turner, bivalve a croissance rapide des sources hydrothermales, oceanique, C.R. Acad. Sci. Paris, 243:63-68.

Fatton, E., Marien, G., Pachiaudi, C., Rio, M., and Roux, M., 1982, Fluctuations de l'activite des sources hydrothermales ocean- iques (Pacifique Est, 21°N) enregistrees lors de la croissance des coquilles de Calyptogena magnifica (Lamellibranche, Vesi- comyidae) par les isotopes stables du carbone et de l'oxygene, C.R. Acad. Sci. Paris, 293:701-706.

Fauchald, K., 1982, A eunicid polychaete from a white smoker, Proc. Biol. Soc. Wash., 95(4):781-787.

Felbeck, H., 1981, Chemoautotrophic potentials of the hydrothermal vent tube worm, Riftia pachyptila Jones (Vestimentifera), Science, 213:336-338

Felbeck, H., Childress, J. J., and Somero, G. N., 1981, Calvin- Benson cycle and sulphide oxidation enzymes in animals from sulphide-rich habitats, Nature 293:291-293.

Francheteau, J., and Ballard, R. D., The East Pacific Rise near

21°N, 13°N and 20°S: Inferences for along-strike variability of axial processes of the Mid-Ocean Ridge, Earth Planet. Sci. Lett., (in press).

Francheteau, J., Needham, D., Juteau, T., and Rongin, C., 1980, Naissance d'un ocean sur la borsale du Pacifique est, CYAMEX, Centre National pour l'Exploitation des Oceans, Paris.

Fretter, V., Graham, A., and McLean, J. H., 1981, The anatomy of the Galapagos Rift limpet, Neomphalus fretterae, Malacologia, 21:337-361.

Galapagos Biology Expedition Participants: Grassle, J. F., Berg, C. J., Childress, J. J., Grassle, J. P., Hessler, R. R., Jannasch, H. W., Karl, D. M., Lutz, R. A., Mickel, T. J., Rhoads, D. C., Sanders, H. L., Smith, K. L., Somero, G. N., Turner, R. D., Tuttle, J. H., Walsh, P. J., and Williams, A. J., 1979, Galapagos '79: Initial findings of a biology quest, Oceanus, 22(2):2-10.

Giere, O., 1981, The gutless marine oligochaete Phallodrilus leukodermatus. Structural studies on an aberrant tubificid associated with bacteria, Mar. Ecol. Prog. Ser., 5:353-357.

Grassle, J. F., 1982, The biology of hydrothermal vents: a short summary of recent findings, MTS Jour., 16(3):33-38.

Harwood, C. S., Jannasch, H. W., and Canale-Parola, E. 1982, Anaerobic spirochaete from a deep-sea hydrothermal vent, Appl. Environ. Microbiol., 44:234-237.

Hessler, R., 1981, Oasis under the sea - where sulfur is the staff of life, New Scient., 10 December, pp. 741-747.

Hiatt, B., 1980, Sulfides instead of sunlight, Mosaic, 11(4): 15-21.

Humes, A., and Dojiri, M., 1980, A siphonostome copepod associated with a vestimentiferan from the Galapagos Rift and East Pacific Rise, Proc. Biol. Soc. Wash., 93(3):697-707.

Jannasch, H. W., and Wirsen, C. O., 1979, Chemosynthetic primary production of East Pacific sea floor spreading centers, Bioscience, 29:592-598.

Jannasch, H., and Wirsen, C., 1981, Morphological survey of microbial mats near deep-sea thermal vents, App. Environ. Microbiol., 41:528-538.

Jones, M. L., 1980, Riftia pachyptila, a new genus, new species, the vestimentiferan worm from the Galapagos Rift geothermal vents (Pogonophora), Proc. Biol. Soc. Wash., 93(4):1295-1313.

Jones, M. L., 1981, Riftia pachyptila Jones: some observations on the vestimentiferan worm from the Galapagos Rift, Science, 213:333-336.

Karl, D., Wirsen, C., and Jannasch, H., 1980, Deep-sea primary production at the Galapagos hydrothermal vents, Science, 207:1345-1347.

Killingley, J. S., Berger, W. H., MacDonald, K. C., and Newman, W. A., 1981, $^{18}O/^{16}O$ variations in deep-sea carbonate shells from the Rise hydrothermal field, Nature, 287:218-221.

Krantz, G. W., 1982, A new species of Copidognathus Trouessart (Acari:Actinedida:Halacaridae) from the Galapagos Rift, Can. Jour. Zool., 60:1728-1731.

Lalou, C., and Brichet, E., 1981, Possibilites de datation des depots de sulfures metalliques hydrothermaux sous-marins par les descendants a vie courte de l'uranium et du thorium, C.R. Acad. Sci. Paris, 293:821-826.

Lalou, C., and Brichet, E., 1982, Ages and implications of East Pacific Rise sulphide deposits at 21°N, Nature, 300:169-171.

Laubier, L., Desbruyères, D., and Chassard-Bouchaud, P., Evidence of sulfur accumulation in the epidermis of the polychaetes Alvinella pompejana from deep-sea hydrothermal vents, A micro-analytical study, Nature, (in press).

Liley, M. D., deAngelis, M. A., and Gordon, L. I., 1982, CH_4, H_2, CO and N_2O in submarine hydrothermal vent waters, Nature, 300:48-50.

Lonsdale, P., 1977, Clustering of suspension feeding macrobenthos near abyssal hydrothermal vents at oceanic spreading centers, Deep-Sea Res., 24:857-863.

Lonsdale, P., Batiza, R., and Simkin, T., 1982, Metallogenesis at sea mounts on the East Pacific Rise, MTS Jour., 16(3):54-61.

Lutz, R. A., 1982, Dissolution of molluscan shells of deep-sea hydrothermal vents, EOS, 63:1014.

Lutz, R. A., Jablonski, D., Rhoads, D. C., and Turner, R. D., 1980

Larval dispersal of a deep-sea hydrothermal vent bivalve from the Galapagos Rift, Mar. Biol., 57:127-133.

Maciolek, N. J., 1981, Spionidae (Polychaeta, Annelida) from the Galapagos Rift geothermal vent, Proc. Biol. Soc. Wash., 94: 826-837.

Malahoff, A., McMurtry, G. M., Wiltshire, J. C., and Yeh, H.-W., 1982, Geology and chemistry of hydrothermal deposits from active submarine volcano Loini, Hawaii, Nature, 298:234-239.

McDonald, K. C., Becker, K., Spiess, F. N., and Ballard, R. D., 1980, Hydrothermal heat flux of the "black smoker" vents on the East Pacific Rise, Earth Planet. Sci. Lett., 48(1980):1-7.

McLean, J., 1981, The Galapagos Rift limpet Neomphalus: relevance to understanding the evolution of a major Paleozoic-Mesozoic radiation, Malacologia, 21:291-336.

Mickel, T. J., and Childress, J. J., 1982, Effects of pressure and temperature on the EKG and heart rate of the hydrothermal vent brachyuran crab, Bythograea thermydron (Brachyura), Biol. Bull., 162:70-82.

Mickel, T. J., and Childress, J. J., 1982, Effects of temperature, pressure and oxygen concentration on the oxygen consumption role of the hydrothermal vent crab Bythograea thermydron (Brachyura), Physiol. Zool., 55:199-207.

Newman, W. A., 1979, A new scalpellid (Cirripedia): a Mesozoic relic living near an abyssal hydrothermal spring, Trans. San Diego Soc. Nat. Hist., 19:153-167.

Normark, W. R., Lupton, J. E., Murray, J. W., Koski, R. A., Clague, D. A., Morton, J. L., DeLaney, J. R., and Johnson, M. P., 1982, Polymetallic sulfide deposits and water column tracers of active hydrothermal vents on the Southern Juan de Fuca Ridge, MTS Jour., 16(3):46-53.

Powell, M. A., and Somero, G. N., 1983, Blood components prevent sulfide poisoning of respiration of the hydrothermal vent tube-worm Riftia pachyptila, Science, 219:297-299.

Pugh, P. R., A review of the Family Rhodaliidae (Siphonophora: Physonectae), Phil. Trans. Roy. Soc. B., (in press).

Rau, G. H., 1981, Hydrothermal vent clam and tube worm $^{13}C/^{12}C$: further evidence of non-photosynthetic food sources, Science, 213:338-340.

Rau, G. H., 1981b, Low $^{15}N/^{14}N$ of hydrothermal vent animals: On-site N_2 fixation and organic nitrogen synthesis?, Nature, 289: 484-485.

Rau, G. H., and Hedges, J. I., 1979, Carbon-13 depletion in a hydrothermal vent mussel: suggestion of a chemosynthetic food source, Science, 203:648-649.

Rhoads, D. C., Lutz, R. A., Revelas, E. C., and Cerrato, R. M., 1981, Growth of bivalves of the deep-sea hydrothermal vents along the Galapagos Rift, Science, 214:911-913.

Rhoads, D. C., Lutz, R. A., Cerrato, R. M., and Revelas, E. C., 1982, Growth and predation activity at deep-sea hydrothermal vents along the Galapagos Rift, Jour. Mar. Res., 40:503-516.

Rise Project Group: Spiess, F. N., Macdonald, K. C., Atwater, T., Ballard, R., Carranza, A., Cordoba, D., Cox, C., Diaz Garcia, V. M., Francheteau, J., Gurerrero, J., Hawkins, J., Haymon, R., Hessler, R., Juteau, T., Kastner, M., Larson, R., Luyendyk, B., Macdougall, J. D., Miller, S., Normark, W., Orcutt, J., and Rangin, C., 1980, East Pacific Rise: Hot springs and geophysical experiments, Science, 207:1421-1433.

Ruby, E. G., and Jannasch, H. W., 1982, Physiological characteristics of Thiomicrospira sp. Strain L-12 isolated from deep-sea hydrothermal vents, Jour. Bacteriol., 149:161-165.

Ruby, E. G., Wirsen, C. O., and Jannasch, H. W., 1981, Chemolithotrophic sulfur-oxidizing bacteria from the Galapagos Rift hydrothermal vents, Appl. Environ. Microbiol., 42:317-324.

Simoneit, B. R. T., and Lonsdale, P. F., 1982, Hydrothermal petroleum in mineralized mounds at the seabed of Guaymas Basin, Nature, 295:198-202.

Smithey, W. M., Jr., and Hessler, R. R., Megafaunal distribution at deep-sea hydrothermal vents: an integrated photographic approach, in: "Underwater Photography for Scientists," (in press).

Southward, A. J., Southward, E. C., Dando, P. R., Rau, G. H., Felbeck, H., and Flügel, H., 1981, Bacterial symbionts and low $^{13}C/^{12}C$ ratios in tissues of Pogonophora indicate unusual nutrition and metabolism, Nature, 293:616-620.

Terwilliger, R. C., Terwilliger, N. B., and Schabtach, E., 1980, The structure of hemoglobin from an unusual deep-sea worm (Vestimentifera), Comp. Biochem. Physiol., 65B:531-535.

Turekian, K. K., and Cochran, J. K., 1981, Growth rate of a vesicomyid clam from the Galapagos Spreading Center, Science, 214:909-911.

Turekian, K., Cochran, J. K., and Nazaki, Y., 1979, Growth rate of a clam from the Galapagos Rise hot spring field using natural radionuclide ratios, Nature, 280:385-387.

Turner, R. D., 1981, "Wood islands" and "thermal vents" as centers of diverse communities in the deep sea, Biologiya Morya, 1: 3-10.

Tuttle, J. H., Wirsen, C. O., and Jannasch, H. W., Microbial activities in the emitted hydrothermal waters of the Galapagos Rift vents, Mar. Biol., (in press).

van Andel, T. H., and Ballard, R. D., 1979, The Galapagos Rift at 86°W: 2. Volcanism, structure, and evolution of the Rift Valley, Jour. Geophys. Res., 84(B10):5390-5406.

van Praet, M., Regime alimentaire des Actinies, Bull. Soc. Zool. France, (in press).

Williams, A. B., 1980, A new crab family from the vicinity of submarine thermal vents on the Galapagos Rift (Crustacea:Decapoda; Brachyura), Proc. Biol. Soc. Wash., 93(2):443-472.

Williams, A. B., and Chase, F. A., 1982, Shrimp of the family Bresiliidae from thermal vents of the Galapagos Rift Crustacea: Decapoda: Caridae), Jour. Crust. Biol., 2(1):136-147.

Williams, P. M., Smith, K. L., Druffel, E. M., and Linick, P. W., 1981, Dietary carbon sources of mussels and tubeworms from Galapagos hydrothermal vents determined from tissue ^{14}C activity, Nature, 292:448-449.

Wittenberg, J. B., Morris, R. J., Gibson, Q. H., and Jones, M. L., 1981, Hemoglobin kinetics of the Galapagos Rift vent tube worm, Riftia pachyptila Jones (Pogonophora: Vestimentifera), Science, 213:344-346.

MICROBIAL PROCESSES AT DEEP SEA HYDROTHERMAL VENTS

Holger W. Jannasch

Woods Hole Oceanographic Institution
Woods Hole, Massachusetts 02543

ABSTRACT

The primary production of organic carbon by chemosynthetic sulfur-oxidizing bacteria has been proposed to provide the base of the food chain for the extensive populations of animals found at hydrothermal vents at depths of about 2600 m. The oxidation of reduced inorganic compounds (such as H_2S, S^o, $S_2O_3^{2-}$, NH_4^+, NO_2^-, Fe^{2+} and possibly Mn^{2+}) as the source of energy for chemosynthesis is equivalent to the role of light in photosynthesis. Reported here is the present state of proof of this hypothesis which includes the work of many collaborating scientists. Epifluorescence microscopy and nucleotide determinations demonstrated substantial bacterial densities in the emitted vent waters. Multi-layered mats of unicellular bacteria were observed, often encased in heavy Mn/Fe deposits, as well as assemblages of *Leucothrix*/*Thiothrix*-like filaments and others resembling trichomes of apochlorotic cyanobacteria. Masses of *Beggiatoa* filaments were found on artificial surfaces deposited near the vents for 10 months. Species of the genera *Thiomicrospira*, *Thiobacillus* and *Hyphomonas* have been isolated and studied in detail. Furthermore, an anaerobically chemosynthetic, extremely thermophilic, methanogenic bacterium was isolated as well as a number of "Type I" methylotrophic bacteria oxidizing methane and methylamine. The gills of bivalves, collected from areas intermittently flushed with H_2S-containing vent water and oxygenated ambient seawater, contained masses of bacteria showing high activities of sulfur metabolism and Calvin-Benson cycle enzymes. Likewise the "trophosome" tissue of the gutless tube worm *Riftia* was found to consist of procaryotic cells exhibiting ATP-generating and CO_2-reducing activity. Thus, three locations of chemosynthetic production are proposed: (1) within the subsurface vent system at elevated temperatures, (2) in microbial mats in the immediate surrounding of the vents, and (3) in various symbiotic associations with invertebrates. It appears that the predominant chemosynthetic production, in combination with the most efficient transfer of organic carbon to the vent animals, occurs via symbiosis.

INTRODUCTION

Chemical and physical alterations of hydrothermal fluid during its rise through the oceanic crust and its contact with oxygenated seawater within complex vent systems provide favorable conditions for a number of microbial processes. Dense populations of sizable organisms in the immediate vicinity of the vents indicated the most important of these processes: in the absence of light, the photosynthetic base of the food chain appears to be replaced by chemosynthetic production of organic carbon. In other words, solar energy normally used by plants is substituted by geothermal energy liberated through the bacterial oxidation of reduced chemical species in the hydrothermal fluids. This terrestrial energy is then used for the chemosynthetic conversion of CO_2 into organic carbon, a process that appears to be highly efficient as indicated by the enormous production and quantity of biomass. Next to this microbial conversion of primary importance for the existence of life at the vents, a number of other aerobic and anaerobic microbial processes have been found or are expected to exist.

The detection and qualitative study of these processes started with macro- and microscopic observations and were followed by pure culture isolations, physiological description and identification of organisms responsible for particular microbial transformations. Quantitative studies consisted of in situ measurements of the physico-chemical conditions, the conversion of added radiolabeled tracer compounds and, finally, laboratory reproductions of the in situ conditions for the experimental re-enactment of particular microbial processes. This presentation of work conducted during the first five years following the discovery of the deep sea hydrothermal vents at the Galapagos Rift ocean floor spreading center includes observations from a recent biology cruise to the 21°N, 109°W site of the East Pacific Rise during April and May of 1982.

IN SITU OBSERVATIONS AND MICROSCOPY

Direct observations from the research submersible ALVIN and carefully directed sampling have been most important for microbiological work. The first observation, highly suggestive of chemosynthesis, was the "milky-bluish" water (Corliss et al., 1979) emitting from the warm (8-23°C) vents (Figure 1). (Figure 1 will be found in the color insert following page 18). It indicated the possibility of both chemical and biological oxidation of hydrogen sulfide. White deposits and iridescent particle suspensions commonly are signs of the formation of colloidal or particulate elemental sulfur at or near H_2S/O_2 interfaces in various natural environments where sulfide appears as a normal corollary of anaerobiosis followed by a bacterial reduction of sulfate.

The contact of reduced sulfur compounds with free oxygen provides a suitable source of energy and the necessary electron sink to a number of well known sulfur oxidizing bacteria. The energy liberated by this biologically mediated oxidation is used for the reduction of carbon dioxide to organic carbon, a process called chemosynthesis (see below). By way of regulating the intracellular pH and by other not yet completely understood mechanisms, the bacteria are able to outcompete the spontaneous chemical oxidation of reduced sulfur compounds.

When the turbid water was collected and filtered in the ship's laboratory through 0.2 μm Nuclepore filters, scanning electron microscopy (SEM) showed (Figure 2) that the particulate matter consisted primarily of microbial cells (Jannasch and Wirsen, 1979; Karl et al., 1980). The abundance of dividing stages may indicate active growth, although the emission of cells into the 2°C ambient seawater may have arrested the possibly high rates of growth taking place within the vent system at unknown optimal temperatures. Larger particles, some of them visible to the naked eye, are, unlike pelagic or sedimentary marine detritus, primarily composed of bacterial cells (Figure 3).

Bacterial numbers counted by epifluorescence microscopy resulted in 10^5 to 10^9 cells/ml in waters sampled as closely as possible to the vent openings (Jannasch and Wirsen, 1979; Corliss et al., 1979) where concentrations of hydrogen sulfide were found in a range of 10 to 160 mmol/liter (Edmond et al., 1979). These cell counts were supplemented by determinations of adenosine 5'-triphosphate (ATP) as another possible measurement of living biomass and the guanosine 5'-triphosphate:ATP ratio as an indicator of growth rate (Karl, 1980). The data (Table 1) show that the amount of ATP in the vent waters was 3 to 4 times greater than that measured in the photosynthetically productive surface waters of the Galapagos Rift region. From this generally high output of bacterial biomass from the vents it is concluded that growth must

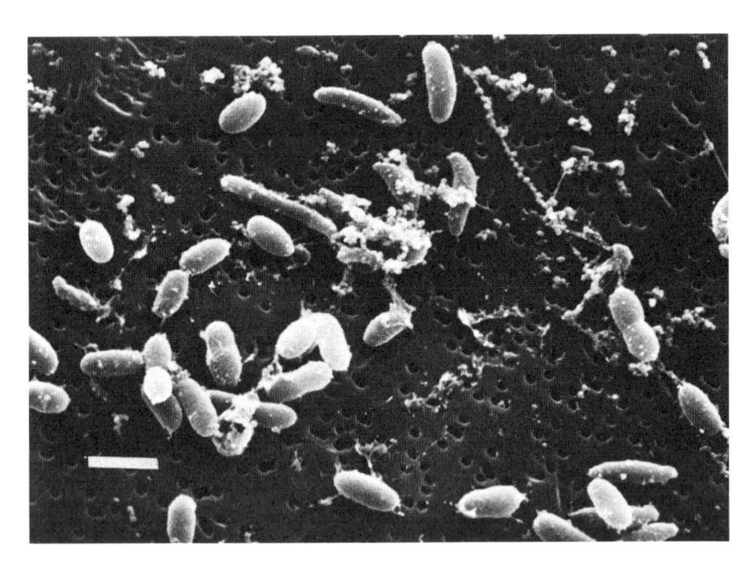

Figure 2. Scanning electron micrograph of suspended material in the milky-bluish vent water on a Nuclepore filter (scale 1 μm; Mussel Bed, Galapagos Rift; from Karl et al., 1980).

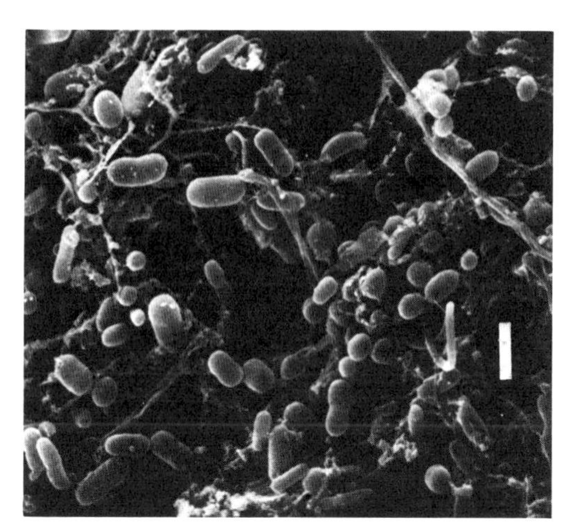

Figure 3. Scanning electron micrograph of particulate matter emitted from vents (scale 1 μm; Mussel Bed, Galapagos Rift; from Karl et al., 1980).

take place in subsurface lava cracks and chambers where the circulation of oxygenated seawater (Lister, 1977 and this conference) comes into contact with the highly reduced hydrothermal fluid, and at any biologically possible temperature above ambient (2.1°C). The above-mentioned abundance of particles, often with a flock- or sheet-like appearance, indicates growth in dense microbial mats on surfaces.

The "warm" vents of the Galapagos Rift area emitting turbid water are commonly surrounded by dense populations of mussels. They appear to be more dependent on a filter-feeding mode of life than the giant clams (Calyptogena magnifica, Boss and Turner, 1980). The nutrition of these clams and the vestimentiferan tube worms (Riftia pachyptila, Jones, 1981), the two other major invertebrate species inhabiting the vent area, is based upon symbiotic associations with chemosynthetic bacteria (see below). Thus, the growth of bacteria within the subsurface vent systems and the emission of biomass represent only one portion of the overall chemosynthetic productivity. It appears that most of this production is lost by dilution into the surrounding seawater, although a number of other filter feeding organisms besides bivalves, such as planktonic Crustacea and invertebrate larvae, will participate in harvesting the bacterial biomass which, thereby, may contribute to the species diversity of the vent populations as a whole. Important in the use of bacterial suspensions as food for the animal community, i.e., the base of the food chain, are the larger particles or cell aggregates (Figure 3) which

TABLE 1

MICROBIAL NUCLEOTIDES AT A HYDROTHERMAL VENT SITE (Galapagos Rift)[a]

Sample	ATP	A_T	GTP:ATP
Surface seawater[b] (50 m)	130 ± 32 ng/l	340 ± 112 ng/l	0.16 ± 0.08
Deep Seawater[b] (2400 m)	1.7 ± 0.3 ng/l	3.4 ± 0.3 ng/l	0.075
"Garden of Eden" vent[b] (2500 m)	491 ± 151 ng/l	1494 ± 553 ng/l	0.86 ± 0.17
"Garden of Eden" vent[c] (2500 m)	1943 ± 1143 ng/g	4248 ± 2031 ng/g	0.89 ± 0.35

[a]Data from Karl et al., 1980 (ATP = adenosine 5'-triphosphate, A_T = total adenylates, GTP:ATP = ratio of guanosine 5'-triphosphate to adenosine 5'-triphosphate).

[b]Filtered.

[c]Settled particles.

facilitate sustenance of organisms with a more coarse filtering apparatus, possibly including the "vent fishes" frequently observed at the Galapagos Rift area head-down in the vent openings.

Many of the warm vents at the Galapagos Rift as well as the 21°N area are characterized by whitish coatings. Viewed microscopically, they reveal a predominance of Thiothrix-like filaments, about 2 μm wide (Figure 4), non-gliding, attached at one end and loaded with intracellular sulfur globules. They appear to predominate in mats collected from micro-environments directly exposed to freshly emitted hydrothermal fluid. Slate panels, originally deposited at the Galapagos Rift vents for the attachment of bivalve larvae and collected after a period of 10 months, were found covered with masses of stringy white material, which consisted almost exclusively of Beggiatoa-like filaments with sheaths and filament widths of up to 20 μm (Figure 5). Recently heavy whitish mats were observed on the sandy bottom of the Guayamas Basin spreading center in the Gulf of California where

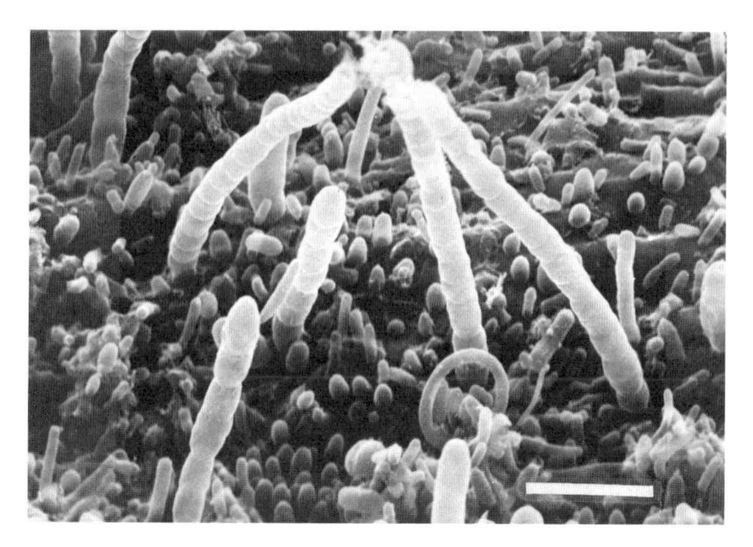

Figure 4. Scanning electron micrograph of microbial mat grown on a clam shell located within the vent water plume, largely Thiothrix-like filaments (scale 5 μm; Rose Garden, Galapagos Rift).

hydrothermal vents are overlayed by sediments (see below). Microscopically these mats consist of Beggiatoa-like filaments with a diameter of up to 160 μm in the collapsed state, and are the largest procaryotic filaments observed so far (Figure 6).

Rusty-brownish deposits occur in the immediate vicinity of the vents indiscriminately covering the surface of mussels, tube worms and lava rocks. SEM revealed densely packed mats partly encrusted with layers of metal oxides (Jannasch and Wirsen, 1981). A KEVEX X-ray defraction spectrum of one of these deposits (Figure 7A) was taken from a sample of periostracum, the keratinaceous protective coating of the mussel shell. The relative abundance of the deposited materials correlates with the mineral todorokite $(Mn, Fe, Mg, Ca, K, Na)_2 (Mn_5O_{12}) 3H_2O$. The fine-grained and poorly crystalline state is characteristic for marine ferromanganese deposits. Manganese and iron oxide-hydroxide complexes occurred in almost equal amounts on glass slides deposited directly in a vent opening for 10 months (Figure 7B). The difference in the Mn/Fe ratios at the two sites reflects the position of the samples in relation to the vents: the further away from the point of hydrothermal fluid emission, the lower the Fe peak due to the lower solubility coefficient of iron as compared to that of manganese (Mottl et al., 1979).

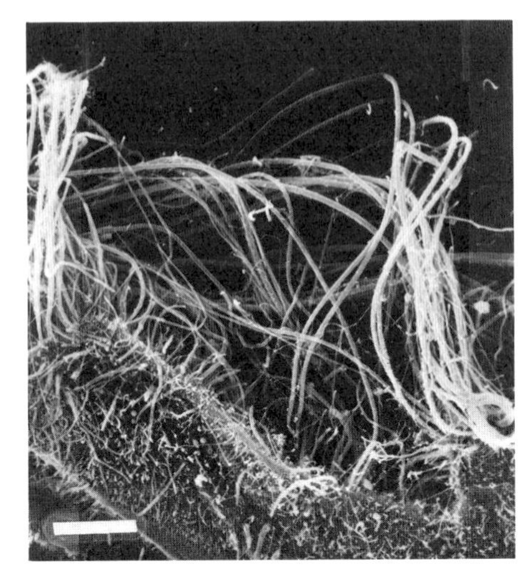

Figure 5. Scanning electron micrograph of Beggiatoa-like filaments (scale 100 μm; Mussel Bed, Galapagos Rift.)

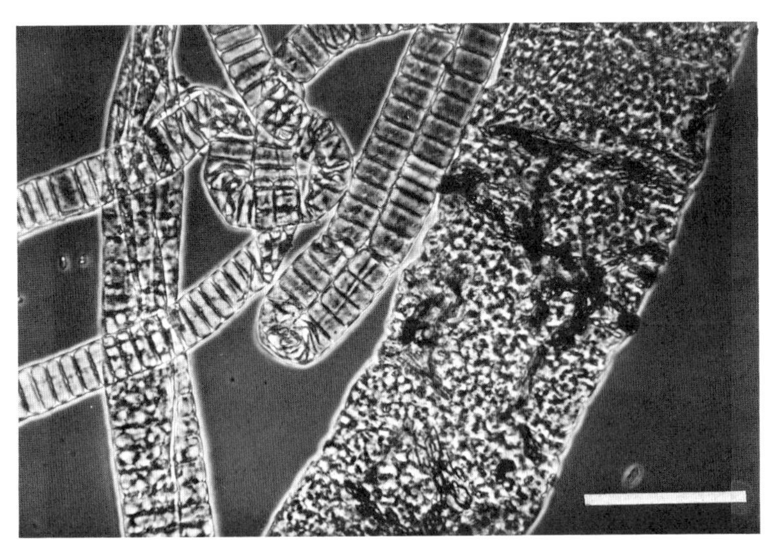

Figure 6. Phase micrograph of Beggiatoa-like filaments (scale 100 μm; Guaymas Basin).

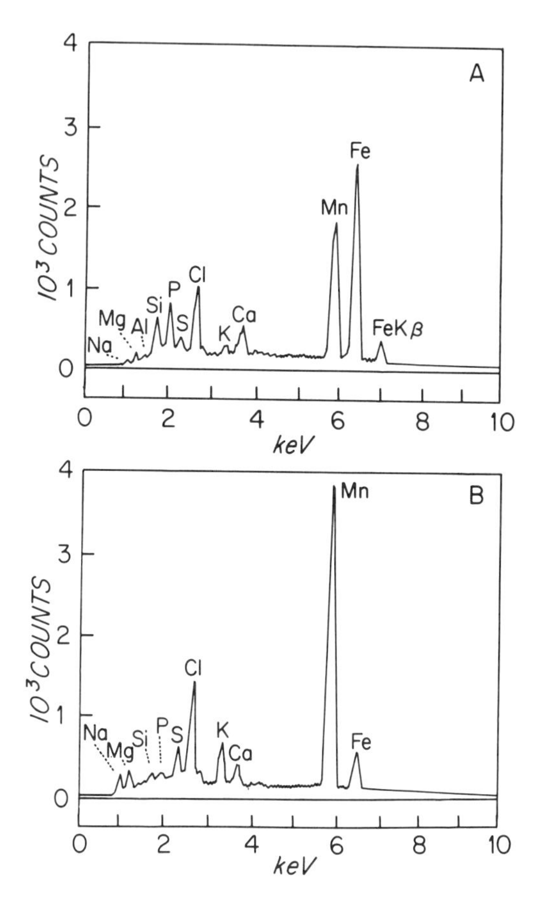

Figure 7. Line drawings from KEVEX X-ray spectrum photos; A: glass slide incubated for 10.5 months in an active fissure at Mussel Bed, Galapagos Rift; B: piece of mussel surface (periostracum) collected at Rose Garden, Galapagos Rift (keV - kilo electron volt; from Jannasch and Wirsen, 1981).

Besides patchy occurrence of Beggiatoa- and Thiothrix-filaments, the rusty-brownish coatings in the vicinity of the vents are characterized by three major types of microorganisms (Jannasch and Wirsen, 1979, 1981, Figure 8): a more or less uniform cover by short rods or spherical cells, less frequent cells of similar shape and size but producing long filamentous appendices of about 0.1 μm diameter, and quite large (about 2 x 20 μm) multicellular cyanobacteria-like trichomes. Sectioning of these mats and transmission electron microscopy (TEM) showed that, with very few exceptions of occasional findings of a protozoan or nematode, all observed cells were procaryotic in structure (i.e., gram-negative cell walls and the absence of a membrane-bound nucleus, Figure 9).

Figure 8. Scanning electron micrograph of microbial mat on the surface (periostracum) of a mussel located within the vent water plume (scale 10 μm; Mussel Bed, Galapagos Rift; from Jannasch and Wirsen, 1979).

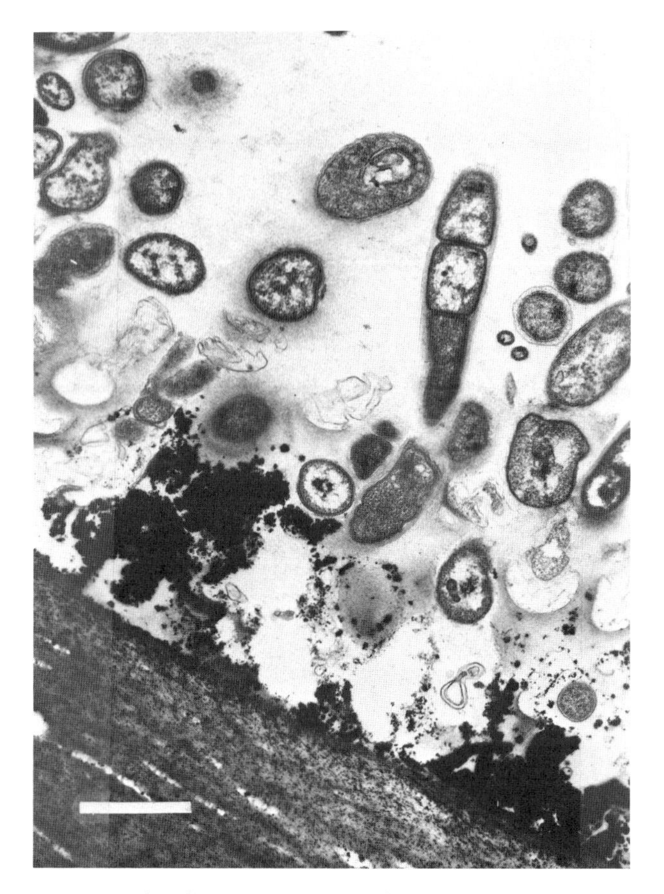

Figure 9. Transmission electron micrograph of a section through a microbial mat on the surface (periostracum) of a mussel located within the vent water plume (scale 1 μm; Mussel Bed, Galapagos Rift).

By their morphological nature, the long slender filaments in Figure 8 could be described as stalks of the genera Hyphomonas or Hyphomicrobium. Later pure culture isolations confirmed this identification. The utilization of amino acids as the sole source of carbon and energy by most of the isolates puts them in the genus Hyphomonas (Moore, 1981). Hyphomicrobium strains have recently been shown to use dimethyl sulfoxide (CH_3-SO-CH_3) and dimethyl sulfide (CH_3-S-CH_3) as sources of carbon and energy (De Bont et al., 1981).

The common occurrence of the cyanobacteria-like structures is still a puzzle. In the absence of light, no pigment formation

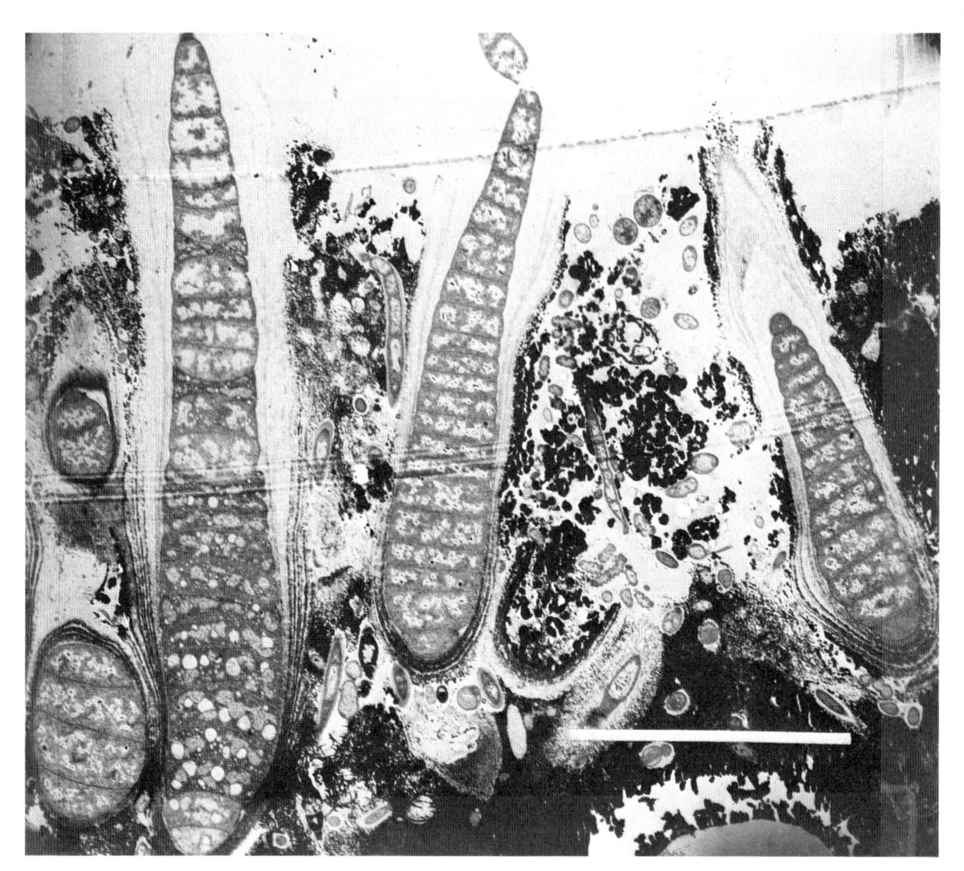

Figure 10. Transmission electron micrograph of a section through the microbial mat pictured in Figure 8 showing the tapered trichomes with metal encrusted sheaths (scale 5 μm; from Jannasch and Wirsen, 1981).

is likely. TEM of sections show the equivalent of "necridia" (sacrificing cells) that lead to the formation of hormogonia (Figure 10). Heterocysts were also observed, indicating the potential capability of nitrogen fixation. The morphological resemblance of this organism with the cyanobacterium Calothrix is much closer than in the photosynthetic/chemosynthetic pair Oscillatoria/Beggiatoa. It is to be expected that in this newly observed analogy the apochlorotic Calothrix is indeed chemosynthetic. The encasement of the trichomes by multi-layered slime sheaths (Figure 10) may provide for a low pH in the immediate surrounding of the cell, thereby making it possible for Fe^{2+} to become a suitable source of electrons. Transmission electron

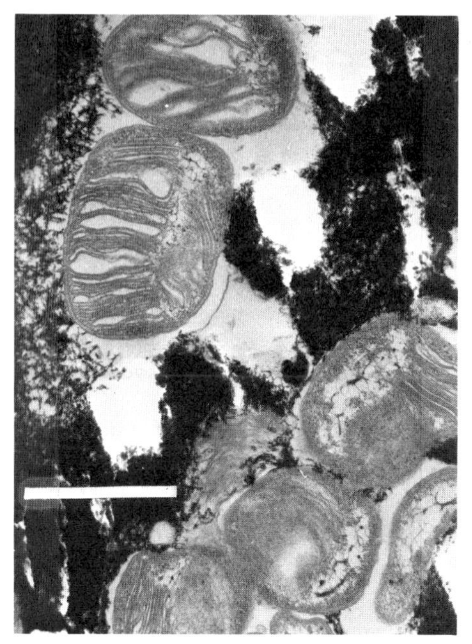

Figure 11. Transmission electron micrograph of part of the section pictured in Figure 9 showing cells with intracellular membrane systems resembling those of "Type I" methylotrophic bacteria (scale 1 μm).

micrographs of microbial mats also revealed groups of cells morphologically resembling the genus Seliberia, an iron bacterium (Schmidt and Swafford, 1981), surrounded by heavy deposits of ferric hydroxide.

Most TEM sections of microbial mats collected in the vent vicinity show cells containing highly developed intracellular membrane systems (Figure 11). This morphological feature is known from methane oxidizing bacteria (Davis and Whittenbury, 1970) and, since methane was found in the hydrothermal fluid, its bacterial consumption at suitable concentrations in an oxic environment is most likely. Indeed methylotrophic (methane and methanol oxidizing) bacteria could be isolated from these mats (see below). Since the bacterial production of methane within the vent system is most likely, according to evidence obtained with mixed populations and pure cultures of methanogenic bacteria (see below), a complete cycling of carbon with hydrogen and possibly hydrogen sulfide as sources of electrons is a distinct possibility.

IN SITU EXPERIMENTS

Microbial transformations can be measured by the in situ inoculation and incubation of vessels which are precharged with the labeled substrate to be converted. Such experiments were done with arrays of six 200 ml syringe-type water bottles each, holding the $^{14}CO_2$ in a few ml, filled with the turbid vent water and left at the site for incubation of one or two days. Parallel vessels contained an organic carbon source or thiosulfate.

The total CO_2 assimilation or chemosynthetic production in the emitted vent waters, freshly mixed with oxygenated ambient seawater of 2-3°C, was indeed minor as compared to the bacterial biomass produced within the subsurface vent system prior to emission (Tuttle et al., 1983). Uptake of acetate and glucose indicated the presence of mixotrophic or facultatively chemolithotrophic bacteria (see below) which was in agreement with the results of the pure culture isolations. The demonstration of certain enzyme activities (ribulose diphosphate carboxylase and phosphenol pyruvate carboxylase) in collected cells as well as in pure culture isolates supports the general notion that these organisms are the chemosynthetic source of the particulate organic carbon in the vent water.

Our data (unpublished) from the recent 21°N cruise demonstrate that maximum amounts of CO_2 assimilation occurred at 25°C, suggesting that chemosynthesis is carried out by a population of mesophilic, not psychrophilic or thermophilic organisms. This result does also agree with data obtained from growth studies on isolated strains. Added thiosulfate increased the final amount of CO_2 assimilation. No oxygen limitation was reached in these experiments.

These endpoint determinations were supplemented by time-course measurements conducted in a pressure-retaining 1 liter sampler (Jannasch et al., 1976). This approach allowed true rate determinations. At 25°C, the approximate temperature at which the samples were collected, a rate of 3-4 x 10^{-3} μmol CO_2/ml/day was measured. Again, the detailed study is not yet published.

Between two cruises to the Galapagos Rift vents, surface materials were left on subsurface buoys in vertical profiles of warm water vent plumes for a duration of 10 months. The surfaces of glass slides, clam shells, plastics and metals were almost equally covered by microbial mats. Ferrous-manganese deposits on glass slides were discussed above (Figure 7B).

MICROBIAL ENRICHMENTS AND ISOLATIONS

Bacterial enrichments are made in order to test for the presence of known or hypothesized metabolic types. To this end, specifically designed enrichment media are inoculated with the natural microbial population to be studied and incubated under special conditions. In order to check for chemosynthetic sulfur-oxidizing bacteria, test tubes with an H_2S-containing agar plug at the bottom overlayed by a sterile seawater medium were allowed to develop an H_2S/O_2 diffusion gradient prior to being inoculated with a piece of mussel periostracum or a sample of hydrothermal fluid. Phenol red was added as a pH indicator. Bacteria grew in a tight band near, but distinctly removed from the liquid/air interface clearly indicating (1) microaerophilic growth characteristics, (2) chemoautotrophic growth (no organic source of energy available), (3) tolerance of certain H_2S concentrations, and (4) production of acid (yellow zone).

The same enrichment principles apply in attempting to find strictly anaerobic, thermophilic, H_2 + CO_2 utilizing methanogenic bacteria and aerobic, methane oxidizing methylotrophic bacteria, etc. Samples of enrichment cultures as well as inoculation material are kept in liquid nitrogen for storage and later use.

The large number of successful enrichment cultures (about 265) from various parts of the Galapagos Rift and 21°N vents indicates the wide distribution of chemosynthetic sulfur bacteria (Ruby et al., 1981). Approximately 70% of these enrichment cultures resulted in base producing strains which oxidize H_2S or thiosulfate to polythionates instead of sulfate:

$$2S_2O_3^{=} + O_2 + H_2O \rightarrow S_4O_6^{=} + 2OH^- \quad \text{base producing sulfur bacteria,} \qquad 1)$$

$$S_2O_3^{=} + 2O_2 + H_2O \rightarrow 2SO_4^{=} + 2H^+ \quad \text{acid producing sulfur bacteria.} \qquad 2)$$

The former outgrew readily the latter when traces of certain organic compounds were added to the medium. This successful competition of the facultatively autotrophic base producers over the obligately autotrophic acid producers appears to be aided by the buffering capacity of seawater. It can be hypothesized that a preferable site for growth of the acid producing bacteria are mats within the subsurface vent systems where densely packed cells may be able to affect the pH of this microenvironment considerably.

Many of these enrichments led to the isolation of pure cultures using common bacteriological techniques such as dilution tubes and agar streaking plates, etc. Out of about 200 isolates from the Galapagos vent site, a large variety of metabolic types was described (Ruby et al., 1981). One of the most prominent genera, the obligate chemoautotrophic Thiomicrospira was characterized in more detail (Ruby and Jannasch, 1982). The organisms are mesophilic, i.e., the incorporation of $^{14}CO_2$ is fastest at 25°C, and barotolerant, i.e., growth and $^{14}CO_2$ incorporation are not affected by pressures of up to 100 atm but are entirely inhibited at 500 atm. At the in situ pressure of the vents, ca. 250 atm, the isolate retained 75% of its 1 atm activity. The optimum growth temperature of this Thiomicrospira isolate was 25°C. These physiological traits are not unlike those of the other known species of this genus (Kuenen and Veldkamp, 1972), isolated from sulfide-rich estuaries.

Enrichments of the chemosynthetic ammonia- and nitrite-oxidizing bacteria were positive (J. B. Waterbury, personal communication) but their numbers estimated by dilution experiments were low. It appears that the chemosynthetic input of organic carbon by nitrification is small whether or not ammonia is generated geothermally or by recycling of organic nitrogen.

If sources of combined nitrogen were omitted from the enrichment media for various metabolic types of bacteria, active growth suggested the potential ability to reduce dinitrogen (N_2). As verified by the acetylene reduction technique, this trait of nitrogen fixation appeared to be widely distributed in the bacterial strains isolated from the vents (unpublished).

Cells resembling Hyphomonas and Hyphomicrobium were frequently observed and later isolated by J. S. Poindexter (personal communication). As mentioned above, the requirement of amino acids as the sole source of carbon appears to put most of them in the former genus (Moore, 1981). An anaerobic spirochaete, the first one isolated from the deep sea, was originally observed with phase-contrast microscopy in acetate containing enrichments from microbial mat material collected at the Galapagos Rift vents. The organism grew extremely slowly, but its purification and isolation was ultimately achieved (Harwood et al., 1982). Manganese-oxidizing bacteria were isolated by H. L. Ehrlich (1982). In comparative experiments it was shown that these isolates were able to oxidize free Mn^{2+} and do not depend on a prior binding of Mn^{2+} to certain specific absorbents as found in similar strains isolated from ferromanganese nodules. The vent cultures also differ from the nodule cultures in that their Mn^{2+} oxidase is inducible and not constitutive. Ehrlich concludes from his studies that manganese oxidizing bacteria do contribute to the primary production at deep sea hydrothermal vents.

Anaerobic microbial enrichments inoculated with water from the 21°N hot vents were found to produce methane and carbon monoxide at about 100°C (Baross et al., 1982). Supporting these observations, an extremely thermophilic methane producing bacterium was recently isolated in pure culture from samples collected at the same site (Leigh and Jones, 1983). When the presence of aerobic methane oxidizing bacteria was indicated by morphological observations (Figure 11), isolations were attempted from a large variety of materials. Pure cultures of Type I methylotrophs, oxidizing methane and methylamine but not methanol, were readily obtained (R. Hanson, personal communication) from samples of microbial mats collected in the near vicinity of the vents, filtered vent water, clam gill tissue and Riftia trophosomes (see below).

AEROBIC MICROBIAL CHEMOSYNTHESIS

Since the chemosynthetic production of organic carbon from CO_2 appears to be the most important biological process at the hydrothermal vents, a general discussion of its definition and ecology is included in this paper.

Chemosynthesis is carried out by chemoautolithotrophic bacteria. "Chemo"-autotrophy designates the use of chemical energy instead of light as in "photo"-trophy. The independence of organic carbon and the use of CO_2 as the sole source of carbon is expressed by the term "auto"-trophic. The term "litho"-trophic indicates that inorganic compounds are serving as sources of electrons (namely H_2, H_2O, $S^=$, $S°$, $S_2O_3^=$, NH_4^+, NO_2^-, Fe^{++} or Mn^{++}). Of the three autolithotrophic processes:

non-oxygenic photosynthesis (bacteria),

$$2CO_2 + H_2S + 3H_2O \xrightarrow{hv} 2[CH_2O] + H_2SO_4 + H_2O; \quad 3)$$

oxygenic photosynthesis (green plants),

$$CO_2 + H_2O \xrightarrow{hv} [CH_2O] + O_2; \quad 4)$$

chemosynthesis (bacteria),

$$CO_2 + H_2S + O_2 + H_2O \rightarrow [CH_2O] + H_2SO_4; \quad 5)$$

the latter requires free oxygen as the necessary electron sink. Chemosynthetic organisms, therefore, can be considered late arrivals from the evolutionary point of view assuming that free oxygen appeared in the early atmosphere only or largely as a product of green plant photosynthesis. This point is made independently of the thought that the vents might indeed be a location with the potential of producing primary life forms such as methanogenic archaebacteria (see below).

On the per-mole basis, the gain of energy obtained by the complete oxidation of hydrogen sulfide via elemental sulfur to sulfate exceeds the energy required for the reduction of CO_2 to a carbohydrate by about 40%. The yields are different for other electron donors, a fact that is important, for example, for the occurrence of nitrification (chemosynthetic oxidation of ammonia and nitrite) or the bacterial deposition of iron and manganese oxides under various microenvironmental conditions. The emphasis on sulfur oxidation as the primary chemosynthetic process at the hydrothermal vents is based merely on the relative abundance of reduced sulfur species and on the results of enrichment and isolation studies.

Under conditions where most of the reduced sulfur occurs as a result of mining operations, its chemical or chemosynthetic oxidation leads commonly to a strong acidification of the environment. Acid mine water drainage and acid rain, after the SO_2^- disposal into the atmosphere, are examples par excellence. The high buffering capacity of seawater prevents acidification in marine environments wherever active chemosynthetic bacteria occur commonly in succession with sulfate reducing bacteria (Figure 13). Growth of acidophilic bacteria may also occur in seawater if the environmental conditions favor the formation of dense bacterial mats.

For practical purposes, the chemosynthetic sulfur oxidizing bacteria can be divided into three categories: (1) the acid producing, obligate autolithotrophic bacteria and (2) the non-acid-producing, facultative autolithotrophic bacteria (both groups of the unicellular thiobacillus-type morphology as in Figures 2 and 3), and (3) the "large", often filamentous, bacteria of the _Beggiatoa_-, _Thiothrix_-, or _Thiovulum_-type (Figures 4, 5 and 6) which are notoriously difficult to grow in culture and few of which have been purified. The non-acidophilic bacteria produce polythionates instead of sulfate (equation 1) slightly raising the pH over that of seawater. The organisms of the category (3) belong to the largest bacteria known commonly reaching cell widths of 5 to 50 μm and, as recently recorded, over 100 μm (Figure 6). Few of the filamentous sulfur-oxidizing bacteria have so far been cultivated. Growth of these organisms is more interface-(H_2S/O_2) dependent than that of the unicellular thiobacillus-type bacteria, a fact that accounts for the difficulties in their cultivation. Globules of elemental sulfur are commonly observed within the cells. They may disappear when the external concentration of H_2S decreases and the internal store of S° is oxidized.

Requirements for organic substrates as additional sources of energy and carbon vary within most groups of chemosynthetic bacteria. While "obligately" autotrophic chemosynthetic bacteria may even be inhibited by the presence of organic carbon, "facultative" autotrophs can switch from the inorganic to an organic carbon source, and "mixotrophic" bacteria are able to use both at the same time. The highly acidophilic thiobacilli tend to belong to the group of obligately chemoautotrophic bacteria. At the other extreme end of the line, the polythionate-producing isolates use little reduced sulfur along with various sources of organic carbon. There are a number of strains known which fit various points on the gradient between these two extremes. From the isolates so far obtained, it is not quite clear at this time which of the various metabolic types of bacteria is responsible for the bulk of sulfur oxidation and aerobic chemosynthetic CO_2 reduction at the hydrothermal vents.

METHANOGENESIS AND METHANE OXIDATION

The bacterial formation of methane from hydrogen and carbon dioxide can be defined as anaerobic chemosynthesis. It commonly occurs in mixed populations in association with fermenting, hydrogen producing bacteria at extremely low redox potentials, as, for instance, in the sediments of organically rich lakes, estuaries or sewage plants.

The emission of methane from deep sea hydrothermal vents is believed to represent its major source in the oceans (Welhan and Craig, 1979). It is generally assumed that this origin of methane is based on a geothermic process. On the other hand, highly reduced conditions within some of the vent systems appear to be conducive also to the biological formation of methane. The subsurface introduction of oxygenated seawater into the rising hydrothermal fluid must be assumed an irregular event. In many cases it can be expected to be insufficient to raise the redox potential beyond the range required for growth of strictly anaerobic methanogenic bacteria. Even warm vents with low flow rates may be non-oxic prior to emission into the oxic ambient seawater. Proof of this fact is difficult to provide since the first mixing of hydrothermic fluid with seawater may occur in places of vent openings inaccessible for sampling. By the same token, strictly anaerobic bacteria may not survive small amounts of oxygenated seawater introduced accidently into the sample. The observation of methane production in enrichment cultures from hot water samples (Baross et al., 1982) as well as the pure culture isolation of an extremely thermophilic methanogenic bacterium

(Leigh and Jones, 1983) with an optimum growth temperature of 86°C, provide strong evidence of more than a potential for methanogenesis in deep sea hydrothermal vents.

If written to include the production of organic cell carbon, bacterial methanogenesis can be expressed as:

$$6H_2 + 2CO_2 \rightarrow (CH_2O) + CH_4 + 3H_2O \qquad 6)$$

Similarly bacterial acetogenesis is carried out by a number of anaerobic chemoautotrophic genera (*Clostridium*, *Acetobacterium*, *Acetogenium*), none of which has yet been isolated from vent waters:

$$10H_2 + 5CO_2 \rightarrow (CH_2O) + 2CH_3COOH + 5H_2O \qquad 7)$$

Acetate has been found to be readily metabolized by a number of facultatively chemoautotrophic sulfur bacteria isolated from warm vents (Ruby et al., 1981), but this fact is no conclusive evidence for the existence of acetate in vent waters, although it is likely to occur.

As indicated by the microscopic observation of cytoplasmic membrane systems characteristic for methylotrophic bacteria (see above, Figure 11) and their subsequent isolation, the oxidation of methane in the bacterial mats surrounding the vent openings is most likely. This aerobic transformation:

$$2CH_4 + 3O_2 + H_2O \rightarrow (CH_2O) + CO_2 + 4H_2O \qquad 8)$$

implies the complete recycling of dissolved organic carbon within vent systems. This methane turnover is complete only in the qualitative sense. Considerable quantities of methane are dispersed in ocean waters escaping the biological oxidation near the vents.

The net of the equations 6) and 8):

$$4H_2 + CO_2 + O_2 \rightarrow (CH_2O) + 3H_2O \qquad 9)$$

corresponds to an aerobic transformation carried out by hydrogen bacteria. Although their occurrence at the vents is likely, they have not yet been isolated.

SYMBIOSIS OF CHEMOSYNTHETIC BACTERIA WITH VENT INVERTEBRATES

The unusually large clams (*Calyptogena magnifica*; Boss and Turner, 1980) and the vestimentiferan tube worms (*Riftia pachyptila*; Jones, 1980) appear to constitute the major portion of total animal biomass at the vents. The nutrition of both of these species was a puzzle to the early observers. As morphological studies showed, the clams are not typical filter-feeders. They are primarily found certain distances away from vent openings where the concentrations of bacterial cells in the surrounding seawater are extremely low or imperceptible. In Riftia the anatomical description revealed the complete absence of the ingestive system, i.e., mouth and gut (Jones, 1980). Its suggested nutrition by the resorbance of organic matter through the epidermis was energetically untenable. Eventually, microscopic (Cavanaugh et al., 1981) and enzymatic (Felbeck, 1981) evidence suggested that the animals harbor actively metabolizing populations of chemosynthetic bacteria in new and not yet fully understood symbiotic associations.

The cavity in the trunk of Riftia, the most extensive section of the worm's length (over 50%), is occupied by the trophosome. It represents a tissue-like material that was detected and named in another pogonophoran genus, Lamellibrachia, but its function was unknown. Its granular appearance resolves in scanning electron micrographs as densely packed spherical bodies (Figure 12) with diameters of 3 to 5 μm. When sectioned and observed by transmission electron microscopy, these structures were identified as procaryotic cells with walls resembling those of Gram-negative bacteria (Cavanaugh et al., 1981).

Figure 12. Scanning electron micrograph of "trophosome" material from Riftia (scale 5 μm).

This trophosome material contains ATP-producing enzymes of the microbial sulfur metabolism (thiosulfate sulfur transferase, APS reductase and ATP sulfurylase) as well as the Calvin-Benson cycle enzymes ribulose biphosphate carboxylase and ribulose 5-phosphate kinase (Felbeck, 1981). A study on the fine structure and physiology of a different pogonophoran genus, Siboglinum, and its trophosomal enzymes confirms the fact that the symbiosis with chemosynthetic bacteria is widespread in this phylum (Southward et al., 1981).

It is believed that the blood system in Riftia acts as the carrier for oxygen as well as hydrogen sulfide. The kinetics of the hemoglobin reaction with oxygen is relatively independent of temperature (Wittenberg et al., 1981), a fact that might relate to the occurrence of these animals in waters changing from 2° to 23°C, and most likely to much higher temperatures at the 21°N vents. Compatible with the high oxygen demand of chemoautotrophic metabolism in a variable environment is the fact that the extracellular hemoglobin has a specifically high oxygen affinity and carrying capacity (Arp and Childress, 1981).

Our recent work on bacterial isolations from trophosome material (in preparation) resulted in a number of strains of sulfur- and methane-oxidizers, only one of them resembling the large spherical cells when cultured in liquid medium. According to the present state of this work, it appears that the trophosome does not represent a nearly pure culture of symbionts but contains a number of bacteria that might (a) either exist in this heterogeneous environment with no or little consequence to the function of the trophosome, or (b) belong to a complex consortium-type, more or less stable, population of symbionts.

Corresponding work on the giant clam, Calyptogena, is yet largely unpublished. New material was collected during the April/May 1982 Biology Cruise to the 21°N spreading area. In the meantime, however, the phenomenon of chemosynthetic symbiosis in marine benthic invertebrates living in sulfide containing sediments has been also found in several shallow water species (Felbeck et al., 1981) and is now studied independently of the much larger vent specimen.

Another possible confirmation of the use of chemosynthetically produced organic carbon for food by the vent animals has been based on data indicating increased stable carbon isotope ratios ($^{13}C/^{12}C$) in their tissues (Rau, 1981; Southward et al., 1981).

DISCUSSION AND CONCLUSIONS

Table 2 provides an overview of the microbial transformations so far under study. Chemosynthesis can be taken as the most important microbial process at the vents because of its role in giving rise to a highly productive population of animals in the absence of light. Although chemosynthesis is long known in microbiology and biochemistry, its more or less sole support of an ecosystem is ecologically novel. The scheme of the interacting carbon and sulfur cycles (Figure 13) demonstrates the reasons why the non-photosynthetic production of organic carbon is generally termed "secondary production" because it depends on a photosynthetic source of carbon for the bacterial reduction of sulfate. If, however, the source of hydrogen sulfide is non-biological, the chemosynthetic production of organic carbon becomes a primary

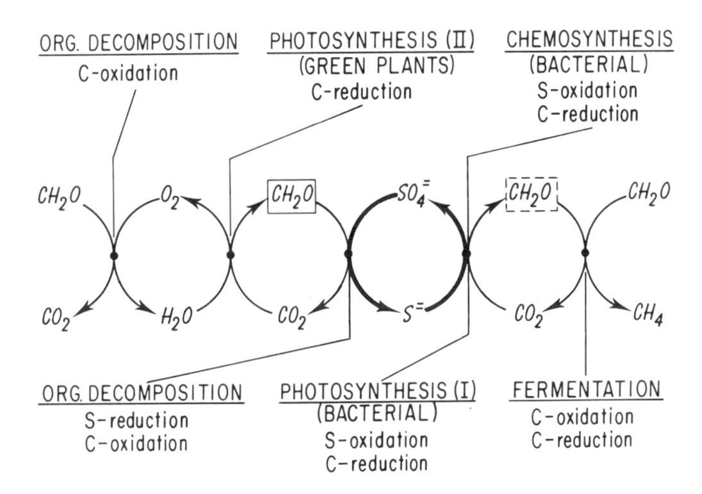

Figure 13. Partial scheme of the carbon/sulfur cycle and types of microbial transformations. Primary photosynthetic production of organic carbon (solid square) normally leads to secondary chemosynthetic production (dashed square). At the vents, this chemosynthetic production may be termed primary because the source of $S^=$ is not related to another biological process but to the geothermal (abiotic) reduction of sulfur (from Jannasch, 1983).

TABLE 2

MICROBIAL PROCESSES AT DEEP SEA HYDROTHERMAL VENTS

1. Chemosynthetic reduction of CO_2 to organic carbon by

   a) sulfur-oxidizing bacteria ($S^=$, S°, $S_2O_3^=$),

   b) symbiotic associations of sulfur-oxidizing procaryotes with various invertebrates,

   c) other aerobic chemoautotrophic bacteria (H_2, NH_4^+, NO_2^-, Fe^{++} and Mn^{++} oxidizers),

   d) anaerobic and possibly highly thermophilic, chemoautotrophic bacteria (methanogens, acetogens).

2. Oxidation of methane by free-living bacteria and possibly also in symbiosis with invertebrates.

3. Nitrogen fixation in various of the above-mentioned types of microorganisms.

4. Heterotrophic microbial processes that occur as a result of chemosynthesis (including sulfate reduction based on the oxidation of organic compounds as well as H_2) and the possible microbial involvement in the formation or in transformations of indigenous hydrocarbons.

process (Jannasch and Wirsen, 1979). This terminology may be challenged on the grounds that the necessary electron acceptor, free oxygen, is also a product of photosynthesis. The terms primary and secondary production, however, are commonly based on the source of energy, not on the sink of energy. In photosynthesis the source of energy is solar, while in chemosynthesis at the vents it is terrestrial (geothermal).

In the fictional event of the sun's radiation being blocked from reaching the earth's surface, photosynthetically supported life would cease to exist while the vent ecosystems, sustained by aerobic chemosynthesis, would persist as long as the store of free oxygen in seawater would permit. Anaerobic chemosynthesis could be assumed to prevail after the total consumption of free oxygen. Vice versa, the vents have been speculated as being suitable sites for the origin of life and the first appearance of certain bacteria now phylogenetically grouped under the archaebacteria (see below).

As an alternative explanation to the chemosynthetic support of life at the vents, the hypothesis has been advanced (Lonsdale, 1977; Enright et al., 1981) that the food supply for the vent populations is related to the formation of thermo-convective cells. Warm water rising over a vent would produce a bottom current which collects and carries particulate food materials toward the vents. Not to be exhausted within a relatively short time, this system must rely on a more or less direct connection to photosynthetically productive surface waters.

Data on the distribution of dissolved manganese, however, show that even the emissions of the very hot vents (350°C) at the 21°N, 109°W area with the maximum flow rates of 2 m/sec do not rise higher than the bottom 10% of the 2600 m water column (Lupton et al., 1980). Furthermore, vents with low flow rates (less than 2 cm/sec) and small temperature gradients (maximally 23°C) and plumes of a few m in diameter, consistently exhibit much higher concentrations and total quantities of animals than the hot vents.

If photosynthesis is assumed to be the primary food source for the vent communities, simple calculations from the data presented by Karl et al. (1980, Table 1) show that the required particulate material produced by a considerable area of surface water must funnel its way down to the mussel or clam beds in a highly concentrated form with virtually no loss by decomposition. Besides being an oceanographically improbable feature, no macroscopical or microscopical evidence supports this possibility.

Finally, the distribution of animals in well separated populations around small vents, often not more than a few meters apart, indicates their dependence on a tightly localized food source. The population size of those animals that depend on symbiosis with chemosynthetic bacteria, and the distance of these populations from the point of hydrothermal fluid emission, appear to be related to a favorable concentration of sulfide in the seawater. One extreme is represented by the "Rose Garden" vent site at the Galapagos Rift where the sulfide appears to be emitted over a large bottom area at low flow rates causing growth of a dense forest of tube worms over a relatively uninterrupted layer of mussels. At the "Black Smokers" of the 21°N vent site, on the other hand, the sulfide concentrations in the emitted hot water are extremely high (Spiess et al., 1980), possibly at toxic levels. Growth of the giant clams and tube worms (mussels are absent from this area) takes place in greater distance from these vents. Commonly mussels as well as clams and, more rarely, few individual tube worms, are found along the cracks and furrows between the boulders of pillow lava. At such sites, whitish deposits often indicate minute vents that seem to support small numbers of sizable specimens.

Three major sites of chemosynthetic production can be identified: microbial subsurface mats and emission from the vents, microbial mats within and surrounding vent openings, and symbiotic associations with vent invertebrates. Growth within and very near an opening of a vent system is favored by the elevated temperatures. The rate of CO_2 uptake in turbid vent water incubated at 25°C (the temperature measured in a large warm water pocket at the 21°N vent site) was 1.5 to 2.0 orders of magnitude higher than that determined at 3°C (data to be published). Other new and yet unpublished results concern considerable enzymatic (ribulose biphosphate carboxylase) activity in microbial mats surrounding some vents.

Quantitatively the symbiotic associations appear to represent the most important site giving rise to the predominating populations of clams and tube worms representing up to 90% of animal biomass at the vents. The highly increased efficiency in the use of the available energy source, hydrogen sulfide and possibly methane, by localizing the chemosynthetic production within the animals themselves, is a striking evolutionary accomplishment. Nothing quite comparable exists on land or shallow marine waters where light as the predominant source of energy led to very different types of food chains. Only recently, and after the discovery of the symbiosis between chemosynthetic bacteria and invertebrates at deep sea vents, have similar systems been found in coastal waters (Cavanaugh et al., 1981; Felbeck et al., 1981; Southward et al., 1981). However, here chemosynthesis represents a "secondary" production since hydrogen sulfide is supplied by bacterial sulfate reduction (Figure 13). An exception are warm water springs found in shallow coastal waters.

The newly discovered vent sites in the Guaymas Basin, located in the center of the Gulf of California, presents a situation very different from those of the Galapagos Rift and 21°N sites. At a depth of about 2000 m, only 90 km from either shore, the vents are overlayed by sediments penetrated by the hydrothermal discharges (Lonsdale et al., 1980). Recent and fossil diatomaceous oozes are evidence for a considerable photosynthetic input. A thermogenic formation of hydrocarbons as a result of a high conductive heat flow within these sediments appears to be another characteristic of these complex hydrothermal vent systems (Simoneit and Lonsdale, 1982). Microbiological observations so far have been limited to the above mentioned heavy layers of whitish mats on large segments of sediment (Figure 6).

Much interesting speculation has been advanced with respect to the abiotic formation of organic compounds in deep sea hot vents rendering the latter preferred sites for the origin of life on earth (Corliss et al., 1981). Apart from this aspect, the possible geothermal production of organic compounds under high pressure and temperatures would certainly aid facultatively chemoautotrophic as well as heterotrophic microorganisms entering the vents. Data on the presence and concentrations of organic compounds in vent waters are now forthcoming after the recent use of an in situ pump system specially designed to prevent sample contamination with bottom seawater. There will be certain difficulties in distinguishing abiologically produced dissolved organic compounds from those produced by chemosynthetic bacteria within the vent systems, unless studies on stable carbon isotope fractionation will be able to solve the problem. The capacity of acidophilic bacterial vent isolates to excrete up to 9% of the total assimilated organic carbon in dissolved form has been documented by Ruby et al. (1981), and the occurrence of acetogenic bacteria (see above) is a definite possibility.

There is a potential for pyrophosphate to be formed within the vents, an energy-rich compound that has been proposed as an evolutionary precursor to adenosine 5'-triphosphate (Lipmann, 1966). This inorganic pyrophosphate has recently been shown to serve as a source of energy for growth of certain sulfate reducing bacteria (Liu et al., 1982).

The idea of looking for unique groups of microorganisms in the vents was fostered by the work on "archaebacteria" (Balch et al., 1979). This group of organisms, comprising methanogens as well as acidophilic, thermophilic and halophilic bacteria, are characterized by their unique nucleotide sequence in the 16S ribosomal RNA structure. Work is in progress studying the usefulness of nucleotide sequence analyses by recombinant DNA techniques to clone the 16S ribosomal RNA genes in order to assess the species composition and phylogenetic relations within the microbial vent populations in the absence of enrichment procedures.

Very recently a number of new thermo-acidophilic bacteria have been isolated (*Thermoproteus*, *Desulfurococcus* and *Thermococcus*) from terrestrial and shallow marine hot springs (Zillig et al., 1981 and 1982). They show the biochemical characteristics of the archaebacteria and exhibit growth optima within a pH range of 4.5 to 5.5 at temperatures of 85 to 95°C. Some of them reduce elemental sulfur with hydrogen to hydrogen sulfide. Since their carbon source, however, is organic they have to be described metabolically as chemo-heterolithotrophic. Heinen and Lauwers (1981) and Stetter et al. (1982) have grown bacteria beyond 100°C at the corresponding pressures, Stetter et al.

reporting no growth below 80°C, optimal growth at 105°C and weak growth still at 110°C in a yet undescribed organism isolated from a submarine solfatara field. Freshly formed and collected sulfide deposits from the chimneys of 21°N hot smokers showed, next to the above-mentioned production of methane (Baross et al., 1982), appreciable concentrations of ATP and rates of RNA and DNA synthesis when incubated at 21°, 50° and 90°C (Karl et al., in preparation).

With respect to the uniqueness of the vent microorganisms, the following considerations can be made at the present state of our knowledge. Most of the aerobic, mixotrophic, facultative or obligate chemoautotrophic bacteria so far isolated represent strains similar to existing metabolic types. This may not be surprising since deep sea conditions do not destroy the viability of those microorganisms which are continuously introduced from surface waters (ZoBell, 1968; Jannasch and Wirsen, 1983). Likewise high barotolerance is known to be widely distributed among microorganisms isolated from deep as well as shallow waters (Marquis and Matsumura, 1978). On the other hand, uniquely high growth rates for chemoautotrophic bacteria were reported by Leigh and Jones (1983) for a methanogenic vent isolate with a 30 min doubling time at 86°C, and by Nelson et al. (in preparation) for a new vent isolate of the genus Thiomicrospira with a 70 min doubling time at 36°C. The ability to fix nitrogen appears to be widely distributed among the vent chemoautotrophs (unpublished), and pressure as well as temperature adaptations are to be expected.

According to the finding of highly pressure adapted, heterotrophic deep sea bacteria in certain niches of high nutrient concentrations, such as the intestinal tract of benthic invertebrates (Yayanos et al., 1979; Deming et al., 1981), similar isolates may be found in microbial vent populations living at ambient temperature of 2.1°C. In addition to the common occurrence of high-pressure and low-temperature adaptations in deep sea bacteria, a combination of the former with thermophilic characteristics may also be expected in vent isolates. While thermophilic, methanogenic and acetogenic bacteria have been reported from other environments, growth of such vent isolates at temperatures substantially beyond 100°C aided by the elevated pressures might indeed be unique for this particular deep sea environment.

ACKNOWLEDGEMENTS

Much of the newly reported work in this paper was done, where not specifically mentioned, in close cooperation with C. O. Wirsen and D. C. Nelson. Thanks for invaluable assistance in the SEM and TEM work are due to E. Seling and J. B. Waterbury. I am also indebted to the submersible's pilots R. Hollis, G. Ellis and R. Brown as well as the entire crews of DSRV ALVIN and R/V LULU. The research was supported by the National Science Foundation grants OCE80-24253 and OCE81-17560. Contribution No. 5217 of the Woods Hole Oceanographic Institution and No. 7 of the 21°N Biology ("Oasis") Expedition, 1982.

LITERATURE CITED

Arp, A. G. and Childress, J. J., 1981, Blood function in the hydrothermal vent vestimentiferan tube worm. Science 213: 342-344.

Balch, W. E., Fox, G. E., Magrum, L. J., Woese, C. R. and Wolfe, R. S., 1979, Methanogens: reevaluation of a unique biological group. Microbiol. Rev. 43:260-296.

Baross, J. A., Lilley, M. D. and Gordon, L. I., 1982, Is the CH_4, H_2 and CO venting from submarine hydrothermal systems produced by thermophilic bacteria? Nature 298:366-368.

Boss, K. J. and Turner, R. D., 1980. The giant white clam from the Galapagos Rift *Calyptogena magnifica* species novum. Malacologia 20:161-194.

Cavanaugh, C. M., Gardiner, S. L., Jones, M. L., Jannasch, H. W. and Waterbury, J. B., 1981. Procaryotic cells in the hydrothermal vent tube worm *Riftia pachyptila* Jones: possible chemoautotrophic symbionts. Science 213:340-341.

Corliss, J. B., Dymond, J., Gordon, L. I., Edmond, J. M., von Herzen, R. P., Ballard, R. D., Green, K., Williams, D., Bainbridge, A., Crane K. and van Andel, T. H., 1979, Submarine thermal springs on the Galapagos Rift. Science 203:1073-1083.

Corliss, J. B., Baross, J. A. and Hoffman, S. E., 1981, An hypothesis concerning the relationship between submarine hot springs and the origin of life on earth. Oceanol. Acta No. SP, 59-69.

Davis, S. L. and Whittenbury, R., 1970, Fine structure of methane and other hydrocarbon utilizing bacteria. J. Gen. Microbiol. 61:227-232.

De Bont, J. A. M., van Dijken, J. P. and Harder, W., 1981. Dimethyl sulphoxide and dimethyl sulphide as a carbon, sulphur and energy source for growth of *Hyphomicrobium* S. J. Gen. Microbiol. 127:315-323.

Deming, J. W., Tabor, P. S. and Colwell, R. R., 1981, Barophilic growth of bacteria from intestinal tracts of deep-sea invertebrates. Microb. Ecol. 7:85-94.

Edmond, J. M., Corliss, J. B. and Gordon, L. I., 1979, Ridge crest-hydrothermal metamorphism at the Galapagos spreading center and reverse weathering, in: "Deep Drilling Results in the Atlantic Ocean: Ocean Crust," M. Talwani, C. Harrison and D. Hayes eds., Amer. Geophys. Union, Washington, D.C., pp. 383-390.

Ehrlich, H. 1982. Manganese oxidizing bacteria from a hydrothermally active area on the Galapagos Rift. Ecol. Bull. 35. In press.

Enright, J. T., Newman, W. A., Hessler, R. R. and McGowan, J. A., 1981, Deep-ocean hydrothermal vent communities. Nature 289: 219-221.

Felbeck, H. 1981. Chemoautotrophic potentials of the hydrothermal vent tube worm, Riftia pachyptila (Vestimentifera). Science 213: 336-338.

Felbeck, H., Childress, J. J. and Somero, G. N., 1981, Calvin-Benson cycle and sulfide oxidation enzymes in animals from sulfide-rich habitats. Nature 293:291-293.

Harwood, C. S., Jannasch, H. W. and Canale-Parole, E., 1982, An anaerobic spirochaete from deep sea hydrothermal vents. Appl. Environ. Microbiol. 44:234-237.

Heinen, W. and Lauwers, A. M., 1981, Growth of bacteria at 100°C and beyond. Arch. Microbiol. 129:127-128.

Jannasch, H. W., 1983, Interactions between the carbon and sulfur cycles in the marine environment, in: "The Major Biochemical Cycles and Their Interactions," B. Bolin and R. Cook, eds., Wiley, New York. In press.

Jannasch, H. W. and Wirsen, C. O., 1979, Chemosynthetic primary production at East Pacific sea floor spreading centers. BioScience 29:592-598.

Jannasch, H. W. and Wirsen, C. O., 1981, Morphological survey of microbial mats near deep-sea thermal vents. Appl. Environ. Microbiol. 41:528-538.

Jannasch, H. W., Wirsen, C. O. and Taylor, C. D., 1976, Undecompressed microbial populations from the deep sea. Appl. Environ. Microbiol. 32:360-367.

Jones, M. L., 1980, Riftia pachyptila, n. gen., n. sp., the vestimentiferan worm from the Galapagos Rift geothermal vents (Pogonophora). Proc. Biol. Soc. Wash. 93:1295-1313.

Jones, M. L., 1981, Riftia pachyptila Jones: observations on the vestimentiferan worm from the Galapagos Rift. Science 213: 333-336.

Karl, D. M., 1980, Cellular nucleotide measurements and applications in microbial ecology. Microbiol. Rev. 44: 739-796.

Karl, D. M., Wirsen, C. O. and Jannasch, H. W., 1980, Deep-sea primary production at the Galapagos hydrothermal vents. Science 207:1345-1347.

Kuenen, J. G. and Veldkamp, H., 1972, Thiomicrospira peloptila gen. n., sp. n., a new obligately chemolithotrophic colorless sulfur bacterium. Ant. van Leeuwen. 38:241-256.

Leigh, J. A. and Jones, W. J., 1983, A new extremely thermophilic methanogen from a submarine hydrothermal vent. Am. Soc. Microbiol., 83rd Ann. Meetg., New Orleans.

Lipmann, F., 1966, The Origins of Prebiological Systems, "Mir" (Moscow), pp. 261-271.

Lister, C. R. B., 1977, Qualitative models of spreading center processes, including hydrothermal penetration. Tectonophysics 37:203-218.

Liu, C. L., Hart, N. and Peck, H. D., 1982, Inorganic pyrophosphate: energy source for sulfate reducing bacteria of the genus Desulfotomaculum. Science 217:363-364.

Lonsdale, P., 1977, Clustering of suspension-feeding macrobenthos near abyssal hydrothermal vents at oceanic spreading centers. Deep Sea Res. 24:857-863.

Lonsdale, P. F., Bischoff, J. L., Burns, V. M., Kastner, M. and Sweeney, R. E., 1980, A high-temperature hydrothermal deposit on the seabed at a Gulf of California spreading center. Earth Planet. Sci. Lett. 49:8-20.

Lupton, J. E., Klinkhammer, G., Normark, W., Haymon, R., Macdonald, K., Weiss, R. and Craig, H., 1980, Helium-3 and manganese at the 21°N East Pacific Rise hydrothermal site. Earth Planet. Sci. Lett. 50:115-127.

Marquis, R. E. and Matsumura, P., 1978, Microbial life under pressure, in: "Microbial Life in Extreme Environments," D. J. Kushner, ed., Academic Press, New York, pp. 105-158.

Moore, R. L., 1981, The genera Hyphomicrobium, Pedomicrobium, and Hyphomonas, in: "The Prokaryotes," M. P. Starr et al., eds., Springer Verlag, Berlin, pp. 480-487.

Mottl, M. J., Holland, H. D. and Corr, R. F., 1979, Chemical exchange during hydrothermal alteration of basalt by seawater - II. Experimental results for Fe, Mn, and sulfur species. Geochim. Cosmochim. Acta 43:869-884.

Rau, G. H., 1981, Hydrothermal vent clam and tube worm $^{13}C/^{12}C$: further evidence of non-photosynthetic food source. Science 213:338-340.

Ruby, E. G. and Jannasch, H. W., 1982, Physiological characteristics of Thiomicrospira sp. L-12 isolated from deep sea hydrothermal vents. J. Bacteriol. 149:161-165.

Ruby, E. G., Wirsen, C. O. and Jannasch, H. W., 1981, Chemolithotrophic sulfur-oxidizing bacteria from the Galapagos Rift hydrothermal vents. Appl. Environ. Microbiol. 42:317-324.

Schmidt, J. M. and J. R. Swafford. 1981. The genus Seliberia. in: "The Prokaryotes," M. P. Starr et al., eds., Springer Verlag, Berlin, pp. 516-519.

Simoneit, B. R. T. and Lonsdale, P. F., 1982, Hydrothermal petroleum in mineralized mounds at the seabed of Guaymas Basin. Nature 295:198-202.

Southward, A. J., Southward, E. C., Dando, P. R., Rau, G. H., Felbeck, H. and Flugel, H., 1981, Bacterial symbionts and low $^{13}C/^{12}C$ ratios in tissues of Pogonophora indicate unusual nutrition and metabolism. Nature 293:616-620.

Spiess, F. N., MacDonald, K. C., Atwater, T., Ballard, R., Carranza, A., Cordoba, D., Cox, C., Diaz Garcia, V. M., Francheteau, J., Guerrero, J., Hawkins, J., Haymon, R., Hessler, R., Juteau, T., Kastner, M., Larson, R., Luyendyk, B., Macdougall, J. D., Miller, S., Normark, W., Orcutt, J. and Rangin, C., 1980, East Pacific Rise: Hot springs and geophysical experiments. Science 207:1421-1433.

Stetter, K. O., 1982, Ultrathin mycelia-forming organisms from submarine volcanic areas having an optimum growth temperature of 105°C. Nature 300:258-260.

Tuttle, J. H., Wirsen, C. O. and Jannasch, H. W., 1983, Microbial activities in the emitted hydrothermal vent waters of the Galapagos Rift vents. Mar. Biol. In press.

Welhan, J. A. and Craig, H., 1979, Methane and hydrogen in East Pacific Rise hydrothermal fluid. Geophys. Res. Lett. 6:829.

Wittenberg, J. B., Morris, R. J., Gibson, Q. H. and Jones, M. L., 1981, Hemoglobin kinetics of the Galapagos Rift vent tube worm Riftia pachyptila Jones (Pogonophora, Vestimentifera). Science 213:344-346.

Yayanos, A., Dietz, A. S. and Van Boxtel, R., 1979, Isolation of a deep-sea barophilic bacterium and some of its growth characteristics. Science 205:808-810.

Zillig, W., Schnabel, R., and Tu, J., 1982, The phylogeny of Archaebacteria, including novel anaerobic thermoacidophiles in the light of RNA polymerase structure. Naturwiss. 69:197-204.

Zillig, W., Stetter, K. O., Schafer, W., Janekovic, D., Wunderl, S., Holz, I. and Palm, P., 1981, Thermoproteales: a novel type of extremely thermoacidophilic anaerobic archaebacteria isolated from Icelandic solfataras. Zbl. Bact. Hyg., I. Abt. Orig. C2:205-227.

ZoBell, C. E., 1968, Bacterial life in the deep sea. Bull. Misaki Mar. Biol. Inst. (Kyoto) 12:77-96.

PRIMARY CONSUMERS FROM HYDROTHERMAL VENTS ANIMAL COMMUNITIES

D. DESBRUYERES and L. LAUBIER

Centre Océanologique de Bretagne
B.P. 337
29273 Brest Cédex - France

ABSTRACT

This paper reviews the present knowledge on the primary consumers from the hydrothermal vent community discovered in three different sites of the eastern Pacific (namely Galapagos ridge, East Pacific Rise at 21° N and 13° N). They display uniform ecological structure and zoological composition. Four major species of primary consumers dominate in term of biomass : the giant tube worm *Riftia pachyptila*, the large white clam *Calyptogena magnifica*, an undescribed mytilid musel and the Pompeii worm *Alvinella pompejana*. These species as well as some others primary consumers of minor importance are strictly linked with active hydrothermal vents, as shown by their spatial microdistribution surrounding the vents. In a way, the primary consumers of the hydrothermal community can be considered as having an r-type ecological strategy ? Large dispersal capabilities have been deduced in the case of the Galapagos mussel from the morphology of its larval shell. Actively swimming stages could also provide dispersal capability to the Pompeii worm.

Growth rates of bivalves have been studied using natural radionuclide shell content and direct growth measurement. Both methods show very high growth rates, from 1 to 4 centimeters per year as an average.

The luxuriance of the primary consumers from the hydrothermal community, as originally proposed in a preliminary hypothesis, comes from their ability to utilize the food source produced by chemosynthetic bacteria. Several approaches have been used including enzymological assays, histological investigations and radiogeochemical methods. At present, the most elaborate situation is probably the case of Riftia pachyptila, which possesses a special organ containing symbiotic chemoautotrophic bacteria. The energy transfer from chemosynthetic bacteria to primary consumers can also occurs in simpler ways in the case of other species.

At the boundary of the hydrothermal community, different types of ordinary deep-sea primary consumers are also found. Their abnormally high density can easily be explained by a hypothetical advective mechanism concentrating food particles.

The existing data are not sufficient to discuss the origin and antiquity of the strictly adapted primary consumers from hydrothermal vents, and future investigations in other suitable fast spreading centers would be of great interest.

INTRODUCTION

The discovery of luxuriant benthic animal colonies clustered around hydrothermal vents on the Galapagos ridge at 2 500-2 600 meters depth has been a tremendous surprise for marine biologists (Lonsdale, 1977). On a scale of a few hundred square meters around the vents, few species of huge size thrive, composing a true biogenic reef. The zoological composition of these communities is relatively constant within the three sites already known occurring on oceanic basalts on the Galapagos ridge and on the East Pacific Rise at 21° N and 11°-13° N[(1)]. The probably strangest species, Riftia pachyptila Jones is present at the three stations. This tube-worm belongs to the order Vestimentifera and the phylum Pogonophora. These animals are found to overcome 1.5 meters in length. They build sturdy cylindrical tubes which are basally closed. The posterior part of the tube lies in an iridescent water with temperature ranges from 8 to 13° C, while the ambient sea-water temperature is nearly 2° C. These tubes are clustered, forming real bushes in which several smaller organisms are living : different species of gasteropods, the most frequent being the limpets, polychaetous annelids (hesionids, polynoids, ampharetids, spionids, dorvilleids) as well as amphipods and decapods.

Two large bivalves live in the vicinity of these Riftia colonies (Calyptogena magnifica and an undescribed mytilid mussel). The giant white clam Calyptogena magnifica Boss and Turner (fam. Vesicomyidae), whose length can exceed 25 cm, occurs on the Galapagos site and on the EPR at 21° N, but has not been found at

(1) A similar community occurring on soft sediments has been recently discovered and sampled in the Guaymas basin, Gulf of Baja California (Dr. F. Grassle, pers. comm.).

11°-13° N where 28 active sites have been explored(1). The so-called mussel, an unnamed Mytilid, is very common at the Galapagos site where it creates real mussel-beds on basalt pillows and fissures. It also lives attached to vestimentiferan tubes, and measures up to 15 cm long. The "mussel" also occurs on the EPR, at 11°-13° N, in sparse numbers. In most of these mussels live one or two polynoid polychaetes. The mussels are generally attached to each other and to the rocks by strong byssal threads ending in greatly enlarged attachment areas.

On the two EPR hydrothermal sites, the two types of smokers (white smokers with water temperature from 70° to 150° C and black smokers > 300° C) are frequently partially covered by clusters of parchment-like tubes built by two species of polychaetous annelids belonging to the subfamily alvinellids. These animals live at the lower edge of a strong temperature gradient (from 250° C in the inner part of the tube-mass to 20° C at the surface with an accuracy of 0.1°). The tubes are open and the gills of the worm appear at the surface of the colony. The first species discovered at 21° N and nicknamed the Pompeii worm by the submersible divers has been described as _Alvinella pompejana_ ; this worm frequently emerges from its tube and has been observed swimming actively in the hydrothermal discharge. _Alvinella pompejana_ can reach nearly 10 cm in length, and exhibits two different morphological stages. _Alvinella_ has been collected at 21° N and 11-13° N. The second species more recently discovered, _Paralvinella grasslei_ Desbruyères and Laubier, occurs on the Galapagos ridge, on the EPR, at 11-13° N and 21° N, and in the Guaymas basin. The tube-building activity of alvinellids might contribute to the collapse of the hydrothermal chimneys, as suggested by some tube masses capping dead chimneys, resulting in the so-called "snow-ball" formation. Actinarians populations are present among the tubes, living at temperatures ranging from 12° to 20° C.

At the foot of the black smokers, metallic sulfide particles are deposited and are inhabited by a rich and little diversified fauna. The meiofauna is largely composed of harpacticoid copepods and polychaete larvae. On the basalts, few cirripeds of the family Scalpellidae are attached (_Neolepas zevinae_ Newman). In the boundary of the hydrothermal community, a crown of filter-feeding serpulids polychaetes marks the outer limit of the primary consumers strictly associated with the hydrothermal activity. Amongst these few vent species, whose biomass can reach several tens of kilograms per square meter (fresh weight), live some carnivorous organisms : a zoarcid fish, a crab belonging to a

---

(1) Dead shells of _Calyptogena magnifica_ have also been recorded on deep-tow pictures near Easter Island, on the EPR at 20° N (Dr. R. Ballard, pers. comm.).

family new to science, Bythograea thermydron. A galatheid crab belonging to the genus Munidopsis is rather abundant throughout the hydrothermal environment : its role in the food-chain is not clearly known.

SETTLEMENT AND ECOLOGICAL STRATEGY

The exploration of a given hydrothermal field of the EPR type(1) demonstrates the strong link between the primary consumers and the vents : as a matter of fact, at a distance of a few meters from the vents, the benthic fauna becomes very scarce, and the settlement consists of species usually observed on deep-sea basaltic substrates. This spatial microdistribution suggests the existence of a trophic rather than a physical or chemical dependence between the primary consumers and the hydrothermal vents. This strong dependence has been corroborated by the following observation. In spring 1978, the submersible CYANA discovered a graveyard of the giant white clam Calyptogena magnifica together with empty serpulid tubes on the axial zone of EPR at 21° N, a short distance from massive polymetallic sulfides deposits in the shape of large termit-mounds. These termit-mounds were later recognised as extinct hydrothermal chimneys. Presumably, the "death" of the hydrothermal vent had induced the demise of benthic life around the vent. The life span of a given vent is rather short : dating of hydrothermal polymetallic sulfides from 21° N using the ^{210}Pb have shown their age in the range of 23 to 61 years (Lalou and Brichet, 1981). Moreover, the direct observation of a chimney has revealed large fluctuations of flow within a few hours, and even complete interruption of the discharge (Van Praet, pers. comm.). The survival of the hydrothermal community is strictly related to an event transient with time and discontinuous with space.

Those unstable environments are occupied by primary consumers with the following features : they are able to settle quickly, they grow fast and reproduce early, and they can disperse easily to find another suitable environment when the conditions deteriorate, by means of their larvae or juveniles (Grassle and Grassle, 1974).

In a way, the primary consumers of the hydrothermal community could be considered as having an r-type ecological strategy.

(1) In the Galapagos area, the hydrothermal fluid seeps from cracks and fissures without polymetallic sulfide deposition.

Typical r-type species have fast growth rate, precocious sexual maturation, very large dispersal capability (through larval production or adult migration). Such species are adapted to unstable, short-life environments. They are very rare in the deep-sea, with the exception of wood-boring species (Turner, 1973) and species recolonizing azoïc sediments (Grassle, 1977 and Desbruyères et al., 1980). Ordinarily, r-type species have a small size. On the contrary, the primary consumers associated with hydrothermal vents have an unusual large size.

Dispersal capacities of organisms can be due to the presence of long pelagic larval life or to the existence of an adult dispersal phase. The existence of long life pelagic larvae has been deduced from the morphology of the larval shell for the Galapagos mussel (Lutz et al., 1980). Vitellus contents in molluscs ovocytes have direct relationships with the larval shell morphology. The prodissoconch I larva, which is the first ontogenetic stage of larval shell following the trocophor, has a small size in planktotrophic larva species. The prodissoconch II well developed in planktotrophic larva species (up to 400 μm) does not display shell ornamentation except growth lines. According to Lutz et al. (1980), comparison with another mytilid *Modiolus modiolus* which have similar larval shell morphology suggests a possible pelagic life of 3 to 4 weeks for the Galapagos mussel. Moreover, it has been established that settlement of the larva of *Mytilus edulis* can be delayed when a suitable environment is lacking. The hypothesis that the presence of adequate physico-chemical and biological conditions prevailing around the hydrothermal vents could induce the metamorphosis of pelagic larvae and their settlement, has been proposed. Although Lutz et al. (1980) underline the scarcity of pelagic larvae for abyssal molluscs, Bouchet (1976) gave evidence of long pelagic life larvae in a common deep-sea prosobranch gastropod from the North Atlantic, *Benthonella tenella*. The small size of prodissoconch I of the Galapagos mussel (95 μm) suggests low yolk content. A long pelagic life implies a stay in rich waters ; the migration to the photic layer seems improbable, because it would lead to a large dissemination with little probability of finding another hydrothermal site. A demersal larval stage within the nepheloid layer seems more probable with the larvae feeding on suspended particulate matter. The advective currents created by the vent (Enright et al., 1981) could concentrate larvae living in the nepheloid layer. Also, current measurements (Hekinian, unpublished data) conducted at 20 meters above the bottom, in the axial zone of the EPR at 11°-13° N show a NNW to SSE direction, generally along the main rift strike. Larvae can be passively carried by currents up to the vents where they can be concentrated by advective processes, the metamorphosis being adequately delayed. An increase of temperature or of particulate organic matter content could induce the settlement.

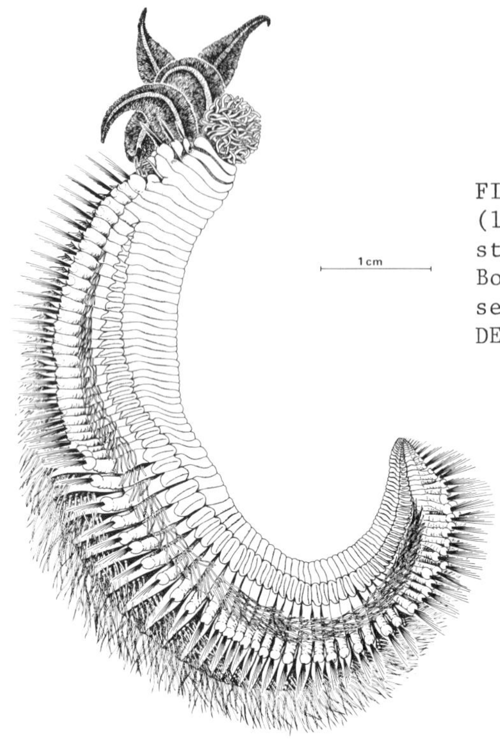

FIGURE 1 : *Alvinella pompejana* (lateral view), mostly tubicolous stage with abruptly tapering end. Body thick and plump with 91 setigerous segments (from DESBRUYERES and LAUBIER, 1980).

Gonadic maturation in the Galapagos mussel does not seem to be cyclic ; in the same individual, all gamete stages are present, from small ovogonies to pedunculate ovocytes, together with atretic gonia (Le Pennec, pers. comm.). This observation suggests a continuous production of ovocytes, which is the usual case for deep-sea animals. In the limpet-shaped gastropod Neomphalus fretterae, Mc Lean (1981) has suspected that the development occurs within a demersal egg-mass up to the veliger stage. Juveniles are benthic, and have been found in cracks and byssal threads of mussels. At the end of postprotoconch II, the animals becomes more sedentary. The colonization of new hydrothermal sites by this species is difficult to understand.

New hydrothermal vent colonization could also take place by means of actively swimming adults drifted along the rise axis by general bottom currents. In the case of Alvinella pompejana, slight morphological differences led us to the hypothesis of two different stages (figures 1 and 2), the first one being able to swim actively in the water, the second one being more sedentary (Desbruyères and Laubier, 1980). No direct evidence of this hypothesis has yet been given, but the existence of adult swimming forms in polychaetes associated with reproduction processes is well known in several families (e.g. nereids and syllids). Such successive stages do not

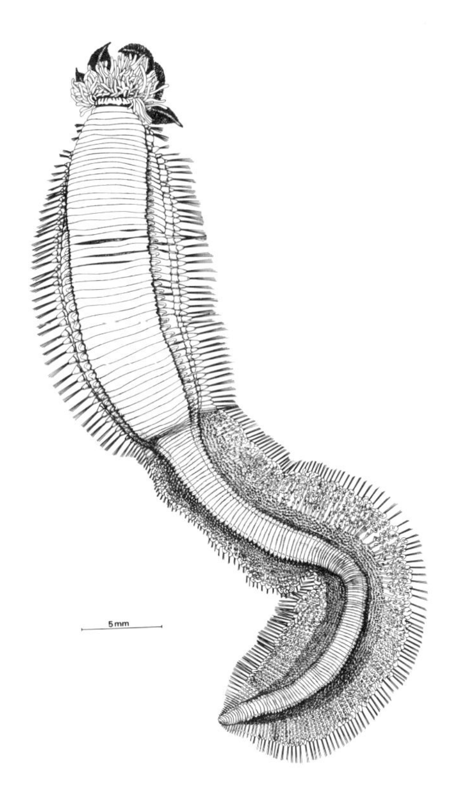

FIGURE 2 : *Alvinella pompejana* (ventral view) mostly swimming stages with regularly tapering posterior part. Body with 122 to 196 setigerous segments. The body grows narrow between 49th and 54th setigerous segments and notopods are modified, longer with digitiform branches.

occur in Paralvinella grasslei, the second and more primitive form of alvinellid. Also numerous benthic larvae have been collected recently in the sulfide particles deposited at the foot of the hydrothermal chimneys by 11°-13° N. Paralvinella grasslei, originally found in the Galapagos ridge, is now known from all hydrothermal sites, including the sedimentary site at Guaymas basin, and the propagation strategy of this species is not yet explained. Several scientists have recorded during their dives real swarms of large sized worms (Dr. R. Ballard, pers. comm.) which could well be composed of adult polychaetes swimming and drifting along the rise axis by the general circulation. Several species of the hydrothermal community belong to taxa whose larvae are either demersal, on truly benthic (ampharetids, pogonophorans, etc...) and it is difficult at the present state of knowledge to propose a general scheme of the colonisation of a new hydrothermal zone. Numerous adults of an ampharetid polychaete belonging to the genus Amphisamytha (unknown species) have been collected among the plumose abdominal setae and the mouthpieces of the carnivorous crab Bythograea thermydron caught in baited traps. Such photeric transport could well account for the colonisation by benthic adults of new vents through large migrations of this crab. These different means of adult dispersal reduce the energy loss during reproductive processes.

GROWTH RATES

Is the high biomass exhibited in the vent community a sign of a high secondary production ? Till now, it is difficult to answer this question, because growth studies have been restricted to only the two giant bivalves from the Galapagos site. Moreover, we are still lacking a reliable dating method and life span measurement for non-calcified invertebrates. In the case of the bivalves, two different methods have been utilised for growth studies. Calyptogena magnifica has been studied by Turekian and associates (Turekian et al., 1979 ; Turekian and Cochran, 1981) using the measurement of natural radionuclide content. The activity ratio of radionuclide belonging to the same series enables one to determine the time of incorporation of the original radionuclide in the tissue. Using $^{210}Po/^{210}Pb$ (^{238}U series) and $^{228}Th/^{228}Ra$ (^{232}Th series) ratios, these authors aged a series of samples taken along the maximum growth axis of a 12 cm long shell of Calyptogena (figures 3 and 4). The growth rate was 5 cm per year for the first 6 cm, the decreased to 2 to 3 cm per year for the following 2 cm, with an average growth rate of 4 cm per year. Studying the shell microstructure, Fatton and Roux (1981) have shown an alternating

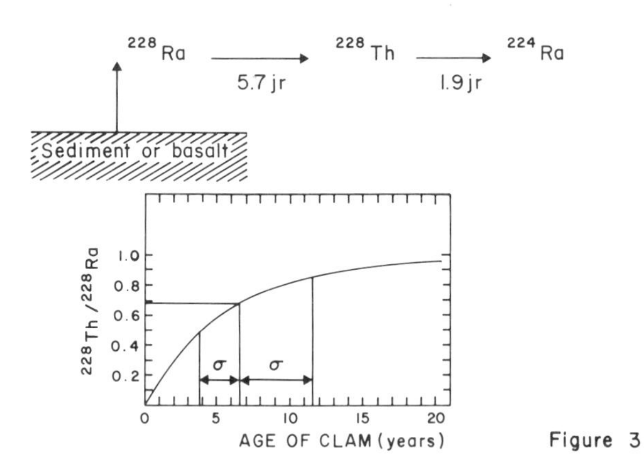

FIGURE 3 : Plot of $^{228}Th/^{228}Ra$ activity ratio for a whole shell of a Vesicomyid clam as a function of age of organism. The change in the average $^{228}Th/^{228}Ra$ activity ratio in a clam depositing a constant mass of calcium carbonate per unit time with a constant $^{228}Ra/Ca$ ratio and no initial ^{228}Th. Study made by TUREKIAN et al. (1979) on *Calyptogena magnifica* collected dead from the Galápagos hot spring field.

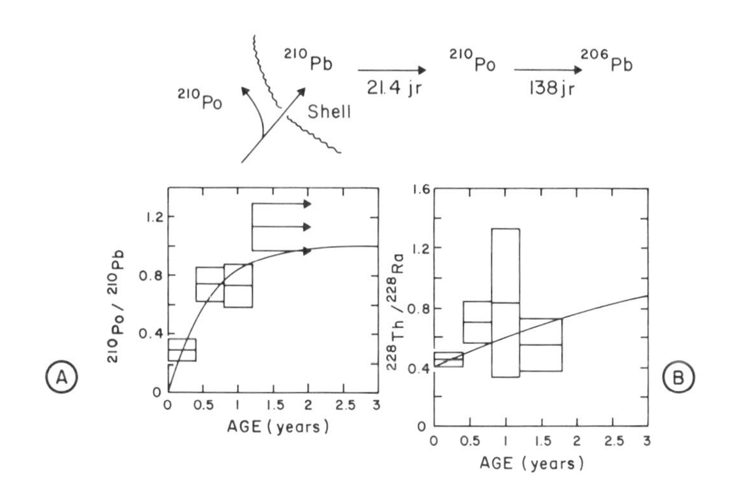

FIGURE 4 : Plots of natural radionuclide ratios as a function of age of a Vesicomyid clam (*Calyptogena magnifica*) from the Galapagos. Samples of the outer layers were made along the axis of maximum growth.
(A) The $^{210}Po/^{210}Pb$ activity ratio growth curve with an initial value of ^{210}Po equal to 0. Continuous deposition is assumed. The last sample is in equilibrium and therefore could be 2 years old or older.
(B) The $^{228}Th/^{228}Ra$ activity plotted as a function of age for sequential layers in the clam shell. The best fit curve for $^{225}Th/^{228}Ra$ growth with time requires an initial activity ratio of 0.4 (from TUREKIAN and COCHRAN, 1981).

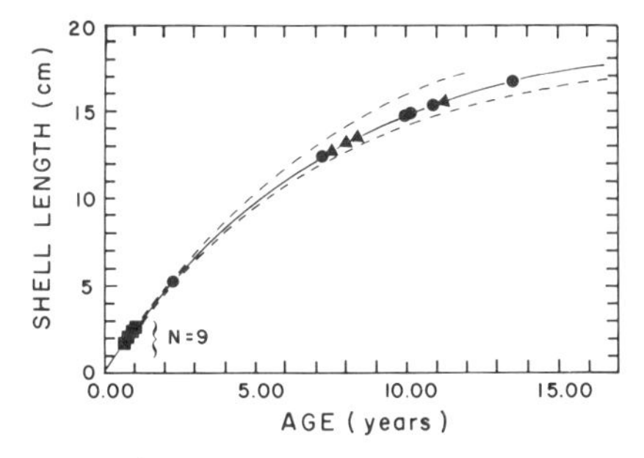

FIGURE 5 : The ontogenic curve for mussels from Mussel Bed based on the mark recapture technique.
Probably marked adult mussels.
Clearly marked adult mussels.
Juvenile mussels from a slide box and bottle rack used for microbiological studies.
The curve was generated from growth curves plotting increase in shell length for mussels of different sizes by use of the Von Bertalanffy growth equation (from RHOADS et al., 1981).

fast and slow growth rate in the same species, which is not related with a seasonal cyclic activity, but which could well depend upon the vent activity fluctuations or the movement of animals. The size of different individuals of the same species does not seem to vary with average ambient temperature as shown by the ratio of stable oxygen isotopes (Fatton et al., 1982). A direct method of growth measurement has been applied to the Galapagos mussel by Rhoads et al. (1981) in the "Mussel bed" sampling station. Shells of living mussels were abraded with a file along their posterior margins with the use of ALVIN manipulator arm. Marked specimens were collected, 294 days later, and the growth was measured between the mark and the shell edge (figure 5). Integrating these data with independant information about juvenile growth rates from young mussels newly attached to glass plates deposited during the first dives led to define growth curves fitting well with the Von Bertalanffy's model. The average growth rate of a 18.4 cm long mussel was of the order of 1 cm per year. This growth rate is similar to shallow water species of mytilids and to the growth rate of Scrobicularia plana from an estuarine environment as determined by mark-recapture method (Bachelet, 1981). The Galapagos mussels have an extended longevity and a continuous growth during the life. Such a growth type where the maximum size is reached belatedly is common in eutrophic low temperature environments (polar gigantism) ; but, contrary to opportunistic species the sexual maturity is delayed, favouring body growth. In the soft bottom deep-sea, data on growth rate are very scarce. Using radiochemical datation (^{228}Ra), Turekian et al. (1975) have estimated the age of the bivalve Tindaria callistiformis from 3 806 meters : individual of 8.4 mm has an age of 100 years. While the accuracy of this method could be discussed (Jumars and Gallagher, in press), one can assume faster growth rates in the hydrothermal community compared with order deep-sea environments. Generally speaking, the mean sizes of deep-sea organisms are also smaller than those of their shallow water relatives.

FOOD SOURCES AND NUTRITION PROCESSES

Growth measurements of both species of bivalves and opportunistic features of the primary consumers are consistent with the importance of hydrothermal community secondary production. The rich biomass of vent communities drastically contrasts the extremely poor biomasses usually encountered in the deep-sea, on basaltic substrates as well as sedimentary ones. Following Sanders and Hessler (1969), most authors postulate that the abundance of food generally controls density and biomass of deep-sea benthos. Few exceptions to this asumption have been observed, for instance in the Namiby abyssal plain where the benthic biomass is almost non existent although the particulate non refractory organic matter content in the sediment is high (our unpublished data). Such an example suggests that limiting factors different from organic sources could intervene (vitamins, oligoelements, genetic factors). The luxuriance of the hydrothermal community has been directly related to the high organic matter content (Londsale, 1977) coming either from chemosynthesis locally achieved by chemolithotrophic bacteria, or from advection of particulate organic matter produced in the photic zone (Enright et al., 1981). Several methods have been proposed to elucidate the origin of the organic matter used by primary consumers of the hydrothermal community. All of them are based upon the radiogeochemical methods : ratios between stable isotopes of carbon and nitrogen and dating of organic matter using the ^{14}C technique.

Based on the fact that the stable isotope carbon ratio ($\delta^{13}C$) of an organic source is not modified significantly through the ingestion and assimilation processes, it is possible to use this ratio as a tracer to identify the origin of the organic matter consumed by vent animals (Rau and Hedges, 1979). The $\delta^{13}C$ of primary producers is mainly determined by metabolic pathways during primary production and by the stable isotop ratio of the inorganic carbon source (CO_2, CH_4...). Incorporation of ^{13}C is favoured by Hatch and Slack cycle (figure 6) whereas incorporation of ^{12}C is favoured by Calvin cycle (Haines and Montague, 1979). Other pathways, such as the bacterial methane oxydation (Barker and Fritz, 1981) seem to favour ^{12}C incorporation. Parameters such as temperature (Fontugne and Duplessy, 1981), pH, CO_2 content also play a minor role on $\delta^{13}C$.

$\delta^{13}C$ measurements on various primary consumers of the hydrothermal community show extreme differences. The $\delta^{13}C$ approximates $- 11^{o}/_{oo}$ in Riftia pachyptila, while it is abnormally low in both bivalves species with values from $- 32$ to $- 34^{o}/_{oo}$ (Rau and Hedges, 1979 ; Williams et al., 1981 ; Rau, 1981). Boulègue et al. (in press) have recorded the occurrence of bacteria-like cells associated with pyrite and iron sulfate on an hydrothermal site of the EPR by 7° N. Those cells which could oxidize pyrite and produce iron sulfate have a very low $\delta^{13}C$ ($- 28.4$ to $- 31^{o}/_{oo}$). The carbon source could come from the ambient sea-water which has an isotopic ratio of nearly $0^{o}/_{oo}$ (Degens, 1969) of from the hydrothermal fluid ($\delta^{13}C > - 4^{o}/_{oo}$ from Craig et al., 1980) : thus, the fractioning effect due to the chemolithotrophic metabolism should be $- 26^{o}/_{oo}$ in that example.

This data argues for incorporation by bivalves of organic matter fixed by chemolithotrophic bacteria. Inversely, relatively high $\delta^{13}C$ of Riftia pachyptila tissue suggests that more than one carbon fixation pathway is of nutritional importance to vent animals (Rau, 1981).

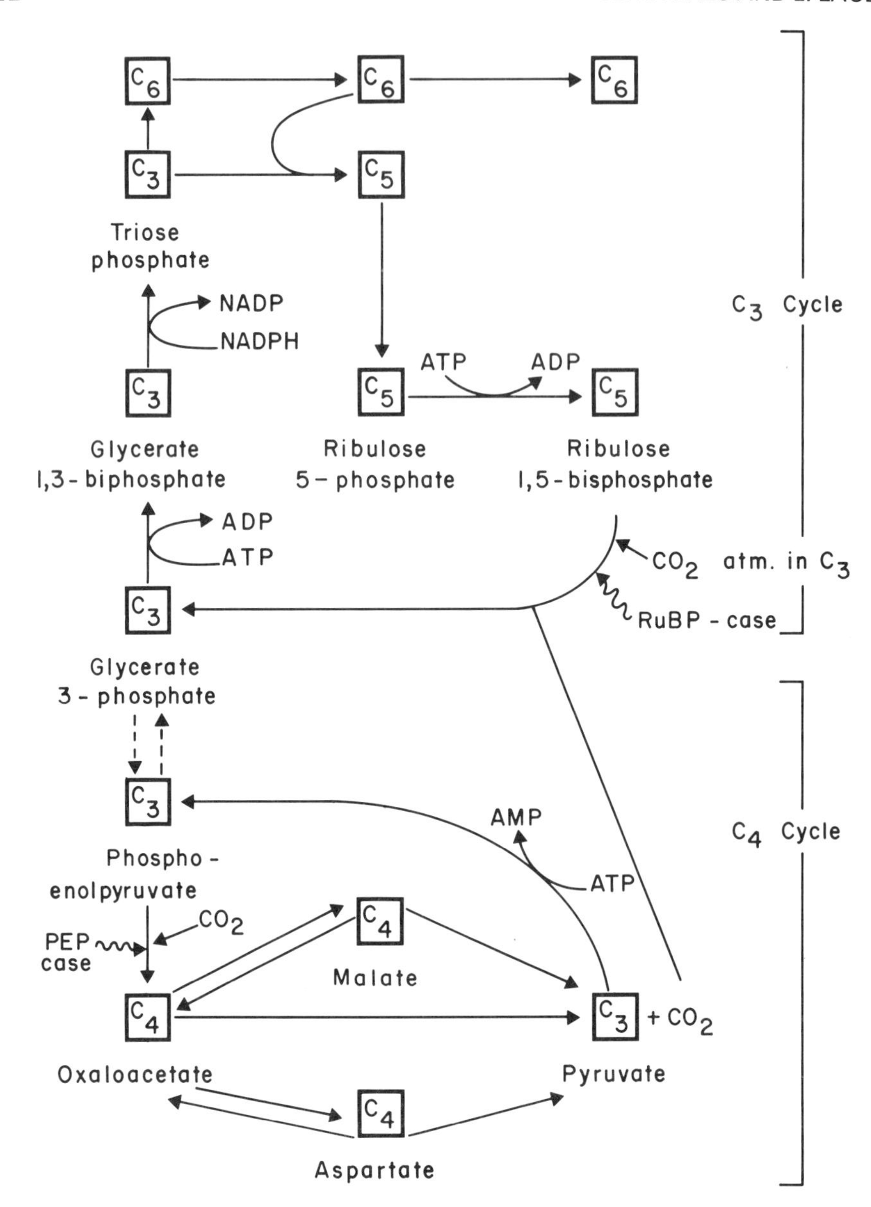

FIGURE 6 : CO_2 reduction cycle following the C_3 and C_4 pathways. In the C_4 species the CO_2 released inside the cell through malate or aspartate decarboxylation is retrieved by the Calvin cycle, which is the only one operating in the C_3 cells (from JACQUES, 1981).

Data dealing with stable nitrogen isotope ratios (Rau, 1981) enable one to conclude that the organic matter from primary consumer tissues has evolved through a short recycling, more probably of local than of pelagic origin.

Arguments in favour of both trophic source hypotheses are numerous (Enright et al., 1981 ; Williams et al., 1981). It remains highly probable that the major process of biological production is the local chemotrophic source. The contribution of advected organic particles provide a complementary food source specially important at the outer limits of the hydrothermal community for filtering organisms (serpulids polychaets). Smith (pers. comm.) has shown by transplantation experiments a direct link between distance from the vents and the elaboration of glycogen reserves in Galapagos mussel tissue. Likewise, the oxygen consumption of these mussels from the middle of the mussel bed (near the vents) is higher than consumption of individuals in the periphery of the bed, where the particulate organic matter is half that in the central area (K. Smith, unpub. data).

The most common primary consumer of the hydrothermal community is the pogonophoran Riftia pachyptila. Like all other representatives of the phylum Pogonophora since their discovery by Caullery (1914), it is devoid of mouth, gut and anus. The most recent theories on feeding habits present the pogonophora as a typical example of nutrition whereby dissolved organic compounds are absorbed through epithelial tissues. However, Southward et al. (1979) experimentally demonstrated, using labelled molecules that the quantities absorbed by an individual are far below its nutritional needs. This shortage would be even larger in the case of Riftia, where the surface/volume ratio is much smaller than those of ordinary species of Pogonophora measuring a few centimeters long.

The discovery of the role of the trophosome in Riftia, massive trunk organ surrounded by the coelomic cavity, enables us to elucidate the feeding ethology of this pogonophoran. The trophosome, when discovered in another vestimentiferan, Lamellibrachia, was considered as a possible reserve source for spermiogenesis or as a detoxicating organ. Several findings demonstrate that the trophosome of Riftia is involve in a symbiotic relationship with chemosynthetic sulfide oxidizing bacteria : significant depositions of sulfur crystals have been observed in the trophosome together with large vascularisation comprising a central vessel and numerous capillaries (Jones, 1981 ; Cavanaugh et al., 1981). Epifluorescent microscopy and TEM and SEM observations show the existence of a huge number of prokaryotic cells in the trophosome (3.7×10^9 cells per g of tissue). Later analyses showed that the prokaryotic cell wall has a GRAM$^-$ reaction. Measurements of nitrogen isotope ratio (figure 7) show that the $\delta^{15}N$ of the

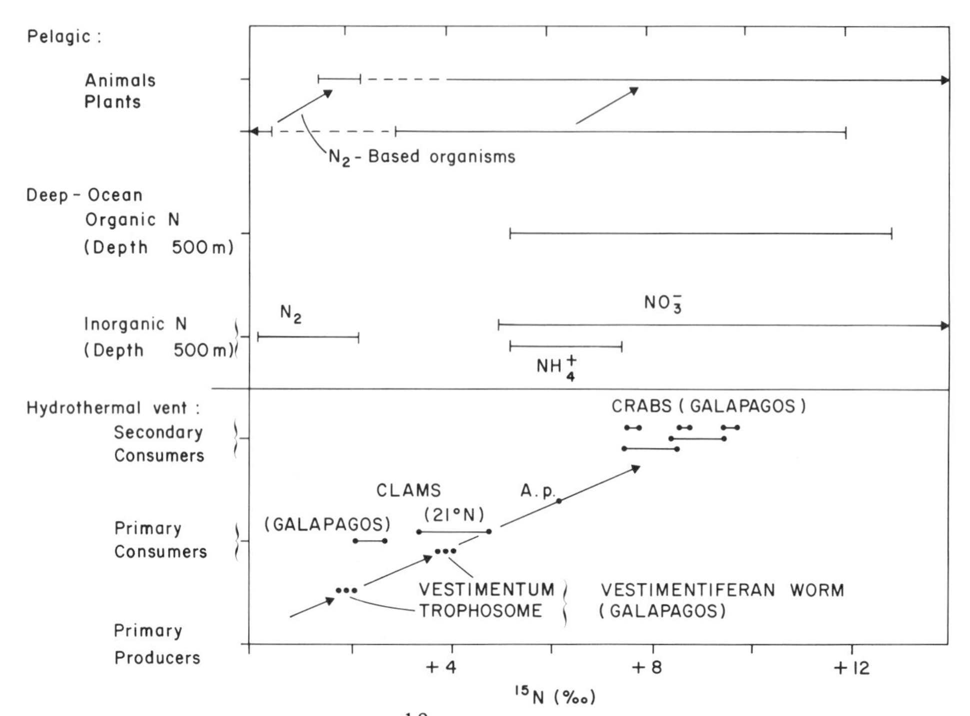

FIGURE 7 : Diagram of the $\delta^{13}N$ ocean nitrogen pools in the Pacific Ocean : the assumed trophic levels of hydrothermal vent animals, based on knowledge of related species, are plotted in ordinate. A.p. : *Alvinella pompejana* (from RAU, 1981).

trophosome is similar to that of primary producers (Rau, 1981). On a biochemical basis, Felbeck (1981) has discovered several enzymes in the trophosome (rhodanese, APS-reductase, ATP-sulfurylase) which can produce biological energy (e.g. ATP) from sulfide oxidation. Two enzymes typical of the Calvin-Benson cycle have also been identified : ribulose 1-5 biphosphate carboxylase and ribulose 5-phosphate kinase. This metabolic pathway using ATP as an energy source enables the animals as symbionts to reduce the CO_2 to organic carbon.

This symbiosis implies that nutrients (CO_2, O_2, sulfides, nitrates...) are carried to the trophosome by the blood. The whole blood of Riftia has a small Bohr effect (a shift of the oxygen equilibrium to the right) when CO_2 concentration increases and pH decreases (Arp and Childress, 1981). That property is more common in transient burrow dwellers than in worms living in a permanent tube. This small Bohr effect may be important for the simultaneous transport of CO_2 and O_2. Carrying capacities of the whole blood for oxygen is also very high compared with other invertebrates (8.4 ± 1.8 ml O_2/100 ml of blood). Hydrogen sulfide is well recognized as being very toxic for mammals : the formation of sulphaemoglobin (which cannot act as a respiratory carrier) is generally considered to be one of the mechanisms of toxicity. Arp and Childress (1981), working on the whole blood of Riftia found no indication of sulphaemoglobin (peak at 623 nm). This compatibility is not unusual in marine worms : Wells and Pankust (1980) demonstrate that in vitro treatment of Abarenicola haemoglobin with sulfide at a neutral pH does not result in any change to the oxygen affinity and is not converted to sulphaemoglobin contrary to similarly treated human haemoglobin. Those blood features do not seem unusual in worms living in low oxygen conditions.

Riftia pachyptila is not the only example of an animal relying almost exclusively (a part of the food may be supplied by direct uptake of dissolved organic compound through integument) on a chemoautotrophic symbiosis. Southward et al. (1981) demonstrated, using similar criteria (e.g. anatomy, natural isotope ratio, enzymology) that such an association existed in several species belonging to another pogonophoran genus Siboglinum. A model of a host-algal symbiont system (Hallock, 1981) predicts that the production of organic matter by the system can be increased by 2-3 orders of magnitude with efficient recycling of nutrients over that of a system comprised of only autotrophs and heterotrophs. Riftia pachyptila is unique because of its large size. Several vestimentiferan species living close to the hydrothermal vents and all other deep-sea Pogonophora do not exceed twenty centimeters in length. Until now, we have no information on Riftia growth rate and life span ; the only technique available seems to be a population survey of a new colonised vent area.

From morphological evidences, other primary consumers seem to possess different feeding habits : both species of alvinellids, as other ampharetids (Fauchald and Jumars, 1979) are surface deposit feeders. They utilize their retractile buccal structure (buccal tentacles and membrane) to seize the food particles. Observations of the buccal tentacle groove show that mineral particles driven to the mouth are coated with bacterial cells. The inner surface of the tubes in which the alvinellids live is also heavily lined with true bacterial mats. Sulfide deposits (crystals) are also seen in the tube. The worm living in such tubes can feed directly upon these bacterial populations using its retractile buccal structures. Gut content studies showed that the diet is mainly composed of mineral particles coated with organic matter and bacterial cells, together with rare centric diatoms. In the case of Alvinella pompejana, the two morphological stages seem to have different relationships with the bacterial population of the inner tube surface. The tubicolous stage exhibits rows of long tubules on the dorsal region which could function in the organic matter exchanges by the increase of the body surface. The swimming stage possesses a modified posterior part with numerous parapodial branches which are highly vascularised and have attached dense tuffs of bacteria. Elementary sulfur deposits have been observed in the epidermis of this region : they can be considered as a testimony to chemical exchanges between the polychaete and the bacteria (Laubier *et al.*, 1983). Other bacterial colonies are present on the dorsal region epidermis, the so-called necrotic area. This region also displays high vascularisation of the epidermis, and could be the site of important nutritional exchanges between the polychaete and associated bacteria. Gut content analyses and histological study give clear indication of a fully functional digestive tract, at least from morphological evidence (Desbruyères *et al.*, submitted). The amount of organic input through the tegument of worm is unknown but might be sufficient to provide for the needs of epidermal and associated tissues, like in most soft-bodied marine invertebrates (Fergusson, 1982). Chemoautotroph epibiotic bacteria, mainly *Thyotrix* (D. Nelson, person. comm.), can act increasing D.O.M. concentration in the worm environment inside the tube and accordingly increasing D.O.M. influx through epidermal tissue (Costopoulos *et al.*, 1979). Important metal deposits occur on the epidermis in the areas where bacteria are abundant. Iron oxides can represent up to 80 % of the tissue weight in the integument of "necrotic" area. Parapodial branches can act as detoxication organs when they reach a high level of metal oxides by autotomic process. The hypothesis of a dominant symbiotic relationship is difficult to justify by comparison with *Paralvinella grasslei*, a related species living in the same part of the hydrothermal community, but completely devoid of bacterial epibiosis.

Another species of ampharetid polychaete belonging to the genus *Amphisamytha* is common in the sulfide deposits ; its buccal structure is typical of the ampharetinae, and its feeding habits is probably that of a surface deposit feeder. On the contrary, the underscribed polynoid living as a commensal on the large mytilid is mainly carnivorous : pieces of mussel mantle and gills are frequent in their gut contents, together with large numbers of bacteria which demonstrates the non-selective feeding habits of this polynoid. Van Praet (in press) has recorded a similar absence of food selectivity in the actinians of hydrothermal chimneys sampled on the 11°-13° N site. Within phagocytic cells of mesenteric filament, fragments of crustaceous muscles and bacteria are present. Disalvo (1971) had previously reported the ability of the related group Scleractinian to phagocyte bacteria. In the case of actinians, it is difficult to distinguish between primary consumer habits and carnivorous feeding, the size of the preys compared with the oral disk size being the main factor in the choice of ingested particles (Sebens, 1981).

The limpet-shaped gastropod is more complex. They are very abundant in the hydrothermal community, not only on basaltic substrates, but also on *Riftia*'s tubes and in sulfide particle deposits at the foot of chimneys. They are zoologically heterogeneous : more than ten different species of limpet-shaped gastropods have been identified in the samples from 11°-13° N (Bouchet, pers. comm.). More than 1 000 individuals were collected during a single dive of the submersible CYANA. The single species already described as _Neomphalus fretterae_ has been collected on the Galapagos site from mytilid shells, _Riftia_ tubes and on rocks (Mc Lean, 1981). The digestive content of this species demonstrates a combination of filtration feeding and grazing. Radiolarian and foraminiferan fragments and crustacean skeletons are found in the gut contents (Fretter _et al._, 1981). The wear of rachidian and lateral teeth of the radula are considered as an additional evidence of the grazer ecology of _Neomphalus_ by McLean. The bacterial mats which settle on all available substrates (Jannasch and Wirsen, 1981) are probably the main source of food for these animals. Bacterial mats occur on basaltic as well as biological substrates. The filter feeding behavior is acquired later as adults. In juveniles, the gill, which ciliaty action drives the filtered particles to the mouth, is uncompletely developed while the jaw is very prominent (Mc Lean, 1981).

Data on the feeding habits of both large bivalvia are very scarce. Morphological studies on _Calyptogena magnifica_ demonstrate the existence of a relatively reduced particle collecting apparatus (labial palps very reduced, alimentary groove absent on the ventral margin of the two half-gills). Gut contents are practically absent, and composed of transparent mucous material (Boss and Turner, 1980). Cavanaugh _et al._, (1981) suggested the occurrence of a symbiotic relationship between _Calyptogena magnifica_ and chemosynthetic bacteria. This hypothesis is confirmed by the positive result of the enzymatic bioassay of the RuBP carboxylase, which intervene in the Calvin-Benson cycle, and by the observation of prokaryote symbiotic cells using T.E.M. (Felbech _et al._, 1981). However it seems difficult to ascertain if the bacteria observed on the gills are symbiotic bacteria or cells being filtered and on the way to the mouth to be ingested. The transport mechanism, if it exists, is also to be analysed.

The anatomy of the mytilid gives evidence of an active and non-selective filtration : the labial palps are extremely developed, their hypertrophied muscles extend to the oesophagus. Observation with SEM of gills of mussels from 11-13° N, prepared with the critical point method, does not show any sign of epibiotic bacteria, but rather the transit along the gill lamella of large sized mineral particles (10 to 100 µm) coated by dense bacterial mats. The latero-frontal and lateral ciliations are absent or reduced, and the attachment disks only remain. This loss of the filtering screen suggests a faint retention of the smallest particles. On the gill frontal grooves, mineral particles coated with bacteria and phytoplanktonic cells are present. However Felbeck _et al._ (1981) found Calvin-Benson enzymes in the vent mussel gill tissue. This data demonstrates a chemoautotrophic activity probably due to the presence of symbiotic bacteria embedded in the gill. The quantitative contribution of that chemoautotrophic activity to the nutrition of the mussel is unknown, although probably important.

CONCLUSION

Less than six years since its discovery, the deep-sea hydrothermal community has been observed in four different locations far away from each other, and displays very uniform ecological structure and zoological composition. The strange primary consumers living in the vicinity of the vents as originally proposed in a preliminary hypothesis, are able to utilize the food source produced by bacteria, in a more or less sophisticated way : the most elaborate situation is probably the case of the vestimentiferan tube-worm Riftia, which possesses a special organ containing symbiotic bacteria. Surprisingly, this symbiosis between a pogonophoran and bacteria is not unique. Several small-sized Pogonophora living in low oxygen environments in marine fjords also display symbiotic bacteria within their bodies (Southward et al., 1981). It seems highly probable this explanation will apply to other species of invertebrates previously known for their reduced or completely lacking digestive tracts and their ability to live in more or less reduced shallow-water sediment. In a way, the discovery of the hydrothermal primary consumers gives a key to the solution of related unsolved questions of other marine environments.

The energy transfer from chemosynthetic bacteria to primary consumers is more often performed in a simpler way : it is more or less the case with two large bivalves, the giant white clam and the undescribed mytilid. The ecological ability of the Pompeii worm to tolerate the highest range of temperature in close contact with hydrothermal vents, makes it an appropriate species for the development of several kinds of association with epibiotic bacteria. Other primary consumers from the hydrothermal community, generally much smaller by size, are presently insufficiently known to enable any satisfactory conclusion dealing with their food habits.

Apart from the very high biomass, the very few existing data on some of the primary consumers from the hydrothermal community demonstrate the fast growth rate of the bivalves and the corresponding important secondary production level.

At the boundary of the hydrothermal community, other types of primary consumers are found : they belong to previously known species, and their abnormally high density can easily be explained by a hypothetical advective mechanism which concentrates large amounts of particulate organic matter in the proximity of the vents region.

It would be unwise to discuss with the existing data the origin and antiquity of the strictly adapted hydrothermal primary consumers. Up to now, new taxa vary from new species of previously known genera to new families. Perhaps more surprising is the fact that the variation of intensity of hydrothermal activity as demonstrated by the type and density of vents and the importance of polymetallic sulfide deposits, does not make any significant difference in the community. What we do know now in terms of hydrothermal community variation can be considered as purely local variation from the same pool of species. Distances between each of the known locations is in the order of a thousand nautical miles, which is a limited distance. Future investigations in other suitable fast spreading centers areas would be of great interest and provide a better understanding of this astonishing community. Also, the mud substrate hydrothermal community recently discovered in Guaymas basin could give us some new biological and ecological guide-lines.

ACKNOWLEDGEMENTS

We are most appreciative for the helpful comments made by H. Felbeck, R.R. Hessler and K. Smith who carefully read and corrected this manuscript.

REFERENCES

Arp, A.J. and Childress, J.J., 1981. Blood function in the hydrothermal vent vestimentiferan tube worm. Science, 213 : 342-344.

Bachelet, G., 1981. Application de l'équation de Von Bertalanffy à la croissance du bivalve *Scrobicularia plana* . Cah. Biol. mar., 22 : 291-311.

Barker, J.F. and Fritz, P., 1981. Carbon isotop franctionation during microbial methane oxidation. Nature, 293 (5830) : 289-291.

Boss, K.J. and Turner, R.D., 1980. The giant clam from the Galapagos rift, *Calyptogena magnifica* species novum. Malacologia, 20 (1) : 161-194.

Bouchet, P., 1976. Mise en évidence d'une migration de larves véligères entre l'étage abyssal et la surface. C.R. Acad. Sc. Paris, 283 : 821-824.

Boulègue, J., Pineau, F., Javoy, M., Perseil, E.A. (in press). Bacterial oxidation of pyrite from an East Pacific rise hydrothermal deposit. Nature.

Cavanaugh, C.M., Gardiner, S.L., Jones, M.L., Jannash, H.W. and Waterbury, J.B., 1981. Prokariotic cells in the hydrothermal vent tube worm Riftia pachyptila Jones : possible chemoautrophic symbionts. Science, 213 : 340-341.

Costopoulos, J.J., Stephens, G.C. and Wright, S.H., 1979. Uptake of amino acids by marine Polychaetes under anoxic conditions. Biol. Bull., 157 : 434-444.

Craig, H., Welhan, J.A., Kim, K., Poreda, R. and Lupton, J.E., 1980. Geochemical studies of the 21° N EPR hydrothermal fuids. E.O.S., 61, 992.

Degens, E.T., 1969. Biochemistry of stable carbon isotopes. In Organic Geochemistry, G. Englington and M.J.J. Murphy, Eds. Springer-Verlag, New-York, 1969, 304-329.

Desbruyères, D., Bervas, J.Y. and Khripounoff, A., 1980. Un cas de colonisation rapide d'un sédiment profond. Oceanologica Acta, 3 (3) : 285-291.

Desbruyères, D., Gaill, F., Laubier, L., Prieur, D. and Rau, G.H. (submitted). Unusual nutrition of the "Pompeii worm" (Alvinella pompejana, Polychaetous annelid) from a hydrothermal vent environment : SEM, TEM, ^{13}C and ^{15}N evidence. Mar. Biol.

Desbruyères, D. and Laubier, L., 1980. Alvinella pompejana gen. sp. nov., Ampharetidae aberrant des sources hydrothermales de la ride Est-Pacifique. Oceanologica Acta, 3 (3) : 267-274.

Desbruyères, D. and Laubier, L. (under press). Paralvinella grasslei, new genus, new species of Alvinellinae (Polychaeta, Ampharetidae) from the Galapagos rift geothermal vents. Proc. Biol. Soc. Wash.

Disalvo, I.H., 1971. Ingestion and assimilation of bacteria by two scleractinian coral species. In Experimental Coelenterate biology. H.M. Lenhoff, L. Muscatine and L.V. Davis, University of Hawaii Press. Honolulu ed., 129-136.

Enright, J.J., Newman, W.A., Hessler, R.R. and Mc Gowan, J.A., 1981. Deep-ocean hydrothermal vent communities. Nature, 289, 219-221.

Fatton, E., and Roux, M., 1981. Etapes de l'organisation microstructurale chez Calyptogena magnifica Boss et Turner, bivalve à croissance rapide des sources hydrothermales océaniques. C.R. Acad. Sc. Paris, 293 : 63-68.

Fatton, E., Marien, G., Pachiaudi, C., Rio, M. and Roux, M., 1982. Fluctuations de l'activité des sources hydrothermales océaniques (Pacifique Est, 21° N) enregistrées lors de la croissance des coquilles de Calyptogena magnifica (Lamellibranche, Vesicomydae) par les isotopes stables du carbone et de l'oxygène. C.R. Acad. Sc. Paris, 293 : 701-706.

Fauchald, K. and Jumars, P.A., 1979. The diet of worms : a study of Polychaete feeding guilds. Oceanogr. Mar. Biol. Ann. Rev., 17 : 193-284.

Felbeck, H., 1981. Chemoautotrophic potential of the hydrothermal vent tube worm Riftia pachyptila Jones (Vestimentifera). Science, 213 : 336-338.

Felbeck, H., Childress, J.J. and Somero, G.N., 1981. Calvin-Benson cycle and sulphide oxidation enzymes in animals from sulphide-rich habitats. Nature, 293, 5830 : 291-293.

Ferguson, J.C., 1982. A comparative study of the net metabolic benefits derived from the uptake and release of free amino acids by marine invertebrates. Biol. Bull., 162 : 1-17.

Fontugne, M.R. and Duplessy, J.C., 1981. Organic carbon isotopic fractionation by marine plankton in the temperature range - 1 to 31° C. Oceanologica Acta, 4 (1) : 85-90.

Grassle, J.F., 1977. Slow recolonisation of deep-sea sediment. Nature, 265, 5595 : 618-619.

Grassle, J.F. and Grassle, J.P., 1974. Opportunistic life histories and genetic systems in marine benthic Polychaetes. J. Mar. Res., 32, 2 : 253-284.

Haines, E.B. and Montague, C.L., 1979. Food sources of estuarine invertebrates analyzed using $^{13}C/^{12}C$ ratios. Ecology, 60 (1) : 48-56.

Hallock, P., 1981. Algal symbiosis : a mathematical analysis. Mar. Biol., 62 : 249-255.

Jacques, G., 1981. Approche physiologique de la production primaire pélagique. Océanis, 7 (5) : 511-530.

Jannash, H.W. and Wirsen, C.O., 1979. Chemosynthetic primary production at East Pacific sea-floor spreading center. Bio-Science, 29 (10) : 592-598.

Jannash, H.W. and Wirsen, C.O., 1981. Morphological survey of microbial mats near deep-sea thermal vents. Appl. Environ. Microbiol., 41 (2) : 528-538.

Jones, M.L., 1981. Riftia pachyptila, new genus, new species, the vestimentiferan worm from the Galapagos rift geothermal vents (Pogonophora). Proc. Biol. Soc. Wash., 93 (4) : 1295-1313.

Jumars, P.A. and Gallagher, E.D. (under press). Deep-sea community structure : three plays on the benthic proscenium. In Ecosystem processes in the deep ocean, W.G. Ernst and J. Morin, Eds., Prentice-Hall, Englewood Cliffs, N.J. U.S.A.

Lalou, C. and Brichet, E., 1981. Possibilités de datation des dépôts de sulfures métalliques hydrothermaux sous-marins par les descendants à vie courte de l'uranium et du thorium. C.R. Acad. Sc. Paris, 293 : 821-824.

Laubier, L., Desbruyères, D. and Chassard-Bouchaud, C. (under press). Evidence of sulfur accumulation in the epidermis of the polychaete Alvinella pompejana from deep-sea hydrothermal vents. A microanalytical study. Mar. Biol. Progr. ser.

Lonsdale, P., 1977. Clustering of suspension-feeding macrobenthos near abyssal hydrothermal vents at oceanic spreading centers. Deep-Sea Res., 24 : 857-863.

Lutz, R.A., Jablonski, D., Rhoads, D.C. and Turner, R.D., 1980. Larval dispersal of a deep-sea hydrothermal vent bivalve from the Galapagos rift. Mar. Biol., 57 : 127-133.

Mc Lean, J.H., 1981. The Galapagos rift limpet Neomphalus : Relevance to under-standing the evolution of a major paleozoic-mezozoic radiation. Malacologia, 21 (1-2), 291-336.

Newman, W.A., 1979. A new scalpellid (Cirripedia) : a Mesozoic relic living near an abyssal hydrothermal spring. Trans. San Diego Soc. Nat. Hist., 19 (11), 153-167.

Rau, G.H., 1981. Low $^{15}N/^{14}N$ in hydrothermal vent animals : ecological implications. Nature, 289 (5797) : 484-485.

Rau, G.H., 1981. Hydrothermal vent clam and tube worm $^{13}C/^{12}C$: further evidence of non-photosynthetic food sources. Science, 213 : 338-340.

Rau, G.H. and Hedges, J.I., 1979. Carbon-13 depletion in a hydrothermal vent mussel : suggestion of a chemosynthetic food source. Science, 203 : 648-649.

Rhoads, D.C., Lutz, R.A., Revalas, E.P. and Cerrato, R.M., 1981. Growth of bivalves at deep-sea hydrothermal vents along the Galapagos rift. Science, 214 : 911-933.

Sanders, H.L. and Hessler, R.R., 1969. Ecology of the deep-sea benthos. Science, 163 : 1419-1424.

Sebens, K.P., 1981. The allometry of feeding, energetics and body size in three anemones species. Biol. Bull., 161 : 152-171.

Southward, A.J., Southward, E.C., Brattegard, T. and Bakke, T., 1979. Further experiments on the value of dissolved organic matter as food for Siboglinum fiordicum (Pogonophora). J. mar. biol. Ass. U.K., 59 : 133-148.

Southward, A.J., Southward, E.C., Dando, P.R., Rau, G.H., Felbeck, H. and Flugel, H., 1981. Bacterial symbionts and low $^{13}C/^{12}C$ ratios in tissues of Pogonophora indicate unusual nutrition and metabolism. Nature, 293 (5834) : 616-620.

Turekian, K.K., Cochran, J.K., Kharkar, D.P., Cerrato, R.M., Vainys, J.R., Sanders, H.L., Grassle, J.F. and Allen, J.A., 1975. Slow growth rate of a deep-sea clam determined by ^{228}Ra chronology. Proc. Nat. Acad. Sci. U.S.A., 72 (7) : 2829-2832.

Turekian, K.K., Cochran, J.K. and Nazaki, Y., 1979. Growth rate of a clam from the Galapagos rise hot spring field using natural radionuclide ratios. Nature, 280 : 385-387.

Turekian, K.K. and Cochran, J.K., 1981. Growth rate of a Vesicomyd clam from the Galapagos spreading center. Science, 214 : 909-911.

Turner, R.D., 1973. Wood-boring bivalves, opportunistic species in the deep-sea. Science, 180 : 1377-1379.

Van Praet, M. (under press). Régime alimentaire des Actinies. Bull. Soc. Zool. France.

Wells, R.M.G. and Pankhurst, N.W., 1980. An investigation into the formation of sulphide and oxidation compounds from the haemoglobins of the lugworm Abarenicola affinis (Ashworth). Comp. Biochem. Physiol., 664 : 255-259.

Williams, P.M., Smith, K.L., Druffel, E.M. and Linick, T.W., 1981. Dietary carbon sources of mussels and tubeworms from Galapagos hydrothermal vents determined from tissue ^{14}C activity. Nature, 292 : 448-449.

THE DISTRIBUTION AND COMMUNITY STRUCTURE OF MEGAFAUNA

AT THE GALAPAGOS RIFT HYDROTHERMAL VENTS

Robert R. Hessler and William M. Smithey, Jr.

Scripps Institution of Oceanography

La Jolla, California 92093 USA

ABSTRACT

The distributions of the twenty-two megafaunal species at the Galapagos Rift hydrothermal vents vary markedly with respect to the discharging warm water. Vent associated water temperature ranged to 14.72°C, substantially above the 2.01°C ambient temperature of the area. Because it is a conservative property, temperature is a general index of vent-water quality. Some animals (the vestimentiferan, limpets, clam, a shrimp, an anemone, and for the most part, the mussel) are limited to the mouths of vents, where the temperature is several degrees above ambient. Others (serpulid worm, a second anemone, galatheid crab, turid gastropod) are abundant around the vents, but avoid the vent openings and so never experience much more than a degree above ambient. A third group (the siphonophore, brachiopod, a third anemone, enteropneust, a shrimp, ophiuroid) remains at the periphery of the vent field where temperature is at most a few tenths of a degree above ambient. Some mobile species (vent fish, brachyuran crab, galatheid crab, amphipods) are most abundant at vent openings but range even into nonvent terrain. Among the taxa that are peripheral or at least avoid vent openings are species which also live in the vast nonvent milieu, but most vent field species are endemic. Conversely, most members of the nonvent environment are absent from vent fields. While vents are obviously a source of abundant nutrition, most deep-sea animals are probably not adapted to the elevated temperature and/or unusual chemistry. Some may be inhibited by interference competition. Those that are totally excluded must be especially sensitive because dilution at the periphery is high.

Chemoautotrophic bacteria form the base of the food chain. The largest portion of metazoan biomass thrives through symbiosis with an incorporated chemoautotrophic bacterial flora; these animals are most closely associated with vent openings. Others feed on suspended bacteria ejected from the vents, those that have settled out, or bacteria growing as a film on the substratum. Vent fields possess a well-developed plankton, but the extent to which they form an intermediate link is not known. Nor do we know the amount of photosynthetically derived plankton and detritus that is contributed via the thermally induced convection cell. The top of the food chain consists of scavengers, mostly malacostracan crustaceans, some of whom combine deposit feeding with carnivory. Oddly, fish are not important at this level.

INTRODUCTION

The Galapagos spreading center is the first place where deep-sea hydrothermal vents were investigated at close hand. The initial suspicion that they might be populated with a special fauna came from Deep Tow photographs made in 1976 (Lonsdale, 1977), but the full community, with its high standing crop and wonderous morphologies, was first seen in 1977 by geologists and chemists using Angus and Alvin (Corliss et al., 1979). News of the 1977 discovery resulted in a multi-investigator biological expedition to the site in 1979 (Grassle et al., 1979).

Hessler's role on this cruise was to characterize the megafauna (animals large enough to be recorded in photographs) and document its distribution. Such information supplies much basic data that is useful in biogeographic studies and analyses of vent aut- and synecology. It is therefore a natural complement to the studies of taxonomy, life histories, physiology and microbiology which took place at the same time.

The present paper provides a general description of nearly all the megafaunal taxa seen at the vent fields; a few rare ones are omitted. These distributions are summarized for each of the vent fields we visited, and the vent fauna is compared with the adjacent nonvent community. The reasons for these differences are discussed, and finally, distributional information is combined with what else is known about the biology of the animals to construct a food web for the community.

METHODS

All data were collected using the research submersible Alvin. Three cruise legs in January-February and November-December, 1979, allowed 23 dives of up to seven hours bottom time per dive. Each dive accommodated two scientific observers of varying interest. All dives had multiple tasks, but no individual programs had time allotted on every dive.

Distributional data for the megafauna were collected by direct observation and photography. Observations were recorded on voice tape. Photographs were taken either through view ports using 35mm, hand-held cameras (single lense reflex or stereo) by automatic external survey cameras, or our "arm-stereo" camera held by the starboard mechanical arm. The arm-stereo camera yielded especially clear close-up photographs which were taken with 28mm plane port corrected lenses one meter from the subject. The cameras of the stereo pair were 32.5mm apart. Distance was determined with a one-meter wand that incorporated a temperature sensor at its tip.

The temperature sensor provided information on the extent to which animals in the picture were exposed directly to vent water. If the temperature were ambient (2.01°C), vent water was not involved. Elevated temperatures need more careful interpretation because they should be related to the temperature of the vent water prior to exiting. For example, a temperature of 3°C results from much less dilution if the emerging vent water were at 8°C than at 15°C. About 400 arm-stereo photographs were taken, but unfortunately, the temperature sensor was not always working.

Hessler participated in seven dives, visiting all four vent sites. The observations in this paper are primarily his, although he profited from the observations of others where possible. In addition to the 400 arm-stereo photographs, 520 hand-held single lense reflex photographs (taken by many observers), 340 hand-held stereos, and nine 800 shot camera surveys constitute the photographic data base of this paper.

Some correlations for which there was insufficient data in the Galapagos study were clarified during our recent expedition to the vents on the East Pacific Rise at 21°N (Spiess et al., 1980). These will be utilized where necessary.

No maps of the vents we visited can be made. The terrain was too rough for surveys like that of Grassle et al. (1975), and Alvin's cameras and lighting were not oriented for higher altitude transects like the Angus surveys in Crane & Ballard (1980). Further, the navigation was too uncertain to map our wanderings while closer range photography was done. As a result, there exists only the most general idea about the interrelationship of various sections within any vent field. Finally, the photography is not unbiased. Arm-stereo photographs required the submarine to have settled, and the terrain limited where this was possible. Also, most photographs were taken of subjects of interest, usually biological, and not in a random way. For all these reasons, a calculation of total or average standing crop at a vent is impossible (Smithey & Hessler, in press). Arm-stereo photographs can be quantified, but the most they can give us is maximum densities that were photographed (Table 3).

Equipment and methods of photographic analysis are treated more fully in Smithey & Hessler (in press).

LOCATION

Three vent fields received several visits: Garden of Eden (4 dives), Musselbed (11 dives) and Rose Garden (7 dives) (Table 1). A fourth vent, here called Small Fry, was seen briefly once while searching for Musselbed. Garden of Eden was one of five vents visited by the physical scientists in 1977 and is described by Crane & Ballard (1980). Musselbed and Rose Garden were discovered on our 1979 expedition. Because vents are easily missed, there are probably more. The lineation of these nearly equatorial sites is essentially east-west, and approximately 20 km separates the extremes. Rose Garden is approximately 7,800 m from Musselbed, and the latter is 2,700 m from Garden of Eden.

The physical setting of the vents is described in van Andel & Ballard (1979), and Crane & Ballard (1980). The water chemistry is covered by Corliss et al. (1979), and Edmond et al. (1979a,b).

Table 1. Depth and coordinates of vents visited on the 1979 biological expedition.

Vent	Coordinates	Depth (approx. m)
Rose Garden	00°48.247'N; 86°13.478'W	2450
Musselbed	00°47.894'N; 86°09.210'W	2490
Small Fry	unknown, but near Musselbed	2495
Garden of Eden	00°47.692'N; 86°07.739'W	2485

DISTRIBUTIONS

Because the hydrothermal effluent is the driving force governing the existence of life at vents, the communities are characterized by a diffuse zonation centered on vent openings. Accordingly, the following descriptions of species distributions will begin with species found at the warmest vent openings and continue through those at cooler fissures, to intervening rock surfaces, and finish at the periphery of the vent field (Table 2). In reality, this zonation is not clear-cut; the untidy distribution of major and minor vent openings, combined with complexities in topography result in patchiness which is often difficult to explain. The vent megafauna is listed in Table 2. Undescribed species are identified to the lowest taxonomic level currently available. Additional photographs illustrating vent fauna distributions are found in Corliss & Ballard (1977), Grassle et al. (1979), and Ballard & Grassle (1979).

Vestimentifera

Riftia pachyptila (Jones, 1981) is only found at active vent openings, where it usually grows in clusters (Figs. 1,2,5,7). (Figures 1-6 will be found in the color insert following page 18.) Their density ranges from as few as one to dense stands of thousands. The largest thickets, seen at Rose Garden, stand two meters high and run many meters in length. The base of the tube is always hidden, either because it is attached down in the vent opening, or because of an overgrowth of other organisms. The tubes generally tend to parallel each other, but may form a more tangled pattern. For the most part, they are erect, but in cases where the density of individuals is sparse, they may be recumbent. The only part of the animal itself which emerges from the tube is the lamellate, red obturaculum, and the path and length of the tubes is such that the obturacula are usually exposed at the surface of the thicket where they are frequently clumped.

The growth form of thickets is easily interpreted in terms of nutritional needs. As an obligate chemoautotroph (Cavanaugh et al., 1981; Felbeck, 1981; Cavanaugh, 1983), a vestimentiferan requires exposure of its absorptive organ, the obturaculum, to H_2S coming from the vent and O_2 from ambient bottom water. The CO_2 which is also required could come from either source. To achieve this, the obturaculum must remain on the periphery of dense thickets, where ambient and vent waters mix. Within the thicket there may not be sufficient ambient water to allow survival of an animal whose obturaculum is placed there. Conversely, an obturaculum which protrudes too far may not be exposed to the vent water itself. The generally erect form of most thickets should be a response to the rising of the lighter vent water, and the clustering of obturacula could be in response to preferential channels of vent water flowing through the thicket. The reclined orientation seen in some sparse colonies is probably related to a weak flow of vent water which forces the animal to keep its obturaculum close to the vent crack in order to obtain a proper mixture.

Archaeogastropoda

Three limpets were visible in the photographs. Only one, *Neomphalus fretterae* (McLean, 1981), has been described. All three

Table 2. Megafaunal taxa of the Galapagos Rift hydrothermal vents. Abbreviations: Ca, carnivore; Ch, hemoautotroph; D, deposit feeder; indet., indeterminate; Ms, strongly mobile; Mw, weakly mobile; Nf, near part of vent field, but not in vent openings; Nv, nonvent; P, peripheral part of vent field; Sc, scavenger; Se, sessile; Su, suspension feeder; V, vent opening. Where capitals are not used in an abbreviation, it indicates lesser importance.

Major taxon	Name	Locality	Feeding type	Mobility	Common name
Vestimentifera	*Riftia pachyptila*	V	Ch	Se	Giant tube worm
Gastropoda	*Neomphalus fretterae*	V	Su, D	Se, mw	Limpet
Gastropoda	2 limpet species indet.	V	D	Ms	Limpet
Bivalvia	*Calyptogena magnifica*	V	Ch, Su?	Mw	Giant clam
Bivalvia	Mytilidae species indet.	V, nf	Ch, Su?	Mw	Mussel
Actinaria	Species indet.	V	Su	Se	Large, plumose, translucent anemone
Vertebrata	?*Diplacanthopoma* sp.	V, Nf, p	Ch?	Ms	Vent fish
Crustacea	*Bythograea thermydron*	V, Nf, p, nv	Sc, Ca	Ms	Brachyuran crab
Crustacea	*Alvinocaris lusca*	V	Sc	Ms	Vent shrimp
Annelida	Polynoidae species indet.	V	Ca, Sc?	Ms	Scale worm
Crustacea	Amphipoda, several kinds	V, Nf, p	Sc, Ca?	Ms	Amphipods
Actinaria	Species indet.	V, Nf, p	Su	Se	Small, translucent anemone
Annelida	Serpulidae species indet.	Nf, P	Su	Se	Featherduster worm
Crustacea	*Munidopsis* species indet.	Nf, p, nv	Sc	Ms	Galatheid crab, squat lobster
Crustacea	Natantia species indet.	Nf, P	Sc	Ms	Small, red shrimp
Gastropoda	Turidae species indet.	Nf, P	Ca	Ms	Whelk
Siphonophora	Species indet.	P	Su	Se, ms?	Dandelion; rhodaliid siphonophore
Brachiopoda	Inarticulata, species indet.	P, Nv?	Su	Se	Brachiopod
Actinaria	Species indet.	P, Nv?	Su	Se	Amber-scaled anemone
Enteropneusta	Species indet.	P, Nv	Su?	Mw	Spaghetti, acorn worm
Crustacea	?*Nematocarcinus* species indet.	P, Nv	Sc	Ms	Long-legged shrimp
Echinodermata	Ophiuroidea, species indet.	p, Nv	D	Ms	Thin, white ophiuroid; brittle star

Major taxon	Name	Locality	Feeding type	Mobility	Common name
Vestimentifera	*Riftia pachyptila*	V	Ch	Se	Giant tube worm
Gastropoda	*Neomphalus fretterae*	V	Su,D	Se,mw	Limpet
Gastropoda	2 limpet species indet.	V	D	Ms	Limpet
Bivalvia	*Calyptogena magnifica*	V	Ch,Su?	Mw	Giant clam
Bivalvia	Mytilidae species indet.	V,nf	Ch,Su?	Mw	Mussel
Actinaria	Species indet.	V	Su	Se	Large, plumose, translucent anemone
Vertebrata	?*Diplacanthopoma* sp.	V,Nf,p	Ch?	Ms	Vent fish
Crustacea	*Bythograea thermydron*	V,Nf,p,nv	Sc,Ca	Ms	Brachyuran crab
Crustacea	*Alvinocaris lusca*	V	Sc	Ms	Vent shrimp
Annelida	Polynoidae species indet.	V	Ca,Sc?	Ms	Scale worm
Crustacea	Amphipoda, several kinds	V,Nf,p	Sc,Ca?	Ms	Amphipods
Actinaria	Species indet.	V,Nf,p	Su	Se	Small, translucent anemone
Annelida	Serpulidae species indet.	Nf,P	Su	Se	Featherduster worm
Crustacea	*Munidopsis* species indet.	Nf,p,nv	Sc	Ms	Galatheid crab, squat lobster
Crustacea	Natantia species indet.	Nf,P	Sc	Ms	Small, red shrimp
Gastropoda	Turidae species indet.	Nf,P	Ca	Ms	Whelk
Siphonophora	Species indet.	P	Su	Se,ms?	Dandelion; rhodaliid siphonophore
Brachiopoda	Inarticulata, species indet.	P,Nv?	Su	Se	Brachiopod
Actinaria	Species indet.	P,Nv?	Su	Se	Amber-scaled anemone
Enteropneusta	Species indet.	P,Nv	Su?	Mw	Spaghetti, acorn worm
Crustacea	?*Nematocarcinus* species indet.	P,Nv	Sc	Ms	Long-legged shrimp
Echinodermata	Ophiuroidea, species indet.	p,Nv	D	Ms	Thin, white ophiuroid; brittle star

species were limited to vent openings or their immediate proximity. They were seen on the rock walls of the vent opening, on mussels or the tubes of vestimentiferans. None were seen beyond the throat of the vent.

Neomphalus fretterae often forms aggregations so tightly packed on the walls of the vent that there is no space between individuals (Fig. 7). The largest specimens rested at the center, and individual size diminished toward the periphery. The white shell of this species is radially grooved and irregular. Its rim not only fits the irregular contour of the substrate, but conforms to that of adjacent individuals. The total effect is reminiscent of a chemical incrustation. Clearly, the animal does not move about much. Anatomical studies (Fretter et al., 1981) suggest some mobility and grazing activity, as well as the ability to suspension feed (McLean, 1981). It is likely that suspension feeding dominates in the clusters. N. fretterae occurs sparsely on vestimentiferan tubes, but was not found on other organisms.

The other two limpet species (Figs. 2,6) cannot be separated reliably in photographs; one is black, and the other translucent, but the latter may have either a dark oxide covering or transmit the color of the rock underneath. Both species occur on rock, vestimentiferan tubes and mussels. They may be abundant, but are always well separated from neighboring limpets, and their shell margin is regular. These observations suggest that the two species wander during their regular activities and are probably deposit feeders.

Calyptogena

The vesicomyid bivalve Calyptogena magnifica (Boss & Turner, 1980) easily catches the eye by virtue of its large, white shell (Figs. 1,3,4,8,9). It was seen at all vent areas, and concentrations of its shell were found even in areas where vents were not identified, indicating vent fields that had expired (Corliss & Ballard, 1977). Indeed, except at Rose Garden, most specimens were dead (Fig. 8). As a result, the animal appeared more common than was actually the case. This is in contrast to the large fields of living clams seen at 21°N on the East Pacific Rise (Ballard & Grassle, 1979; pers. obs.).

Both living animals and dead shells were concentrated in clefts between lava pillows or cracks in lava sheets. Rarely, a dead shell would be found lying on the unbroken rock surface, but never a living animal. The situations at Rose Garden and Musselbed encompass the range of conditions of Calyptogena's distribution.

At Rose Garden (Fig. 1,3,4), the dominant environment for Calyptogena was a broad, flat, relatively unbroken expanse. There

Fig. 7. Garden of Eden. Vent opening encrusted with Neomphalus. A few Bythograea are concealed among the bases of the vestimentiferans, and one ?Diplacanthopoma hovers just beyond. (Photo by A. Giddings)

Fig. 8. Garden of Eden. Serpulids and translucent anemones, with a few Bythograea and Munidopsis. No vent openings can be seen in this view. (Photo by R. Hessler)

were no major fissures, but the surface was crossed by a number of straight cracks up to a few inches wide. Some ran into the main thicket of vestimentiferans and mussels, and disappeared under the clot of animals. Typically, these cracks were well populated with *Calyptogena* (Figs. 1,3). Often living individuals were oriented with the anterior end headed down into the crack. Some cracks were only wide enough to accommodate a single line of clams; others allowed two or more abreast. In places, gaps interrupted the line of clams.

For the most part, living individuals dominated the cracks. Dead shells were more abundant on the rock adjacent to the cracks. *Calyptogena* has a long, flexible foot and can adjust its position actively. As animals oriented themselves in the cracks, it would be natural for dead shells to be knocked out of the way.

Nearer to the main vestimentiferan/mussel thicket, mussels become more abundant. Where the two species were in direct contact, the clam would tend to be below, often lying on its side (Fig. 1). Clams were in direct contact with the main thicket, tucked under the mass of mussels that often made the thicket's fringe.

At Musselbed, living *Calyptogena* were far less abundant, but dead shells were numerous (Fig. 8). Typically a crevasse between pillows would be covered with shells lying in a jumble or lying flat and packed more closely. Among these might be a few living individuals poked down into the substrate. Here, mussels were commonly associated with the clams, either singly or in clumps; again, *Calyptogena* would be underneath.

We saw several beds where all the clams were dead. The general distribution of shells with respect to general topography was the same as with living animals. The relative age of the bed could be discerned from whether the valves were still articulated, whether any organic structures remained, degree of breakage, or dissolution and amount of associated sediment. At one extreme were fresh, articulated shells loosely packed and in a variety of orientations. Such relatively recent situations were common in portions of the vent field. The oldest dead clam field, to the best of our knowledge, was not in the immediate vicinity of any active vent. There were no living vent organisms in the area. All shells were disarticulated and broken or with their thin, central areas dissolved away. They were lying flat, mostly convex side up and closely packed down. There was a heavy dusting of sediment over the whole area. Killingley et al. (1980) estimate a whole shell will dissolve in approximately 25 years.

An impression gained by all who dove on the site was that individuals of *Calyptogena* were smaller at Rose Garden. Elsewhere,

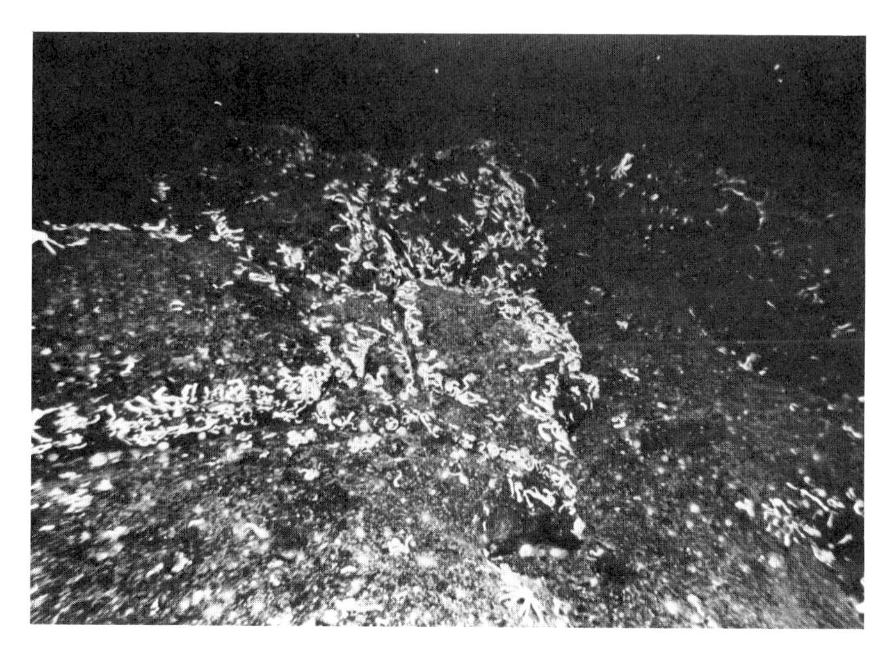

Fig. 9. Musselbed. Siphonophores in the peripheral zone. The dead clams and mussels suggest this is an extinguished vent. Temperature 2.15°C. (Photo by R. Hessler)

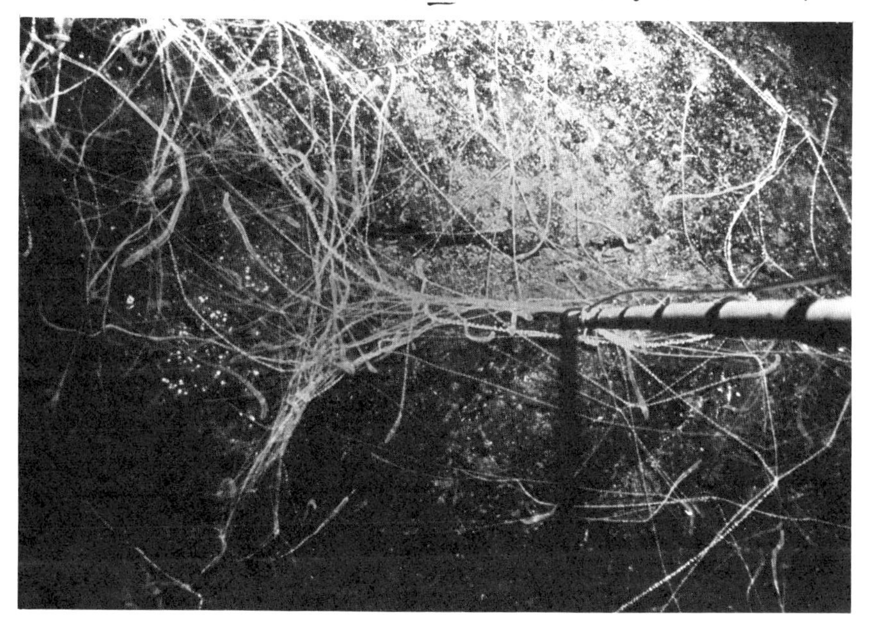

Fig. 10. Garden of Eden. Enteropneusts on peripheral pillow basalt. Temperature 2.08°C. (Photo by R. Hessler)

living and dead tended to be of a relatively uniform, large size. Our experience at 21°N tells us smaller individuals probably resided underneath. Therefore, the hypothesis that clam populations are the result of a single pulse of colonization (Corliss et al., 1979) may not be correct.

Mytilidae

The distribution of this undescribed mussel is a result of its apparent preference for vent water together with its ability to exist where not much is present, and its facility for attaching to steep surfaces through use of byssal threads. Musselbed, the vent area where mussels were the dominant life form, shows their range of distribution clearly (Figs. 5,6,9).

Here, the main source of exiting water was a complex cluster of openings in a nearly vertical surface of jumbled pillow basalt. The mussels occurred in dense clumps, often many individuals thick, with animals attached to each other. We estimate densities as high as 312 ind./m^2, with a biomass of approximately 10.1 kg/m^2. Smoky vent water could be seen emerging around these clumps. Here and there small clusters of vestimentiferans were interspersed. Often the mussel shells or vestimentiferan tubes had numerous byssal thread bases attached to them, giving evidence of the mussels' mobility. Nearby, other clumps were lying in the bottom of declivities between pillows or stuffed tightly into cracks. In this situation, they were often interspersed among living and dead Calyptogena (Fig. 8); whether the mussels were clumped or single, they rested on top. Finally, in this region (never far from vent cracks) individual mussels might be lying on top of unbroken pillows.

While vestimentiferans attracted most notice at Rose Garden, mussels probably had an equivalent standing crop (Figs. 1-3). The interstices of the lower half or third of each vestimentiferan thicket was completely filled with mussels, so that this region of the thicket was truly a solid mound of life (Table 3). Many mussels were attached to the higher, free-standing portion of vestimentiferan tubes (Fig. 2). At the bottom of the thickets, there was often a densely packed fringe of mussels. Cracks near the thickets were dominated by mussel clumps rather than clams (Fig. 1). In some places a few twisted vestimentiferans protruded from large mussel clumps, and clams fringed the clump below. Even in the region somewhat further away, clusters of mussels might dominate a crack. Wherever mussels occurred at cracks, a few scattered individuals might be seen on the adjacent unbroken lava among the anemones (Fig. 3).

Table 3. Maximum numerical density of selected species as seen with the arm-stereo camera. This is not necessarily the highest density the animals achieve.

Taxon	Site	Density .125 m^2 (measured)	Density 1.0 m^2 (extrapolated)
Riftia	Rose Garden	22	176
Mytilid	Musselbed/Rose Garden	39	312
Serpulid	Garden of Eden	135	1080
Neomphalus	Garden of Eden	68	544
Other limpets	Garden of Eden	77	616
Small, translucent anemone	Garden of Eden	57	456
Alvinocaris	Rose Garden	14	112
Brachiopod	Rose Garden	21	168
Enteropneust	Garden of Eden	34	272

Actinarians

Describing the distribution of anemones is complicated because, as with the archaeogastropod limpets, there are more than one species that cannot be discriminated reliably from photographs. Unlike the limpets, however, the different anemone species may have differing distributions. In the present discussion, we limit ourselves to types that are common in the vent area. There are others, usually quite large, that were seen individually only a few times and were too poorly documented to discuss. One of these is memorable because a large individual was videotaped with its long tentacles streaming in the current of a vent. Its tolerance of higher temperatures suggests that it probably does not occur away from vents, making it a unique example of a rare vent species.

The common anemones are all of modest size. They appear in elevated concentrations from the margin of the vent field to the vent openings themselves. One is characterized by amber-colored papillae in rows on its column and elongate form, even when contracted. The others are more translucent and when contracted are much shorter than broad. Some are pigmentless, while others are light pink. Papillae, if present, are scattered. Of the colorless anemones, one type is larger and has more numerous tentacles.

The larger, pigmentless, more numerously tentacled anemone is seen frequently at vent openings, growing on vestimentiferan tubes or on rock. Often they are tucked tightly into crevasses, much as with mussels, with which they may be associated. Neomphalus sometimes encrusts the vent wall above them.

The smaller translucent anemones may be seen anywhere in the vent area (Figs. 1,3,4,9,11). Sometimes they occur peripherally, but most typically, they are found in intermediate environments. The largest display is at Rose Garden, where they dominate the rock flats adjacent to the main thickets (Figs. 1,3,4). Individuals of every size are mixed together in densities as high as 456 ind/m^2. These concentrations extend undiminished up to the edge of the vestimentiferans or the clams and mussels in the cracks that crossed the flats. These anemones are also abundant at Garden of Eden, where the topography was much more irregular (Fig. 9). Here, individuals showed especially clearly a general tendency to occupy topographic lows as well as flats. Unlike serpulids, they seem to avoid ridges and crests. Anemones and serpulids (below) are frequently mixed, but tend to display inverse abundances.

The elongate anemones with rows of papillae are found mainly in peripheral locations and never in high densities. They are associated with enteropneusts, siphonophores, brachiopods and low densities of serpulids (Fig. 11). They have also been seen with the nonvent holothurians and xenophyophorians.

Vertebrata

The only fish commonly seen at vent fields has been tentatively identified as the bythitid ?*Diplacanthopoma* (Cohen & Haedrich, in press). This is uncertain because it has never been caught.

?*Diplacanthopoma* might be seen throughout the vent area, but is clearly concentrated at the most active vent openings, such as with vestimentiferan thickets (Fig. 7), or the main area of Musselbed (Fig. 5) . Here, they would swim down into the vent, to the extent allowed by other associated organisms. In the vents, they would hover, heads angled downward, with their tails undulating slowly. Often, several would be next to each other so that their movement appeared to be in unison. Otherwise, ?*Diplacanthopoma* swam slowly about. They were never seen purposefully in contact with the bottom, even in vents, except when stationary in clumps of organisms.

Bythograea

The brachyuran crab *Bythograea thermydron* (Williams, 1980) occurs throughout the vent area and even beyond, but their peak abundance is unquestionably on animal clumps in the vent openings themselves, particularly vestimentiferan thickets (Figs. 2,5-7,9). Here they usually nestle down among the tubes (Fig. 2), often in the throat of the vent, but also climb up individual tubes. Where mussels are mixed among the vestimentiferans, as at Rose Garden, *Bythograea* is equally abundant on the mussels, except at the

Fig. 11. Garden of Eden. Detail of pillow lava. Numerous unknown tubes, anemones, and brachiopods (particularly lower center) are shown. Temperature 2.14°C. (Photo by R. Hessler)

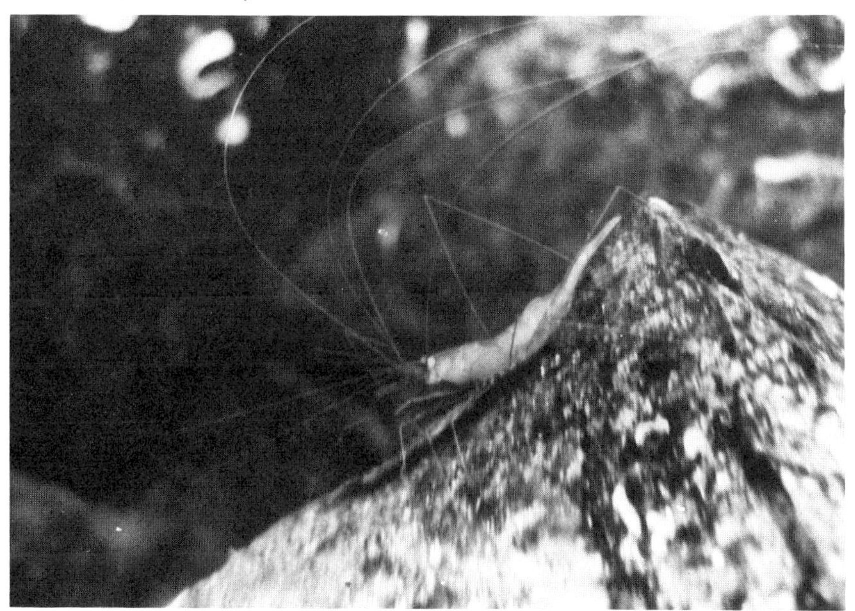

Fig. 12. Peripheral to Musselbed. ?*Nematocarcinus* on pillow basalt.

margins, where fewer crabs are sitting upon the pure bivalve fringe.

Bythograea also is present in elevated abundance at vents where species other than vestimentiferans dominate. At Musselbed they clambered among the mussels and into the cavities between them (Figs. 5,6). In the rare instances of vents dominated by limpets, Bythograea might be seen in the vent opening. In all of these cases, crabs would frequently be sitting on the rock adjacent to the vents.

Specimens in ones and twos occur elsewhere in the vent field: on bare rocks or rocks dominated by serpulids and anemones. They are found occasionally in peripheral areas where siphonophores or brachiopods may be abundant, or where nonvent taxa might be found. Rarely, they are even seen in barren terrain away from vent fields.

Alvinocaris

The bresiliid caridean shrimp Alvinocaris lusca (Williams & Chace, 1982) is present for the most part only at vent openings, usually residing on the larger sessile organisms (Figs. 2,6). Its highest densities were seen in the large vestimentiferan/mussel thickets at Rose Garden (Fig. 2), where closeup photographs usually show numbers of them on either kind of sessile animal. Here an occasional individual might also be seen on the adjacent rock, in one case on neighboring ledges with serpulids and anemones.

Alvinocaris was also found on vestimentiferans at Garden of Eden, but at Musselbed they were found in mussel encrustations, even where vestimentiferans were not present (Fig. 6). Here, small numbers of shrimp were sometimes seen on mussel clumps away from the main bed, even under conditions where dead clams were present. Rarely, an individual was present where vent conditions were barely detectable.

Polynoidae

A pink polynoid polychaete is abundant at vent openings, in association with vestimentiferans, mussels, clams, limpets and the usual other vent organisms. It crawls on the rock walls, mussel shells or less commonly, vestimentiferan tubes.

Amphipoda

Some amphipods are marginally large enough to be seen and photographed. They were noticed throughout the vent field, sometimes swimming, but because they are usually cryptic, their real abundance is much greater than what was seen in undisturbed circumstances; while sampling clumps of mussels or vestimentiferans, many might swim out. No patterns can be deduced because several species are involved.

Serpulidae

These abundant tubicolous polychaetes have not yet been identified. Their distribution is difficult to describe because they are found in so many situations, and while some patterns seem apparent, we are aware of exceptions to most of them.

From some approaches upon entering a vent field, serpulids are the first animals seen in any concentration. At Garden of Eden (Fig. 9), they are abundant in the same region as the generally peripheral siphonophores (below). They are also found in close proximity to vent openings. At Musselbed, patches of them live on the vertical wall amongst the clumps of mussels at the main vent opening (Fig. 5). For all practical purposes, they are not found directly in the vent opening, but even this occurs rarely; at Garden of Eden, a concentration of them extended down into a cleft where they were finally replaced by limpets. In another case at an apparently peripheral region, serpulids were concentrated at the bottom of a cleft, much as a vent species would be. No temperatures were taken in such situations to verify the amount of vent activity.

The densest concentrations of serpulids are found on steeper surfaces. Often, but by no means invariably, the slopes of a pillow will be heavily encrusted while the top is barren or only lightly populated (Fig. 9). As a result, they are mostly found in irregular topography; they are abundant in a pillow region adjacent to the large thickets at Rose Garden, but are nearly absent from the extensive flats. In many places, the concentration of individuals also decreases going down into nonvent clefts. On complex surfaces with ridges separated by narrow clefts, the serpulids tend to be concentrated on the ridges and absent from the intervening troughs.

A striking pattern that is frequently seen is an abruptly dense concentration along the edge of a rock, such as a sharp crest or the broken edge of a collapse structure (Fig. 9). A small ridge on an otherwise flat surface may bear numerous serpulids, which are abruptly absent from the adjacent plane.

Individuals of a cluster tend to be the same size, but frequently smaller individuals occupy the edges. Areas where all the individuals are dead are not uncommon, particularly near patches of dead Calyptogena. Where a serpulid patch butts against a cluster of mussels, there is frequently a border of unoccupied rock. Perhaps interference from the movement of the mussels keeps serpulids from settling. Serpulids are usually found in pure stands, but also occur in reduced numbers mixed with translucent anemones.

Galatheidae

An unidentified species of the galatheid crab Munidopsis is as broadly distributed as the brachyuran crab Bythograea, with one important difference--it is virtually absent from vent openings or on clumps of sessile animals living in vent openings. Its greatest concentrations are on pillows with a healthy growth of serpulids (Fig. 9), and rock surfaces adjacent to animal-occluded vent openings. The rocks around the main mussel-filled vent at Musselbed are a good example (Fig. 5). Generally, Munidopsis is less common on surfaces dominated by anemones; at Rose Garden, it was relatively uncommon on the anemone flats adjacent to the main vent system, compared to pillows with serpulids.

Munidopsis also lives in more peripheral portions of vent areas. They are found in small numbers near small vents with moderate or small clumps of mussels and living clams, or where dead clam shells are abundant. They appear where serpulids are sparsely present, along with siphonophores and brachiopods. Finally, they are seen occasionally in nonvent regions in association with the sparse nonvent fauna.

Natantia indet.

A very small shrimp (not yet identified) with red viscera occurs broadly, but sporatically through the vent field. It is most common on rocks with serpulids near secondary vent cracks occupied by mussels, but it was also seen as peripherally as with elongate anemones and dead Calyptogena shells, and as centrally as on rock by a major mussel vent.

Turidae

The distribution of whelks is opposite that of the limpets. They are never in the vent throat itself, although they come close to it (Fig. 3). At Rose Garden, they were not uncommon on the flat basalt area that was rich in anemones right up to the margin of the vestimentiferan/mussel thicket. At Musselbed, whelks were seen one-third meter or less from a vent crack with mussels in it. They also occur at the periphery of a vent, in the vicinity of siphonophores and even enteropneusts. Throughout their range, they are associated with serpulids and anemones. The strong development of the foot suggests that individuals move about actively.

Siphonophora

The rhodaliid siphonophore (Pugh, in press) may be found in abundance, and even dominates the macrofauna locally (Fig. 8). They are most abundant toward the periphery of a vent field, but not so far out as with the center of concentration of enterpneusts (below). Unlike other sessile taxa, even at the height of concentration, they are always well separated from each other.

The majority of individuals are attached in low areas--general topographic lows, the clefts between lava pillows, or in collapse chambers. This may well be caused by their mode of attachment. Siphonophores are positioned in the water column a few centimeters above the bottom. They are held there by many long, radiating attachment filaments. These are fragile (only a little current stirred up by Alvin would break them loose), so that exposed individuals might be in danger of detachment. Occasionally, however, individuals are seen near the tops of rocks.

Siphonophores may be found some distance from vent outflows, in areas where the nonvent fauna begins to dominate--with holothurians and xenophyophorians (large, mud agglutinating protozoans), for example. They may rarely come close to a vent opening. One individual was seen in a cleft next to a crack with a concentration of healthy mussels. Most commonly this species is associated with brachiopods, serpulids or anemones, although the last two usually occupy rocks above the attachment sites of the siphonophores.

Brachiopoda

No brachiopods were collected, so that detailed identification is not possible. They appear to be Inarticulata of the family Discinidae or Craniidae and are easily recognized by their translucent nacreous luster, somewhat irregular discoid shape, subcentral umbo, and recumbent orientation (Fig. 11).

None was seen near vent openings. Indeed, they were always found in areas remote from direct influence, on horizontal or steeply sloping surfaces, even in the clefts between lava pillows. The most typical associates are anemones and an unidentified agglutinated tube builder. Occasional enteropneusts, siphonophores and isolated mussels are also associates. Brachiopods are not not found with living serpulids, but can be found in conjunction with dead ones or near dead Calyptogena.

Enteropneusta

A still undescribed species of enteropneust, aptly called "spaghetti" by those who first saw it, attracts immediate attention because of the characteristic aggregations in which it lives. Typically they lie draped over the tops of rocks in dense profusion (Fig. 10). Concentrations may be so great that the rock is completely hidden, and the covering is many animals deep. The general impression is somewhat like that of cobwebs because along the vertical and lower undercutting portions of the rock, the animals dangle freely in the water or arc over to other surfaces. Often it is the anterior end of the animal that is dangling freely.

Unquestionably, the animals prefer high, exposed surfaces. In places where a slab of lava has tilted and cracked, the enteropneusts upon it are concentrated along the ridge formed by the broken edge. These animals also occur at low densities and even as isolated individuals, where they are occasionally seen on open upper surfaces that are not necessarily topographic highs, but never in depressions.

Enteropneusts only appear towards the periphery of vent areas. Indeed, at vents where they are present, they are the last indication of the presence of a vent before entering the adjacent nonvent terrain. Water temperature is always ambient. Various members of the nonvent or peripheral vent fauna may be nearby, but are never found on the same rocks where enteropneusts are aggregated in abundance.

No pattern is perfect. One photograph at Musselbed shows three small individuals on a dead mussel shell in a peripheral crevasse. Other mussels are in the clump, and most are dead. It was not possible to tell whether any were living. As will be discussed, enteropneusts are also found in nonvent regions.

?Nematocarcinus

Of all the mobile animals that were seen commonly in vent areas, this long-legged, red caridean shrimp is the only one that showed greatest abundance at the margins of the vent region (Fig. 12). Indeed, it does not occur anywhere near the vents themselves. Typical associates among the vent fauna are the siphonophore, enteropneust, brachiopod and amber-scaled anemone. They also associate with dead *Calyptogena* and sparse mussels or serpulids. ?*Nematocarcinus* was not uncommon in nonvent terrain, associated with nonvent organisms. Usually they occurred in ones or twos, but occasionally in higher densities. Approximately 10 were seen once in a small collapse pit that also contained siphonophores.

Ophiuroidea

A thin, white ophiuroid was seen at peripheral portions of the vent field, sometimes in concentrations above that of the nonvent region. It was associated with siphonophores, *Calyptogena* shells or solitary mussels.

Vent Field Temperatures and Faunal Overview

At the hydrothermal vents, temperature is both a critical environmental parameter and an index of a complicated interplay of the 2.01°C ambient bottom water with the chemically unique and nutrient-rich vent water. It is clear that some of the fauna live preferentially in areas of less diluted vent water, while others prefer water with only a very minor vent-originated component.

Water exiting from a Galapagos vent is estimated to have been heated to 350° in its journey through the earth's crust, yet the warmest temperature actually recorded at the vents was 17°C (Corliss et al., 1979). We recorded a range of elevated temperatures from slightly above ambient to 14.72°C.

The mixing process resulting in the cooling of the 350°C water to ambient has several components. The largest drop occurs within the plumbing of the vents. Our high of 14.72°C was reached by plunging a temperature probe at least 30 cm into a vent choked with *Riftia*, so that the temperature was unlikely to have been influenced by surface mixing. The decrease is a result of subsurface mixing within the "leaky" plumbing of the vent combined with cooling to surrounding rocks.

After leaving the mouth of a vent, mixing is rapid. In cases where water flow is unimpeded by organisms, the temperature is reduced to a few tenths of a degree above ambient within a meter of the vent opening. Dense aggregations of fauna can modify the mixing process. Temperatures within a clump of *Riftia* are definitely higher than would be the case at the same elevation were the thicket not there, because the aggregation of tubes acts as a porous chimney. Mussels clogging the main vent at Musselbed experienced temperatures several degrees above ambient (once as high as 12°C) on the vent-facing (upstream) side, while the surface that faced away from the vent was exposed to water as low as a few tenths of a degree above ambient.

The maximum temperature recorded varied little among vents, with Rose Garden registering 14.72°C, Garden of Eden 13.00°C and Musselbed around 12.00°C. Perhaps more telling than this was the amount that the vent water warmed the ambient water in the area of the site. This effect is a combined function of both the maximum temperature and the volume of water spewed out at each vent. With the temperature probe at a height of 1.5 to 2.5 meters above the bottom, a transect over Rose Garden recorded a high of 2.32°C, Garden of Eden, 2.35°C and Musselbed, 2.10°C. The extent and shape of this regional anomaly must be influenced by several factors. Local currents and tidal flux undoubtedly have an effect. Also, as evidenced at Musselbed, the extent, type and density of fauna may influence the rapidity of mixing. We know nothing of the variation in temperature with time or volume of the vent emissions.

Because of convection lifting most of the diluted vent water upward, vent influence degrades quickly from areas directly bathed by warm water to those lateral to the source (Fig. 13). The distribution pattern of each species with respect to the exiting vent water is much the same at the four vents, as seen from the treatment of individual taxa. In some cases the species closely reflect the abrupt changes in water quality while other species distributions may transcend them.

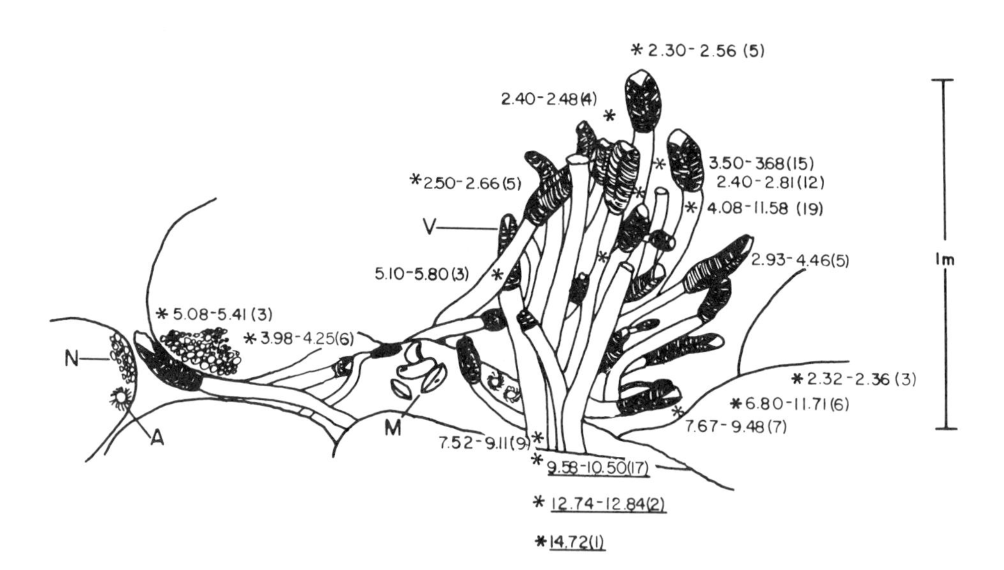

Fig. 13. Diagram of the temperature field around a vent dominated by vestimentiferans. This is a composite of many observations. Temperature (°C) is given as a range; number of observations is indicated parenthetically. Underlined numbers indicate measurements down in the vent opening. Symbols: A, plumose anemone; M, mytilid, some with limpets; N, Neomphalus; V, vestimentiferan.

Most easily categorized are those animals found at vent openings, in water of 4-15°C. *Riftia*, *Alvinocaris* and the large, plumose anemone are limited to this situation. The mytilid, *Bythograea*, ?*Diplacanthopoma*, *Neomphalus* and the other limpets for the most part are found in these conditions. While *Calyptogena* is occasionally found at such temperatures, it is more numerous lining cracks with exiting water a few tenths of a degree above ambient. Within vestimentiferan thickets, the temperature grades from above 12°C at the base to around 3.5°C at the obturacula. Mussels clogging vents at Musselbed were bathed in water in excess of 8°C on the surface facing the flow. This fell to around 3°C on the side away from the flow.

Other animals live lateral to vents, but close enough to be within the influence of vent waters, where the temperature is just a few tenths to a full degree above ambient. Dominant are serpulids (up to 2.94°C) and small, translucent anemones (up to 2.68°C). Living among them are the whelk and the small red shrimp. Mussels are found in this area in temperatures up to 2.72°C. The galatheid crab is most often in these areas.

The siphonophore, enteropneust, brachiopod and ?*Nematocarcinus* are found in areas sufficiently peripheral that the temperature is usually ambient, but occasionally slightly above. The maximum temperature was 2.24°C for the siphonophore, 2.11°C for the enteropneust, 2.10°C for the brachiopod and 2.15°C for ?*Nematocarcinus*. All these values are within 0.2°C of ambient.

Animals we have classified as nonvent are those that were never observed in water of elevated temperature. These will be discussed shortly.

Vents as a Whole

While the distribution pattern of species remains consistent between vent fields, the relative abundance may vary.

Rose Garden (Figs. 1-4). This vent field displays the highest standing crop that has been found in the Galapagos system, primarily due to the huge thickets of vestimentiferans and mussels. The largest thickets are confined to a single fissure which is manifested by the smoky water percolating out through the animals. Other, smaller thickets are abundant; they are dominated more by mussels with decreasing size. Clams are a minor component by comparison, forming a fringe of the mussel clumps, but gradually coming to dominate smaller fissures away from the main thickets. Brachyuran crabs and vent shrimp are abundant in the larger thickets. The two mobile limpets and the large translucent anemone are not uncommon.

Much of the area surrounding the main vents, particularly to the south, is flat and unbroken, except for occasional cracks clogged with bivalves. These surfaces are dominated by dense concentration of the small translucent anemone, but with brachiopods locally abundant toward the outer edges. Further away from the central thickets, this gives way to dense stands of serpulids.

The more jumbled pillow basalts that abutt on the central thickets, especially to the northwest have anemones closest to the thickets, but these are shortly replaced by serpulids, which extend nearly to the margin of the vent field. Galatheids are particularly abundant on these pillows.

At the limits of the vent field occur a few patches of siphonophores which are not extensive and many small clumps of enteropneusts.

?*Diplacanthopoma* and *Neomphalus* are uncommon here, perhaps because of the domination by vestimentiferans and mussels leaving few open surfaces or spaces exposed to vent water.

Musselbed (Figs. 5,6,8,12). This vent field contrasts strongly with Rose Garden, both topographically and in relative positions of its fauna. It is located in a very jumbled terrain of pillow basalt. The main vent openings are on a steep slope. Here dense beds of mussels dominate, but tufts of vestimentiferans poke out here and there. Immediately adjacent rocks have high concentrations of serpulids.

?*Diplacanthopoma* is abundant in the smoky vent water. Crabs are common--*Bythograea* more on the vent organisms, *Munidopsis* on the adjacent serpulids. As elsewhere, they also occur in more peripheral areas.

At least one small vent is dominated by *Neomphalus*.

Surrounding the incline bearing the main vents are flatter areas with numerous beds of dead clams in varying states of decalcification, with an intermixture of a few living clams and mussels, many of them dead. Serpulids cover the intervening pillows in moderate densities. Here and there in peripheral regions are modest patches of siphonophores, and a few patches of enteropneusts occur at the limits of the vent field. At least one of these was spectacularly large.

Garden of Eden (Figs. 7,9,10,11). The terrain of this vent field is entirely pillow basalt. Vent water flow seemed less than at Rose Garden or Musselbed, as indicated by the amount of vent-opening faunal clusters. Only a few modest tufts of vestimentiferans were seen. Well-developed rock encrustations of *Neomphalus* were associated with each of these, interspersed with openings having *Neomphalus* alone. ?*Diplacanthopoma* was abundant. Mussels were not common, mainly appearing in small numbers in a few feeble vent cracks. Small clam shells were seen, but no living individuals.

Extensive fields of serpulids, including many very dense concentrations butted directly up to the vents. There were also healthy patches of small, translucent anemones, some of the patches being quite large. More peripherally were dense concentrations of siphonophores.

The crabs *Munidopsis* and *Bythograea* were both common throughout the area, the former particularly so, with aggregations even peripheral to the concentrations of sessile suspension feeders.

Small Fry. This small site (radius 10 m or less) was covered with pillow basalt. No discrete vent was seen, yet ?*Diplacanthopoma* was common, particularly swimming out of a deep, crater-like depression. The dominant sessile organisms were siphonophores, with a sparse admixture of serpulids. Galatheids were common, and a few *Bythograea* were present. Enteropneusts covered some pillows in the vicinity. None of the usual sessile vent-opening organisms were seen. It is possible that we missed the main vent, but that does not explain why ?*Diplacanthopoma* was so common where we saw it.

General

Clearly, while the participants are the same, the communities differ distinctly in dominance at different vents. For example, the importance of mussels varies strikingly between Rose Garden and Garden of Eden, and living *Calyptogena* appeared to be missing from the latter locality. Crane & Ballard (1979) give other examples involving different vents.

We have insufficient knowledge of the natural history of the animals and physical conditions at the vents to be able to identify the causes, but some reasonable possibilities may be suggested. Exiting vent water chemistry is known to vary in proportion to the amount of subsurface mixing (Corliss et al., 1979; Edmond et al., 1979a,b), so that waters of different temperatures represent different chemical milieus. To a minor extent, different vent fields can display variation in abundance of some chemicals, presumably because of differing rock regimes (Corliss et al., 1979). These differences in chemistry and temperature should affect the conditions for growth of the microflora, both free-living and symbiotic, as well as affecting the animals directly. The rate of vent water flow and bacterial production will vary, and differences in topography, which affect currents, will influence the path, residence time and diffusion of the plume. The size and abundance of particles emitted from the vent varies, as is visually obvious; vent water may be clear or smoky, and with or without large particulates. The bacterial content of the particulates is not likely to be constant. All these variables could influence metazoan distribution, but there are still no concrete data for testing whether correlations exist.

At the Galapagos site in general, vestimentiferans experienced higher temperatures than those recorded for clams or mussels. However, at Clam Acres at the 21°N site on the East Pacific Rise, tufts of vestimentiferans grow in the same cracks as clams (a combination not seen at the Galapagos), and both experience 15°C. It may mean that volume of flow is equally important. Vestimentiferans may require higher flow to bathe their generally elevated obturacula, while bivalves can make do with lesser flow because they can nestle down into the water more effectively. Growing in a reclined position is only a limited solution for vestimentiferans because physical interference by bivalves must offer problems. Thus, inadequate flow at any one spot may explain low abundance of vestimentiferans at Musselbed and Garden of Eden.

If ?Diplacanthopoma needs access to vent openings, its low abundance at Rose Garden may result from the generally clogged condition of vents there.

The success of sessile animals in the vent field away from vents must be related to the volume and quality of vent water of the field as a whole. The general ambient temperature anomaly is a measure of total vent flow and is lowest at Musselbed, where this sessile fauna is least well developed.

These are examples of possible causes for variation in the vent community. The reasons for most differences await clarification. Indeed, the distribution of many species is puzzlingly patchy within a vent field; for example, the reason for the presence of serpulids on one pillow and absence on the adjacent one remains to be revealed.

Dispersal does not seem to be a factor in this issue. Except for the apparent absence of Neomphalus at Rose Garden, all species were seen at all three of the well-studied vent fields; Small Fry was not explored well enough to be sure of its fauna. Living clams were absent from Garden of Eden, but shells prove they had been there. This shows that at Galapagos, distances of a kilometer more or less are not a discernable barrier to the vent fauna. Therefore, if the faunas vary, ecological causes are more likely.

Corliss et al. (1979) have suggested the vent faunas vary in a successional sequence that reflects the aging process of the vent itself. Crane & Ballard (1980) propose the cycle propagates from east to west. That is, the western vent fields are youngest. In some ways, the fauna supports the Crane & Ballard hypothesis. The western-most vent Rose Garden seems to have the strongest flow of warm water, as documented by the luxuriance of the vent-opening fauna, and the generally smaller size of *Calyptogena* there might be construed as an indication of recent colonization. However, Garden of Eden (eastern-most) did not have the large fields of dead clams to indicate it had ever been more active. Perhaps it was always a more modest vent system. In general, local peculiarities of the physical setting dominate over the influence of vent age.

Relation to Nonvent Fauna

So far, we have devoted attention to distributions within vent fields. There remains the question of the extent to which the vent fauna is endemic. Answering this is difficult because of the paucity of observations away from vents. At those times when we did visit nonvent areas, we cruised, which could result in overlooking smaller organisms. Thus, with many taxa, the possibility of vent endemism must be viewed cautiously.

Taxa which are most clearly limited to vent areas are those which live at discharge openings: *Riftia*, *Calyptogena*, the mytilid, the limpets, the large plumose anemone and ?*Diplacanthopoma*. None of these were observed away from vent fields, and except for the mytilid and fish, they were always directly at vent openings. Exceptional is the crab *Bythograea*, which was seen rarely completely away from vents (Mickel & Childress, 1982a; pers. obs.); these occurrences seem best interpreted as expatriates--animals that accidentally wandered from the vent area and are perishing or living so marginally that reproduction is impossible. The absence of deep bathyal records of brachyurans prior to the discovery of the vents (Balss, 1955; Zarenkov, 1969; Hessler & Wilson, in press) shows we have little reason to expect *Bythograea* to be a regular part of the rocky deep-sea fauna.

Routine members of the general vent-field fauna that were not seen away from vents include the serpulid, the most common translucent anemone, the siphonophore, brachiopod and whelk. The serpulid, whelk and siphonophore are so noticeable that their presence would probably have been detected. Thus, they are likely to be vent endemics. The anemone and brachiopod easily could have been missed. Inarticulate brachiopods of that sort are typical of hard bottoms in the deep sea (Zezina, 1965, 1969). It could well be an example of higher standing crop resulting from the enhanced vent nutrition.

As already mentioned, the enteropneust and ophiuroid are successful inhabitants of the nonvent bottom. ?*Nematocarcinus* and

Munidopsis were seen several times away from the vent field. The abundant records of these two crustacean genera in the deep sea (Zarenkov, 1969), and their aversion for situations with elevated temperatures at the vents suggest they are nonvent taxa profiting from vent productivity.

Only two species of nonvent fish (both of the macrourid genus Coryphaenoides) showed somewhat elevated abundance near the vent field (Cohen & Haedrick, in press). But taking the fish fauna as a whole, species diversity is lower near vents than away from them (ibid.).

The rest of the nonvent fauna does not even penetrate the vent field, that is, the region near vents where species show elevated abundance. Principal among these are the holothurians and many kinds of typical deep-sea anthozoan coelenterates including gorgonians, antipatharians, hydroids, hydrocorallines and actinarians. Other invertebrates with this distribution were asteroids, hexactinellid sponges, xenophyophorians (rhizopod protozoans) and other animals we were not able to identify.

The dominant pattern, then, of the distribution of animals which thrive on hard bottoms away from vents is absence from the vent field. The importance of this pattern is accentuated by the high taxonomic level at which it operates (nearly all echinoderms, most coelenterates, sponges). At this stage of investigation, one can only speculate about causes.

The most reasonable possibility is that most nonvent taxa are intolerant of physical conditions engendered by the vents. Perhaps they cannot survive elevated temperatures or aspects of the water chemistry. In both cases, unusual sensitivity is suggested because the thermally driven convection cell lifts vent water away from the bottom (Corliss et al., 1979; Lipton et al., 1980), so that except in the immediate vicinity of vent openings, vent field temperatures are at most a few tenths of a degree above ambient, indicating considerable dilution of vent-water chemicals. Hydrogen sulfide levels should be even lower than what would result from dilution because of oxidation on contact with oxygen-rich ambient water. Temperature has not been recorded in such places over time, and it is possible that occasionally, warmer boluses would impinge. (Oxygen isotope ratios in Calyptogena [Killingley et al., 1980; Fatton et al., 1981] suggest variation in vent discharge, but do not take into account the possibility of movement of the clam.) Still, because of the efficiency of mixing, it is unlikely that the temperature would be elevated by even a degree. The concentration of potentially toxic substances has not been measured at the bottom away from vent openings. We lack more than anecdotal information on the tolerance of typical deep-sea organisms to temperature fluctuation. Deep-sea organisms in nonvent regions live under thermally unvarying circumstances and therefore have been under no selective pressure for tolerance to temperature change. How much variation is acceptable in such cases, and for how much time? It is not simply a question of the latitude of single metabolic pathways, which tend to react more coarsely, but the synergistically complex functioning of the whole organism.

The Food Web

A reasonable approximation of the food web (Fig. 14) at hydrothermal vents can be derived from consideration of behavior of living animals, their physiology, biochemistry, anatomy and distribution, as well as knowledge of the functional morphology and natural history of related taxa. The primary source of nutrition is surely the vent water, with chemoautotrophic bacteria using its chemically reduced constituents as an energy source to synthesize protoplasm (Jannasch & Wirsen, 1979, 1981; Rau & Hedges, 1979; Karl et al., 1980; Felbeck, 1981; Cavanaugh et al., 1981; Rau, 1981a,b; Felbeck & Somero, 1982). The intense concentration of metazoans in the mouth of vent openings is a good indication of this. The bacteria are utilized by the community in three major ways.

They may grow symbiotically in close conjunction with the animal, as best seen with *Riftia*, which brings the essential inorganic constituents of vent and ambient water to the bacterial culture within it (Felbeck, 1981; Cavanaugh et al., 1981). *Calyptogena* and the mytilid also have associated bacteria, in this case in their gills (Felbeck & Somero, 1982). All three taxa show stable isotope ratios that indicate a local food source (Rau & Hedges, 1979; Rau, 1981a,b).

Their distribution correlates nicely with this life style. Because dilution of vent water is so complete at the bottom, the concentration of reduced ions will be too low except in the vent openings. Accordingly, *Riftia* and *Calyptogena* are limited to vents, and that is the only place the mytilid is abundant. Further, mytilids growing away from vents are less well nourished (Hiatt, 1980 quoting K. L. Smith).

That mussels can live away from directly outwelling vent water at all points toward an ability to nourish themselves in another way. Both the mussel and *Calyptogena* have a well-developed digestive tract (unlike *Riftia*), and it is possible that a portion of their diet is based on suspension feeding (see below). To date, gut-content studies have not been adequate to clarify this point.

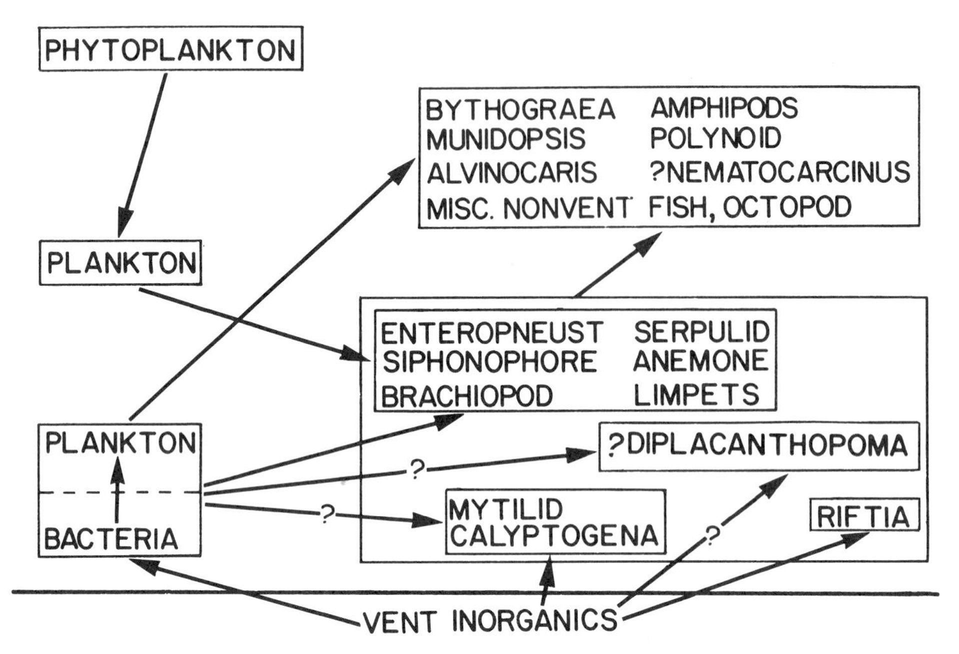

Fig. 14. Food web of vent megafauna. Linkages that are likely to be minor are not included.

?Diplacanthopoma also may be a chemoautotroph, as shown by its behavior. Its unwillingness to approach bait suggests it is not a scavenger. Nor was it ever seen to feed (Cohen & Haedrich, in press). Its frequent hovering in the vent throat without touching the rock walls and without any apparent movements of its mouth implies some significant purpose. Chemoautotrophy is quite possible. In this connection, it is important to note that we have seen the hovering behavior even when the vent water appeared quite clear, implying a paucity of particles of filterable size. The growing list of animals that independently evolved the ability to utilize chemoautotrophic bacteria demonstrates that this may not be a difficult ability to acquire.

While it has never been rigorously quantified, it is obvious that the greatest portion of vent biomass (75%) is made up of these symbiotically chemoautotrophic taxa.

A second source of nutrition is encrusting bacteria (Jannasch & Wirsen, 1979, 1981). Because the most flourishing bacterial growth depends on vent water, this food source is developed primarily at vent openings, either on rock or organisms. The two limpet species that move about on rocks, bivalves and vestimentiferans likely feed on this resource. The vent-water adapted crustaceans (*Alvinocaris*, *Bythograea*) also partake; *Bythograea* has been observed to sweep surfaces with its maxillipeds, and at 21°N we saw *Munidopsis* feed on bacterial overgrowth.

Finally, bacteria grow suspended in the subsurface vent water and also surely break loose from surfaces and are washed out. High concentrations of bacteria have been measured in vent water, and much of it is in aggregations that are large enough to be filtered by suspension feeders (Karl et al., 1980). Those suspension feeding animals that live in the vent throat (*Neomphalus*, anemones) may well live primarily off this food source.

The greatest cause of uncertainty about the precise nature of food for suspension feeders is our scant understanding of the vent plankton assemblage. Observers at vents on the East Pacific Rise at 21°N noted plankters living in the vent throat. Subsequently we saw the same thing in photographs from the Galapagos vents. Vent-mouth suspension feeders may also be feeding on these.

The plume of vent water rises away from the bottom because of its lower specific gravity, carrying living and nonliving particulates. This was observed on several occasions a number of meters up. These particles slowly rain down on the adjacent terrain. Thus, vent bacteria are even available to suspension and deposit feeders living in the immediate vicinity. An observable plankton fauna also occupies this plume and associated water. Amphipods are not uncommon swimming here, and one could see many plankters too small to identify. Once again, there is no certainty about the extent to which the benthic fauna away from a vent opening is feeding directly on bacteria or indirectly by eating plankton and plankton byproducts.

Further, some portion of the plankton is likely to be derived from the outlying nonvent region, having been carried in by the centripetal flow of the thermal convection cell (Enright et al., 1981). These provide a nutritional resource whose ultimate origin is derived from the normal, sunlight-powered photosynthetic food chain.

Serpulids, anemones, brachiopods, the siphonophore and the enteropneust are the obvious megafaunal suspension feeders relying on this whole complex, suspended and settling resource, while *Munidopsis*, ?*Nematocarcinus*, amphipods and wandering *Bythograea* are the dominant deposit feeders. Enteropneusts are normally considered deposit feeders (Hyman, 1959), but their intensely aggregated draping over elevated surfaces, often with their anterior ends hanging freely, makes suspension feeding more likely. Utilization of dissolved organics was suggested by one anonymous reviewer.

The distribution of the siphonophore, brachiopod, enteropneust, elongate anemone and ?Nematocarcinus are not fully explained by this senario because it would predict that their concentration should be highest toward the vent openings. Instead, their highest concentrations are more peripheral. The siphonophore may be inhibited by the fragileness of its attachment to the bottom (Pugh, in press). This would exclude it from regions of higher current or concentrations of larger mobile organisms, such as the galatheid. Its preference for sheltered low spots is consistent with the suggestion.

Interference by other organisms may limit the other taxa to peripheral areas, but this does not seem likely for any but the enteropneust. As already suggested, they are possibly more sensitive to elevated temperatures or vent-water chemistry. Finally, it may indicate a shift in food preferences. The enteropneust is not totally limited to vent sites. Dive 985 cruised extensively over nonvent terrain where more than once enteropneusts were sighted, and in each case they were concentrated on the crest of irregular ridges, once at the lip of a deep gorge. We interpret this as indicating a preference for places where current intensification exists. If this is so, the presence of enteropneusts at vents may be stimulated by the convection cell drawing food in from the outlying nonvent region (Enright et al., 1981). But need for current does not explain why they are not closer to the vents, where current would be stronger. It would be helpful to emplace sedimentation traps in the vicinity of peripherally concentrated taxa to see what kinds of food are available there.

So far, we have emphasized primary consumer activities and plankton feeding. Higher in the food chain are scavengers--those omnivorous animals that feed on deposited animal byproducts (carcasses, exuveae, feces, etc.), smaller living animals, and even pieces of much larger animals in addition to benthic microflora. At the Galapagos vents, crustaceans dominate this category: Munidopsis, Bythograea, Alvinocaris, ?Nematocarcinus and amphipods. The polynoid polychaete also probably belongs here. A few nonvent fish which appear in the vent region at somewhat higher concentrations (Cohen & Haedricn, in press) also fall in this category, but are not significant factors. Of all these, only Bythograea actually has been documented as a carnivore (Mickel & Childress, 1982b; pers. obs.). Carnivory has been attributed to Alvinocaris (Jones, 1981; Williams & Chase, 1982), but Childress was cited erroneously on this point (Childress, pers. comm.).

This food web suffers from neglecting the macrofauna (Sanders, in prep.) and smaller taxa. These surely play a major role in detritus feeding. A further weakness is that this scheme is based more on knowledge of the activities of homologous taxa than is desirable, but concrete data on most of the vent fauna have not yet been accumulated. While gut contents are difficult to analyze, they might be useful in specific instances; inspection of potential chemoautotrophs would be especially valuable. Particularly vexing is the void in information on plankton. Settling trap deployments would do much to reveal the kinds of particulates available in different parts of the vent field. The plankton fauna must be analyzed before we can hope to have a balanced picture of trophic structure of vent communities.

ACKNOWLEDGMENTS

The list of people whose willing help made this work possible is long. It includes the scientific members of the 1979 expedition, the companies of three ships (Giliss, New Horizon, Alvin/Lulu), the National Geographic Society film crew and the Angus team. The extraordinary efforts of our expedition leader, Fred Grassle, and the Alvin/Lulu Deep Submergence Group deserve special mention. Clifford Keller and Leslie Snider scored numerous photographs, and Ted Bayer, Austin Williams, Jim McLean, Ruth Turner and Daphne Dunn helped with identifications. Jim Childress and Tom Mickel contributed data for mussel biomass calculations. Suggestions by Ken Smith and two anonymous reviewers improved the manuscript very much. We give our heartfelt thanks to them all. This work was supported by Grants OCE78-10460 and OCE80-24901 of the National Science Foundation. It is 42 in the series of papers stemming from the Galapagos Rift Biology Expedition.

REFERENCES

Ballard, R. D., and Grassle, J. F. 1979, Return to oases of the deep (strange world without sun). Natl. Geogr. 156:680-703.

Balss, H., 1955, Decapoda:Okologie, Bronns, H. G., and Klassen, U., Ordnungen des Tierreichs, Bd. 5, Abt. 1, Buch 7, Lief. 10:1285-1367.

Boss, K. J., and Turner, R. D., 1980, The giant clam from the Galapagos Rift, Calyptogena magnifica species novum, Malacologia, 20:161-194.

Cavanaugh, C. M., Gardiner, S. L., Jones, M. L., Jannasch, H. W., and Waterbury, J. B., 1981, Prokaryotic cells in the hydrothermal vent tube worm Riftia pachyptila Jones: Possible chemoautotrophic symbionts, Science, 209:340-342.

Cavanaugh, C. M., 1983, Symbiotic chemoautotrophic bacteria in marine invertebrates from sulphide-rich habitats, Nature, 302:58-61.

Cohen, D. M., and Haedrich, R. L., in press, The fish fauna of the Galapagos thermal vent region, Deep-Sea Res.

Corliss, J. B. and Ballard, R. D. 1977, Oases of life in the cold abyss, Natl. Geogr., 152:441-454.

Corliss, J. B., Dymond, J., Gordon, L. I., Edmond, J. M., von Herzen, R. P., Ballard, R. D., Green, K., Williams, D., Bainbridge, A., Crane, K., and van Andel, T. H., 1979, Submarine thermal springs on the Galapagos Rift, Science, 203:1073-1083.

Crane, K., and Ballard, R. D., 1980, The Galapagos Rift at 86°W: 4. Structure and morphology of hydrothermal fields and their relationship to the volcanic and tectonic processes of the rift valley, J. Geophys. Res., 85(B3):1443-1454.

Edmond, J. M., Measures, C., McDuff, R. E., Chan, L. H., Collier, R., Grant, B., Gordon, L. I., and Corliss, J. B., 1979a, Ridge crest hydrothermal activity and the balances of the major and minor elements in the ocean: The Galapagos data, Earth Planet. Sci. Lett., 46:1-18.

Edmond, J. M., Measures, C., Mangum, B., Grant, B., Sclater, F. R., Collier, R., Hudson, A., Gordon, L. I., and Corliss, J. B., 1979b, On the formation of metal-rich deposits at ridge crests, Earth Plant. Sci. Lett., 46:19-30.

Enright, J. T., Newman, W. A., Hessler, R. R., and McGowan, J. A., 1981, Deep-ocean hydrothermal vent communities, Nature, 289:219-221.

Fatton, E., Marien, G., Pachiaudi, C., Rio, M., and Roux, M., 1981, Fluctuations de l'activite des sources hydrothermales oceaniques (Pacifique Est, 21°N) enregistrees lors de la croissance des coquilles de Calyptogena magnifica (Lamellibranche, Vesicomyidae) par les isotopes stables du carbone et de l'oxygene, C. R. Acad. Sci. Paris, 293(serie III):701-706.

Felbeck, H., 1981, Chemoautotrophic potential of the hydrothermal vent tube worm, Riftia pachyptila Jones (Vestimentifera), Science, 209:336-338.

Felbeck, H., and Somero, G. N., 1982, Primary productivity in deep-sea hydrothermal vent organisms: Roles of sulfide-oxidizing bacteria, Trends Biochem. Sci., 7:201-204.

Fretter, V., Graham, A., and McLean, J. H, 1981, The anatomy of the Galapagos Rift limpet, Neomphalus fretterae, Malacologia, 21:337-361.

Grassle, J. F., Berg, C. J., Childress, J J., Grassle, J. P., Hessler, R. R., Jannasch, H. J., Karl, D. M., Lutz, R. A., Mickel, T. J., Rhoads, D. C., Sanders, H L., Smith, K. L., Somero, G. N., Turner, R. D., Tuttle, J. H., Walsh, P. J., and Williams, A. J., 1979, Galapagos '79: Initial findings of a deep-sea biological quest, Oceanus, 22(2):1-10.

Grassle, J. F., Sanders, H. L., Hessler, R. R., Rowe, G. T., and McLellan, T., 1975, Pattern and zonation: a study of the bathyal megafauna using the research submersible Alvin, Deep-Sea Res., 22:57-481.

Hiatt, B., 1980, Sulfides instead of sunlight, Mosaic, 11(4):15-21.
Hessler, R. R., and Wilson, G. D., 1983, The origin and biogeography of malacostracan crustaceans in the deep-sea, in "Evolution, Time, and Space: The Emergence of the Biosphere," R. W. Sims, J. H. Price, and P. E. S. Whalley, eds., Systematics Association Special Vol. 23:227-254.
Hyman, L. H., 1959, "The Invertebrates V: Smaller Coelomate Groups," McGraw-Hill, New York.
Jannasch, H. W, and Wirsen, C. O., 1979, Chemosynthetic primary production at East Pacific sea floor spreading center, BioSci., 29:592-598.
Jannasch, H. W., and Wirsen, C. O., 1981, Morphological survey of microbial mats near deep-sea thermal vents, Appl. Environ. Microbiol., 41:528-538.
Jones, M. L., 1981, Riftia pachyptila, a new genus, new species: The vestimentiferan worm from the Galapagos Rift geothermal vents (Pogonophora), Proc. Biol. Soc. Wash., 93:1295-1313.
Karl, D. M., Wirsen, C. O., and Jannasch, H. W., 1980, Deep-sea primary productivity at the Galapagos hydrothermal vents, Science, 207:1345-1347.
Killingley, J. S., Berger, W. H., Macdonald, K. C., and Newman, W. A., 1980, 0-18/0-16 variations in deep-sea carbonate shells from the Rise hydrothermal field, Nature, 288:218-221.
Lonsdale, P., 1977, Clustering of suspension-feeding macrobenthos near abyssal hydrothermal vents at oceanic spreading centers, Deep-Sea Res., 24:857-863.
Lupton, J. E., Klinkhammer, G. P., Normark, W. R., Haymon, R., Macdonald, K. C., Weiss, R. F., and Craig, H., 1980, Helium-3 and manganese at the 21°N East Pacific Rise hydrothermal site, Earth Planet. Sci. Lett., 50:115-127.
McLean, J. H., 1981, The Galapagos Rift limpet Neomphalus: Relevance to understanding the evolution of a major Paleozoic-Mesozoic radiation, Malacologia, 21:291-336.
Mickel, T. J., and Childress, J. J., 1982a, Effects of temperature, pressure and oxygen concentration on the oxygen consumption rate of the hydrothermal vent crab Bathyograea thermydron (Brachyura), Physiol. Zool., 55:199-207.
Mickel, T. J., and Childress, J. J., 1982b, Effects of pressure and temperature on the EKG and heart rate of the hydrothermal vent crab Bathyograea thermydron (Brachyura), Biol. Bull., 162:70-82.
Pugh, P. R., in press. Benthic siphonophores: A review of the family Rhodaliida, Phil. Trans. Roy. Soc. London.
Rau, G. H., 1981a, Hydrothermal vent clam and tube worm C-13/C-12: Further evidence of non-photosynthetic food sources, Science, 209:338-340.
Rau, G. H., 1981b, Low N-15/N-14 in hydrothermal vent animals: Ecological implications, Nature, 289:484-485.

Rau, G. H., and Hedges, J. I., 1979, Carbon-13 depletion in a hydrothermal vent mussel: Suggestion of a chemosynthetic food source, Science, 203:648-649.

Smithey, W. M., Jr., and Hessler, R. R., in press, Megafaunal distribution at deep-sea hydrothermal vents: An integrated photographic approach, in: "Underwater Photography for Scientists," Oxford University Press, London.

Spiess, F. N., Macdonald, K. C., Atwater, T., Ballard, R., Carranza, A., Cordoba, D., Cox, C., Diaz Garcia, V. M., Francheteau, J., Guerrero, J., Hawkins, J., Haymon, R., Hessler, R., Juteau, T., Kastner, M., Larson, R., Luyendyk, B., Macdougall, J. D., Miller, S., Normark, W., Orcutt, J., and Rangin, C., 1980, East Pacific Rise: Hot springs and geophysical experiments, Science, 207:1421-1433.

van Andel, and Ballard, R. O., 1979, The Galapagos Rift at 86°W: 2. Volcanism, structure, and evolution of the rift valley, J. Geophys. Res., 84:5390-5406.

Williams, A. B., 1980, A new crab family from the vicinity of submarine thermal vents on the Galapagos Rift (Crustacea: Decapoda: Brachyura), Proc. Biol. Soc. Wash., 93:443-472.

Williams, A. B, and Chace, F. A., Jr., 1982, A new caridean shrimp of the family Bresiliidae from thermal vents of the Galapagos Rift, J. Crust. Biol., 2:136-147.

Zarenkov, N. A., 1969, Decapoda, in "Biology of the Pacific Ocean, Part II, The Deep-Sea Bottom Fauna," L. A. Zenkevich, ed., Vol. 7:79-82, U.S. Naval Oceanogr. Office, Washington, D. C. (translation).

Zezina, O. N., 1965, Distribution of the deepwater brachiopod Pelagodiscus atlanticus (King), Oceanology, 5:127-131.

Zezina, O. N., 1969, Brachiopoda, in "Biology of the Pacific Ocean, Part II," The Deep-Sea Bottom Fauna," L. A. Zenkevich, ed., Vol. 7:100-102, U.S. Naval Oceanogr. Office, Washington, D. C. (translation).

APPENDIX 1

LANDMARKS IN STUDIES OF HYDROTHERMAL PROCESSES AT SEAFLOOR SPREADING CENTERS

The process of seafloor spreading was hypothesized (Holmes, 1931; Hess, 1962; Dietz, 1961) and verified by application of the magnetic polarity-reversal time scale (Cox *et al.*, 1963) to interpretation of the sequence of seafloor magnetic anomalies (Morley, unpublished manuscript; Vine and Matthews, 1963; Vine and Wilson, 1965; Pitman and Heirtzler, 1966) and dating of samples recovered in transects across the seafloor by the Deep Sea Drilling Project (Maxwell *et al.*, 1970).

Seafloor spreading centers were incorporated into the theory of plate tectonics as divergent plate boundaries where lithosphere (oceanic crust and upper mantle) is generated by the process of seafloor spreading (McKenzie and Parker, 1967; Morgan, 1968; Le Pichon, 1968; Isacks *et al.*, 1968).

The existence of subseafloor hydrothermal convection systems in ocean basins was inferred from the presence of the components of such systems at oceanic ridges comprising magmatic heat sources to drive convection, seawater as a fluid medium, and fractured volcanic rocks as a permeable solid medium (Elder, 1965; Deffayes, 1970).

The hypersaline hydrothermal solutions in the vicinity of the Atlantis II Deep of the Red Sea appeared as an anomaly on the hydrographic data of the Swedish Deep Sea Expedition at a water sampling station of the research vessel ALBATROSS while transiting the Red Sea in 1948 (Bruneau *et al.*, 1953). The anomaly was not recognized at that time. The Red Sea represents an early stage of opening of an ocean basin about a seafloor spreading center.

The combination of stratified high temperature and high salinity solutions ponded in certain basins along the axis of the Red Sea was recognized in hydrographic data of water sampling stations made by ships transiting the Red Sea between 1963 and 1965 during the International Indian Ocean Expedition (Charnock, 1964; Miller, 1964; Swallow and Crease, 1965).

Metalliferous sediments were found in certain basins along the axis of the Red Sea and their origin related to the metal-rich hypersaline hydrothermal solutions present (Miller *et al.*, 1966; Hunt *et al.*, 1967; Bischoff, 1969).

Anomalously high contents of Fe, Mn, Cu, Cr, Ni, and Pb were identified in sediments of the East Pacific Rise and attributed to hydrothermal activity alternatively related to volcanism (Skornyakova, 1964; Arrhenius and Bonatti, 1965), or to exhalations from mantle-level magmatic processes (Boström and Peterson, 1966).

The first sample of solid hydrothermal material reported from an oceanic ridge was recovered in the form of a fractionated ferromanganese encrustation from a seamount located off the axis of the East Pacific Rise (Bonatti and Joensuu, 1966).

From the vantage point of Iceland, Pálmason (1967) suggested that hydrothermal circulation may play a significant role in the distribution of oceanic heat flow, and Talwani *et al.* (1971) attributed low heat flow measured at the axis of the Reykjanes Ridge to dissipation of heat by convective circulation of seawater.

An excess of the primordial inert isotope ^{3}He was detected at mid-depth in the Pacific Ocean and attributed to mantle degassing at sites along the crest of the East Pacific Rise (Clarke *et al.*, 1969), providing the basis for use of ^{3}He as an index of hydrothermal discharge (Craig *et al.*, 1975).

On the basis of the mineralogy of altered basalts, Hart (1970) suggested that much of the basalt-seawater interaction occurs by low-temperature reaction near the rock-seawater interface, but some basalt-seawater reaction takes place at depths of up to 5 km within the oceanic crust.

Recognition of the widespread occurrence of hydrated metamorphosed oceanic crust and upper mantle led to realization that a large volume of water was required to react with the rocks (Melson *et al.*, 1968; Miyashiro *et al.*, 1971; Christensen, 1972); analysis of oxygen isotopes indicated that a low $\delta\,^{18}O$ isotopic source such as seawater had interacted with the hydrated rocks (Muehlenbachs and Clayton, 1972; Spooner *et al.*, 1974).

Various lines of evidence from crustal sections interpreted as oceanic crust exposed on land (ophiolites) and in ocean basins were consolidated into a unified model for subseafloor metamorphism, heat and mass transfer at seafloor spreading centers (Spooner and Fyfe, 1973).

The observation that theoretical values of conductive heat flow calculated from models of heat produced by the generation of lithosphere at oceanic ridges (McKenzie, 1967; Le Pichon and Langseth, 1969; Sclater and Francheteau, 1970) exceeded the average measured values of conductive heat flow (Langseth and Von Herzen, 1970) led to the interpretation that the major portion of heat dissipation at oceanic ridges must be due to convective heat transfer by deeply circulating hydrothermal fluids (Lister, 1972).

The observed discrepency between theoretical and average measured conductive heat flow was used to estimate the global large magnitude of heat transfer by hydrothermal convection at seafloor spreading centers (Williams and Von Herzen, 1974; Wolery and Sleep, 1976).

A basal layer of metalliferous sediments was identified overlying basaltic basement in the Pacific Ocean Basin similar to the sediment forming at the crest of the East Pacific Rise, which demonstrated that metalliferous sediments have formed more or less continuously related to hydrothermal activity during seafloor spreading (von der Borch and Rex, 1970).

"Volcano-exhalative" and hydrogenous metal-bearing components of East Pacific Rise sediments were distinguished on the basis of radiogenically determined metal accumulation rates and distinctive lead isotopic composition (Bostrom, 1970; Bender *et al.*, 1971; Dasch *et al.*, 1971).

The observation that transition metals were depleted in the slowly cooled interior relative to the rapidly quenched exterior of extrusive basalts from the Mid-Atlantic Ridge was the basis for the inference that metals were mobilized from the interior by dissolution as chloride complexes in seawater that produced ore-forming hydrothermal solutions adequate to form metallic minerals observed at seafloor spreading centers (Corliss, 1971).

Discovery of the first active submarine hydrothermal field on an oceanic ridge in an open ocean basin by the NOAA Trans-Atlantic Geotraverse (TAG) project in 1972, the TAG Hydrothermal Field on the Mid-Atlantic Ridge, confirmed that hydrothermal processes could act more or less continuously from early to advanced stages in the opening of an ocean basin about a seafloor spreading center represented, respectively, by the Red Sea and the Atlantic Ocean (Rona, 1973; Rona and R. B. Scott, 1974; M. R. Scott *et al.*, 1974; R. B. Scott *et al.*, 1974; Rona, 1980).

The first direct measurements of the magnitude and gradient of near-bottom water temperature anomalies with the characteristics of hydrothermal discharge at an oceanic ridge were made in 1973 at the TAG Hydrothermal Field on the Mid-Atlantic Ridge (Rona *et al.*, 1974, 1975) and at the Galapagos Spreading Center (Williams *et al.*, 1974).

Stockwork-type copper-iron sulfides were recovered from oceanic crust exposed in large-offset ridge-ridge transform fault segments of major fracture zones in the Atlantic Ocean (Bonatti *et al.*, 1976) and Indian Ocean (Dmitriev *et al.*, 1970; Rosanova and Baturin, 1976), and penetrated in a drillhole through the upper portion of oceanic crustal Layer 2 (Deep Sea Drilling Project Hole 504B south of the Costa Rica Rift; Anderson *et al.*, 1982).

Laboratory experiments reacting basalt with seawater in closed containers simulating pressures and temperatures of subseafloor hydrothermal systems produced elemental exchanges analogous to those determined for oceanic crust from studies of basalt alteration and indicated chemical conditions for generating ore-forming hydrothermal solutions (Bischoff and Dickson, 1975; Hajash, 1975).

Simultaneous measurements of increased temperature and excess ^3He in a near-bottom water sample recovered at the Galapagos Spreading Center in 1976 confirmed the existence of hydrothermal discharge from an oceanic ridge (Weiss *et al.*, 1977).

The first direct observation of active submarine hydrothermal vents at an oceanic ridge and the discovery of a biological community associated with the vents was made diving with DSRV ALVIN at the Galapagos Spreading Center in 1977 where four groups of vents discharging hydrothermal solutions at temperatures up to 17°C were investigated (Corliss *et al.*, 1979). Clams in the vent biota were photographed by the Scripps Deep Tow instrumentation system (Lonsdale, 1977).

Global fluxes of certain elements extrapolated from their concentrations in dilute hydrothermal solutions sampled at the Galapagos Spreading Center were determined to equal or exceed input of those elements to the ocean by rivers transporting material derived from weathering of continents (Corliss *et al.*, 1979; Edmond *et al.*, 1979).

The global magnitude of convective heat flux at seafloor spreading centers calculated at between 4.0 and 6.4×10^{19} cal y^{-1} from the discrepency between theoretical heat flow and average measured conductive heat flow (Williams and Von Herzen, 1974; Wolery and Sleep, 1976) was experimentally verified on the basis of extrapolation of the measured ratio between ^3He and water temperature in hydrothermal effluent recovered at the Galapagos Spreading Center to the average oceanic flux of ^{3}He (Jenkins et al., 1978).

Investigation of biological communities at active hydrothermal vents at the Galapagos Spreading Center and the East Pacific Rise at latitude 21°N revealed that bacteria capable of metabolizing hydrogen sulfide in the hydrothermal effluents were the primary producers at the base of the food chain, and that the chemosynthetic products are ultimately derived from internal heat energy of the Earth independent of photosynthesis (Corliss et al., 1979; Jannasch and Wirsen, 1979; Karl, 1980).

The chemosynthetic bacterial processes at oceanic hydrothermal vents were proposed as an alternate to photosynthetic processes in shallow seas as an environment favorable for the evolution and maintenance of early forms of life (Corliss et al., 1981).

Massive sulfide deposits in the form of mounds exposed on the seafloor were first directly observed and sampled at an oceanic ridge in 1978 by a team of French, American and Mexican scientists diving with the French submersible CYANA near the axis of the East Pacific Rise at latitude 21°N (CYAMEX Scientific Team, 1979).

Hydrothermal effluents with the properties of high-temperature (c. 350°C) end member solutions (acidic, reducing, metal-rich) predicted on the basis of extrapolation of mixing curves of low-temperature effluents (c. 10°C) previously sampled at the Galapagos Spreading Center (Edmond et al., 1979) were recovered at "black smokers" discharging from massive sulfide chimneys surmounting mounds at the axis of the East Pacific Rise at latitude 21°N by a team of scientists diving with DSRV ALVIN in 1979 (RISE Project Group, 1980) within hundreds of meters of the relict massive sulfide mounds observed in 1978 (CYAMEX Scientific Team, 1978).

The existence of high-temperature hydrothermal solutions (c. 300°C) at slow-spreading oceanic ridges comparable to such solutions at intermediate- to fast-spreading oceanic ridges was indicated by several lines of evidence from the Mid-Atlantic Ridge comprising temperature determinations from oxygen isotopes in hydrothermal quartz vugs (Rona et al., 1980), chemistry of fluid inclusions in quartz crystals (Delaney et al., 1983), and the presence of sedimentary layers enriched in metals (Cu, Fe, Zn) derived from discharge of high-temperature solutions (Shearme et al., 1983).

Anomalies in conductive heat flow (Lawver and Williams, 1979) and in ^{3}He in near-bottom water (Lupton, 1979) led to the discovery of high-temperature hydrothermal activity associated with mineral deposits (Lonsdale et al., 1980) and hydrocarbons (Simoneit and Lonsdale, 1982) in the Guaymas Basin of the Gulf of California representative of the early stage of opening of an ocean basin about an intermediate-rate spreading center.

Sites at seafloor spreading centers in back-arc basins behind volcanic island arcs in the western Pacific were found to be loci of hydrothermal mineralization (Bertine and Keene, 1975; Bonatti et al., 1979) and of ongoing hydrothermal discharge (Horibe et al., 1982).

A relict massive sulfide body apparently formed of coalesced mounds surmounted by chimneys with overall external dimensions comparable in size to economically interesting massive sulfide deposits in volcanogenic rocks on land was found at a site at the faulted margin of the axial zone of the Galapagos Spreading Center (Malahoff, 1982).

The occurrence of two-phase subseafloor hydrothermal convection systems involving separation liquid and vapor phases at oceanic water depths was inferred from the measurement of anomalous salinities in fluid inclusions within secondary alteration minerals in basalts recovered from certain sites on oceanic ridges (Jehl, 1975; Vanko and Batiza, 1982; Delaney et al., 1983; Cosens, 1983).

The first drillhole to attain penetration of one kilometer into oceanic crust (Deep Sea Drilling Project Hole 504B on the intermediate spreading rate Costa Rica Rift) encountered a sheeted dike complex supporting the analogy between oceanic crust and certain ophiolites, and measured variations in permeability related to the horizontally layered structure of oceanic crust previously determined in seismic models (Anderson et al., 1982; Becker et al., 1982).

Observations of the distribution of metal-enriched sediments on the seafloor (Boström et al., 1969), and of geochemical tracers from hydrothermal sources at mid-depth in the water column (^{3}He; Lupton and Craig, 1981) support the interpretation that buoyant hydrothermal effluents at the crest of the East Pacific Rise discharge into a water mass characterized by higher temperature on an isopycnal surface at mid-depth in the water column (Reid, 1982) and dynamically drive that water mass (Stommel, 1982).

REFERENCES

Anderson, R. N., Honnorez, J., Becker, K., Adamson, A. C., Alt, J. C., Emmermann, R., Kempton, P. D., Kinoshita, H., Laverne, C., Mottl, M. J., and Newmark, R. L., 1982, DSDP Hole 504B, the first reference section over 1 km through Layer 2 of the oceanic crust, Nature, 300:589-594.

Arrhenius, G. O. S., and Bonatti, E., 1965, Neptunism and volcanism in the ocean, in: "Progress in Oceanography," M. Sears (Editor), Pergamon Press, London, 3:7-22.

Becker, K., Von Herzen, R. P., Francis, T. J. G., Anderson, R. N., Honnorez, J., Adamson, A. C., Alt, J. C., Emmermann, R., Kempton, P. D., Kinoshita, H., Laverne, C., Mottl, M. J., and Newmark, R. L., 1952, In-situ electrical resistivity and bulk porosity of the oceanic crust, Costa Rica Rift, Nature, 300:594-598.

Bender, M., Broecker, W., Gornitz, V., Middel, U., Kay, R., Sun, S.-S, and Biscaye, P., 1971, Geochemistry of three cores from the East Pacific Rise, Earth Planet. Sci. Ltrs., 12:425-433.

Bertine, K. K., and Keene, J. B., 1975, Submarine barite-opal rocks of hydrothermal origin, Science, 18:150-152.

Bischoff, J. L., 1969, Red Sea geothermal brine deposits, in: "Hot Brines and Recent Heavy Metal Deposits of the Red Sea," E. T. Degens and D. A. Ross (Editors), Springer-Verlag, New York, pp. 348-401.

Bischoff, J. L., and Dickson, F. W., 1975, Seawater-basalt interaction at 200°C and 500 bars: implications for origin of seafloor heavy-metal deposits and regulation of seawater chemistry, Earth Planet. Sci. Ltrs., 25:385-397.

Bonatti, E., and Joensuu, O., 1966, Deep-sea iron deposit from the South Pacific, Science, 154:643-645.

Bonatti, E., Kolla, V., Moore, W. S., and Stern, C., 1979, Metallogenesis in marginal basins: Fe-rich basal deposits from the Philippine Sea, Mar. Geol., 32:21-37.

Boström, K., 1970, Submarine volcanism as a source for iron, Earth Planet. Sci. Ltrs., 9:348-354.

Boström, K., and Peterson, M. N. A., 1966, Precipitates from hydrothermal exhalations on the East Pacific Rise, Econ. Geol., 61:1258-1265.

Boström, K., Peterson, M. N. A., Joensuu, O., and Fisher, D. E., 1969, Aluminum-poor ferromanganoan sediments on active oceanic ridges, Jour. Geophys. Res., 74:3261-3270.

Bruneau, L., Jerlov, N. G., and Koczy, F. F., 1953, Physical and chemical methods, Swedish Deep-Sea Expedition Report, v. III, Physics and Chemistry, No. 4, Appendix, Table 1, Physical and Chemical Data, pp. XXIX, Station 254.

Charnock, H., 1964, Anomalous bottom water in the Red Sea, Nature, 203:591.

Christensen, N. L., 1972, The abundance of serpentinites in the oceanic crust, Jour. Geol., 80:709-719.

Corliss, J. B., 1971, The origin of metal-bearing submarine hydrothermal solutions, Jour. Geophys. Res., 76:8128-8138.

Corliss, J. B., Dymond, J., Gordon, L. I., Edmond, J. M., Von Herzen, R. P., Ballard, K., Green, K., Williams, D., Bainbridge, A., Crane, K., and van Andel, T. H., 1979, Submarine thermal springs on the Galapagos Rift, Science, 203:1073-1083.

Corliss, J. B., Baross, J. A., and Hoffman, S. E., 1981, Submarine hydrothermal systems: a probable site for the origin of life, Oceanologica Acta, Supplement to v. 4, pp. 59-69.

Cosens, B. A., 1983, Initiation and collapse of active circulation in a hydrothermal system at the Mid-Atlantic ridge, 23°N, Jour. Geophys. Res. (in press).

Cox, A., Doell, R. R., and Dalrymple, G. B., 1963, Radiometric dating of geomagnetic field reversals, Science, 140:1021-1023.

Craig, H., Clarke, W. B., and Beg, M. A., 1975, Excess ^{3}He in the deep water on the East Pacific Rise, Earth Planet. Sci. Ltrs., 26:125-132.

CYAMEX Scientific Team (Francheteau, J., Needham, H. D., Choukroune, P., Juteau, T., Seguret, M., Ballard, R. D., Fox, P. J., Normark, W. R., Carranza, A., Cordoba, D., Guerrero, J., and Rangin, C.), 1979, Massive deep-sea-sulfide ore deposits discovered on the East Pacific Rise, Nature, 277:523-528.

Deffayes, K. S., 1970, The axial valley: a steady-state feature of the terrain, in: "The Megatectonics of Continents and Oceans," H. Johnson and B. L. Smith (Editors), Rutgers Univ. Press, Camden, New Jersey, pp. 194-222.

Delaney, J. R., Mogk, D. W., and Mottl, M. J., 1983, Quartz-cemented, sulfide-bearing greenstone breccias from the Mid-Atlantic Ridge: samples of a high-temperature upflow zone, Science (in press).

Dietz, R. S., 1961, Continent and ocean basin evolution by spreading of the seafloor, Nature, 190:854-857.

Dmitriev, L. V., Barsukov, V. L., and Udintsev, G. B., 1970, Oceanic rift zones and problems of ore formation, Geokhimiya, 8:937.

Edmond, J. M., Measures, C., McDuff, R. E., Chan, L. H., Collier, R., Grant, B., Gordon, L. I., and Corliss, J. B., 1979, Earth Planet. Sci. Ltrs., 46:1-18.

Elder, J. W., 1965, Physical processes in geothermal areas, in: "Terrestrial Heat Flow," W. H. K. Lee (Editor), Amer. Geophys. Union Geophys. Monograph Series, 8:211-239.

Hajash, A., 1975, Hydrothermal processes along mid-ocean ridges, Contrib. Mineral. Petrol., 53:205-226.

Hart, R. A., 1973, A model for chemical exchange in the basalt-seawater system of oceanic Layer 2, Canadian Jour. Earth Sci., 10:799-816.

Holmes, A., 1931, Radioactivity and earth movements, Geol. Soc. Glasgow Trans., 18:559-606.

Horibe, Y., Kim, K-R., and Craig, H., 1982, Deep ocean hydrothermal vents in Mariana Trough, Fifth Int. Conf. on Isotope Geol., Nikko, Japan.

Hunt, J. M., Hayes, E. E., Degens, E. T., and Ross, D. A., 1967, Red Sea: detailed survey of hot brine areas, Science, 156:514-516.

Isacks, B., Oliver, J., and Sykes, L. R., 1968, Seismology and the new global tectonics, Jour. Geophys. Res., 73:5855-5899.

Jehl, V., 1975, Abstract in Fluid Inclusion Research, Proceedings of COFFI, E. Roedder, Editor, 8:75.

Jenkins, W. J., Edmond, J. M., and Corliss, J. B., 1978, Excess ^{3}He and ^{4}He in Galapagos submarine hydrothermal waters, Nature, 272:156-158.

Langseth, M. G., Jr., and Von Herzen, R. P., 1970, Heat flow through the floor of the world oceans, in: "The Sea," A. E. Maxwell (Editor), V. 4, Part I, Wiley-Interscience, New York, pp. 299-352.

Lawver, L. A., and Williams, D. L., 1979, Heat flow in the central Gulf of California, Jour. Geophys. Res., 84:3465-3478.

Le Pichon, X., 1968, Seafloor spreading and continental drift, Jour. Geophys. Res., 73:3661-3697.

Le Pichon, X., and Langseth, M. G., Jr., 1969, Heat-flow from the mid-ocean ridges and seafloor spreading, Tectonophysics, 8:319-344.

Lister, C. R. B., 1972, On the thermal balance of a mid-ocean ridge, Royal Astron. Soc. Geophys. Jour., 26:515-535.

Lonsdale, P. F., 1977, Clustering of suspension-feeding macrobenthos near abyssal hydrothermal vents at oceanic spreading centers, Deep-Sea Res., 24:857-863.

Lonsdale, P. F., Bischoff, J. L., Burns, V. M., Kastner, M., and Sweeney, R. E., 1980, A high-temperature hydrothermal deposit on the seabed at a Gulf of California spreading center, Earth Planet. Sci. Ltrs., 49:8-20.

Lupton, J. E., 1979, Helium-3 in the Guaymas Basin: evidence for injection of mantle volatiles in the Gulf of California, Jour. Geophys. Res., 84:7446-7452.

Malahoff, A., 1982, A comparison of the massive submarine polymetallic sulfides of the Galapagos Rift with some continental deposits, Mar. Tech. Soc. Jour., 16:3:39-45.

Maxwell, A. E., Von Herzen, R. P., Hsu, K. J., Andrews, J. E., Saito, T., Percival, S. F., Milo, E. D., and Boyce, R. E., 1970, Deep-sea drilling in the South Atlantic, Science, 168:1047-1059.

McKenzie, D. P., 1967, Some remarks on heat flow and gravity anomalies, Jour. Geophys. Res., 72:6261-6273.

McKenzie, D. P., and Parker, R. L., 1967, The North Pacific: an example of tectonics on a sphere, Nature, 216:1276-1280.

Melson, W. G., Thompson, G., and van Andel, T. H., 1968, Volcanism and metamorphism in the Mid-Atlantic Ridge, 22°N, Jour. Geophys. Res., 73:5925-5941.

Miller, A. R., 1964, Highest salinity in the world ocean? Nature, 203:590-591.

Miller, A. R., Densmore, C. D., Degens, E. T., Pocklington, R., and Jokela, A., 1966, Hot brines and recent iron deposits in deeps of the Red Sea, Geochim. Cosmochim. Acta, 30:341-359.

Miyashiro, A., Shido, F., and Ewing, M., 1971, Metamorphism in the Mid-Atlantic Ridge near 24° and 30°N, Royal Soc. Phil. Trans. (London), 268:589-603.

Morgan, W. J., 1968, Rises, trenches, great faults, and crustal blocks, Jour. Geophys. Res., 73:1959-1982.

Morley, L. W., 1963, An explanation of magnetic banding in ocean basins, in: "The Oceanic Lithosphere, The Sea," 1981, C. Emiliani (Editor), v. 7, Wiley, New York, pp. 1717-1719.

Muehlenbachs, K., and Clayton, R. H., 1972, Oxygen isotope geochemistry of submarine greenstones, Canadian Jour. Earth Sci., 9:471-478.

Pálmason, G., 1967, On heat flow in Iceland in relation to the Mid-Atlantic Ridge, in: "Iceland and Mid-Ocean Ridges," S. Bjornsson (Editor), Soc. Sci. Islandica, 38:111-117.

Pitman, W. C., III, and Heirtlzer, J. P., 1966, Magnetic anomalies over the Pacific-Antarctic Ridge, Science, 154:1164-1171.

Reid, J. L., 1982, Evidence of an effect of heat flux from the East Pacific Rise upon the characteristics of the mid-depth waters, Geophys. Res. Ltrs., 9:381-384.

RISE Project Group (Speiss, F. N., Macdonald, K. D., Atwater, T., Ballard, R., Carranza, A., Cordoba, D., Cox, C., Diaz Garcia, V. M., Francheteau, J., Guerrero, J., Hawkins, J., Haymon, R., Hessler, R., Juteau, T., Kastner, M., Larson, R., Luyendyk, B., Macdougall, J. D., Miller, S., Normark, W., Orcutt, J., and Rangin, C.), 1980, East Pacific Rise: hot springs and geophysical experiments, Science, 207:1421-1444.

Rona, P. A., 1973, New evidence for seabed resources from global tectonics, Ocean Management, 1:145-159.

Rona, P. A., and Scott, R. B., 1974, Convenors, Symposium: axial processes of the Mid-Atlantic Ridge, EOS, Trans. Amer. Geophys. Union, 55:292-295.

Rona, P. A., McGregor, B. A., Betzer, P. R., and Krause, D. C., 1974, Anomalous water temperatures over Mid-Atlantic Ridge crest at 26°N, Amer. Geophys. Union Trans., EOS, 55:293.

Rona, P. A., McGregor, B. A., Betzer, P. R., Bolger, G. W., and Krause, D. C., 1975, Anomalous water temperatures over Mid-Atlantic Ridge crest at 26° north latitude, Deep-Sea Res., 22:611-618.

Rona, P. A., Boström, K., and Epstein, S., 1980, Hydrothermal quartz vug from the Mid-Atlantic Ridge, Geology, 8:569-572.

Rona, P. A., 1980, TAG Hydrothermal Field: Mid-Atlantic Ridge crest at latitude 26°N, Jour. Geol. Soc. London, 137:385-402.

Rozanova, T. V., and Baturin, G. N., 1971, Hydrothermal ore shows on the floor of the Indian Ocean, Oceanology, Academy of Sciences of USSR, translation by Amer. Geophys. Inst. 11(6):847-879.

Sclater, J. G., and Francheteau, J., 1970, The implications of terrestrial heat-flow observations on current tectonic and geochemical models of the crust and upper mantle of the earth, Geophys. Jour., 20:509-542.

Scott, M. R., Scott, R. B., Rona, P. A., Butler, L. W., and Nalwalk, A. J., 1974, Rapidly accumulating manganese deposit from the median valley of the Mid-Atlantic Ridge, Geophys. Res. Ltrs., 1:355-358.

Scott, R. B., Rona, P. A., McGregor, B. A., and Scott, M. R., 1974, The TAG Hydrothermal Field, Nature, 251:301-302.

Simoneit, B. R. T., and Lonsdale, P. F., 1982, Hydrothermal petroleum in mineralized mounds at the seabed of Guaymas Basin, Nature, 295:198-202.

Shearme, S., Cronan, D. S., and Rona, P. A., 1983, Geochemistry of sediments from the TAG Hydrothermal Field, Mid-Atlantic Ridge at latitude 26°N, Mar. Geol. (in press).

Skornyakova, I. S., 1965, Dispersed iron and manganese in Pacific Ocean sediments, Inter. Geol. Rev., 7:2161-2174.

Spooner, E. T. C., and Fyfe, W. S., 1973, Sub-seafloor metamorphism, heat and mass transfer, Contrib. Mineral. Petrol., 42:287-304.

Spooner, E. T. C., Beckinsdale, R. D., Fyfe, W. S., and Smewing, J. D., 1974, O^{18} enriched ophiolitic metabasic rocks from E. Liguria (Italy), Pindos (Greece), and Troodos (Cyprus), Contrib. Mineral. Petrol., 47:41-74.

Stommel, H., 1982, Is the South Pacific helium-3 plume dynamically active? Earth Planet. Sci. Ltrs., 61:63-67.

Swallow, J. C., and Crease, J., 1965, Hot salty water at the bottom of the Red Sea, Nature, 205:165-166.

Talwani, M., Windisch, C. C., and Langseth, M. G., Jr., 1971, Reykjanes Ridge crest: a detailed geophysical study, Jour. Geophys. Res., 76:473-517.

Vanko, D. A., and Batiza, R., 1982, Plutonic rocks from the Mathematicians seamounts fossil ridge, East Pacific, EOS, Trans. Amer. Geophys. Union, 63:472.

Vine, F. J., and Matthews, D. H., 1963, Magnetic anomalies over ocean ridges, Nature, 199:947-949.

Vine, F. J., and Wilson, J. T., 1965, Magnetic anomalies over a young ocean ridge off Vancouver Island, Science, 150:485-489.

von der Borch, C. C., and Rex, R. W., 1970, Amorphous iron oxide precipitates in sediments cored during Leg 5, Deep Sea Drilling Project, in: "Initial Reports of the Deep Sea Drilling Project," D. A. McManus et al. (Editors), v. 5, Government Print Office, Washington, D.C., pp. 541-544.

Weiss, R. F., Lonsdale, P., Lupton, J. E., Bainbridge, A. E., and Craig, H., 1977, Hydrothermal plumes in the Galapagos Rift, Nature, 269:600-603.

Williams, D. L., and Von Herzen, R. P., 1974, Heat loss from the Earth: new estimate, Geology, 2:327-328.

Williams, D. L., Von Herzen, R. P., Sclater, J. G., and Anderson, R. N., 1974, The Galapagos Spreading Center: lithospheric cooling and hydrothermal circulation, Royal Astron. Soc. Geophys. Jour., 38:587-608.

Wolery, T. J., and Sleep, N. H., 1976, Hydrothermal circulation and geochemical flux at mid-ocean ridges, Jour. Geol., 84:249-275.

Photograph of authors and other participants at Cambridge University during the NATO Advanced Research Institute, "Hydrothermal Processes at Seafloor Spreading Centers," convened 5-8 April 1982.

Front Row: Left to Right

C. Lalou, D.S. Cronan, G. Thompson, K.C. Macdonald, R. Hekinian, P.A. Rona, T.H. van Andel, K.K. Turkenian, M.J. Mottl, J. Edmond

Middle Row: Left to Right

H. Craig, H.P. Taylor, Jr., B.J. Skinner, A.H.F. Robertson, V. Stefánsson, R.D. Ballard, R.N. Anderson, J. Boyle, H. Jannasch

Back Row: Left to Right

J. Francheteau, C.R.B. Lister, B.E. Parsons, R.J. Rosenbauer, A. Fleet, F.J. Grassle, R. Hessler

Photograph of participants at Cambridge University during the NATO Advanced Research Institute, "Hydrothermal Processes at Seafloor Spreading Centers," convened 5-8 April 1982.

Front Row: Left to Right

H. Craig, D.S. Cronan, J. Francheteau, C.R.B. Lister, G. Thompson, K.C. Mcdonald, F. Machado, P.A. Rona, J. Honnorez, R.F. Dill, R.D. Ballard, N.A. Ostenso, R. Hessler, H. Thiel, F. Grassle

Second Row from Front: Left to Right

J. Verhoef, R. Whitmarsh, V. Stefánsson, B.E. Parsons. T. Juteau, G.A. Gross, H.P. Taylor, Jr., F. Albarede, H. Jannasch, E. Bonatti, K. Crane, J. Lydon, I.D. MacGregor, E.R. Oxburgh

Third Row from Front: Left to Right

R. Hekinian, B.J. Skinner, C. Mevel, L. Widenfalk, R. Bowen, H. Bougault, T.H. van Andel, J.R. Cann, R.J. Rosenbauer, D.T. Rickard, A. Malahoff, S.P. Varnavas, M.J. Mottl

Fourth Row from Front: Left to Right

K. Brooks, J.W. Elder, B. Stuart, K. Gunnesch, A. Fleet, H.T. Papunen, A.H.F. Robertson, S.A. Moorby, J. Boyle, C. Lalou, V. Ittekkot

Top Row: Left to Right

K.K. Turekian, J. Hertogen, J.A. Pearce, J. Edmond, S.D. Scott, D.B. Duane, A.S. Laughton, H-W. Hubberton, R. Chesselet, R.L. Chase

INDEX